AMPHIBIANS AND REPTILES OF NEW MEXICO

William G. Degenhardt   Charles W. Painter   Andrew H. Price

Illustrations by Clay M. Garrett

Foreword by Roger Conant

AMPHIBIANS AND REPTILES OF NEW MEXICO

UNIVERSITY OF NEW MEXICO PRESS   ALBUQUERQUE

© 1996 by the Univesity of New Mexico Press
First Edition
All rights reserved.

Paperback ISBN 0-8263-3811-9

PRINTED IN THE UNITED STATES OF AMERICA

10 09 08 07 06 05   1 2 3 4 5 6

Share with Wildlife is a non-profit fund dedicated to the conservation of all wildlife in New Mexico. Administered by the New Mexico Department of Game and Fish, Share with Wildlife is supported entirely by voluntary contributions—most received through the state income tax check-off. Share with Wildlife funds research, habitat protection, wildlife rehabilitation, and public education, with an emphasis on projects benefiting nongame wildlife, including state and federal endangered species.

LIBRARY OF CONGRESS CATALOGING-IN-PUBLICATION DATA

Degenhardt, William G. (William George), 1926–

The amphibians and reptiles of New Mexico /
William G. Degenhardt, Charles W. Painter,
Andrew H. Price; illustrations by Clay M. Garrett.—1st ed.
   p. cm.
   Includes bibliographical references and index.

ISBN 0-8263-1695-6

1. Amphibians—New Mexico—Indentification.
2. Reptiles—New Mexico—Indentification.
I. Painter, Charles W. (Charles Wilson), 1949–   .
II. Price, Andrew H. (Andrew Hoyt), 1951–   .
III. Title.
   QL653.N6D44   1996
   597.6'09789—dc20
                               95–41784
                                 CIP

Designed by Sue Niewiarowski

To my late wife, Paula Braden Degenhardt, who was not only a very loving wife and friend, but also served as an assistant and field companion during the many years of data collection for this book.

My great regret is that she cannot see the result.

William G. Degenhardt

For my mother, Leah F. Painter, who never discouraged my childhood interest in herpetology, and for Brenda and Ashley who lost much valuable family time as this book was being prepared.

Charles W. Painter

For Jack McCoy and Joe LaPointe, who placed me firmly on the road to whatever professional success I enjoy and who always believed in me, and for my son, Alec, who had the great good fortune of being born a native New Mexican.

Andrew H. Price

CONTENTS

Foreword	xiii
Preface	xv
Acknowledgments	xvii
A PHYSIOGRAPHIC SKETCH OF NEW MEXICO	1
A BRIEF HISTORY OF HERPETOLOGY IN NEW MEXICO	7
A CHECKLIST OF THE AMPHIBIANS AND REPTILES	11
A KEY TO THE TADPOLES AND SALAMANDER LARVAE	15
A KEY TO THE SALAMANDERS	17
Family Ambystomatidae—Mole salamanders	19
Ambystoma tigrinum Tiger salamander	20
Family Plethodontidae—Lungless salamanders	24
Aneides hardii Sacramento mountain salamander	25
Plethodon neomexicanus Jemez Mountains salamander	27
A KEY TO THE TOADS AND FROGS	31
Family Pelobatidae—Spadefoots	35
Scaphiopus couchii Couch's spadefoot	36
Spea bombifrons Plains spadefoot	38
Spea multiplicata New Mexico spadefoot	40
Family Leptodactylidae—Tropical Frogs	43
Eleutherodactylus augusti Barking frog	44
Family Bufonidae—Toads	46
Bufo alvarius Colorado River toad	47
Bufo boreas Western toad	49
Bufo cognatus Great Plains toad	51

Bufo debilis Green toad — 53
Bufo microscaphus Southwestern toad — 55
Bufo punctatus Red-spotted toad — 57
Bufo speciosus Texas toad — 59
Bufo woodhousii Woodhouse's toad — 61
Family Hylidae—Treefrogs — 64
Acris crepitans Northern cricket frog — 65
Hyla arenicolor Canyon treefrog — 67
Hyla eximia Mountain treefrog — 69
Pseudacris triseriata Western chorus frog — 71
Family Microhylidae—Narrowmouth Toads — 74
Gastrophryne olivacea Great Plains narrowmouth toad — 75
Family Ranidae—True Frogs — 78
TABLE 1 Characteristics Used in the Identification of Five Species of Leopard Frogs (*Rana* spp.) — 79
Rana berlandieri Rio Grande leopard frog — 79
Rana blairi Plains leopard frog — 81
Rana catesbeiana Bullfrog — 83
Rana chiricahuensis Chiricahua leopard frog — 85
Rana pipiens Northern leopard frog — 87
Rana yavapaiensis Lowland leopard frog — 89

A KEY TO THE TURTLES — 93
Family Chelydridae—Snapping Turtles — 95
Chelydra serpentina Snapping turtle — 96
Family Emydidae—Box and Water Turtles — 99
Chrysemys picta Painted turtle — 100
Pseudemys gorzugi Western river cooter — 102
Terrapene ornata Ornate box turtle — 104
Trachemys gaigeae Big Bend slider — 107
Trachemys scripta Slider — 109
Family Kinosternidae—Mud Turtles — 113
Kinosternon flavescens Yellow mud turtle — 114
Kinosternon sonoriense Sonoran mud turtle — 116
Family Trionychidae—Softshell Turtles — 119
Trionyx muticus Smooth softshell — 120
Trionyx spiniferus Spiny softshell — 122

A KEY TO THE LIZARDS — 125

   Family Crotaphytidae — Collared and Leopard Lizards — 131

    Crotaphytus collaris Collared lizard — 132

    Gambelia wislizenii Leopard lizard — 135

   Family Phrynosomatidae — Zebratail, Earless, Spiny, Tree, Side-blotched, and Horned Lizards — 138

    Callisaurus draconoides Zebratail lizard — 139

    Cophosaurus texanus Greater earless lizard — 141

    Holbrookia maculata Lesser earless lizard — 145

    Phrynosoma cornutum Texas horned lizard — 148

    Phrynosoma douglasii Short-horned lizard — 151

    Phrynosoma modestum Roundtail horned lizard — 154

    Phrynosoma solare Regal horned lizard — 157

    Sceloporus arenicolus Sand dune lizard — 159

    Sceloporus clarkii Clark's spiny lizard — 161

    Sceloporus graciosus Sagebrush lizard — 163

    Sceloporus jarrovii Yarrow's spiny lizard — 166

    Sceloporus magister Desert spiny lizard — 170

    Sceloporus poinsetti Crevice spiny lizard — 173

    Sceloporus scalaris Bunch grass lizard — 175

    Sceloporus undulatus Prairie lizard — 178

    Sceloporus virgatus Striped plateau lizard — 183

    Urosaurus ornatus Tree lizard — 185

    Uta stansburiana Side-blotched lizard — 189

   Family Gekkonidae — Geckos — 194

    Coleonyx brevis Texas banded gecko — 195

    Coleonyx variegatus Western banded gecko — 197

    Hemidactylus turcicus Mediterranean gecko — 199

   Family Teiidae — Whiptails — 201

    TABLE 2 Characteristics of Whiptail Lizards — 203

    Cnemidophorus burti Canyon spotted whiptail — 205

    Cnemidophorus dixoni Gray-checkered whiptail — 207

    Cnemidophorus exsanguis Chihuahuan spotted whiptail — 209

    Cnemidophorus flagellicaudus Gila spotted whiptail — 211

    Cnemidophorus grahamii Checkered whiptail — 212

    Cnemidophorus gularis Texas spotted whiptail — 216

Cnemidophorus inornatus Little striped whiptail — 218
Cnemidophorus neomexicanus New Mexico whiptail — 221
Cnemidophorus sexlineatus Six-lined racerunner — 223
Cnemidophorus sonorae Sonoran spotted whiptail — 226
Cnemidophorus tigris Western whiptail — 227
Cnemidophorus uniparens Desert grassland whiptail — 231
Cnemidophorus velox Plateau striped whiptail — 233
Family Scincidae—Skinks — 236
Eumeces multivirgatus Many-lined skink — 237
Eumeces obsoletus Great Plains skink — 239
Eumeces tetragrammus Four-lined skink — 241
Family Anguidae—Alligator Lizards — 244
Elgaria kingii Madrean alligator lizard — 245
Family Helodermatidae—Venomous Lizards — 247
Heloderma suspectum Gila monster — 248

A KEY TO THE SNAKES — 251

Family Leptotyphlopidae—Blind Snakes — 255
Leptotyphlops dulcis Texas blind snake — 256
Leptotyphlops humilis Western blind snake — 258
Family Colubridae—Colubrids — 260
Arizona elegans Glossy snake — 261
Bogertophis subocularis Trans-Pecos rat snake — 262
Coluber constrictor Racer — 264
Diadophis punctatus Ringneck snake — 267
Elaphe guttata Corn snake — 269
Gyalopion canum Western hooknose snake — 271
Heterodon nasicus Western hognose snake — 273
Hypsiglena torquata Night snake — 276
Lampropeltis alterna Gray-banded kingsnake — 278
Lampropeltis getula Common kingsnake — 280
Lampropeltis pyromelana Sonoran mountain kingsnake — 282
Lampropeltis triangulum Milk snake — 284
Liochlorophis vernalis Smooth green snake — 286
Masticophis bilineatus Sonoran whipsnake — 288
Masticophis flagellum Coachwhip — 289
Masticophis taeniatus Striped whipsnake — 291
Nerodia erythrogaster Plainbelly water snake — 293

Pituophis melanoleucus Bullsnake, Gopher snake	295
Rhinocheilus lecontei Longnose snake	298
Salvadora deserticola Big Bend patchnose snake	299
Salvadora grahamiae Mountain patchnose snake	301
Senticolis triaspis Green rat snake	303
Sonora semiannulata Ground snake	305
Tantilla hobartsmithi Southwestern black-headed snake	307
Tantilla nigriceps Plains black-headed snake	309
Tantilla yaquia Yaqui black-headed snake	311
Thamnophis cyrtopsis Blackneck garter snake	312
Thamnophis elegans Western terrestrial garter snake	314
Thamnophis eques Mexican garter snake	317
Thamnophis marcianus Checkered garter snake	320
Thamnophis proximus Western ribbon snake	322
Thamnophis radix Plains garter snake	324
Thamnophis rufipunctatus Narrowhead garter snake	326
Thamnophis sirtalis Common garter snake	328
Trimorphodon biscutatus Lyre snake	332
Tropidoclonion lineatum Lined snake	334
Family Elapidae—Elapids	337
Micruroides euryxanthus Western coral snake	338
Family Viperidae—Vipers	340
Crotalus atrox Western diamondback rattlesnake	341
Crotalus lepidus Rock rattlesnake	344
Crotalus molossus Blacktail rattlesnake	346
Crotalus scutulatus Mojave rattlesnake	348
Crotalus viridis Western rattlesnake	351
Crotalus willardi Ridgenose rattlesnake	353
Sistrurus catenatus Massasauga	356
AMPHIBIANS AND REPTILES OF QUESTIONABLE OCCURRENCE IN NEW MEXICO	359
LIST OF SCIENTIFIC AND COMMON PLANT NAMES USED IN THE TEXT	363
LIST OF MUSEUM SYMBOLIC CODES	365
Glossary	367
Literature Cited	373
Index	421
CONVERSION TABLE	431

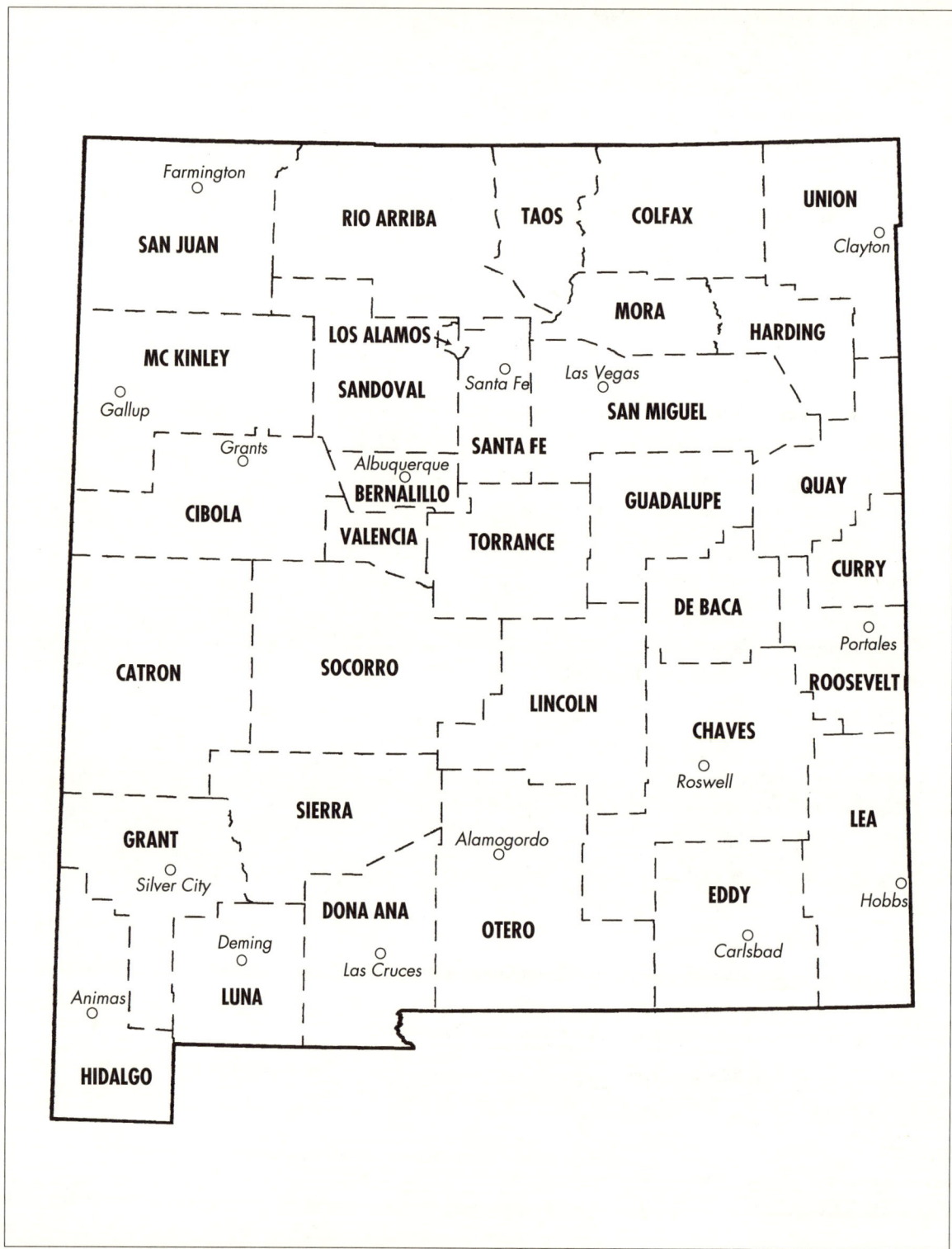

Figure 1 Map of New Mexico with counties and select cities mentioned in the text.

FOREWORD

New Mexico, fourth largest of the lower forty-eight states, is home to at least 123 species (166 including subspecies) of amphibians and reptiles. This scholarly volume is based on decades of field work, an intensive search of the scientific literature, and the studies of three long-time residents. It brings together an exhaustive compendium about virtually everything known, at the time of going to press, concerning the state's salamanders, toads, frogs, lizards, snakes, and turtles. Because it has been so carefully assembled, this book will remain the chief authority on the herpetology of New Mexico for a long time to come.

The illustrations, 134 of them in color, numerous drawings, and 123 carefully prepared and detailed distribution maps, serve to identify these various kinds of animals and to give the reader a quick overview of where, in New Mexico, each has been found to date.

Work on this long-awaited book began when William G. Degenhardt arrived at the University of New Mexico in 1960. He quickly envisioned a survey of the cold-blooded vertebrates (excluding the fishes) of the state, and he was instrumental in building up the large and important study collection deposited in the Museum of Southwestern Biology in the Biology Department on campus. With one of his students, James L. Christiansen, Degenhardt published a review on the "Distribution and Habitats of the Turtles of New Mexico," in the *Southwestern Naturalist* in 1974. It seemed likely at the time that papers on the herpetofauna of New Mexico would probably appear in print piecemeal, one group at a time.

Another of Degenhardt's students, Charles W. Painter, became herpetologist for the New Mexico Department of Game and Fish Endangered Species Program and, through his efforts, came the impetus and funding for a single, all-inclusive volume.

Andrew H. Price was long associated with New Mexico State University, and he did extensive field work, especially in the southern counties. He also specialized on lizards and was thus the logical choice to prepare the text and maps for that diverse, widespread, and abundant group of reptiles.

As thorough as this volume is in 1995, it is inevitable that much more will eventually be discovered and recorded about the herpetology of New Mexico. Although many parts of the state have become accessible by road during the past few decades, huge areas remain to be explored herpetologically. There are, for example, two great Wilderness Areas, the Gila and the Pecos; there are more than 13,000,000 acres of land under the control of the Bureau of Land Management; and the topography varies from many mountains ranging from 10,000 to more than 13,000 feet in elevation to the lowlands of the Chihuahuan Desert. What will be found in the future? The authors mention possibilities and outline many problems that remain to be studied.

New Mexico is a dry state. Proportionately it has less fresh water than any of the others, and much of it lies within man-made impoundments, the largest of which are Elephant Butte and Caballo reservoirs. It is a wonder that so many kinds of turtles and semiaquatic natricine snakes live within the borders of The Land of Enchantment.

Over a period of two decades I have watched, here in Albuquerque, the steady progress made on this major contribution to the herpetological literature. I can appreciate the enormous amount of hard work and dedication that provided not only the data base, but also the preparation of the text and illustrations. Here is a worthy and valuable addition to the herpetology literature for New Mexico.

Roger Conant
Department of Biology
University of New Mexico
Albuquerque

PREFACE

New Mexico, The Land of Enchantment, is blessed with an interesting and diverse herpetofauna. We recognize 123 species, an assemblage of 3 salamanders, 23 frogs and toads, 10 turtles, 41 lizards, and 46 snakes, which include one lizard and perhaps one frog introduced into the state. When all subspecies are included, there is a total of 166 taxa. In response to increased public interest in nongame wildlife resources, we endeavored to produce a treatise on the state's amphibians and reptiles that is useful to students, herpetologists, resource managers, and the public at large who are interested in the great diversity of New Mexico's natural heritage. We hope this book stimulates further study and conservation of the state's herpetofauna.

This book was a collaborative effort; however, each of us took primary responsibility for certain sections. William G. Degenhardt contributed sections on the turtles and snakes; Charles W. Painter contributed sections on the amphibians, venomous reptiles, and most of the species listed as threatened or endangered by the New Mexico Department of Game and Fish or by the United States Fish and Wildlife Service; Andrew H. Price contributed sections on the lizards. With multiple authors, uniformity in writing style is difficult to achieve, and although we tried for consistency we were not always successful. We hope this does not detract from the purpose of the book.

For widespread taxa, we did not attempt a thorough literature review but selected information from references that generally applied to New Mexico populations. We have tried to provide standard bibliographic data for each species including the actual dates of publication. Where the year of publication differs from the imprint date, we have so indicated by placing the publication date (that is, the date of distribution) in brackets after the imprint date. We have attempted to determine these dates by going to the original publications or by referring to indexes produced for this purpose. The index to the *Journal and Proceedings of the Philadelphia Academy of Sciences* (Nolan, 1913) has been especially helpful. We consider the journal numbers, as well as volume and pages, to be useful and have included these whenever available. The systematic arrangement of the family accounts follows Frost (1985) for amphibians, Iverson (1992) for turtles, Estes et al. (1988) for lizards, and Schmidt (1953) for snakes. Individual species accounts are arranged alphabetically within the family accounts. The scientific and common names most often used are those in the recent field guides and standardized lists of common and scientific names though we question certain usage on grammatical or other grounds. Deviations, such as retention of the name *Trionyx* for the softshell turtles and usage of the generic name *Liochlorophis* for the smooth green snake, are explained in the appropriate species accounts. We have also deviated from "standard" common names (Collins, 1990) where we considered them inappropriate (e.g., Yaqui black-headed snake for Yaqui blackhead snake, which is both bad grammar and politically incorrect).

We have not personally examined all of the more than 55,000 specimens from New Mexico that are widely scattered in major museums and universities across the United States. We examined, or had collection curators examine, those specimens we thought were misidentified or had questionable locality data.

The distribution maps were hand-plotted on a base map of New Mexico then electronically scanned into a Macintosh computer. Using Aldus Freehand software, each species was individually layered on the base map and then reproduced as a single map. Each dot covers an area with an approximate diameter of 4.8 km so that a given dot may contain several closely adjacent locality records. Unless otherwise noted, all specimen photographs were taken by Charles W. Painter. We tried to provide photographs of specimens from New Mexico although in a few cases, the rare, secretive, or perhaps extirpated species were photographed in adjacent states. Most habitat photographs were taken by Don MacCarter.

ACKNOWLEDGMENTS

We are grateful for the assistance and support of many people. Each made an extra effort to help out, and we appreciate their valuable contributions. Through the dedication and perseverance of Barney R. Tomberlin, James N. Stuart, Tony A. Snell, and Mark W. Doles we were provided specimens for most of the photographs in this book. Ronn Altig wrote the key to the tadpoles of New Mexico. Ted L. Brown and James N. Stuart provided us with much unpublished data from their field notes and personal collections. Mark Hakkila, Alice Marshall, and Mary Koch provided maps of New Mexico that made plotting records on the distribution maps much easier. Thanks to Randy D. Jennings and Norman J. Scott, Jr., who have worked hard to understand the distribution of leopard frogs in New Mexico, for allowing us to reproduce their leopard frog distribution maps. Mark Jordan, Jon Klingel, Kerry Mower, Richard Pfaff, Anne Rice, and Heidi Solper helped solve many of our computer problems. Owen J. Sexton allowed use of his laboratory and library while Degenhardt was in St. Louis. Heide and Howard Snell worked hard to computerize the distribution maps. Bob Wilson edited both early and late stages of the manuscript.

The majority of specimens that form the basis of the distribution maps are accessioned into the herpetology collection at The University of New Mexico Museum of Southwestern Biology (MSB). Since the beginning of that collection, many faculty members and former graduate students have contributed significant numbers of specimens from New Mexico. Foremost among these is the late William J. Koster whose early collections account for the majority of specimens accessioned during the formative years of the collection. Other noteworthy individuals include Robert D. Aldridge, John S. Applegarth, Michael A. Bogan, Terence P. Boyle, Ted L. Brown, James L. Christiansen, Roger Conant, Arthur E. Dunham, Susan M. Federer, James S. Findley, Eugene D. Fleharty, Richard B. Forbes, Julianne N. Green, Michael R. Gruner, Richard B. Halley, Laurence M. Hardy, James S. Jacob, Randy D. Jennings, Clyde J. Jones, Kirkland L. Jones, John E. Kimmons, Joseph Lavandoski, Joseph M. Lewandowsky, Ronald V. Lucchino, Bruce E. Miller, Roger Mongold, Steven P. Platania, Norman J. Scott, Jr., Robert C. Semmler, Anthony P. Sena, James N. Stuart, Richard A. Sugerman, Douglas P. Reagan, Robert P. Reynolds, James A. White, Stephen R. Williams, Donald E. Wilson, Bruce D. Woodward, and John W. Wright.

Grants from the National Science Foundation, New Mexico Department of Game and Fish, United States Army Corps of Engineers, United States Fish and Wildlife Service, United States Forest Service, and United States Park Service to one or more of us supported field work during the many years of data collection. Many grants to Degenhardt from the Research Allocations Committee of The University of New Mexico assisted his field activity over the years. Continued support for the Museum of Southwestern Biology by the Department of Biology at UNM was important in developing and sustaining the large collections that form the basis for this work.

We thank the many landowners who, throughout many

years, have given us permission to search for specimens or conduct studies on their private ranches. Most specifically we thank Eugene Catanzaro, Ed Elbrock, Drum and Terry Hadley, Adeline Hill, Jim Nance, George Pendleton, Ricky Pierce, and Robert Scholes.

Personnel at the Earth Data Analysis Center (EDAC) provided the vegetative, topographic, and county maps. We specifically thank Michael Inglis and Benny Campos for their help.

For a variety of favors and help in many ways we thank Hank Adams, Chris Anderson, Laurie Angel, Dale Belcher, Martha Bogert, the late Charles M. Bogert, William Bromberg, Ila Bromberg, Douglas Burkett, Howard Campbell, Steve Corn, Mark Doles, Lee Fitzgerald, Billy Gorum, Mark Hakkila, Wally Hausman, Toby Hibbitts, Patricia Holbrook, John Hubbard, Mariah Hughes, the late Dane Johnson, Mark Jordan, Larry Kamees, Jim Krupa, Letitia Peirce, Leland Pierce, Alicio Pino, Cecil Schwalbe, John Sherman, Don Sias, Richard Snearly, Tony Snell, Lex Snyder, Eddie Stegall, Dave Stricker, Jim Stuart, Gary Swinford, Barney Tomberlin, Michael L. Treadaway, Mike Vermillion, Roland Wauer, Brenda Williams, Larry David Wilson, and John Woodward.

Funding for this book was provided by a generous grant from the New Mexico Department of Game and Fish Share with Wildlife Program and Division of Public Affairs. We thank John Crenshaw, Don MacCarter, and Claire Tyrpak for their assistance. We are also indebted to a private donor who generously helped support publication of this book and to the taxpayers of New Mexico who support the Share with Wildlife Program. We hope they are all proud of this book.

Many individuals reviewed all or a portion of this manuscript. We sincerely appreciate the helpful comments of the following individuals: Kraig Adler (history); Marilyn J. Altenbach (plethodontid salamanders); Ralph W. Axtell (lizards); Royce E. Ballinger (lizards); Daniel D. Beck (Gila monster and rattlesnakes); Charles J. Cole (lizards and black-headed snakes); Roger Conant (entire manuscript); James R. Dixon (amphibians, lizards, venomous reptiles, and endangered species); Neil B. Ford (garter snakes); Fred R. Gehlbach (entire manuscript); John B. Iverson (turtles); Randy D. Jennings (leopard frogs and treefrogs); Thomas R. Jones (tiger salamander); Douglas A. Rossman (garter snakes and water snake); Damon T. Salceies (gray-banded kingsnake); Norman J. Scott, Jr. (amphibians, lizards, turtles, venomous reptiles, and endangered species); James F. Scudday (entire manuscript); Michael E. Seidel (turtles); Richard A. Seigel (all snakes); Wade C. Sherbrooke (horned lizards); Hobart M. Smith (entire manuscript); Nancy L. Staub (plethodontid salamanders); James N. Stuart (entire manuscript); Brian K. Sullivan (toads and treefrogs); Laurie J. Vitt (lizards); Robert G. Webb (entire manuscript); Bruce D. Woodward (toads).

To all of the curators and curatorial assistants who provided specimen records and verified specimens, we extend a special thanks. We particularly thank Robert P. Reynolds (USNM) for so carefully and diligently providing data on the large number of type specimens from New Mexico that are accessioned into the herpetology collections at the U.S. National Museum. For their time and many courtesies the following individuals (with institution symbolic codes) are gratefully acknowledged: Richard G. Zweifel and Charles J. Cole (AMNH); Edmond V. Malnate and J. E. Cadle (ANSP); Terry C. Maxwell (ASNHC); James P. Collins and Steven M. Norris (ASU); R.W. Van Devender and Matthew Rowe (ASUC); Stanley E. Trauth (ASUMZ); Craig Guyer (AUM); B.T. Clarke (BMNH); Cynthia A. Ramotnik (BS/FC); Robert Powell (BWMC); Wilmer W. Tanner and Skip Skidmore (BYU); Ronald Vasile (CA); Frank J. Deckert, Felix Hernandez III, and Edward M. Chamberlin (CACA); Jens V. Vindum (CAS); Clarence J. McCoy (deceased) and Ellen J. Censky (CM); Alan de Queiroz and Roxanna Normark (CU); Karl E. Krumke (DMNH); Ernie A. Liner, Private Collection (EAL); A. L. Gennaro (ENMU); Hymen Marx (FMNH); William Stanley and Paula M. Guthrie (HSU); Kevin S. Cummings (INHS); John O. Whitaker, Jr. (ISU); John B. Iverson, Private Collection (JBI); Joseph T. Collins, Adrian Nieto, and John E. Simmons (KU); John W. Wright (LACM); Douglas A. Rossman, David Cannatella, Jeff Boundy, Van Wallach, and David A. Good (LSUMZ); Laurence M. Hardy (LSUS); Billy J. Davis (LTU); Jose P. Rosado (MCZ); J. R. Choate (MHP); Max A. Nickerson

ACKNOWLEDGMENTS

and Gary S. Casper (MPM); Howard L. Snell, Paul A. Stone, Derrick W. Sugg, Alexis Schuler, and Allan J. Landwer (MSB); Dennis Parmley (MSUM); William H. Gutzke and Duane Cagle (MSUMZ); Harry W. Greene (MVZ); William M. Palmer (NCSM); Neil H. Douglas (NLU); Joseph L. LaPointe (deceased), Paul W. Hyder, and Geoffrey C. Carpenter (NMSU); Charles C. Carpenter, Laurie J. Vitt, and Janalee P. Caldwell (OMNH); Stanley F. Fox and George R. Cline (OSUS); Scott M. Moody (OUVC); Dennis R. Paulson (PSM); Ralph W. Axtell, Private Collection (RWA); Sherman A. Minton, Private Collection (SAM); Gregory K. Pregill (SDSNH); Ronald A. Brandon (SIUC); Jeff Boundy (SJSU); Bryce C. Brown (SMBU); J. Whitfield Gibbons (SREL); James F. Scudday (SRSU); Allan H. Chaney (TAIC); James R. Dixon (TCWC); David Cannatella and Tod W. Reeder (TNHC); John S. Mecham and Robert D. Owen (TTU); Harold A. Dundee (TU); James M. Walker (UADZ); Charles H. Lowe, Jr., and George Bradley (UAZ); Museum Staff (UCC); Hobart M. Smith (UCM); Robert Dubos and Frank J. Dirrigl, Jr. (UCS); Samuel S. Sweet and W. Bryan Jennings (UCSB); David L. Auth (UF-FSU); Benjamin Kaufman, Linda E. Maxson, Steven D. Sroka, and Thomas Uzzell (UIMNH); Arnold G. Kluge and Gregory Schneider (UMMZ); Dean E. Metter (UMOC); Patricia W. Freeman (UN); Michael D. Stuart (UNCA); Thomas H. Fritts, Robert P. Reynolds, and Steve W. Gotte (USNM); Jonathan A. Campbell, John Darling, and Carol K. Malcolm (UTA); Carl S. Lieb and Robert G. Webb (UTEP); Frank A. Iwen (UWZM); Bruce J. Hayward and Sandy Kruse (WNMU); Flavius C. Killebrew (WTSU); Willard D. Hartman and Fred C. Sibley (YPM).

We deeply appreciate the efforts by the staff of the University of New Mexico Press during the production of this book. David V. Holtby, the associate director, has given us continual help and encouragement from the very early stages of our work. Susan V. Niewiarowski designed the book and assisted in perfecting the artwork, maps, and photographs.

Finally we wish to thank all of those students of herpetology who have worked in New Mexico. We apologize to anyone who may have been overlooked in these acknowledgments.

A PHYSIOGRAPHIC SKETCH OF NEW MEXICO

Amphibians and reptiles are ectotherms, therefore their daily and seasonal activity along with their zoogeography are directly governed by temperature and moisture. Abiotic factors also exert profound influences on landforms and vegetation patterns, and thus on the substrates on which these animals live, the foods available for them to eat, and how they escape their enemies. The following account provides a backdrop on climate, geology, and vegetation for understanding the distribution and natural history of the herpetofauna of New Mexico, and depends heavily upon Maker et al. (1978), Morafka (1977), Bloom (1978), Dott and Batten (1981), Williams (1986), Van Devender (1986), Chronic (1987), and Dick-Peddie (1993) for content. Interested readers are invited to consult these sources and references therein for details.

Understanding the earth as a dynamic, evolving physical system of great antiquity, a concept first postulated by James Hutton in 1795 and conclusively demonstrated by the evidence gathered since then, is a key to understanding the current distribution of the herpetofauna of New Mexico. Energy is supplied to this open, steady-state system by solar radiation and the earth's own internal heat, the latter derived from cooling of its molten interior and radioactive decay. The earth's surface is divided into about 12 lithographic plates that are being driven apart along oceanic ridges and together along oceanic trenches by thermal convection in the earth's mantle. About 35% of this surface is continental crust, largely low-density rocks that "float" above the mantle, and the continents have been dragged erratically across the face of the planet by plate movements for at least 3.5 billion years. The currently familiar configuration of the globe began to develop about 200 million years ago with the breakup of the supercontinents Laurasia and Gondwanaland. Rock is added to the continents from sea-floor spreading and subduction at plate margins and from volcanic activity. The continental surface is constantly being sculpted into the mountains, plains, valleys, soils, and other landforms we see by climatic forces like wind and rain. The weather patterns these forces comprise are caused by the earth spinning on its tilted axis and orbiting the sun. Climate and landforms have a profound influence on global vegetation patterns, all of which in turn help determine the kinds of animals that inhabit a region.

A primary reason for the enchantment cast on those who live in and visit New Mexico is its landforms, which, as one might infer, have also undergone many changes. The oldest exposed rocks in New Mexico are Precambrian and occur in the cores of many of the state's mountain ranges, from the Taos Mountains (two billion years old) to the Franklin Mountains (one billion years old). Most of New Mexico was covered by shallow seas during the Paleozoic Era, 570–240 million years ago. Extensive limestone deposition occurred during this interval, and outstanding examples such as the Sacramento and Guadalupe mountains and associated karsts, salt deposits, and soils of southeastern New Mexico remain. The conspicuous sandstones and shales of the northern part of the state resulted from fluvial erosion and deposition during much of the Mesozoic Era, 240–100 million years ago.

Most of the coal deposits in New Mexico were formed between 100 million and 63 million years ago when another shallow sea repeatedly advanced and retreated from the east, supporting vast swampy habitats in which lush vegetation and associated dinosaurs flourished. The beginning of the Cenozoic Era 63 million years ago coincided with the initiation of the Laramide Orogeny, which formed the Rocky Mountains and the Colorado Plateau, modern landscapes in northern New Mexico. Most of the mountain ranges in the southwestern portion of New Mexico and many in the central part of the state resulted from prodigious volcanic activity, which began about 40 million years ago and lasted for 20 million years, associated with the collision of the Pacific and North American lithographic plates. Less extensive volcanic activity occurring until a few hundred years ago has left its imprint virtually everywhere throughout the state in landforms such as the Grants and Carrizozo malpais, Capulin Mountain, and Mount Taylor. Crustal weakening and longitudinal extension beginning about 30 million years ago created the Rio Grande Rift through which that river currently flows, and associated fault-block mountains and basins such as the Tularosa Basin in the southern half of the state. This area is still seismically active, as the junior author can remember awakening early one morning in Las Cruces to the peculiar and unnerving motion caused by the earth adjusting itself along a nearby fault. The ebb and flow of North American Pleistocene glaciation beginning about 2 million years ago, when most modern species or species groups of the continent's herpetofauna were already extant, etched the final surface veneer on New Mexico's physiography. At the zenith of the last glacial advance about 20,000 years ago the Sangre de Cristo Mountains were covered in alpine tundra and grassland, the Plains of San Augustin, the Estancia and Otero basins, and the Animas Valley contained lakes up to 512 km^2 in extent, spruce and fir trees covered what is now Albuquerque and piñon and pine trees were dominant in Carlsbad and Las Cruces. The Chihuahuan Desert did not then exist in New Mexico (Murray, 1957; Van Devender and Everitt, 1977).

A pattern of complex intermingling of regional biogeographic components is shared by New Mexico's flora and herpetofauna. The biota east of the Rio Grande Rift are derived from an eastward extension of the Great Plains. Rocky Mountain components extend into the northern half of the state and are also isolated at higher elevations in the south, and Chihuahuan Desert biota extend into southern New Mexico at lower elevations. A mosaic of biota from the latter two regions and Sierra Madrean and Sonoran Desert components exist in the southwestern corner of the state. Lower elevations of the San Juan Basin and surrounding areas in northwestern New Mexico share biota derived from the Great Basin. Several species of amphibians and reptiles in New Mexico with statewide distributions that appear at first glance to obscure these conventions, for example *Bufo woodhousii*, *Crotaphytus collaris*, *Sceloporus undulatus*, *Cnemidophorus tigris*, *Pituophis melanoleucus*, and *Crotalus viridis*, have well-marked geographic races that conform to the general biogeographic pattern.

The current climate and physiography of New Mexico is one of spectacular contrasts. The state covers about six degrees each of latitude and longitude. The Continental Divide meanders through the western tier of counties; the San Juan, Zuni, and Gila rivers flow to the Gulf of California whereas the Rio Grande, Pecos, and Canadian rivers flow to the Gulf of Mexico. Vast grassy plains stretching from the east meet the Rocky Mountains in the north and the Sierra Madre Occidental extending northward from Mexico in the south. The northern mountains include Wheeler Peak in Taos County, the highest point in the state at 4011 m. The Rio Grande and the Pecos River leave the state at 1148 m and 861 m, the latter the lowest point in New Mexico. The summit of Sierra Blanca in Otero County, the southernmost glaciated peak in the continental United States at 3627 m, is 32 km east of the floor of the Tularosa Basin, which at 1333 m contains some of the most arid lands in the United States.

New Mexico's high altitude (mean elevation 2436 m) leads to high insolation; the relatively less dense air heats up faster during the daytime (and cools faster at night). Temperature varies with latitude and altitude; mean temperature decreases about 1°C for every degree of latitude northward and about 3°C for every 300 m increase in altitude. New Mexico is a dry state; 90% of its surface receives less than 500 mm of rainfall annually. The state's continental position on the eastern aspect of the Rocky Mountain–Sierra Madre Occidental cordillera means that Pacific weather systems lose much of their moisture as rising air cools, precipitating on higher elevations at the expense of interior basins and lowlands. Average annual rainfall is 374 mm at Reserve and 482 mm at Chama, for instance, compared to 208 mm at both Albuquerque and Las Cruces. Moisture-laden air from the Gulf of Mexico, moving inland during the summer, similarly rises and cools; what precipitation it does bring falls on the eastern plains. Most of the precipitation in New Mexico occurs as convective summer thunderstorms of local high intensity and short duration. The intense heat at ground level causes air to rise explosively, expand and cool rapidly, and dump its moisture content in a very short time. Much of this moisture runs off as characteristic flash floods or sheet flow or is lost through evaporation. The annual rate of evaporation is 2–10 times greater than the average rainfall in the state. In addition, the angle at which the sun's rays strike the earth results in higher daytime temperatures and evaporation rates on south-facing slopes than north-facing slopes at a given altitude and latitude. Soils at high altitudes and latitudes are generally leached, well-developed, and acidic, whereas soils at low altitudes and latitudes are generally not leached, are not well-developed, and are neutral or alkaline. The pattern of life zones reflected in New Mexico's vegetation communities is a result of this climatic and landscape diversity. Relatively xeric-adapted plants grow at lower altitudes/latitudes, and relatively mesic-adapted plants grow at higher altitudes/latitudes.

Probably the most characteristic vegetation type in New Mexico is Desert Scrubland, which covers about 58,848 km^2 or 19% of the surface area of the state. This type can be subdivided into five categories, each with a suite of characteristic plant species. Montane Scrub occurs as small, isolated pockets distributed throughout the state, dominated by *Cercocarpus montanus*. Plains-Mesa Sand Scrub, dominated by *Quercus havardii* and *Artemisia filifolia*, occurs extensively on deep sands in the southeastern corner of New Mexico, and along the Rio Grande Valley, where the former species does not occur and the latter is the dominant. Great Basin Desert Scrub, dominated by *Artemisia* spp., *Atriplex* spp., and *Sarcobatus vermiculatus*, occurs extensively throughout the San Juan Basin and along the Taos/Rio Arriba county line. Chihuahuan Desert Scrub occurs throughout the southern tier of counties and is dominated by *Larrea tridentata* and *Flourensia cernua*. Amphibians and reptiles commonly associated with Desert Scrubland in New Mexico include *Scaphiopus couchii*, *Bufo punctatus*, *B. speciosus*, *Eleutherodactylus augusti*, *Coleonyx brevis*, *Gambelia wislizenii*, *Cophosaurus texanus*, *Phrynosoma modestum*, *Sceloporus arenicolus*, *S. graciosus*, *S. magister*, *Cnemidophorus gularis*, *C. tigris*, *C. uniparens*, *Bogertophis subocularis*, *Hypsiglena torquata*, *Masticophis taeniatus*, *Rhinocheilus lecontei*, *Sonora semiannulata*, *Trimorphodon biscutatus*, *Crotalus atrox*, and *C. scutulatus*.

Grassland is another important vegetation type in New Mexico, covering 125,088 km^2 or about 40% of the state. This type can be subdivided into Montane Grassland, Plains-Mesa Grassland, and Desert Grassland. Desert Grassland, covering about 54,589 km^2 (17%) throughout southern and western New Mexico, is considered by many ecologists to be an ecotone between grasslands and scrublands and shares many of the dominant plant species of the latter type. The dominant grass is *Bouteloua eriopoda*, but this category is highly variable seasonally and geographically in its floristic composition. Plains-Mesa Grassland extends over much of the eastern third of the state, and occurs in parts of west-central New Mexico as well. *Bouteloua gracilis* is a dominant species throughout, and co-dominates locally with *Agropyron smithii*, *Aristida* spp., *Bouteloua curtipendula*, *Buchloe dactyloides*, *Hilaria* spp., *Oryzopsis hymenoides*, or *Stipa* spp. Montane Grassland occurs in small patches between 2700–3500 m throughout the state, and is characterized by various bunch grasses *(Danthonia, Deschampsia, Festuca, Koeleria, Muhlenbergia, Poa)*, rushes *(Juncus)*, sedges *(Carex, Cyperus)*, and a variety of common forbs. Amphibians and reptiles commonly associated with grasslands in New Mexico include *Spea bombifrons*, *S. multiplicata*, *Bufo boreas*, *B. cognatus*, *B. debilis*, *Pseudacris triseriata*, *Gastrophryne olivacea*, *Rana blairi*, *Terrapene ornata*, *Holbrookia maculata*, *Sceloporus scalaris*, *Eumeces obsoletus, C. grahamii, C. sexlineatus, Coluber constrictor, Elaphe guttata, Gyalopion canum, Heterodon nasicus, Liochlorophis vernalis, Thamnophis radix, Crotalus viridis,* and *Sistrurus catenatus.*

A third major vegetation type in New Mexico is Woodland-Savanna, covering 71,823 km² (23%) of the state. There are two subdivisions, Coniferous-Mixed Woodland and Juniper Savanna, and they occur throughout the state except for the southeastern quarter. Juniper Savanna occurs as an ecotone at the upper elevational extent of grassland communities where edaphic conditions permit. The dominant plant species, in addition to grasses and/or shrubs, is *Juniperus monosperma,* although *J. deppeana, J. erythrocarpa, J. osteosperma,* or *J. scopulorum* may co-dominate or replace *J. monosperma* locally. Most woodlands in New Mexico are dominated by *Pinus edulis* and *Juniperus monosperma.* Other tree species which may co-dominate locally are *Pinus aristata, P. discolor, Juniperus* spp., *Quercus* spp., *Arbutus xalapensis, Arctostaphylos pungens,* and *Garrya wrightii.* Amphibians and reptiles often associated with woodlands and savannas in New Mexico include *Bufo punctatus, Elgaria kingii, Phrynosoma douglasii, Sceloporus clarkii, S. virgatus, Urosaurus ornatus, Eumeces tetragrammus, Cnemidophorus burti, C. exsanguis, C. flagellicaudus, C. sonorae, Diadophis punctatus, Lampropeltis pyromelana, Salvadora grahamiae, Senticolis triaspis, Thamnophis elegans, Crotalus molossus,* and *C. willardi.*

The fourth major vegetation type in New Mexico is Coniferous Forest, which covers about 52,750 km² (17%) of the state and is subdivided into montane and subalpine components. Subalpine Coniferous Forest occurs generally above 3000 m and is concentrated in north-central New Mexico. Dominant tree species are *Picea engelmannii* and *Abies lasiocarpa.* Montane Coniferous Forest occurs over wide areas in northern, western, and south-central New Mexico from 2100–3000 m, grading into Coniferous Woodland at lower elevations. The dominant tree species by descending altitude are *Pseudotsuga menziesii, Abies concolor, Picea pungens,* and *Pinus ponderosa,* with *Pinus arizonica, P. flexilis, P. leiophylla,* and *P. strobiformis* locally dominant and *Populus tremuloides* present throughout as a disturbance species. Amphibian and reptile species characteristic of this vegetation type in New Mexico are comparatively few, but include the endemic plethodontid salamanders *Aneides hardii* and *Plethodon neomexicanus* as well as *Bufo microscaphus, Hyla eximia, Phrynosoma douglasii,* and *Thamnophis elegans* at lower elevations. Although not a distinct classification as the preceding, riparian vegetation communities are an important correlate to the distribution of several amphibian and reptile species in New Mexico. Riparian areas range from perennial rivers and streams to dry arroyos that only carry water during major rainfalls and playas that hold such runoff intermittently. Many plant species (i.e., *Celtis reticulata, Chilopsis linearis, Brickellia* spp., *Eleocharis* spp., *Equisetum* spp., *Scirpus* spp.) are restricted to riparian situations, others (i.e., *Quercus* spp., *Prosopis* spp., *Fallugia paradoxa, Rhus* spp., *Sporobolus* spp.) grow elsewhere but not to the same size and/or density. All amphibians in New Mexico except *Aneides hardii, Plethodon neomexicanus,* and *Eleutherodactylus augusti* require temporary or permanent water for breeding. All turtles in the state except *Terrapene ornata* are aquatic or semiaquatic, and all except *Kinosternon flavescens* and *T. ornata* do not wander far from water. Several snakes are largely riparian in New Mexico, including *Nerodia erythrogaster, Thamnophis cyrtopsis, T. eques, T. marcianus, T. proximus, T. rufipunctatus,* and *T. sirtalis.*

The Spanish conquest and 400 years of subsequent settlement activities have caused profound changes in the vegetation patterns of New Mexico. The great riparian gallery forests of cottonwood (*Populus* spp.), walnut (*Juglans* spp.), sycamore (*Platanus wrightii*), and other trees were cut down. Farming and livestock grazing and the interruption of the natural hydrogeological cycle in floodplains through dam-building and water diversion prevented these species from reseeding themselves and allowed the establishment of alien species like Russian olive *(Elaeagnus angustifolia)* and salt cedar *(Tamarix* spp.). Extensive livestock grazing combined with fire suppression during the past 150 years have converted much of the state's savannas, grasslands, and desert-grasslands into piñon-juniper and juniper-sagebrush woodlands and desert-scrublands, while promoting the invasion and establishment of additional exotic species like Russian thistle *(Salsola kali)*. Most of these changes are naturally irreversible because the topsoils have eroded away, seed banks have been lost, and the hydrological characteristics of the remaining surfaces are unsuitable for the reestablishment of native grasses. Some species of New Mexico's herpetofauna (e.g., *Cnemidophorus tigris, C. uniparens, Cophosaurus texanus, Crotalus atrox*) have probably benefitted from these changes whereas others (e.g., *Cnemidophorus inornatus, Holbrookia maculata, Crotalus viridis*) probably have not (see discussion in Jones, 1981). Such large-scale habitat changes acting alone or synergistically with other anthropogenic factors like the introduction and establishment of exotic species (i.e., fish, bullfrogs) and their pathogens, pollution, habitat acidification, and depletion of ozone in the upper atmosphere have recently been implicated in regional declines or extinctions of populations of several wetland herpetofaunal species (Hayes and Jennings, 1986; Schwalbe and Rosen, 1988; Jennings and Scott, 1991; Dunson et al., 1992; Carey, 1993; Grant and Licht, 1993; Blaustein et al., 1994a,b). New Mexican species affected may include several species of *Rana, Bufo boreas, Thamnophis eques,* and *Kinosternon sonoriense*. Clearly humans are a significant factor shaping the earth's physiography, and the consequences for New Mexico's herpetofauna remain to be determined.

Color maps presenting topographic and vegetation data together with eleven color photographs of vegetation types are found following page 204.

A BRIEF HISTORY OF HERPETOLOGY IN NEW MEXICO

The first herpetologists in New Mexico were undoubtedly the Native Americans. Amphibians and reptiles were significant totems in their everyday and spiritual lives (e.g., Whipple, 1854 [1856]; Kennerly, 1856; Henderson and Harrington, 1914), and several species were seasonally utilized for food. Various forms were iconographed by the Anasazi and Mimbres cultures, and many of the Puebloan groups (Zia, Zuni, and Acoma among them) as well as the Apache, Navajo, and the Ute incorporated snakes, especially rattlesnakes, into their rituals (Klauber, 1972; C. Raish, pers. comm.). The first clear written reference to the herpetofauna of New Mexico by Europeans was of rattlesnakes made by members of the Spanish colonial expedition led by Father Augustín Rodríguez and Francisco Sánchez Chamuscado in 1581 (Klauber, 1972). New Mexico remained herpetologically *terra incognito* until well into the 19th century. Following the War of 1812, the U.S. Army created a division of topographical engineers, in part to explore the western frontier. The first exploring expedition under this aegis to enter what is now New Mexico was led by Major Stephen H. Long to the Rocky Mountains (James, 1823). A small detachment of this expedition spent about two weeks of August 1820 traversing the Mora and Canadian river valleys in the northeastern corner of the state. The only herpetological references made during this portion of the journey were to the ubiquitous rattlesnake *(Crotalus viridis)* and to two forms of "orbicular lizards" *(Phrynosoma cornutum* and *P. douglasii)*, said to be quite common.

Rising tensions between the United States and Mexico provided the next opportunity for an addition to the printed record of New Mexico herpetology. In August and September 1845, Lieutenant James W. Abert of the newly formed Corps of Topographical Engineers spent about two weeks exploring the Canadian River valley from Raton Pass to the Texas border and commented on the abundant rattlesnakes *(C. viridis)* associated with prairie dog colonies along the way (Abert, 1846). The following year both Lieutenant Abert and First Lieutenant William H. Emory accompanied the military expedition of Colonel Stephen W. Kearny against Mexico which entered New Mexico along the Santa Fe Trail on August 6, although Lieutenant Abert was delayed approximately 5 weeks in Colorado by illness. During this expedition, both men recorded only meager herpetological observations (Abert, 1848; Emory, 1848). First Lieutenant Emory noted the abundance of "*Agama cornuta*" (actually *P. douglasii*) near the junction of the Vermejo and Canadian rivers on 9 and 11 August and, continuing on with Colonel Kearny to California, an encounter on October 15 with a reddish *C. atrox* near Las Palomas on the Rio Grande. Lieutenant Abert remained in northern New Mexico, visiting various sites including El Rito, Acoma, and Socorro, leaving New Mexico in mid-January 1847. He noted "great numbers of horned lizards" *(P. douglasii)* near Lamy on 29 September, and he obtained a specimen of a "singular lizard" *(Crotaphytus collaris?)* between Bernalillo and the mouth of the Jemez River on October 13. The disposition of this specimen is unknown. In 1846, Adolphus Wislizenus, an emigrant Swiss naturalist, embarked on one of the first privately funded natural history explorations of the West. He commented on the general abundance of lizards on his sojourn down the Santa Fe Trail and the Rio Grande Valley between 23 June and 8 August 1846 (Wislizenus, 1848).

The decade prior to the Civil War marked the birth of New Mexico as a center of herpetological interest. Following the end of the Mexican War in 1848, several military expeditions were dispatched to survey the international boundary with Mexico and to locate the best route for a transcontinental railroad. Spencer Fullerton Baird, Assistant Secretary of the nascent Smithsonian Institution, arranged to have naturalists attached to many of these expeditions (Dall, 1914; Adler, 1989), and the material they collected and returned to Washington revealed the richness of the herpetofauna of the American Southwest. Sixty-one (37%) of the 166 taxa of amphibians and reptiles that occur in New Mexico were described from specimens obtained by these expeditions. The first, led by Captain Lorenzo Sitgreaves, explored the Zuni River and the Little Colorado River to its junction with the Colorado River and beyond, traveling over 900 miles from Santa Fe to Fort Yuma between 15 August and 24 November 1851 (Sitgreaves, 1853). The expedition traveled up the Rio Puerco and Rio San José from the Rio Grande to Laguna Pueblo, and thence westward via Acoma Pueblo, Inscription Rock, and Zuni Pueblo to the Zuni River, leaving New Mexico along the course of the Zuni River on 25 September. The United States and Mexican Boundary Survey, led by Major William H. Emory, consisted of a series of separate expeditions conducted between 1851 and 1855 to determine the international boundary following the treaty of Guadalupe Hidalgo in 1848 and the Gadsden Purchase in 1853 (Emory, 1857, 1859). In 1853, Congress authorized surveys to find the best route for a transcontinental railroad from the Mississippi River to the Pacific Coast. The results of these surveys were published as a 12-volume treatise with multiple authors between 1854 and 1859 (e.g., Pope, 1854 [1855]). The expedition led by Lieutenant A. W. Whipple entered New Mexico along the Canadian River on 21 September 1853, traveled up Pajarito Creek and across Arroyo Cuerbito and the Gallinas River to the Pecos River at Anton Chico on 27 September and, crossing there, reached Albuquerque via Galisteo and Peña Blanca on 5 October. On 8 November, the expedition traveled south to Isleta Pueblo, crossed the Rio Grande and followed the approximate route of the Sitgreaves expedition, leaving New Mexico on 28 November. Whipple (1854 [1856]) commented in general terms several times on the success of the zoological collecting by members of his party. Another expedition, led by Lieutenant John G. Parke, crossed the Peloncillo Mountains into New Mexico through Stein's Pass on 6 March 1854 and traveled the old Boundary Commission Road past Fort Webster and Cooke's Spring, reaching the Rio Grande in the vicinity of Leasburg on 12 March. After a few days' rest at Fort Fillmore, a short excursion was made along the present route of I-10 to the vicinity of Deming, returning on 21 March. Parke (1855) noted that many of his specimens were lost because the containers in which they were stored leaked, a common fate not shared by contemporary bird, mammal, or botanical collections. A third expedition, led by Brevet Captain John Pope, departed eastward from Doña Ana on 12 February 1854 and, passing through Soledad Canyon in the Organ Mountains on 14 February, traveled via the Hueco, Alamo, and Cornudas mountains through Guadalupe Pass to the Delaware River and down the latter to its junction with the Pecos River on 8 March. Pope (1854 [1855]) noted on 7 March the expedition "killed a rattle-snake (the first we have yet seen) on a hill near camp. It was put in spirits and carried along." A small detachment was sent eastward on 9 March when it traveled south and crossed into Texas.

Relatively little herpetological activity took place in New Mexico during the near-century following the establishment of territorial boundaries in 1863 and prior to World War II. A series of government-sponsored expeditions to explore the geographic and geological features west of the 100th meridian, led by First Lieutenant George M. Wheeler, U.S. Army Corps of Engineers, took place from 1871 to 1874. The results were published in a 6-volume treatise, one of which contained the zoology of the expeditions. Two chapters (Coues, 1875; Yarrow, 1875) treated New Mexican herpetology, and Yarrow (1875) reported 1 salamander, 8 anurans, 21 lizards, 17 snakes, and 1 turtle from the state. Other published references during this period (Cope, 1883b, 1896; Garman; 1887; Townsend, 1893; Stone and Rehn, 1903; Bailey, 1905; MacBride, 1905; Ellis, 1917) provided general information or cursory notes on specimens collected in the state. Personnel associated with the U.S. Biological Survey (1896–1939) procured a number of specimens, and Bailey (1913) listed 76 species (3 salamanders, 7 anurans, 31 lizards, 32 snakes, and 3 turtles) in New Mexico and provided the first extensive statewide discussion of the ecological characteristics and zoogeography of individual species. Van Denburgh (1924) followed with a checklist containing 85 species (2 salamanders, 12 anurans, 30 lizards, 36 snakes, and 5 turtles).

The founding of New Mexico State University in 1880 and the USDA's Jornada Experimental Range nearby in 1912, along with the University of New Mexico in 1889, resulted in a concentration of herpetological activity in the Rio Grande Valley (i.e., Cockerell, 1896; Herrick et al., 1899; Little and Keller, 1937). Other noteworthy accounts of limited focus during this period included Ruthven (1907), who spent a month in Alamogordo and collected between White Sands (National Monument) to the west and Cloudcroft in the nearby Sacramento Mountains. He provided extensive ecological annotations along with photographs of representative habitats for the 2 species of anurans, 13 lizards, 2 snakes, and 1 turtle he encountered. Mosauer (1932) spent three weeks in Dark Canyon of the Guadalupe Mountains, and provided observations on 1 salamander, 1 anuran, 7 lizards, and 3 snakes.

World War II and the advent of global post-war tensions changed the face of New Mexico. Major military and supporting civilian installations were built, and existing roadways were paved or new ones built to transport personnel and materials between them as well as to and from other states. As a result, more of the landscape became accessible to natural history exploration. The routine use of motorized vehicles made it easy to get to many previously unexplored places as well as providing a direct mechanism for collecting specimens (e.g., Klauber, 1939b; Campbell, 1953, 1956). New Mexico's population increased substantially in postwar decades (Williams, 1986), as did visitation by amateur and professional herpetologists from around the United States. Research programs or research collections in herpetology were established at New Mexico colleges during this period. The arrival of William J. Koster at the University of New Mexico in 1938 was especially noteworthy. Koster was an ichthyologist, although in keeping with his broadly based Cornell training he was an excellent naturalist, and due to his association with A. H. Wright had a special interest in herpetology. His collections of amphibians and reptiles, although incidental to his fish collections, resulted in over 5000 carefully preserved and stored specimens. His publications in herpetology included first state records for a number of species. James S. Findley, trained at the University of Kansas, arrived in 1955 and became the second vertebrate biologist on the University of New Mexico faculty. Findley was primarily a mammalogist although he and his students collected numerous amphibians and reptiles and published in herpetology. William G. Degenhardt arrived from Texas A. & M. University in 1960, resulting in the development of the division of herpetology and undergraduate and graduate programs. During this same time period other programs in herpetology developed at New Mexico State University (James R. Dixon, Walter G. Whitford, Joseph L. LaPointe), Western New Mexico University (Bruce J. Hayward) and Eastern New Mexico University (A. L. "Tony" Gennaro). This activity resulted in a substantial increase in the knowledge of the distribution of New Mexico's herpetofauna (Fig. 15), as well as in the number of publications similar to that seen in other western states and Mexico (e.g., Webb, 1970b; Dixon, 1987; Carpenter and Krupa, 1989; Smith and Smith, 1973, 1976, 1979 [1980], 1993). Many of these citations are in the individual species accounts; additional notes from this period include Bugbee (1942) and Bragg and Dundee (1949 [1950]). Noteworthy among the contributions that focused on regional herpetofaunas within the state during this period are Lewis (1950; Organ Mountains and adjacent Tularosa Basin, 34 species), Harris (1963; San Juan Basin, 21 species), Gehlbach (1965; Zuni; Zuni Mountains, 30 species), Jones (1970; Chaco Canyon National Monument, 18 species), Mecham (1979; Guadalupe Mountains, 50 species) and

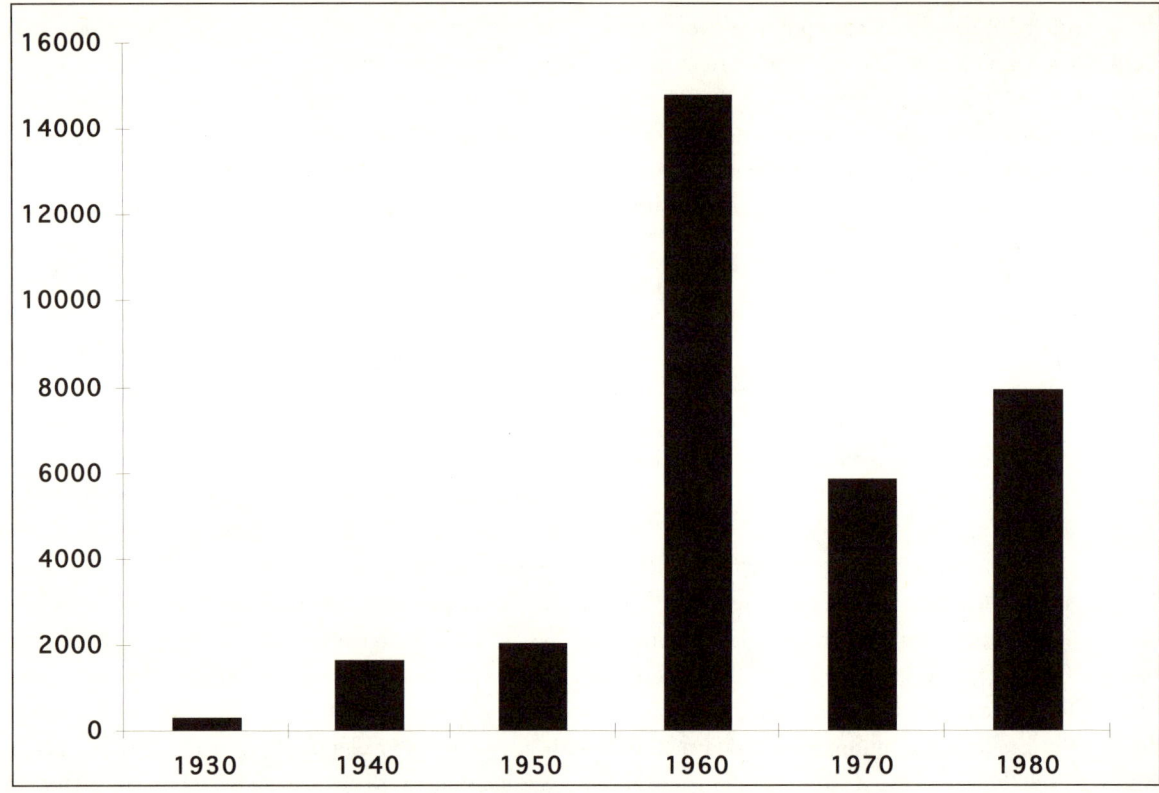

Figure 15 *Number of specimens added to The University of New Mexico Herpetology Collection by decade.*

Best et al. (1983; Pedro Armendariz lava field, Sierra and Socorro counties, 26 species). Important studies of taxonomic assemblages within New Mexico during this period included Zweifel (1968b), Creusere and Whitford (1976), and Woodward (1982c, 1983, 1987a) for anurans; Degenhardt and Christiansen (1974) for turtles; Tanner (1975), Whitford and Creusere (1977), Gehlbach (1979), Creusere and Whitford (1982), and Baltosser and Best (1990) for lizards; and Price and LaPointe (1990) for snakes.

The efforts of many of the explorers and naturalists who have come before us were indeed heroic, and the authors of this book can truly be said to be standing on the shoulders of giants in presenting our view of the herpetology of New Mexico. Part of the enchantment which is New Mexico lies in the rich and unique diversity of amphibians and reptiles contained within its borders. New Mexicans are the prime stewards of this natural diversity and can contribute much to the understanding of life on this planet by continuing to encourage the study of herpetology in our state. We are hopeful that this book will contribute to our collective efforts to maintain the beauty and integrity of New Mexico for human generations yet to come.

A CHECKLIST OF THE AMPHIBIANS AND REPTILES OF NEW MEXICO

ORDER CAUDATA — Salamanders

Family Ambystomatidae — Mole salamanders

Ambystoma tigrinum mavortium Baird, 1850.
Barred tiger salamander

Ambystoma tigrinum nebulosum Hallowell, 1852 [1853]. Arizona tiger salamander

Family Plethodontidae — Lungless Salamanders

Aneides hardii (Taylor, 1941).
Sacramento mountain salamander

Plethodon neomexicanus Stebbins and Riemer, 1950.
Jemez Mountains salamander

ORDER ANURA — Frogs and Toads

Family Pelobatidae — Spadefoots

Scaphiopus couchii Baird, 1854. Couch's spadefoot
Spea bombifrons (Cope, 1863). Plains spadefoot
Spea multiplicata (Cope, 1863). New Mexico spadefoot

Family Leptodactylidae — Tropical Frogs

Eleutherodactylus augusti latrans (Cope, 1880).
Eastern barking frog

Family Bufonidae — Toads

Bufo alvarius Girard *in* Baird, 1859. Colorado River toad
Bufo boreas Baird and Girard, 1852. Boreal toad
Bufo cognatus Say *in* James, 1823. Great Plains toad
Bufo debilis insidior Girard, 1854. Western green toad
Bufo microscaphus microscaphus Cope, 1866 [1867]. Arizona toad
Bufo punctatus Baird and Girard, 1852.
Red-spotted toad
Bufo speciosus Girard, 1854. Texas toad
Bufo woodhousii australis Shannon and Lowe, 1955.
Southwestern Woodhouse's toad
Bufo woodhousii woodhousii Girard, 1854. Woodhouse's toad

Family Hylidae — Treefrogs

Acris crepitans blanchardi Harper, 1947. Blanchard's cricket frog
Hyla arenicolor Cope, 1866. Canyon treefrog
Hyla eximia Baird, 1854. Mountain treefrog
Pseudacris triseriata triseriata (Wied-Neuwied, 1838).
Western chorus frog
Pseudacris triseriata maculata (Agassiz, 1850).
Boreal chorus frog

Family Microhylidae — Narrowmouth Toads

Gastrophryne olivacea (Hallowell, 1856 [1857]).
Great Plains narrowmouth toad

Family Ranidae — True Frogs

Rana berlandieri Baird, 1859.
Rio Grande leopard frog
Rana blairi Mecham, Littlejohn, Oldham, Brown and, Brown, 1973. Plains leopard frog
Rana catesbeiana Shaw, 1802. Bullfrog
Rana chiricahuensis Platz and Mecham, 1979.
Chiricahua leopard frog
Rana pipiens Schreber, 1782. Northern leopard frog
Rana yavapaiensis Platz and Frost, 1984. Lowland leopard frog

ORDER TESTUDINES — Turtles

Family Chelydridae — Snapping Turtles

Chelydra serpentina serpentina (Linnaeus, 1758).
Common snapping turtle

Family Emydidae — Box and Water Turtles

Chrysemys picta bellii (Gray, 1831). Western painted turtle
Pseudemys gorzugi Ward, 1984. Western river cooter
Terrapene ornata luteola Smith and Ramsey, 1952.
Desert box turtle
Terrapene ornata ornata (Agassiz, 1857).
Ornate box turtle

Trachemys gaigeae (Hartweg, 1939). Big Bend slider
Trachemys scripta elegans (Wied-Neuwied, 1838).
Red-eared slider

Family Kinosternidae — Mud Turtles
Kinosternon flavescens flavescens (Agassiz, 1857).
Yellow mud turtle
Kinosternon sonoriense sonoriense LeConte, 1854.
Sonoran mud turtle

Family Trionychidae — Softshell Turtles
Trionyx muticus muticus LeSueur, 1827.
Midland smooth softshell
Trionyx spiniferus emoryi (Agassiz, 1857).
Texas spiny softshell
Trionyx spiniferus hartwegi (Conant and Goin, 1948).
Western spiny softshell

ORDER SQUAMATA — Lizards and Snakes

SUBORDER SAURIA — Lizards

Family Crotaphytidae — Collared and Leopard Lizards
Crotaphytus collaris auriceps Fitch and Tanner, 1951.
Yellowheaded collared lizard
Crotaphytus collaris baileyi Stejneger, 1890.
Western collared lizard
Crotaphytus collaris collaris (Say *in* James, 1823).
Eastern collared lizard
Crotaphytus collaris fuscus Ingram and Tanner, 1971.
Chihuahuan collared lizard
Gambelia wislizenii punctata (Tanner and Banta, 1963).
Pale leopard lizard
Gambelia wislizenii wislizenii (Baird and Girard, 1852).
Longnose leopard lizard

Family Phrynosomatidae — Zebratail, Earless, Spiny, Tree, Side-blotched, and Horned Lizards
Callisaurus draconoides ventralis (Hallowell, 1852).
Arizona zebratail lizard
Cophosaurus texanus scitulus (Peters, 1951).
Southwestern earless lizard
Holbrookia maculata approximans Baird, 1858 [1859].
Speckled earless lizard
Holbrookia maculata elegans Bocourt, 1874.
Western earless lizard
Holbrookia maculata maculata Girard, 1851.
Northern earless lizard
Holbrookia maculata ruthveni Smith, 1943.
Bleached earless lizard
Phrynosoma cornutum (Harlan, 1825).
Texas horned lizard
Phrynosoma douglasii hernandesi (Girard, 1858).
Mountain short-horned lizard
Phrynosoma modestum Girard in Baird and Girard, 1852.
Roundtail horned lizard
Phrynosoma solare Gray, 1845. Regal horned lizard
Sceloporus arenicolus Degenhardt and Jones, 1972.
Sand dune lizard
Sceloporus clarkii clarkii Baird and Girard, 1852.
Sonoran spiny lizard
Sceloporus graciosus graciosus Baird and Girard, 1852. Northern sagebrush lizard
Sceloporus jarrovii jarrovii Cope *in* Yarrow, 1875.
Yarrow's spiny lizard
Sceloporus magister bimaculosus Phelan and Brattstrom, 1955.
Twin-spotted spiny lizard
Sceloporus magister cephaloflavus Tanner, 1955.
Orangeheaded spiny lizard
Sceloporus poinsetti poinsetti Baird and Girard, 1852.
Crevice spiny lizard
Sceloporus scalaris slevini Smith, 1937.
Bunch grass lizard
Sceloporus undulatus consobrinus Baird and Girard, 1853.
Southern prairie lizard
Sceloporus undulatus cowlesi Lowe and Norris, 1956.
White Sands prairie lizard
Sceloporus undulatus elongatus Stejneger, 1890.
Northern plateau lizard
Sceloporus undulatus erythrocheilus Maslin, 1956.
Red-lipped prairie lizard
Sceloporus undulatus garmani Boulenger, 1882.
Northern prairie lizard
Sceloporus undulatus tedbrowni Smith, Bell, Applegarth, and Chiszar, 1992. Mescalero sand dunes prairie lizard
Sceloporus undulatus tristichus Cope *in* Yarrow, 1875.
Southern plateau lizard
Sceloporus virgatus Smith, 1938. Striped plateau lizard
Urosaurus ornatus levis (Stejneger, 1890). Canyon tree lizard
Urosaurus ornatus linearis (Baird, 1859). Lined tree lizard
Urosaurus ornatus schmidti (Mittleman, 1940).
Big Bend tree lizard
Urosaurus ornatus wrighti (Schmidt, 1921). Northern tree lizard
Uta stansburiana stejnegeri Schmidt, 1921.
Desert side-blotched lizard
Uta stansburiana uniformis Pack and Tanner, 1970.
Colorado side-blotched lizard

Family Gekkonidae — Geckos
Coleonyx brevis Stejneger, 1893. Texas banded gecko
Coleonyx variegatus bogerti Klauber, 1945.
Tucson banded gecko
Hemidactylus turcicus (Linnaeus, 1758). Mediterranean gecko

Family Teiidae — Whiptails
Cnemidophorus burti stictogrammus Burger, 1950.
Giant spotted whiptail

CHECKLIST OF AMPHIBIANS AND REPTILES

Cnemidophorus dixoni Scudday, 1973.
Gray-checkered whiptail

Cnemidophorus exsanguis Lowe, 1956.
Chihuahuan spotted whiptail

Cnemidophorus flagellicaudus Lowe and Wright, 1964.
Gila spotted whiptail

Cnemidophorus grahamii Baird and Girard, 1852.
Checkered whiptail

Cnemidophorus gularis gularis Baird and Girard, 1852.
Texas spotted whiptail

Cnemidophorus inornatus gypsi Wright and Lowe, 1993.
Little white whiptail

Cnemidophorus inornatus heptagrammus Axtell, 1961.
Trans-Pecos striped whiptail

Cnemidophorus inornatus juniperus Wright and Lowe, 1993.
Woodland striped whiptail

Cnemidophorus inornatus llanuras Wright and Lowe, 1993.
Plains striped whiptail

Cnemidophorus neomexicanus Lowe and Zweifel, 1952.
New Mexico whiptail

Cnemidophorus sexlineatus viridis Lowe, 1966.
Prairie lined racerunner

Cnemidophorus sonorae Lowe and Wright, 1964.
Sonoran spotted whiptail

Cnemidophorus tigris gracilis Baird and Girard, 1852.
Southern whiptail

Cnemidophorus tigris marmoratus Baird and Girard 1852.
Western marbled whiptail

Cnemidophorus tigris reticuloriens Vance, 1978.
Eastern marbled whiptail

Cnemidophorus tigris septentrionalis Burger, 1950.
Northern whiptail

Cnemidophorus uniparens Wright and Lowe, 1965.
Desert grassland whiptail

Cnemidophorus velox Springer, 1928.
Plateau striped whiptail

Family Scincidae — Skinks

Eumeces multivirgatus epipleurotus Cope, 1880. Variable skink

Eumeces obsoletus (Baird and Girard, 1852). Great Plains skink

Eumeces tetragrammus callicephalus Bocourt, 1879.
Mountain skink

Family Anguidae — Alligator Lizards

Elgaria kingii nobilis Baird and Girard, 1852.
Arizona alligator lizard

Family Helodermatidae — Venomous Lizards

Heloderma suspectum suspectum Cope, 1869.
Reticulate Gila monster

SUBORDER SERPENTES — Snakes

Family Leptotyphlopidae — Blind Snakes

Leptotyphlops dulcis dissectus (Cope, 1896).
New Mexico blind snake

Leptotyphlops humilis segregus Klauber, 1939.
Trans-Pecos blind snake

Family Colubridae — Colubrids

Arizona elegans elegans Kennicott *in* Baird, 1859.
Kansas glossy snake

Arizona elegans philipi Klauber, 1946.
Painted desert glossy snake

Bogertophis subocularis subocularis (Brown, 1901).
Trans-Pecos rat snake

Coluber constrictor flaviventris Say *in* James, 1823.
Eastern yellowbelly racer

Coluber constrictor mormon Baird and Girard, 1852.
Western yellowbelly racer

Diadophis punctatus arnyi Kennicott, 1859.
Prairie ringneck snake

Diadophis punctatus regalis Baird and Girard, 1853.
Regal ringneck snake

Elaphe guttata emoryi (Baird and Girard, 1853).
Great Plains rat snake

Elaphe guttata meahllmorum Smith, Chiszar, Staley, and Tepedelen, 1994. Southern plains rat snake

Gyalopion canum Cope, 1860. Western hooknose snake

Heterodon nasicus kennerlyi Kennicott, 1860.
Mexican hognose snake

Heterodon nasicus nasicus Baird and Girard, 1852.
Plains hognose snake

Hypsiglena torquata jani (Dugès, 1865).
Texas night snake

Hypsiglena torquata loreala Tanner, 1944.
Mesa Verde night snake

Lampropeltis alterna (Brown, 1901).
Gray-banded kingsnake

Lampropeltis getula splendida (Baird and Girard, 1853).
Desert kingsnake

Lampropeltis pyromelana pyromelana (Cope, 1866 [1867]).
Arizona mountain kingsnake

Lampropeltis triangulum celaenops Stejneger, 1902 [1903].
New Mexico milk snake

Liochlorophis vernalis blanchardi (Grobman, 1941).
Western smooth green snake

Masticophis bilineatus bilineatus Jan, 1863.
Sonoran whipsnake

Masticophis flagellum piceus (Cope, 1892).
Red coachwhip

Masticophis flagellum testaceus (Say *in* James, 1823).
Western coachwhip

Masticophis taeniatus taeniatus (Hallowell, 1852).
Desert striped whipsnake

Nerodia erythrogaster transversa (Hallowell, 1852).
Blotched water snake

Pituophis melanoleucus affinis Hallowell, 1852.
Sonoran gopher snake

Pituophis melanoleucus deserticola Stejneger, 1893.
Great Basin gopher snake

Pituophis melanoleucus sayi (Schlegel, 1837). Bullsnake

Rhinocheilus lecontei lecontei Baird and Girard, 1853.
Western longnose snake

Rhinocheilus lecontei tessellatus Garman, 1883.
Texas longnose snake

Salvadora deserticola Schmidt, 1940.
Big Bend patchnose snake

Salvadora grahamiae grahamiae Baird and Girard, 1853.
Mountain patchnose snake

Senticolis triaspis intermedia (Boettger, 1883).
Green rat snake

Sonora semiannulata Baird and Girard, 1853.
Ground snake

Tantilla hobartsmithi Taylor, 1936 [1937].
Southwestern black-headed snake

Tantilla nigriceps Kennicott, 1860. Plains black-headed snake

Tantilla yaquia Smith, 1942. Yaqui black-headed snake

Thamnophis cyrtopsis cyrtopsis (Kennicott, 1860).
Western blackneck garter snake

Thamnophis elegans arizonae Tanner and Lowe, 1989.
Arizona garter snake

Thamnophis elegans vagrans (Baird and Girard, 1853).
Wandering garter snake

Thamnophis eques megalops (Kennicott, 1860).
Mexican garter snake

Thamnophis marcianus marcianus (Baird and Girard, 1853).
Checkered garter snake

Thamnophis proximus diabolicus Rossman, 1963.
Arid land ribbon snake

Thamnophis radix haydenii (Kennicott *in* Cooper, 1860).
Western plains garter snake

Thamnophis rufipunctatus (Cope *in* Yarrow, 1875).
Narrowhead garter snake

Thamnophis sirtalis dorsalis (Baird and Girard, 1853).
New Mexico garter snake

Trimorphodon biscutatus lambda Cope, 1885 [1886].
Sonoran lyre snake

Trimorphodon biscutatus vilkinsonii Cope, 1885 [1886].
Texas lyre snake

Tropidoclonion lineatum (Hallowell, 1856 [1857]). Lined snake

Family Elapidae — Elapids

Micruroides euryxanthus euryxanthus (Kennicott, 1860).
Arizona coral snake

Family Viperidae — Vipers

Crotalus atrox Baird and Girard, 1853.
Western diamondback rattlesnake

Crotalus lepidus klauberi Gloyd, 1936. Banded rock rattlesnake

Crotalus lepidus lepidus (Kennicott, 1861).
Mottled rock rattlesnake

Crotalus molossus molossus Baird and Girard, 1853.
Blacktail rattlesnake

Crotalus scutulatus scutulatus (Kennicott, 1861).
Mojave rattlesnake

Crotalus viridis cerberus (Coues, 1875).
Arizona black rattlesnake

Crotalus viridis nuntius Klauber, 1935.
Hopi rattlesnake

Crotalus viridis viridis (Rafinesque, 1818).
Prairie rattlesnake

Crotalus willardi obscurus Harris and Simmons, 1976.
New Mexico ridgenose rattlesnake

Sistrurus catenatus edwardsii (Baird and Girard, 1853).
Desert massasauga

A KEY TO THE TADPOLES AND SALAMANDER LARVAE OF NEW MEXICO

By Ronn Altig

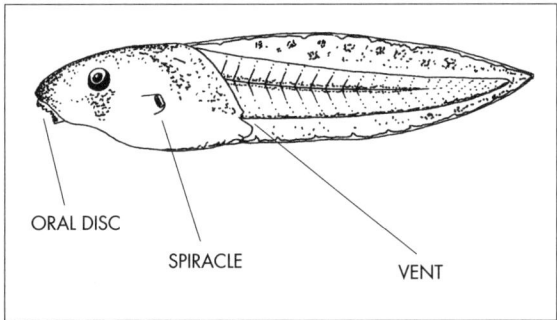

Structural features of generalized tadpole.

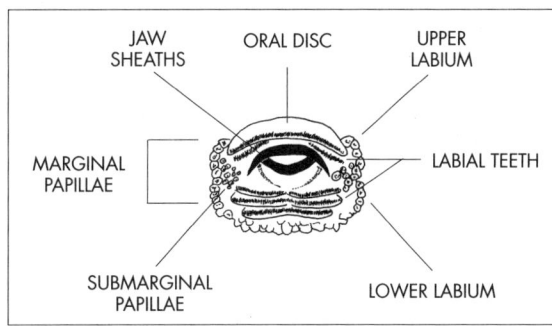

Structural features of generalized tadpole mouthparts.

1. Larvae found within egg capsules in terrestrial sites (e.g., under logs, among rock talus, in burrows, etc.); miniature salamander or frog exists without a free-living larval stage   2
1. Free-living larvae found in either lentic or lotic aquatic sites; metamorphosis produces a terrestrial salamander or frog well after hatching   4
2. Small gills not noticeable without magnification; tail with flimsy musculature and wide, membranous fins; if development is advanced, embryo looks like a small frog with its hind legs tucked beneath body. (Leptodactylidae)   *Eleutherodactylus augusti*
2. Gills large but very transparent; tail looks like typical salamander tail; if development is advanced, embryo looks like a small salamander curved laterally in egg capsules (Plethodontidae) 3
3. Collected above 2130 m in Jemez Mountains; Los Alamos, Rio Arriba, and Sandoval counties   *Plethodon neomexicanus*
3. Collected above 2400 m in Capitan, White, or Sacramento mountains; Lincoln and Otero counties   *Aneides hardii*
4. Large external gills present at head-body junction; body elongate with four legs (except at stages near hatching); mouth large and extends backward to a point beyond the eye; salamander larvae (Ambystomatidae)   *Ambystoma tigrinum*
4. Large external gills absent; body globular; hind legs visible throughout much of development, but forelegs develop beneath the skin and are not visible until the approach of metamorphosis; mouth small and surrounded by a disc of fleshy tissue; frog and toad tadpoles   5
5. Keratinized jaw sheaths and labial teeth absent; hemispherical oral flaps pendant over mouth; spiracle mid-ventral on belly; eyes lateral; body rounded in dorsal view, depressed in lateral view (Microhylidae)   *Gastrophryne olivacea*
5. Keratinized jaw sheaths and labial teeth present; oral flaps absent but oral disc surrounding mouth present; spiracle on left side of body; eyes lateral or dorsal; body not rounded in dorsal view, may be somewhat depressed in lateral view   6
6. Narrow or wide dorsal gap in marginal papillae present, ventral gap absent; nostrils not large; eyes dorsal or lateral, vent medial or dextral   7

6. Wide dorsal and ventral gaps in marginal papillae; nostrils large for size of tadpole; eyes dorsal; vent medial. (Bufonidae) *Bufo* 9

7. Dorsal gap in marginal papillae wide; spiracle definitely on left side; eyes dorsal or lateral; tooth row A-1 long; vent dextral 8

7. Dorsal gap in marginal papillae narrow (width of 3–6 papillae); spiracle low on left side; eyes dorsal; tooth row A-1 much shorter than A-2; vent medial (Pelobatidae) *Spea* and *Scaphiopus* 16

8. Eyes dorsal; lateral margin of oral disc emarginate; marginal papillae large (Ranidae) *Rana* 18

8. Eyes dorsal (*Acris, Hyla*, part) or lateral (*Hyla*, part and *Pseudacris*); lateral margin of oral disc not emarginate; marginal papillae small (Hylidae) 23

9. Labial tooth row formula 2/2 *Bufo debilis*

9. Labial tooth row formula 2/3 10

10. Dorsum of body intensely black 11

10. Dorsum of body not intensely black, often lightly pigmented; tail often patterned dark on light 13

11. Tail musculature black on top, white ventrally *Bufo woodhousii*

11. Tail musculature totally black in lateral view 12

12. Medial gap in row A-2 about 7% of length of lateral part of row; found statewide except extreme north-central region; to about 1.5 cm total length *Bufo punctatus*

12. Medial gap in row A-2 about 30% of length of lateral part of row; found in small area in north-central Rio Arriba County; to about 2.5 cm total length *Bufo boreas*

13. P-3 short, much shorter than P-2 *Bufo cognatus*

13. P-3 about same length as P-2 14

14. Lateral parts of tail muscle unicolored; extreme southwestern part of state (Hidalgo County only) *Bufo alvarius*

14. Lateral parts of tail muscle with some contrasting dark pattern 15

15. In lateral view snout is abruptly rounded; Catron, Grant, Sierra, and Socorro counties in southwestern part of state *Bufo microscaphus*

15. In lateral view, snout slopes uniformly; Chaves, Eddy, and Lea counties in southeastern part of state *Bufo speciosus*

16. Margin of upper jaw sheath forms uniform arc; jaw musculature not enlarged 17

16. Margin of upper jaw sheaths with a median cusp; enlarged jaw musculature visible through lightly pigmented skin cannibal morphs–*Spea bombifrons* and *S. multiplicata*

17. Black knob on roof of mouth *Spea bombifrons* and *S. multiplicata*

17. Black knob on roof of mouth absent *Scaphiopus couchii*

18. Body dorsum and tail greenish with discrete black dots; jaw sheaths narrow; nostrils minute; *catesbeiana* group *Rana catesbeiana*

18. Body and tail with mottling and blotches but not as above; jaw sheaths wide and robust; nostrils not minute; *Rana pipiens* group 19

19. Oral disc width 22% or more of body length *Rana yavapaiensis*

19. Oral disc width less than 22% of body length 20

20. Belly musculature prominent, obscures intestine *Rana chiricahuensis*

20. Belly musculature not prominent to not visible; gut coil easily visible 21

21. Tail patterned with discrete dark and pale spots or sharply defined reticulations enclosing pale blotches *Rana berlandieri*

21. Tail pattern obscure or mottled with indistinct blotches 22

22. Body unspotted; tail muscle exceptionally wide; iris in life without a cross-like pattern *Rana blairi*

22. Body usually spotted; tail muscle narrow; iris in life with cross-like pattern *Rana pipiens*

23. Labial tooth row formula (LTRF) 2/2; tail tip typically black; spiracular tube long and free from body for most of length *Acris crepitans*

23. LTRF 2/3; tail tip not black; spiracular tube not long and free from body 24

24. P-3 as long as P-2 and longer than the transverse length of the upper jaw sheath; dorsal fin not much higher than the plane of the upper body surface; collected in backwater areas of rocky streams *Hyla arenicolor*

24. P-3 considerably shorter than P-2 and the transverse length of the upper jaw sheath; dorsal fin distinctly higher than the plane of the upper body surface; collected in temporary pools and flooded grassy areas 25

25. Marginal papillae uniserial midventrally on lower labium *Hyla eximia*

25. Marginal papillae biserial midventrally on lower labium *Pseudacris triseriata*

A KEY TO THE SALAMANDERS OF NEW MEXICO

1. No external gills present   2
1. External gills present (Fig. 1); tail with high tail fin; aquatic   *Ambystoma tigrinum* (larval stage)
2. Toes of adpressed limbs do not overlap (Fig. 2)   3
2. Toes of adpressed limbs overlap   *Ambystoma tigrinum*
3. Costal grooves (Fig. 3) number 14–15; hind limbs with 5 toes; Sacramento Mountains, SE New Mexico   *Aneides hardii*
3. Costal grooves number 18–19; 5th toe on hind foot absent or greatly reduced; Jemez Mountains, north-central New Mexico   *Plethodon neomexicanus*

Figure 1

Figure 2

Figure 3

FAMILY AMBYSTOMATIDAE   *Mole Salamanders*

The family Ambystomatidae contains about 34 species in two genera. They occur from southern Alaska and Canada throughout the United States and most of Mexico. They are among the largest of the terrestrial salamanders, often with bright color patterns in sharp contrast with the dark ground color. They are squat-bodied with smooth skin, a laterally compressed tail, and well-developed limbs. In most species, the eggs are laid in water and hatch into free-swimming larvae called waterdogs that are commonly used for fish bait. Two species of *Ambystoma* are known to reproduce by parthenogenesis. The adults are fossorial, spending almost their entire life underground. They are rarely seen on the surface except during the breeding season. New Mexico has one genus with one species and two subspecies.

Some authorities recognize the family Dicamptodontidae, with two genera, *Dicamptodon* and *Rhyacotriton*, as distinct from this family (Lanza et al., 1992).

AMBYSTOMA TIGRINUM (Green, 1825)   *Tiger Salamander*

See Color Plate 1.

Type: Jacob Green described *Ambystoma tigrinum* in 1825 with the type locality given as "near Moore's town in New Jersey." No type specimen is known to exist.

Distribution: The natural range of *Ambystoma tigrinum* extends from the east to west coasts of North America, and from southern Canada to Puebla, Mexico. It is absent as a native from most of the Great Basin, most of the Pacific Coast, the Mojave and Colorado deserts, the Appalachian region, and south Florida (Stebbins, 1985). (See Remarks.) In New Mexico, *A. tigrinum* occurs statewide from 900–3355 m wherever suitable habitat is available.

Description: This large, distinctively marked salamander has four stout, well-developed legs. There are four fingers and five toes that are flattened and pointed and without claws. There are no scales covering the moist smooth skin. The broad head has dorsally placed small eyes. There is a prominent gular fold. The lateral body muscles have a segmental arrangement, forming distinct vertical grooves. The dorsal color varies from brownish black to shiny black, grading to light gray ventrally. There are usually yellowish bars or spots on the dark background in the lowland forms, although montane individuals may be unicolored or nearly so.

Non-transformed (neotenic) adults and larvae have elongate, laterally compressed bodies and tails, three pairs of external plume-like gills and well-developed tail fins. The mouth of the larva is broad as it is in the adult. The overall color may be related to substrate color and water transparency (Fernandez and Collins, 1988). Maximum total length of terrestrial adults is about 345 mm (Smith and Reese, 1968). Neotenic adults may reach 400 gm (Snow, 1978) and 385 mm TL (Smith, 1949; Larson, 1974).

Similar Species: With its distinctive coloration and body form, *A. tigrinum* should not be confused with any other amphibian in New Mexico. The aquatic larvae may be distinguished from tadpoles by the plumose gills and four legs.

Systematics: Six or seven subspecies are currently recognized (Collins, 1990; Gehlbach, 1967a). Two of these, *A. t. mavortium* Baird, 1850 and *A. t. nebulosum* Hallowell, 1852 [1853] occur in New Mexico. *Ambystoma t. mavortium* was described as *A. mavortia* from a specimen collected by Dr. Wislizenus, with "New Mexico" given as the type locality. *Ambystoma t. nebulosum* was described as *A. nebulosum*, also with "New Mexico" given as the type locality. ("New Mexico" actually referred to the Territory of New Mexico prior to the division of Arizona and New Mexico.) Hallowell (1853) later restricted the type locality of *A. nebulosum* to "San Francisco Mountain" (= San Francisco Peaks, Coconino County, Arizona). Dunn (1940) first used the trinomials.

Collins et al. (1980) discussed the confusion encountered in the classification of the subspecies of *A. tigrinum* and questioned the taxonomic reality of a single, widely distributed, polytypic species. Pierce and Mitton (1980), in a study of the patterns of allozyme variation of the subspecies *A. t. mavortium* and *A. t. nebulosum* along the Front Range in Colorado, revealed significant genetic divergence between these two forms. In contrast to Pierce and Mitton (1980), Jones and Collins (1992) sampled the same taxa along a contact zone in central New Mexico and found no fixed allelic differences between the two subspecies. They found that limited gene flow apparently does occur. They suggested further study of the phylogenetic relationships among all races of *A. tigrinum* before taxonomic revisions in this group are attempted.

Habitat: *Ambystoma tigrinum* occupies a wide variety of habitats provided there is nonflowing water nearby for breeding. In New Mexico, specimens are known to occur from desert scrub at elevations of 900–1500 m to spruce-fir forest at 2500–3000 m (Delson and Whitford, 1973). Bragg (1941) reported specimens from 3355 m elevation near the Taos County line. Temporary rain pools, stock ponds, fish hatcheries, beaver ponds, and deep clear lakes may be used for breeding. Tom R. Jones (pers. comm.) found larvae in the intermittent pools of a creek in the Gila National Forest, and Woodhouse (1854) reported larvae from a large spring near the Zuni Pueblo. Bodies of water with predatory fish or crawfish may occasionally be used for breeding although the eggs and larvae may soon be eliminated. Aquatic vegetation may or may not be present. In optimal habitat, aquatic tiger salamanders can be very abundant with hundreds of individuals in the same pond. During dry periods, the terrestrial adults remain underground in rodent burrows or take shelter within large logs, under rocks, boards, or other surface debris. Wet or rainy weather may stimulate terrestrial activity.

Ambystoma tigrinum does not seem to have been seriously threatened by urbanization or agricultural activity. They are found within large metropolitan areas and in areas of intense farming, provided pollution levels are not too high.

Behavior: Adult *A. tigrinum* are primarily burrowing animals and are above ground in great numbers only during the wet weather in early spring when they migrate to breeding ponds. Individuals usually estivate in the ground or in rodent burrows when their breeding ponds dry up. Webb (1969) found 10 subadult salamanders buried 15–30 cm deep within the mud cracks when a cattle tank near Las Cruces dried up.

Whitford and Massey (1970) reported the responses of larval *A. tigrinum* to thermal and oxygen gradients in a natural situation in south-central New Mexico. They found that in the winter when water temperatures were higher in the shallow water during the day, the salamanders moved to the periphery of the pond and remained there. As temperatures in the shallow water cooled in the early evening, the salamanders returned to deep, warmer areas in the middle of the pond. In the absence of thermal and oxygen gradients, movements of salamanders during the daylight hours were not confined to one particular area of the pond and were greater than on days when gradients were established.

Rose and Armentrout (1976) reported cannibal morphs of *A. tigrinum* that have disproportionally large heads, wide mouths, and enhanced development of vomerine teeth. Pedersen (1993) compared skull growth in cannibal and non-cannibal *A. tigrinum* morphs. Experiments by Collins and Cheek (1983) suggested that larval density stimulated expression of these cannibalistic traits. Pfennig et al. (1991a) suspected that certain pathogens may also affect the incidence of cannibalism. The cannibal morphs they studied contained higher levels of pathogenic bacteria and parasitic nematodes, which were likely acquired from the conspecific victims they consumed. These pathogens may cause the death of the cannibals. Thus, Pfennig et al. (1991a) suggested that intraspecific predation may be costly enough to constrain an organism's tendency to produce certain morphological traits. This may help explain why cannibal morphs are rare. Cannibalism is nevertheless an adaptive advantage to this species because the increased food availability allows more rapid growth and earlier metamorphosis (Lannoo and Bachmann, 1984).

Growth rates are dependent primarily on temperature and food availability. In high-elevation populations of *A. tigrinum* in the Rocky Mountains, temperature of the aquatic habitat determined the body size and timing of metamorphosis (Bizer, 1978).

Reproduction: Males and females at low elevations likely reproduce at one year of age (Bishop, 1941), although at high elevations in New Mexico, individuals are small and not sexually mature during their first year (T. R. Jones, pers. comm.). The courtship pattern of *A. tigrinum* is similar to that of most ambystomatids, with such activity taking place in the water. Breeding activity is induced by rising temperatures combined with saturation of the ground from snow melt and/or spring rains. With these environmental cues, large numbers of tiger salamanders emerge from their wintertime underground retreats and migrate to breeding pond sites. During this migration, they can become entrapped in ditches, wells, or gas and water meter boxes. The breeding season is short with the males arriving at the pond first, followed by the females. In *A. tigrinum*, there is no actual physical contact between the sexes during the transfer of sperm. As the male first becomes aware of a potential mate, he approaches her, nudges and rubs the uncaptured female, and then forcefully pushes her through the water to a spermatophore. Males usually deposit several spermatophores at scattered localities and then may repeat the courtship sequence several times (Duellman and Trueb, 1986). The female secures the sperm by picking up the head portion of the spermatophore between the lips of her cloaca.

The aquatic eggs of *A. tigrinum* are subject to predation primarily by aquatic invertebrates, especially in temporary pond situations. Aquatic predators include leeches, caddisflies, (Dalrymple, 1970), and adult and larval *A. tigrinum*. There are also various egg predators that invade and consume these eggs. Salamander larvae are subject to heavy predation in permanent ponds by crawfish and fishes, and in temporary ponds by insects such as diving beetles and dragonfly larvae (Holomuzki, 1986). Coyotes and bobcats are known to feed on transforming larvae in drying temporary ponds (Webb and Roueche, 1971). Other vertebrate predators include *Rana catesbeiana*, *Thamnophis* spp., turtles, gulls, wading birds, and raccoons. Anderson et al. (1971) reported only 3.3% survival of the eggs and larvae of *A. tigrinum* studied in New Jersey. They felt that such high mortality may be characteristic of pond-dwelling amphibians of the temperate region.

Among salamanders, fecundity is highest in species that deposit their eggs in water, much lower in those having terrestrial eggs, and lowest in those that give birth to living young. The largest clutches in salamanders are those produced by salamanders of the genus *Ambystoma* (Duellman and Trueb, 1986). For example, Rose and Armentrout (1976) estimated that more than 5,000 eggs are laid by adult *A. tigrinum* in west Texas. After being fertilized, eggs are laid singly or in rows or small clusters on vegetation or debris 5–25 cm below the surface of the water. Eggs may be laid from mid-March to mid-August in the plains, and generally later at higher elevations (Hammerson, 1982). Eggs hatch 2–5 weeks after being laid, taking longer at higher elevations (Sexton and Bizer, 1978). Gehlbach (1965) reported that larval development may take as long as two years in certain populations in northwest New Mexico. According to Powers (1903, 1907) and Reese (1969) size at transformation averages 130–140 mm. However, Roth et al. (1991) described a recently transformed individual from northwestern Colorado (UCM 56294) that was 36 mm SVL and 57 mm TL and retained no trace of external gills or gill clefts. This is the smallest known specimen of *A. tigrinum* at which complete loss of the external branchial apparatus is known.

Reproduction in New Mexico of *A. t. mavortium* appears temporally flexible, so that breeding, although typically in summer, may occur at any time when sufficient water is present. *Ambystoma t. nebulosum* appears to breed only in the spring (Jones and Collins, 1992). Readers are referred to Webb and Roueche (1971) for a discussion of embryonic development and larval growth rates of an isolated population of *A. tigrinum* in south-central New Mexico.

Food Habits: Larger larvae, especially the cannibalistic morphs, will eat almost any animal they can catch and swallow, including snails, leeches, earthworms, crawfish, tadpoles, amphibian eggs, and mice (Hammerson, 1982; Collins, 1982). Painter (1985) reported small unidentified fish and a scorpion as food items. Dodson and Dodson (1971) reported various aquatic insect larvae, including water boatmen, backswimmers, dragonflies, mayflies, caddisflies, mosquitoes, midges, and true bugs. The diet of the cannibalistic morphs studied by Holomuzki and Collins (1987) contained 84% conspecific larvae. However Pfennig (pers. comm. *in* Fellman, 1995) suggested that these cannibal morphs avoid eating their brothers, sisters, and first cousins. Ontogenetic changes in the diet of *A. tigrinum* larvae have been reported by Brophy (1980) and Holomuzki and Collins (1987). As the larvae grow there is a significant increase in the variety of foods taken. The small intestine is about 25% shorter in adults than in the larvae (Tilley, 1964).

Remarks: *Ambystoma tigrinum* larvae are often referred to as waterdogs and are commonly used as fish bait. As a result, numerous introductions of various subspecies have been made in New Mexico. Dundee (1988) discussed this problem in Louisiana.

Harte and Hoffman (1989) investigated the possible effects of acidic deposition on a Rocky Mountain population of *A. tigrinum*. Their experimental population declined by 65% over seven years. They implicated acidic deposition for this decline, although they stated that natural population fluctuation may also be responsible. They found *A. tigrinum* eggs to have an LD-50 at pH 5.6.

Thompson and Jones (1992) reported on the occurrence of paedomorphic cave-dwelling *A. tigrinum* in southeastern New Mexico, just north of Roswell in Chaves County. Maldonado-Koerdell and Firschein (1947) were the first to establish the occurrence of the subspecies *A. t. nebulosum* east of the Rio Grande. Gehlbach (1967a) reviewed this species.

FAMILY AMBYSTOMATIDAE

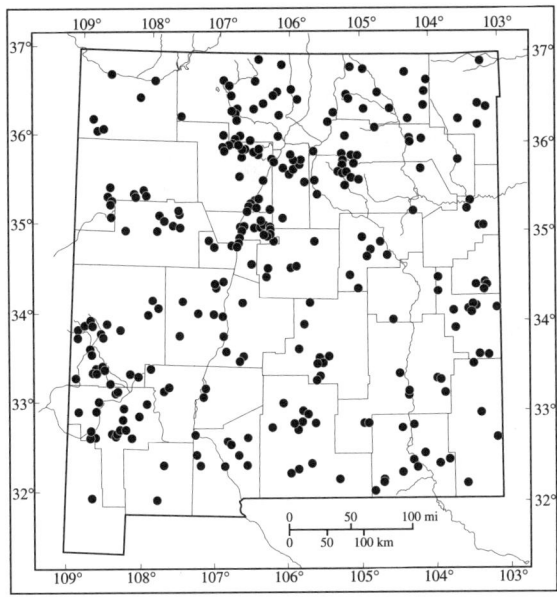

Distribution of
Ambystoma tigrinum
in New Mexico.

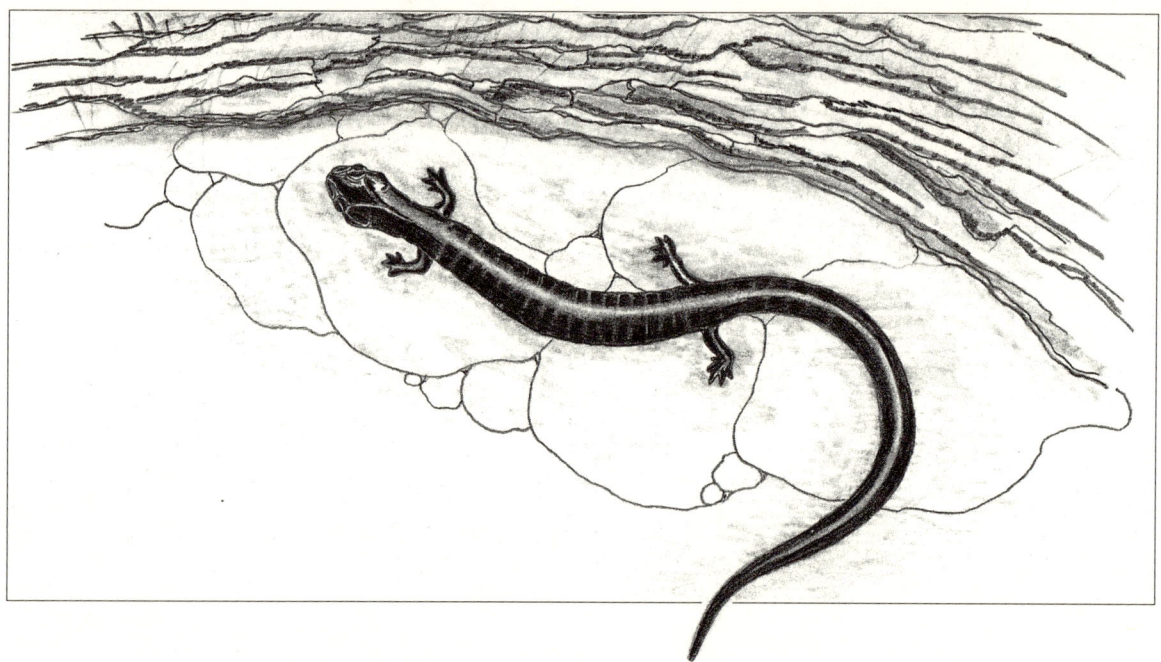

Lungless Salamanders

FAMILY PLETHODONTIDAE

Lungless salamanders in the family Plethodontidae make up the largest family in the order Caudata and include about 60% of the known living species of salamanders. About 240 species in 30 genera are found in the Americas from Nova Scotia and extreme southeastern Alaska to central Bolivia and eastern Brazil. A few species are European, inhabiting Sardinia, the southeast of France, and Italy from the Maritime Alps to the central Apennines (Lanza et al., 1992).

Two endemic plethodontid salamanders occur in New Mexico. These unique salamanders lack lungs, they breath through mucous membranes in the mouth and throat and through the skin. The grape-like clusters of eggs are laid in rotted logs or deep within crevices in the soil. After hatching, the juveniles are small replicas of the adults. Both species in New Mexico are found on isolated mountain ranges, and both have been the subject of much concern as increasing commodity demands are placed on the shrinking coniferous forests of the Southwest. They are currently listed as Notice of Review species by the U.S. Fish and Wildlife Service (USDI, 1994) and as endangered by the New Mexico Department of Game and Fish (NMGF, 1990).

ANEIDES HARDII (Taylor, 1941) — *Sacramento Mountain Salamander*

See Plate 2.

Type: Edward H. Taylor (1941a) described *Aneides hardii* as *Plethodon hardii* from an adult male collected by D. E. Hardy on 29 June 1940. The holotype is FMNH 100103. The type locality is "Sacramento Mountains at Cloudcroft (9000 ft.), New Mexico."

Distribution: *Aneides hardii* is endemic to south-central New Mexico where it is found at high elevations in the Capitan, White, and Sacramento mountains of Lincoln and Otero counties. Schad et al. (1959) are given credit as the first to report *A. hardii* from the Capitan and White mountains, although Findley (1959), perhaps unaware of the earlier report by Schad et al. (1959), also reported a collection of 20 specimens that he thought were the first from the Capitan Mountains. Findley's specimens were collected 2–3 July 1958 whereas the specimens reported by Schad et al. were collected 6–9 August 1958. *Aneides hardii* is locally abundant only where essential microhabitat characteristics are available.

Description: *Aneides hardii* is blackish brown to brown dorsally and is often heavily mottled with greenish gray to bronze; the venter is light in color. Tail length averages slightly shorter than the body length. There are 14–15 costal grooves. The limbs are short with 2–4.5 costal folds between the toe tips of the adpressed limbs. The digits are short and only slightly expanded distally. The head is somewhat triangular when viewed from above, with the temporal region slightly swollen by the jaw musculature, especially in mature males. Mature males average 48.4 mm SVL; females 44.5 mm. Hatchlings resemble the adults and first appear on the surface at approximately 15 mm SVL.

Similar Species: In New Mexico, only *Plethodon neomexicanus* is similar to *A. hardii*. These species are easily separated on the basis of geography; *P. neomexicanus* is endemic to the Jemez Mountains of north-central New Mexico. *Plethodon neomexicanus* has 18–20 costal grooves and 7.5–8.5 intercostal folds between the toe tips of the adpressed limbs.

Systematics: No subspecies of *Aneides hardii* are recognized. Lowe (1950) first assigned this species to the genus *Aneides*. The three isolated populations show little morphological (Schad et al., 1959), skeletal (Wake, 1965), or genetic (Pope and Highton, 1980) variation.

Habitat: *Aneides hardii* is generally associated with Douglas fir and spruce at elevations from 2400–3570 m where it is usually found under large woody debris or rocks. Moir and Smith (1970) found *A. hardii* at an elevation of 3570 m in the vicinity of Sierra Blanca Peak in the White Mountains. This represents the only known occurrence of any North American salamander in tundra habitat. Dominant overstory species in *A. hardii* habitat include Douglas fir and white fir with lesser amounts of Engelmann spruce and southwestern white pine. Rocky Mountain maple, gooseberry, and oceanspray share the understory with seedling conifers and downed logs in various stages of decay. Common components of the ground cover are usually various low-growing herbs, fine woody debris, and moss. There are usually limestone rocks and boulders exposed on the surface (Scott and Ramotnik, 1992).

Ramotnik and Scott (1988) analyzed the habitat of *A. hardii* by comparing transects with and without salamanders. They found these transects to differ in several respects: transects with salamanders occurred at significantly higher elevations, on shallower slopes, and had higher numbers of spruce and lower numbers of pine than transects without salamanders. An analysis of size classes of fir and spruce revealed that transects with salamanders had significantly higher densities of large fir and all size classes of spruce. Weigmann et al. (1980) provided additional habitat data for this species. Ninety-five percent (589/620) of *A. hardii* discovered by Ramotnik and Scott (1988) were distributed among four cover classes as follows: coarse woody debris (64%), rocks (22%), bark (10%), and fine woody debris (4%). Additionally, 14 salamanders were found under aspen logs and 17 were above or below the surface litter.

Behavior: *Aneides hardii* is active on the surface from mid-June to mid- to late September, although most activity occurs during summer rains in July and August. Other times of the year are spent underground. Staub

(1986) found that marked animals moved an average of 22.7 m. There was great variation in the distances moved by individual salamanders; two were recaptured in their original log, while others moved between 4–50 m. Whitford (1968) found the critical thermal maximum to be only 33.3°C and suggested that this low temperature reflected the low ambient temperatures characteristic of its microhabitat. Mean body temperatures of 19 *A. hardii* tested in laboratory experiments by Carey (1988) averaged 11.3°C (6.1–15.9).

Staub (1993) related the presence of scarring on the body to intraspecific agonistic behavior in the genus *Aneides*. In *A. hardii*, the frequency of scarred individuals was only 4% (6 of 145 individuals examined), thus indicating very little apparent aggressive behavior in wild individuals.

Reproduction: Reproduction in *A. hardii* was extensively studied by Williams (1972, 1976, 1978). Mature males have an average length of 48.4 mm (39–59) SVL. Males are larger than females and mature in two years at about 39 mm. Mature females average 44.5 mm (36–53) SVL. Females reach maturity at about 36 mm SVL and two years of age, although an additional year is required for the ova to reach sufficient size for oviposition. Oviposition in *A. hardii* occurs every three years (Williams, 1978).

Two reproductive classes of adult female *A. hardii* are discernible based on average ova diameter. In Class I, females would not lay eggs during the current summer. Ova diameter ranged from 1.2 mm in late June to 2.6 mm in early September. The number of maturing eggs averaged 10.1 mm, with about five in each ovary. In Class II, females are considered gravid and would mate and oviposit their eggs by the beginning of the next active season. Ova diameter ranged from 2.4 mm in late June to 4.3 mm in early September. The number of maturing eggs averaged 7.7, with equal numbers in each ovary. A third class of females may be present. These are the females presumed to be underground or within logs with their egg clutches. These females would move into Class I the next summer.

In males, primary spermatocytes are probably formed first in spring and are found until late July. Secondary spermatogonia are first formed in late July and are present until the following spring. Spermatids are probably formed in late spring and are found until early September. Spermatozoa are first formed in mid-July and are probably transferred to the vas deferens in early fall (Williams, 1978).

Mating occurs in early June, and probably takes place underground. *Aneides hardii* produces one of the smallest clutches of eggs known among salamanders (Staub, 1986). Lowe (1950) found three eggs in a large, moist, decomposing Douglas fir log on a steep north-facing slope in August. Schwartz (1955) found 10 eggs and an accompanying female on 12 August. The eggs were each suspended by a stalk from the upper surface of a cavity in a Douglas fir log and had an average diameter of 8.1 mm (7.4–9.1 mm). Johnston and Schad (1959) found four egg clutches, one on 14 July and three on 27 July. The clutches contained six, four, four, and one eggs with an average egg diameter of 6.4 mm. Staub (1986) summarized clutch data from the literature and her personal observations. In 19 clutches, egg number ranged from 1–10 and averaged 5.9 eggs per clutch; nine of 19 clutches were discovered in September. On 15 June 1987, eight juvenal salamanders (15–16 mm SVL) were found under a piece of bark near Wofford Lookout Tower in the Sacramento Mountains (M. J. Altenbach, pers. comm.). Staub (1986) reported an individual 14.9 mm SVL taken in September. We assume these salamanders represent the approximate size at hatching.

Food Habits: Various arthropods and other invertebrates are eaten, including annelid worms, mollusks, collembolans, beetles, ants, and small wasps. Johnston and Schad (1959) reported the following percentages of the total food items taken from a sample of 16 salamanders: 40.2% ants, 30.9% beetles, 9.7% spiders, 8.4% mites, 4.4% mollusca, 3.2% unidentified material, 1.3% orthopterans, 1.3% wasps, and 0.6% hemipterans.

Remarks: Because of the presumed threats from various silvicultural activities, primarily logging, *Aneides hardii* is listed as a Category 2, Notice of Review species by the U.S. Fish and Wildlife Service (USDI, 1994), and as endangered by the New Mexico Department of Game and Fish (NMGF, 1990). Staub (1986) reported that logged areas had significantly fewer salamanders than unlogged areas and that the logged areas were significantly hotter and had less ground and canopy cover than the unlogged sites. Ramotnik and Scott (1988) suggested that intensive logging, slash removal, and burning may reduce or eliminate populations of *A. hardii*. They suggested that long-term, intensive studies of salamander populations throughout the logging cycle are necessary to provide information needed to make informed management decisions regarding this species. Scott and Ramotnik (1992) reported on such studies before and after logging of mixed-conifer habitat and found no detectable short-term (0.5–3.5 years) effect on populations of *A. hardii*. They stated that although these populations persist through the first intensive cutting, they may not survive a repeated 10-year logging cycle. More frequent cutting may destroy the present subterranean habitat structure that is vital to the salamander's survival. Interested readers are referred to Ramotnik and Scott (1988) and Scott and Ramotnik (1992) for further discussion on the persistence of amphibian populations after habitat alteration.

Wake (1965) reviewed this species.

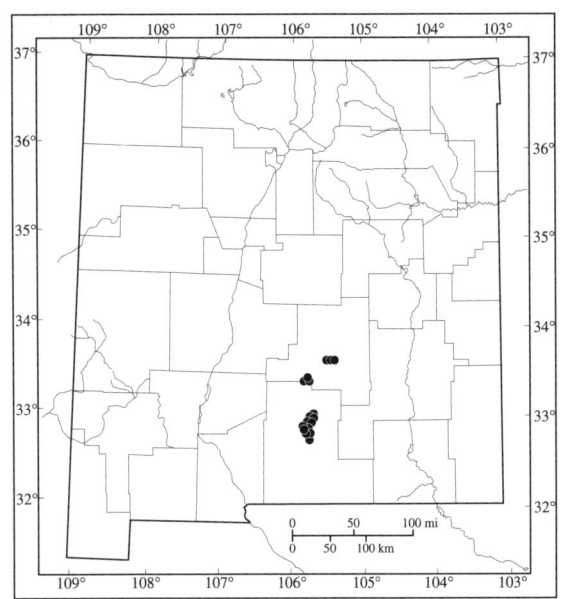

Distribution of Aneides hardii *in New Mexico.*

PLETHODON NEOMEXICANUS Stebbins and Riemer, 1950   *Jemez Mountains Salamander*

See Plate 3.

Type: The holotype of *Plethodon neomexicanus* is MVZ 49033, an adult male collected by Robert C. Stebbins on 14 August 1949 from "12 miles west and 4 miles south of Los Alamos, 8750 +/- feet, Sandoval County, New Mexico." The first specimen of *Plethodon neomexicanus* was actually collected in the Jemez Mountains by Junius Henderson and identified as *Spelerpes multiplicatus* (= *Eurycea multiplicata*) (Bailey, 1913). This locality in the Jemez Mountains was recorded as a westward extension of its range (Dunn, 1926). Stebbins and Riemer (1950) correctly reidentified the specimen, USNM 42921, as *P. neomexicanus*.

Distribution: *Plethodon neomexicanus* is endemic to north-central New Mexico where it is found only in the Jemez Mountains in portions of Los Alamos, Sandoval, and Rio Arriba counties. It is locally common only in areas where essential microhabitat exists.

Description: *Plethodon neomexicanus* is uniformly dark brown above, with occasional fine gold stippling dorsally. The venter is sooty gray, being lighter on the chin and on the underside of the tail. The head may be slightly wider than the body, especially in sexually mature males. The body form is slender and elongate, with 18–20 costal grooves. There are 7.5–8.5 costal grooves between the toe tips of the adpressed limbs. The fifth toe is much reduced, projecting only slightly beyond the foot web. The mental gland of males is not evident. Brodie and Altig (1967) discussed morphological variation in *P. neomexicanus*; Dwyer and Hanken (1990) reported on the limb skeletal variation. Total length of 296 specimens with undamaged tails collected during 1992 and 1994 averaged 82.3 mm (30.9–134.4) (our unpubl. data).

Similar Species: In New Mexico, only *Aneides hardii* is similar to *P. neomexicanus*. These species are easily separated on the basis of geography; *A. hardii* is endemic to the Sacramento, Capitan, and White mountains of south-central New Mexico. *Aneides hardii* has 14–15 costal grooves and 2–4.5 intercostal folds between the toe tips of the addressed limbs.

Systematics: No subspecies of *Plethodon neomexicanus* are recognized.

Habitat: *Plethodon neomexicanus* is generally found on loose rocky soils between 2200 and 2900 m. They occur in and under rotting coniferous logs or under rocks on both flat areas and steep slopes. They are rarely observed on the surface or encountered under bark, surface litter, or aspen logs. The habitat is coniferous forest dominated by Douglas fir, blue spruce, Engelmann spruce, ponderosa pine, and white fir with occasional aspen, Rocky Mountain maple, New Mexico locust, oceanspray, and various shrubby oaks. Ramotnik and Scott (1988), investigating the habitat requirements of *P. neomexicanus*, found that transects along which they found salamanders occurred on significantly steeper slopes and at lower elevations than transects lacking salamanders. Ninety-six percent (141/147) of the salamanders encountered were distributed among four cover classes as follows: coarse woody debris (68%), rocks (27%), and fine woody debris (1%); no specimens were encountered under bark.

Behavior: *Plethodon neomexicanus* spends most of its life underground; surface activity depends upon warm temperatures and favorable moisture conditions, which are necessary for cutaneous respiration. It seldom leaves the shelter of rotted logs or rocks, although darkness and sufficient moisture may enable it to travel short distances overland. Whitford (1968) found the critical thermal maximum to be only 33.5°C and suggested that this low temperature reflects the low ambient temperatures characteristic of its microhabitat.

Reproduction: Reproduction in *P. neomexicanus* was extensively studied by Williams (1976, 1978) and much of the following information, unless otherwise credited, is based on his work.

Females are larger than males with sexual maturity reached at 3–4 years in females and 3 years in males. Sexually mature males average 55.2 mm SVL (47–63 mm); females 56.2 mm SVL (49–67 mm). Two reproductive classes of *P. neomexicanus* are discernible based on average ova diameter. In Class I, females would not lay eggs during the current summer. Ova diameters ranged from 1.1 mm in mid-June to 2.1 mm in mid-September. The number of maturing eggs averaged 7.7, with about four in each ovary. In Class II, females are considered gravid and would mate and oviposit their eggs by the beginning of the next active season. Ova diameter ranged from 1.9 mm in early July to 3.6 mm in late August. The number of maturing eggs averaged 7.8, with equal numbers in each ovary. A third class of females may be present. These are the females presumed to be underground with their egg clutches. These females would move into Class I the next summer.

Spermatogenesis begins at the posterior end of the testes and progresses slowly anteriorly. Secondary spermatogonia are formed during late fall and are present until late July. Primary spermatocytes are found from early June to late August, and spermatids are found from mid-July to mid-September. Spermatozoa are first formed in late August and fill the entire testes by early fall. *Plethodon neomexicanus* has sperm in the vasa deferentia during the entire activity season from June through September.

Brodie and Altig (1967), through examination of the gonads, stated that individuals under 50 mm SVL were juveniles. Oviposition probably occurs in the spring but may take place between August and the following spring; hatching commences by early August. Courtship and deposition of eggs have not been observed for *P. neomexicanus* in its natural habitat, although Williams (1976) reported one clutch of seven eggs laid in the laboratory. It is likely that the eggs are laid beneath the soil surface in interstitial spaces between fractured rocks, in rotted root channels, or in the burrows of rodents or large invertebrates. Juvenal *P. neomexicanus* 18–20 mm SVL have been found active on the surface as early as mid- to late July. We assume these salamanders represent the approximate size at hatching.

Food Habits: Little is known of the food habits of *P. neomexicanus*. Reagan (1972) examined the stomachs and intestines of 39 adults and reported the following distribution of food items by percent occurrence: ants 77%; lepidopteran larvae 18%, beetle larvae 13%, mycetophilid flies 10%, staphylinid beetles 8%, pseudoscorpions, tenebrionid beetles, tipulid flies, and termites 5% each, and annelid worms, mites, carabid beetles, and snails 3% each. The specimens Reagan (1972) collected during the late summer contained a greater variety of prey than those collected earlier in the season. He speculated this was due to an increase in the types of prey available rather than a shift in dietary preference. Specimens examined by us have contained earthworms, collembolans, mites, spiders, centipedes, crickets, ants, beetles, and flies.

Remarks: *Thamnophis elegans* is abundant in the habitat of *P. neomexicanus* and may often prey upon these salamanders. On 31 July 1992, we discovered a *T. elegans* 279 mm SVL that had consumed an adult *P. neomexicanus* 59.8 mm SVL.

Since 1980, increases in timber harvest by the U.S. Forest Service and changes in timber practices have prompted concern about the effects of logging on these salamanders (Scott et al., 1987; U.S. Fish and Wildlife Service, 1986, 1987). Most of the range occurs on National Forest lands, and the close association of *P. neomexicanus* with mixed coniferous forests makes them vulnerable to some forest management practices (Ramotnik and Scott, 1988). The species should be protected from unregulated collecting and from the destruction of potential and occupied habitat on federal, state, and private lands.

Plethodon neomexicanus is listed as endangered by the New Mexico Department of Game and Fish (NMGF, 1990) and as a Category 2, Notice of Review species by the U.S. Fish and Wildlife Service (USDI, 1994). Altenbach (1992) provided a complete annotated bibliography. Williams (1973) reviewed this species.

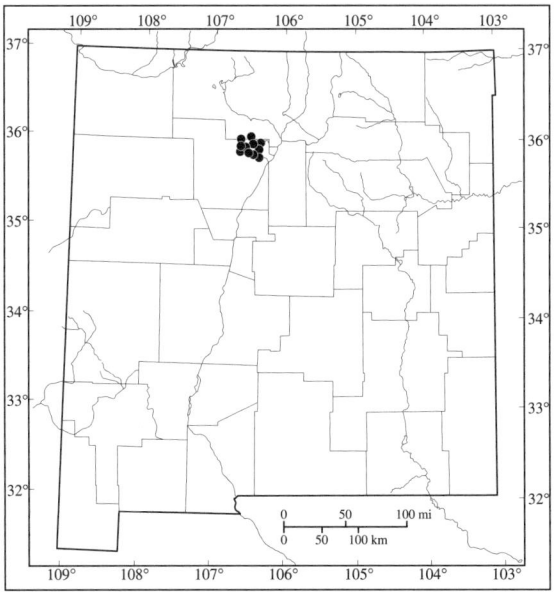

Distribution of
Plethodon neomexicanus
in New Mexico.

A KEY TO THE TOADS AND FROGS OF NEW MEXICO

1. Parotoid glands absent; skin smooth and moist, lacking warts, may have small glandular bumps   2
1. Parotoid glands present (**FIG. 1**); dry warty skin (Bufonidae)   6
2. Tympanum absent; snout distinctly pointed; toes unwebbed; a transverse fold of skin across head in back of eyes (**FIG. 2**) (Microhylidae)   *Gastrophryne olivacea*
2. Tympanum absent or not; snout rounded or only somewhat pointed; toes unwebbed or not; lacking a fold of skin across head in back of eyes   3
3. Toes unwebbed; circular fold of skin forming large disc on venter (**FIG. 3a**); prominent, pointed tubercles on undersides of joints of toes; juveniles greenish or blackish with fawn-colored band across the back (Leptodactylidae)   *Eleutherodactylus augusti*
3. Toes webbed; dorsolateral folds present or not (**FIG. 3b**); juveniles similar to adults   4
4. Pupil vertically elliptical (vertically oval to round when widely dilated) (**FIG. 4a**); single metatarsal tubercle (**FIG. 4b**) (Pelobatidae)   14
4. Pupil round; metatarsal tubercles lacking; dorsolateral folds present or absent   5
5. Dorsolateral ridges lacking (Hylidae)   16
5. Dorsolateral ridges present or at least curving down behind the tympanum (Ranidae)   19
6. Cranial crests present   7
6. Cranial crests absent or obscure   10
7. Parotoid glands tapering posteriorly, large, and as long as head (**FIG. 7**); size of adults small (38–51 mm); head flattened; cranial crests curve closely around eyelids when present   *Bufo debilis*
7. Parotoid glands not as long as head   8

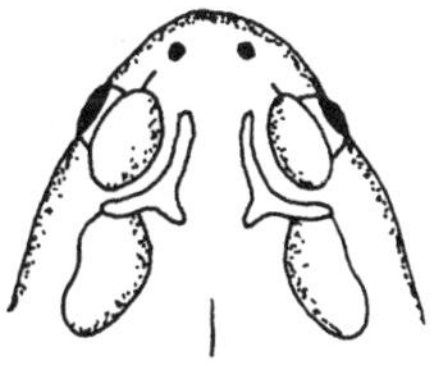

Figure 1

Figure 2

Figure 3a

Figure 3b

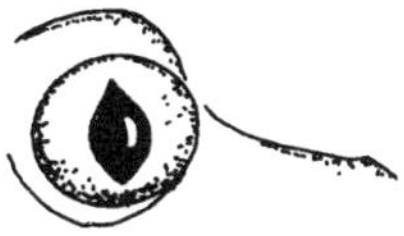

Figure 4a

Figure 4b

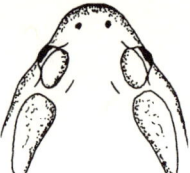 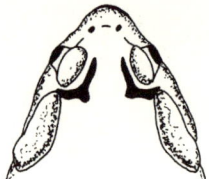

Figure 7   Figure 8

Figure 9a   Figure 9b

Figure 10   Figure 11

Figure 13a   Figure 13b

Figure 15a   Figure 15b

8. Large warts on dorsal surface of femur and/or wart or warts on tibia; size large (76–152 mm); skin usually smooth and of uniform color (olive or brown); tympanum about the same size as eye; long parotoid glands touch cranial crests and extend diagonally downward (**FIG. 8**); (juveniles may lack glands and have small spots on the back) *Bufo alvarius*

8. Not as above 9

9. Usually well-defined boss on snout; cranial crests diverging posteriorly from boss; parotoid glands almost triangular (**FIG. 9a**); usually with large dorsal and lateral dark blotches; venter unspotted *Bufo cognatus*

9. No boss on snout; interorbital crests parallel or sometimes absent; parotoids about twice as long as wide, touch cranial crests (**FIG. 9b**); vertebral stripe present; venter with dark blotches or spots on chest and upper sides; males with dark throat *Bufo woodhousii*

10. Parotoid glands small, same size or smaller than upper eyelid, almost perfectly round (**FIG. 10**) *Bufo punctatus*

10. Parotoid glands not as above 11

11. Vertebral stripe present; oval parotoid gland about the size of eye (**FIG. 11**); fold of skin on tarsus; adults to 100 mm *Bufo boreas*

11. Vertebral stripe absent 12

12. Size small (38–51 mm); eyes widely spaced (1½ to 2 times width of eyelid between) head flattened; ground color (in life) green to yellowish with small black spots *Bufo debilis*

12. Size larger (63 mm); eyes closer together (about same width of eyelid between) 13

13. Inner (large) metatarsal tubercle elongate and sharp and edged with black; smaller metacarpal tubercle not prominent; short oval parotoid glands less than twice as long as wide (**FIG. 13a**) *Bufo speciosus*

13. Inner metatarsal tubercle rounded and not sharp-edged; usually brown in color; smaller metacarpal tubercle usually prominent; oval parotoid glands twice as long as wide (**FIG. 13b**) *Bufo microscaphus*

14. Metatarsal tubercle elongate or sickle-shaped; tympanum distinct; interorbital space wide, the width of an eyelid equal to or slightly less than the interorbital space; ground color greenish, often with coarse mottling of dark color *Scaphiopus couchii*

14. Metatarsal tubercle rounded or wedge-shaped, not sickle-shaped; interorbital space narrow, the width of an eyelid 1⅓ to 1½ times interorbital distance 2

15. Interorbital mass of bone (boss) present, making a prominent bump just anterior to the eyes (**FIG. 15a**) *Spea bombifrons*

15. Interorbital mass of bone (boss) absent (**FIG. 15b**) *Spea multiplicata*

16. Toe tips slightly expanded; rear of thigh with dark stripe or elongated, ragged band; dark triangle between the eyes *Acris crepitans*

16. Toe tips slightly or conspicuously expanded; rear of thigh lacking dark, longitudinal stripes 17

17. Toe tips only slightly expanded (less than one-half the diameter of the tympanum) (**FIG. 17a**); usually three dark stripes down the back, may be reduced or broken; light line along upper lip
Pseudacris triseriata

17. Toe tips broad, at least one-half or more the diameter of eyelid (**FIG. 17b**); coloration not as above 18

18. Dark eyestripe present; anterior dorsal surface of body usually unmarked; bright green in life *Hyla eximia*

18. Dark eyestripe absent; scattered spots and blotches of darker color present dorsally; considerable yellow or orange color in femoral region; dark bar below eyes; skin moderately rough
Hyla arenicolor

19. Dorsolateral fold curving down behind tympanum (**FIG. 19a**); tympanum large; body lacking spots or with small irregular spots; blotches absent or inconspicuous on dorsal surface of thighs *Rana catesbeiana*

19. Dorsolateral fold present and well-developed at least on anterior and middle of body; tympanum small, not conspicuously sexually dimorphic; body with prominent spotted pattern, spots variable in size but usually large; blotches conspicuous on posterior surface of thighs 20

20. Posterior surface of thigh spotted, most spots not fused into elongate blotches or reticulations or if so only on ventral margin of posterior surface and most of thigh spotted 21

20. Posterior surface of thigh reticulated; light or dark markings fuse into network or irregular elongate blotches 22

21. Posterior thigh with dark spots on lighter ground color (**FIG. 21a**); dorsal spots frequently with light halos; dorsolateral folds continuous and well-developed posteriorly *Rana pipiens*

21. Posterior thigh with small white spots on dark ground color at least on proximal portion; posterior thigh conspicuously tuberculate (**FIG. 21b**); head relatively broad; dorsal spots small, numerous, and lacking halos; dorsolateral folds discontinuous and often poorly defined posteriorly *Rana chiricahuensis*

22. Dorsal spots with light halos; adult males with well developed vestigial oviducts; body robust; lower lip frequently mottled; reticulation on posterior thigh bold and contrasting (**FIG. 22**)
Rana berlandieri

22. Dorsal spots without light halos; adult males usually lacking vestigial oviducts; body less robust; lower lip variable but tending to lack mottling; reticulation on posterior thigh somewhat diffuse, less contrasting than above 23

23. Spot usually present on head anterior to eyes; posterior thigh reticulated with more light than dark coloration; reticulation often with indistinct margins (**FIG. 23a**) *Rana blairi*

23. Spot usually absent on head anterior to eyes; posterior thigh reticulated with more dark than light coloration, reticulation with distinct margins (**FIG. 23b**) *Rana yavapaiensis*

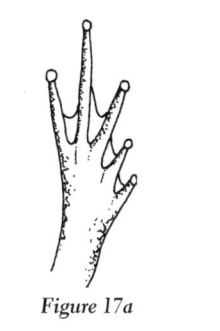
Figure 17a

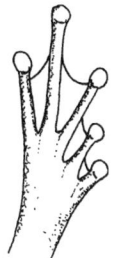
Figure 17b

Figure 19a

Figure 21a

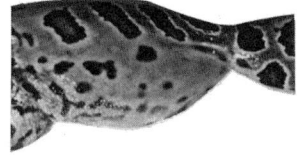
Figure 21b

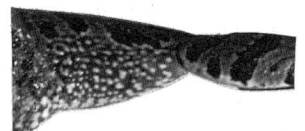
Figure 22

Figure 23a

Figure 23b

FAMILY PELOBATIDAE *Spadefoots*

The family Pelobatidae has fewer than 90 species in nine genera. Seven genera are found in Asia, from Pakistan and western China to the Philippines and through Indonesia to the Greater Sunda Islands in the Malay Archipelago. Their habitats vary from forested tropical lowlands and uplands to Himalayan peaks. The European genus, *Pelobates*, occurs in some areas of Mediterranean northwestern Africa, and western Asia (Zweifel, 1992). Two genera, *Spea* and *Scaphiopus*, occur in North America where they are referred to as spadefoots or spadefoot toads. Three, possibly four, species of these interesting toad-like burrowing frogs occur in New Mexico. They are well adapted to arid conditions, being explosive breeders with short-duration, high-density breeding aggregations that form during periods of summer thundershowers. Breeding in these ephemeral desert pools constrains the life cycle into precise synchrony with a very temporary aquatic environment. Sound or vibration, such as that caused by rainfall or possibly thunder, is apparently a primary cue for emergence in these frogs. The males are the first to arrive at flooded areas where they set up loud breeding choruses to attract the females. All spadefoot toads in New Mexico have smooth skins and darkened horny spades on their hind feet. All members of this family have vertical pupils, and all have free-living aquatic tadpoles.

SCAPHIOPUS COUCHII Baird, 1854 — *Couch's Spadefoot*

See Plate 4.

Type: Spencer Fullerton Baird (1854) gave the type locality as "Coahuila and Tamaulipas," which was later restricted to "Matamoros, Tamaulipas" Mexico by Smith and Taylor (1950b). Probable syntypes are USNM 3713, 3714, and 3715, collected by D. N. Couch in 1853, although these were reported lost or destroyed by Kellogg (1932). In the original description, Baird (1854) neglected to cite the USNM catalog numbers of the syntypes. Cope (1863), in revising the genus, removed USNM 3714 from the supposed syntypes and transferred it to his new species, *Scaphiopus rectifrenis*. In the same paper, Cope designated a specimen, USNM 3713, obtained by D. N. Couch at Matamoros, Tamaulipas, Mexico, as the basis for the original diagnosis of *S. couchii*.

Distribution: *Scaphiopus couchii* occurs from central Texas and adjacent Oklahoma to extreme southeastern California, south to the tip of Baja California and to Nayarit, Zacatecas, and Querétaro, Mexico. It is absent from the highlands of western Mexico. An isolated colony occurs in southeastern Colorado (Conant and Collins, 1991). In New Mexico, *S. couchii* occurs at 900 m to about 1800 m nearly statewide except for the northern and west-central counties.

Description: *Scaphiopus couchii* is recognized by the dark, irregular, dorsal mottling on a yellowish or greenish ground color. There is a dark, elongated, sickle-shaped tubercle found at the base of each hind foot. There is no boss between the eyes, and the eyelids are as wide as the space between them. The venter is whitish, without dark spotting. Adults are generally larger than other spadefoot toads in New Mexico and may reach 90 mm SVL. The male has less well-defined dorsal markings than the female. Stroud (1949) discussed a pale form of *S. couchii* from White Sands National Monument. The call is a groaning bleat, suggestive of a goat or sheep. Each bleat is relatively long, lasting 0.5–1 second (Conant and Collins, 1991).

The mature tadpole is approximately 18–24 mm TL. The body is slightly wider than the head. The ground color is coppery bronze with golden spots or sheens and much iridescence. The upper half of the tail musculature is spotted, although most of the lower half and tail tip are unspotted. The upper and lower tail fins are transparent but finely stippled with small black melanophores which are most abundant in the upper fin. The labial tooth row formula is usually 2/4 or 4/4, rarely 5/4, 3/4, or 5/5 (Stebbins, 1962).

Similar Species: *Scaphiopus couchii* is similar to the other spadefoots in New Mexico, although it differs in having a dorsum of yellow or greenish background color and in the absence of a bony boss between the eyes. The metatarsal tubercle is elongate and sickle-shaped rather than wedge-shaped as in *Spea bombifrons* and *S. multiplicata*.

Systematics: No subspecies of *S. couchii* are recognized. We follow Conant and Collins (1991) and retain this form in the genus *Scaphiopus*. See discussion in the *Spea bombifrons* account.

Habitat: *Scaphiopus couchii* is probably the most xeric adapted of all the North American anurans. In southeastern California, Mayhew (1962) reported periods as long as three years without sufficient rainfall to stimulate emergence. The physiological tolerance of *S. couchii* to long periods of dormancy has been studied by McClanahan (1967). In New Mexico, *S. couchii* occupies arid grasslands and areas grown to creosotebush and mesquite, where soils are sandy and well drained. They are occasionally found in the irrigated agricultural lands of the larger river valleys and are often abundant on desert roadways after summer thundershowers. Stroud (1949) found that *S. couchii* may bury themselves in gypsum sands at the base of the dunes at White Sands National Monument.

Behavior: *Scaphiopus couchii* is largely nocturnal. It spends most of its life buried in the soil and emerges only during spring and summer rains. Wright and Wright (1949) observed an individual retreat into a packrat burrow and speculated that it may aestivate in rodent burrows during dry periods.

Reproduction: As with many desert anurans, breeding activity in *S. couchii* usually occurs during the summer rains in temporary rain-filled depressions. Dimmitt and Ruibal (1980a) studied the environmental factors that may influence emergence and found that low frequency sound and/or vibration, such as caused by rainfall or thunder, is a primary cue for emergence. In their laboratory experiments, increasing the soil temperature and wetting the soil to less than saturation failed to break dormancy in the absence of a sound stimulus. Woodward (1984b) reported that more than 90% of the breeding generally occurs on the first night following pond formation. Among three choruses reported by Sullivan (1989), average breeding period duration was only 1 day. Woodward (1987a) reported adult female *S. couchii* from southern New Mexico averaged approximately 3,310 eggs, although he noted high levels of variation in clutch sizes within breeding aggregations (Woodward, 1987b).

Tadpoles of *S. couchii* have the most rapid rate of development of any North American anuran. The eggs hatch in only 15 hours at 30°C, although at 10°C 82 hours are required for hatching (Justus et al., 1977). Tadpoles metamorphose at approximately 15–20 mm SVL. Ted Brown (pers. comm.) collected 10 newly transformed young, approximately 20 mm SVL, on 15 July near Artesia in Eddy County. Woodward (1982b) investigated sexual selection and mating patterns in *S. couchii* in southern New Mexico. He found nonrandom mating (with larger males mating disproportionately often) in one of four *S. couchii* breeding aggregations. Wells (1977) proposed that the most likely explanation for the nonrandom mating observed in explosive breeders, such as *S. couchii*, is male-male interactions, essentially in the form of scramble competition for females.

Food Habits: *Scaphiopus couchii* is a generalized arthropod predator, concentrating primarily on ground-dwelling species. Punzo (1991) examined 367 specimens from west Texas and found that beetles, orthopterans, ants, spiders, and termites comprised 82.1% of the overall diet. He found no significant differences in the diet as a function of sex or season. Arthropods with well-known chemical defenses, such as blister beetles, velvet ants, stink bugs, and millipedes usually are not included in the diet. Dimmitt and Ruibal (1980b) studied this species in southeastern Arizona and southwestern New Mexico and determined that it is capable of eating enough food in one meal to last it an entire year. They found that *S. couchii* has a diet composed primarily of termites and that it may consume 55% of its body weight in a single feeding. The high energy content of termites and their simultaneous emergence with *S. couchii* during the first summer rains, probably makes the presence of termites essential to the survival of spadefoot toads in the desert.

Remarks: Skin secretions from *S. couchii* may be toxic to some humans. Blair (1947) reported, "While handling the toads the night of August 27, the writer found that scratches on his hands became intensely painful. Subsequently he experienced incessant sneezing and discharge from the eyes and nostrils. *Scaphiopus couchii* is believed to be the toad at fault, since Mr. C. M. Bogert related several instances in which handling of this species produced like symptoms. However, there is good reason to believe that the secretions of other species of *Scaphiopus* may have a similar effect."

This species was reviewed by Wasserman (1970a).

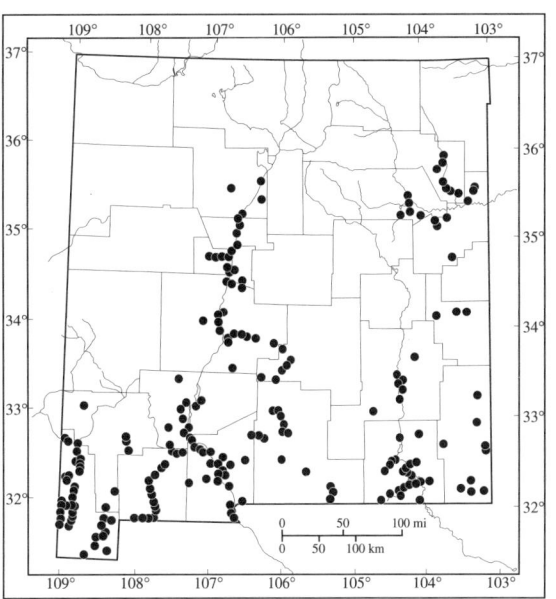

Distribution of Scaphiopus couchii *in New Mexico.*

SPEA BOMBIFRONS (Cope, 1863) — *Plains Spadefoot*

See Plate 5.

Type: Edward Drinker Cope (1863) gave the type locality, with syntypes, as "Fort Union, on Missouri River, lat. 48° N., from Mr. E. J. Denis, (Smithsonian, No. 3704.) On Platte River, 200 miles west Fort Kearney, from W. S. Wood, of Lieut. Bryan's Expedition (Smithsonian, No. 3520.) Llano Estacado Texas, Capt. Popes' Exped. Coll., (Smithsonian, No. 3703.)" The type locality was later restricted to Fort Williams [= Fort Union], North Dakota, USA, by Schmidt (1953). No collecting dates are given (Frost, 1985).

Distribution: *Spea bombifrons* occurs from southwestern Manitoba to southern Alberta and south to Chihuahua. It follows the Missouri River valley eastward across Missouri. Disjunct populations occur in extreme southern Texas and northeastern Mexico and in northwestern Arkansas and south-central Colorado (Conant and Collins, 1991). In New Mexico, *S. bombifrons* occurs statewide at 900 m to approximately 2200 m where favorable habitat exists.

Description: Adult *S. bombifrons* have a pronounced bony boss between the eyes. The eyes are large and have vertically elliptical pupils. The skin is smooth but may have small tubercles on the dorsum that are yellowish or reddish. The overall coloration is grayish or brownish, often with a greenish tinge. The darker markings on the back are brown or gray. There is a wedge-shaped spade on the hind foot. The eyelids are wider than the space between them. The call is a short rasping bleat repeated at intervals of ½–1 second, or a rather low-pitched rasping snore. Each trill lasts ½–¾ second (Conant and Collins, 1991).

The tadpole is medium sized and deep bodied. The body is broadest just behind the eyes. Live specimens are uniformly tan or brown. The eyes are positioned dorsally. The nostrils are small and closely spaced. The tail fins are clear and of medium height, with the tail musculature delineated with pigment. The oral disc is round and is mostly surrounded with small, unpigmented, marginal papillae. There is a small median gap on the anterior labium in the papillary fringe. The labial tooth row formula may range from 2/4 to 4/6. The upper jaw is cuspate; the lower jaw is notched. The total length is 40–48 mm (Johnson, 1987).

Similar Species: *Spea bombifrons* is easily mistaken for the similar *S. multiplicata*. Newly metamorphosed toadlets or specimens that have been in preservative for long periods of time are particularly difficult to distinguish. *Spea bombifrons* possesses a frontal boss that is absent in *S. multiplicata*. When sympatric, however, these species may hybridize (Simovich, 1994), and this character is not always reliable. *Scaphiopus couchii* is easily separated from *S. bombifrons* because of the dorsal greenish-yellow coloration and dark mottling. Also, the width of the eyelid of *S. couchii* is equal to or only slightly less than the interorbital distance. The metatarsal tubercle is wedge-shaped as in *S. multiplicata* rather than elongate and sickle-shaped as in *S. couchii*.

Systematics: No subspecies of *Spea bombifrons* are currently recognized. Cope (1866b) erected the genus *Spea*, which he separated from *Scaphiopus* based on the co-ossification of the skin and the cranium. He presented the following couplets in his 1866 key: "Derm involved in the cephalic ossification, which is complete . . . *Scaphiopus*" and "Derm distinct from cranium, which is only ossified superiorly in two superciliary bars . . . *Spea*." Later it was suggested that *Spea* was a subgenus of *Scaphiopus* (e.g., Wasserman, 1970b), although several workers (Bragg, 1944, 1945; Blair, W. F., 1955; Kluge, 1966; Tanner, 1989) provided morphological, life history, ecological, and skeletal evidence to support the retention of *Spea* as a distinct genus. We follow the latter authors.

Hughes (1965) collected spadefoot toads from Lubbock County, Texas, representing *S. bombifrons* and *S. hammondii* (= *S. multiplicata*). He compared the bony protuberance ("boss") of these two forms by measuring the frontoparietal bone of the skull and suggested that, although they frequently hybridize, they should be maintained as separate species and not subspecies as earlier suggested by Wright and Wright (1949). Wasserman (1970b) compared karyotypes of *S. bombifrons* and *S. hammondii* (= *S. multiplicata*) from New Mexico and found them "virtually indistinguishable." Wiens and

Titus (1991), in a phylogenetic analysis of the genus *Spea*, suggested that although *S. bombifrons* and *S. multiplicata* do frequently hybridize, they are the most distantly related species in *Spea*.

Habitat: Primarily a grasslands species, *S. bombifrons* generally avoids river bottoms or wooded areas. Aquatic habitats are used only for breeding.

Behavior: *Spea bombifrons* usually remains hidden in burrows during the day and emerges only at night, especially after hard summer rains. As all spadefoots in New Mexico, *S. bombifrons* is an accomplished burrower, burying itself by backing into the loose soil with a sideways shuffling of the rear feet. It may burrow to any depth necessary to remain moist. It is known to use gopher or spotted ground squirrel burrows as underground retreats (Hammerson, 1982).

Reproduction: Like other spadefoot toads in the arid Southwest, *S. bombifrons* has an opportunistic, relatively short, and high-density mating pattern. It is dependent upon heavy rainfall to stimulate reproductive activity and to provide suitable habitat for egg deposition. *Spea bombifrons* may not breed every year due to the localized nature of the summer thundershowers in New Mexico. Bragg (1950c) observed 111 breeding congresses of *S. bombifrons* in Oklahoma. The earliest date he recorded was 16 March, and the latest was 18 July. Woodward (1984b) reported that in most breeding aggregations of *S. bombifrons* at least 90%, and usually 100% of the mating occurs on one night. Of seven choruses reported by Sullivan (1989), average breeding period duration was 1.8 days.

Male *S. bombifrons* usually call from the edge of a pool while sitting on the bank or floating in the pond. Woodward (1987a) found that adult female *S. bombifrons* from southern New Mexico averaged approximately 1,600 eggs, although he noted high levels of variation in clutch parameters within breeding aggregations. At higher environmental temperatures *S. bombifrons* has extremely rapid development, the eggs hatching in only 20 hours at 30°C, although at 10°C 84 hours are required for development to hatching (Justus et al., 1977). King (1960) reported on a population of *S. bombifrons* from Oklahoma that required only 13–14 days from egg laying to metamorphosis. The rate of hatching and transformation is regulated by the water temperature, food supply, and the amount of dissolved oxygen. Ted L. Brown (pers. comm.) collected 5 newly transformed young, approximately 20 mm SVL, in Albuquerque on 5 July. Tadpoles metamorphose at approximately 18–22 mm SVL.

Tadpoles of *S. bombifrons* (and all other New Mexico spadefoots) may occur as two environmentally induced, trophic morphs: a large, rapidly developing carnivorous morph and a smaller, more slowly developing omnivorous morph (Pfennig et al., 1991b). Carnivorous morphs can be induced to develop by feeding the normally omnivorous tadpoles live fairy shrimp. This morph determination is also reversible. When studying both trophic morphs under experimental lab conditions, Pfennig (1992) found the carnivorous morphs survived better in highly ephemeral artificial ponds because they developed faster; the omnivorous morphs survived better in longer-duration artificial ponds because their larger fat reserves enhanced postmetamorphic survival. Pfennig (pers. comm. *in* Fellman, 1995) suggested that, unless stressed for food, these cannibal morphs avoid eating their brothers and sisters.

Forester (1973) demonstrated the distinctness of the mating calls between the morphologically similar *S. bombifrons* and *S. multiplicata*.

Food Habits: Food habits of *S. bombifrons* have not been studied in New Mexico, although Whitaker et al. (1977) found that the major foods in northeastern Colorado consisted of adult moths, caterpillars, carabid beetles, and various other small arthropods.

Remarks: On the basis of mating call analysis, Pierce (1976) indicated that two distinct call types of *S. bombifrons* are evident and suggested that the fast-mating-call form from the southwestern portion of the species' range may represent a cryptic species. Sattler (1980) however, found no electrophoretic differences between these two mating-call variants of *S. bombifrons*.

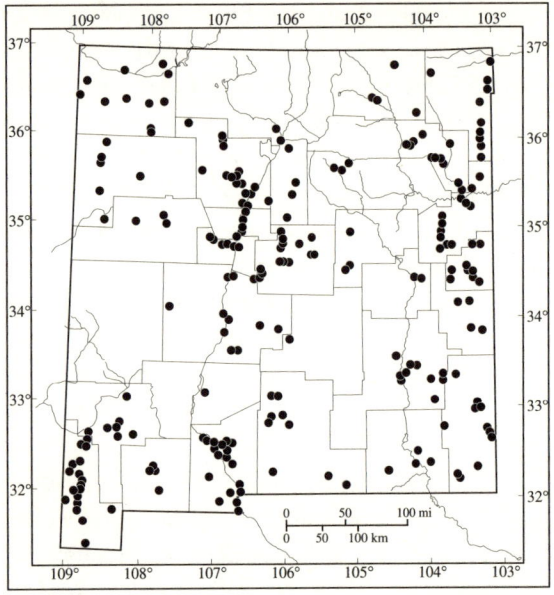

Distribution of Spea bombifrons in New Mexico.

SPEA MULTIPLICATA (Cope, 1863) *New Mexico Spadefoot*

See Plate 6.

Type: Edward Drinker Cope (1863) gave the type locality as "Valley of Mexico." The holotype, USNM 3694, was sent to the Smithsonian Institution by Mr. Jno. Potts. The date of collection is unknown (Cochran, 1961).

Distribution: *Spea multiplicata* ranges from western Oklahoma and central Texas to Arizona and far south into Mexico (Conant and Collins, 1991). In New Mexico, *S. multiplicata* occurs statewide from 900 m to about 2600 m where favorable habitat exists.

Description: *Spea multiplicata* lacks a boss between the eyes and the eyelids are wider than the space between them. The dorsal color is uniformly brown or dark gray with small dark spots or blotches and red-tipped tubercles scattered over the dorsum. No dorsolateral stripes are evident. There is a short wedge-shaped spade on each hind foot. The iris is slightly variegated and appears pale copper colored. Adults may reach 65 mm SVL. The call is a vibrant, metallic trill, like running a fingernail along the stiff teeth of a large comb. Each trill lasts about ¾–1.5 seconds (Conant and Collins, 1991).

The tadpole body is broadest just behind the eyes, tapering gradually posteriorly and sharply anteriorly. The snout is short. The tail is about 1⅓–1¼ times the head-body length. The depth of the tail musculature at its base is about ½–⅓ the depth of the body. The greatest width of the tail is near the midpoint, with the tail fins at that point being about equal in height to the width of the tail musculature. The dorsal fin originates well posteriorly on the body. The eyes are close together and well up on the head. The anus is medial, emerging in the base of the ventral fin. The spiracle is low on the left side, below the lateral axis of the body. The labial tooth row formula is 2.3, 2/4, 3/3, 3/4, 4/4, or 5/4 (Stebbins, 1962).

Similar Species: *Spea multiplicata* is easily mistaken for the similar *S. bombifrons*. Newly metamorphosed toadlets or specimens that have been in preservative for long periods of time are difficult to distinguish. *Spea multiplicata* lacks the frontal boss of adult *S. bombifrons*, although these two species frequently hybridize (Simovich, 1994) and this character is not always reliable. *Scaphiopus couchii* is easily separated from *S. multiplicata* because of the dorsal greenish-yellow and dark mottling. Also, the width of an eyelid of *S. couchii* is equal to or slightly less than the interorbital distance. The metatarsal tubercle is wedge-shaped as in *S. bombifrons* rather than elongate and sickle-shaped as in *S. couchii*.

Systematics: Currently there are no recognized subspecies of *S. multiplicata*; however, based on skeletal morphology, Tanner (1989) assigned the name *Spea hammondii stagnalis* to this species. Smith (1978) recognized *S. multiplicata* as a subspecies of *S. hammondii*. Brown (1976) and Sattler (1980) provided evidence to support the recognition of *S. multiplicata* as being morphologically and biochemically distinct from *S. hammondii*.

Habitat: *Spea multiplicata* is found in a wide variety of habitats, including grasslands, sagebrush flats, semiarid shrublands, river valleys, and agricultural lands. It is often found on roads after summer thundershowers.

Behavior: *Spea multiplicata* is largely nocturnal and secretive. During the summer rainy season *S. multiplicata* may be found hidden under surface objects. Like other spadefoots in New Mexico, they usually occupy underground burrows they dig in the soft earth with their specialized hind feet, which are equipped with sharp-edged spades. During nocturnal wanderings in search of breeding sites or prey they are often seen on the roadways. When molested, these toads may secrete a musty skin toxin that smells somewhat like raw peanuts. This toxin can irritate the sensitive membranes of the eyes and nose of those who rub their face after handling a spadefoot. Roger Conant (pers. comm.) when he first encountered a large breeding chorus of this spadefoot in Mexico in 1960, suffered a violent, hour-long nasal and lachrymal reaction. He had held two or three males close to his nose to test the raw peanut odor he had read about.

Reproduction: Spadefoot breeding is closely associated with the summer monsoon rains that fill playa lakes and cause the rapid formation of pools in low-lying areas. Dimmitt and Ruibal (1980a) studied the environmental factors that may influence the emergence of these toads and found that low-frequency sound and/or vibration, such as caused by rainfall or thunder, is a primary cue for emergence. James N. Stuart (pers. comm.) noted two males of this species calling from within the drying mud bottom of a stock pond in the Animas Valley as a thunderstorm occurred approximately one km away. One male and one female were located 10–15 cm beneath the surface.

The monsoon rains normally occur during mid-July but, during dry years, *S. multiplicata* may not breed until later in the season. In Bernalillo County, we have observed males initiate breeding (although unsuccessful) as late as 13 August at an elevation of 2226 m. Of seven choruses reported by Sullivan (1989), the average breeding period duration was only 1.6 days. Woodward (1987a) found adult female *S. multiplicata* from southern New Mexico averaged approximately 1,070 eggs, although he noted high levels of variation in clutch size within breeding aggregations. The males usually call while floating on the surface of the water. The vocal sac appears as a dark, heavily pigmented area on the throat of the male. During amplexus, the eggs are fertilized by the male as they are laid. The eggs are laid in cylindrical masses that are then attached to submerged aquatic vegetation or debris and hatch in as little as 42–48 hours. The tadpoles undergo metamorphosis in about three weeks, and the toadlets emerge from the drying pond and disperse. Brown (1976) reported 25 sexually mature males collected at a breeding chorus near Rodeo with a mean SVL of 46.9 mm. Forester (1973) demonstrated the distinctness of the mating calls between the morphologically similar *S. multiplicata* and *S. bombifrons*.

Food Habits: *Spea multiplicata* is a generalized arthropod predator, concentrating on ground dwelling species. Punzo (1991) examined 293 specimens from west Texas during the spring and summer of 1988 and found that beetles, orthopterans, ants, spiders, and termites comprised 93.8% of their total diet. He found no significant differences in diet by sex or season. Dimmitt and Ruibal (1980b) found that the diet of *S. multiplicata* in southwestern New Mexico consisted of 72% termites and 22% beetles. Although *S. multiplicata* occasionally feeds on centipedes and scorpions, other arthropods with well known chemical defenses such as blister beetles, velvet ants, stink bugs, and millipedes are usually avoided. Dimmitt and Ruibal (1980b) suggested that *S. multiplicata* may require seven feedings before it has accumulated the fat reserves necessary to survive 12 months. MacKay et al. (1990) reported tadpole shrimp, fairy shrimp, and conspecific tadpoles eaten by tadpoles of *S. multiplicata*. There seems to be a high degree of dietary overlap among the spadefoot toads of New Mexico.

Remarks: Much of the literature for this species is found under the name *Scaphiopus hammondii*. Bragg (1955–56, 1965) presented a good overall view of spadefoot biology.

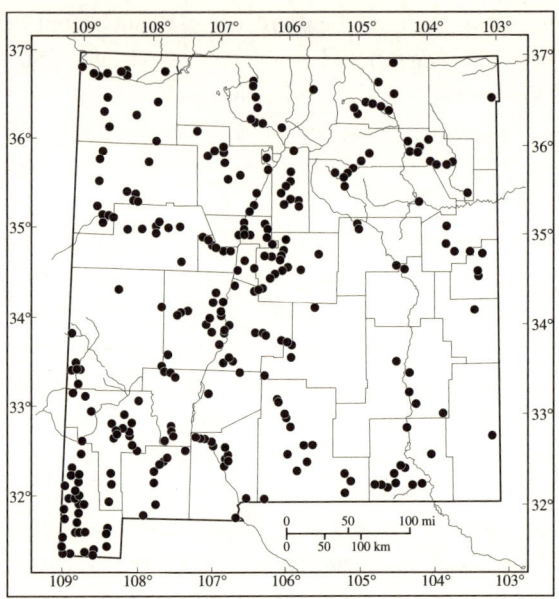

Distribution of Spea multiplicata in New Mexico.

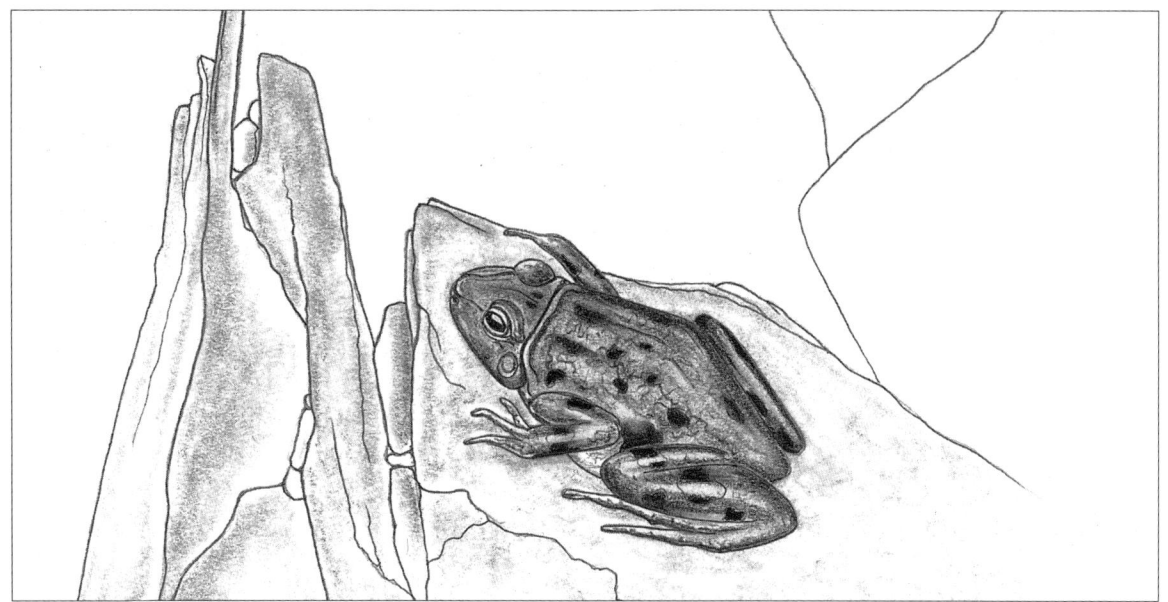

FAMILY LEPTODACTYLIDAE *Tropical Frogs*

The family Leptodactylidae is a diverse assortment of almost 800 species of frogs in 52 genera, inhabiting South America, the Caribbean Islands, Central America, and Mexico (Zweifel, 1992). A great diversity of habitat preferences, morphology, reproductive mode, and life-style occurs in this large family. Seven frogs in this family occur in the United States; five are native and two are introduced. Only a single species occurs in New Mexico. The barking frog, *Eleutherodactylus augusti*, inhabits limestone caves and crevices and rodent burrows in southeastern New Mexico. Reproduction in this little-known frog is unique among New Mexico anurans. The eggs are laid on land, metamorphosis takes place entirely in the egg, and the young emerge as fully formed small froglets.

ELEUTHERODACTYLUS AUGUSTI (Dugès in Brocchi, 1878 [1879])   *Barking Frog*

See Plates 7A, 7B.

Type: Alfredo Dugès (1869) first mentioned this species, which was later described as *Hylodes augusti*. Dugès *in* Brocchi (1878 [1879]) formally described the species and gave the type locality as "Guanajuato (Mexique)." According to Smith and Necker (1943), "The original description was apparently based upon 3 cotypes (= syntypes), none of them now extant in [the Alfredo Dugès Museum in] Guanajuato. They are replaced by another specimen, marked "type" by Dugès, 'trouvè à plus de 40m. de profondeur dans la mine du Cedro, Guanajuato (Juin).' This specimen was designated as a neotype." Smith and Taylor (1948), however, stated that a skeleton in the Dugès Museum is the type specimen. This agrees with the statement by Mocquard (1899) that Dugès had sacrificed the type for the sake of a prepared skeleton (Zweifel, 1967).

Distribution: *Eleutherodactylus augusti* ranges from southeastern Arizona and southeastern New Mexico to central Texas, and south through central and western Mexico to the Isthmus of Tehuantepec (Frost, 1985). In New Mexico, *E. augusti* is known only from isolated localities in Chaves, Eddy, and Otero counties at elevations between 900–1200 m. Populations are well established at Bottomless Lakes State Park, near Carlsbad, and southeast of Whites City. Price (1986b) reported the first specimen from Otero County. Bezy et al. (1966) reported on the second known specimen from Arizona. Milstead (1960) considered this species to be relict in the Chihuahuan Desert, stating it exists there as disjunct populations that are considerably removed from the main portion of the range.

Description: The distinctive *E. augusti* has a broad head, chunky body, and short limbs which present a toad-like appearance, although the skin is smooth with no warts. There is an intertympanic dermal fold. The slender toes lack webbing and have tips that are slightly expanded. The smooth venter is light colored and has a distinct abdominal disc. Dorsal coloration of the adults is greenish to brown and is marked with dark blotches. Body size is moderately large; the largest of 13 females from a Texas population was 94 mm SVL, the largest of 35 males, 77.2 mm SVL (Zweifel, 1956).

The young have a broad ivory-colored band across the back that rapidly fades with age. James N. Stuart (pers. comm.) provided the following field notes on a juvenal *E. augusti* collected approximately 11 km SE of Whites City in Eddy County, New Mexico: "Color notes on juv. *Eleutherodactylus*, 16 mm SVL. Light color on dorsum, off white; dark [color] on dorsum, mottled black & brown; dorsum of legs yellow tinged green. Throat white, belly essentially translucent as are hindlimbs. SW Gotte, collector."

The call is explosive, like a barking dog when heard at a distance but more a guttural "whurr" at close range. The single note may occur at regular intervals of 2–3 seconds (Conant and Collins, 1991). The female will often emit a blaring screech when grasped. Jameson (1954) reported that *E. augusti* has a warm-up call, which is voiced crepuscularly and which changes abruptly to the normal bark as darkness falls. He likened a chorus of these frogs to an orchestra tuning up before a concert.

Similar Species: The lack of webbing between the toes, the abdominal disc, and large adult size should separate this species from all other anurans in New Mexico. The distinctive dorsal coloration of the juvenile is diagnostic.

Systematics: Four subspecies are currently recognized (Zweifel, 1967). The eastern barking frog, *E. a. latrans* (Cope, 1880), is the form known to occur in New Mexico, although the western form, *E. a. cactorum* Taylor, 1938 [1939], eventually may be discovered in extreme southwestern Hidalgo County. The type locality of *E. a. latrans* is given as "Helotes, Bexar County, Texas." Possible syntypes (see "Remarks" in Zweifel, 1967) include USNM 10058 (2 specimens), USNM 10529 (2 specimens), USNM 10751–53, and ANSP 10757–58, all collected by G. W. Marnock.

This species often appears in the recent literature as *Hylactophryne augusti*. We follow Lynch (1986) and Duellman (1993) and recognize *Hylactophryne* as a synonym of *Eleutherodactylus*.

Habitat: This unique frog is most often associated with escarpments and other rocky outcroppings where numerous cracks and fissures exist (Zweifel, 1956). In southeastern New Mexico, *E. augusti* is encountered in barren creosotebush flats with numerous and extensive rodent burrows on gypsum soils and in and near limestone and gypsum outcrops.

Behavior: In New Mexico, this nocturnal, secretive frog is more often heard than seen. It will often sit at the mouth of rodent burrows or in rock crevices during rainy weather, and its call can be heard for considerable distances. It is extremely wary and difficult to collect. When approached, it can rapidly disappear into the underground honeycomb of cracks and crevices, and trying to dig one out is usually futile. The occasional individual seen abroad during heavy rainstorms may have been flooded from its burrow or may be a female seeking the calling males. The small black-and-white banded juveniles may be encountered as they disperse from the hatching egg clutch.

Reproduction: Several older works (Cope, 1878; Strecker, 1910; Noble, 1925) reported *E. latrans* to have aquatic development. Although Livezy and Wright (1947) assumed the form to have terrestrial development, the mode of embryological development was accurately described by Jameson (1950).

Valett and Jameson (1961) reported on the embryology of *E. augusti* and outlined several important changes that occurred in the evolution of the genus. These changes include; no evidence of a ventral sucker, internal gills, horny tadpole mouthparts, lateral line system, true operculum, or equatorial cleavages, and an egg tooth that is present at the tip of the snout. Jameson (1950) reported the time from cleavage to hatching to be approximately 25–35 days.

Jameson (1954) suggested the peak of breeding activity occurs in April and May, as indicated by the numbers of individuals calling and the numbers of individuals with eggs. He reported a pair of barking frogs that were observed under one rock for three nights. The male called each night but the pair did not mate, thus suggesting a long and possibly complex mating pattern. When mating does occur, the eggs are laid in a cluster in a small cavity in the soil under a rock. The male remains with the eggs and apparently maintains the necessary moisture conditions by excretion. Jameson (1950) reported a calling male that was discovered under a rock with a cluster of 67 eggs. These were in a soil pocket about 10 cm deep which was filled with mud and eggs.

Reproduction in *E. augusti* is not well known in New Mexico. A clutch of 60 eggs was taken from a preserved female 82 mm SVL (MSB 58175) collected 16 June 1994 by Don Sias in southern Eddy County. On 16 July 1991, Steve Gotte collected a small juvenile (16 mm SVL) (MSB 52340) near the same locality as it moved about on the surface around midnight. Based on our presumed size at hatching, this individual was only a few days old. Barking frogs lay their eggs in crevices or under rocks during the summer rainy season. These eggs hatch if sufficient moisture is available, and the froglets emerge fully formed after metamorphosis is completed within the egg.

Food Habits: The food habits of *E. augusti* have not been studied in New Mexico or elsewhere. They are known to eat camel crickets and land snails (Olson, 1959).

Remarks: *Eleutherodactylus augusti* is often referred to as the robber frog. It was not reported from New Mexico until 5 September 1944 when Dr. W. J. Koster found a juvenal specimen that had taken refuge under his tent during a heavy thunderstorm (Koster, 1946b). The second specimen reported from New Mexico was collected at Bottomless Lakes State Park during May, 1966. Zweifel (1967) reviewed this species.

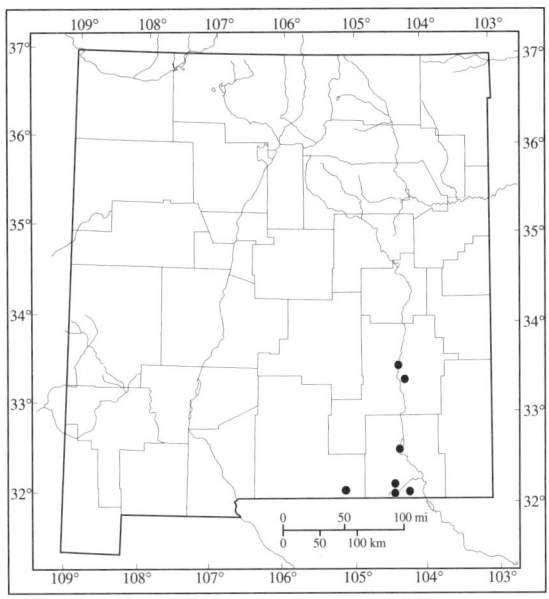

Distribution of Eleutherodactylus augusti *in New Mexico.*

Toads

FAMILY BUFONIDAE

Members of the family Bufonidae are native to temperate and tropical zones, deserts and rain forests, and mountains and prairies. At least one species, *Bufo marinus*, has been introduced into many regions, although the family is not native to the Australian Region, Madagascar, or the Oceanic Region. About 360 species in 31 genera make up this large and diverse group (Zweifel, 1992). In New Mexico, this family is represented by a single genus, *Bufo*, with eight species. All species in New Mexico are squat-bodied, heavy-set, short-legged toads with numerous wart-like glands on the body and legs. Habitats in New Mexico vary from high-mountain beaver ponds to riverbanks to low desert. Most are stimulated to breed by heavy summer thundershowers, although others may breed independently of rainfall. Populations of toads are susceptible to environmental change. Three species in New Mexico are currently listed either as federal Notice of Review species by the U.S. Fish and Wildlife Service (USDI, 1994) or as endangered by the New Mexico Department of Game and Fish (NMGF, 1990).

BUFO ALVARIUS Girard *in* Baird, 1859   *Colorado River Toad*

See Plate 8.

Type: Charles Girard (*in* Baird, 1859) gave the type locality as, "Valley of Gila and Colorado." This type locality was restricted to, "Colorado River bottomlands below Yuma, Arizona" by Schmidt (1953) and later modified to "Fort Yuma, Imperial County, California," by Fouquette (1968). The lectotype is USNM 2572, a female collected by A. Schott.

Distribution: *Bufo alvarius* ranges from extreme southeastern California and southern Arizona to extreme southwestern New Mexico, and south to northwestern Sinaloa, Mexico (Stebbins, 1985). In New Mexico, *B. alvarius* occurs only in southwestern Hidalgo County in the vicinity of Rodeo and in scattered localities in the adjacent Peloncillo Mountains at elevations of 1250–1387 m. *Bufo alvarius* was not collected in New Mexico until the summer of 1961 (Cole, 1962). Peters and McCoy (1978) were the first to report it from the Animas Valley.

Description: *Bufo alvarius* attains a body size of at least 190 mm SVL. The smooth leathery skin is olive to dark brown with scattered low, rounded tubercles. The hind legs are covered with numerous large warts and there is a conspicuous, large, light-colored wart at the angle of the jaw. The large kidney-shaped parotoid glands are longer than the head and extend diagonally downward. A distinct cranial crest curves above each eye. The advertisement call is a deep low-volume honk of 0.5–1 second (Smith, 1978). Fifteen individual calls measured by Blair and Pettus (1954) averaged only 0.7 second (0.6–0.8) in length. (See Reproduction.)

The tadpole, which may reach 57 mm TL, has a somewhat flattened body, which is lightly pigmented and brassy colored. The tail tip is rounded and the musculature of the tail is evenly pigmented with large punctate melanophores. Melanophores are absent from the center of the belly. The jaws are coarsely serrate (Altig, 1970).

Similar Species: The large size and numerous large warts on the hind legs will separate *B. alvarius* from all other toads in New Mexico. The only large toad that occurs sympatrically with *B. alvarius* in New Mexico is *B. cognatus*, which may be distinguished by the dark, often paired dorsal blotches that include many warts, and by the lack of large warts on the hind legs. *Bufo woodhousii* may be distinguished by the middorsal light stripe. *Scaphiopus* and *Spea* may be distinguished by the lack of warts and parotoid glands, and the presence of horny spades on the bottoms of their feet.

Systematics: No subspecies of *Bufo alvarius* are recognized.

Habitat: In New Mexico, *B. alvarius* has been encountered in desert shrub characterized by broad flat expanses of creosotebush and mesquite, in rocky riparian zones grown to cottonwood and sycamore, in muddy stock ponds, and in stock ponds with abundant aquatic vegetation.

Behavior: In New Mexico, *B. alvarius* is rarely seen except during the breeding season when it may be encountered on roadways or around permanent or temporary stock ponds. Much of the time may be spent underground in rodent burrows. Wet weather is not necessary to initiate nocturnal activity, although Stebbins (1951) stated that *B. alvarius* is more aquatic than most other southwestern toads. Large size, a handicap in seeking underground shelter, and smooth skin, subject to drying, are characteristics that may restrict this species to the vicinity of water or rodent burrows where the humidity is high.

Reproduction: Little information regarding the breeding behavior of *B. alvarius* is available. In central Arizona, Sullivan and Malmos (1994) found that breeding activity usually occurred on one night following rainfall events of greater than 25 mm. Males at these breeding aggregations employed two behavioral strategies; some males actively searched for females while others called from shoreline sites or in very shallow water within the pond. Although considerable variation existed, they noted that calling males were significantly larger than those males observed only actively searching for females. Sullivan and Malmos (1994) suggested that male *B. alvarius* may call more frequently when fewer conspecific males are present in a breeding aggregation. The advertisement call may be heard 100–150 m from the breeding site.

Although breeding is typically associated with summer rains, B. alvarius in New Mexico have been observed breeding independently of rainfall. We observed a single male and two amplexing pairs in a permanent stock tank in Guadalupe Canyon in southwestern Hidalgo County on 16 June 1992, prior to the start of the summer monsoon rains. Blair and Pettus (1954) found a large breeding aggregation of roughly 200 individuals in a stock tank in Maricopa County, Arizona, on the evening of 22 July 1953. Eggs are laid in long jelly-coated strings, and they may number several thousand per string. Larval periods appear quite short, lasting no more than a month. Based on sightings by David Pettus of numerous tadpoles in a stock tank in Maricopa County, Arizona, during early October, 1953, the breeding season may be quite long.

Blair and Pettus (1954) suggested that the advertisement call has little importance in the reproductive behavior of this species, stating "it is probably without function in the pattern of reproductive behavior of this species." Based on anatomical evidence, they suggested that the call of B. alvarius is probably evolving out of existence; Inger (1958) agreed. McAlister (1961) examined three males and found a large larynx with well developed vocal cords. He suggested that any loss of call capability must be caused by failure of some other portion of the vocal mechanism.

Sullivan and Malmos (1994) studied variation in the advertisement call, release call, and calling behavior of B. alvarius at breeding aggregations in central Arizona over a 3-year period. The advertisement call is short, approximately 1 KHz in frequency, and around 60 pulses/sec. The release call is a series of pulse groups (ca. 10.6 groups/sec.) lasting about one second. They found advertisement call frequency and duration were not correlated with SVL, nor were they related to body temperature. The release call pulse rate was negatively correlated with body temperature; a phenomenon not previously reported in anurans. They showed that females are indeed attracted to male advertisement calls and that these calls may play an important role in mate selection by females.

Food Habits: Cole (1962) reported on the stomach contents of five B. alvarius taken from the extreme eastern edge of their range and included individuals taken from Hidalgo County, New Mexico. The most commonly encountered prey items included beetles, ants, termites, and solpugids. Numerous other food items have been noted in the diet, including snails, grasshoppers, spiders, centipedes, millipedes, wasps, lizards, mice, and the amphibians Bufo cognatus and Scaphiopus couchii (King, 1932; Gates, 1957). Dr. Thomas Eisner (in Cole, 1962) was impressed by the number of prey items that were protected by sting-mechanisms or defensive secretions. It seems likely that B. alvarius will prey on anything it can overcome.

Remarks: Skin toxins of B. alvarius may paralyze or kill dogs (Musgrave, 1930) and are a potent hallucinogen. Hanson and Vial (1956) discussed the defensive behavior and the effects of the skin toxins on a variety of mammalian predators. Bufo alvarius is listed as endangered by the New Mexico Department of Game and Fish (NMGF, 1990). Fouquette (1970) reviewed this species and reported it to be one of the least known of the North American toads.

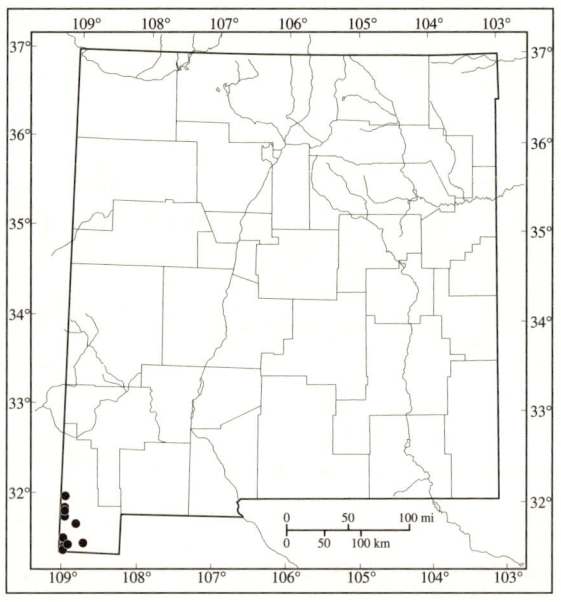

Distribution of Bufo alvarius *in New Mexico.*

BUFO BOREAS Baird and Girard, 1852 — *Western Toad*

See Plate 9.

Type: Baird and Girard (1852b) described *Bufo boreas* and gave the type locality as "Columbia River and Puget Sound." This was later restricted to "vicinity of Puget Sound," by Schmidt (1953). Syntypes are USNM 15467–15470, males collected by C. J. Pickering during May, 1841.

Distribution: *Bufo boreas* occurs in western North America from southern Alaska, through western Canada and to southern Colorado, Utah, Nevada, and northern Baja California, Mexico (Frost, 1985). It is absent from most of the arid Southwest. In New Mexico, *B. boreas* is known only from north-central Rio Arriba County at elevations between 2775 and 3200 m in the San Juan Mountains at Canjilon Lakes, Trout Lakes, and Lagunitas Lakes (Stuart and Painter, 1994). Campbell (1970a) reported an individual collected from a small lake at 3557 m elevation in Rocky Mountain National Park in Colorado.

The distribution and abundance of *Bufo boreas* in Colorado have declined considerably in the last 20 years (Corn, 1993; Corn et al., 1989; Carey, 1993) This is likely true for the limited population in New Mexico. (See Remarks.)

Description: This is a medium-sized toad with cranial crests that are absent or only faintly evident. Warts are large and numerous with particularly large warts on the dorsal surface of the tibia. The well-separated, tear-shaped parotoid glands are about one and one-half the length of the upper eyelids. The overall coloration of *B. boreas* is dark brownish to black, and there is a distinct, whitish vertebral stripe, which may be broken into segments of variable length. This middorsal stripe is usually lacking in juveniles. The tympanum is round and smaller than the eye. The tarsal fold, extending from the inner metatarsal tubercle almost to the heel, is well developed. There is often a well-defined light patch on the lower eyelid. The venter is buff or whitish, with considerable dark spotting on the chest. The underside of the feet may be yellowish, especially in the juveniles. Maximum SVL is usually less than 90 mm.

The tadpole body is broadest between the eye and the spiracle. The eyes are nearer to the lateral margin of the body than to the dorsal midline. The tail fins are narrow, widest near the midpoint of the tail. The greatest height of the dorsal fin is nearly the same height as the musculature at the tail base. The ventral fin is similar in height to the dorsal fin. The anus is medial and emerges in the ventral fin. The spiracle is on the left side of the body near the middle. The general coloration is dull blackish with the underside slightly paler with little or no iridescence. The tail fin is clouded; it is without spots and marked with dense, minute melanic stippling, especially on the dorsal fin. The muscular portion of the tail is black. The labial tooth row formula is 2/3 (Stebbins, 1962).

Similar Species: At high elevations of northern New Mexico the only anuran sympatric with *B. boreas* is *Pseudacris triseriata*. Adults of these two species may be easily separated by the smooth skin and lack of warts in *P. triseriata*. Tadpoles of *B. boreas* are uniformly black and the eyes are located dorsally. *Pseudacris* tadpoles are brown, and the eyes protrude from the dorsolateral edge of the squarish body.

The lack of cranial crests and shape of the parotoid glands will separate *B. boreas* from all other toads in New Mexico. Spadefoots (*Scaphiopus* and *Spea*) may be distinguished by smoother skin, a single metatarsal tubercle, and the vertically elliptical pupil.

Systematics: Three subspecies of *B. boreas* are listed by Collins (1990); only the nominotypical form described by Baird and Girard (1852b) occurs in New Mexico. Steve Corn (pers. comm.), based on work by Anna Goebel, suggested that populations of *B. boreas* in the southern Rocky Mountains may represent an undescribed species.

Habitat: Although reported to prefer beaver ponds, *B. boreas* may be found in high-elevation lakes, slow-moving streams, or low marshy areas. Dominant vegetation around these areas in New Mexico where *B. boreas* has been collected consists of corkbark fir, Engelmann spruce, aspen, willows, and various sedges. Adults often seek cover in high grass surrounding wet areas.

Behavior: *Bufo boreas* is well known for its activity at low temperatures. Adult toads may be active during day or night at air temperatures as low as 3°C. During the day they may sit in contact with the moist substrate in the sun where their body temperatures may be elevated to 30°C or higher for several hours (Carey, 1978). Larvae may exhibit the same behavioral thermoregulatory patterns by remaining immobile in the deeper, colder water at night and moving to shallower, warmer water during the day. Both tadpoles and juveniles tend to aggregate at the warmest site available.

These toads may overwinter in hibernacula. Campbell (1970b) reported such hibernacula may be natural underground chambers with a continuous flow of groundwater beneath the floor of the chamber. Temperatures within these chambers remain near 0°C.

Jones (1978) found toads active throughout the day with a period of reduced activity between 1200–1400. Movements of toads recorded by Jones (1978) averaged 6.7 m/day. Escaping toads usually jump into the water and seek refuge under bottom debris or in mud.

Reproduction: Due to short growing seasons and cold nocturnal temperatures which limit food availability and restrict attainment of high body temperatures necessary for digestion, growth rates of montane populations of boreal toads are very slow. Sexual maturity of males and females occurs within four and six years and at 54 mm and 75 mm SVL (Carey, 1976, 1987). Egg clutches may be laid soon after the first snow melt during mid- to late June with metamorphosis occurring approximately two months after egg-laying, usually during late August or early September. Carey (1987) found an egg clutch at Lagunitas Lakes that was laid on 12 June. Two large females (86 and 92 mm SVL) collected 4 September and 10 August from the same area were found to be gravid (J. N. Stuart, pers. comm.). Jones (1978) found metamorphosing toadlets in early August, suggesting breeding during May or June in north-central New Mexico. *Bufo boreas*, breeding at 3000 m in the Colorado Rockies, laid egg clutches ranging in size from 3,239–8,663 (mean 5,213) (Carey, 1976). In the higher elevations of Colorado, the tadpoles may overwinter beneath the ice and metamorphose during the following summer (Campbell, J. B., 1972).

Male *B. boreas* generally lack a vocal pouch and even in breeding aggregations produce only release calls. Awbrey (1972), however, reported an adult male from California that produced what he interpreted as an advertisement call from a well-developed vocal pouch.

Food Habits: No study of the food habits of *B. boreas* has been carried out in New Mexico, although studies elsewhere indicate that various beetles, wasps, and flies comprise the principal arthropod taxa ingested. It is likely that this species consumes a wide variety of invertebrates including snails, moths, beetles, and spiders. In Boulder County, Colorado, Campbell (1970c) found the diet to include 71% Hymenoptera (Formicidae) and 14.6% Coleoptera.

Remarks: *Bufo boreas* was not discovered in New Mexico until June, 1966 (Campbell and Degenhardt, 1971), and since then numerous studies of the status of the species in New Mexico have been conducted. For unknown reasons, it has declined rapidly and may no longer exist in New Mexico. Jones (1978) estimated a population of 327 toads occurred at Upper Lagunitas Lake during 1978. Jones (1978) noted reproduction and high densities of toads at Lagunitas Lakes, although he observed no reproduction at Trout Lakes or Canjilon Lakes and suggested low densities of toads in these areas. Woodward and Mitchell (1985) surveyed 139 ponds in known or expected *B. boreas* habitat in Rio Arriba County and located adults or tadpoles in only two ponds. They concluded that breeding was confined to Lower Lagunitas Lake where the total adult population was possibly as low as 4–7 individuals and that the previously known populations at Trout and Canjilon lakes had vanished. Later, Carey (1987) found 56 toads at Lagunitas Lakes during 1986. The majority of these were year-old toads (20–30 mm SVL), with only four individuals exceeding 40 mm SVL. She speculated that the presence of large numbers of introduced fathead minnows, *Pimephales promelas*, in the ponds where *B. boreas* lay eggs may decrease the reproductive success of the toad population by causing energetic stress for the larvae. Corn (1993) felt that mortality, even from natural predators such as ravens, may be a serious threat for some of the remaining populations of *B. boreas* in the southern Rocky Mountains. Olson (1989) presented data on raven predation on breeding *B. boreas* in Oregon.

In a letter to the New Mexico Department of Game and Fish, Cynthia Carey stated that she could not find any eggs or toads during a one-day search at Lagunitas Lakes during June 1987. John P. Hubbard and James N. Stuart (pers. comm.) visited the same area on different occasions during July–August 1993 and were also unable to confirm the continued existence of *B. boreas* in New Mexico. Carey (1993) believed the population at Lagunitas Lakes to be extinct. Corn et al. (1989) reported finding boreal toads at only 17% of 59 known localities in southern Wyoming and northern Colorado. Blaustein et al. (1994a), working in western Oregon, demonstrated that hatching success was greater in *B. boreas* embryos that were shielded from UV radiation compared to those that were exposed to UV radiation.

The Rocky Mountain population of *B. b. boreas* is listed as a Category 2, Notice of Review species by the U.S. Fish and Wildlife Service (USDI, 1994) and as endangered by the New Mexico Department of Game and Fish (NMGF, 1990).

Stuart and Painter (1994) presented a review of the distribution and status of *B. boreas* in New Mexico.

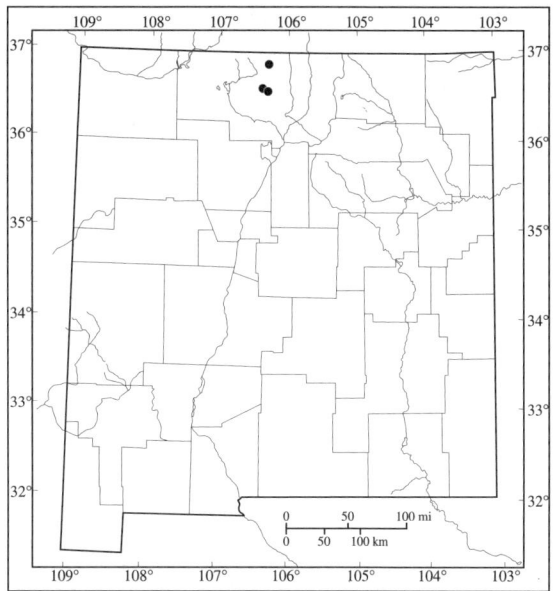

Distribution of Bufo boreas *in New Mexico.*

BUFO COGNATUS Say *in* James, 1823 — *Great Plains Toad*

See Plate 10.

Type: *Bufo cognatus* was described by Thomas Say *in* James (1823). The type locality was stated as "The alluvial fans of the [Arkansas] River," a locality in Prowers County, Colorado. The holotype was deposited in the Philadelphia Museum, although confusion exists about the fate of the type specimen. According to Baird and Girard (1853b), it apparently was destroyed by fire, while Kellogg (1932) reported that the specimen was sold in 1850 and may have been part of the P.T. Barnum "American Museum" that was destroyed by fire in 1865. The collector and date of collection is unknown (Krupa, 1990b).

Distribution: *Bufo cognatus* ranges from central Missouri, western Minnesota, and Iowa westward to central Montana and southeastern California and Nevada, and from southern Manitoba to Alberta southward in Mexico to Aguascalientes and San Luis Potosí (Krupa, 1990b). In New Mexico, *B. cognatus* occurs statewide in arid and semi-arid areas except for the extreme northwestern counties. It is often encountered in the southwestern desert regions of the state, although it ranges from approximately 900 m in Eddy County where the Pecos River leaves the state to at least 1900 m in Colfax County along the Cimarron River.

Description: *Bufo cognatus* has prominent cranial crests that contact the parotoid glands and diverge widely posteriorly. The length of the parotoid glands is distinctly less than the distance between their anterior ends. Dorsal coloration varies from gray, brown, or brownish yellow to the more typical greenish hue. The dorsal skin is covered with numerous uniform small warts and large, dark dorsal blotches bordered by white or light tan, which commonly contain a large number of warts. A middorsal stripe is sometimes present and is typically faint but occasionally distinct. There are two prominent, horny, and usually dark brown to black plantar tubercles on each hind foot. The venter is lightly colored and rarely spotted. Males have a dark throat and a sausage-shaped vocal sac that extends forward and above the snout when inflated. Bragg (1950b) measured 702 adults from Oklahoma; males were 63–103 mm SVL, females 49–112 mm SVL. Krupa (1994) measured 849 breeding adults; 758 males were 56–98 mm SVL; 91 females were 60–115 mm SVL. Conant (1975) gave a record size of 114 mm SVL. Bragg (1958 [1959]) reported a population from central Oklahoma that contained a relatively high frequency of melanistic individuals. The call is a shrill, piercing, metallic trill, sustained and often lasting 20 seconds or more. Call pulse rate is 13–20 per second (Conant and Collins, 1991).

The small tadpoles are totally black but begin to assume a definite color pattern at about 8 mm TL, and at about 25 mm TL (appearance of hindlimbs) the body is mottled brown and gray dorsally and lighter ventrally. The tail fin is highly arched with pigmentation primarily on the dorsal fin. The iris of the eye is golden. The labial formula is 2/3 (Bragg, 1936).

Similar Species: The large blotches bordered with white or light tan should serve to distinguish *B. cognatus* from all other anurans in New Mexico. Calling males may be distinguished by their unique sausage-shaped vocal sac and very long call.

Systematics: No subspecies of *Bufo cognatus* are recognized (Krupa, 1990b). Rogers (1972) compared morphological characteristics of *B. cognatus*, *B. compactilis*, and *B. speciosus*. Sullivan (1990) reported a natural hybrid between *B. cognatus* and *B. punctatus* from central Arizona.

Habitat: *Bufo cognatus* is characteristic of warmer grassland areas in New Mexico, rarely venturing into upland woodlands. Normal habitat includes prairie grasslands, creosotebush desert, mesquite woodland, sagebrush plains, and river bottoms. This toad is relatively common in cultivated or other man-made environments, such as the agricultural areas of the Rio Grande Valley where large choruses are common during the summer rains. It seems to thrive in areas drier than those occupied by most other toads in New Mexico.

Behavior: *Bufo cognatus* is most active during the summer rains when it is primarily nocturnal. At other times this toad remains in underground burrows or retreats; it is often very common on the roadways at night after summer rains. *Bufo cognatus* is a proficient burrower.

Reproduction: James J. Krupa has contributed significantly to the understanding of the breeding biology of *Bufo cognatus* in Oklahoma. Unless credited otherwise, most of the following information is taken from his published research (Krupa 1986a, 1986b, 1988, 1989, 1990a, 1991, 1994). As with most southwestern toads, *B. cognatus* breeds opportunistically with rainfall, exhibiting a relatively short, high-density mating pattern. In 10 choruses reported by Sullivan (1989), average breeding period duration was 2.6 days. Rain-filled ditches or temporary ponds are usually selected as breeding sites, although *B. cognatus* will breed in permanent ponds and irrigated fields as well. Breeding occurs at any time suitable temperature and moisture conditions prevail, although most large breeding choruses are formed during the summer monsoon season. Bragg (1950c) observed 143 breeding congresses of *Bufo cognatus* in Oklahoma. The earliest date he recorded was 28 March, the latest 9 August. Krupa (1986a) reported an earlier breeding date of 20 March. In southwestern New Mexico, large choruses of 200–500 individuals have been found on 12 and 21 July (Brown and Pierce, 1967). We noted a large chorus on 7 June in an irrigated field just east of Bernardo in Socorro County. In eight choruses reported in Sullivan (1989), the mean number of males was 25.2.

Males precede females to breeding sites and, shortly after sunset, set up large choruses to attract the females. In these choruses, male *B. cognatus* exhibit two mating tactics. Whereas most males give advertisement calls to attract the females, other males, termed "satellites," remain silent and position themselves near calling males in an attempt to intercept females. Up to 74% of the males marked during a study in Oklahoma switched between these two tactics, and the percentage of males that exhibited satellite behavior varied from 0–57%. Krupa (1990a) demonstrated the influence of body size and temperature on the advertisement call of *B. cognatus*. Males and females generally do not breed until 60 mm SVL or larger, although occasional males may breed at 56 mm SVL. Breeding females are significantly larger than breeding males. Average duration of amplexus is 13.3 hours. The clutch of an average-sized breeding female (ca. 82 mm SVL) is 11,074 eggs, although clutches may range from 1,342–45,054 eggs with clutch size being directly related to female size. Some females may lay two clutches a year, and many pairs exhibit communal egg-laying, e.g., laying eggs in close proximity to one another at a specific location in a breeding pool. The larval period (from the day eggs are laid to the day when toadlets emerge onto the shore) varies from 18–49 days. Bragg (1937) found that tadpoles of B. cognatus may start metamorphosis within 1–1.5 months after fertilization. The tiny, newly metamorphosed toadlets average 10 mm SVL (Johnson, 1987). Lewis (1950) found gravid females as late as 7 September and metamorphosing tadpoles in late September and early October in south-central New Mexico.

Food Habits: Dimmitt and Ruibal (1980b) found that the diet of *Bufo cognatus* in southwestern New Mexico consisted of 47% ants and 44% termites. Other prey items included beetles, crickets, spiders, centipedes, and mites.

Remarks: *Bufo cognatus* can store as much as 30% of its gross body weight as water in the urinary bladder. The adaptive value of this reserve is tremendous as it permits them to survive long periods of drought, and allows them to forage far from their water source (Ruibal, 1962). Krupa (1990b) reviewed this species.

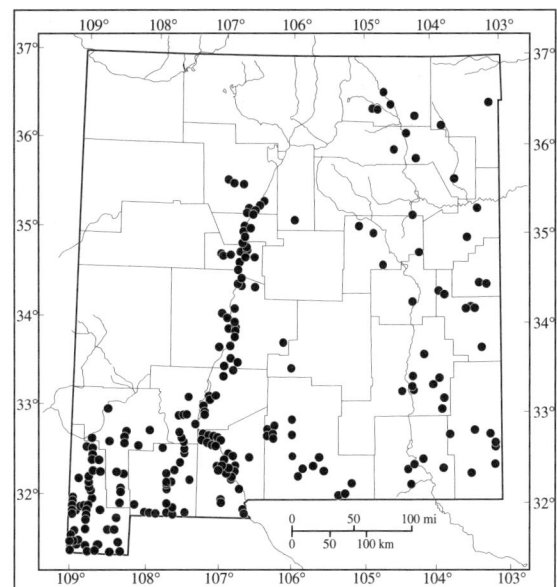

Distribution of Bufo cognatus *in New Mexico.*

BUFO DEBILIS Girard, 1854 — *Green Toad*

See Plate 11.

Type: Charles Girard described *Bufo debilis* in 1854 from specimens, "found in the lower part of the Rio Bravo (Rio Grande del Norte), and in the Province of Tamaulipas." Kellogg (1932), by inference, restricted the type locality to Matamoros, Tamaulipas, Mexico. Flores-Villela (1993) pointed out that both Schmidt (1953) and Frost (1985) erroneously stated that the type locality was restricted to Brownsville, Texas, by Sanders and Smith (1951) although there is no such restriction in that publication. Considerable confusion exists regarding designation of a holotype. Bogert (1962) suggested the holotype may have been collected by Arthur Schott in Tamaulipas, Mexico. Frost (1985) lists USNM 2621 (8 specimens) as the syntypes. No date of collection is given in the original description.

Distribution: *Bufo debilis* ranges from southeastern Colorado and southwestern Kansas to Zacatecas and from southeastern Arizona to eastern Texas. In New Mexico, *B. debilis* is widespread in the eastern and southern counties, generally avoiding the north-central and northwest areas and ranging from approximately 900–1500 m.

Description: The small size and green coloration make this little toad easy to identify. The body and head are noticeably flattened. The head is broader than long, the width a little less than one-half that of the body. Viewed from above, the head is wedge-shaped and sharply truncate at the tip. The cranial crests are reduced and surmounted by a discontinuous series of black-tipped warts. The interorbital space is slightly concave. The parotoid glands are large and elongate, situated obliquely on the shoulders. The tympanum is small and indistinct. There are numerous warts on the dorsal surface, although these are small and less conspicuous than in other *Bufo*. Throat color in males is dusky or black; in females, it is yellow or white. Males are slightly smaller than females, averaging approximately 35.1 mm SVL; females average 36.5 mm (Savage, 1954; Bogert, 1962). The call is a cricket-like trill held at one pitch and lasting 3–7 seconds (Conant and Collins, 1991). The dominant frequency is the highest (3180–3699 cps) of any toad in the Southwest; trill rate is high, 120–128 trills per second at 18–23°C (Blair, 1956).

The tadpole body is rounded, with the spiracle almost halfway up the left side, and the anal tube opening on the right side of the tail fin. The eyes are situated dorsally. The tail fins are only moderately developed, with the tail fin depth/total length ratios varying from approximately 16–26% and averaging 20%. The total length at hatching is approximately 3.1–3.4 mm. Labial tooth row formula is 2/2. When the mouth parts are well developed, there are two rows of teeth anterior to the beak and two rows posterior. The second anterior row is broadly broken in the middle. The third and fourth rows are virtually the same length, although they are complete and shorter than the first row. Oral papillae are few and confined to the corners of the mouth. The abdominal region is black laterally with golden flecks. In the latest stages of development, tadpoles are lightly stippled with melanophores, although, in contrast to the dense black aspect common to many *Bufo* tadpoles, they are relatively transparent even when well grown (Bragg, 1955; Zweifel, 1970).

Similar Species: Considering the small adult size, distinctive greenish coloration, and flattened body, *Bufo debilis* should not be confused with any other anuran native to New Mexico.

Systematics: Two subspecies are currently listed in Collins (1990), although Taylor (1938 [1939]) recognized these as two distinct species, *B. debilis* and *B. insidior*. Bogert and Oliver (1945), noting the close relationship between the two forms, suggested they were probably subspecies. The western green toad, *B. d. insidior*, occurs in New Mexico. This form was originally described as *B. insidior* by Girard (1854) from specimens collected in Chihuahua, Mexico. See Ferguson and Lowe (1969) for a discussion of relationships with *B. kelloggi*, *B. punctatus*, and *B. retiformis*.

Habitat: Most records from New Mexico are below 1300 m elevation. *Bufo debilis* occurs primarily in the lower desert grasslands, or areas grown to mesquite and creosotebush.

Behavior: *Bufo debilis* is largely nocturnal, although occasionally males may be heard calling during the daytime after warm summer rains. Adults are secretive, usually seeking shelter in underground burrows or beneath surface objects including rocks, old lumber, or dry cow dung. They are not usually seen other than during the breeding season.

Reproduction: *Bufo debilis* breeds in shallow temporary rain pools, fishless stock ponds, or along intermittent streams. Arrival at breeding sites coincides with the onset of summer rains and generally takes place during the first week in July, with males usually calling from exposed positions on muddy banks. Other calling males may be found floating throughout the pond or hiding amongst streamside vegetation. Of five choruses reported in Sullivan (1989), average breeding period duration was 2.6 days; mean number of males per chorus was 11.4. Satellite males are unknown in this species; amplexus is initiated when a female approaches a calling male and actually makes physical contact (Sullivan, 1984). We noted breeding on 13–14 June in Guadalupe County. Newly transformed toadlets have been found in Hidalgo County as late as 27 September, although most metamorphosis normally takes place during late July. At the Pierce Ranch in eastern Chaves County, M. W. Doles (pers. comm.) reported calling males first appeared at Lone Wolf Well on 19 July, and newly metamorphosed toadlets emerged from the same pond on 14 August. At emergence, five toadlets averaged 15.4 (14.8–15.9) mm SVL. In southeastern Arizona, Zweifel (1968b) found *Bufo debilis* required temperatures of 18.2–33.8°C for normal development of the eggs and larvae. Within this temperature range, development was notably faster at higher temperatures.

Food Habits: The food habits of *B. debilis* have not been studied in New Mexico or elsewhere. They likely

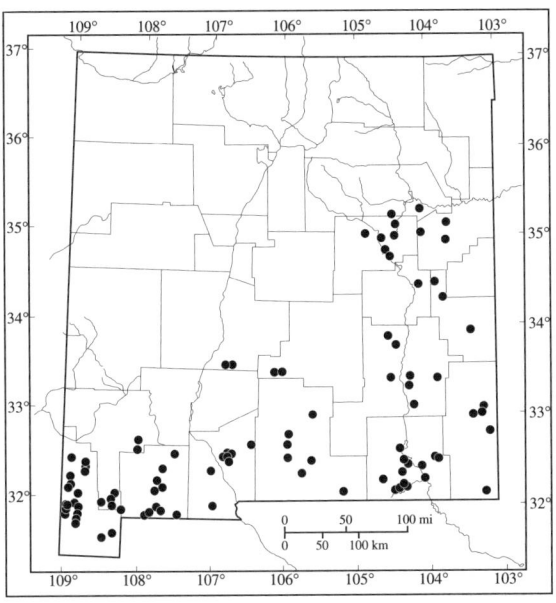

Distribution of Bufo debilis in New Mexico.

consume a large variety of small invertebrates including ants, beetles, and moths.

Remarks: There is little known about the natural history of this toad.

BUFO MICROSCAPHUS Cope, 1866 [1867] *Southwestern Toad*

See Plate 12.

Type: *Bufo microscaphus* was described by E. D. Cope (1866 [1867]). He gave the type locality as "Arizona . . . near the parallel of 35 degrees and along the valley of the Colorado from Fort Mojave to Fort Yuma." This was later restricted to Fort Mohave, Mohave County, Arizona, by Shannon (1949) who designated USNM 4184, an adult female collected by H. B. Möllhausen, as the lectotype. The date of collection is unknown (Price and Sullivan, 1988).

Distribution: *Bufo microscaphus* is found in a series of isolated populations from the Mogollon Plateau of southwestern New Mexico westward to the Colorado and Virgin river basins of northwestern Arizona, southern Nevada, and southwestern Utah. Other forms range west of the deserts in southern California and northern Baja California, Mexico, and at high elevations in the Sierra Madre Occidental of Chihuahua, Durango, and Sonora, Mexico. An isolated population occurs along a headwater of the Río Aguanaval in northwestern Zacatecas, Mexico (Price and Sullivan, 1988). In New Mexico, *B. microscaphus* has been collected in Catron, Grant, and Sierra counties. Fitzgerald and Degenhardt (1986) reported this species from Sierra and Socorro counties, the first localities known from east of the Continental Divide. The species is common along the rivers and lakes of the Gila and San Francisco river drainages. Sullivan (1993) reported on the status of *B. microscaphus* in Arizona.

Description: *Bufo microscaphus* is a small to medium-sized toad with cranial crests weak or absent. The dorsum is covered with small, uniformly sized warts. The parotoid glands are elongate, oval, widely separated, and approximately twice as long as wide. The sacral humps and the anterior half of the parotoid glands and eyelids are usually pale-colored. No pale middorsal stripe is present. The belly is whitish. Breeding males lack a dark throat, and the vocal pouch is pale and round when inflated. Females are larger than males, with adults ranging in size from approximately 50–84 mm body length. Sullivan (1986a) reported on the morphology of hybrids between *B. microscaphus* and *B. woodhousii* in central Arizona. The mating call is a trill lasting 8–10 seconds. At 20°C, the advertisement call pulse rate is around 60 pulses per second, the release call 70 pulses per second (Price and Sullivan, 1988).

The tadpole morphology is typical of other *Bufo* tadpoles in New Mexico. There is an emarginate oral disc present. The spiracle is single, sinistral, and near the longitudinal axis. The tail musculature is light with dark blotches. The dorsum is light colored. Stebbins (1962) described the tadpoles, although he suggested his description was of *B. microscaphus–B. woodhousii* hybrids.

Similar Species: In New Mexico, only *B. woodhousii*, *B. cognatus*, or *B. speciosis* may be confused with *B. microscaphus*. *Bufo woodhousii* may be separated by the presence of a prominent middorsal stripe and parotoid glands, which are in contact with heavy cranial crests. *Bufo cognatus* has cranial crests that join at a boss on the snout and large light-edged (often paired) dark spots on the back. *Bufo speciosus* lacks prominent cranial crests and has oval parotoid glands that are less than twice as long as wide. Juveniles of all of these species are difficult to distinguish. Geography often helps with identification; only *B. woodhousii* and *B. microscaphus* are sympatric in southwestern New Mexico.

Systematics: Three subspecies of *Bufo microscaphus* are recognized (Price and Sullivan, 1988). Only the nominotypical form described by Cope (1866 [1867]), occurs in New Mexico. In the literature, *B. microscaphus* has been confused with *B. boreas*, *B. woodhousii*, and *B. compactilis*. In spite of the hybridization of *B. microscaphus* and *B. woodhousii* reported by Sullivan (1986a), he suggested that the continued recognition as separate species is appropriate given the narrowness of this hybrid zone and the distinctive morphological and behavioral differences.

Habitat: *Bufo microscaphus* is usually associated with permanent ponds or rocky streams with relatively shallow water flowing over sandy or rocky bottoms. It is generally found in unaltered riparian areas grown to sycamore or cottonwood, and in grasslands, piñon-juniper, or ponderosa pine and is less likely to be found along agricultural fields, irrigation ditches, or meandering broad river bottoms than most other species of *Bufo* in New Mexico. In New Mexico, *B. microscaphus* ranges from 1900–2700 m.

Behavior: The harsh desert environment that surrounds the area where these toads live is severely limiting, and they do not move away from permanently watered areas. They are sometimes encountered sitting in shallow river backwater areas at night.

Reproduction: *Bufo microscaphus* has an explosive breeding period, normally breeding in early spring shortly after the snow melt in New Mexico. In Catron County, R. D. Jennings (pers. comm.) observed these toads in amplexus under the ice in March. Unlike many toads in New Mexico, *B. microscaphus* does not depend upon spring or summer rains to stimulate breeding activity. This may be a result of breeding in streams and ponds where there is usually permanent water. Photoperiod and temperature seem to be the cues that stimulate breeding. Sullivan (1992) reported variation in the calling behavior in *B. microscaphus* and found that the pulse rate and duration of the advertisement call were significantly correlated with temperature in a central Arizona population. Active searching or satellite behavior has not been described for *B. microscaphus*.

Dorcas and Foltz (1991), studying the effects of variation in the physical environment on anuran advertisement calling, found *B. microscaphus* called almost exclusively at night and infrequently during rain. They suggested that rain may interfere with call transmission and stated that because calling is energetically expensive and may increase predation risks, the benefits of calling during rain may not be sufficient to offset the costs associated with calling.

In New Mexico, gravid females have been found in early April. Calling and egg laying were observed by Sullivan (1992) as early as 7 February in Arizona. Mature tadpoles have been found in late May in central Arizona (Blair, A. P., 1955) and newly metamorphosed young between 29 June and 2 July in Chihuahua (Riemer, 1955). We found newly metamorphosed young to be abundant at 1952 m elevation in Catron County at Wall Lake on 13 July 1969. The tadpoles metamorphose at approximately 15–20 mm SVL; 8 toadlets, with the tail bud remaining, were collected in southern Catron County on 10 July and averaged 15.3 mm (14.2–16.5) SVL (our unpubl. data). The eggs of *B. microscaphus* are laid in long strings in typical bufonid fashion. Albert P. Blair (1955) counted five clutches that varied from approximately 3,000–4,300 eggs. Males first show secondary sexual characteristics at approximately 46 mm SVL.

Sullivan (1986a, 1993) suggested the breakdown in reproductive isolation between *B. microscaphus* and *B. woodhousii* may be the result of ecological disturbance. He speculated the alteration of riparian habitats associated with the construction of impoundments may lead to continued hybridization if populations of *B. microscaphus* are forced to utilize the aquatic habitats occupied by *B. woodhousii*. Sullivan and Lamb (1988) suggested that the hybrid zones between these species may be unstable and speculated that *B. woodhousii* is replacing *B. microscaphus* along streams in south-central Arizona.

Food Habits: The food habits of *B. microscaphus* have not been studied in New Mexico, although they likely consume a wide variety of small arthropods, including true bugs, beetles, and moths. Five specimens from Utah reported in Tanner (1931) contained a sand cricket, a variety of beetles, a true bug, ants, bees, moth larvae, snails, and plant fragments.

Remarks: *Bufo microscaphus* was first reported from New Mexico by Stebbins (1954), when he mapped (with query) a specimen from Mimbres, Grant County. The existence of this species in New Mexico was not verified until Findley (1964) reported four specimens collected in Catron County.

This species was reviewed by Price and Sullivan (1988).

Bufo microscaphus is listed as a Category 2, Notice of Review species by the U.S. Fish and Wildlife Service (USDI, 1994).

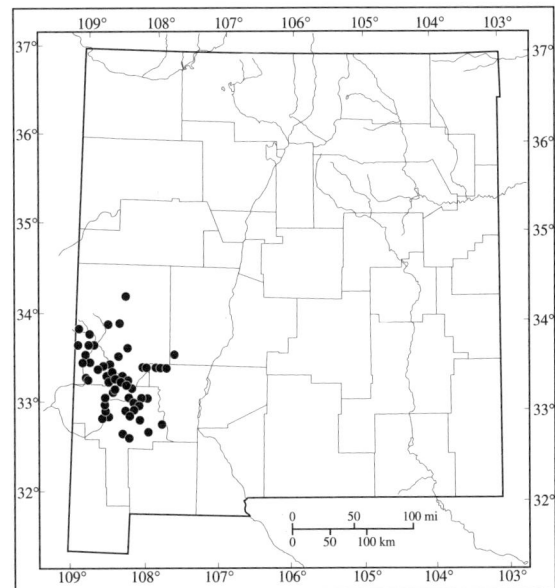

Distribution of Bufo microscaphus in New Mexico.

BUFO PUNCTATUS Baird and Girard, 1852 — *Red-spotted Toad*

See Plate 13.

Type: Spencer Fullerton Baird and Charles Girard (1852b) gave the type locality as "Rio San Pedro of the Rio Grande Del Norte." No type specimen was designated. Syntypes are USNM 2618 (3 specimens), collected by J. H. Clark, although these are apparently lost.

Distribution: *Bufo punctatus* ranges from southwestern Kansas and central Texas to southeastern California, and far south into mainland Mexico and to the southern tip of Baja California (Conant and Collins, 1991). *Bufo punctatus* occurs statewide in New Mexico at elevations ranging from 900–2200 m. It is most commonly encountered in the southern counties and northward throughout the Rio Grande and Pecos river drainages.

Description: *Bufo punctatus* has a short squat body and head that are noticeably flattened. Dorsal coloration is reddish brown to gray with small red spots that are usually on dermal tubercles. The oval parotoid glands are approximately the size of the eye. Cranial crests are weak or absent. The venter is white to cream-colored, although males have a darkened throat. Record body size is 76.2 mm (Conant and Collins, 1991), although 38–63 mm SVL is average. The call is a clear musical trill, high and essentially on one pitch with a duration of 4–10 seconds. The call interval is variable, sometimes longer or sometimes shorter than the call itself (Conant and Collins, 1991).

Bufo punctatus tadpoles are generally black with a

bronze-flecked venter. Larger tadpoles have faint mottlings of lighter color on a blackish background color. The tail fin is translucent with evenly spaced black dots. The tail, including the fins, is about the same height as the greatest body depth. The greatest height of the dorsal and ventral tail fins are similar, and each is the same as the greatest width of the tail musculature. The dorsal tail fin originates well posteriorly on the body and ends at the rounded tail tip. The spiracle is sinistral. The body is ovoid in shape and is broadest behind the midpoint. The eyes are placed well up on the head. The iris is bronze. The labial tooth row formula is 2/3, with a reduced papillary fringe confined to the lower half of the sides of the oral disc (Stebbins, 1951, 1954).

Similar Species: The flattened body shape and the round parotoid glands will distinguish B. *punctatus* from all others in New Mexico. Only B. *debilis* has the same body shape and it has large, triangular parotoid glands and is greenish in color.

Systematics: No subspecies of *Bufo punctatus* are recognized.

See Ferguson and Lowe (1969) for a discussion of the relationships with B. *kelloggi*, B. *debilis*, and B. *retiformis*.

Habitat: *Bufo punctatus* is most often encountered in dry rocky areas at lower elevations, where it typically occurs near desert springs, persistent pools along rocky arroyos, or less commonly, around cattle tanks. The toad's small size and relatively flat shape probably helps it retreat under rocks or into rock crevices during dry weather.

Behavior: *Bufo punctatus* is largely nocturnal, although it may occasionally be found abroad during mornings and early evenings. It is active and alert and is often found under large flat rocks where it seeks shelter during the day. It is often encountered on the roadways at night during summer rains.

Reproduction: *Bufo punctatus* generally calls on exposed boulders or rocks at the water's edge, although males will call from shallow water. Although large choruses are found, small choruses of fewer than 10 males are more common. Of five choruses reported in Sullivan (1989), average breeding period duration was 20 days; mean number of males was 3.0. In southern Arizona and adjacent areas, the males of B. *punctatus* call when air temperatures range from 19–29°C, but seldom when water temperatures are in excess of 27–28°C (Ferguson and Lowe, 1969). We have found B. *punctatus* breeding at 2150 m elevation in Cedro and Sabino canyons in the Manzano Mountains of central New Mexico. Satellite males are unknown in B. *punctatus*, and amplexus is initiated when a female approaches a calling male and makes physical contact (Sullivan, 1984).

The eggs of B. *punctatus* are the most distinctive of any toad in North America. *Bufo* eggs typically are laid in strings, however those of B. *punctatus* are laid singly on the bottom of pools or streams. The eggs are protected by a layer of sticky gelatin and tend to adhere together in masses a single layer deep (Smith, 1950). We have collected eggs and tadpoles on 31 May in Harding County in northeastern New Mexico. The tadpoles metamorphose at approximately 13–18 mm SVL.

Food Habits: Food habits of B. *punctatus* have not been investigated in New Mexico and little information is available from elsewhere. Beetles, bees, bugs, ants, and various other arthropods have been reported in the diet (Smith, 1934, 1950; Little and Keller, 1937; Tanner, 1931). Little and Keller (1937), at the Jornada Experimental Range near Las Cruces, found that adults in May and June generally had empty stomachs.

Remarks: Little is known about the natural history of B. *punctatus*. It seems to have no close relative among United States toads. Blair (1956) reported the call to show striking similarities to the federally endangered B. *houstonensis*.

FAMILY BUFONIDAE

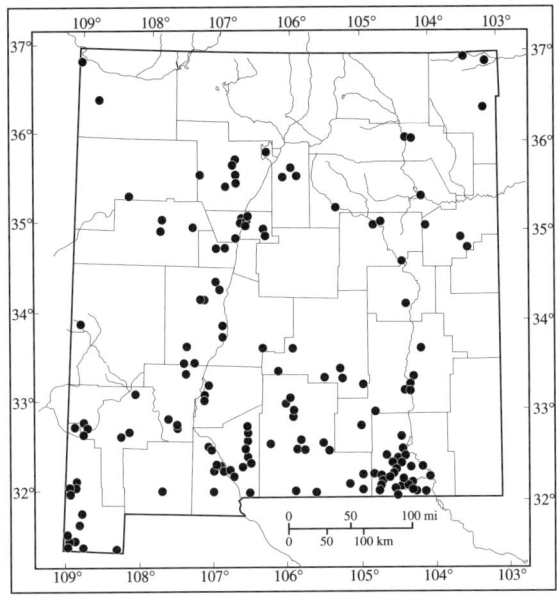

Distribution of Bufo punctatus *in New Mexico.*

BUFO SPECIOSUS Girard, 1854 — *Texas Toad*

See Plate 14.

Type: Charles Girard (1854) designated the type locality of *Bufo speciosus* as "the valley of the Rio Bravo (Rio Grand del Norte), . . . not uncommon in the province of New Leon." (See Remarks.) No type specimen was designated. Syntypes are USNM 2608, Ringgold Barracks (Rio Grande City, Starr County) Texas; USNM 2610, Brownsville, Cameron County, Texas, collected by S. Van Vliet; USNM 2611 and USNM 131559, Pesquieria Grande, Nuevo Leon, Mexico, collected by D. N. Couch during April 1853 (Cochran, 1961).

Distribution: *Bufo speciosus* ranges from northwestern Oklahoma and southern New Mexico across the eastern part of the Mexican Plateau and the Gulf coastal plain of Mexico (Blair, 1972). In New Mexico, this toad occurs only in the southeast in Lea, Chaves, and Eddy counties at elevations between 900–1300 m.

Description: *Bufo speciosus* has short oval parotoid glands less than twice as long as wide. The cranial crests are indistinct or absent. There are many uniformly small warts. The general coloration is olive to grayish brown above with small scattered irregular dark spots; the belly is unspotted. There is no middorsal stripe. On the underside of the foot there are two black tubercles that are sharp-edged with the inner one sickle-shaped. The male has a sausage-shaped vocal sac that extends up and outward, similar to that of *B. cognatus*. The call is a continuous series of loud explosive trills, each ½ second or more in length. The pulse rate is 39–57 per second, hence much more rapid than *B. cognatus* (Conant and Collins, 1991).

The tadpole is lightly pigmented dorsally. The tail musculature is light with dark lateral blotches that tend to form a stripe. The body is elongate oval in dorsal aspect, and is broadest about midpoint. The tail fins are low with the greatest height of the dorsal fin at midpoint. The eyes are well up on the head, and the nostril is located about midway between the eye and the tip of the snout. The spiracle is on the left side of the body. The anus is medial, opening at the edge of the ventral fin close to the juncture with the body. The labial tooth row formula is 2/3. The labial papillae are emarginated medially and in a single row on either side of the mouth (Stebbins, 1951).

Similar Species: *Bufo speciosus* is similar in appearance to *B. punctatus*, *B. woodhousii*, and *B. microscaphus*. *Bufo punctatus* has round parotoid glands and a more flattened body; *B. woodhousii* has prominent cranial crests and usually a distinct middorsal stripe; *B. microscaphus* has a pale bar across the eyes and lacks the sickle-shaped tubercles found in *B. speciosus*.

Systematics: No subspecies of *B. speciosus* are recognized. Although *Bufo speciosus* and *B. compactilis* were once considered subspecies of a single species (Smith, 1947), they are currently treated as distinct species, both within the *B. cognatus* species group (Blair, 1963). Additional supporting evidence for their separation is found in Bogert (1960) and Rogers (1973).

Habitat: *Bufo speciosus* shows a preference for sandy soils, and is generally found in short-grass plains or mesquite savannahs. It is abundant in the mesquite grassland areas along the Black River in southern Eddy County.

Behavior: After metamorphosis, the toadlets remain around the natal pool and disperse only when the pool evaporates or when they reach approximately 2–3 months of age, whichever comes first (Bragg, 1950a). These toads are primarily nocturnal, often encountered on the roadways at night or feeding on arthropods attracted to streetlights.

Reproduction: *Bufo speciosus* breeds only after summer rainfall in almost any standing water, and it may appear in deep muddy cattle tanks as well as roadside ditches or flooded low-lying grassy areas. As with most desert toads in New Mexico, with summer rainfall the males precede the females to the breeding sites and initiate a chorus to attract the females. When amplexus is achieved, the eggs are fertilized by the males and are laid in long strings. Wright and Wright (1949) reported the brown and yellow eggs to be crowded within the jelly tube, being 14–20 eggs per 30 mm. The eggs hatch in about two days and in 40–60 days transform into tadpoles about 12 mm TL.

Regarding breeding of *B. speciosus* in Oklahoma, Bragg (1950a) states "sexually stimulated males congregate in large numbers in a limited area and behave as though trying to attack each other; each actively tries to clasp those near him. Sometimes there are so many thus engaged that a pool may be a seething mass of struggling males. I have seen as many as five males piled on top of one another, each struggling to retain its sexual clasp upon the one next below." Breeding has been recorded in Oklahoma from mid-April through early August (Bragg, 1950b). In Texas, Wiest (1982) reported calling males from mid-April to early May, although he found no females or tadpoles present. In New Mexico, calling males have been recorded on 15 July, gravid females collected on 17–19 July, and tadpoles collected on 15 August. Ted L. Brown (pers. comm.) collected 30 newly transformed young, approximately 15–25 mm SVL, on 14 July in Eddy County.

Food Habits: No dietary studies of *B. speciosus* have been conducted in New Mexico, although they likely feed on a variety of terrestrial and flying arthropods, as do most desert toads.

Remarks: Smith and Taylor (1950b) restricted the type locality to Brownsville, Cameron County, Texas. The description of the tadpole was taken from Stebbins (1951) based on living material collected by A. N. Bragg in Greer County, Oklahoma. *Bufo speciosus* is known to hybridize with *B. cognatus* and *B. woodhousii* (Blair, 1972).

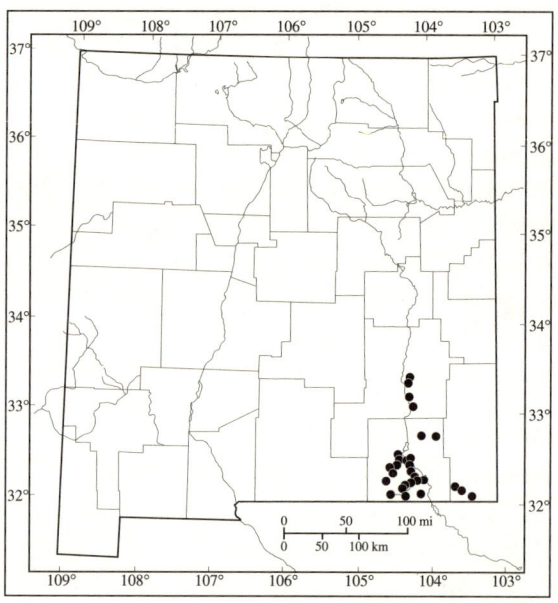

Distribution of Bufo speciosus *in New Mexico.*

BUFO WOODHOUSII Girard, 1854 — *Woodhouse's Toad*

See Plate 15.

Type: Charles F. Girard described *Bufo woodhousii* in 1854, although Hallowell (1852) had earlier described this form under the preoccupied name of *Bufo dorsalis*. The type locality was given by Girard (1854) as "[Territory of] New Mexico, having so far been found in the province of Sonora, and in the San Francisco Mts." This was later restricted to "San Francisco Mountain, Coconino County, Ariz." by Smith and Taylor (1948). The holotype is USNM 2531, collected by S. W. Woodhouse. The date of collection is unknown (Cochran, 1961).

Distribution: *Bufo woodhousii* ranges from northern Montana to Durango, Mexico, and from the Atlantic Coast to southeastern Washington, western Utah, and southeastern California. In New Mexico, *B. woodhousii* is generally found at lower elevations statewide where suitable soils and moisture conditions exist. It is most common at lower elevations in the Rio Grande Valley, although we have collected it at 2150 m in Sabino Canyon in the Manzano Mountains in Bernalillo County and at 2269 m just south of Vallecitos in Rio Arriba County. Elevational range of *B. woodhousii* in New Mexico is approximately 900–2400 m.

Description: *Bufo woodhousii* has dry skin covered with numerous irregular-sized warts with one to several warts found within the irregular dark dorsal spots. The overall coloration may be olive or greenish, and there usually is a prominent light middorsal stripe. The parotoid glands are oblong, approximately twice as long as wide. The cranial crests are prominent and often touch the parotoid glands. The belly is whitish, often with dark flecks on the chest between the forelegs. This is a large toad, often reaching 100–130 mm T. L. Sexual dimorphism is pronounced in this species. Males average smaller than females and often have a dark throat when sexually mature. As in most toads in New Mexico, the first and second digits on the forefeet of males have dark thickened horny patches. Newly transformed toadlets are approximately 12–15 mm SVL. Conant and Collins (1991) describe the call as a nasal "w-a-a-a-h," lasting 1–2.5 seconds and sounding like a sheep bleating in the distance.

The tadpole is heavily pigmented with dark brown or gray to slate. The dorsal musculature of the tail is lighter than the body; the ventral musculature is lighter than the dorsal musculature. The tail fins have a few scattered flecks of pigment, more numerous in the dorsal fin than the ventral fin, which is immaculate. The tadpole is small, with a maximum length of approximately 23 mm. The dorsal tail fin extends to a point less than halfway between the anus and vertical of the spiracle. The spiracle is sinistral, directed backward and upward at an angle of about 35–40°. The anus is medial, opening distinctly higher than the lower edge of the ventral tail fin. The labial tooth row formula is 2/3 (Youngstrom and Smith, 1936; Stebbins, 1951).

Similar Species: *Bufo woodhousii* may be confused with *B. boreas*, *B. microscaphus*, and *B. speciosus*. *Bufo boreas* has small, oval parotoid glands and usually a gland of similar size on the shank. The cranial crests of *B. microscaphus* and *B. speciosus* are indistinct or absent. Newly metamorphosed young are difficult to distinguish.

Systematics: Four subspecies of *Bufo woodhousii* are listed in Collins (1990). Two of these, the Woodhouse's toad, *B. w. woodhousii*, and the southwestern Woodhouse's toad, *B. w. australis*, occur in New Mexico. Specimens from a large part of northern and central New Mexico are considered intergrades. *Bufo w. woodhousii* is the form described by Girard (1854). Shannon and Lowe (1955) described *Bufo w. australis* and gave the type locality as "a damp irrigation ditch within the city limits of Tempe, Maricopa County, Arizona." The holotype, FAS 6817 [currently UIMHN 67056], is an adult male formerly in the private collection of Frederick A. Shannon of Wickenburg, Arizona. Paratypes were designated as follows: Charles H. Lowe, Nos. 3488, 6702–04, 6708, 6711 [currently UAZ 50067–50072] and FAS Nos. 1577–79, 1581–83, and 2918–2937.

Sanders (1987) described *Bufo antecessor*. The type specimen, USNM 25322, is an adult female from "nine miles northwest of Louiston (= Lewiston) Nez Perce County, Idaho." Paratypes include (Ottys Sanders) OS 169744, OS 170341, MVZ 10230, and UCM 6535. The range included portions of northern New Mexico; OS

169744 was collected 25 August 1953 from "Albuquerque, Bernalillo County, New Mexico." Pending further study, we do not recognize this form as a distinct species separate from *Bufo woodhousii*. Collins (1989) suggested the relegation of *B. antecessor* to synonymy with *B. woodhousii*.

Habitat: In New Mexico, *B. woodhousii* is generally confined to relatively mesic areas in the vicinity of streams and river valleys, often in agricultural areas and river floodplains. It is typically associated with larger river systems and other sources of permanent water, although it is occasionally found in more xeric habitat. We found an individual in San Miguel County on Glorieta Mesa where the habitat is ponderosa pine, grassland savannah. This toad seems partial to areas of sandy, loose soils.

Behavior: *Bufo woodhousii* generally lives near permanent water or floodplains, resting in shallow burrows during the day, feeding at dusk and after dark. It may be found feeding in gardens and beneath streetlights on warm summer evenings.

Reproduction: The breeding season may last from one to several weeks. Of 10 choruses reported by Sullivan (1989), average breeding period duration was 19.8 days. Bragg (1950c) observed 147 breeding aggregations of *Bufo w. woodhousii* in Oklahoma. The earliest date he recorded was 21 March, the latest 4 September. Gehlbach (1965) suggested that populations in northwestern New Mexico have a biannual breeding regime, corresponding to spring and summer peaks in precipitation. Currently, we have no data to support his suggestion.

Woodward (1982a) found that *B. woodhousii* in New Mexico breeds in temporary and permanent ponds over a 40–50 night breeding period from April to June. Only a small portion of the breeding aggregation is present on any night, and different males and females continue to arrive and leave the pond throughout the breeding season. Individual males spend about two nights in the breeding pond, females about one night (Woodward, 1984a); thus males spend on average almost twice as many nights in the pond as females. Bragg (1941) found tadpoles in flowing streams around Las Vegas in June, but was unable to find any in stock tanks. Bragg (1941) stated that *Bufo woodhousii* breeds more or less independently of rainfall at Las Vegas, New Mexico. Woodward (1987a) found adult female *B. woodhousii* from southern New Mexico averaged approximately 10,500 eggs. Krupa (1995) reported a clutch of 28,493 eggs laid 29 May by a 101 mm SVL *B. woodhousii* from Oklahoma. Eggs are laid in single long strings and may or may not be attached to vegetation. We encountered breeding choruses on 29 May in Harding County and on 13 June in Guadalupe County. Ted L. Brown (pers. comm.) collected newly transformed young, approximately 15–20 mm SVL, on 25 June in Albuquerque.

Woodward (1982b) investigated sexual selection and mating patterns in *B. woodhousii* in southern New Mexico. He presented several hypotheses that have been suggested as explanations for the nonrandom mating patterns observed in anurans. Male *B. woodhousii* were observed to wrestle intraspecifically and to attempt to displace one another from amplexus (Woodward, 1982b) suggesting that male-male interactions may play a role in causing the observed nonrandom mating patterns observed in southern New Mexico among mixed anuran breeding aggregations. Of five choruses reported in Sullivan (1989), the mean number of males was 5.4.

Food Habits: No detailed dietary studies of *B. woodhousii* are available from New Mexico. Smith (1934) reported bees, beetles, insect larvae, spiders, and ants being eaten by this toad in Kansas, and pointed out the economic importance of this species consuming large numbers of insects, sometimes eating as much as two-thirds of its own weight in a single day. Stebbins (1951) provided a list of arthropods recorded in the diet of *B. woodhousii*, including sowbugs, scorpions, centipedes, spiders, and a variety of insects.

Remarks: Woodward and Mitchell (1990) reported *Thamnophis marcianus*, raccoons, and skunks preying on *B. woodhousii* at breeding ponds in Albuquerque. Stuart (1991a) documented the occurrence of partial albinism in tadpoles of a north-central New Mexico population of Woodhouse's toad.

FAMILY BUFONIDAE

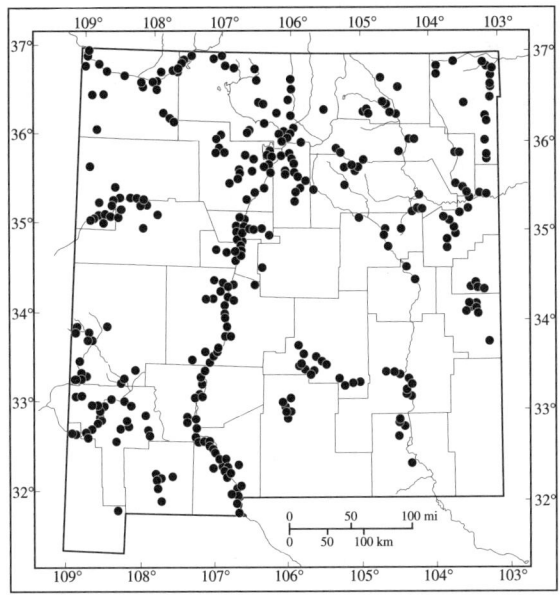

Distribution of
Bufo woodhousii
in New Mexico.

Treefrogs **FAMILY HYLIDAE**

This diverse and widespread treefrog family has over 680 species in 40 genera. More than 500 species live in the Americas, especially in the tropics. There are more than 140 species in the Australia–New Guinea area, and about 12 closely related species in the genus *Hyla* are distributed across temperate regions of Eurasia, scattered from Spain to Japan. One species occurs in northwestern Africa (Zweifel, 1992). Some of the most remarkably colored frogs belong to this family. Three genera and four or five species occur in New Mexico. All have free swimming tadpoles and all are aquatic to varying degrees. In New Mexico, they occur along healthy riparian corridors, woodland meadows and ponds, and flooded roadside ditches. These long-legged frogs are powerful jumpers. Some species have the toe tips expanded into small adhesive discs.

ACRIS CREPITANS Baird, 1854 — *Northern Cricket Frog*

See Plate 16.

Type: Baird (1854) did not designate any type material. The type locality was "northern states generally" although that was later restricted to Albany, Albany County, New York, by Smith and Taylor (1950b). Duellman (1970) noted that this restriction was in error, stating that the locality is outside of the known range of the species. Schmidt (1953) restricted the type locality to "Potomac River at Harper's Ferry, West Virginia." Smith et al. (1995b) refuted this restriction and gave the type locality as "The Locusts, near Oyster Bay, Nassau Co., Long Island, New York". No neotype has been designated.

Distribution: *Acris crepitans* occurs throughout most of the United States east of the Rocky Mountains, and into southern Ontario (Pt. Pelee), Canada, and south to northern Coahuila, Mexico (Frost, 1985). Frost (1983a) reported a specimen (KU 13473) from Arizona "10 mi. E Douglas" collected in 1905 by F. H. Snow. Based on the historic marshy habitat at this site, Frost (1983a) believed this record to be reliable but stated that the species may be extinct in Arizona. In New Mexico, *A. crepitans* have been found only in the southeast corner of the state, where they are often abundant along the permanent springs and rivers of the Pecos River drainage of Eddy and Chaves counties. There is a single locality reported from near old Fort Sumner in DeBaca County, and *A. crepitans* may occur along the Pecos River between Roswell and Fort Sumner. Although most collecting localities for *A. crepitans* in New Mexico range from 900–1100 m, the species may range as high as 1275 m in DeBaca County. Milstead (1960) considered *A. crepitans* a relict species in the Chihuahuan Desert, existing as apparently disjunct populations considerably removed from the main portion of the range.

Description: This is a small frog, with a maximum SVL of about 38 mm. The head is narrow with a pointed snout and there is a dark triangle between the eyes, with the apex of the triangle pointing posteriorly. The hind legs are very long with the heel reaching the snout when the leg is turned forward. The hind feet are fully webbed, each with five toes terminating in small, poorly developed toe pads. Small irregular warts cover the dorsal surface, somewhat suggestive of toad skin. The lower belly and sides are light-colored and also coarsely granular. The dorsal background color may be green, gray, light tan, or dark brown, and often there is a medial stripe of brown, tan, or green. The throat is heavily mottled in the males, less so or immaculate in females, and unmarked in the young. Post-metamorphic juveniles resemble the adults. The voice resembles two pebbles being clicked together, slowly at first but picking up speed and continuing for 20–30 or more beats (Conant and Collins, 1991). Ryan and Wilczynski (1991) demonstrated significant variation in the advertisement call among populations of *A. crepitans* in the western and eastern parts of its range.

The tadpole body is pear-shaped, and depressed dorsoventrally with a long, low-finned tail. The tail may have a black tip or may be plain-colored. The eyes are positioned dorsolaterally and the nostrils are large. Up to one-half of the spiracle tube is free from the body wall. The intestinal coil is visible through the skin of the belly. Tail fins are faintly mottled, with the dorsum of the tail musculature banded. The oral disc is slightly emarginate at the sides of the jaw and the labial tooth row formula is 2/2. Total length is approximately 10 mm at hatching and 35 mm at transformation (Johnson, 1987; Conant and Collins, 1991).

Similar Species: *Acris crepitans* can easily be distinguished from most New Mexico anurans by its small adult size, longitudinal dark stripe on the rear surface of the thigh, long legs with fully webbed toes with small, poorly developed toe tips, and the dark triangle between the eyes. *Pseudacris triseriata* has a white stripe on the upper lip and dark stripes down the back and does not occur sympatrically with *Acris*.

Systematics: Three subspecies are recognized (Collins, 1990); *A. crepitans blanchardi* Harper, 1947 occurs in New Mexico. This subspecies was named for Frank Nelson Blanchard, a herpetologist at the University of Michigan from specimens collected at "Smallen's Cave, at Ozark, Christian County, Missouri." The holotype is CM 26607, an adult male collected by Charles E. Mohr on 9 June 1938. It was originally described as *Acris gryllus blanchardi*.

Habitat: *Acris crepitans* prefers low, sunny, marshy areas along rivers, streams, and desert springs where there is abundant vegetation. They are non-arboreal, aquatic-margin frogs that generally do not frequent open water. Unless pursued, they venture into the water away from the shore line only when mats of algae are present on the surface.

Behavior: These alert, long-legged little frogs avoid predators by a series of quick erratic hops. After jumping in the water, they reverse directions and swim underwater toward the shelter of the bank or seek mats of algae where their mottled coloration tends to conceal them. *Acris crepitans* are rarely observed climbing in bushes or plants. Although specimens have been collected in New Mexico as late as 11 November, most are observed during the spring and summer, mid-April through late August.

Caldwell (1982) found polymorphism in the tail color of *A. crepitans* tadpoles. Working in Kansas, she reported that populations of tadpoles in ponds usually had large numbers of individuals with black tail tips, whereas tadpoles in lakes and creeks were mostly plain-tailed. This polymorphism of tail tip color among habitat types was correlated with the type of predators commonly occurring within each habitat. Caldwell (1982) suggested the black tail tip of the pond form functioned as a deflection mechanism to avert the attack of larval dragonflies.

Reproduction: Sexual maturity in *A. crepitans* occurs at less than one year of age (Burkett, 1984). In Kansas, males emerge from their winter retreats while temperatures are still highly variable and begin calling during mid-March and call until mid-September, increasing their call rates with increasing temperatures. These frogs call by day, as well as at night. Females arrive at the breeding sites and breeding takes place between late April and mid-June. Each female deposits 200–400 2–3 mm eggs on aquatic vegetation. Eggs may be laid singly or in small clusters. Burkett (1984) speculated that some females may lay two clutches of eggs during each annual breeding season. In Texas, larvae are present in ponds between early May and early August (Wiest, 1982). Near Austin, Texas, Pyburn (1958) noted breeding activity in early February. He observed calling males on 5 February at 21°C water temperature and 22.2°C air temperature. Breeding activity increased considerably in March, and reached its peak in April and May. Newly metamorphosed frogs began appearing in May, and increased rapidly in numbers in June. In Colorado, newly metamorphosed *A. crepitans* have been observed in early July (Hammerson, 1982). Bragg (1950c) observed 144 breeding choruses of *Acris crepitans* in Oklahoma. The earliest date he recorded was 12 March, and the latest was 17 August. In southeastern New Mexico, males may be heard calling day and night throughout the late spring and summer.

Food Habits: No dietary studies are available from New Mexico, although *A. crepitans* probably eats a wide variety of small terrestrial invertebrates captured near the water's edge. Burkett (1984) studied the diet of *A. crepitans* in Kansas and found that beetles, true bugs, leaf hoppers, flies, collembolans, and hymenopterans make up most of the diet. During postmetamorphic growth, *A. crepitans* ingests increasingly larger prey items as well as a broader spectrum of prey.

Remarks: The Latin derivation of *Acris* is "grasshopper" and is appropriately applied to these frogs because of their capacity for such prodigious leaps. The specific name *crepitans* is Latin for "rattling," and apparently refers to the call of this species (Duellman, 1970). Burkett (1984) suggested that *A. crepitans* is a useful subject for population studies since it is generally abundant and tends to form separate and distinct populations.

FAMILY HYLIDAE

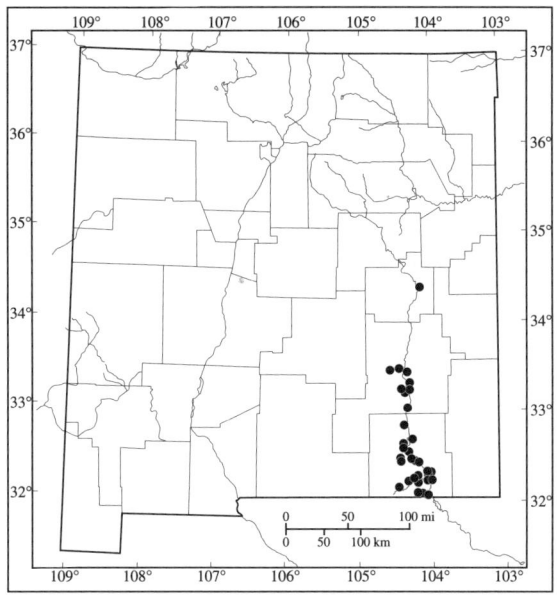

Distribution of Acris crepitans *in New Mexico.*

HYLA ARENICOLOR Cope, 1866   *Canyon Treefrog*

See Plate 17.

Type: Baird (1854) first described *Hyla arenicolor* as *Hyla affinis*, a name previously occupied by *H. affinis* Spix, 1824. Cope (1866b) later substituted the name *Hyla arenicolor*. Baird (1854) gave the type locality as "Northern Sonora," which was restricted to the Santa Rita Mountains, Arizona, by Smith and Taylor (1950b), and later restricted to Peña Blanca Springs, 10 mi. (16 km) NW Nogales, Santa Cruz County, Arizona, by Gorman (1960). Syntypes include USNM 11410 (5 specimens), collected by J. H. Clark. Gorman (1960) designated USNM 11410a as the lectotype, the other specimens in the series becoming paralectotypes.

Distribution: *Hyla arenicolor* occurs from southern Utah and extreme western Colorado to Oaxaca, Mexico. Disjunct populations occur in the Chisos and Davis mountains and canyons of Trans-Pecos Texas, and in northeastern New Mexico (Conant and Collins, 1991). In New Mexico, *H. arenicolor* has been found from about 1200–2500 m. It is most commonly found in Grant and Catron counties, although specimens have been found at other scattered localities west of the Rio Grande, and in Taos, Santa Fe, Harding, and Doña Ana counties.

Description: *Hyla arenicolor* has the toe tips expanded into large, wide, adhesive discs. The well-camouflaged body is plump with a warty skin and dorsal markings that vary from dark or medium brown to greenish. When exposed to sunlight these frogs often become a uniform grayish white. There is a dark-edged, light spot under the eye. The concealed surfaces of the hind limbs are orange-yellow. Maximum TL reaches 57 mm (Conant and Collins, 1991). Stebbins (1962 and references therein) described the call as the "ba-a-a" of a slightly hoarse lamb, like that of a goat, or like the quack of a duck.

The relative tail length of the tadpole increases with body length, with tail length/total body length ratios varying from approximately 53–65%. The relative height of the tail fin also increases with growth, with tail fin depth/body length ratios varying from 48–68%. Transformation takes place at approximately 38 mm total length. The eyes are slightly displaced dorsally and cannot be seen from below. Dark melanophores are well developed and scattered across the body, giving the ground color of the body and tail musculature a dark brown hue. Overall color changes to a golden brown as the tadpole matures. There is a considerable amount of geographic variation in the amount of pigmentation in the tail fins. Pigmentation varies from large, dark, intense blotches to the absence of a distinct pattern on the tail musculature and weak development of melanin in the tail fins. The oral disc is mostly surrounded by a double row of papillae, which are absent anteriorly, and the sides are not emarginate. There are five rows of labial teeth, two upper and three lower. The second upper row of teeth is narrowly interrupted along the midline. The two upper rows and first two lower rows are similar in length, and the third lower row is only slightly shorter than the others (Zweifel, 1961).

Similar Species: In New Mexico, only *Hyla eximia* has similarly expanded toe tips. *Hyla arenicolor* can be separated from *H. eximia* on the basis of color. *Hyla eximia* is bright green with a brown lateral stripe extending from the snout posteriorly to the shoulder.

Systematics: No subspecies of *Hyla arenicolor* are recognized.

Habitat: *Hyla arenicolor* is closely associated with areas of large boulders and rock outcrops along wooded canyon streams. Individuals may also be found in high mountain talus slopes where sufficient moisture can be found deep within the interstitial spaces of the talus. Individuals have been found at least a kilometer from a permanent water source at 2500 m in talus slopes on the highest peaks in the Animas Mountains of southern Hidalgo County and also in high-elevation rocky slopes near the Springtime Canyon campground in the San Mateo Mountains in Sierra County (our obs.).

Behavior: These small frogs may be observed clinging to boulders a short distance above the water. They appear to "sun" on large boulders adjacent to mountain pools, returning to nearby water periodically to rehydrate. They often call during daylight hours while hidden in small spaces among rocks.

Reproduction: *Hyla arenicolor* breeds in rocky streams, in potholes in the solid rock of canyon bottoms, and in rain pools on top of rock cliffs. Gehlbach (1965) suggested that *H. arenicolor* in the Zuni Mountains of northwestern New Mexico may have two or more breeding seasons coinciding with separate peaks of precipitation.

A chorus of 20–30 males was found on 17 May in a canyon in the Mimbres Mountains in Sierra County. The males were calling from the vertical, smooth rock walls at and above the waterline and on the level shoreline near the rock walls. Metamorphosing tadpoles and newly metamorphosed froglets were found in early September in the Peloncillo Mountains in Hidalgo County (J. N. Stuart, pers. obs.). We observed a breeding chorus along the East Fork Jemez River on 21–23 June. The males started calling just before dark from along the stream in flat, grassy areas. *Pseudacris triseriata* were heard in the same chorus.

Zweifel (1961) noted breeding activity in the Chiricahua Mountains of southeastern Arizona during late spring or very early summer, when calling males were heard after the summer rains started. Gravid females were found on 28 June and 1 July. Newly hatched tadpoles were observed on 18 July, and Zweifel suggested that oviposition took place on 12–13 July. The period from oviposition to metamorphosis was between 50 to 60 days. The SVL of newly metamorphosed froglets observed by Zweifel in the Chiricahua Mountains was about 15 mm, those of Gehlbach (1965) from the Zuni Mountains averaged 24.7 mm.

Food Habits: Food habits of *H. arenicolor* are unknown in New Mexico. It is likely they feed on a variety of invertebrates including beetles, ants, bugs, and caterpillars. Painter (1985) found beetles, true bugs, caddisflies, and annelid worms in specimens collected from the Gila and San Francisco river drainages.

Remarks: Frost (1985) and Collins (1990) erroneously cited the date of description of this species as "Cope, 1886" instead of "Cope, 1866."

FAMILY HYLIDAE

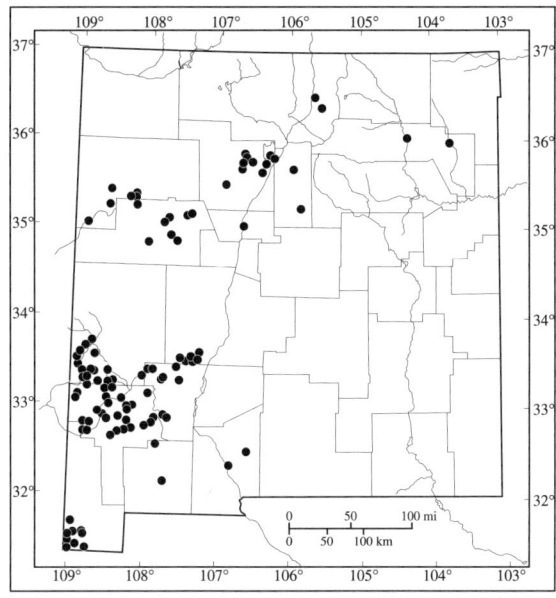

Distribution of Hyla arenicolor in New Mexico.

HYLA EXIMIA Baird, 1854 — *Mountain Treefrog*

See Plate 18.

Type: Baird (1854) designated the type locality of *Hyla eximia* as "Valley of Mexico" (Distrito Federal, Mexico). (See Remarks.) Syntypes are USNM 3248 (2 specimens) collected by William Rich in 1853.

Distribution: *Hyla eximia* ranges from the mountains of central Arizona and southwestern New Mexico, south in the Sierra Madre Occidental in northwestern Mexico, and throughout the southern part of the Mexican Plateau, the Sierra Madre Oriental, and Cordillera Volcanica in central Mexico. It is found from 900–2900 m elevation (Duellman, 1970). In New Mexico, it is known only from the higher elevations (approximately 2000–2750 m) of Catron and Sierra counties where it appears to be common in its limited habitat and not rare, contrary to the suggestion of Price and Johnson (1978b). The status of this species in McKinley County (USNM 8508) needs clarification. *Hyla eximia* does not range into Texas as indicated by Chapel (1939).

Description: *Hyla eximia* is a beautiful little treefrog, bright green in color with a dark line extending from the snout posteriorly through the eye onto the side of the body where it may be broken up into several segments. The posterior surfaces of the femur and groin are orange or gold with a greenish tinge. The throat of males is dull greenish and tan, that of females whitish. Males are 24–44 mm SVL, females 24–48 mm (Wright and Wright, 1949). Duellman (1970) gave the maximum length of females as 36.2 mm and males as 35 mm, although Renaud (1977) and Sullivan (1986b) reported different samples of 20 calling males from Baker Lake in Coconino County, Arizona, that ranged from 31–45 mm SVL and 37–47 mm SVL. The mating call is a rough clink, like metal striking metal, repeated 1–3 times a second (Smith, 1978). Sullivan (1986b) included an audio-spectrogram of the advertisement call.

Tadpole tail length ranges from about half to more than two-thirds of the total length, with the relative length increasing as the tadpole grows. The depth of the tail fin decreases relative to increasing body size, with tail fin depth/body length ratios varying from approximately 55–65%. Newly hatched tadpoles are 4.9–5.2 mm in total length; metamorphosis takes place at approximately 38 mm. The eyes are slightly displaced dorsally and cannot be seen from the ventral view. Dark melanophores are scattered over the body and base of the tail. Ventral surfaces anterior to the abdomen are largely free of melanophores and no pattern is evident in the pigmentation of the body, although there may be a faint dark lateral line. The oral disc is mostly surrounded by a double row of papillae, which are absent anteriorly and the sides are not emarginate. There are five rows of labial teeth, two upper and three lower. The second upper row is broken above the beak (Zweifel, 1961).

Similar Species: The expanded toe tips, dark lateral stripe, and distinctive green coloration should separate this frog from all others in New Mexico. Only *Hyla arenicolor* has similarly expanded toe tips, but it lacks the dark lateral stripe extending through the eye to the groin.

Systematics: No subspecies of *Hyla eximia* are recognized. Prior to 1970, this species was referred to as *Hyla wrightorum* (Taylor, 1938b). Duellman (1970) presented evidence for the synonymy of *H. eximia* and *H. wrightorum* based on the absence of any distinctive morphological characters and on the general similarity of mating calls. However, in an unpublished Ph.D. dissertation, Renaud (1977) found that *H. eximia* and *H. wrightorum* are distinct in their morphology (primarily size) and advertisement calls and, in light of their allopatric distributions, should warrant recognition as separate species. Sullivan (1986b) agreed with this assessment. The systematic relationship of these species has not been resolved.

Habitat: In New Mexico, these small frogs occur at high elevations, generally above 2000 m, and are found in coniferous forests of ponderosa pine and Douglas fir. They occur along small streams, in wet meadows and cienegas, and in temporary roadside ditches. Chapel (1939) reported that these frogs ascend to considerable heights in trees. In Hidalgo, Mexico, they have been found in bromeliads growing on pine trees (Duellman, 1970).

Behavior: In experimental field exclosures where habitat complexity and ratio of predator and prey were manipulated, Sredl and Collins (1992) found that *Ambystoma tigrinum* larvae strongly affected the survival of *H. eximia* tadpoles. When salamanders were absent, survival was nearly five times greater than when salamanders were present.

Reproduction: Sullivan (1986b) found the advertisement call frequency in *H. eximia* to be significantly correlated with male snout-vent length, although males found in amplexus in a central Arizona breeding aggregation were not significantly larger than nonmating males. Hence, he suggested that the mating success of male *H. eximia* appears unrelated to size. Breeding in *H. eximia* occurs sporadically over a few weeks during the summer in temporary and permanent ponds. Such breeding choruses usually last only 2–3 days, although some males may remain at the site for an additional 2–4 days (Chapel, 1939). In a small roadside pond in Catron County, Randy Jennings (pers. comm.) observed breeding on 28 July, tadpoles on 6 September, and newly metamorphosed froglets on 4 October 1988. In east-central Arizona, Zweifel (1961) encountered a large chorus of *H. eximia* in a dirt stock tank on 23 June. Eggs were present the following day and he speculated they were laid on 22 June. The small loose clusters of eggs were attached to rushes in shallow water. Sullivan (1986b) reported breeding aggregations on the night of 15 July and 21 July in central Arizona in Gila and Coconino counties. In central Arizona, Sredl and Collins (1992) identified two distinct types of breeding habitat, one was a long duration site, the other an ephemeral site. At these sites, breeding success depended on the relative risks of predation and desiccation. Breeding commenced with the summer rains in early July, and larvae metamorphosed in late August. On 23 July, Sredl and Collins (1992) found eggs of *H. eximia* and tadpoles 4 mm total length at prestage 25 (Gosner, 1960). These eggs and tadpoles were contained in field enclosures and metamorphosed frogs were observed between 16 September and 16 October. Chapel (1939) reported newly metamorphosed froglets to be 10–13 mm SVL.

Food Habits: Very little is known of the food habits of *H. eximia* in New Mexico or elsewhere. Chapel (1939) examined the stomach contents of seven specimens from west-central Arizona. He found beetles, spiders, earthworms, flies, grass particles, and four specimens of the bark beetle, *Ips*. The beetles are well known pests in pine forests and Chapel (1939) speculated that *H. eximia* may help control these insects.

Remarks: Smith and Taylor (1950b) restricted the type locality to Coyoacán, Distrito Federal, Mexico. Very little is known about the natural history of this species in

New Mexico. Vest (1992) speculated that *H. eximia* is fairly toxic and reported two cases of severe burning of the eyes after handling this frog. Randy Jennings (pers. comm.) reported the same sensation caused by specimens from New Mexico.

Wright and Wright (1949) and Stebbins (1954) reported two females, 37.5 and 42 mm SVL, collected at Santa Fe on 18 June 1874, by H. W. Henshaw (now USNM 9338). It is likely these specimens were collected elsewhere and shipped from Santa Fe to the U.S. National Museum in Washington, D.C.

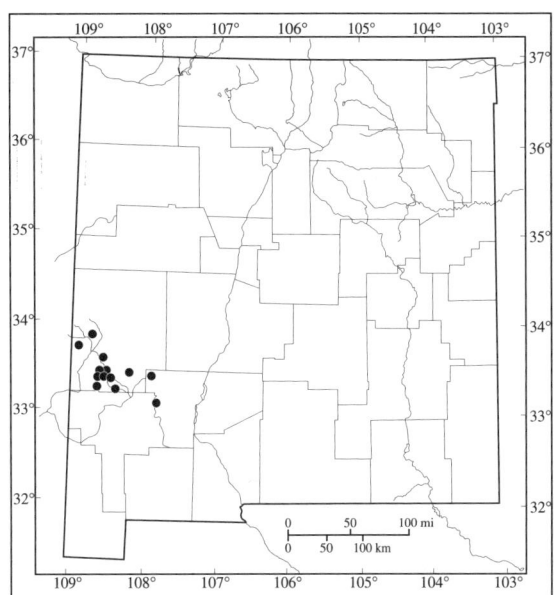

Distribution of Hyla eximia *in New Mexico.*

PSEUDACRIS TRISERIATA (Wied-Neuwied, 1838) *Western Chorus Frog*

See Plate 19.

Type: Maximilian Wied-Neuwied first described this species as *Hyla triseriata* in 1838 in the first part of Volume I (the second part was published in 1839). The holotype was lost. The type locality was given as "Mount Vernon, Ohio River, Indiana." Harper (1955) restricted the type locality "to the area between Rush Creek and Big Creek along the route from New Harmony to Mt. Vernon in Posey County, Indiana."

Distribution: *Pseudacris triseriata* is strictly North American, ranging from the Gulf of Mexico to New York and southern Ontario in the east and from central Arizona almost to the Arctic Circle in the west (Conant and Collins, 1991). In New Mexico, *Pseudacris triseriata* occurs mostly in the northern counties, and south in the Rio Grande Valley to northern Socorro County. Somewhat disjunct populations occur in the Gila National Forest, the Zuni Mountains, and along the San Juan River. Because of the arid landscape and lack of additional specimens, the continued existence of a population at Lordsburg (LTU 22549–50) needs confirmation. *Pseudacris triseriata* ranges from approximately 1280–3050 m in New Mexico.

Description: This small frog ranges from 19–39 mm SVL. The dorsal pattern is extremely variable, ranging from three solid stripes or partially coalesced spots that form stripes to the complete lack of a dorsal pattern. There is a white supralabial stripe extending the length of the lip and a dark stripe extending from the nostril, through the eye, above the forelimb to the groin. The hindfoot has greatly reduced webbing and toe pads. The dorsal background color may be light tan, grayish, or dark brown, with occasional individuals having a greenish wash. The spotted or striped pattern is brown, olive, or grayish and is darker than the background color. The throat is heavily mottled with dark pigment in the males, less so or immaculate in females, and unmarked in the young. The venter is immaculate or with dark spots on the chest and throat. Males average slightly smaller than females. The call is a vibrant, regularly repeated "crreek" or "prreep" (roll the r's) speeding up and rising in pitch toward the end (Conant and Collins, 1991). It is often likened to the sound made by running a fingernail over the small teeth of a pocket comb.

Total length of the tadpole is 2.6–3.0 cm with a body/tail length ratio of 1:2. The intestinal coil is visible. The tail musculature is unicolored with the dorsal tail fin higher than the ventral fin. The tail fin is clear or only weakly pigmented. The oral disc is emarginate at the jaw edges and ventrally. The third posterior labial tooth row (if present) is shorter than the first and second rows. The labial tooth formula is 2/2 or 2/3 (Conant and Collins, 1991). The young tadpoles appear squarish and tend to swim near the surface of the water.

Similar Species: The small size, smooth skin, only slightly expanded toe tips, and the white stripe on the upper lip will separate *Pseudacris triseriata* from all other frogs in New Mexico. *Acris crepitans* lacks the white labial stripe, has a rough warty skin, and has a longitudinal dark stripe on the posterior surface of the thigh; *Hyla arenicolor* is larger, lacks the white labial stripe, and has prominent toe pads.

Systematics: As currently recognized (Conant and Collins, 1991), *P. triseriata* consists of one wide-ranging species with four subspecies. Platz and Forester (1988) and Platz (1989), citing biochemical and morphological data and mating call analysis, suggested recognition of four distinct species. Platz (1989) recognized a broad contact zone between *P. t. triseriata* and *P. t. maculata*, extending as far west as Fort Collins, Colorado, and as far south as Lawrence, Kansas. The taxonomic status of *P. triseriata* in New Mexico is not fully resolved.

We follow Conant and Collins (1991) and recognize *P. t. triseriata*, the nominotypical form, and *P. t. maculata* (Agassiz, 1850) in New Mexico. *Pseudacris t. maculata* was described as *Hylodes maculatus* and the type locality given as the north shore of Lake Superior, which was later restricted to the vicinity of Sault Ste. Marie by Schmidt (1953). There are two syntypes (MCZ 38). No date of collection or collector is provided by P. W. Smith (1956).

Habitat: *Pseudacris triseriata* is found in a variety of habitats in New Mexico. These range from high mountain lakes and wet meadows to roadside ditches, shortgrass prairie, playa lakes, and flooded fields. These frogs tend to be diurnal in early spring and fall and nocturnal during warm spring and summer months. They hide under surface debris, in thick vegetation, or in burrows when inactive.

Behavior: *Pseudacris triseriata* is shy and secretive and is seldom seen or heard except during the breeding season. Almost 70% of the records available from New Mexico were collected from March–May. A large percentage of those taken at other times during the year were from choruses at higher elevations.

Reproduction: *Pseudacris triseriata* is among the earliest of all New Mexico amphibians to commence breeding. On 8 May, Jack Herring (pers. comm.) heard a breeding chorus during a light snow storm at 2623 m in the Jemez Mountains of north-central New Mexico. At lower elevations in the central Rio Grande Valley, they often begin calling in late February or early March, depending upon early spring temperatures. They often call during the day as well as at night. James N. Stuart (pers. obs.) found *P. triseriata* calling in shallow playa lakes in Colfax County on 18 July along with *Spea bombifrons* and *Bufo cognatus*. Bragg (1941) found them breeding on 6 August near Las Vegas. They breed in shallow and often very temporary bodies of water, with large aggregations found in flooded grassy meadows, shallow weedy borders of stock ponds, or irrigated fields. Adults may breed at less than one year of age (Duellman and Trueb, 1986). However, in montane populations of Colorado, males breed at two years, and females at three years (Spencer, 1964; Matthews, 1971). When reproductive traits of montane and lowland populations of *P. triseriata* are compared, there are significant differences in mean size of eggs, hatchling tadpoles, and adult females, although the mean number of eggs does not differ significantly. The larger ova, hatchling tadpoles, and adult females and the more rapidly developing ova in montane populations are useful in averting the limitations of shorter growing seasons in this cooler environment (Pettus and Angleton, 1967). *Pseudacris triseriata* may lay two clutches of eggs per season at lower elevations. The tadpoles metamorphose at approximately 13–15 mm SVL. Recently transformed specimens collected at about 2806 m on 13 July and reported by Gehlbach (1965) were 13 and 14 mm SVL.

Food Habits: Food habits of *P. triseriata* have not been studied in New Mexico. In Colorado, Christian (1976) found they feed upon a variety of flies, springtails, mites, true bugs, spiders, beetles, and ants.

Remarks: The systematics of this group is in need of further study.

FAMILY HYLIDAE

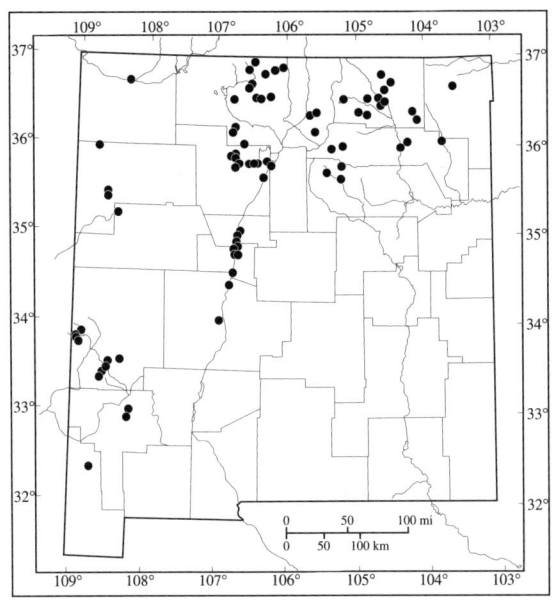

Distribution of
Pseudacris triseriata
in New Mexico.

Narrowmouth Toads

FAMILY MICROHYLIDAE

The family Microhylidae occurs in the Americas from the southern United States to Argentina, equatorial and southern Africa, eastern India and Sri Lanka, and through Southeast Asia to New Guinea and northern Australia. More than 300 species in 65 genera are recognized, the largest number of genera in any family of frogs (Zweifel, 1992). Only one species occurs in New Mexico, where it is known only from south-central Luna County. The Great Plains narrowmouth toad is easily recognized by its small, wide body with a narrow pointed head. It feeds primarily on ants and termites.

GASTROPHRYNE OLIVACEA (Hallowell, 1856 [1857])   *Great Plains Narrowmouth Toad*

See Plate 20.

Type: *Gastrophryne olivacea* was originally described as *Engystoma olivaceum* by Hallowell (1856 [1857]), who did not designate a type locality. The type locality was later restricted to "Kansas, Geary Co., Ft. Riley" by Smith and Taylor (1950b) and to "vicinity of Lawrence, Kansas" by Schmidt (1953). Although no type specimen was designated, ANSP 2745, a female collected by Dr. Hammond in Kansas, is likely the holotype. No date of collection was provided (Nelson, 1972a). (See Remarks.)

Distribution: *Gastrophryne olivacea* occurs from southeastern Nebraska and Missouri, south to Tamaulipas and San Luis Potosí and west through most of Texas and northern Mexico to west-central Chihuahua and northeastern Durango. A large, disjunct western population occurs from extreme south-central Arizona, south along the Pacific Coast of Mexico to Nayarit. Isolated records are from west Texas and southwestern New Mexico (Conant and Collins, 1991). In New Mexico, *G. olivacea* is known only from south-central Luna County near the U.S.-Mexican border (Degenhardt, 1986; Stuart, 1992b). Based on the distribution in Colorado (Hammerson, 1982), Texas (Dixon, 1987), and Oklahoma (Black and Sievert, 1989), it may occur in Union and Lea counties as well.

Description: *Gastrophryne olivacea* has a small plump body with smooth skin. There is a fold of skin just behind the pointed head. There is no visible tympanum. The dorsum is uniformly grayish, slate-colored, or olive with small black spots scattered over the back and hind legs. The belly is unmarked, except for a darkened vocal sac in the males during the breeding season. The digits lack webbing and there is a single metatarsal tubercle on each hind foot. Sexually mature males have a dark, distensible throat pouch. Females are slightly larger than males, with females reaching 42 mm, males 37 mm SVL. The call lasts 1–4 seconds and consists of a distinct very short "peep" followed by a buzz like that of an angry bee (Conant and Collins, 1991).

The tadpoles are easily recognized by the soft mouth disc instead of horny jaws and a single spiracle that opens mid-ventrally rather on the side as in other New Mexican anurans. Stuart (1992b) described the tadpoles from New Mexico as being grayish-tan dorsally, with a mottled gray-and-white venter. The tail tip is dark. Nelson and Cuellar (1968) described the tadpole and compared it to tadpoles of the closely related genus *Hypopachus*.

Similar Species: With the small plump body shape, short limbs, and distinctly pointed snout, *G. olivacea* should not be confused with any other species. Juvenal specimens of *Scaphiopus*, *Spea*, and *Bufo* have spades on the soles of the hind feet. Juvenal *Eleutherodactylus augusti* also lack webbing between the toes; however, they are distinctly marked by a broad, cream-colored band across the back.

Systematics: No subspecies of *Gastrophryne olivacea* are currently listed by Collins (1990). Nelson (1972b) discussed the range of variation in the dorsal and limb patterning in *G. olivacea* and concluded that subspecific recognition is unwarranted since no distinct geographic break is seen in characteristics used to separate previously recognized taxa. Two subspecies, *G. o. olivacea* and *G. o. mazatlanensis* (Taylor, 1943) had been recognized earlier (Chrapliwy, et al., 1961; Nelson, 1972a) and Stebbins (1985) continued to recognize these subspecies. This species is referred to *Microhyla* in much of the older literature. Based on morphological evidence, De Carvalho (1954) resurrected the generic name *Gastrophryne* for the American species formerly included in *Microhyla*.

Habitat: All specimens of *G. olivacea* known from New Mexico have been collected from low-lying, flooded roadside ditches between elevations of 1281–1342 m in overgrazed desert scrub dominated by mesquite, creosotebush, and various arid-land grasses. Much of this area has been converted to agriculture through irrigation with groundwater. Smith (1950) stated that *G. olivacea* is often found in mesquite flats that are devoid of rocks.

Behavior: This small frog is shy and secretive, spending most of its time hidden beneath surface objects or in rodent burrows. Surface activity is usually nocturnal and limited to times when the soil is wet, mainly during heavy rains and immediately afterward. In a wooded area in northeastern Kansas, Fitch (1956b) found that adults may remain in an area with a radius of only a few hundred feet for months or years. *Gastrophryne olivacea* in Kansas was found to be active over a temperature range of at least 16–37.6°C. They tolerate high temperatures that would be lethal to many other kinds of amphibians (Fitch, 1956b).

Stuart (1992b) noted distinctive behavior of the *G. olivacea* tadpoles he encountered in rainwater pools in southern New Mexico. These tadpoles would hang motionless at the water surface for several minutes at a time, a behavior not observed in other tadpoles in the area.

Hunt (1980) described an interesting mutualistic relationship between *G. olivacea* and tarantulas. The narrowmouth toad is known to live in the burrow of the tarantula and seek protection under the belly of the tarantula when threatened by potential predators. The toad in turn helps keep the nest free of ants which could pose a problem for an unprotected tarantula.

Reproduction: Although calling males had been observed in south-central Luna County, reproduction in *Gastrophryne olivacea* had not been described in New Mexico until Stuart (1992b) reported finding calling males and, later in the same pond, tadpoles. Calling males have been noted in New Mexico on 27 June, 25 and 30–31 July, and 30–31 August; in each case activity occurred shortly after heavy rainfall. Males were observed calling in a vertical posture from within or immediately adjacent to emergent grass clumps in turbid water 15–35 cm deep. These calling toads are often extremely well hidden with only the head protruding above the water surface.

A lot of 13 tadpoles (MSB 54896) at stages 35–38 (Gosner, 1960) were taken on 16 August, 16–17 days after calling males were observed at Stuart's (1992b) study site. Since no tadpoles were found on 30–31 July when the pool was newly formed nor on 5 September when the pool was mostly dried up, Stuart's (1992b) observations are consistent with reports that the species is able to complete the larval period within 30 days (Collins, 1982).

Breeding normally takes place after heavy summer rains, and Fitch (1956b) suggested that in Kansas, precipitation of at least five cm within a few days is necessary to bring forth large breeding choruses. With smaller amounts of precipitation, only stragglers or small aggregations may arrive at breeding ponds. The egg mass is laid as a surface film; one clutch from Texas reported by Wright and Wright (1949) contained 645 eggs. These eggs hatch within two days and newly metamorphosed young are 15–16 mm SVL. Sexual maturity is reached within 1–2 years (Collins, 1982).

Amplexus in *Gastrophryne olivacea* is termed "glued" by Duellman and Trueb (1986) and is described by Fitch (1956b). The males possess specialized secretory cells in the dermis of the venter that provide an adhesive substance for attachment to the back of the female. This adhesion may have survival value. These frogs are shy and at the least disturbance at the breeding pond they respond by vigorous attempts to escape and hide. This adhesion of the pair may prevent separation or it may serve to prevent displacement of a clasping male by a rival male.

Food Habits: Once aptly called the "ant-eating" frog, *G. olivacea* feeds almost exclusively on ants. No detailed dietary studies are available from New Mexico, although Stuart (1992b) reported finding ant remains in the feces. Freiburg (1951) examined the stomachs of 52 specimens from northeastern Kansas and found mostly ants and some beetle fragments.

Remarks: We have chosen 1857 as the publication date for the original description in spite of the fact that all previous authors (Schmidt, 1953; Nelson, 1972a; Collins, 1990) have used 1856. Roughly half of the papers presented to the Philadelphia Academy in 1856 and published in the *Proceedings* were received by subscribers in 1857 (Nolan, 1913). Since Hallowell's paper appeared on pages 238–253 and there were only 322 pages published in the *Proceedings* that year, it seems logical to assume that this paper must have been received by the subscribers in 1857.

Stuart (1992b) reported a distinctive behavior of tadpoles. He observed tadpoles hanging motionless at the surface of the water for several minutes at a time. Tadpoles of *Spea* in the same pool, however, were more active and usually not visible in the turbid water except when they came briefly to the surface for air.

Although Fitch (1956b) indicated that the subspecies *G. o. mazatlanensis* occurred in New Mexico, the presence of *G. olivacea* was not verified in the state until a

FAMILY MICROHYLIDAE

single calling adult male was collected from a temporary pool under a railroad trestle in Luna County on 27 June 1986 by W. G. Degenhardt. Earlier, the species had been located on 25 July 1977 by E. Stegall (pers. comm.) who tape-recorded the call of a single adult male in an intermittent pond immediately northeast of Hermanas.

Gastrophryne olivacea is listed as endangered by the New Mexico Department of Game and Fish (NMGF, 1990). Nelson reviewed the genus (1973) and the species (1972a).

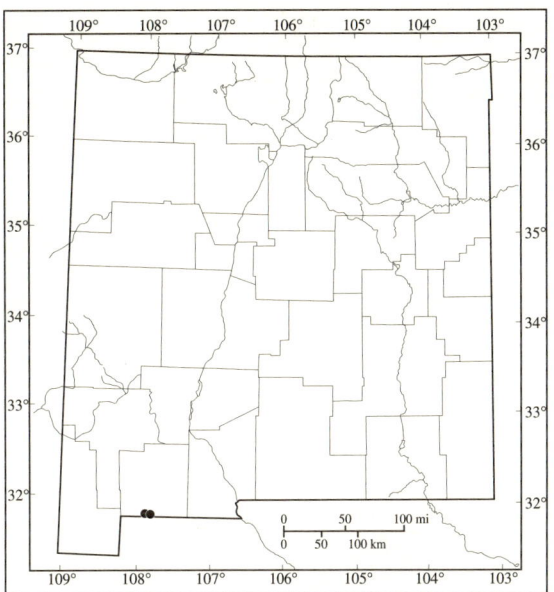

Distribution of Gastrophryne olivacea *in New Mexico.*

True Frogs   FAMILY RANIDAE

Ranid frogs have the widest distribution of any family of frogs. They occur in North America (including Alaska), Central America, and northern South America, in Europe and across Asia south of the Arctic Circle, through the East Indies to New Guinea, in the extreme north of Australia and the Fiji Islands, and in most of Africa and Madagascar (Zweifel, 1992). The genus *Rana*, the sole representative of the family in New Mexico, includes more than two-thirds of the 650-odd species credited to the family. At least six species of this diverse genus occur in New Mexico. These are the familiar frogs typically living in and on the margins of streams and lakes. The long and powerful legs are smooth-skinned and have extensive webbing between the long thin toes. In all cases, the eggs are laid in water and there is a tadpole stage that may last from several weeks to two years. The bullfrog is classified as a game species in New Mexico.

In many parts of the world, populations of frogs, toads, and salamanders are dwindling, and some have disappeared from their native habitats completely. (See Blaustein and Wake, 1995; Blaustein et al., 1994a,b,c; Pechmann and Wilbur, 1994; Wake 1991, for a discussion of this problem.) In the Southwest, notable declines in two species of leopard frogs have led to their listing as Notice of Review species by the U.S. Fish and Wildlife Service (USDI, 1994) and as endangered by the New Mexico Department of Game and Fish (NMGF, 1990).

	BERLANDIERI	BLAIRI	CHIRICAHUENSIS	PIPIENS	YAVAPAIENSIS
DORSOLATERAL FOLDS	DISCONTINUOUS, INSET POSTERIORLY	DISCONTINUOUS, INSET POSTERIORLY	DISCONTINUOUS, INSET INDISTINCT POSTERIORLY	CONTINUOUS INSET	DISCONTINUOUS, INDISTINCT POSTERIORLY
THIGH PATTERN	BOLD, DARK RETICULATIONS	FUZZY RETICULATIONS	DARK BACKGROUND WITH WHITE TUBERCLES FORMING "SALT & PEPPER" PATTERN	DARK SPOTS	VARIABLE RETICULATION
SPOTS ON THE NOSE ANTERIOR TO THE EYE	USUALLY 0	1 OR MORE	AT LEAST 1, FREQUENTLY MORE	USUALLY 1 LARGE SPOT	USUALLY 0
SUPRALABIAL STRIPE	INDISTINCT ANTERIOR TO EYE	USUALLY WELL-DEFINED	INDISTINCT ANTERIOR TO EYE	VARIABLE	INDISTINCT ANTERIOR TO EYE
TYMPANUM SPOT	VARIABLE, USUALLY ABSENT	DISTINCT	ABSENT	VARIABLE	USUALLY ABSENT
VESTIGIAL OVIDUCTS IN MALES	PRESENT	PRESENT	USUALLY PRESENT	VARIABLE	ABSENT
CONSPICUOUS LIGHT HALOS SURROUNDING DORSAL SPOTS	PRESENT, USUALLY FAINT; RARELY ABSENT	USUALLY ABSENT; IF PRESENT THEN FAINT	USUALLY ABSENT; IF PRESENT THEN FAINT	PRESENT; RARELY ABSENT	USUALLY ABSENT; IF PRESENT THEN FAINT

Table 1 Characteristics used in the Identification of Five Species of Leopard Frogs (Rana spp.) in New Mexico.

RANA BERLANDIERI Baird, 1859 *Rio Grande Leopard Frog*

See Plate 21.

Type: Baird (1854) gave the type locality as "Southern Texas generally." The syntypes include MCZ 155 (2 specimens), USNM 3293 (9 specimens), and USNM 131513, which was recorded as a syntype by Cochran (1961), who noted that two specimens of USNM 3293 were exchanged to MCZ in 1879. The lectotype (USNM 131513), designated by Pace (1974), is an adult male collected at Brownsville, Cameron County, Texas, by Capt. S. Van Vliet. The date of collection is unknown (Cochran, 1961).

Distribution: *Rana berlandieri* ranges from central and west Texas and extreme southeastern New Mexico south into Mexico (Conant and Collins, 1991). In Mexico, *R. berlandieri* occurs in the eastern states and as far south as Nicaragua, although D. M. Hillis (*in* Frost, 1985) notes that populations south of southern Veracruz and Oaxaca are of questionable taxonomic status. It has been recently introduced in Arizona and is established at numerous sites along the lower Colorado and Gila rivers (Platz, 1991). In New Mexico, *R. berlandieri* is known only from the lower Pecos River drainage in Eddy County at elevations between 900–1450 m.

Description: *Rana berlandieri* has well-developed dorsolateral folds that are discontinuous posteriorly and displaced medially. The supralabial stripe does not typically extend anterior to the eye. The dorsal spots often have faint, light-colored halos. There is usually a lack of spots on the head anterior to the eyes, although rarely there is one head spot or mottling in this area. The venter is cream colored, with mottling on the chin of older individuals. The posterior surface of the thigh has well defined, dark, and highly contrasting reticulations. There are tubercles or ridges on the side of the body and dorsal surface of the shank. Males exhibit paired, external vocal sacs and possess prominent vestigial oviducts. Males are smaller than females, with the mean body length in males 64.4 mm SVL, females 73.5 mm (Fritts et al., 1984). The mating call is a short, guttural rattle with 13 or more pulses per second (Conant and Collins, 1991).

The tadpole of *Rana berlandieri* has large dark and pale spots on the tail fin. The dark spots may coalesce to form a coarse dark reticulation enclosing these spots. The tail fin is of medium depth and forms 65–80% of the body length. About half of all large tadpoles have three rows of upper labial teeth. The oral papillae are mostly unpigmented and relatively long and moderately dense lateral to the beak. The iris is pale gold in color with dorsal, lateral, and ventral dark spots (Scott and Jennings, 1985).

Similar Species: *Rana berlandieri* is most similar to *R. yavapaiensis*. *Rana berlandieri* can be distinguished from *R. yavapaiensis* by its greater propensity for green coloration on the body, its highly contrasting reticulated thigh pattern, and the presence of vestigial oviducts in males. Although *Rana blairi* is the only other leopard frog that occurs in Eddy County, it has not been found sympatrically with *R. berlandieri*. Characters used to separate *R. berlandieri* from all other leopard frogs in New Mexico are presented in Table 1.

Systematics: No subspecies of *Rana berlandieri* are currently recognized, although based on the color of the groin, Smith (1978) suggested that two subspecies may occur in the United States. He presented no name except for the nominotypical subspecies. Hillis et al. (1983) assigned this species, along with *R. yavapaiensis*, to the *berlandieri* species group of Mexican leopard frogs. Frost (1982) removed *R. forreri* from the synonymy of *R. berlandieri*.

Habitat: In New Mexico, *R. berlandieri* is generally found in clear, flowing streams or permanent pools in intermittent streams that originate from springs (Jennings, 1987); it occasionally also occurs in stock tanks (Fritts et al., 1984). Individuals collected from sites without permanent water probably represent fortuitous collections of dispersing individuals during wet periods (Jennings and Scott, 1991). Jennings and Hayes (1994) stated that a key habitat component for *R. berlandieri* is the presence of holes or burrows that metamorphosed frogs can utilize as refugia.

Behavior: Primarily nocturnal, *R. berlandieri* seeks shelter under rocks and in thick streamside vegetation during the day. When startled, it is quick to leap into the water where it buries itself under mud or debris. Platz (pers. comm. *in* Jennings and Hayes, 1994) stated that *R. berlandieri* seems to thrive in habitats that have limited numbers of introduced bullfrogs and may actually be replacing them in some situations.

Reproduction: *Rana berlandieri* has an extended reproductive season that includes spring, summer, and fall, although large numbers of egg masses and small to medium tadpoles found during April suggest a peak in activity during early spring. Adult frogs have been heard calling between April and August, and egg masses are found as early as mid-April through early July. There are no data to suggest eggs are laid any later than August. James N. Stuart (pers. obs.) found adult *R. berlandieri* calling and observed several egg masses in Dark Canyon in Eddy County on 22–23 March. The egg masses were 7–9 cm across and were attached to emergent vegetation in 9–15 cm deep, quiet water along the stream. Tadpoles may overwinter (Scott and Jennings, 1985; Jennings, 1987). Tadpoles metamorphose at approximately 30–32 mm SVL. Bragg (1950c) observed 168 breeding choruses of *Rana berlandieri* in Oklahoma. The earliest date he recorded was 9 February, and the latest was 27 September.

Food Habits: Food habits of *R. berlandieri* have not been studied in New Mexico, although it probably feeds on a variety of insects and invertebrates.

Remarks: The distribution map for this species is modified after Jennings and Scott (1991). Platz (1991) reviewed this species. Frost (1985) and Collins (1990) erroneously gave the year of this taxon as "Baird, 1854" although the original description appears in Baird (1859).

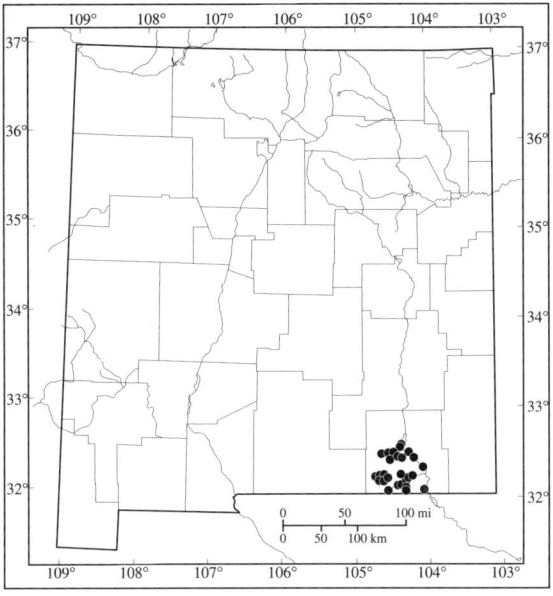

*Distribution of
Rana berlandieri
in New Mexico.*

RANA BLAIRI Mecham, Littlejohn, Oldham, Brown, and Brown, 1973   *Plains Leopard Frog*

See Plate 22.

Type: The holotype of *Rana blairi* is UMMZ 131690, an adult male collected "1.6 km. W New Deal, Lubbock Co., Texas" on 6 August 1971 by Charles Everett and described by Mecham et al. (1973).

Distribution: *Rana blairi* ranges from western Indiana to southeastern South Dakota and eastern Colorado and south to central Texas. Isolated colonies occur in Illinois, New Mexico, and southeastern Arizona (Conant and Collins, 1991). Brown (1992) provided a detailed description of the range. In New Mexico, *R. blairi* ranges from about 1000–2250 m, and is widely scattered east of the Rio Grande, primarily in the area drained by the Pecos, Canadian, and Dry Cimarron rivers. There are isolated populations in Sierra County along the Rio Grande near Truth or Consequences and in Rio Arriba County near El Rito. The continued existence of *R. blairi* near El Rito needs confirmation. Although not currently sympatric with *R. berlandieri*, historical ranges of these two species overlapped.

Description: *Rana blairi* may be recognized by dorsolateral folds that are well developed and discontinuous with the posterior ends displaced medially. These posterior segments may rarely appear continuous or broken but not inset. The upper lip has a distinct white line that is usually well defined anterior to the eye. There is usually a light spot in the center of the tympanum. The venter is ivory or cream-colored, frequently with some yellow posteriorly, as well as in the groin and on the proximal part of the thigh. There is often mottling on the throat. The posterior surface of the thigh has dark, usually indistinct, reticulations on a light background; this pattern rarely appears spotted or weakly reticulate. The area around the cloaca and the posteroventral surface of the thigh is densely covered with tubercles. The dorsal spots may or may not be surrounded by light colored halos. There is usually a spot on the head anterior to the eyes and the spots on the body are often arranged in loosely defined longitudinal rows. The area between the spots is buff, pale brown, or sometimes dull green. Males are smaller than females, with the mean SVL in males 64.4 mm, in females 75.5 mm (Fritts et al., 1984). The mating call is usually 2–3 distinctly spaced, abrupt guttural notes delivered at a pulse rate of about three per second (Conant and Collins, 1991).

The tadpole of *R. blairi* is the palest of the leopard frog tadpoles in New Mexico. The tail has fine indistinct mottling. As the tadpole approaches metamorphosis, the distal half of the tail tends to darken. The tail fin is deep, 70–90% of the body length. There is at least part of a third row of upper labial teeth present in about half of all large specimens. This row usually consists of 1–3 labial teeth on either or both sides of the beak. There are three lower rows of labial teeth. The oral papillae are normally unpigmented, relatively small, and densely packed lateral to the beak. The iris is medium-gold in color and lacks dorsal or ventral dark spots (Scott and Jennings, 1985).

Similar Species: *Rana blairi* is most likely to be confused with *R. pipiens* and *R. yavapaiensis*. It can be distinguished from *R. pipiens* by its discontinuous dorsolateral folds, the reticulated thigh pattern, and halos surrounding the dorsal spots that are absent or very faint. *Rana blairi* can be distinguished from *R. yavapaiensis* by the presence of spots on the nose anterior to the eyes, the presence of a spot on the tympanum, and a complete supralabial stripe. *Rana blairi* is sympatric with *R. pipiens* in Colfax, Mora, San Miguel, and historically, Rio Arriba counties, and also in the lower Rio Grande Valley in Sierra and Socorro counties. Its range abuts that of *R. berlandieri* in Eddy County. Characters used to separate *R. blairi* from all other leopard frogs in New Mexico are presented in Table 1.

Systematics: No subspecies of *R. blairi* are recognized. Hillis et al. (1983) assigned this species to the *pipiens* species group of temperate leopard frogs. In New Mexico, *R. blairi* is known to hybridize with *R. pipiens* at three localities including along the Cimarron River near Springer, along the Mora River just downstream of the confluence with Wolf Creek, and along the lower Rio Grande just north of Elephant Butte Reservoir. The first extensive data for *R. blairi* were provided by McAlister (1962) while this species was still confused with *R. pipiens*.

Habitat: *Rana blairi* inhabits a variety of temporary and permanent aquatic habitats, including streams, ponds, irrigation ditches, pools in rocky canyons, and rivers. They are generally found in playa lakes, grasslands, and river marshes. Scott and Jennings (1985) commonly found tadpoles in muddy tanks and rivers.

Behavior: *Rana blairi* may be somewhat terrestrial during part of the year. Adults are often found near bodies of water during the spring and summer breeding season, or during winter hibernation. The remainder of the year, they follow irrigation waters into cultivated fields, or they find low-lying moist spots in green meadows. When caught by predators, this species issues a loud explosive distress call. *Rana blairi* seems to have a marked tendency to leap away from water rather than toward it to escape predators (pers. obs.).

Reproduction: Pace (1974) reported breeding from early February to early October. In Arizona, egg masses have been recorded in March, May, and August with tadpoles present from late March to early June and from August to October (Frost and Platz, 1983). Large clutches of eggs are attached to vegetation in shallow water. Collections of small and medium-sized tadpoles in April probably indicates breeding in New Mexico as early as March or perhaps February. In the Truth or Consequences area, *R. blairi* has been observed to breed from May to August (J. N. Stuart, pers. comm.) The tadpoles may overwinter (Gillis, 1975; Scott and Jennings, 1985). Tadpoles metamorphose at approximately 28–30 mm SVL.

Food Habits: Food habits of *R. blairi* have not been studied in New Mexico, although it probably feeds on a variety of insects and invertebrates. Hammerson (1982) found a specimen in Colorado that had a stomach full of grasshoppers and Creel (1963) captured a large individual (114 mm SVL) in Texas consuming a bat (*Pipistrellus subflavus*) with only one of its wings protruding from the frog's mouth.

Remarks: The distribution map for this species is modified after Jennings and Scott (1991). Brown (1992) reviewed this species. The specific name honors the late Dr. W. Frank Blair, professor of zoology at the University of Texas at Austin.

FAMILY RANIDAE

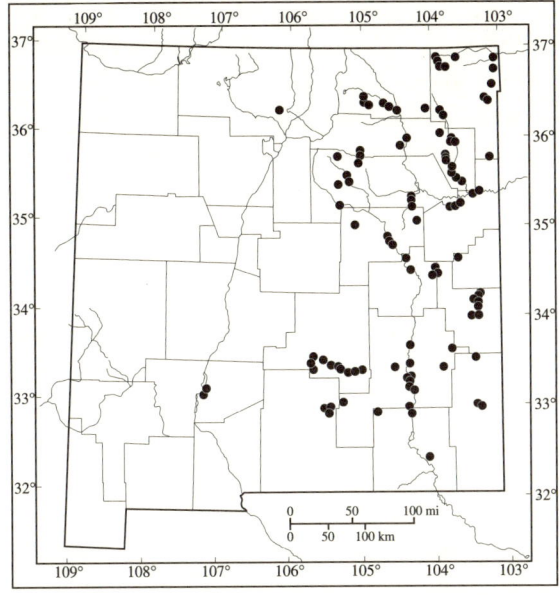

*Distribution of
Rana blairi
in New Mexico.*

RANA CATESBEIANA Shaw, 1802 — *Bullfrog*

See Plate 23.

Type: George Shaw (1802) gave the type locality of *Rana catesbeiana* as "North America" although that was later restricted to "vicinity of Charleston, South Carolina" by Schmidt (1953). A type specimen is not known to exist. The collector and date of collection are unknown (Frost, 1985).

Distribution: One of the most widely distributed amphibians on the North American continent, *R. catesbeiana* originally ranged from Nova Scotia to central Florida, west to the Rocky Mountains. The historic western limits of its range are hopelessly confused because of its introduction into a large number of localities as far west as British Columbia and California. *Rana catesbeiana* has also been introduced into many other countries, including Mexico, Cuba, and Jamaica (Conant and Collins, 1991). In New Mexico, *R. catesbeiana* occurs statewide at lower elevations (approximately 900–2100 m) wherever suitable freshwater habitats exist.

Description: *Rana catesbeiana* is easily recognized by the short glandular fold that curves around the dorsal edge of the tympanum, stopping just above the forearm. There are no dorsolateral ridges on the body. The dorsum is green or brown and is often roughened by numerous very small tubercles. Larger individuals may possess a mottled, somewhat melanistic pattern. The venter is white or yellow and is often mottled with varying amounts of pigment. The long, powerful, and delicious hind legs may be banded or blotched. Males have a single gular vocal pouch and the throat is pale to bright yellow. Throat color in females is similar to the venter. The tympanum of the male is conspicuously larger (up to 2x as large) than the eye diameter, that of the female equal to or smaller than the eye. Large adults may reach over 203 mm in body length; newly transformed young are approximately 35–45 mm SVL. The familiar call is a deep, sonorous "jug-o-rum."

The tadpole may be quite large, 78–121 mm TL. It is usually a drab olive green with numerous fine black pinhead-sized specks scattered throughout the tail fin and on the dorsal surface. The labial tooth row formula is 2/3 or 1/3 with the second anterior row very short (Johnson, 1987).

Similar Species: The dorsolateral fold that curves around the tympanum will distinguish this species from all other ranid frogs in New Mexico. Adults may be distinguished by their large size and the lack of large, round dorsal spots as seen in the leopard frogs.

Systematics: No subspecies are currently recognized, although Dundee and Rossman (1989) suggested *Rana catesbeiana* may represent two species.

Habitat: Generally associated with permanent freshwater habitats such as the state's larger rivers and lakes, this frog may appear in stock ponds or roadside ditches after heavy rains. It also occurs in irrigation ponds in otherwise arid terrain. Plant growth in *R. catesbeiana* habitat is often dense and emergent. In New Mexico, it is usually not found above 2100 m elevation. In Gunnison County, Colorado, Hammerson (1982) found *R. catesbeiana* thriving at 2743 m where warm spring waters enabled them to reproduce in an area that was otherwise uninhabitable.

Behavior: Startled juvenal *R. catesbeiana* usually squawk as they leap into the water. When disturbed, they retreat to deeper water with a series of long leaps and considerable splashing. Ferguson et al. (1968) observed *R. catesbeiana* making movements of 137–275 m during the summer. They suggested that individuals make sporadic appearances at small isolated pools, and sites near large bodies of water may be alternately occupied and vacated for several days or weeks. After nocturnal rains, they noted occasional one-night excursions of up to 159 m beyond the home range.

In New Mexico, *R. catesbeiana* often occupies ephemeral stock tanks that are several miles from any permanent water source. It is unknown how these individuals colonize these ponds or where they seek shelter when the ponds dry up, although it is likely that they estivate in the mud or in nearby rodent burrows until summer rainfall has refilled these tanks. Often there is a single chorusing male present at these sites and it is unlikely that many of them successfully reproduce, although J. N. Stuart (pers. comm.) observed newly metamorphosed juveniles in ephemeral ponds in Luna County near Hermanas in late August.

Rana catesbeiana usually spends the daylight hours hiding under vegetation or overhanging banks or sitting in the shallow water near shore. As nightfall approaches, they may leave these retreats and establish calling sites along the bank or on floating logs.

Reproduction: During the breeding season, adult males establish territories and aggressively exclude conspecific males. Unlike many North American anurans, in which members of a given population attain peak reproductive condition synchronously and mating occupies only a brief period of days or weeks, males remain reproductively active throughout much of the summer. Each female, however, is sexually receptive for only a short time, and the long breeding season results from the great variation in the dates at which different females reach this condition (Emlen, 1968). Woodward (1987a) found that on the average adult females in southern New Mexico contained approximately 11,200 eggs. In New Mexico, eggs are laid during late spring and summer.

Food Habits: *Rana catesbeiana* is a voracious and opportunistic "sit-and-wait" predator, eating almost anything it can overcome and swallow. Bury and Whelan (1984) published a list of known food items, which includes such varied items as vegetation, leeches, salamanders, snakes, alligators, turtles, birds, bats, and small rodents. Clarkson and deVos (1986) reported a western diamondback rattlesnake, a softshell turtle, and a muskrat as food items. Beringer and Johnson (1995) included a young mink that was taken from a bullfrog collected in Missouri. In New Mexico, the most commonly eaten prey seems to be nocturnal insects, including moths and beetles, and snails and crawfish. Randy Jennings (pers. comm.) reported an adult *Ambystoma tigrinum* eaten by an adult *R. catesbeiana* from the Rio Grande Valley in Valencia County. Conspecific juveniles and tadpoles may be important food items in high-density populations in New Mexico (Stuart and Painter, 1993a).

Toads (Bufonidae) are rarely found in the diet of *R. catesbeiana* and are generally thought to be unpalatable. Brown (1974) reported that the parotoid gland secretions of adult *B. woodhousii* and *B. valliceps* can immobilize large *R. catesbeiana*. Tucker and Sullivan (1975) observed *R. catesbeiana* swallowing adult *B. woodhousii* and *B. americanus* and quickly regurgitating them. However, Stuart (1995a) collected two large adult *R. catesbeiana* from a pond in southern Luna County. There was a large chorus of toads in the pond and the *R. catesbeiana* contained partly digested adult *Bufo cognatus* and *B. debilis* in their stomachs.

FAMILY RANIDAE

Remarks: In New Mexico, *R. catesbeiana* is widely introduced west of the Continental Divide. The earliest specimen known from New Mexico is in the collection of the Museum of Comparative Zoology at Harvard University (MCZ 1917). This specimen was reported collected from Santa Fe in 1885, although Santa Fe may actually have served only as a shipping point. It is unknown if *R. catesbeiana* is native to New Mexico.

In some cases predation by this large frog may be the cause of localized extinctions of associated wetland herpetofaunas (Moyle, 1973; Hammerson, 1982; Rosen and Schwalbe, 1988). Rosen and Schwalbe (1988) and Schwalbe and Rosen (1988) hypothesized that predation by *R. catesbeiana* caused the decline of a native wetland herpetofauna in southeastern Arizona, including *Thamnophis eques*, *Rana yavapaiensis*, and *Kinosternon sonoriense*. Hayes and Jennings (1986) reviewed data relating to several factors that have been suggested as causing the decline of ranid frogs in the western North America. Conant (1977 [1978]) reported his observations on the result of a Mexican governmental boondoggle translated as "frogs for the tables of Mexico." Bullfrogs introduced into the lower portions of the Río Casas Grandes were present in large numbers, and they had completely extirpated all the riparian vertebrates, frogs, snakes, and small turtles, that normally would be expected along that stream.

Rana catesbeiana is protected in New Mexico and a fishing license is required to take them during the limited hunting season. In view of their detrimental effect on other wildlife, this protection is inconsistent with good conservation practice.

Bury and Whelan (1984) reviewed much of the literature on this species.

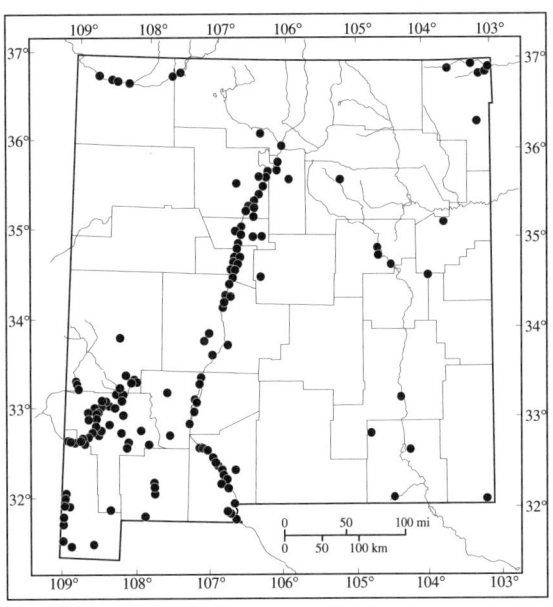

Distribution of Rana catesbeiana *in New Mexico.*

RANA CHIRICAHUENSIS Platz and Mecham, 1979   *Chiricahua Leopard Frog*

See Plate 24.

Type: *Rana chiricahuensis* was described from "Herb Martyr Lake (elev. 1768 m), 6 km west of Portal, Coronado National Forest, Cochise County, Arizona." The holotype is AMNH 100372, an adult male collected on 10 September 1971 by James E. Platz. Paratypes include AMNH 100373–100382 and UMMZ 150049–54, 150187–88.

Distribution: *Rana chiricahuensis* occurs as disjunct populations in Arizona, New Mexico, and Sonora and Chihuahua, Mexico. Populations in the northern part of the range are found in montane habitats of the Mogollon Rim of central and eastern Arizona, and adjacent mountains and foothills of western New Mexico. This part of the range is separated from populations along the southern borders of Arizona and Sonora, Mexico, and New Mexico by broad expanses of scrub desert with elevations below 1000 m. Another series of populations is distributed southward in Chihuahua along the eastern base of the Sierra Madre Occidental. The species occurs from 1000–2600 m (Platz and Mecham, 1984).

In New Mexico, *R. chiricahuensis* is known from the southwestern portion of the state and is most abundant in the Gila and San Francisco river drainages. The Rio Grande drainage is occupied by these frogs only in Alamosa Creek in Socorro County and Cuchillo Negro

Creek in Sierra County. Other localities include the Mimbres River drainage of Grant and Luna counties and the numerous stock tanks and intermittent creeks of southwestern Hidalgo County, including those in the Animas and Peloncillo mountains.

Description: *Rana chiricahuensis* is the most heavily proportioned of all the leopard frogs in New Mexico. Posteriorly, the dorsolateral folds are discontinuous and inset medially or indistinct and poorly defined. The supralabial stripe is absent anterior to the eye. The dorsum is moderately ornamented with blunt tubercles between the dorsolateral folds. The venter is dull whitish or yellowish, often with gray mottling on the throat and chest. The posterior surface of the thigh is highly tuberculate, with a dark ground color and white spots corresponding to tips of tubercles. Light-colored halos are usually lacking around the dorsal spots. There is usually at least one spot on the head anterior to the eyes. Males are smaller than females, with the mean body length in males 64.3 mm SVL, females 76.9 mm (Fritts et al., 1984). Adult SVL is usually 57–95 mm (Platz and Mecham, 1984), although Scott (1992) recorded a female 107 mm. The mating call is a long snore-like trill with an unusually high pulse repetition (approximately 34 pulses/sec at 22°C) and pulse number (28–68 per call). The call is repeated intermittently and is typically offered as a single note lasting between 1–2 seconds, depending upon temperature (Platz and Mecham, 1984). The call of populations from New Mexico has not been described. This species is known to call underwater (N. J. Scott, pers. comm.) similar to *R. subaquavocalis* (Platz, 1993).

This is our darkest leopard frog tadpole. The tail has numerous dark blotches which may coalesce. The tail fin is of medium depth, 60–80% of the body length. There are one or two rows of labial teeth on the upper labium. The unpigmented oral papillae are relatively small and sparse in the area lateral to the beak. The iris is dark in color with lateral, dorsal, and ventral dark spots (Scott and Jennings, 1985). Tadpoles show considerable variation in several morphological traits depending upon the type of aquatic habitat in which they occur (Jennings and Scott, 1993).

Similar Species: *Rana chiricahuensis* is the most distinctive of the leopard frogs in New Mexico. It is unique in having prominent white spots on a dark ground color on the tuberculate posterior surface of the thigh. *Rana chiricahuensis* is most likely to be confused with *R. yavapaiensis*. It can be distinguished from *R. yavapaiensis* by the presence of spots on the nose anterior to the eyes, its distinctive thigh pattern, green coloration on the dorsum, and the presence of vestigial oviducts in males. *Rana chiricahuensis* is sympatric with *R. yavapaiensis*, and the range abuts that of *R. pipiens* in northern Catron County. Characters used to separate *R. chiricahuensis* from all other leopard frogs in New Mexico are presented in Table 1.

Systematics: No subspecies of *Rana chiricahuensis* are recognized. Populations of *Rana chiricahuensis* from Hidalgo County may represent one or more undescribed species (N. J. Scott and R. D. Jennings, pers. comm.). Hillis et al. (1983) assigned this species to the *montezumae* species group of Mexican leopard frogs.

Habitat: *Rana chiricahuensis* is found in a variety of aquatic habitats including thermal springs and seeps, stock tanks, wells, intermittent rocky creeks, and mainstream river reaches. They are the most aquatic of New Mexico leopard frogs.

Behavior: These shy, nocturnal frogs are quick to seek shelter when approached. During the day they usually rest hidden among the vegetation surrounding their aquatic habitat and are quick to enter the water when approached. Nocturnal activity may take them farther from the bank, or they may be observed on exposed mats of algae or other floating aquatic vegetation.

Populations of *R. chiricahuensis* may exhibit seasonal fluctuations in relative abundance. Overall abundance increases with the metamorphosis of tadpoles in August and September and is lowest from December through March. Throughout the year, frog activity generally increases as nocturnal water temperatures increase (Jennings, 1990).

Reproduction: In New Mexico, populations of *R. chiricahuensis* occurring in thermally stable habitats may be reproductively active throughout the year, with tadpoles growing continuously during the winter months. Jennings (1988, 1990) found reproductive activity throughout the entire year at Alamosa Warm Springs in Socorro

County where the surface temperature of the water remains above 16°C. At a nearby stock tank with a varying temperature regime, reproduction was noted only during late April through May and from mid-August through late September. The time required to pass through the larval stage may be much more rapid at sites with relatively warm water temperatures than at sites where water temperature is strongly affected by ambient air temperature. The time from hatching to metamorphosis is 2–3 months at Alamosa Warm Springs while it may be 8–9 months at sites that experience marked drops in temperature over the winter (Jennings, 1988). Tadpoles metamorphose at approximately 35–40 mm SVL.

In the Chiricahua Mountains of southeastern Arizona, near the type locality, eggs of *R. chiricahuensis* have been found from early February through early September (Zweifel, 1968b).

Food Habits: The food habits of *R. chiricahuensis* have not been studied in New Mexico, although, as with all leopard frogs, it likely eats a wide variety of insects and other arthropods.

Remarks: Prior to formal description, Platz (1976) discussed the biochemical and morphological variation in this species under the name "southern form." The distribution map for this species is modified after Jennings and Scott (1991). This species is in decline throughout its range in New Mexico and Arizona. Scott (1992) visited eight known collection localities on the Gray Ranch in southern Hidalgo County and was unable to find *R. chiricahuensis* at any of these localities. The Chiricahua leopard frog is listed as a Category 2, Notice of Review species by the U.S. Fish and Wildlife Service (USDI, 1994). Platz and Mecham (1984) reviewed this species.

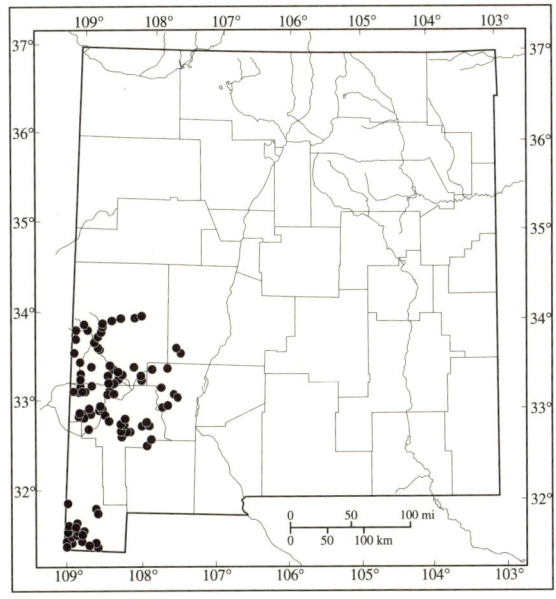

Distribution of Rana chiricahuensis *in New Mexico.*

RANA PIPIENS Schreber, 1782 — *Northern Leopard Frog*

See Plate 25.

Type: The original description of *Rana pipiens* (Schreber, 1782) was based on a specimen sent to H. Schreber from New York by Dr. Johann D. Schoepf, a doctor with the Royal Brandenberg Troops in America during the Revolutionary War. The type locality was given as "New York and Raccoon Landing, Gloucester County, New Jersey" and was later restricted to "White Plains, New York," by Schmidt (1953). The date of collection is unknown. In the original description, a holotype was not identified. Pace (1974) later designated UMMZ 71365, from "Fall Creek, Etna, Tompkins County, New York" USA, as the neotype.

Distribution: *Rana pipiens* ranges from southern Quebec west to the extreme southern District of Mackenzie, and south to Pennsylvania and Kentucky in the East with isolated records in Maryland and West Virginia. It occurs west to the Pacific states, and south, to Nevada, Arizona, New Mexico, and Texas with disjunct colonies in eastern Colorado and southern New Mexico along the Rio Grande south of Caballo Reservoir. It has been introduced into western Newfoundland and extensively in the West, particularly in California. There is an introduced, disjunct colony in Labrador (Conant and Collins, 1991). In New Mexico, *R. pipiens* is known from about 1120–

3050 m in a large area in the northern and western part of the state and along the entire length of the Rio Grande Valley except southern Elephant Butte and northern Caballo reservoirs. Although recent records are lacking from much of the Rio Grande Valley, the species persists in the Isleta–Los Lunas area.

Description: *Rana pipiens* has continuous dorsolateral folds that are typically not broken or inset posteriorly. The supralabial stripe is variable, being well-defined, faint, or absent anteriorly depending upon the population being studied. The dorsal spots usually have faint or prominent light colored halos. There is usually one spot on the head anterior to the eyes. Few or no tubercles are on the dorsal and lateral body surfaces. The venter is white or cream colored. The posterior surface of the thigh has dark spots or blotches, typically with the spots not forming a reticulate pattern. Males appear to be only slightly smaller than females. Mean SVL in males is 68.3 mm and in females, 74.2 mm (Fritts et al., 1984). The mating call is a long, deep rattling snore interspersed with clucking grunts that may be single or of two or more syllables. The call duration is more than a second (usually three) and the pulse rate is about 20/sec (Conant and Collins, 1991).

The tadpole of *Rana pipiens* has coarse indistinct mottling on the tail. The distal half of the tail tends to darken approaching metamorphosis. The tail fin is of medium depth, 65–80% of the body length. There are usually two rows of labial teeth on the upper labium; rarely a partial third row is present. The unpigmented oral papillae are medium-sized and sparse lateral to the beak. The iris is gold in color with dorsal, lateral, and ventral dark spots (Scott and Jennings, 1985).

Similar Species: *Rana pipiens* is most likely to be confused with *R. blairi*. It can be distinguished from *R. blairi* by its continuous dorsolateral folds that are not broken posteriorly and inset medially, (a character that separates *R. pipiens* from all other leopard frogs in New Mexico), the absence of spots on the tympanum, and a thigh pattern of discrete dark spots on a light ground color. *Rana pipiens* is sympatric with *R. blairi* in Colfax, Mora, San Miguel, and historically, Rio Arriba counties, and also in the lower Rio Grande Valley in Sierra and Socorro counties. Its range abuts that of *R. chiricahuensis* in northern Catron County. Characters used to separate *R. pipiens* from all other leopard frogs in New Mexico are presented in Table 1.

Systematics: No subspecies of *R. pipiens* are recognized. Hillis et al. (1983) assigned this species, along with *R. blairi*, to the *pipiens* species group of temperate leopard frogs. In New Mexico, *R. pipiens* is known to hybridize with *R. blairi* at three localities, including along the Cimmaron River near Springer, along the Mora River just downstream of the confluence with Wolf Creek, and along the lower Rio Grande just north of Elephant Butte Reservoir.

Rana pipiens is the oldest name that has been assigned to the leopard frogs of the United States, including at one time frogs from Washington to Maine, from California to Florida, and from Canada to Panama (Fritts et al., 1984; Hillis, 1988). This widespread usage of a single name for all leopard frogs in the United States led to considerable confusion in the earlier literature of this group.

Habitat: *Rana pipiens* is generally associated with streams and rivers, although lakes, marshes, and irrigation ditches are also occupied. Much of the river valley habitat of these frogs has been modified by human activities, including draining of wetlands, channelization and damming of rivers, and the development of irrigation systems (Scott and Jennings, 1985).

Behavior: *Rana pipiens* rests near the water's edge and quickly leaps into or away from the water when alarmed. It may forage long distances from the water during wet periods.

Reproduction: Scott and Jennings (1985) reported eggs and small tadpoles of this species from April–July and September–October in New Mexico. In Colorado, males begin calling on warm, sunny days in March or April. Calling usually wanes in April but low-elevation frogs sometimes call during May or early June. Females begin laying eggs a few days after calling begins, and most females in the plains region lay their eggs by mid-April (Hammerson, 1982). Corn and Livo (1989) investigated the reproduction of *R. pipiens* in Colorado and Wyoming. They examined 68 egg masses that averaged 3,045 eggs (range 645–6277). Mean hatching success ranged from

70–99%. Egg masses are regularly attached to emergent vegetation. The initial breeding activity observed by Corn and Livo (1989) was related more to temperature than to precipitation. Oviposition followed the onset of male chorusing by two or three days and corresponded to periods of warm weather. Males continued calling for about two weeks after the last egg masses were deposited. Breeding among high-elevation populations of *R. pipiens* begins later in the year than low-elevation populations. Hahn (1968) observed metamorphosing leopard frogs on 2 August at 3200 m in southern Colorado. Tadpoles metamorphose at approximately 30–35 mm SVL.

Food Habits: Food habits of *R. pipiens* in New Mexico are unknown, but they undoubtedly feed on a wide variety of invertebrate prey. Drake (1914) examined 209 specimens from Ohio and found that almost 90% of the prey consisted of various insects and spiders. Breckenridge (1970) reported a large adult that had eaten a small garter snake.

Remarks: Corn and Fogleman (1984) discussed the extinction of montane populations of *R. pipiens* in Colorado. The distribution map for this species is modified after Jennings and Scott (1991).

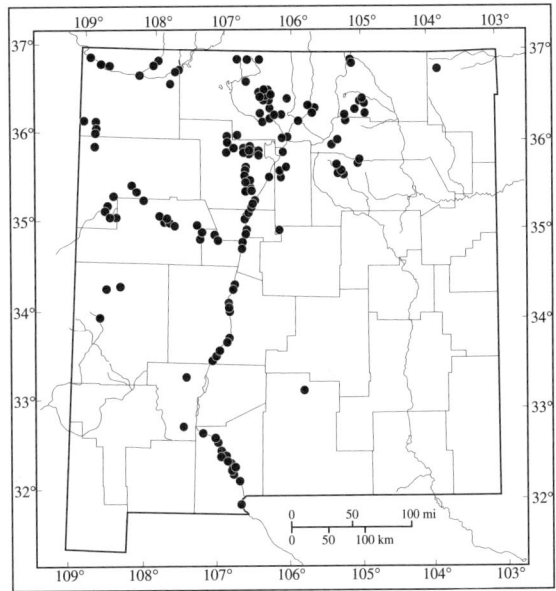

Distribution of Rana pipiens *in New Mexico.*

RANA YAVAPAIENSIS Platz and Frost, 1984 — *Lowland Leopard Frog*

See Plate 26.

Type: The holotype of *Rana yavapaiensis*, AMNH 117623, is an adult male collected from "Tule Creek (elev. 670 m), 34°00', 112°16', Yavapai Co., Arizona." on 25 August 1971 by James E. Platz. Paratypes include AMNH 117633–639 and UMMZ 174839–44.

Distribution: *Rana yavapaiensis* occurs in western New Mexico, Arizona, and in adjacent parts of Sonora, Mexico. Specimens have also been collected in Imperial County, California, and the Virgin River drainage of Arizona, Nevada, and Utah (Platz, 1988; R. D. Jennings, pers. comm.). In New Mexico, *R. yavapaiensis* occurs from 1128 m to about 1700 m and is known only from western Catron, Grant, and Hidalgo counties in the southwestern corner of the state. Most localities are in the San Francisco and Gila river drainages, although specimens have been taken in western Hidalgo County in Double Adobe Creek, Pine Canyon, and Guadalupe Canyon. Despite numerous attempts to locate additional individuals, the last specimen observed in New Mexico was seen in Guadalupe Canyon during April 1985. During May through September 1986, Jennings (1987) visited 12 historical collection sites and 14 new sites considered likely habitat and was unable to locate *R. yavapaiensis*.

Description: The posterior dorsolateral folds of *R. yavapaiensis* are discontinuous and inset medially or are poorly defined. The supralabial stripe is absent anteriorly. The dorsal spots usually lack light colored halos and there are no spots on the head anterior to the eyes. There are few or no tubercles on the dorsal and lateral body surfaces. The venter is cream colored with a yellow wash on the groin and posterior venter. The posterior surface of the thigh is reticulated with various degrees of dark and light coloration (Fritts et al., 1984). Adult males have a body length of 46–72 mm SVL and females 53–87 mm. The advertisement call is a series of short chuckles, the first of these longer than the 6–15 that follow. These notes may last from 3–8 seconds depending upon the total number of notes and the temperature. The pulse rate is low (12 pulses/sec at 24°C) and the pulse number per note varies, decreasing from approximately 11 pulses in the first note to 3–4 in the last of a series (Platz, 1988).

The tadpole of *Rana yavapaiensis* has sparse, discrete, small, dark blotches on the tail. The tail fin is shallow, 50–65% of the body length. There are two or three rows of labial teeth on the upper labium. The oral papillae are small and very sparse lateral to the beak. Many of the oral papillae have dark pigment on the tips. The iris is dark in color with dorsal, lateral, and ventral dark spots (Scott and Jennings, 1985). Tadpoles of this species have not been collected in New Mexico.

Similar Species: *Rana yavapaiensis* is likely to be confused with *R. berlandieri*, *R. blairi*, and *R. chiricahuensis*. It can be distinguished from *R. berlandieri* by its diffusely reticulated thigh pattern and the absence of vestigial oviducts in males; from *R. blairi* by the absence of spots on the nose anterior to the eyes, and the absence of vestigial oviducts in males; and from *R. chiricahuensis* by the absence of spots on the nose anterior to the eyes, the absence of vestigial oviducts in males, the reticulated thigh pattern, and the frequent lack of green coloration on the dorsum. Only *R. chiricahuensis* is found sympatrically with *R. yavapaiensis* in New Mexico. Characters used to separate *R. yavapaiensis* from all other leopard frogs in New Mexico are presented in Table 1.

Systematics: No subspecies of *Rana yavapaiensis* have been recognized. Hillis et al. (1983) assigned this species, along with *R. berlandieri*, to the *berlandieri* species group of Mexican leopard frogs.

Habitat: In New Mexico, most populations occupy small streams and rivers, springs, and associated pools at low elevations in scrub desert localities. In Arizona, they are most abundant where pools are deep enough to provide a safe retreat from predators (Platz, 1988). In southwestern New Mexico and the adjacent Sierra San Luis of Sonora, Mexico, individuals have been observed in a variety of aquatic situations, including along rocky rivers and small intermittent creeks. Along these streams, *R. yavapaiensis* seems to be concentrated where springs enter the stream, near debris piles that had collected around small, streamside trees and shrubs, and near deep pools associated with root masses of large trees growing along the bank (Jennings, 1987; our obs.).

Behavior: Due to the scarcity of observations, very little is known about the behavior of *R. yavapaiensis* in New Mexico. In the Sierra San Luis of Sonora, Mexico, they were observed to seek shelter in the streamside vegetation or in the root masses of downed trees along small creeks. Jennings and Hayes (1994) included a photograph of *R. yavapaiensis* in a defensive posture.

Populations of *Rana yavapaiensis* may be especially susceptible to events such as severe floods and droughts. These catastrophic, 100-year floods may displace leopard frog populations and destroy or alter habitats so that recolonization may take several years.

Reproduction: The breeding chronology of *R. yavapaiensis* is little known in New Mexico, although in Arizona there seems to be an early and late season peak in breeding activity. Egg masses and newly hatched tadpoles have been found in late February to late April and during October (Frost and Platz, 1983; Collins and Lewis, 1979). At a site near Pima, Arizona, we have observed calling males as well as newly laid egg masses and amplecting pairs of adult frogs on 31 January 1992. Jennings (1987) collected post-metamorphic frogs on 25 August along Cañon Bonito in the Sierra San Luis in Sonora, Mexico. Metamorphosis occurs at 25–29 mm in length (Platz, 1988).

Food Habits: No studies of the food habits of *R. yavapaiensis* have been reported, although they probably eat a wide variety of insects and other invertebrates.

Remarks: Platz and Platz (1973) and Platz (1976), under the description "Lowland form," discussed the distribution in Arizona and the biochemical and morphological variation in *R. yavapaiensis*. The distribution map for this species is modified after Jennings and Scott (1991). *Rana yavapaiensis* is listed as a Category 2, Notice of Review species by the U.S. Fish and Wildlife Service (USDI, 1994) and as endangered by the New Mexico Department of Game and Fish (NMGF, 1990). Platz (1988) reviewed this species. Jennings and Hayes (1994) reviewed the decline of this species in the desert southwest.

FAMILY RANIDAE

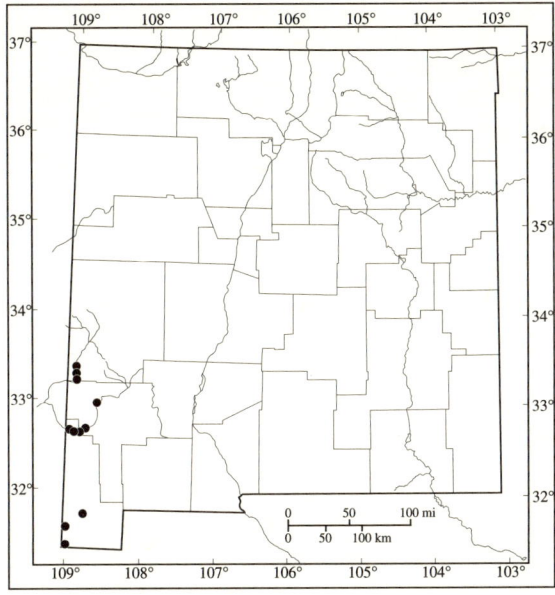

*Distribution of
Rana yavapaiensis
in New Mexico.*

A KEY TO THE TURTLES OF NEW MEXICO

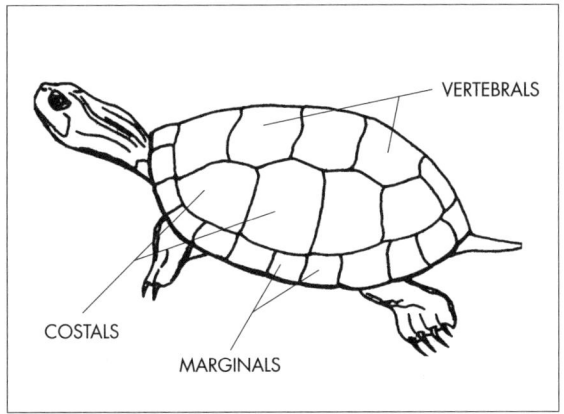

Dorsal aspect of a turtle showing scutes of the carapace.

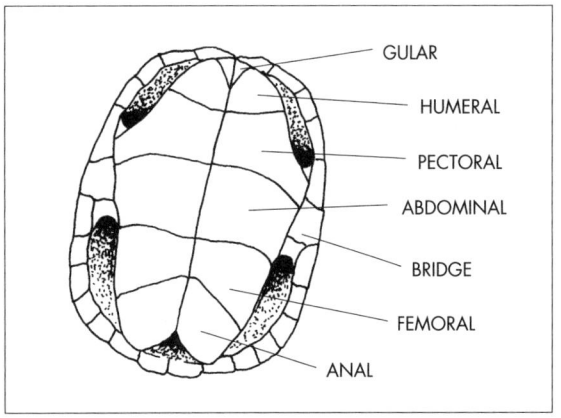

Ventral aspect of a turtle shell showing scutes of the plastron.

1. Carapace covered with numerous horny plates — 3

1. Carapace covered with leathery skin (Trionychidae) — 2

2. Carapace smooth to the touch, not sandpapery, and without tubercles on its anterior surface (**FIG. 2a**); nasal septum without lateral ridges projecting into nostrils (**FIG. 2d**) — *Trionyx muticus*

2. Carapace rough and sandpapery in adults, with tubercles on its anterior surface (**FIG. 2b**); nasal septum with lateral ridges projecting into nostrils (**FIG. 2c**) — *Trionyx spiniferus*

3. Plastron reduced and cross-shaped (**FIG. 3**); soft underparts exposed; head massive and tail longer than one-half the carapace length (Chelydridae) — *Chelydra serpentina*

3. Plastron, head, and tail not as above — 4

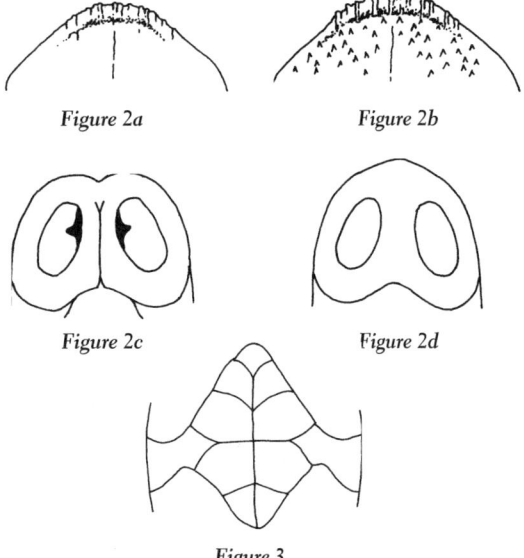

Figure 4a

Figure 4b

4. Hind feet elephant-like and not webbed (**FIG.** 4a) (Testudinidae) *Gopherus agassizii*

4. Hind feet not elephant-like and more or less webbed (**FIG.** 4b) 5

5. Plastron with 11 scutes (one gular) and two prominent hinges bordering the abdominal scutes (**FIG.** 5a) (Kinosternidae) 6

5. Plastron with 12 scutes (two gulars) and one or no prominent hinges (**FIG.** 5b) (Emydidae) 7

Figure 5a

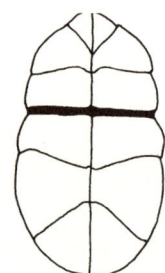
Figure 5b

6. Ninth marginal noticeably higher than eighth (**FIG.** 6a); color brown to olive or green; chin and throat yellowish *Kinosternon flavescens*

6. Ninth marginal not greatly higher than eighth (**FIG.** 6b); color olive-brown to dark brown; mottled pattern on head and neck *Kinosternon sonoriense*

7. Plastron with prominent single hinge *Terrapene ornata*

7. Plastron without prominent hinge 8

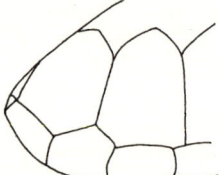
Figure 6a

Figure 6b

8. Carapace not serrated posteriorly; no vertebral keel (**FIG.** 8a) *Chrysemys picta*

8. Carapace serrated posteriorly; vertebral keel present though often weakly represented (**FIG.** 8b) 9

9. Second costal scute light with C-shaped marking directed posteriorly; no prominent patches of red on side of head behind eye; upper jaw notched in front *Pseudemys gorzugi*

9. Markings on second costal scute not C-shaped; prominent patches of red or yellow on side of head behind eye; upper jaw not notched in front 10

Figure 8a

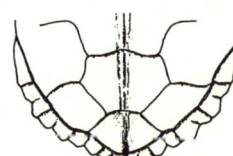
Figure 8b

10. Usually a broad reddish stripe behind the eye without a distinct black border (**FIG.** 10a) *Trachemys scripta*

10. A large black-bordered orange spot on the side of the head with a second smaller spot directly behind the eye (**FIG.** 10b) *Trachemys gaigeae*

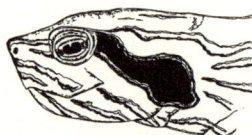
Figure 10a

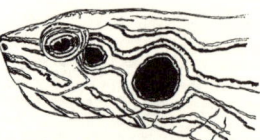
Figure 10b

FAMILY CHELYDRIDAE *Snapping Turtles*

The family Chelydridae ranges from southern Canada to Ecuador and contains only two living genera, each with a single species: *Chelydra serpentina* and *Macroclemys temminckii*. These turtles have large heads equipped with powerful jaws. The massive carapace is strongly serrated posteriorly and has three longitudinal keels. It is connected to a reduced, cross-shaped hingeless plastron by only a narrow bridge. The tail has a dorsal row of prominent saw-toothed scales and is as long or longer than the carapace. These turtles are highly aquatic and spend considerable time lying on the bottom, only rarely coming ashore to bask or lay eggs. They are omnivorous and feed on a variety of plant and animal matter. When handled, they may emit a musk as potent as that of the mud turtles. Snapping turtles are among the largest and most ferocious freshwater turtles. They can strike with astonishing speed and their strongly hooked jaws are capable of tearing flesh. The massive, well-developed limbs have long claws that can produce deep lacerations. These are dangerous turtles and should be handled with caution. Only one representative of this family occurs in New Mexico.

CHELYDRA SERPENTINA (Linnaeus, 1758) — *Snapping Turtle*

See Plate 27.

Type: A type specimen was not designated in the original description. However, for the holotype, Iverson (1992) stated, "Originally in NRM; lost according to Andersson (1900:4, 23); however, NRM GA 49 is apparently the holotype [Wallin, pers. comm.]." The type locality, given as the "Warmer Regions" (North America), was restricted to "New Orleans, La." by Smith and Taylor (1950b). Schmidt (1953) later restricted the type locality to "vicinity of New York City," but gave no reason for doing so. It is difficult to envision Schmidt's locality as a "warmer region."

Distribution: *Chelydra serpentina* is found in southern Canada and the United States east of the Continental Divide, and from southeastern Mexico to Colombia and Ecuador. Populations, probably introduced, are found in Arizona, California, Nevada, and Utah. In New Mexico, *C. serpentina* is widespread in the eastern part of the state and is largely associated with the major drainage systems of the Pecos, Canadian, and Dry Cimarron rivers. A smaller, perhaps introduced, population is present in the Rio Grande drainage of central New Mexico (see Remarks). Most records of *C. serpentina* from New Mexico are below 1400 m (Degenhardt and Christiansen, 1974), however, Shields and Lindeborg (1956) recorded specimens from the upper Pecos River drainage at 2050 m. There is a specimen (MSB 44378) from Cimarron Creek in Colfax County from approximately the same elevation. *Chelydra serpentina* occurs throughout eastern Colorado at elevations below 1650 m (Hammerson, 1982).

Description: *Chelydra serpentina* is a large, heavy-bodied turtle usually attaining an adult size of 230–490 mm CL. During 1990–93, 35 individuals collected in New Mexico averaged 221.1 (24–370) mm CL (our unpubl. data). The large, pointed head has eyes that can be seen from above, and powerful jaws that the turtle is quick to use in defense. The long tail is about the length of the carapace with a row of large sawtooth scales along the top. The carapace has three rows of prominent keels in young turtles, but these are reduced with age and may be difficult to discern in old adults. The plastron is greatly reduced and has very narrow bridges. The dorsum varies from light brown to almost black in color. The venter is off-white or cream-colored. Since *C. serpentina* seldom basks, the carapace is often covered with algae and mud.

Similar Species: The massive head and serrated tail plus the very small plastron will distinguish *C. serpentina* from all other New Mexican turtles.

Systematics: Four subspecies are recognized (Gibbons et al., 1988; Iverson, 1992). *Chelydra s. serpentina* occurs in New Mexico.

Habitat: Quiet permanent waters with aquatic vegetation are preferred by snapping turtles, but we have collected them under less than optimal conditions. Degenhardt and Christiansen (1974) stated that they tend to avoid moving waters which are devoid of vegetation in New Mexico. A large individual captured in the Rio Grande, however, was found in a sandy-bottomed channel of moving water without vegetation. Snappers are present in the Pecos River channel below Brantley Dam, which is devoid of vegetation at present. Another large specimen, estimated to be 13–22 kg, was found walking down a street in Clovis, an eastern plains town far from a major drainage system or even a river tributary. However, this turtle could have been an escaped or a released individual. The digestive turnover time (ingestion to defecation) of *C. serpentina* is significantly faster than most other North American freshwater turtles; this is considered to be an adaptation for living in a cool, benthic habitat (Parmenter, 1981) where turnover time would normally be longer due to slowed metabolism.

Behavior: *Chelydra serpentina* is one of our most aquatic turtles, spending most of its time on the bottom, often partially buried in mud or debris. It is seldom seen basking out of water but may float near the surface, possibly taking advantage of the warmer surface water and sunshine in order to raise its body temperature. In the central and eastern United States, snapping turtles are often seen crossing roads or moving on land, sometimes great distances from water. These sightings are rare in New Mexico, probably because of our dry and sunny climate as well as reduced turtle numbers. Ernst (1968) reported that snapping turtles lose water more rapidly than other freshwater turtles he tested. Hutchison et al. (1966) found them to be less tolerant of high temperatures, having a mean critical thermal maximum of 39.46°C in a sample of eight that he tested. Acclimation to higher temperatures may increase the critical thermal maximum (Williamson, et al., 1989). Collins (1982) reported that snapping turtles become more active at night when they forage for food. In Ontario, Obbard and Brooks (1981) recorded most activity in morning or early evening. In his laboratory study of juveniles, Graham (1978) also found most activity in morning and evening with least activity during midday and nocturnal hours. In New Mexico, we have taken them in sardine-baited traps checked at intervals during the day and also in those traps left in the water overnight.

These turtles overwinter on the bottom of permanent, preferably shallow waters, buried in the mud, under bottom debris, or under overhanging banks. The scarcity of suitable hibernacula may be one of the factors limiting aquatic turtle numbers in New Mexico. Snapping turtles may become active at water temperatures as low as 5°C (Ernst et al., 1994) however feeding does not normally occur at temperatures below 15°C (Obbard and Brooks, 1981).

Chelydra serpentina is vigorous in its own defense and there is danger of a severe bite. The jaws are large and powerful, the neck is long, and the turtle is capable of striking with remarkable speed. Small snappers can be safely carried by the tail but larger ones should be grasped by one or both hind legs in order to avoid injury to the turtle. It is probably unnecessary to suggest that the plastron should be facing the handler's body in order to avoid surprises!

Reproduction: Much of our knowledge of the reproductive biology of *C. serpentina* has been summarized in general works on turtles such as Carr (1952), Ernst and Barbour (1972, 1989), and Ernst et al. (1994). Male snapping turtles mature at about 15 cm PL and at 4–5 years of age. Females mature at about 17 cm PL and probably need a minimum of six years to attain this size in New Mexico. In the northernmost populations, growth is slowed because of a shorter yearly active period, and a larger body size is needed to attain sexual maturity.

Courtship and mating take place mostly in late spring to early summer, but may occur anytime between April and November when the turtles are active. Males will usually mount the females with little preliminary behavior. However, displays between male and female individuals have been described (Legler, 1955; Taylor, 1933) that were probably involved with courtship. Sperm may remain viable in females for several years.

Nesting occurs from May–September, usually peaking in June, but the time may vary because of weather conditions and the size of the female. Larger females tend to nest earlier. The female digs a flask-shaped hole 7–18 cm deep, in the open, and as much as several hundred meters from water. The spherical, tough-shelled eggs are about 28 mm in diameter. The egg number is correlated with female size, but the mean is near 30. Record clutch sizes of 104 and 109 have been reported in a Nebraska population (Packard et al., 1990).

The incubation time of 55–125 days is dependent on temperature. Hatchlings usually emerge from August – October. Delayed emergence until the following spring has been reported (Ernst, 1966) but the neonates in the nest may succumb to temperatures below –2°C in frozen soil (Packard et al., 1993). Since sex is determined by temperature (Yntema, 1976), there be may different temperatures within a large nest resulting in different sexes in a single clutch. In the laboratory incubation of eggs, Bobyn and Brooks (1994) found that a temperature of 24.8°C during the critical period of sex determination produced predominantly males (51 of 62), whereas 21.2°C and 29.1°C produced mostly females (85 of 101 and 42 of 43). There is normally one clutch per year, but individuals may skip years. Mammalian predation on nests is very high in populations studied elsewhere (Congdon et al., 1987). The effects of predation, and the tendency for some females to forego nesting in a given year, are offset by a maximum age of at least 47 years in this species. In an extensively studied Ontario population, females mature in 17–19 years, are larger in size, and produce larger clutches (Galbraith et al., 1989).

Food Habits: The snapping turtle is omnivorous, feeding on a large variety of both living and dead animal matter, and a variety of aquatic plants. Individuals appear to feed throughout the day and night. It is doubtful that snapping turtles have any effect on game fish or waterfowl populations in New Mexico.

Remarks: With few exceptions *Chelydra serpentina* cannot be considered abundant anywhere in New Mexico. Koster (1946a) examined specimens from the Pecos River and was the first to report this species from New Mexico. It is rare in the Rio Grande drainage where the status of the population needs clarification. A large *C. serpentina* was taken from the Rio Grande at Albuquerque in 1981 and photographed but the turtle later escaped. Stuart and Painter (1988) reported on two Rio Grande specimens in the University of New Mexico collection taken in 1978 and 1987. More recently, at least four additional occurrences in the Rio Grande have been documented.

Ernst et al. (1994) have summarized most of the extensive published life history information on *C. serpentina*. This genus has been reviewed by Ernst et al. (1988) and the species by Gibbons et al. (1988).

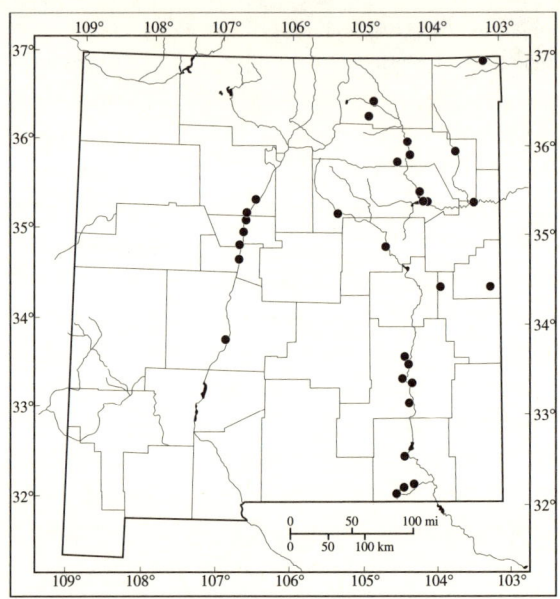

Distribution of Chelydra serpentina *in New Mexico.*

FAMILY EMYDIDAE *Box and Water Turtles*

The family Emydidae represents the most diverse group of turtles known, with representatives of the 33 genera and 91 species occurring on all continents except Australia and Antarctica. Some authorities recognize two subfamilies, Batagurinae and Emydinae (Ernst and Barbour, 1989). There are four genera with five species in New Mexico.

These turtles are distinguished by having a full bony shell covered with horny plates, and in some genera, well-developed hinges on the plastron, allowing for complete closure of the shell. Most have well-developed limbs with webbed feet. Although some species are terrestrial and have evolved a tortoise-like appearance, most are semiaquatic and inhabit an array of wetland areas, including coastal lagoons, rivers, lakes, and swamps. A large variety of plant and animal matter is eaten. These turtles are confirmed baskers and spend hours basking on logs, overhanging limbs, or floating debris.

CHRYSEMYS PICTA (Schneider, 1783) — *Painted Turtle*

See Plate 28.

Type: No type specimen or locality was given in the original description. Iverson (1992) says of the type material, "possibly in the MZUS according to R. Bour (pers. comm.)." Mittleman (1945) designated "Lancaster, Pennsylvania" as the type locality for the species.

Distribution: This species is found in southern Canada, much of the United States (absent in Florida and Nevada; introduced in California), and extreme northern Chihuahua, Mexico. In New Mexico, principal populations are present in the Pecos, Rio Grande, and San Juan river systems, but other populations are present in lakes and ponds some distances from these systems.

Description: The painted turtle reaches an adult CL of 80–180 mm, but large females may occasionally exceed 230 mm. During 1989–92, 37 individuals collected in New Mexico averaged 117.1 (78–171) mm CL (our unpubl. data). The skin is olive with yellow stripes. The carapace is brownish to olive with a network of light lines forming a reticulate pattern over the dorsal scutes. The marginals have light vertical bars and are usually red-edged. The smooth carapace is not keeled and is not serrated posteriorly. The plastron is yellow, and usually has a large central dark figure with branches extending laterally along the sutures. Males have slightly elongated claws on their forefeet. Males also differ from females in that they have longer and thicker tails with the anal opening nearer the tip rather than closer to the body. Young have an indistinct vertebral stripe and a weak dorsal keel.

Similar Species: *Trachemys* spp. usually have no red markings on the shell, and the rear margin of the carapace is serrated (sawtooth) rather than smooth. *Pseudemys gorzugi* also has a notched upper jaw but the rear of the carapace has a serrated margin. *Kinosternon* spp. have a double hinged plastron and lack red markings.

Systematics: Four subspecies are recognized (Ernst, 1971). The single subspecies found in New Mexico, *C. p. bellii* Gray, 1831, was named from a type in the Royal College of Surgeons Museum, England. That specimen was destroyed by bombing in World War II. The type locality was not stated, but was designated as "Manhattan, Kans." (Riley County, Kansas) by Smith and Taylor (1950b).

Habitat: The painted turtle typically inhabits permanent waters such as slow-moving portions of rivers, lakes, marshes, and ponds. It is also occasionally found in semipermanent waters such as irrigation ponds and ditches accessible by short overland excursions. In New Mexico, habitats vary and may have clear or turbid water, sandy, silty, or muddy bottoms, and may or may not support permanent vegetation. Other turtles occurring at the same localities are *Chelydra serpentina*, *Pseudemys gorzugi*, *Trachemys scripta*, *T. gaigeae*, *Trionyx spiniferus*, and *Kinosternon flavescens*. Degenhardt and Christiansen (1974) described in detail some specific collecting sites in New Mexico. The highest elevation collection sites within the state are 1700 m in the Rio Grande and 1600 m in the San Juan River, and the lowest elevation is about 900 m in the Pecos River near the Texas border (see Remarks).

Behavior: Painted turtles are active during the day and remain inactive at night, usually on the bottom or on submerged objects. Much of the day is spent basking in the warmed surface water, on the bank, or most often on logs or other emergent debris away from the shoreline. Many individuals may make use of a choice basking spot at the same time. This wary and agile species is often detected first by the splash when entering the water, and shortly thereafter by a head breaking the surface some distance away. Thermoregulation by basking is especially important when the water is cold and it is necessary to elevate body temperatures for feeding, digestion, and reproduction.

In a Pennsylvania population, painted turtles have been found hibernating as deep as 46 cm in the mud at the bottom of permanent waters (Ernst and Barbour, 1972). However, the usually anoxic conditions present in this environment may prevent long-term burial. Although Ultsch (1988) presented data that show his *Chrysemys p. bellii* sample was better able to compensate for anoxic conditions than both *C. p. dorsalis* and *Sternotherus odoratus*, long-term exposure to these conditions (31 days) resulted in the death of all. Possibly, periodic exits from the anoxic conditions throughout the winter dormancy period is sufficient to rid turtles of their lactate load and to allow survival. Painted turtles usually emerge in March or April and retreat again in October, but annual activity varies with seasonal temperatures. Of 331 New Mexico collection records, the earliest date is 12 March at LaJoya Waterfowl Management Area (Socorro County), and the latest is 5 November in Alameda (Bernalillo County). Ted L. Brown observed adults basking on logs in a canal in Bosque del Apache National Wildlife Refuge on 3 November and 28 February. This longer than expected activity period may be possible because of the ability of New Mexico C. picta to maintain high metabolic rates over a broad range of ambient temperatures (Seidel, 1977).

Reproduction: Reproductive traits vary markedly between populations of Chrysemys picta. Iverson and Smith (1993) have tabulated most of the published information on this subject. Their data showed differences between subspecies as well as differences between populations within subspecies. In C. p. bellii, age at maturity is reached later and the size is larger in northern populations. Frazer et al. (1993) presented evidence suggesting that warmer and longer growing seasons result in more rapid attainment of sexual maturity in male Chrysemys picta. In New Mexico (below 1200 m), Christiansen and Moll (1973) found that males matured as early as the third year, and were able to mate in the spring of the following year, at 80–90 mm PL. They also found that New Mexico females do not usually mature until at least 2 years later than males, and their smallest mature female was 132 PL (139 mm CL). Moll (1973) reported that age at maturity varied with latitude and that Wisconsin populations were older at maturity than those in Louisiana and Tennessee. These latitudinal differences, presumably resulting from seasonal temperatures, may be present between the northern Upper Rio Grande and San Juan River populations and the southern Pecos River population.

Courtship and mating in New Mexico probably follow the pattern of other populations, taking place in spring and fall. Most nesting is from May until mid-July, with most females producing two or three clutches per year (Iverson and Smith, 1993). Nests are flask-shaped, average about 125 mm deep, and are in sandy or loamy soil. Egg number varies from 4–20 (commonly 5–8) in C. p. bellii and increases with female size. Eggs are white to cream color and about 20 x 32 mm in size. Incubation time is 65–80 days. Most hatchlings overwinter in the nest. In northern Idaho, Lindeman (1991) found that all hatchlings overwintered in the nest; the time from oviposition to emergence ranged from 292–348 days in seven nests with complete data. Survivorship was low (between 21% and 33% for 193 eggs) in spite of the lack of predation at his study site. Even though the eggs have flexible shells, they are more resistant to drying than those of squamates, and have a decreased incubation time when subjected to warmer and dryer conditions (Packard et al., 1982). Sex is temperature-dependent in this species. In general, lower incubation temperatures produce more males whereas higher temperatures produce more females; however, C. picta has two threshold temperatures for sex determination and females are also produced at the both the higher and lower thresholds. Average egg size and hatchling size are positively correlated with female size (Lindeman, 1991). Also, hatchlings can survive subfreezing temperatures while overwintering in shallow subterranean nests by avoiding freezing (supercooling) instead of by tolerating freezing (Packard and Packard, 1993a; Storey et al., 1988). In the laboratory, Packard et al. (1989) also found that hatching success was lower at high temperatures and on dry substrates, and that the embryos in large eggs have a better chance than those in small eggs for surviving in stressfully dry environments. The studies by Packard and Packard (1993b) and Cagle et al. (1993) supported earlier suggestions that moisture in the environment is of major importance to developing embryos. These studies are applicable to New Mexico in that they suggest that larger females, which produce larger eggs and therefore larger hatchlings, should have higher reproductive success in the warm and dry environments found here.

Food habits: These turtles are omnivorous and opportunistic, showing a preference for animal foods when available, but younger ones tend to be more carnivorous than older individuals. Both higher and lower plants and animals as well as carrion are eaten. Sight and smell are used for food detection, and they can be readily trapped using sardines as bait. We have had most success with daytime trap sets. Feeding activity is temperature dependent and was found to occur from 18–32° C in a Missouri marsh (Kofron and Schreiber, 1987). Since digestive turnover time is long in this species (Parmenter, 1981), basking is necessary for efficient food utilization.

Remarks: Although *C. picta* is one of the most studied turtles, little life history work has been done in New Mexico. The San Juan population, with no other sympatric aquatic species, and subject to cold winters and a short activity season, should be compared with the population in the lower Pecos, where there are five aquatic turtles in sympatry, milder winters, and a long season.

A turtle, believed to be *C. picta*, was observed basking on the river bank about 5 river km above Pilar (ca. 1842 m), Taos County (Stuart, 1995b). This locality is about 50 river km upstream from the Española collection site reported by Degenhardt and Christiansen (1974).

This species has been reviewed by Ernst (1971).

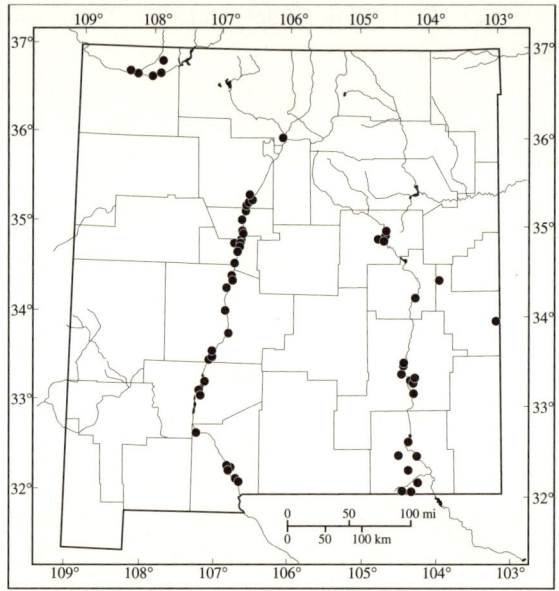

Distribution of Chrysemys picta *in New Mexico.*

PSEUDEMYS GORZUGI Ward, 1984 — *Western River Cooter*

See Plate 29.

Type: The type locality is "from 3 ½ mi W. Jiménez, Río San Diego, Coahuila, Mexico, 850 feet altitude." The holotype is KU 39986, an adult female shell, skull, and skeleton collected by P. S. Chrapliwy on 19 June 1952.

Distribution: *Pseudemys gorzugi* is found in the lower Rio Grande and Pecos River drainages from Tamaulipas, Nuevo León, and Coahuila, Mexico and south Texas, through west Texas, to southeast New Mexico. It occurs from sea level to around 1100 m elevation. In New Mexico, the western river cooter is restricted to the lower Pecos River drainage, ranging as far north as the lower reaches of Brantley Reservoir and including the entire length of the Black and Delaware rivers. There is a questionable record for Bitter Lakes National Wildlife Refuge in Chaves County (Bundy, 1951). Recent attempts to locate this species there have been unsuccessful.

Description: The western river cooter is a large turtle. Of 238 specimens collected in New Mexico, 97 females averaged 195.3 (80–285) mm CL and 1139.1 (85–3100) gm; 141 males averaged 152.3 (89–232) mm CL and 496.2 (112–1675) gm (our unpubl. data).

The carapace is an elongate oval in outline, highest at the middle and widest behind the midline. It has a serrated rear margin and is ornately marked with yellow and black lines and blotches. The second costal scute bears a conspicuous C-shaped marking. The plastron is mostly unmarked, although in the juveniles there may be thin dark lines along the seams. There are yellowish green stripes on the head and neck with a large blotch of similar color on each side of the head. The upper jaw has a prominent medial notch bordered on each side by tooth-like cusps. The crushing surfaces of both jaws bear a cluster of well-developed denticulations. The legs and exposed skin are marked with red, yellow, and black, especially on the webbing between the toes where there are black half-moon markings on a red background. Ward (1984) described additional skeletal and scute characteristics. Old adult males may become melanistic.

Sexual dimorphism is pronounced in New Mexico

populations of *P. gorzugi*, with adult males having long, straight foreclaws and long, thick tails, with the anal opening beyond the carapacial margin. In females, the anal opening is positioned beneath the posterior marginals. Males average smaller (152.3 mm CL; 496.2 grams; n = 141) than females (195.3 mm CL; 1139.1 grams; n = 97) (Painter, 1991a; our unpubl. data).

Similar Species: In its restricted range in southeastern New Mexico, the western river cooter may be mistaken for *Chrysemys picta* or *Trachemys scripta*. These species may be separated from *P. gorzugi* by the smooth rear margin of the carapace and the ornately marked plastron in *C. picta* and by the prominent patch of red on each side of the head in *T. scripta*.

Systematics: Most of the literature on the western river cooter in New Mexico is found under the name *P. floridana texana* or *P. concinna texana*. The current name came into usage when Ward (1984) elevated *P. texana* to full species rank and described specimens from the Rio Grande and Pecos river drainages as *P. c. gorzugi*. Ernst and Barbour (1989) chose not to recognize *P. c. gorzugi* and retained the name *P. concinna texana*. Ernst (1990b) elevated the Rio Grande and Pecos river populations to full species rank, *P. gorzugi*, citing the lack of gene exchange and allopatry from other *P. concinna* populations as justification. Seidel (1994) presented morphometric and biochemical evidence in support of this arrangement. We follow Ernst (1990b) and Seidel (1994) and retain the name *P. gorzugi* for populations of the western river cooter in New Mexico.

Because the species does not occur in the Rio Grande in New Mexico, we suggest the common name "western river cooter" for this species instead of Rio Grande river cooter as listed in Collins (1990) or Rio Grande cooter as listed in Ernst et al. (1994).

Habitat: In New Mexico, *P. gorzugi* occurs mostly in riverine habitat, although there is a single, verified sight record from Willow Lake near Malaga. It tends to avoid riffles, usually being confined to the larger, deeper pools found along the Pecos, Black, and Delaware rivers. The presence of aquatic vegetation for foraging and cover is desirable but not necessary. Suitable substrate varies from muddy, sandy, or rocky to large areas of algae-covered limestone. Elevation records in New Mexico extend from 900 m to about 1100 m.

Pseudemys gorzugi is most often associated with *Chrysemys picta*, *Trachemys scripta*, *Chelydra serpentina*, and *Trionyx spiniferus*.

Behavior: A confirmed basking species, this turtle is often seen basking at the water's surface or on logs, overhanging vegetation, or muddy banks. It is very shy and quick to retreat underwater when approached. It is most active during daylight hours and spends the night resting underwater. When handled, it will attempt to bite. Preliminary data from radiotelemetry studies have shown this species to be a rather sedentary turtle in the pool-shallow run habitat of the Black River in south-central Eddy County. Marked turtles remained in the same small river reach with maximum movements of approximately 300 m downstream from the point of release (our unpubl. data). We have found these turtles active at water temperatures of 12°C in early December in the Black River.

Reproduction: Although not studied in New Mexico, the ritualized courtship of river cooters in Florida was well described by Marchand (1944). In New Mexico, recently plundered nests and gravid females have been observed in late May. On 23 May we collected a large female, 24 cm straight carapace length, in the upper reaches of the Black River in Eddy County. She was X-rayed (Gibbons and Greene, 1979) and found to contain nine oviductal eggs; injection of oxytocin (Ewert and Legler, 1978) induced deposition of all nine eggs. These eggs were oblong in shape and averaged 42 mm x 31 mm. Hatching occurred after 70 days and four hatchlings averaged 35.3 mm curved carapace length, and 33.9 mm straight carapace length. Average weight was 10.1 grams, 3 days posthatching. On 17 August, four hatchlings found along the upper Black River in Eddy County averaged slightly larger than those artificially incubated, with the curved carapace length 39.3 mm, straight carapace length 37.8 mm, and average weight 11.3 grams. These hatchlings had caked-on mud and an egg tooth and presumably had recently emerged from a nest chamber and were dispersing to the water. We do not know how sex is determined in this species, however, sex is temperature dependent in the closely related species, *P. concinna* and *P. floridana*.

Hatchlings may overwinter in the nest cavity. Seven were discovered on 14 April along the Pecos River near Carlsbad as they emerged from the nest cavity (John Sherman, pers. comm.) and a single hatchling, with the egg tooth present, was discovered on 27 October. It is unknown if western river cooters in New Mexico produce multiple clutches each year.

Food Habits: Detailed food habits of *P. gorzugi* are unknown in New Mexico, although it is likely they include a variety of aquatic plants, invertebrates, and vertebrates. Western river cooters in New Mexico are easily trapped in hoop nets baited with fresh fish, sardines, watermelon, or lettuce as bait, although this does not appear to be true elsewhere. Legler (1958 [1959]) examined one specimen from "Blue Springs" and found the stomach and gut filled with "finely chopped vegetable matter." Recently collected individuals often pass fecal matter composed of vegetable matter, particularly green algae. One fecal mass examined contained fragments of crawfish. In captivity, hatchlings will feed on dead *Gambusia*, crawfish, lettuce, spinach, and various aquatic plants collected from the wild.

Remarks: During the early seventies, this species was considered to be in the genus *Chrysemys* by some authors (Ernst and Barbour 1972; Conant, 1975) but it was retained in the genus *Pseudemys* by Degenhardt and Christiansen (1974).

The western river cooter was first reported from New Mexico by Bundy (1951) who cited a specimen from Bitter Lakes National Wildlife Refuge in Chaves County under the name *P. floridana texana*. We cannot locate the specimen on which this record is based and we question its identification. Legler (1958 [1959]) reported a second specimen (KU 15929) taken at Blue Springs in Eddy County. The western river cooter is listed as endangered by the New Mexico Department of Game and Fish (NMGF, 1990). Ernst (1990b), in his review of the species, suggested that the biology of the species is poorly known and a study of its behavior and ecology would be a valuable contribution to the literature on freshwater turtles.

This turtle is named after George R. Zug, curator of amphibians and reptiles, National Museum of Natural History, Smithsonian Institution.

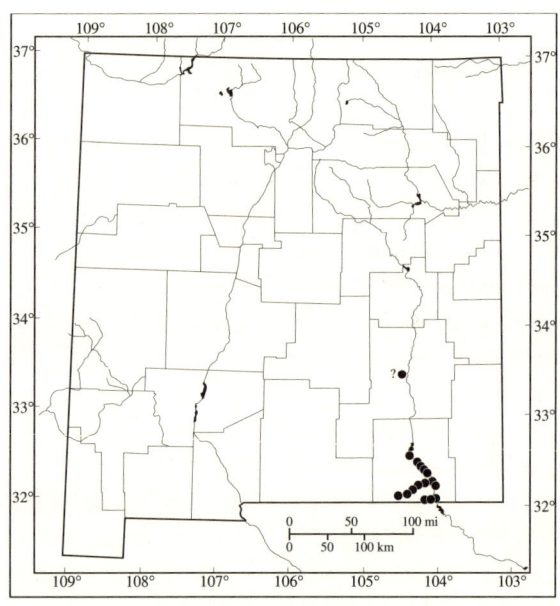

Distribution of Pseudemys gorzugi *in New Mexico.*

TERRAPENE ORNATA (Agassiz, 1857)   *Ornate Box Turtle*

See Plate 30.

Type: Agassiz used six syntypes (USNM 7541, 7547, 7692, 7862, 131837, and MCZ 1536) in naming this species, stating neither date nor collector. The type locality was originally described as "from the upper Missouri . . . and from Iowa." Smith and Smith (1979) corrected earlier restrictions, designating MCZ 1536 as the lectotype, and restricting the type locality to "Burlington [Des Moines County], Iowa."

Distribution: The ornate box turtle is found from southern Wyoming to western Indiana, south to Louisiana and southeastern Arizona into north-central Mexico. The range is discontinuous in the northeast. In New Mexico, this turtle is absent in the northwestern portion of the state, but is otherwise widespread below 2100 m. The northernmost record in the Rio Grande Valley from Santa Fe County may be a release.

Description: This moderately small terrestrial species may reach 15 cm CL, but most mature turtles measure 10–13 cm CL. Females grow larger than males. Skin color is dark brown to reddish-brown with conspicuous yellow to orange spotting. Jaws are yellow. The head may be greenish in some individuals. The carapace and plastron are dark brown to reddish-brown with radiating light lines, or occasionally rows of light spots, on each scute. There may also be a dorsal light stripe. In the southern portions of the state and extending up the Rio Grande Valley, the light lines are more numerous and narrower, the shell has a more yellow appearance, and there is a tendency for turtles to lose their patterns and become yellowish, greenish, or straw-colored with age. There is a single front hinge on the broad plastron. There are four (occasionally three) toes on the hind foot. Eye color is bright red in males and yellowish-brown to reddish-brown in females. The first toe on the hind foot of males is thickened and turned in to assist in grasping the shell of the females during copulation.

Similar Species: No other New Mexico turtle has a plastron with only a single hinge.

Systematics: Both recognized subspecies, the nominotypical subspecies, and *T. o. luteola* Smith and Ramsey, 1952 are found in New Mexico. In the description of *T. o. luteola*, the authors used TCU 1280, an adult male, "secured October 22–24, 1950, 17 miles south of Van Horn, Culberson County, Texas, by W. Elton Smith," as the holotype.

Habitat: Box turtles are not dependent on free water, so they can occupy a wide range of habitats. They are most abundant, however, in grasslands with soils suitable for burrowing. They do not normally occupy dense woodlands, steep, rocky, mountain slopes, or elevations above 2100 m. However, one turtle was found crossing a gravel road at 2200 m in piñon-juniper-ponderosa woodland in the Manzanita Mountains of Bernalillo County. This turtle could be an escapee; a well-weathered shell, however, was found nearby on a 2300 m ridge, and non-rocky soils are available in the vicinity.

Behavior: These shy turtles can quickly withdraw into their shells when threatened, and the hinged plastron, snapping jaws, and copious urine seem to discourage most predators. The turtles usually emerge from their retreats with the morning sun, and after a short basking period begin foraging. They usually remain active until mid-morning, then retreat to their burrows, shade under vegetation, or other shelter to escape the heat. They may emerge again in late afternoon. The daily activity period is lengthened by cloudy or rainy weather; in the driest southern portions of the state it is often restricted to rainy periods. Norman Scott (pers. comm.) reported that *T. o. luteola* in the Sevilleta National Wildlife Refuge, Socorro County, in the summer is active well before sunrise and retires before mid-morning; it is also active around sunset. Although active during the day as much as climatic conditions allow, these turtles in most habitats are seldom seen or collected other than when crossing roads. In sandy areas, fresh tracks can often be followed to a retreat.

On 21 August 1947, C. H. Lowe (*in* Norris and Zweifel, 1950) observed four turtles using enlarged burrows on a banner-tailed kangaroo rat mound in Socorro County. Tom Garrison (pers. comm.) observed a desert box turtle inside a banner-tailed kangaroo rat mound southeast of Belen in Valencia County on 20 July 1985, and J. L. Iverson (pers. comm.) reported that *T. o. ornata* also uses banner-tailed kangaroo rat burrows in western Nebraska. All *T. o. luteola* studied by Nieuwolt (1993) in the Sevilleta National Wildlife Refuge (Socorro County) were never found to dig their own burrows but used those dug by kangaroo rats. Thus at least some populations of *T. ornata* use existing excavations rather than dig their own; however they are able and do indeed dig their own burrows. Metcalf and Metcalf (1970) presented detailed observations on *T. ornata* burrows and burrowing procedure in a Kansas population.

Many reports document the use of water for drinking and soaking. Charles H. Lowe (*in* Norris & Zweifel, 1950) observed at least seven or eight variously sized ornate box turtles in a muddy pond in Guadalupe County, and commented on their excellent swimming ability. Blair (1976) reported pond use in Texas, and Gehlbach (1956) observed similar activity in New Mexico. Webb

(1970b) saw an Oklahoma turtle that was startled in shallow water of a small pond in a rocky stream and which submerged to a depth of about six inches and wedged itself among rocks on the bottom. Although some of these turtles were submerged or swimming below the surface, most ornate box turtles swim only when necessary, and then on the surface, seeming reluctant to enter deep water (Legler, 1960b). John B. Iverson (pers. comm.) says that in western Nebraska, after emerging from hibernation in early spring, box turtles will frequently migrate from their natural home range to water, "tank up," and then return to their natural home range up to as much as a km away. The annual activity period of *T. ornata* extends from March or April to October or November throughout the species' range (Ernst and Barbour, 1989). Based on about 300 collection dates in New Mexico, activity extends at least from April through October with only four records in April and 21 in October. It could be argued that this type of information coincides roughly with the field activity of herpetologists, and therefore is somewhat biased. However, observations based on "yard turtles" in Albuquerque, and data gathered by Blair (1976), using over 800 sightings in Texas, support these data. Length of the actual hibernation period is variable and primarily dependent on temperatures in the fall or spring of a given year. Data from Grobman (1990), based on seven years of study in St. Louis, Missouri, suggest that warming soil temperatures in the spring control emergence. Turtles entering hibernation may dig their own burrows, modify those already dug by other animals, or use piles of leaves or other surface litter if available. On the Sevilleta National Wildlife Refuge in Socorro County kangaroo rat (*Dipodomys spectabilis*) burrows are used almost exclusively for hibernation (Nieuwolt, 1993).

Reproduction: Based on Blair's (1976) 22-year study in central Texas, sexual maturity is attained near 100 mm CL at about 7–8 years of age. Mating most often occurs in the spring on emergence from hibernation, but may take place anytime during the activity period.

Legler (1960b) found nesting from early May to mid-July, and most frequently in June, but it may be delayed if conditions are not suitable. A second but smaller clutch is possible. Ernst and Barbour (1972) summarized the known reproductive activities of *T. ornata* and unless otherwise cited our information is taken from that source. The flask-shaped nests are 50–60 mm deep in soft well-drained soil in open areas. Clutch size ranges from 2–8; the smaller clutches usually have larger eggs. In a Socorro County population of *T. o. luteola*, Nieuwolt (1993) X-rayed gravid females and found that clutch size was smaller, was correlated with female size, and varied from 1–4 (mean = 2.68) in 77 clutches. Nieuwolt found gravid females in May, June, July, and in one year, as late as August. Egg retention for more than 30 days was common and one turtle was recorded carrying eggs for 50 days. The ability to retain eggs until good laying conditions are available is important in the locally unpredictable climate of southwestern habitats. Nieuwolt also found that the eggs were larger in size than reported for other populations and there was no evidence for more than one clutch a year. In the three years of her study no turtle was found gravid for three consecutive years, but it is possible that a very early laying may have occurred in some females resulting in an unrecorded clutch. The ellipsoidal eggs have granular, somewhat brittle, white shells and measure 31–41 x 20–26 mm. Incubation time is dependent on temperature but averages about 70 days at 28°C. Females use care in choosing a nest site, lay the eggs at night, and camouflage the nest well. In *Terrapene*, sex is temperature-dependent; lower incubation temperatures produce more males whereas higher temperatures produce more females (Vogt and Bull, 1982). Some details and techniques of captive breeding are given by Wagner (1995).

Food Habits: These turtles are omnivorous, usually showing a preference for animal foods. However, at times certain plant foods may be eaten solely, and mulberry is mentioned by both Legler (1960b) and Blair (1976) as a favorite. Feeding tends to be opportunistic; this may explain why an abundant food, such as insects, is an important staple in the diet. The list of foods eaten is long, and includes such diverse items as dead mammals, birds and eggs, reptiles (including dead box turtles), spadefoot toad tadpoles (but not adults), crayfish, pillbugs, earthworms, carrion, spiderwort, cactus fruits and stems, melons, and invertebrates found in cow dung, along with some of the dung itself. We have supplemented the diet of our "backyard turtles" with canned dog food, and M. L. Riedesel (pers. comm.) uses moistened parrot food pellets as turtle food.

Food items are grasped by the jaws and swallowed whole or torn apart with help from the front claws. All feeding activity occurs during the daytime, using sight to locate and smell to identify food items.

Remarks: We use ornate box turtle (Ward, 1978; Iverson, 1992) rather than western box turtle (Collins, 1990) for the common name of the species. The genus was reviewed by Ernst and McBreen (1991) and the species by Ward (1978).

Box turtles are reputed to live for very long periods of time, and Legler (1960b) suggested 50 years as a possibility. Long-term studies in Texas (Blair, 1976) and Kansas (Metcalf and Metcalf, 1985) presented no evidence that these turtles live much past three decades under natural conditions.

The distribution, intergradation, and other relationships of the two subspecies in New Mexico are unclear and merit further study.

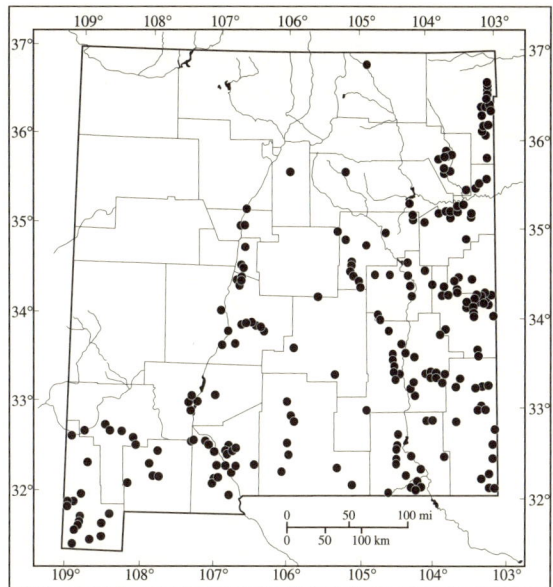

Distribution of Terrapene ornata *in New Mexico.*

TRACHEMYS GAIGEAE (Hartweg, 1939) — *Big Bend Slider*

See Plate 31.

Type: The holotype, UMMZ 66472, is a female collected by Helen T. Gaige during the summer of 1928 at "Boquillas, Rio Grande River, Brewster County, Texas."

Distribution: The Big Bend slider is found in the Rio Grande drainage in southern New Mexico, Texas, and Mexico and the Río Concho drainage (a tributary of the Rio Grande) in Chihuahua, Mexico.

In New Mexico, it has been found only in the southern half of the Rio Grande. Here, specimens have been taken between 1280–1410 m at LaJoya Waterfowl Management Area, Bosque del Apache National Wildlife Refuge, Elephant Butte and Caballo reservoirs, and in deeper stretches of the Rio Grande between these reservoirs.

Description: This is a medium-sized turtle with adults usually measuring 120–220 mm CL. During 1990, 12 males collected from Elephant Butte Reservoir averaged 149.7 (119–199) mm CL and 475.4 (250–800) gm; 10 females averaged 214.4 (181–248) mm CL and 1320.5 (880–2125) gm (our unpubl. data). Stuart et al. (1993) reported a maximum size of 257 mm CL for a female (shell only) collected below Elephant Butte Reservoir, Sierra County. The skin is green to olive brown with light stripes. There is a prominent black-bordered red, orange, or yellow ovoid spot on the side of the head, with another smaller spot usually present between it and the eye. The carapace is olive or brownish with numerous orange curved lines, some of which form ocelli. There is a weak keel on the carapace, and the rear margin is serrated. The plastron is yellow with a dark median figure that may be disrupted longitudinally. With increasing age, melanism may obscure much of the normal pattern, especially in males, although this is less often seen in specimens from New Mexico. Young have a lighter, more contrasting color pattern, often with spots and blotches having light centers.

Similar Species: *Trachemys scripta* is most similar but it has an elongate wide red, orange, or yellow stripe or blotch rather than black-bordered spots on the head, and the carapace has straight light lines rather than curved lines and ocelli. *Chrysemys picta* has a smooth, keelless carapace without serrations on the posterior margin. *Pseudemys gorzugi* has a notched upper jaw, and the second costal scute has C-shaped markings.

Systematics: We consider *Trachemys gaigeae* monotypic because it is sympatric with *T. scripta* and hybridization in New Mexico populations is uncommon. In the opinion of M. E. Seidel (pers. comm.; Stuart, 1995b), hybridization occurs only where *T. s. elegans* has been introduced in the Rio Grande. Seidel believes that the natural (historical) ranges of these species probably did not overlap. We realize this is a simplistic treatment of a complex problem. Some authors (Legler, 1990; Iverson, 1992) suggested retaining *T. gaigeae* as a subspecies of *T. scripta* until the relationships within the neotropical complex are better understood. Iverson (1992) also summarized the principal points concerning the systematics of this difficult group. Ward (1984) treated *T. gaigeae* as a subspecies of *T. nebulosa.* Ernst (1992b) discussed this problem and considered *T. gaigeae* specifically distinct from *T. scripta elegans*, but suggested that the relationship to other subspecies of *T. scripta* is unclear. Other information indicates that these two forms are morphologically and genetically distinct (Francis Rose, pers. comm.; our unpub. data).

Habitat: Early collections of *T. gaigeae* were almost solely from rivers (Carr, 1952; Legler, 1960a). Degenhardt and Christiansen (1974) reported two specimens from large lakes and ponds adjacent to the Rio Grande on Bosque del Apache National Wildlife Refuge, Socorro County. Other non-river records are from man-made artificial ponds near the Rio Grande in Big Bend National Park, Texas, a nearly permanent side channel of the Rio Grande south of Elephant Butte Lake in New Mexico, and a large flooded oxbow lake of the Rio Grande in the LaJoya Waterfowl Management Area in New Mexico. Stuart (1995b) described turtle habitats in his Rio Grande study sites. He found that *T. gaigeae* occurred most commonly in permanent or near-permanent ponds with substantial growth of vegetation and a moderate accumulation of dead organic material. Other turtles found associated with *T. gaigeae* in New Mexico are *T. scripta, Chrysemys picta,* and *Trionyx spiniferus.*

Behavior: This turtle is diurnally active and probably very similar to *T. scripta* in general behavior. In Chihuahua, Mexico, Legler (1960a) reported surface basking chiefly in the morning and late afternoon at locations where out-of-water basking sites were scarce. Our observations in New Mexico indicate activity at least from April to October. Gary Stolz (pers. comm.) observed sliders, believed to be *T. gaigeae,* basking on mud banks at Bosque del Apache National Wildlife Refuge in early February 1993. More recently, J. N. Stuart (pers. comm.) has observed both basking and terrestrial activity from February to November at the same refuge. These winter observations were during warmer than usual seasonal temperatures.

Reproduction: Using measurements of carapace length, Legler (1960a) proposed that most males attain sexual maturity at 105–115 mm, whereas most females are 170–180 mm when mature. The smallest mature male he measured was 115 mm; the smallest female, 169 mm. Another male, 103 mm in length, had partial development of secondary sexual characteristics.

The courtship pattern of *T. gaigeae* differs from that of *T. scripta.* It does not include the frontal face stroking, and is more similar to that of *T. s. taylori*, with pursuit from the rear and possible biting (Ernst, 1992b).

Legler (1960a) reported four females with oviductal eggs, numbering 11, 9, 7, and 6, at carapace lengths of 202, 186, 194, and 169 mm. This small sample indicates correlation of clutch size with female body size. A female (MSB 50523) collected in Socorro County on 8 June contained 10 shelled eggs that appeared ready for laying. The method of sex determination is unknown, however, in the closely related species, *T. scripta,* sex is temperature dependent.

Food Habits: Carr (1952) believed *T. gaigeae* to be carnivorous based on the "character of the bodies of water from which it has been collected." Legler (1960a) examined stomach contents of "several" and found only aquatic vegetation. Digestive tracts of several New Mexico adults contained filamentous algae and possibly *Potamogeton* (J. N. Stuart, pers. comm.). *Trachemys scripta* undergoes an ontogenetic change in diet, with juveniles more carnivorous and adults more herbivorous (see *T. scripta* account). We suspect that *T. gaigeae* is omnivorous and opportunistic, with food habits similar to *T. scripta*. They are attracted to traps set with fresh fish, canned sardines, or lettuce.

Remarks: This species has been referred to both the genera *Chrysemys* and *Pseudemys* in the recent past.

More life history information is needed on this turtle. The relationship with *T. scripta* should be determined and the sympatric occurrence of both species in the Rio Grande Valley affords an excellent opportunity for further study. James N. Stuart is presently studying the middle Rio Grande population (Stuart, 1995b).

This species has been reviewed by Legler (1990), Ernst (1992b), and Iverson (1992).

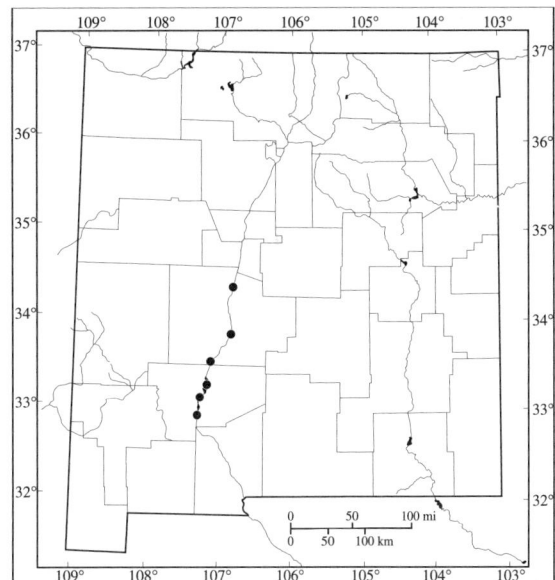

Distribution of Trachemys gaigeae *in New Mexico.*

TRACHEMYS SCRIPTA (Schoepff, 1792)   *Slider*

See Plate 32.

Type: No type or type locality was designated for *Trachemys scripta*, but Schmidt (1953) designated "Charleston, South Carolina," as the type locality.

Distribution: This species is found from the eastern and south-central United States to Brazil and Argentina, but the range is disjunct in the southwestern United States and farther south. In New Mexico, *T. scripta* is most common in the Canadian and Pecos river drainages, but is also present in small numbers in the Rio Grande, where it was almost certainly introduced.

Description: This is a medium to large turtle; adults usually measure 120–200 mm CL and large females reach nearly 300 mm CL. During 1989–92, 196 males collected in the lower Pecos River in Chaves and Eddy counties averaged 145.7 (40–210) mm CL and 496.1 (100–1299) gm; 152 females averaged 175.3 (30–264) mm CL and 888.4 (80–2700) gm (our unpubl. data). The skin is green to dark olive brown with yellow stripes. There is a conspicuous enlargement of the stripe extending posteriorly from the eye. This enlargement may be completely or partially separated from the stripe and is usually red or orange, occasionally yellow. The carapace has a weak keel, usually has faint longitudinal "wrinkles," and a serrated posterior edge. The carapace color is normally olive to brown with yellow stripes and bars, but dark spots, blotches, or bars may obscure some of the basic pattern. The plastron is yellow with roundish black markings, each marking usually contained within a scute. With increasing age, particularly in males, melanin obscures much of the normal pattern on soft body parts as well as carapace and plastron. Old males are typically olive colored with wide dark areas that are particularly evident along the sutures. The young are very brightly marked with an absence of these dark markings. Males tend to be smaller and darker than the females. The males have elongate slender claws and long thick tails, and the anal opening is posterior to the edge of the carapace.

Similar Species: *Trachemys gaigeae* usually has an isolated black-bordered orange or yellow spot, rather than a line, which broadens to include a red patch on the head; it often has a median branched dark figure on the plastron, rather than paired discrete solid dark blotches. *Pseudemys gorzugi* has a prominent greenish-yellow blotch behind the eye, and the second costal scute has a C-shaped marking. *Chrysemys picta* has a smooth carapace with a non-serrated posterior margin.

Systematics: Ernst (1990a) listed 14, Iverson (1992) 16, and Legler (1990) 17 subspecies of *Trachemys scripta*. *Trachemys s. elegans* (Wied-Neuwied, 1838) occurs in New Mexico (see Remarks). We consider *T. gaigeae* to be a full species although some consider that form to be a subspecies of *T. scripta* (e.g. Legler, 1990) and Ward (1984) treats it as a subspecies of *T. nebulosa*. The original description for *T. s. elegans* stated no type or locality, but Wied-Neuwied (1865) listed "Fox Rivers bei New-Harmony" [Posey County, Indiana] as the type locality.

Habitat: *Trachemys scripta* is primarily an inhabitant of permanent wetlands. Aquatic vegetation, a soft bottom, still or slow-moving water, and depths from 1–2 m are general habitat requirements (Morreale and Gibbons, 1986). Water temperatures between 25–29°C are optimal (Brattstrom, 1965; Parmenter, 1980; Spotila et al., 1990). In New Mexico, both the Canadian and Pecos river drainages have many wetlands with suitable habitat that support populations. In the Rio Grande drainage, suitable habitat is limited primarily to waters found in the larger lakes and refuges, since most of the water in the main river channel is rerouted for irrigation, or stored in these lakes and refuges. Degenhardt and Christiansen (1974) reported the upper elevational limit as 1380 m at Santa Rosa in the Pecos drainage. Since then, a specimen, possibly a released or escaped pet, was collected in a riverside drain across the Rio Grande from Albuquerque at 1500 m elevation. Applegarth (1982) collected sliders in shallow portions of Conchas Lake in the Canadian River arm, and Seidel (1975b) collected one about 60 km farther upstream. On three different occasions we trapped upstream from the area Seidel collected but found only *Chelydra serpentina* and *Trionyx spiniferus*.

Behavior: This turtle is shy, gentle, and agile. It quickly disappears below the water surface at the approach of potential danger, taking refuge in aquatic vegetation or the soft bottom. It feeds and basks mostly during the daytime, and is inactive below the surface at night. Preferred basking sites are logs or rocks away from the shoreline, but turtles may float at the surface or use banks if these preferred sites are not available. Movement overland from one habitat site to another is common in eastern populations of this species. In the dry New Mexico climate, only limited overland migration probably occurs, explaining the presence of populations in small isolated ponds. Some of these ponds are spring-fed and higher than the nearest permanent river valley populations. Migrations uphill seem illogical for purposes other than females hunting nest sites or possibly males hunting females. Gibbons (1990) suggested that turtles may be able to detect the presence of other bodies of water visually, by some mechanism involving the properties of reflected light.

The subject of home ranges in this species has been summarized by Schubauer et al. (1990). It appears that turtles of all species living in larger bodies of water tend to have large home ranges. However, this principle may not apply to turtles living in smaller bodies of water when overland travel is a major component of the home range. *Trachemys scripta* does travel overland, and home ranges as large as 104 ha have been reported (Schubauer et al., 1990). Results from marking and recapturing New Mexico turtles are not yet available, but we expect little overland travel and that turtles have larger home ranges in larger bodies of water. Also, males tend to have larger home ranges than females in this species, despite their smaller size.

In all parts of New Mexico, temperatures remain below freezing for extended periods, making hibernation or winter dormancy necessary. Sliders probably overwinter buried in the bottom of permanent bodies of water. Ultsch et al. (1984) have demonstrated in the laboratory that emydid turtles especially are able to survive under the anoxic conditions found in aquatic substrates. The 140 collection dates for *T. scripta* in New Mexico extend from 10 April to 24 October. However, little collecting has been done in March, since waters are usually still too cold for feeding, and therefore the usual trapping methods using bait are unsuccessful.

Reproduction: Data from New Mexico are lacking, but there is much published information from populations elsewhere. Gibbons and Greene (1990) found that in male *T. s. scripta*, size appears to be more important than age in the determination of sexual maturity. In females, however, age is apparently more critical, although the variability among individuals is extreme. They found that most males in South Carolina attain maturity at plastron lengths of 90–110 mm. Males reach these sizes normally during their third to fifth years, but in a warm productive pond, growth is more rapid, and mature size may be reached in the second year. Most females mature at six to eight years of age and from 160 to over 210 mm PL in these populations. Mating occurs most often in the spring, at which time ovulation occurs and fertilization is possible. However, sperm are produced and available throughout the active season and may be retained for later use by the female.

Oviposition occurs after ovulation, when the eggs are shelled and environmental conditions are suitable. Some females may not lay eggs in a given year, whereas in others, an additional clutch may be deposited within 2–4 weeks of a preceding one (Gibbons and Greene, 1990). Nests are usually flask-shaped and placed in carefully selected locations, but no nesting information is available from New Mexico. Clutch size is correlated with body size and ranges from 2–25. Eggs are oval, average 22 x 36 mm in size, and are flexible-shelled. Sex is determined by nest temperatures. Males are produced at lower and females at higher temperatures (Bull et al., 1982). Incubation requires 65–75 days, but the hatchlings usually overwinter in the nest cavity. We do not know what effect New Mexico's more arid climate has on this tendency of hatchlings to remain in the nest.

Food Habits: Data from New Mexico are lacking. Elsewhere, sliders are omnivorous and opportunistic and consume a diverse diet of aquatic plants, invertebrates, and to a lesser extent, vertebrate food matter. Although preference is for animal foods at all ages and sizes, there is an ontogenetic shift from animal to plant foods; the effect is that juveniles are largely carnivores and adults largely herbivores, although given the choice, turtles of all ages and sizes prefer animal foods.

These turtles feed during the day and use both sight and smell to locate food. Temperature has pronounced effects on all phases of feeding behavior and physiology. For a comprehensive coverage of feeding ecology, see Parmenter and Avery (1990).

Remarks: *Trachemys scripta* is one of the most studied turtles in the world, however little ecological information is available for New Mexico populations. Sliders in New Mexico occur at high elevations, in an arid climate, and at the westernmost edge of the species' natural range. Simple extrapolation of data from eastern *T. scripta* to populations in this state are therefore questionable. This species has been referred to both the genera *Chrysemys* and *Pseudemys* in the recent past.

More material is needed from the Rio Grande in order to further clarify the relationships between *T. scripta* and *T. gaigeae*. James N. Stuart is presently studying the middle Rio Grande population (Stuart, 1995b).

Because of a long-established and lucrative pet trade, juveniles of this species, primarily *Trachemys scripta elegans*, have become established in a great many localities in the eastern and midwestern United States well out of their normal range, and even in Europe. Roger Conant (pers. comm.) saw young *T. s. elegans* for sale in a pet shop in Hong Kong. James N. Stuart (pers. comm.) recently collected a melanistic male slider (MSB 58206) from Bosque del Apache National Wildlife Refuge which M.E. Seidel identified as *T. s. scripta*. Unlike *T. s. elegans*, we doubt that *T. s. scripta* has become established here.

This species has been reviewed by Ernst (1990a), Legler (1990), and Iverson (1992).

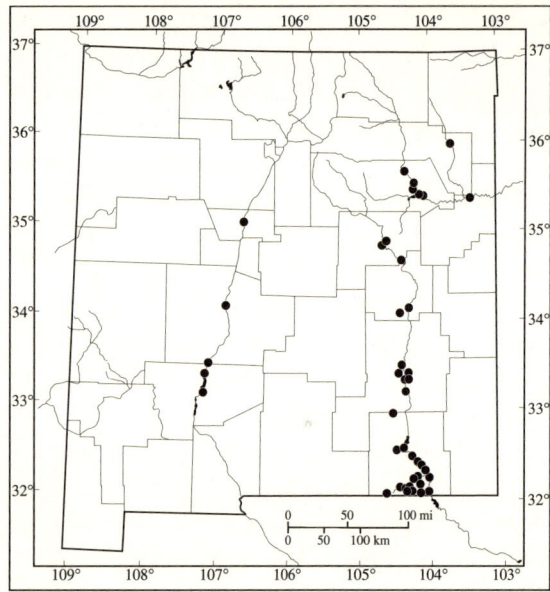

Distribution of Trachemys scripta *in New Mexico.*

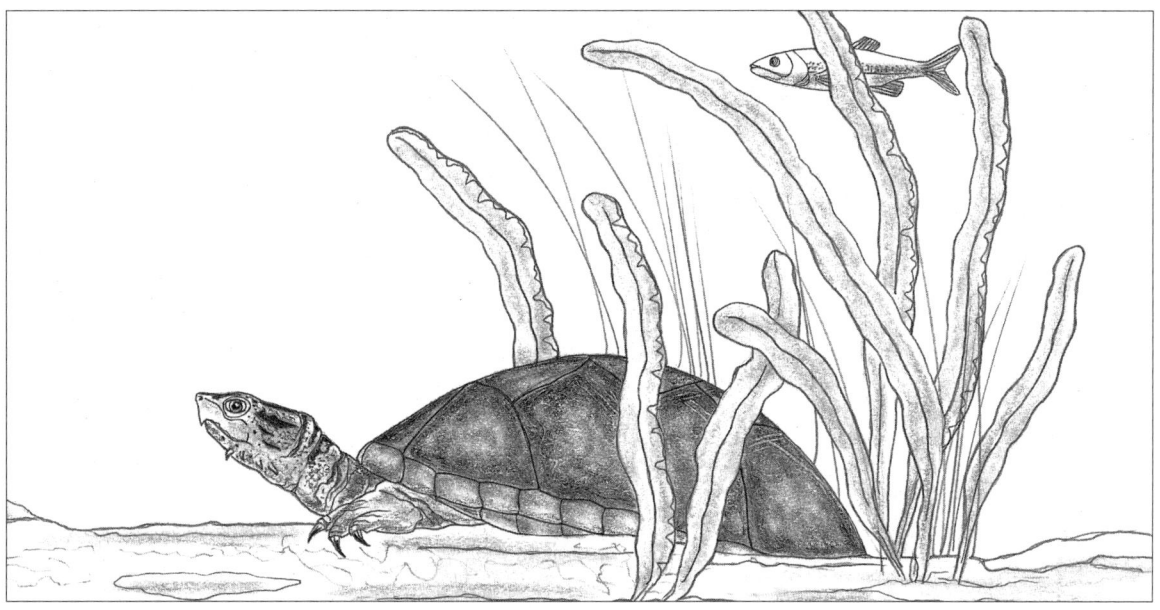

FAMILY KINOSTERNIDAE *Mud and Musk Turtles*

These small turtles average 15–20 cm shell length, although they range from approximately 11–38 cm. There are 22 species in 3–4 genera that are confined to North and South America, from southeastern Canada to central Argentina. Many are characterized by the habit of releasing a strong musky smell when captured; some species are referred to as "stinkpots." The musk is produced by glands at the junction of skin and bridge and not at the base of the tail as in snakes. All species have a solid carapace covered by strong, occasionally overlapping scutes. The plastron may have one or two hinges, allowing some species to completely close their shells. The eggs are elongated with brittle shells. Most are omnivorous, including mollusks, arthropods, annelids, fish, carrion, and aquatic plants in their diet. Two subfamilies, Staurotypinae and Kinosterninae, have been considered separate families.

KINOSTERNON FLAVESCENS (Agassiz, 1857) — *Yellow Mud Turtle*

See Plate 33.

Type: Agassiz (1857) originally listed five syntypes from five different localities with no date or collector information. Iverson (1978) considered only one of these, USNM 50, to have reasonably precise locality data. This specimen is a male and fits the description and figure for the species published by Agassiz. Iverson (1978) proposed that this specimen be designated as the lectotype, and restricted the type locality to Rio Blanco, near San Antonio, Texas.

Distribution: *Kinosternon flavescens* is found primarily in the Great Plains from Nebraska south into Mexico to Veracruz and then northwest to northern Mexico and the southwestern deserts of the United States.

Kinosternon flavescens is most common and widespread in the eastern one-third of New Mexico, but it also occupies suitable habitat in the southwestern portion of the state. With few exceptions, populations are found below 1500 m elevation. (See Remarks.)

Description: The yellow mud turtle is the smallest New Mexico turtle. Adult males are larger than females and reach 165 mm CL. During 1989–93, 36 males collected from the Pecos River drainage of southeastern New Mexico averaged 111.9 (77–141) mm CL; 28 females averaged 95.2 (72–128) mm CL (our unpubl. data). The skin is gray or grayish olive overall, usually with a bright yellow or cream throat and lower jaw. Some head mottling may be present in older individuals, especially males. The carapace is olive. The plastron is yellow to brownish, and hinged both front and back.

The ninth marginal is higher than the eighth and is peaked (except in young). Supraorbital ridges form a narrow crescent over each eye. There are finger-like barbels on the chin and throat. Mature males also differ from mature females in having a longer and thicker tail tipped with a horny nail, a larger head with a more strongly hooked upper jaw, a shorter and slightly more concave plastron, and longer and more curved claws.

Similar Species: *Kinosternon sonoriense* is most similar but differs in having a strongly mottled head, the ninth marginal not higher than the eighth, a slightly longer and higher domed carapace, and a more extensive plastron. *Chelydra* has a very small, cross-shaped, unhinged plastron. *Chrysemys*, *Trachemys*, and *Pseudemys* have large unhinged plastrons. *Terrapene* has a plastron with a single front hinge.

Systematics: Three subspecies are recognized by Berry and Berry (1984) and four by Christiansen and Iverson (1993). One of these, *Kinosternon f. flavescens*, the yellow mud turtle, is found in New Mexico.

Habitat: The yellow mud turtle inhabits arid to semiarid grasslands and open woodlands, usually on nonrocky soils. Soils suitable for digging are important, since most of this turtle's annual cycle is spent buried on land. During its aquatic activity, this turtle will make use of many different habitats, but shows a preference for quiet waters with muddy or sandy bottoms. Since these turtles are poor swimmers, shallow waters are preferred. In streams and rivers, or ponds near these, *Kinosternon flavescens* may be associated with *Chelydra serpentina*, *Trionyx spiniferus*, *T. muticus*, *Chrysemys picta*, *Pseudemys gorzugi*, *Trachemys gaigeae*, and *T. scripta*. On land, *Terrapene ornata* is a cohabitant. In New Mexico, the yellow mud turtle is usually the sole turtle inhabitant of temporary waters. Elevation is limited to near 1500 m, and the turtle reported from Capulin (Union County) at about 1950 m is probably a release. For a more detailed discussion of habitats in New Mexico see Degenhardt and Christiansen (1974).

Behavior: Yellow mud turtles are secretive and shy. They seldom attempt to bite, but may expel small quantities of musk when handled. Although mating, drinking, and most feeding occur in water, this turtle spends more time on land during an average year, and especially during hot and/or dry periods. In Oklahoma, Mahmoud (1969) estimated the annual activity period for *K. flavescens* to be 140 days, excluding estivation and hibernation. Semmler (1979) found activity as early as 13 May and as late as 2 November in Harding County, New Mexico, indicating an activity cycle of at least 172 days. The actual time of activity, however, is much shorter because there are intervening periods when turtles are dormant. Long

(1985) found that the yellow mud turtle maintains the highest carcass lipid index (CLI = 0.45) of all reported turtles and suggested this stored fat may provide the needed energy during the long dormant periods. Some *K. flavescens* can estivate for up to 24 months (Rose, 1980). Besides movement out of the ponds for hibernation in winter, estivation in summer, and drought-forced migration, individual turtles routinely leave and return to ponds throughout the annual activity period. Semmler (1979) found that there was no significant difference between permanent and temporary ponds in these turtles' movements. In all but two instances, turtles were found to return to the same ponds that they left. Iverson (1991) found that *K. flavescens* in the Nebraska Sandhills always returned to the same sandhills for estivation, nesting, or overwintering. These studies indicate a strong homing tendency in this species. Most daily aquatic activity took place from sunrise to sunset, and trapping at night never yielded turtles in Semmler's (1979) study. We have also taken most of our specimens in daytime trap sets. Turtles bask mostly by floating in the warm surface water or burying in the mud at the bottom of shallow water, where they may breathe by extending the head to the surface. Terrestrial movement occurs primarily in morning or evening, but rainy or overcast conditions may extend or further stimulate this activity.

With the onset of cooler weather and shorter days, yellow mud turtles leave the water to dig suitable burrows in which to spend the winter. These turtles are not freeze-tolerant, and even overwintering hatchlings must dig below the frost line for the winter (Iverson, 1991). One or more trial burrows may be dug before a suitable site is discovered, and the burrows are often made under shrubs or other forms of cover for added protection. The early burrows made by the turtle are usually not as deep as the final permanent one. We doubt that this species normally hibernates underwater. In the fall of 1978, fourteen *K. flavescens* and two *K. hirtipes* were placed in a walled-in artificial pond at the University of New Mexico. This pond freezes at the surface but never to the bottom. All *K. flavescens* died, but the two *K. hirtipes* survived that winter. Other turtles, such as *Chelydra* and *Trachemys*, have survived winters in this pond on other occasions. Seidel and Reynolds (1980) report that *K. flavescens* is more resistant to evaporative water loss than *K. hirtipes*. Mahmoud (1969), in comparing temperature tolerances of Oklahoma kinosternid turtles, found *K. flavescens* to be more resistant to warmer, but less so to colder temperatures.

Reproduction: Male *K. flavescens* mature at about 80 mm CL (Mahmoud, 1967), and as early as six or seven years of age (Christiansen and Dunham, 1972). The youngest mature male from New Mexico studied by Christiansen and Dunham (1972) was six years old. Females mature at least one year later than males at 90 mm CL (Iverson, 1991). Maturity in both sexes varies from 6–16 years and is correlated with latitude, with a longer time required northward. Furthermore, maturity is primarily size- rather than age-related. In the Nebraska Sandhills, Iverson (1991) found that the average age of females producing their first clutches was eleven years. A warm wet spring is optimal for normal ovulation and nesting. Under poor conditions the female may forego nesting in that year. This will help insure successful reproduction in the following year, with a predicted relatively greater clutch mass (Iverson, 1991). Iverson (1991) also suggested that by not nesting in a cool year, the chance of producing a clutch with few females is avoided, since sex determination is temperature controlled in this species and cooler incubation temperatures produce more males.

Iverson (1990, pers. comm.) reported that nesting female yellow mud turtles in the Nebraska Sandhills dig down to an average soil depth of 13 cm (to top of the carapace). While buried the female excavates the nest below her plastron and deposits 3–10 elliptical hard-shelled eggs at an average depth of 20 cm. Bladder water is apparently released during this process, helping to moisten the surrounding soil. The female remains in position over the nest for a few hours to many days. Iverson (1990) recorded at least 39 days as the longest time, but some females were still in nests when he left his study area in mid-July. Some of these females may not return to the water until the following spring. This tendency of some females to remain in the nest with their eggs is the first described instance of possible parental care for a turtle in the field, and may help to protect the eggs against some predators. Recently, Tuma (1993), working in Henry County, Illinois, found that some females may divide clutches of eggs between two nests. He suggested that multiple nesting may increase nest survivorship by spreading the risk of total clutch failure due to predation. Nest predation is severe in most populations; common predators include snakes (especially *Heterodon*), skunks, raccoons, and small rodents. Iverson (1991) found that egg mass and clutch size were both correlated with female body size. In 19 New Mexico clutches, Iverson (1991) reported egg size as 16.9 x 28.8 mm. Eggs hatch in the fall, but hatchlings overwinter in the soil and do not migrate to the water until the following spring (Christiansen et al., 1985; Iverson, 1990). Sex is temperature dependent with higher temperatures producing all females and lower temperatures producing mostly males (Vogt et al., 1982; Ewert and Nelson, 1991). For a detailed account of life and demography of this species, see Iverson (1991).

Food Habits: Yellow mud turtles are omnivorous. They feed on a large variety of both living and dead animal matter and a smaller percentage of aquatic plant material (Mahmoud, 1968). Most feeding activity seems to take place during daylight hours. Punzo (1974c), in a Texas study, reported that this turtle actively feeds in both aquatic and terrestrial habitats. He found that turtles feed in the water primarily during the day but often forage on land in the early morning and evening hours, when cooler temperatures prevail. In a laboratory study, D. Moll (1979) reported that while buried in moist sand, the Illinois mud turtle fed on earthworms that had burrowed into the sand after they had been released on the surface. He suggested that turtles may do some feeding during dormant periods if prey items enter their burrows. During June 1994, we observed a large adult male feeding on newly metamorphosed *Rana berlandieri* in a weed-choked stream at Light Spring in Eddy County near Whites City.

Remarks: Much has been learned about this interesting turtle during the last decade, and a number of studies, some long term, are presently being conducted (Christiansen and Iverson, 1993).

The specimen from northern Sandoval county (MSB 49112) likely represents an escaped or liberated individual. This species was last reviewed by Christiansen and Iverson (1993).

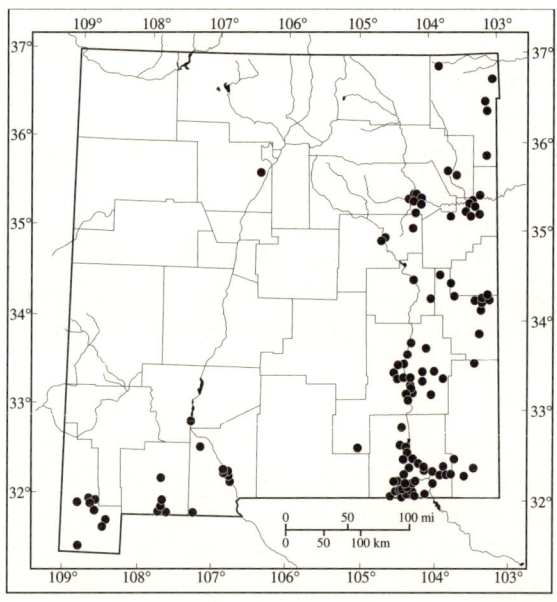

Distribution of Kinosternon flavescens *in New Mexico.*

KINOSTERNON SONORIENSE LeConte, 1854 — *Sonoran Mud Turtle*

See Plate 34.

Type: The holotype, a male collected by LeConte's son with no date specified, was apparently lost from the Philadelphia Academy of Natural Sciences. The type locality was given as, "Tucson, in Sonora," which today is in Pima County, Arizona.

Distribution: The Sonoran mud turtle is found primarily in Arizona and southwestern New Mexico, south through Sonora and eastern Chihuahua, Mexico. Within this area, it is associated with three major drainages, the Colorado River, the Río Yaqui, and the Río Casas Grandes. The more northern population is within the Colorado River drainage and that to the south is in the Río Yaqui drainage. In New Mexico, the Sonoran mud turtle has been taken only in southern Catron, western Grant, and Hidalgo counties in the southwestern corner of the state.

Description: This is a small turtle with adults reaching 175 mm CL in females and 155 mm CL in males. A series of 46 unsexed individuals collected on 14 October 1994 from a single locality in the southern Peloncillo Mountains in Hidalgo County averaged 106.8 (57.2–152.7) mm CL and 180 (32–469) gm (Paul Stone, pers. comm.). The skin is dark gray overall with cream colored mottling on the head and neck that tends to form at least one pair of stripes extending back from the eye. The head pattern is most pronounced in males. The carapace is brown to olive with darker seams. The double-hinged plastron is usually yellow to brownish with darker seams, but varying amounts of plastral melanism may be present. The eighth and ninth marginals are the same height, and the tenth is higher. The chin and throat barbels are large, elongate, and distinct. Mature males differ from mature females in having a longer and thicker tail tipped with a horny nail, and a shorter, more concave plastron.

Similar Species: *Kinosternon flavescens* differs in having the ninth marginal plate higher then the eighth, and little, if any, mottling on the head. *Chelydra serpentina* has a very small unhinged plastron. *Chrysemys picta*, *Trachemys* spp., and *Pseudemys gorzugi* all have unhinged plastrons. *Terrapene ornata* has a plastron with a single front hinge.

Systematics: Two subspecies have been recognized (Iverson, 1981). Only the nominotypical subspecies occurs in this state.

Habitat: The Sonoran mud turtle is more closely associated with water than the yellow mud turtle. Permanent streams, springs, ponds, or lakes in woodland are preferred. These waters are usually clear with rocky or sandy bottoms and aquatic vegetation, however, the water occasionally may be quite turbid and devoid of aquatic vegetation. We took one series from a small spring-fed pool measuring about 1.8 x 3 m in surface and 1.8 m deep near Cliff in Grant County. The water was crystal clear, cool, and the rocky bottom easily seen. No turtles or any kind of movement was discernible; however, two traps baited with sardines and placed on the bottom for two hours yielded 13 turtles. Four turtles were collected by hand in Hidalgo County by walking along a small shallow stream and inspecting the bottom. *Trionyx spiniferus* has been taken in the same river systems. To our knowledge, *K. flavescens* and *K. sonoriense* have not been taken together, but they may coexist in ponds in southwestern New Mexico.

Behavior: Sonoran mud turtles may expel small amounts of musk when handled. They are secretive and spend most of their time on the bottom, except when basking at the surface, or less often on the bank. Terrestrial activity is rare compared to the yellow mud turtle. However, terrestrial activity is indicated by their occasional occurrence in temporary ponds, and Kauffeld (1943a) found two walking on a road near an irrigation ditch at 0700 in mid-July.

Activity cycles vary primarily in response to temperature and light intensity (Hulse, 1974b). During the cooler portions of the year, turtles are active during the daytime unless they are hibernating. During the warmer months, activity is nocturnal or crepuscular except on cloudy and cool days. Deeper pools or ponds may remain cooler and darker at the bottom, allowing for more daytime activity. These turtles hibernate underwater in the mud bottoms of ponds and streams or along shorelines in recesses produced by rocks and stumps (Hulse, 1974b). These turtles are found at a range of elevations from below 1200 m to just over 2000 m. At elevations of 1850 m or less, turtles may remain active year round. We expect that our New Mexico populations at higher elevations are probably inactive for 2–4 months of the year.

Reproduction: The following information is based primarily on Hulse's (1974b) study of this species in Arizona unless cited otherwise. Mature males were at least 76 mm CL and five years old in streams at low elevations (600 m), and 94 mm CL and eight years old at a higher elevation (1200 m) stream. The latter stream differed in being colder and deeper. Mature females were 93 mm CL at 8–9 years and 130 mm CL at 11 plus years from the same two stream sites. However, in his study of 10 different Arizona populations, Rosen (1987) determined that the age at first ovulation is very close to six years, and size may vary from 86–137 mm PL. Rosen also suggested that age and not size determines maturity in female Sonoran mud turtles. Courtship and mating occur in the water in early spring when the water temperature reaches at least 19°C. The male retains sperm throughout the year, permitting early spring mating. Small ovarian follicles are found in the female throughout the year, enlarge in the spring, and are large enough for ovulation by May, June, or July. At the low elevation site (600 m), Hulse (1974b) found that oviductal eggs were present from June through September, and Rosen (1987) found shelled eggs as early as 23 May. Ovulation in this species seems to be later than for other North American kinosternids.

Eggs are hard-shelled, white, elliptical, and smooth. The mean egg size is about 31 x 17 mm; however, egg width varies among populations and is related to the size of the female pelvic opening (Rosen, 1987). Most hard-shelled eggs are resistant to water loss, but Hulse (1974b) reported eggs of this species are not resistant and will desiccate in a few days unless kept in damp soil. Clutch size varies from 1–11 eggs/clutch and means among Arizona populations studied by Rosen (1987) varied from 3.09–8.12. Rosen (1987) reported that in the populations he studied, females mature at six years and may produce three or four clutches per year. Five juveniles were collected from a water-filled tinaja in a southern Hidalgo County canyon bottom 25 September 1993 (Mark Doles, pers. comm.). They measured 29, 29, 31, 43, and 45 mm CL. The umbilical yolk sac scar was very evident on the three smallest turtles. Rosen (1987) estimated minimum hatchling size as about 25 mm CL. Rosen's data also indicate hatchlings first emerge in August, and emergence continues into at least late September and probably into late October to late December, depending on elevation. Overwintering is not proven, but Rosen believed that it is likely at the highest elevations. No other information on nesting behavior, nests, incubation time, and where the young overwinter is available.

Food Habits: The Sonoran mud turtle is basically carnivorous but may eat plant material, both algae and higher plants, in waters with a depauperate benthic fauna. Insect larvae and snails are the most common animal food items, but other aquatic invertebrates, fish, frogs, and tadpoles are also taken.

The turtles forage by slowly crawling along the bottom of the pond or stream, both in dense vegetation and on open bottom. Their acute senses of taste and smell aid the turtles in locating food. They will periodically surface for air, and the time intervals between surfacings are inversely correlated with water temperature and degree of activity. Feeding time shifts from diurnal to nocturnal as the cooler water temperatures of spring progress into the warmer ones of summer. Details of feeding behavior and a list of food items based on stomach content analyses of Arizona turtles are found in Hulse (1974a).

Remarks: This species has been reviewed by Iverson (1976, 1981) and Smith and Smith (1979).

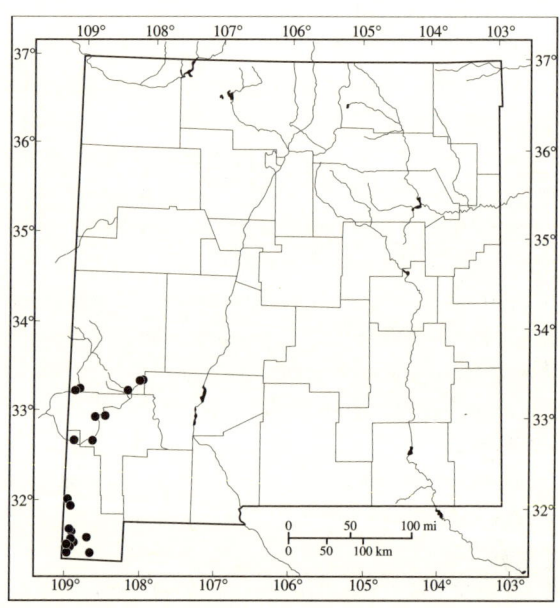

Distribution of Kinosternon sonoriense *in New Mexico.*

FAMILY TRIONYCHIDAE   *Softshell Turtles*

This is a large and diverse family of highly specialized turtles, occurring in Africa, Asia, the Indo-Australian archipelago, and North America. Three species in one genus, *Trionyx*, occur in North America. They are characterized by a round, flat appearance; an extremely long neck and head ending in an elongate, tubular snout; a carapace and plastron that lack scutes and is covered by a leathery skin; and extensive webbing on the paddlelike limbs, each bearing three claws. The highly vascularized pharyngeal and cloacal linings are able to extract oxygen from the water, thus allowing these turtles to remain submerged for extended periods of time. Most are strictly carnivorous, feeding on mollusks, crawfish, aquatic insects, worms, frogs, and fish. These turtles are often seen basking on muddy banks and sandbars. The eggs are spherical and brittle, 2.5–3.5 cm in diameter. Hatchlings average 3.2–5.1 cm with females growing much larger than the males and, unlike most turtles, sex is independent of incubation temperature. Large softshell turtles have a lightning fast strike and can inflict a painful bite. They also use their strong, heavy claws for defense and should be handled carefully.

TRIONYX MUTICUS LeSueur, 1827 — *Smooth Softshell*

See Plate 35.

Type: Five syntypes (MNHN 564, 4143, 7977, 8813, and 8814) were used in the original description. Webb (1962) designated MNHN 8813, a female dried carapace and plastron obtained by LeSueur in August 1827, as the lectotype. All specimens came from "Newharmony, sur le Wabash" (Wabash River, New Harmony, Posey County, Indiana).

Distribution: *Trionyx muticus* ranges through the central United States from western Pennsylvania to North Dakota, and then south to northeastern New Mexico and east to the Florida Panhandle. The smooth softshell enters northeastern New Mexico only in the Canadian River drainage and has been taken to about 1300 m elevation.

Description: *Trionyx muticus* is the smallest American softshell. Females reach 36 cm CL and are larger than males, which reach 21 cm CL. The leathery carapace is round, smooth, keelless, and lacks tubercles on the anterior edge. The carapace ground color of olive to orange-brown is marked with darker spots, short streaks, or blotches; mature females have a lichen-like blotched pattern. The white or gray unmarked plastron is almost translucent and the underlying bones can often be seen through the skin. Postocular stripes extend from the eye onto the neck and have thick black borders which are absent in adult males. There are no ridges on the nasal septum. Males have long thick tails and foreclaws longer than hindclaws; females have shorter and thinner tails and hindclaws longer than foreclaws.

Similar Species: In New Mexico, *Trionyx spiniferus* has more distinct dark-bordered spots (ocelli) on the carapace, a ridged nasal septum, and a gritty surface on the carapace of males.

Systematics: Webb (1973a) recognized two subspecies. Only the nominotypical form, *T. m. muticus*, is found in New Mexico. Meylan (1987), in a cladistic study based on osteological data, considered the North American *Trionyx* generically different from those in the Old World, and suggested using the old Rafinesque (1832) name, *Apalone*, for the New World species. A few recent authors have followed Meylan's lead. Webb (1990a) in his review of the genus considered Meylan's classification an important first step but total acceptance yet premature. Ernst et al. (1994) follow Webb (1990a) and use *Trionyx*. Pritchard (in David, 1994) stated that "Several genera retained by Meylan (1987) are but artifacts of cladistics and do not correspond to the current acceptance of "genus level" in turtle [*sic*]." Until further study, we follow Webb (1990a) and Ernst et al. (1994) by retaining *Trionyx* as the generic name for the New Mexico species.

Habitat: The smooth softshell is primarily a river turtle, but is sometimes found in lakes, impoundments, or ditches along watercourses. Plummer (1977a) studied this species in Kansas and found extensive use of sandbar habitats, especially by juveniles and males. A soft, sandy or silty bottom is the most important habitat requirement for this species. In New Mexico, only seven turtles at five localities have been collected. All localities were in rivers or shallow portions of Conchas Lake where rivers were entering. Water clarity varied from clear to muddy, the substrate varied from sandy to gravel or rock, and the water was slow-moving. Water temperature where softshells were trapped by Applegarth (1982) varied from 24–30°C. Other New Mexico species collected nearby, but never in the same traps, were *T. spiniferus*, *Chelydra serpentina*, *Kinosternon flavescens*, and *Trachemys scripta*.

Behavior: The smooth softshell is alert and wary, so its presence is often unsuspected by its human neighbors. It is shy and withdraws its head completely when captured. Occasional individuals may try to bite, and since they have a long neck and are capable of striking rapidly, they should be handled with some care. This species is the most aquatic of the American softshells and seldom moves overland, although it is extremely agile and can move with surprising speed on land. As would be expected of a river dweller, this streamlined turtle can swim against strong currents with ease, and is reported to have overtaken and captured a small brook trout (Cahn, 1937).

Turtles may bask on sandy banks with the head facing the water, or on emergent rocks, logs, or debris at the water surface. They may also bask in warm shallow water, buried in sand or silt, where the head can be extended to the surface for breathing.

Plummer and Shirer (1975) studied movement patterns in a Kansas river. They found that although there was a tendency for turtles to remain in a limited area for short periods of time, and even sometimes to return after displacement, turtles "seem not to confine their activities to small areas over long periods and females ordinarily do not return to a specific home range after nesting."

A number of reports attest to nocturnal activity in softshell turtles (Webb, 1962). Our trapping records suggest daytime activity, since most turtles were taken from daytime trap checks, rather than from traps left overnight. Plummer and Shirer (1975) noted daytime movement in balloon-tagged *T. muticus*.

Most annual activity occurs from April to October but is temperature dependent. Plummer (1977a) recorded conspicuous activity with water temperatures of 13.5–15°C in October. The annual activity period in New Mexico is probably shorter, due to the colder water of Conchas Lake and the Canadian River drainage. The consensus of most authors suggests hibernation on the bottom of water bodies under a shallow covering of substrate.

Reproduction: Most males reach 80 mm CL (57.6 mm PL) in their fourth year when they mature. Females tend to mature when larger and older, averaging 140 mm CL (100 mm PL) in their ninth year (Plummer, 1977b). Plummer (1977c) described mating behavior, and reported both spring and late summer sexual activity.

Females may lay two or more clutches per season, averaging 11 (3–33) eggs each, but both clutch number and size are directly related to the size of the female. They dig shallow nests, 15–23 cm in diameter, in open areas of sandbars and sandy banks. Eggs are spherical with thick, brittle, white shells. Incubation time varies with temperature, but is usually 65–77 days. Hatchlings average about 25 mm CL and double in length by the end of their first year. Females tend to grow faster than males. In Plummer's (1976) Kansas study, nesting success was affected primarily by water level fluctuations and nest inundation. Webb (1962) attributed most nest mortality of softshells to predation and summarized published reports on the subject. These data are from other populations, however, and may not apply in New Mexico.

Food Habits: The smooth softshell is carnivorous and the small amount of plant materials ingested is probably mostly accidental. Webb (1962) summarized the literature accounts and added his own observations about prey items. He listed a wide variety of aquatic invertebrates, especially arthropods, as important foods, but fishes, amphibians, reptiles, and carrion are eaten when available. Larger turtles select larger prey items resulting in a dietary shift with age. In a detailed dietary study of a Kansas population, Plummer and Farrar (1981) found sexual dietary differences that were due mainly to differences in microhabitat. They found that females forage primarily in stable microhabitats in deep water, whereas males forage at the shallower interface between terrestrial and aquatic environments. Sit-and-wait diurnal hunting, using their excellent eyesight, is probably successful and used most. They locate carrion by olfaction, and therefore they can be trapped with sardines and dead fish.

Remarks: More work needs to be done in the Canadian River system, upstream from Conchas Lake, in order to better understand the status of *T. muticus* in New Mexico. Not only is this area at the westernmost edge of the range, but it is the highest known elevation for the species.

The species was reviewed by Webb (1973a), and the genus by Webb (1990a).

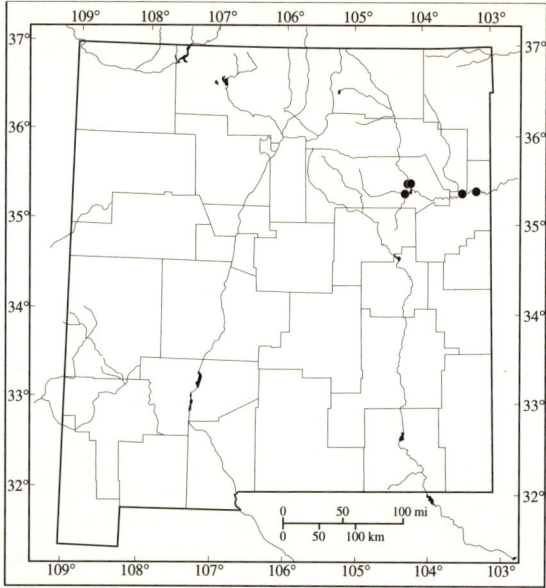

Distribution of Trionyx muticus in New Mexico.

TRIONYX SPINIFERUS LeSueur, 1827 — *Spiny Softshell*

See Plate 36.

Type: Eight syntypes (MNHN 1949, 6957, and 8807–8812) were apparently used in the original description of the species. Webb (1962) designated MNHN 8808, a large stuffed female, as the lectotype. Webb (1962) suggested that MNHN 1949 and 6957 were actually syntypes for *T. ocellatus* LeSueur, 1827, now a synonym of *T. spiniferus*, and he regarded MNHN 8808 as the lectotype. The type locality is "Newharmony, sur le Wabash" (Wabash River, New Harmony, Posey County, Indiana). The collector was Charles LeSueur, and the date of collection was sometime in 1826 or 1827.

Distribution: *Trionyx spiniferus* is found largely east of the Rocky Mountains from southeastern Canada to Montana and south into northern Mexico. Other populations, mostly introduced, are found scattered in the southwestern United States. In New Mexico, *T. spiniferus* occurs naturally in the drainages of the Cimarron, Canadian, Pecos, and Rio Grande rivers below 1600 m elevation. The Gila River population was probably a human introduction (Miller, 1946).

Description: This is a medium-sized *Trionyx*, with females reaching 470 mm CL, and males reaching 365 mm CL. During 1989–91, 97 males collected in the Pecos River and Rio Grande drainages of southern New Mexico averaged 156.9 (108–361) mm CL; 51 females averaged 215.1 (110–367) mm CL (our unpubl. data). The body is compressed (dorsoventrally flattened), and the shell lacks cornified epidermal scutes. The leathery carapace is round, keelless, and usually has tubercles on the anterior edge. The surface of the carapace in adult males (and often subadults) has a rough "sandpapery" surface. The carapace is olive or tan, with a pattern of either white dots (Rio Grande, Pecos, and Gila rivers) or dark spots or ocelli (Canadian and Cimarron rivers), and a marginal dark line. These markings are most distinct in juveniles and males. The white or cream unmarked plastron is almost translucent, and the underlying bones can often be seen through the skin. Postocular stripes are present but may be indistinct. Ridges are present on the nasal septum. Males have a long thick tail with the anal opening near the tip, and retain the juvenal pattern of small ocelli or white dots with distinct borders. Females tend to lose the juvenal pattern (indistinctly bordered ocelli) and develop an overall blotched appearance.

Similar Species: *Trionyx muticus* never has a "sandpapery" carapace, the anterior edge of the carapace lacks tubercles, and ridges on the nasal septum are absent.

Systematics: Webb (1973b) recognized six subspecies of *Trionyx spiniferus*. Two subspecies are found in New Mexico: *T. s. hartwegi* (Conant and Goin, 1948) and *T. s. emoryi* (Agassiz, 1857). The holotype of *T. s. hartwegi* is UMMZ 95365, an adult male collected by Robert Young at Wichita, Sedgwick County, Kansas. The type material (a series of syntypes) of *T. s. emoryi* was discussed by Webb (1962), who designated USNM 7855 as the lectotype from the "lower Rio Grande of Texas, near Brownsville." We follow Webb (1990a) and Ernst et al. (1994) in our use of *Trionyx* over *Apalone*. See explanation in the *T. muticus* account.

Habitat: Webb (1962) documented the occurrence of the spiny softshell turtle in virtually all varieties of permanent water. In Iowa, Williams and Christiansen (1981) reported that *T. spiniferus* occurs throughout the rivers, as well as in permanent and temporary ponds near rivers. Our collections in New Mexico indicate that this turtle is primarily a river dweller, although it also occurs in lakes and impoundments and irrigation ditches near rivers and streams. In lake populations, it seems to prefer shallow areas, where beaches are available, or where streams enter. Deep portions of lakes with steep rocky shorelines distant from beaches have never produced softshell turtles in our trapping. New Mexico stream populations are subject to the wide seasonal fluctuations in size, depth, and turbidity of these steams. When runoff is low, streams may become clear and shallow. At these times, softshells will retreat to the largest available pools, sometimes less than a foot deep. A soft bottom is essential, and although aquatic vegetation is often present, it seems to be less important. Populations of *T. s. emoryi* in New Mexico appear to be thriving in pools (adjacent to the Pecos River) which have relatively high salinity (0.8%) (Seidel, 1975a).

In many parts of the country, softshells often move overland. In the arid New Mexico climate, these turtles are prone to desiccation, and are therefore less likely to move far from permanent water to use newly constructed or seasonal ponds. Other turtles that have been taken in the same waters with *T. spiniferus* in New Mexico are *Chelydra serpentina*, *Kinosternon flavescens*, *K. sonoriense*, *Trachemys scripta*, *T. gaigeae*, *Pseudemys gorzugi*, *Chrysemys picta*, and *Trionyx muticus*. Microhabitat preferences of most of these other species are very different, and less restricted.

Behavior: *Trionyx spiniferus*, like *T. muticus* is highly aquatic and spends little time on land compared to other New Mexico turtles. They are powerful swimmers and are extremely agile on land. They are capable of rapidly striking with their long and extensible neck, but some individuals are very shy and do not attempt to bite. Extreme wariness is characteristic, explaining why their presence is often unsuspected by local human residents. Stuart and Clark (1990) described defensive posturing in *T. spiniferus*. They noted that posturing was most pronounced in large females, but overall was less distinctive than that seen in other turtles such as *C. serpentina*.

These turtles commonly bask on debris in the water, or more often on a bank or sandbar. Basking may serve a variety of needs such as thermoregulation or retarding the growth of epiphytes or epizoites. When basking on banks, the head is usually facing the water, allowing rapid escape from potential predators. Another favored method for warming is aquatic basking; the turtle will bury itself in the soft bottom of warm shallow pools and extend the long neck to the surface for breathing. With any nearby disturbance, the neck is retracted and the turtle is effectively hidden. Graham and Graham (1991) described burying behavior in sand. The turtle "bulldozes" into the sand, and then by a "waggle" or "shimmy" of its rear end, causes the sand to settle smoothly over the carapace, hiding the turtle completely. Another benefit of burying is that it literally scrapes off or prevents the attachment of shell symbionts. Captive softshells exposed to bright light, but prevented from burying, have rapidly developed a growth of algae on the carapace. Swimming or floating in the warm surface water on sunny days is another thermoregulatory method used by many turtles, and softshells are no exception.

Reproduction: Webb (1962) and Ernst and Barbour (1989) have summarized most of the information on reproduction in this species. Male *T. s. hartwegi* mature from 9.6–10.5 cm PL (13.2–14.5 mm CL), and *T. s. emoryi* matures from 8.2–9.0 cm PL (11.3–12.4 mm CL).

Sexually mature males can be recognized, since the anus extends beyond the posterior edge of the carapace. Female maturity may occur from 18–20 cm PL (25–28 mm CL) and all females larger than 22 cm PL (30 mm CL) are considered mature.

Mating normally takes place in April or May. Legler (1955) described courtship between a male *T. spiniferus* and a female *T. muticus*. Both species evidently have similar courtship patterns, making interspecific mating possible in syntopic populations, but hybrids are not known to occur.

Nesting may extend from May to August but normally occurs in June or July. Nests are flask-shaped and 10–25 cm deep. Eggs number from 3–39, are spherical or nearly so, white, brittle, and about 24–32 mm in diameter. Several successive clutches may be produced in a single year, and E. O. Moll (1979) reported that four or five are possible in Illinois. Miller et al. (1989) reported a record clutch size of 39 in a Logan County, Colorado, nest. Fitch (1985) reported an increase in clutch size northward, but this may be a result of more clutches per annum in southern populations. Eggs hatch from August to October. A minimum incubation time of 60 days has been suggested. The time seems dependent primarily on temperature, but moisture may also have an effect. In cooler habitats, eggs or young may overwinter in the nest. Sex determination is apparently independent of the incubation temperature in this species (Vogt and Bull, 1982).

Food Habits: The spiny softshell is carnivorous under normal environmental conditions, and the few plant materials ingested are probably accidental. Webb's (1962) summary of literature accounts, along with his own data, lists a wide variety of aquatic invertebrates, especially crayfish, as the most common food items. Other foods eaten are fishes, amphibians, reptiles, and carrion. We have fed captive softshells strips of beef and chicken, as well as grasshoppers, mealworms, and a variety of insects.

These turtles have excellent eyesight and the use of sit-and-wait hunting is probably most successful. Olfaction serves to locate carrion, and has allowed us to trap them using canned sardines and dead fish as bait. However, these baits often attract crayfish and fishes, which in turn may also serve as an attractant.

Remarks: Where *T. spiniferus* is sympatric with *T. muticus*, niches of the two species have been shown to differ fundamentally yet show considerable overlap (Williams and Christiansen, 1981). In New Mexico, these species are sympatric in the Canadian River drainage, and are at the extreme edge of their geographic ranges where competition could be severe. Studies here should be rewarding.

Painter (1993a) reported a large (312 mm CL) xanthic adult female softshell from the Rio Grande near Radium Springs. The species has been reviewed by Webb (1973b) and the genus by Webb (1990a). Ernst et al. (1994) have summarized the large body of published information on this species.

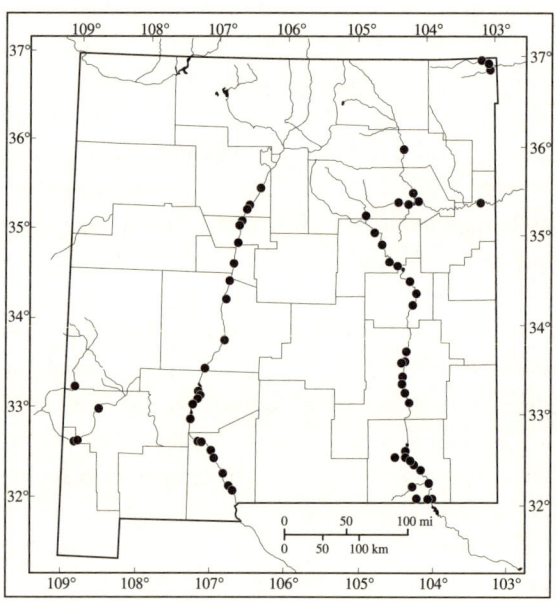

Distribution of Trionyx spiniferus *in New Mexico.*

A KEY TO THE LIZARDS OF NEW MEXICO

1. A well-defined lateral fold along both sides of the body in which the scales are granular (**FIG. 1**); both dorsal and ventral scales large and overlapping (Anguidae) *Elgaria kingii*
1. Lateral fold as above absent; dorsal and ventral scales not as above 2
2. Moveable eyelids absent or, if present, eyes with vertically elliptical pupils (**FIG. 2**); skin on ventral surface thin, translucent (Gekkonidae) 3
2. Moveable eyelids present and pupils round; ventral surface scaled, opaque 5
3. Moveable eyelids absent; dorsal surface with conspicuously enlarged tubercles; expanded toepads present (**FIG. 3**) *Hemidactylus turcicus*
3. Moveable eyelids present; dorsal surface without enlarged tubercles; expanded toepads absent 4
4. Scale series carrying the preanal pores in males (corresponding scales in females usually enlarged and sometimes pitted) divided at the midline by one or more small scales (**FIG. 4a**); usually 6 or fewer pores; tip of cloacal bones in males broad and flat (**FIG. 4b**) *Coleonyx brevis*
4. Scale series carrying the preanal pores in males (or corresponding scales in females) continuous across the midline apex (**FIG. 4c**); usually 7 or more pores; tip of cloacal bones in males narrow and pointed (**FIG. 4d**) *Coleonyx variegatus*
5. Dorsal and ventral body scales cycloid (**FIG. 5**) and equal in size, smooth, and shiny (Scincidae) 6
5. Dorsal and ventral body scales not cycloid and not equal in size, not smooth and not shiny 8
6. Dorsal and lateral scale rows parallel (**FIG. 6a**); longitudinal light stripes present or, if not, ground color olive brown to black without white spots on the labial scales 7
6. Lateral scale rows are oblique to and trend upwards posteriorly with respect to the dorsal scale rows (**FIG. 6b**); ground color of juveniles and subadults black, with white spots on the labial scales; dorsal color of adults yellow to greenish-gray with scale margins edged with black; longitudinal stripes absent *Eumeces obsoletus*

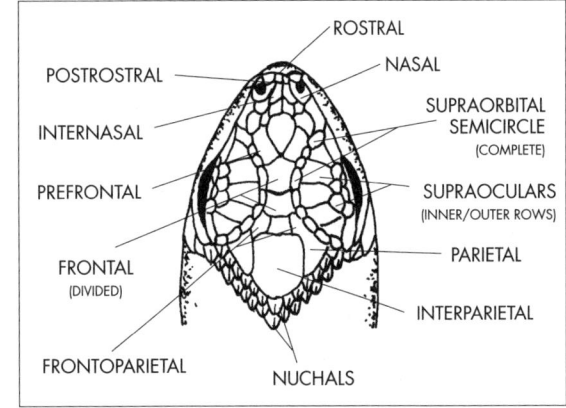

Dorsal aspect of typical lizard head showing scales.

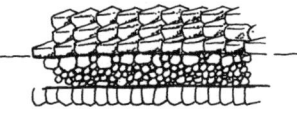

Figure 1

Figure 2

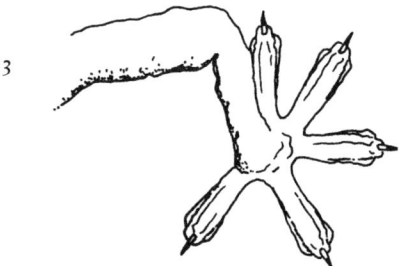

Figure 3

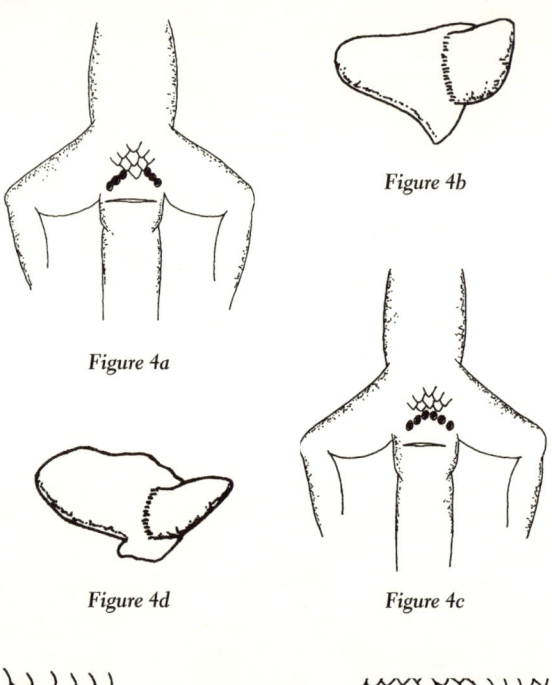

Figure 4a
Figure 4b
Figure 4c
Figure 4d

Figure 5

Figure 6a

Figure 6b

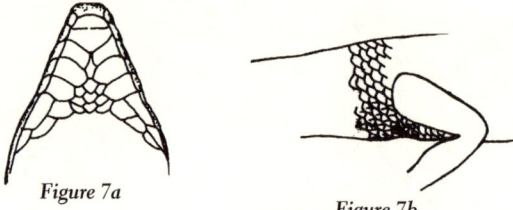
Figure 7a
Figure 7b

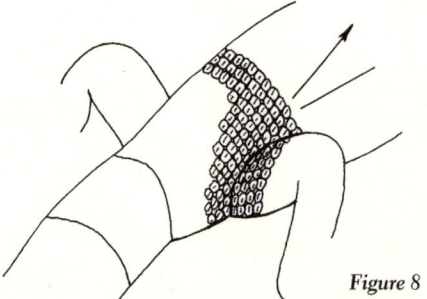
Figure 8

7. Longitudinal light stripes absent or, if present, dorsolateral pair confined to the 3rd scale row from the midline; the light lateral lines, if present, pass through the ear openings; light Y-shaped mark on the dorsal surface of the head absent; postmental scale divided by a transverse suture (**FIG. 7a**) and keeled lateral postanal scales (**FIG. 7b**) absent *Eumeces multivirgatus*

7. Longitudinal dorsolateral light stripes confined to the 4th scale row on either side of the midline; light Y-shaped mark on the dorsal surface of the head present; postmental scale entire or, if divided, the lateral light stripes pass above the ear opening; keeled lateral postanal scales present *Eumeces tetragrammus*

8. Head scales small and beadlike, usually slightly separated from each other; fourth toe on hind foot not greatly elongate, only slightly, if at all, longer than adjacent toes; ventral scales squarish, in transverse but not longitudinal rows (**FIG. 8**); femoral or preanal pores absent (Helodermatidae)
Heloderma suspectum

8. Head scales unequal in size; fourth toe on hind foot elongate, longer than adjacent toes; ventral scales various but not squarish and not arranged only in transverse rows; femoral or preanal pores present 9

9. Dorsal body scales granular; ventral scales large, quadrangular, arranged in 8 longitudinal rows; enlarged preanal scales present (Teiidae) 29

9. Dorsal body scales various, sometimes granular but usually elongate, keeled, or pointed; ventral scales not quadrangular, smaller to slightly larger than dorsal scales, and not arranged in definite longitudinal rows; enlarged preanal scales absent (Iguania) 10

10. Ear opening present and interparietal scale distinctly smaller in size; spines or projecting bony ridge on head absent; no auricular scales (**FIG. 10a**) projecting over ear opening (Crotaphytidae) 11

10. Ear opening present or absent; if present, then interparietal scale distinctly larger than or equal in size or, if smaller, then spines or projecting bony ridge present on head or auricular scales (**FIG. 10b**) project over the ear opening (Phrynosomatidae) 12

11. Black collar bands present *Crotaphytus collaris*
11. Black collar bands absent *Gambelia wislizenii*

12. Head with bony spines or elevated ridge projecting posteriorly 13
12. Head lacking bony spines or ridge as above 16

13. Lateral fringe scales along edge of abdomen absent; 4 short, widely separated horns of equal length at the posterior margin of the head shield (**FIG. 13**) *Phrynosoma modestum*

13. Lateral fringe scales along edge of abdomen present 14

14. A single row of lateral fringe scales along the edge of the abdomen 15

14. Two rows of lateral fringe scales along the edge of the abdomen; a distinct light longitudinal middorsal stripe present; a single pair of elongate, widely spaced, median occipital horns present (**FIG. 14**) *Phrynosoma cornutum*

15. A single pair of short, thick occipital horns separated by a deep, wide median notch between occipital horns on back of neck (**FIG. 15a**); enlarged gular scales absent
Phrynosoma douglasii

15. Two pairs of large, elongate occipital horns with their bases in contact, forming a continuous row along the back of the head (**FIG. 15b**); enlarged gular scales present
Phrynosoma solare

16. Supralabial scales overlap and posterior edges are diagonal, not vertical (**FIG. 16a**); mental scale small, as small as or smaller than adjacent lower labial, and bordered by a median postmental scale (**FIG. 16b**)   17

16. Supralabial scales do not overlap and the edges between them are vertical (**FIG. 16c**); mental scale large and not bordered by a median postmental scale (**FIG. 16d**)   19

17. Ear openings present   *Callisaurus draconoides*
17. Ear openings absent   18

18. Tail flat in cross-section; broad black bands present on the ventral surface of the tail; black marks present on the sides of the body placed far back, just in front of the hind limb insertions (**FIG. 18a**)   *Cophosaurus texanus*

18. Tail rounded in cross-section; ventral surface of the tail immaculate; black marks present on the sides of the body placed well forward, at or just anterior to the midpoint between the forelimb and hindlimb insertions (**FIG. 18b**)
Holbrookia maculata

19. A distinct and complete transverse gular fold present (**FIG. 19a**); dorsal body scales may be keeled and imbricate, but are not mucronate (**FIG. 19b**)   20

19. Transverse gular fold (**FIG. 19c**) absent; dorsal body scales keeled, imbricate, and mucronate (**FIG. 19d**)   21

20. Supranasal scales (**FIG. 20a**) absent; subcaudal scales keeled to within 5 mm of cloacal aperture and mucronate; keels of caudal scales extend entire length of scale, producing a series of parallel longitudinal ridges surrounding the tail (**FIG. 20b**); vertebral and paravertebral scale rows greatly enlarged and bordered by smaller scales (**FIG. 20c**); blue to black axillary spot absent; blue to black ventrolateral band present in males
Urosaurus ornatus

20. Supranasal scales (**FIG. 20d**) present; subcaudal scales never keeled except on posterior one-third of tail; keels of caudal scales do not extend entire length of scale, and parallel longitudinal ridges on tail absent (**FIG. 20e**); vertebral and paravertebral scale rows not greatly enlarged; blue to black axillary spot present in both sexes   *Uta stansburiana*

21. Lateral scale rows arranged parallel to longitudinal dorsal scale rows (**FIG. 21a**)   *Sceloporus scalaris*

21. Lateral scale rows arranged obliquely with respect to longitudinal dorsal scale rows, and trend upward posteriorly (**FIG. 21b**)   22

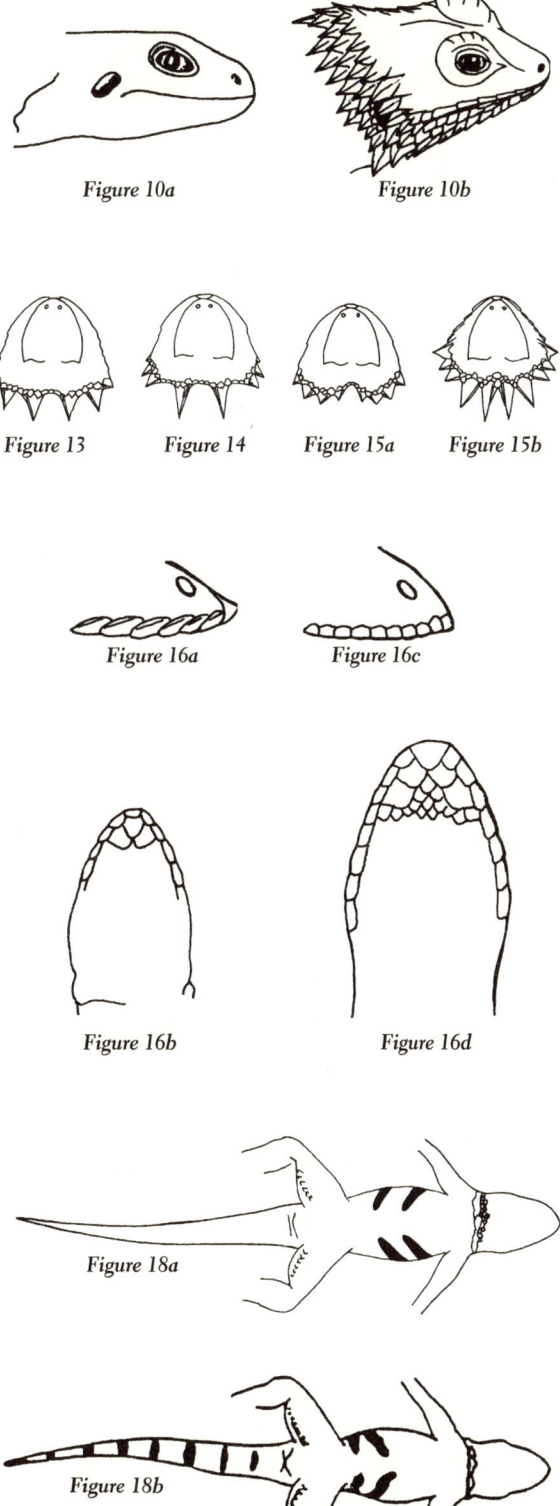

Figure 19a

Figure 19b

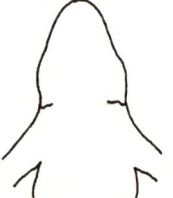
Figure 19c

Figure 19d

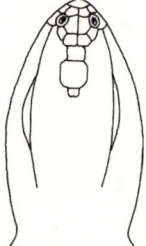
Figure 20a

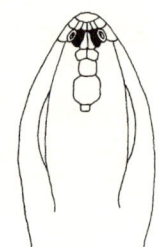
Figure 20d

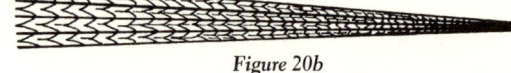
Figure 20b

Figure 20e

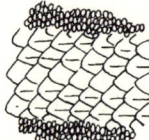
Figure 20c

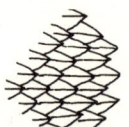
Figure 21a

Figure 21b

22. Supraocular scales large and elongate, in a single longitudinal row; semicircle formed by circumorbital scales incomplete posteriorly, allowing supraocular scales to be in contact with parietal and frontoparietal scales (**FIG. 22a**)   23

22. Supraocular scales smaller, in 1 or 2 rows; circumorbital scales completely separate the supraocular scales from the median head scales (**FIG. 22b**)   24

23. Forearms distinctly barred; sublabial scales usually separated from the mental by a series of smaller scales (**FIG. 23a**); 3 or 4 relatively short and rounded scales along the anterior margin of and projecting posteriorly over the ear opening, the uppermost the largest; 2 or more occipital scales bordering each parietal scale posteriorly (**FIG. 23b**)   *Sceloporus clarkii*

23. Forearms indistinctly barred or lack markings altogether; sublabial scales in contact with mental (**FIG. 23c**); 5 to 7 elongate and mucronate scales along the anterior margin of and projecting posteriorly over the ear opening, with the median scales the largest; a single enlarged occipital scale bordering each parietal scale posteriorly (**FIG. 23d**) *Sceloporus magister*

24. Black nuchal collar present   25
24. Black nuchal collar absent   26

25. Dorsal body scales large, strongly keeled and mucronate, with the distal ends projecting freely; dorsal color pattern consisting of bands, especially on tail; black nuchal collar complete, with both anterior and posterior white borders

Sceloporus poinsetti

25. Dorsal body scales small, weakly keeled, and nonmucronate, the distal ends do not project freely; dorsal banding pattern absent; black nuchal collar edged with white posteriorly only, and often interrupted or notched medially *Sceloporus jarrovii*

26. Scales on posterior surface of the thigh very small, unkeeled, nonimbricate, often granular; 9 or more scales separating the femoral pore series medially   27

26. Scales on posterior surface of the thigh larger, keeled, and imbricate; 7 or fewer scales separating the femoral pore series medially   28

27. Distinct dorsal pattern consisting of paired longitudinal series of dark brown blotches, sometimes fused to form two irregular dark stripes, each bordered by light stripes on either side; males with prominent lateral blue belly patches; 48–60 scales around body   *Sceloporus graciosus*

27. Dorsal pattern absent or consisting of paired faint longitudinal grayish-brown bands; prominent lateral blue belly patches in males absent; 41–52 scales around midbody

Sceloporus arenicolus

28. Lateral blue belly patches, midventral dark brown line, and dark gray or black patch in lateral nuchal fold absent in both sexes; dark lines across the top of the head between the eyes absent; femoral pore scales notched on posterior margins, and femoral pores located within the notches (**FIG. 28a**)

Sceloporus virgatus

28. Lateral blue belly patches and dark gray or black patch in lateral nuchal fold present at least in males; midventral dark brown line present at least posteriorly in some populations; 1 to 3 dark lines across the top of the head between the eyes; notching of the posterior margins of the scales containing the femoral pores absent (**FIG. 28b**)   *Sceloporus undulatus*

29. Dorsal pattern of clearly defined longitudinal light stripes, with or without light spots present in the dorsal fields between stripes, or dorsal pattern with many light spots present, superimposed upon less well-defined longitudinal light stripes   32

29. Dorsal pattern of poorly defined and separated longitudinal light stripes, often broken and connected transversely by light bars creating a checkered or barred appearance   30

30. Mesoptychial scales granular, grading smoothly into anterior throat scales (**FIG. 30a**); dorsal pattern of lines and bars obscure and diffuse; ventral surface with charcoal-gray to black pigment suffusion at least on the throat and chest
Cnemidophorus tigris

30. Mesoptychial scales abruptly enlarged and not grading smoothly into anterior throat scales (**FIG. 30b**); dorsal pattern of lines and bars bold and distinct; ventral surface immaculate except for a few black flecks   31

31. Dorsal pattern finely reticulated, consisting of 10–14 longitudinal stripes broken by frequent fusions across adjacent dark fields; known only from Hidalgo County
Cnemidophorus dixoni

31. Dorsal pattern boldly reticulated, consisting of 6–10 longitudinal stripes broken by frequent fusions across adjacent dark fields; not known from Hidalgo County
Cnemidophorus grahamii

32. Spots present in dark fields between longitudinal light stripes   33

32. Spots absent from dark fields between longitudinal light stripes   38

33. Supraorbital semicircles separate, at least partially, the second supraocular scale from the frontal (**FIG. 33a**); mesoptychial and postantebrachial scales granular (**FIG. 33b**); spots absent from the paravertebral dorsal fields; vertebral stripe noticeably zigzag in outline   *Cnemidophorus neomexicanus*

33. Supraorbital semicircles do not extend beyond the anterior margin of the third supraocular scale (**FIG. 33c**); mesoptychial and postantebrachial scales abruptly enlarged (**FIG. 33d**); some spotting usually present in paravertebral dorsal fields; vertebral stripe, if present, straight in outline   34

34. More than 94 dorsal granules around midbody; longitudinal light stripes often broken up or lost altogether in older individuals; ventral surface of males immaculate
Cnemidophorus burti

34. Less than 95 dorsal granules around midbody; longitudinal light stripes present throughout life; ventral surface of males pigmented   35

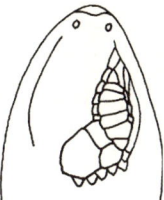
Figure 22a

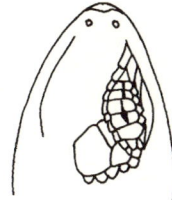
Figure 22b

Figure 23a

Figure 23c

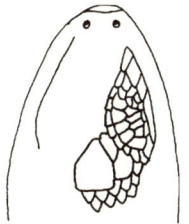
Figure 23b

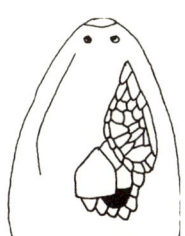
Figure 23d

Figure 28a

Figure 28b

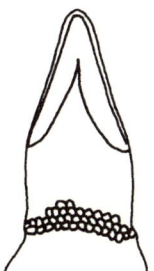
Figure 30a

Figure 30b

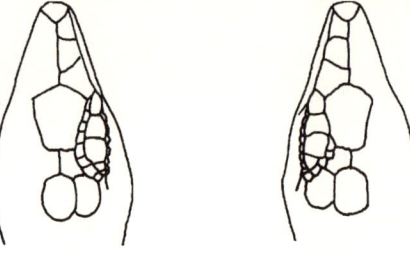

Figure 33a *Figure 33c*

Figure 33b *Figure 33d*

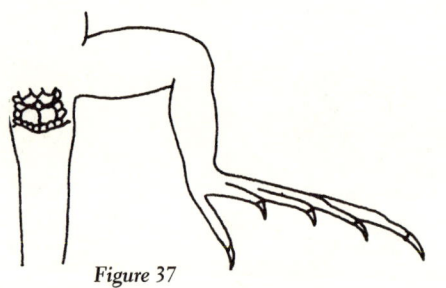

Figure 37

Figure 39

35. Seven or eight complete longitudinal light stripes; paravertebral stripes separated anteriorly by more than 10 scales
Cnemidophorus gularis

35. Six complete longitudinal light stripes; paravertebral stripes separated anteriorly by less than 10 scales   36

36. Dorsal spots large and distinct, arranged randomly or in multiple rows within dorsal fields; dorsal spots present within paravertebral fields and on neck and head; longitudinal stripes fade and dorsal spots expand, some involving the stripes, with age   *Cnemidophorus exsanguis*

36. Dorsal spots small and indistinct, arranged in 1 or 2 rows within dorsal fields; dorsal spots absent from paravertebral fields and usually absent from head and neck; longitudinal stripes remain distinct throughout life   37

37. Dorsal spots relatively few, arranged in 1 or 2 rows, and often touch or lie within the longitudinal light stripes; usually only 2 enlarged preanal scales (**FIG. 37**); pattern on dorsal surface of hind limbs indistinct   *Cnemidophorus flagellicaudus*

37. Dorsal spots relatively numerous, and not involving the longitudinal light stripes; usually 3 or more enlarged preanal scales; pattern on dorsal surface of hind limbs distinct
Cnemidophorus sonorae

38. Mesoptychial scales abruptly enlarged and angular in outline; postantebrachial scales abruptly enlarged or not   39

38. Mesoptychial scales only slightly enlarged, circular in outline, and grade smoothly into the surrounding scales on the throat; postantebranchial scales granular, circular in outline, only slightly if at all larger than adjacent scales; at least the tail and frequently all body surfaces suffused with bright blue coloration in life   *Cnemidophorus inornatus*

39. Sublabial scales usually absent or, if present, rarely more than 6 on each side of the head, extending only to the middle of the fourth infralabial (**FIG. 39**); usually more than 3 enlarged preanal scales; if more than 6 sublabials or less than 4 enlarged preanals, then postantebrachials abruptly enlarged and angular in outline and there are 7 complete longitudinal light stripes
Cnemidophorus velox

39. More than 6 sublabial scales; enlarged preanal scales 3 or less   40

40. Postantebrachial scales granular; hind legs conspicuously short; 7 or 8 complete longitudinal light stripes; color of head and anterior body bright green in life and conspicuously different from remainder of body   *Cnemidophorus sexlineatus*

40. Postantebrachial scales abruptly enlarged and angular; hind legs not conspicuously short; 6 complete longitudinal light stripes; color of head and anterior body not bright green and not conspicuously different from remainder of body
Cnemidophorus uniparens

FAMILY CROTAPHYTIDAE *Collared and Leopard Lizards*

This is a very small group of two genera and six or seven species (Collins, 1990; Zug, 1993) recently elevated to family rank in the taxonomic revision of the metataxon Iguania by Frost and Etheridge (1989). It is a strictly North American group, ranging from the Mississippi River in Missouri, westward to the Pacific Ocean, and southward from eastern Oregon to northern Mexico. These are large (90–150 mm SVL), alert, pugnacious lizards active during the day. Their bodies are covered with small scales, and they have well-developed legs and long tails. They occupy a variety of open desert to semi-arid habitats, often associated with rocks. All species are oviparous. Both genera, each represented by a single species, occur in New Mexico.

CROTAPHYTUS COLLARIS (Say *in* James, 1823)   *Collared Lizard*

See Plate 37.

Type: The species was originally described from a specimen collected near the confluence of the Verdigris and Arkansas rivers, Muskogee County, Oklahoma, on 5 September 1820. Webb (1970b) restricted the type locality to "near Colonel Hugh Glenn's Trading Post on the east bank of the Verdigris River, about two miles above its confluence with the Arkansas River, Wagoner County, Oklahoma." Say *in* James (1823) stated that a specimen was deposited in the museum of the Academy of Natural Sciences of Philadelphia but, if so, that specimen is now lost (Malnate, 1971). According to Webb (1970b), specimens from Long's expedition went to the Peale Museum in Philadelphia and subsequently on to other museums where they were either lost or destroyed. It is probable that the type specimen was stolen by three deserting soldiers along with most of the other scientific specimens and notes collected during Major Long's expedition to the Rocky Mountains and subsequently lost (James, 1823).

Distribution: *Crotaphytus collaris* occurs from the Mississippi River south of St. Louis westward across parts of the Ozark Plateau of southern Missouri and northern Arkansas into eastern Colorado, throughout Texas north and west of the Balcones Escarpment, and westward to the Arizona-California border. It occurs south throughout much of Sonora and on Isla Tiburón in the Sea of Cortéz (Murphy and Ottley, 1984). It extends into southwestern Arizona south of the Gila River, south of the Grand Canyon, and up the Colorado River basin into southeastern Utah and southwestern Colorado. It ranges southward throughout New Mexico and the Chihuahuan Desert to northern Zacatecas and San Luis Potosí. In New Mexico, *C. collaris* occurs nearly statewide at elevations of 900–2750 m. It is probably absent from the high mesas and mountains of the northern part of the state, extending into these regions only along watercourses.

Description: This is a large species with a maximum SVL of 118 mm (our unpubl. data). Males are slightly larger than females; Gehlbach (1965) measured males averaging 97.9 mm (90–105) and females averaging 93.7 mm (87–101) SVL. The head is large and distinctly set off from the body by a narrow neck, and the head of males is broader and more muscular. Body scales are minute and not imbricate. The head scales posterior to the eyes are small and irregularly shaped, and there is usually a complete series of scales comprising the supraocular semicircles which meet and occasionally are fused medially. A transverse gular fold is present. The ventral scales are slightly larger than the dorsals, imbricate, and sometimes keeled. There are 14–27 femoral pores on each leg, and males generally have slightly enlarged postanal scales.

The two black bands around the neck are the characteristic pattern feature of this species. One or both may be incomplete along the dorsal midline. Hatchlings have a distinctly banded dorsal pattern, with narrow transverse light yellow-tan bands alternating with wider dark brown ones. The latter bands are composed of dark spots interspersed with and connected by a lighter reticulate pattern (Fitch, 1956a; Ingram and Tanner, 1971). As individuals age, the banding pattern fades and is replaced by numerous small white spots scattered throughout the dorsal surface. The ground color varies geographically throughout New Mexico. In the San Juan Basin, lizards have light green bodies and bright yellow heads, the yellow extending posteriorly to or just past the second collar and ventrally onto the throat. The yellow and green colors may meet on the gular patch in males. Lizards from most of the remainder of the state have dark green bodies and, if there is yellow on the head, it does not extend posterior to the rear margin of the orbit, and never onto the throat. *Crotaphytus collaris* from the Rio Grande Valley south of the Mogollon Rim have a brown dorsum with no trace of green, and a light to cream colored head with no trace of yellow. The front legs are unpatterned and the hind legs are similar to the rest of the body. The lining of the throat is black. Gravid females develop red or orange spots or bars superimposed on the ground color along the sides of the body and neck.

Similar species: *Gambelia wislizenii* lacks black bands around the neck. Spiny lizards of the genus *Sceloporus* lack black collar bands or have one broad complete black collar, and have larger, distinctly keeled, imbricate, and sometimes mucronate dorsal scales.

Systematics: Six subspecies are recognized; four occur in New Mexico including the nominotypical subspecies. *Crotaphytus c. auriceps* was described by Fitch and Tanner (1951) based on an adult male (KU 29934), collected from "3½ mi. NNE Dewey, west side of the Colorado River, Grand County, Utah" by Henry S. Fitch on 28 June 1950. *Crotaphytus c. baileyi* was originally described as a full species by Stejneger (1890) based on a subadult male (USNM 15821) as the type, collected by C. Hart Merriam and Vernon Bailey from the "Painted Desert, Little Colorado River, Arizona" on 26 September 1889. *Crotaphytus c. fuscus* was described by Ingram and Tanner (1971) based on an adult male (BYU 16970), collected from "6.5 mi. N. and 1.5 mi. W. of Chihuahua City, Chihuahua, Mexico" by Wilmer W. Tanner on 21 July 1960.

Habitat: *Crotaphytus collaris* is typically found in level or hilly rocky terrain in a variety of vegetative communities, including piñon-juniper woodlands, sagebrush, desert shrublands, and desert riparian gallery forests or riparian grasslands (Harris, 1963; Gehlbach, 1965; our obs.). They are abundant on lava fields around the state (Best et al., 1983; our obs.) and are generally absent from flat plains, sandy soils, and other similar habitats unless there are rocky outcrops or riparian zones within them.

Behavior: New Mexico specimens have been collected between 7 March (Grant Co.) and 2 November (Doña Ana Co.). *Crotaphytus collaris* emerges from hibernation in late April or early May upon the onset of warm temperatures. Adults generally enter winter hibernation by the end of August; hatchlings may remain active for up to 6 weeks longer (Fitch, 1956a). They maintain active body temperatures between 37° and 40°C, which are normally considerably above ambient air temperatures and usually seek shelter when the latter approach the lower active body temperature threshold (Fitch, 1956a). This species has a critical thermal maximum temperature near 43°C (Dawson and Templeton, 1963).

Adult male *C. collaris* are highly territorial (Fitch, 1956a; Fox and Baird, 1992) and exclude other males by stereotyped displays or outright aggression. Females are less territorial but tend to occupy more or less exclusive home ranges. These lizards may be highly aggressive when molested and can bite hard.

Individual *C. collaris* are able to run at high speeds on their hind legs directly from a resting position (Snyder, 1952). The body is held off the ground at a 45° angle, and the front legs are held partially flexed and basically still. The stride can be three times the body length of individual lizards at top speed.

Reproduction: In western Texas, most female *C. collaris* reproduce during their first year after hatching (Ballinger and Hipp, 1985). The smallest female found with yolked follicles in early spring was 64 mm SVL, but such individuals grow to 70 mm SVL or more prior to oviposition. Females emerge from hibernation in late March, follicular development begins by early April, mating takes place in April and May, and ovulation occurs from late April through mid-June (Fitch, 1956a; Andre and MacMahon, 1980; Ballinger and Hipp, 1985). Reproductive activity is over by mid-July, and hatchlings appear in August and September. Fat body cycles parallel this activity.

Clutch size is proportional to body size; females add approximately 1 egg per 5 mm increment in SVL. Females 2 years and older may produce 2 clutches a year (Fitch, 1956a; Ballinger and Hipp, 1985). In the latter study, females smaller than 76 mm SVL produced an average clutch of 5.25 eggs. Older females (> 80 mm SVL upon spring emergence) probably produce a first clutch of 9 and a second clutch of 7 eggs. An average clutch size of 5 eggs was reported by Andre and MacMahon (1980) from Utah and Parker (1973b) from southern New Mexico. Clutch weight may represent as much as 45% of female body weight prior to oviposition (Fitch, 1956a). Fitch (1956a) recorded incubation times of 57 to 94 days depending on weather. Hatchlings are about 40 mm SVL and may grow relatively rapidly, 1 mm or so a day for the first week, 10–15 mm SVL prior to hibernation (Fitch, 1956a). Hatchlings from southern Doña Ana County appear in late August through early September, are 34–42 mm SVL, and exhibit similar growth rates prior to hibernation (our unpubl. data).

The minimum recorded size for males at sexual maturity is 83 mm SVL (Parker, 1973b). Male testes develop rapidly upon emergence from hibernation in late March, increase in weight by an order of magnitude by April and May, and possess mature spermatozoa through the end of June in west Texas and southern New Mexico (Ballinger and Hipp, 1985; Parker, 1973b). Regression of testes begins in June, and they essentially attain pre-reproductive condition by winter dormancy beginning in September. Fat bodies are essentially depleted by the peak of reproductive activity in May and June, but are reconstituted during the remainder of the activity season. Fat bodies at the beginning of hibernation are about twice the weight of those upon spring emergence.

Individual *C. collaris* may live to be 7 years old (Fitch, 1956a). Our unpublished data from southern Doña Ana County suggest a natural life span exceeding 10 years.

Food Habits: The bulk of the diet of *C. collaris* throughout the year consists of grasshoppers, except during the spring when adult lepidopterans, coleopterans, and hymenopterans are also important (Burt, 1928; Blair and Blair, 1941; Hotton, 1955; Fitch, 1956a; McAllister, 1985). Best and Pfaffenberger (1987) reported beetles (79%), grasshoppers (mostly acridids, 74%), ants and wasps (52%), spiders (40%), hemipterans (27%), asilid flies (21%), and butterflies and moths (16%) as the most important food items by frequency of occurrence in a population from central New Mexico. Lizards of other species (*Cnemidophorus* spp., *Phrynosoma modestum*, *Sceloporus* spp.) are also significant components of the diet (Cockerell, 1896; Burt, 1928; Little and Keller, 1937; Fitch, 1956a; McAllister, 1985; Best and Pfaffenberger, 1987). The latter also listed one snake (*Sonora semiannulata*) as prey. Spiders and caterpillars are also eaten. Herbivory has also been reported for this species (Pack, 1923a; Knowlton and Thomas, 1934; Banta, 1960).

Raptors (*Accipiter* spp., *Circus cyaneus*, *Buteo* spp., *Falco* spp.) and snakes (*Coluber constrictor* and *Masticophis* spp.) are reported as predators on this species (Fitch, 1956a).

Remarks: Lewis (1949, 1951), Fitch (1956a), Gehlbach (1965) and Best et al. (1983) have reported melanistic individuals from lava flows in New Mexico. We have collected *C. collaris* in the same places as Best et al. (1983) and have noted that many apparently melanistic individuals return to a lighter, more typical dorsal coloration after capture, suggesting that this trait is much more labile in this species than otherwise indicated.

Unlike many other species of lizards, *Crotaphytus collaris* does not readily lose its tail (Fitch, 1956a). The tail of lizards is an important organ in maintaining balance during high-speed locomotion, and therefore in this species the advantages of such locomotion may outweigh benefits from tail autotomy.

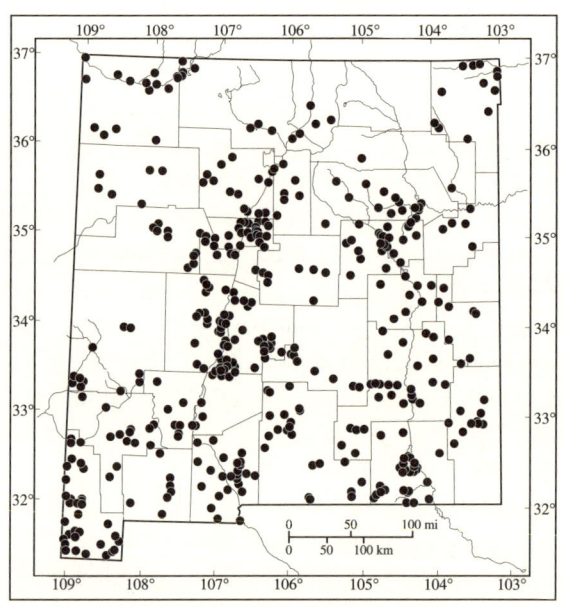

Distribution of Crotaphytus collaris *in New Mexico.*

GAMBELIA WISLIZENII (Baird and Girard, 1852) *Leopard Lizard*

See Plate 38.

Type: The original description (Baird and Girard, 1852a) did not specify any type material. The lectotype (USNM 2685) is an adult female, obtained by Col. J. D. Graham in May or June 1851 during the U.S. and Mexican boundary surveys from "between San Antonio and El Paso del Norte." Considerable confusion exists concerning the provenance, and indeed the existence, of the original material upon which the description was based. We refer the reader to Tanner and Banta (1963), whom we follow in this account.

Distribution: *Gambelia wislizenii* occurs throughout the Great Basin, Mojave, and Sonoran deserts, southward from southeastern Oregon and southern Idaho to northern Sonora, Mexico, and the length of the Baja California peninsula except for the southern tip. It also occurs on Isla Tiburón in the Sea of Cortéz (Murphy and Ottley, 1984). It ranges eastward into the Chihuahuan Desert of northern Chihuahua, Mexico, and down the Rio Grande to the Big Bend region of Texas and adjacent Coahuila. There are apparently disjunct populations in central Coahuila (Cuatro Ciènegas) and northwestern Durango and adjacent Coahuila (Bolsón de Mapimí). It also ranges up the Colorado River basin to the western edge of Colorado.

In New Mexico, most records are from the Rio Grande Valley as far north as the vicinity of Bernalillo, Sandoval County, and in southwestern New Mexico south of the Mogollon Rim to the Arizona border. It also occurs in the Tularosa Basin. It appears to be rare in northern New Mexico (Bragg and Dundee, 1949 [1950]); there are isolated records southeast of Santa Fe and in the Pecos River drainage below Pecos, as well as farther downstream near the Texas border. A single record from the San Juan Basin confirms the presence of the species in the northwestern corner of the state. Further distributional work on this species is needed.

Description: *Gambelia wislizenii* is a large species with a maximum SVL of 132 mm (Conant and Collins, 1991). Females are larger than males; Mitchell (1984) measured males averaging 95.8 mm SVL and weighing 25.8 gm, and females averaging 110.3 mm SVL and 47.0 gm. The head is large and distinctly set off from the body by a narrow neck. Body scales are minute and not imbricate. The head scales are small and granular, and there are no supraorbital semicircles. There are generally 3–5 rows of small scales in the interorbital area. There is a single preocular scale, a long subocular plate, and three postocular scales, all of which may be keeled. One or more transverse gular folds are present. The ventral scales are larger than the dorsals. They are imbricate and sometimes keeled. There are 14–25 femoral pores on each leg. Males have enlarged postanal scales.

The basic ground color of the dorsal surface of the body is tan, grayish-brown, or dark brown and varies geographically. The color pattern of hatchlings consists of up to 11 transverse narrow light bands on the dorsal surface of the body, with the anterior one in the occipital region and the posterior one between the hind legs (Tanner and Banta, 1963; McCoy, 1967). These bands continue onto the tail as light rings. The top of the head has a series of contrasting transverse black and white bars that extend down either side to the throat. There are up to six longitudinal rows of as many as nine reddish-brown spots each on the back, arranged in bilaterally symmetrical pairs. The spots are largest in the row on either side of the midline, and the transverse rows of spots produced by this arrangement alternate with the light-colored transverse bars. As *G. wislizenii* grows this pattern fades and becomes obscured as new dorsal spots are added irregularly to the dorsal pattern and the black-and-white head pattern disappears altogether except on the throat. The normal ground color of ovulating and gravid females is supplanted by a suffusion of red-orange coloration on the sides of the face, body, and ventral surface of the tail. This change is temporary and under the control of hormones associated with the reproductive cycle (Medica et al., 1973).

Similar species: *Crotaphytus collaris* has two black neck bands. All spiny lizards (*Sceloporus*) have larger, distinctly keeled, imbricate, and sometimes mucronate dorsal scales, and lack the transverse spotted and banded dorsal pattern.

Systematics: Five subspecies are recognized (Smith and Smith, 1976; Collins, 1990); two occur in New Mexico. *Gambelia w. wislizenii* occurs throughout most of the state. *Gambelia w. punctata* (Tanner and Banta, 1963) occurs in northwestern New Mexico. The holotype (BYU 20928) is an adult female collected by Wilmer W. Tanner from "the Yellow Cat Mining District approximately 10 miles south of U.S. Highway 50–6, Grand County, Utah" on 28 June 1961.

Habitat: Mitchell (1984) found this species much more abundant on rocky, well-drained substrates than on sandy soils in southeastern Arizona, although L. J. Vitt (pers. comm.) found it common in sandy areas on the Willcox Playa. McCoy (1967) reported that leopard lizards prefer friable sandy or sandy-clay soils with frequent clumps of woody vegetation such as *Sarcobatus vermiculatus*, *Rhus trilobata*, and *Artemisia* spp. in southwestern Colorado; in northern New Mexico, it is often associated with rocks. In southern New Mexico and western Texas, *G. wislizenii* is invariably found in sandy flatlands, loose sandy basins, or low, gently rolling sand dunes, all with sparse vegetative cover of such characteristic plants as *Prosopis glandulosa*, *Larrea tridentata*, *Acacia* spp., *Quercus havardii*, *Gutierrezia sarothrae*, *Yucca glauca* and *Ephedra* spp. (Tinkle, 1959; Milstead and Tinkle, 1969; Whitford and Creusere, 1977; our obs.). Elevational limits in New Mexico, range from 900 m in Eddy County to about 2100 m in Santa Fe County. These lizards include packrat middens as shelters. Other lizards found sympatrically include *Phrynosoma cornutum*, *Uta stansburiana* and *Cnemidophorus tigris*.

Behavior: Adult *G. wislizenii* are active in southwestern Colorado from mid-May to early August (McCoy, 1967) while hatchlings remain active through September. Adults and hatchlings first appear in May and early August in west Texas (Tinkle, 1959). Daily activity in Colorado is bimodal during the summer; most above ground activity occurs prior to 1300 when soil temperatures are below 50°C and then again in the evening (McCoy, 1967). Creusere and Whitford (1982) found daily activity to be bimodal in southern New Mexico, with peak activity around 1000 and 1800. The activity season in southwestern New Mexico extends from April to October (Baltosser and Best, 1990). New Mexico specimens have been collected between 16 March (Bernalillo Co.) and 24 October (Doña Ana Co.). Mean body temperature for 183 active lizards was 37.4°C (Parker and Pianka, 1976). Individual lizards are able to withstand supercooling to –6.7°C (Lowe et al., 1971). These lizards can be active foragers depending on season and year, and may move up to 2 m/min while foraging (Pietruszka, 1986). Juvenal males may disperse distances of 1–2 km (Parker and Pianka, 1976]. The species apparently is not territorial.

Gambelia wislizenii can be aggressive when molested, and are capable of inflicting a painful bite if handled. Aggressive behavior is temperature-dependent (Crowley and Pietruszka, 1983), and at normal active body temperatures the response to danger is to flee at high speeds, often running bipedally (Snyder, 1952). At body temperatures below 26°C defensive behavior shifts to threat postures and attempts to bite. *Gambelia wislizenii* may emit a high-pitched squeal when threatened or agitated, especially at body temperatures below 20°C (Smith, 1974; Crowley and Pietruszka, 1983).

Reproduction: In Colorado, females do not attain sexual maturity until their 3rd activity season at an age of 22 months at body sizes between 80 and 90 mm SVL (McCoy, 1967); females elsewhere are sexually mature at 95 mm SVL (Parker and Pianka, 1976). One annual clutch averaging 7.3 eggs is laid between late May and early July, and larger females produce larger clutches. This species produces small eggs relative to other lizards of similar body size (Tinkle and Hadley, 1975). Hatchlings appear in August and are 38–46 mm SVL (McCoy, 1967; Parker and Pianka, 1976). Individual females may live to be eight years old, but most reproduction in a population is accomplished by females three or four years old. Communal nesting may occur.

Males are reproductively mature by the time they reach 85 mm SVL (Mitchell, 1984). Spermatogenesis is well underway by mid-May in southern Arizona, and testes are completely regressed by the end of June.

Food Habits: Grasshoppers form a dietary staple in *G. wislizenii*. Beetles, butterflies and moths, bees, and wasps are also eaten (Pack, 1922; Hotton, 1955; McCoy, 1967; Milstead and Tinkle, 1969; Parker and Pianka,

1976; Mitchell, 1984). Mitchell (1984) reported that sphecid wasps comprised 24% by frequency of the prey eaten in southeastern Arizona over two seasons. The common name of this species may derive from its propensity to include lizards of other species as a significant component of the diet; reported lizard prey in New Mexico includes *Cnemidophorus tigris* (Lewis, 1951), *Holbrookia maculata* (Gehlbach, 1956), and *Cnemidophorus* spp. and *Sceloporus undulatus* (Little and Keller, 1937). McCoy (1967) and Milstead and Tinkle (1969) also reported *Uta stansburiana* as prey in Colorado and Texas. Other lizards recorded in the diet at various locations throughout the range include *Callisaurus draconoides*, *Sceloporus graciosus*, *S. undulatus*, and *Phrynosoma* spp., *Uta stanburiana*, and *Cnemidophorus tigris* (Stejneger, 1893; Pack, 1922; McCoy, 1967; Parker and Pianka, 1976). Cannibalism has been documented by Stejneger (1893) and Montanucci (1967).

Remarks: We follow Banta (1971) in the citation of the original description of this species.

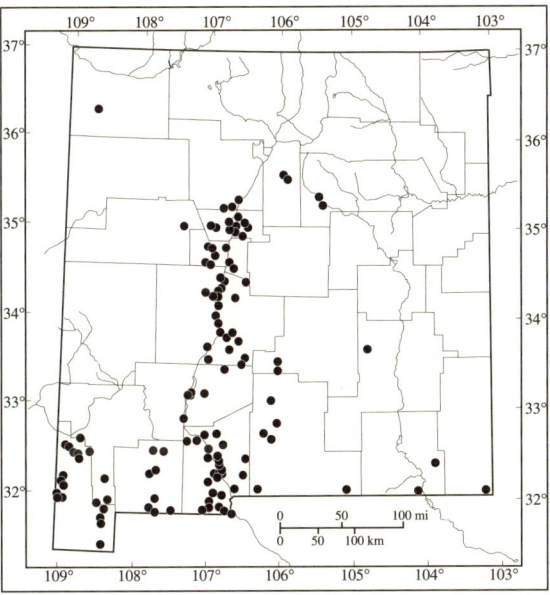

Distribution of
Gambelia wislizenii
in New Mexico.

Zebratail, Earless, Spiny, Tree, Side-blotched, and Horned lizards

FAMILY PHRYNOSOMATIDAE

This is a small New World family of 10 genera and over 100 species (Conant and Collins, 1991; Zug, 1993), recently elevated to family rank by Frost and Etheridge (1989). It occurs from coast to coast and on a number of islands, and ranges from southern Canada to Panama. They range in size between 45–175 mm SVL and come in a variety of body plans, from slender, long-tailed lizards of the genus *Urosaurus* to the spine-covered, squat-bodied horned lizards of the genus *Phrynosoma*. They occupy habitats from deserts to rain forests and are largely diurnal and terrestrial in habits, although arboreal and saxicolous members occur. Many species exhibit well-developed, stereotypical display behaviors that signal social and reproductive status. Most species lay eggs but some give birth to live young. Seven genera and 18 species occur in New Mexico.

CALLISAURUS DRACONOIDES Blainville, 1835   *Zebratail Lizard*

See Plate 39.

Type: A subadult male (MNHN 812) from "California," restricted to Cape San Lucas, Baja California [del Sur, Mexico] by Smith and Taylor (1950b), collected by P. E. Botta sometime during 1827–1829.

Distribution: This species occurs from the vicinity of Pyramid Lake in northwestern Nevada southward throughout the Mojave and Sonoran deserts of Arizona and California, the extreme southwestern corner of Utah, and the Mexican state of Sonora. It occurs along the coastal lowlands of western Mexico to the southern border of Sinaloa (Hardy and McDiarmid, 1969) and throughout the peninsula of Baja California, as well as on a number of islands in the Sea of Cortéz (Murphy and Ottley, 1984). In New Mexico, this species is found only on the western flank of the Peloncillo Mountains in the vicinity of San Simon Cienega, Hidalgo County, between 1180–1250 m. Stejneger (1893), commenting on the type locality of "New Mexico" (which included Arizona at the time) for *C. d. ventralis*, doubted the authenticity of reports from New Mexico. It was first reported from the state by Lowe (1955a).

Description: *Callisaurus draconoides* is a moderate-sized lizard; adult males are 67–99 mm SVL and females are 63–87 mm SVL (Pianka and Parker, 1972). This species is morphologically adapted for living on sandy or loose, friable soils. The legs are relatively long and the body thin. External ear openings are present. The dorsal scales are small and the ventral scales larger; all are smooth and grade into each other laterally. The dorsal head scales are numerous, uneven in size, and smaller than the interparietal. The nasal openings are placed at the end of the canthus rostralis, and open upward. There is a large subocular scale. The upper labials are large and keeled. The posterior margin of each one, which is flared and overlaps the next scale, extends backward diagonally with respect to the line of the mouth. The lower labials are similar morphologically but less pronounced. The mental scale is much reduced and is smaller than the adjacent lower labials. A gular fold is present. There are 14–23 femoral pores on each leg, and males have enlarged postanal scales.

The dorsal ground color is light gray to gray-brown, speckled throughout with numerous minute cream to yellow spots and flecks. A longitudinal series of small dark-brown spots extends posteriorly from the shoulder on either side of the midline. These spots are distinct in juveniles but tend to disappear in adults; they approach each other and fuse posteriorly on the body and onto the tail, where they form the dorsal component of the dark tail rings. There is a distinct horizontal black line on the posterior surface of the thigh, bordered above and below with white. The dorsal limb surfaces are indistinctly barred. Several black bars extend vertically from the upper labials to meet the central dusky area of the throat. The ventral surface of the tail is distinctly barred with black, and these bars fuse laterally with the dorsal tail rings. There are two prominent black bars on each side of the body that begin laterally anterior to midbody. The anterior bar extends vertically onto the ventral surface of the body, whereas the posterior bar extends ventrolaterally into the groin region. In males these bars are enclosed within bright blue patches; in females the blue coloration is absent and the bars themselves are less distinct. During the breeding season, males may develop a greenish hue on the sides of the body, and the light areas of the throat may develop a yellowish or pinkish wash. In gravid females, the dewlap becomes pink and an orange-pink coloration develops along the sides of the body between the limbs, suffusing the dorsum until the dorsal pattern is obscured (Clarke, 1965).

Similar species: Both *Cophosaurus texanus* and *Holbrookia maculata* lack ear openings; the latter species lacks black bars on the underside of the tail. *Uta stansburiana* has only one dark blue or black mark on the side of the body just behind the front legs, and the upper labials do not overlap.

Systematics: Ten subspecies have been recognized (Smith and Smith, 1976; Collins, 1990). *C. d. ventralis* occurs in New Mexico. This taxon was originally described by Hallowell (1852) as a different species in its own genus. The holotype, USNM 2670, is an adult male, based on the original description, collected from "New

Mexico" by Dr. S.W. Woodhouse during the exploration of the Zuni and Colorado Rivers, sometime during 1851. The type locality was erroneously restricted by Smith and Taylor (1950b) to Tucson, [Pima County], Arizona (see Remarks).

Habitat: *Callisaurus draconoides* is almost always found active in open sandy, gravelly, and occasionally rocky areas with little or no vegetative cover (Pianka and Parker, 1972; Smith et al., 1987). It is associated with open ground within *Prosopis-Acacia* (Vitt and Ohmart, 1977a) or *Larrea tridentata* (Lowe, 1955a) vegetation communities. In New Mexico, this species occurs abundantly on relatively open ground, dominated by vegetation such as *Larrea tridentata, Prosopis glandulosa, Gutierrezia sarothrae, Flourensia cernua,* and *Zinnia acerosa* (Baltosser and Best, 1990). It occurs sympatrically with *Cophosaurus texanus* 1.1 km S. Cienaga Peak in Hidalgo County.

Behavior: Pianka and Parker (1972) reported the average body temperature of 352 active lizards as 39.1°C, with lizards in the shade, when first sighted, averaging one degree higher than those in the sun. Lizards do not emerge until mid-morning in the spring, but become active as early as 0730 in the summer. Smith et al. (1987) reported a bimodal diurnal activity period, prior to 1200 and again in the evening, for this species in Mohave County, Arizona. The average preferred body temperature was 38.2°C. Clarke (1965) recorded a mean PBT of 39.2°C (range 34.2–41.6) with a preferred activity range of 39.0–41.6°C. This species is sedentary and spends as little as 1.5% of its daily activity periods in motion in southern California (Anderson and Karasov, 1981). Males are significantly more active than females. Anderson and Karasov (1981) recorded active body temperatures of 40.4°C, at which times field metabolic rates were 30% above maintenance levels and only 23.5% of the total energetic field costs were allocated to movement. This species is active from March to October in New Mexico, with occasional individuals active into November on warm days (Baltosser and Best, 1990). The 28 New Mexico specimens available to us were collected between 2 April and 4 October.

Territorial displays in this species, including head bobbing, leg flexion, and lateral compression of the body, are given by both sexes. These displays are innate, as Clarke (1965) observed them in hatchlings five minutes old. These lizards also frequently curl their tails over their heads as part of their display behavior, exposing the black-and-white banded pattern on the ventral surface. This "tail-wagging" behavior may serve a predator-deterrence function by alerting potential predators to the fact that they have been seen by the displaying lizards (Hasson et al., 1989). These lizards depend upon crypsis and then flight to escape predators. Bulova (1994) demonstrated that escape behavior in this species is significantly influenced by complex interactions among a number of environmental variables, including air temperature, vegetative cover, wind speed, and time of day. *Callisaurus* were more wary than sympatric *Cophosaurus* at San Simon Cienega in Hidalgo County.

Reproduction: In *C. draconoides,* the minimum size at sexual maturity is probably 60 mm SVL for both sexes (Smith et al., 1987) and is achieved during the reproductive season following first hibernation. Mating by older lizards begins in April or May, and females contain oviductal eggs from May through August. Individual females can produce more than one clutch, and clutch size is significantly correlated with body size (Smith et al., 1987). Mean clutch size is 4–5 (1–8) eggs (Pianka and Parker, 1972; Vitt and Ohmart, 1977a; Packard et al., 1980; Smith et al., 1987). Hatchlings may appear from mid-July through the end of November (Smith et al., 1987); the timing of reproductive events in this species is significantly tied to short-term cycles in environmental productivity.

Packard et al. (1980) reported an incubation period of 30–32 days for this species. As is the case for many species of reptiles, the eggs gain water during incubation and die if the soil/nest site in which they are laid is too dry. Favorable hydric conditions, on the other hand, lead to slightly longer incubation times and larger hatchlings, which implies that nest-site selection by female lizards is an important life-history trait.

Male lizards emerge in March and spermatogenesis proceeds until testes reach maximum size in April and May (Vitt and Ohmart, 1977a) or July and early August (Pianka and Parker, 1972). Male reproductive activity continues through the summer, but by August testicular regression is advanced and spermatogenesis is markedly reduced. Fat body weights peak sharply in July and decrease thereafter (Vitt and Ohmart, 1977a).

Food Habits: *Callisaurus draconoides* is an opportunistic, sit-and-wait insectivore, and the diet can vary with the season and location studied. Hotton (1955) reported bees and wasps, robberflies (Asilidae), and caterpillars to be the most important components of the diet, followed by grasshoppers and spiders. Beetles were occasionally taken, and ants rarely so. Pianka and Parker (1972) found adult beetles (30% by volume) and grasshoppers (19% by volume) to be important components of the diet. Grasshoppers were an important and staple food item throughout the season during a study in Mohave County, Arizona (Smith et al., 1987), followed by lepidopterans and dipterans as available.

Remarks: Sitgreaves' (1853) expedition, during which the holotype of *C. d. ventralis* was collected, traveled nowhere near Tucson, Arizona, the restricted type locality given by Smith and Taylor (1950b). An examination of the map accompanying the Sitgreaves report clearly shows the expedition following the Little Colorado River to just north of present day Winslow and then proceeding westward along a line near and to the south of present day Flagstaff and Kingman to the Colorado River near the California/Nevada border. The expedition traveled south along the Colorado River from that point to Fort Yuma in California, and the holotype was probably collected somewhere along this southern journey.

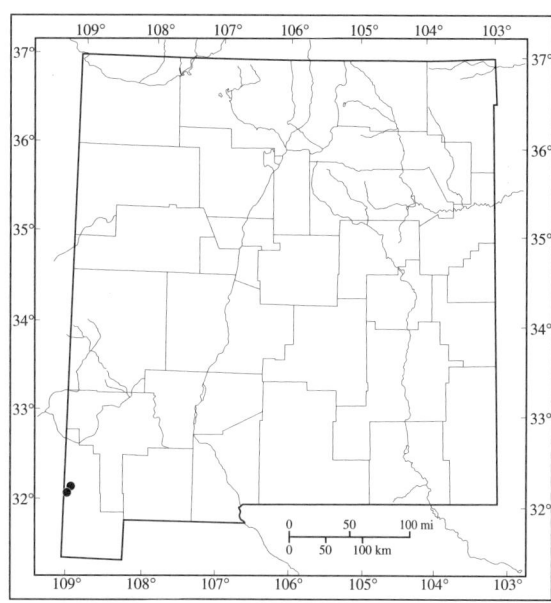

Distribution of Callisaurus draconoides in New Mexico.

COPHOSAURUS TEXANUS Troschel, 1850 [1852]   *Greater Earless Lizard*

See Plate 40.

Type: The type specimen, an adult male from "der deutschen Colonie Neubraunfels an der Guadalupe im westlichen Texas, unter 28° Nordl. Br.," was in the University of Bonn Museum, which was totally destroyed during World War II. Peters (1951) designated a neotype (UMMZ 100811), an adult male, collected from the northeastern edge of the city of New Braunfels, Comal County, Texas, by John Werler on 11 October 1949.

Distribution: *Cophosaurus texanus* is widespread in Texas west of the Balcones Escarpment northward and westward to the Red River, the Caprock Escarpment, and the Canadian River. It ranges westward through Trans-Pecos Texas, southern New Mexico, and along the southern margin of the Mogollon Rim into north-central Arizona. Southward in Mexico it ranges from northeastern Sonora throughout the Chihuahuan Desert as defined by Morafka (1977) to northern Zacatecas and San Luis Potosí, and western Tamaulipas.

The distribution in New Mexico as reflected by specimen records is curious. Locality records largely conform to the regional distribution of thermal, light-colored soils (Maker et al., 1978), and it occurs throughout the Chihuahuan Desert in the southern part of the state. The species ascends the two major rivers traversing this region; along the Rio Grande to the vicinity of Belen and along the Pecos to the vicinity of Sumner Lake on the DeBaca-Guadalupe county line. It occurs along a number of tributary drainages to these main streams, as well as others such as the Gila and San Francisco rivers in the western part of the state. Whether the records around Conchas Lake in San Miguel County represent a natural distribution suggesting stream capture by the Canadian River drainage, or an introduction, remains an interesting dilemma at this point, perhaps to be clarified if additional specimens are found eastward into the Texas Panhandle. The paucity of records in the middle part of the Pecos River, along with the Otero Mesa and surrounding areas south of the Sacramento Mountains, is puzzling, and we attribute this to lack of collecting effort or collector bias rather than the real absence of this species. In New Mexico, *C. texanus* ranges from 900 m to about 2100 m.

Description: *Cophosaurus texanus* is a moderate-sized lizard, with a maximum SVL of 88 mm for males and 78 mm for females. Males and females from a population in Big Bend National Park averaged 77 mm and 64 mm SVL (Howland, 1992). During 1987–92, 1,481 specimens collected at Antelope Pass in Hidalgo County averaged 55.8 (16–80) mm SVL and 6.5 (0.3–22) gm. Four hundred sixty-four adult males averaged 68.9 (60–80) mm SVL and 10.5 (4.5–22) gm; 334 adult females averaged 62 (55–77) mm SVL and 7.9 (3.2–13.5) gm (our unpubl. data). The legs are relatively long and the body robust. There are no external ear openings. The dorsal scales are small and flat, slightly larger along the midline. The ventral scales are slightly larger, and grade into the dorsal scales laterally. The head scales are small and uneven in size; some supraoculars are slightly enlarged. The nasal openings are placed at the end of and slightly medial to the canthus rostralis, and open upward. There are seven upper labials, which are large and keeled, overlap slightly, flare at the free edges, and project diagonally backward. The lower jaw is flat and fits flush underneath the upper labial scales. A gular fold is present. There are 10–19 femoral pores on each leg, and males have enlarged postanal scales.

The dorsal ground color is generally light gray with a longitudinal series of prominent paired black vertebral spots, which often extend onto the tail to form diffuse transverse bands. Ground color may conform somewhat to local substrate color (Cagle, 1950). The sides of the body exhibit a reticulate pattern composed of a black ground color with numerous red, yellow, and orange spots throughout. This reticulation extends onto the sides of the head, and the throat is blackish or with irregular transverse black bars extending slightly backward. The dorsum of males may be suffused with a bright pinkish hue anteriorly and bright green posteriorly during the spring breeding season (our obs.; J. F. Scudday, pers. comm.). There is a pair of black vertical bars on either side of the body just anterior to the insertion of the hind limbs. These bars are embedded in an area of blue or green coloration that is extensive and brilliantly hued in males, and that extends onto the ventral surface from each side and almost come into contact. In females, these bars are small and seldom extend onto the ventral surface or are reduced to spots, and the surrounding coloration, when present, is dark blue. The ventral surface of the tail in both sexes has distinct black transverse bars, which are most prominent on the posterior half. These bars may fuse laterally with the dorsal tail bands. Gravid females develop a pink dewlap and an orange-pink coloration along the sides of the body between the limbs (Clarke, 1965). This coloration may suffuse the dorsal surface until the pattern thereon is obscured.

Similar species: *Callisaurus draconoides* has ear openings. *Holbrookia maculata* lacks black bars on the underside of the tail. In addition, both have the lateral black bars beginning anterior to the midbody rather than in the groin region. *Uta stansburiana* has only one black mark on the side of the body just behind the front legs, and the upper labials do not overlap.

Systematics: Three subspecies of *C. texanus* have been described (Peters, 1951). The infraspecific taxonomy of this species in New Mexico bears investigation. Peters (1951) considered specimens from the southeastern part of the state intergrades between the eastern nominotypical subspecies and the western subspecies *C. t. scitulus* (Peters, 1951). Conant and Collins (1991) considered only the latter form to occur in New Mexico. The type specimen (UMMZ 100818), an adult male, was collected "on rocks at the mouth of a small arroyo entering Cañada del Oro, 16 miles north of Tucson, Pima County, Arizona" by Howard K. Gloyd on 12 August 1930.

Habitat: *Cophosaurus texanus* prefers gravelly to rocky substrates with scattered to moderate shrubby vegetative cover, particularly along floodplains, arroyo edges, and similar topographic features, at elevations between 250–1545 m (Peters, 1951; Nickerson and Mays, 1969 [1970]; Whitford and Creusere, 1977; Smith et al., 1987; our unpubl. data). Whitford and Creusere (1977) and Price et al. (1993) estimated densities of 10–20 and 5–15 lizards/hectare during separate four-year periods near Las Cruces, New Mexico, and Howland (1992) reported a density of 107 lizards/hectare in Big Bend National Park, Texas.

Behavior: New Mexico specimens have been collected between 3 March (Grant Co.) and 22 October (Sierra Co.). The activity season extends from late March through the end of September in southern New Mexico, with adults generally disappearing by mid-August. We have taken adult lizards in pitfall traps during late snows in April. Creusere and Whitford (1982) found lizards active throughout the day in southern New Mexico. Smith et al. (1987) observed a bimodal daily activity period during summer in Arizona; lizards were active between 0900–1300 and again after 1700. The preferred body temperature (PBT) for active lizards averaged 38.5°C. Barbault (1977) observed lizards active throughout the day (between 0700–1900), with most activity between 0900–1700. The body temperature of 49 active lizards averaged 36.7°C. Clarke (1965) measured a mean PBT of 38.3°C (31.8–40.9°C) with a preferred activity range of 38.1–40.9°C. Individual lizards are able to withstand supercooling to –6.65°C (Lowe et al., 1971).

This species has a characteristic territorial display, given by both sexes, which includes head bobbing, leg flexion, and lateral body compression. These behaviors are innate, as Clarke (1965) observed them in 5-minute old hatchlings. These lizards also frequently curl their tails up over their heads as part of their display behavior, exposing the black-and-white banding pattern on the ventral surface. These lizards depend upon crypsis and then flight to escape predators. Bulova (1994) demonstrated that escape behavior in this species is significantly influenced by complex interactions among a number of environmental variables, including air temperature, vegetative cover, wind speed, and time of day. *Cophosaurus* were less wary than sympatric *Callisaurus* at San Simon Cienega in Hidalgo County. Tail-curling displays are often given by these lizards when pursued by potential predators. These lizards lose their tails readily, but there are trade-offs; lizards which lose their tails suffer a significant decrease in speed compared to lizards with intact tails (Punzo, 1982).

Reproduction: Smith et al. (1987) reported that both sexes matured after their first hibernation, females at 55 mm SVL. The reproductive season lasted from April through August in this Arizona population, and individual females produced more than one clutch. Clutch size averaged 3.55 (2–5) eggs and was not related to body size. Female lizards in Texas become sexually mature at 10–12 months and a size of 50–55 mm SVL (Ballinger et al., 1972; Howland, 1992). Older females produced up to four clutches during the reproductive season, which lasts from early May through the end of August, and produced their first clutch earlier in the season than did younger females. Older females also produced larger clutches than did younger females, adding approximately one egg for each four mm increment in SVL (Ballinger et al., 1972). Gravid females were observed between 11 May and 4 June at Elephant Butte Reservoir in Sierra County, New Mexico (Sugg et al., 1995), with oviposition ending by mid-July.

Eggs are about twice as long as wide, and may represent up to 22% of an individual's body weight prior to oviposition (Ballinger et al., 1972). The average clutch size for 70 clutches from a population in west-central Texas was 6.10 eggs (Ballinger et al., 1972), whereas the average from six different locations in Big Bend National Park ranged from 2.82–4.20, with an overall range of 2–9 eggs (Howland, 1992). The average size of four clutches from Sierra County, New Mexico, was 3.25 eggs (Sugg et al., 1995). The incubation period may be about 50 days (Cagle, 1950). Hatchlings (average size 20–25 mm SVL) first appear during July in Texas and New Mexico (Ballinger et al., 1972; Parker, 1973b; Howland, 1992; our obs.). Males grow faster than females after their first winter until adulthood, whereupon growth rates between the sexes equalized. This discrepancy is due in part to differences in the timing of energy allocation to growth and reproduction between the sexes (Sugg et al., 1995), with females diverting energy to reproduction earlier in their lives.

The fat body cycle in female *C. texanus* parallels the reproductive cycle (Ballinger et al., 1972). Fat bodies are largest upon spring emergence from hibernation; they decline to minimal values from late May through early July, and then increase prior to hibernation.

Males mature at a minimum size of 60 mm SVL following their first hibernation (Parker, 1973b; Smith et al., 1987). Parker (1973b) reported spermatozoa in the reproductive tracts of males collected between May and July, but not in males collected in August and September. Testes reached maximum size in mid-May following April emergence of lizards in a population in west-central Texas (Schrank and Ballinger, 1973). Fat bodies were essentially depleted by the beginning of June, and began to be replenished in July after spermiogenesis ceased. Testicular regression is complete by August.

Sugg et al. (1995) reported that only two of 115 marked females (1.7%) and 11.2% of males survived to their second reproductive season. We have the following data for 725 lizards (398 males, 327 females) of known age marked during a 7-year period near Las Cruces: 272 (155 males, 117 females) did not survive to sexual maturity, 271 (135 males, 136 females) survived to their first reproductive season, 134 (71 males, 63 females) survived to their second reproductive season, 39 (29 males, 10 females) survived to their third reproductive season, six (all males) survived to their fourth reproductive season, and three (two males, one female) survived to their fifth reproductive season.

Food Habits: *Cophosaurus texanus* is an opportunistic insectivore. Grasshoppers are the primary food source throughout the year (Hotton, 1955; Smith et al., 1987). Spiders, butterflies, and moths are next in importance, followed by bees and wasps, caterpillars, and leafbugs (Lygaeidae and Miridae). Beetles and ants are occasionally taken. Barbault and Maury (1981) reported the following prey categories (by volume) to vary in importance during different seasons over a 3-year period: lepidopteran larvae (4.6%–44.2%), ants (0.3%–23.5%), other hymenopterans (0.9%–32.2%), hemipterans (1.0%–41.3%), dipterans (0.2%–32.9%), and orthopterans (5.4%–23.3%). Maury (1995) found spiders, hymenopterans, and lepidopteran larvae to be the most important prey items over a 2-year period at the same site in Durango, Mexico, 10 years later. She found pronounced effects of prey availability on seasonal and annual variation in prey consumed by this species during the period encompassed by these two studies, which in turn was correlated with the effect of precipitation on primary productivity. Prey size is significantly correlated with lizard body size and lizard jaw length (Maury, 1995).

Whiting et al. (1991) reported a hatchling *C. texanus* being preyed upon by a lycosid spider. Known predators in New Mexico include *Crotaphytus collaris*, *Masticophis flagellum*, and *Hypsiglena torquata*.

Remarks: Ramsey (1948) found five adult female lizards in the same hibernaculum during December. Goldberg and Bursey (1992b) found helminth parasites in 11% of the lizards they examined from southern Arizona.

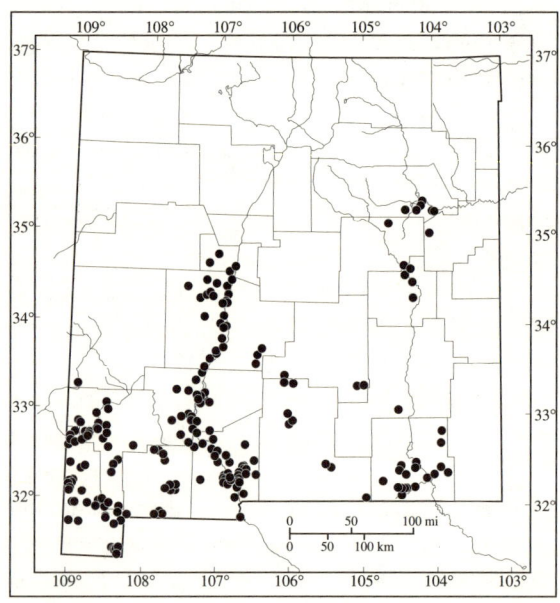

Distribution of Cophosaurus texanus *in New Mexico.*

HOLBROOKIA MACULATA Girard, 1851 — *Lesser Earless Lizard*

See Plate 41.

Type: No type material was designated in the original description (Girard, 1851), and none subsequently. The material used was collected by William Tappan from "opposite Grand Island, Platte River, [Buffalo County, Nebraska]" on 28 June 1848.

Distribution: *Holbrookia maculata* ranges widely throughout the Great Plains of the western United States east of the Rocky Mountains, from southern South Dakota southward to central Texas, and westward into Arizona. It occurs southward between the eastern and western axes of the Sierra Madre to the Mexican states of Jalisco, Guanajuato, and San Luis Potosí (Smith and Taylor, 1950a), with the eastern margin circumscribed by the limits of the Chihuahuan Desert as defined by Morafka (1977). Farther westward, it occurs from southern Arizona southward through the Sonoran Desert and at least to the southern border of Sinaloa along the coastal lowlands (Hardy and McDiarmid, 1969). The species occurs in suitable habitat virtually throughout New Mexico, with the exception of the high northern plateaus in Rio Arriba and Taos counties. The population on the white sands (White Sands National Monument) of Doña Ana and Otero counties appears to be isolated from other populations (Dixon, 1967). Elevational range is from approximately 950–2200 m.

Description: This is a moderate-sized lizard, with a maximum SVL of 75 mm for males and 70 mm for females. Gehlbach (1965) recorded adult females between 41–57 mm SVL in northwestern New Mexico, and Sena (1978) recorded a maximum SVL of 60 mm for a population in eastern New Mexico. During 1992–95, 1061 *H. maculata* collected on the Mescalero Sands in Chaves County averaged 52 (21.3–69.5) mm SVL; 542 adult males averaged 55.3 (44.1–63.8) mm SVL and 4.7 (2.2–6.9) gm; 409 adult females averaged 55.3 (45.2–69.6) mm SVL and 5 (2.3–9.1) gm (L.W. Gorum, unpubl. data). This species is morphologically adapted to live on and beneath the surface of sandy or loose, friable soils. The legs are relatively short and the body stout. The head is broad, very short, and convex. There are no external ear openings. The tail is relatively short, except in specimens from extreme southwestern New Mexico. The dorsal scales are small and flat, the ventral scales are larger, and the two series grade into each other laterally. The head scales are small and uneven in size. The nasal openings are placed at the end of and medial to the canthus rostralis, and open upward. The upper labials overlap, are flared at their free ends, and are separated by diagonal furrows. A complete row separates the supralabials from the subocular scales. A gular fold is present. There are 7 to 16 femoral pores on each leg, and males have enlarged postanal scales.

The dorsal ground color in local populations tends to match the prevailing substrate color (Gehlbach, 1965; our obs.), and varies from almost pure white on the white sands (White Sands National Monument) of Doña Ana and Otero counties to dark brown elsewhere. Lizards from east of the Pecos River tend to be boldly marked with a broad light longitudinal middorsal band and a narrower dorsolateral light stripe extending from the eye to the base of the tail. The intervening darker fields and those laterally are often heavily speckled, and have a longitudinal series of dark brown blotches edged posteriorly with white. The dorsolateral series is most distinct, and often extends onto the tail. In lizards west of the Pecos River the light middorsal band usually remains distinct, but the remainder of the dorsal pattern may disappear entirely, leaving numerous light spots and/or flecks on the dorsal ground color. All lizards of this species have a pair of vertical black bars, sometimes edged with blue, on either side about midbody. These bars may be relatively indistinct in females. The ventral surfaces of both sexes are creamy white, although the throats of males may be suffused with gray or manifest a mottled or barred pattern. There are no markings on the underside of the tail. Gravid females develop an orange-pink to bright crimson coloration that permeates the dorsal ground color (Clarke, 1965; Dixon, 1967).

Similar Species: Both *Cophosaurus texanus* and *Callisaurus draconoides* have distinctive black bars on the underside of the tail, and the latter species also has ear openings. *Uta stansburiana* has only one black mark on the side of the body just behind the front legs, the upper labials do not overlap, and ear openings are present.

Systematics: The phylogenetic relationships of populations of this species are badly in need of revision. Ten subspecies are recognized (Smith and Smith, 1976; Collins, 1990), two or three of which occur in New Mexico in addition to the nominotypical subspecies. The long-tailed form in southwestern New Mexico has been referred to *H. m. elegans* Bocourt, 1874, although this may not be a valid taxon [see Ruthven (1907), Barbour (1921), Schmidt (1922), and Smith (1935) for a flavor of the debate]. It was originally described as a distinct species. The type specimen, an adult female (MNHN, number unknown), was collected from "Mazatlan, [Sinaloa] (Mexique)," collector and date of collection unknown.

Holbrookia m. approximans was originally described as a full species by Baird (1858 [1859]), who gave the type locality as "lower Rio Grande" (and "Tamaulipas" in *Reptiles of the Boundary*, published in 1859). The collector was Lt. D. N. Couch, a member of the United States Army expedition led by Major William H. Emory, Corps of Topographical Engineers, to survey the U.S. and Mexican boundary in 1854 and 1855. No type material was originally designated. Axtell (1958) designated an adult male specimen (USNM 2664A) as the lectotype and restricted the type locality to near Alamo de Parras or Parras de la Fuente, 25°25'N, 102°11'W, Coahuila, Mexico.

Holbrookia m. ruthveni was described by Smith (1943) based on a series of 11 lizards "from White Sands, about twelve miles southwest of Alamogordo, [Otero County], New Mexico." The holotype (FMNH 29452), an adult female, was collected by Wilfred H. Osgood on 26 May 1938.

Habitat: This is a species adapted to relatively level terrain with sparse, low-lying vegetative cover and loose, friable soils (Jameson and Flury, 1949; Harris, 1963; Ballinger and Jones, 1985). In northwestern New Mexico, Gehlbach (1965) reported this species in short-grass or mixed-grass habitats between 1758–2182 m where *Bouteloua gracilis* is dominant or codominant with other grasses such as *B. hirsuta* or *Hilaria jamesii* and sparse shrub cover of *Chrysothamnus nauseosus*, *Gutierrezia sarothrae*, and *Yucca glauca*. In eastern New Mexico, Gennaro (1972) reported that this species inhabits relatively level terrain adjacent to sand dunes, with sparse vegetation such as *Aristida longiseta*, *Artemisia filifolia*, *Cenchrus pauciflorus*, *Gutierrezia sarothrae*, *Helianthus* sp., *Munroa squarrosa*, and *Sporobolus cryptandrus*. This lizard can be abundant in prairie dog towns (Clark et al., 1982).

Spring and summer densities of 20.1 and 18.1 lizards/hectare have been reported for a population in western Nebraska (Jones and Ballinger, 1987). This population declined over a 7-year period from 25 to less than ten per hectare, possibly due to changes in vegetation structure on the study site. Home range sizes of 0.06 and 0.1 hectares for females and males have been reported by Gennaro (1972); Jones and Droge (1980) reported similar values (0.04 and 0.1) in western Nebraska.

Behavior: Clarke (1965) recorded a mean preferred body temperature of 35.7°C (27.4–40.8) with most lizards active between 33.2°C and 40.8°C. On White Sands National Monument, Dixon (1967) found body temperatures of 176 active lizards ranging between 34.5°C and 39°C from June through August. Lizards began seeking shade when soil temperatures reached 36–38°C and became inactive when soil temperature reached 40°C, resulting in a bimodal activity period during most summers, with lizards active in the morning and again in late afternoon or early evening. Sena (1978) measured the average body temperature of 174 active lizards near Portales, Roosevelt County, as 34.3°C. Mean body temperature varied significantly over the active season, ranging from 28.0°C in March to 38.6°C in August, and was significantly correlated with air and shaded substrate temperatures. Lizards basked at body temperatures at or below 29°C and foraged at body temperatures between 30–39°C. Lizards were reluctant to leave shade when ambient air temperature exceeded 39°C, and were observed to climb into vegetation at such temperatures. Gravid females may be active under otherwise unsuitable thermal conditions, and avoid lethal high temperatures by climbing off the substrate and orienting their bodies parallel to the sun's rays (Axtell, 1960). The mean critical thermal maximum for this species has been recorded at 46.5°C (Sena, 1978).

These lizards are active from early to mid-April through mid-October in western Nebraska (Jones and Ballinger, 1987), and can be active as early as mid-March and as late as mid-November. New Mexico specimens have been collected between 5 March (Grant Co.) and 21 November (DeBaca Co.). Individuals are very often active in the open, far from any vegetation, and will take shelter by burying themselves in the sand. Nocturnal activity has been noted by Woodin (1953) in southern Arizona, and we collected an alert individual on a paved road just before midnight on 21 July 1985. This species exhibits characteristic territorial display behavior, given by both sexes, which includes head bobbing, leg flexion, and lateral compression of the body. This behavior is innate, as Clarke (1965) observed it in hatchlings five minutes old.

Reproduction: Tinkle et al. (1970) reported 45 mm SVL as the minimum size at maturity for females in western Texas. Gennaro (1974) reported minimum size at maturity for male and female lizards of this species in eastern New Mexico as 44 and 45 mm SVL achieved the year following hatching. Individual female reproductive effort may vary with age (Jones and Ballinger, 1987). Clutch size increases approximately by one egg for each 5 mm increment in SVL. Average clutch sizes of 2.98, 4.70, 5.27, and 6.1 eggs have been reported (Tinkle et al., 1970; Jones and Ballinger, 1987). Clutch weight may represent 23% of female body weight prior to oviposition (Parker, 1973b).

Gehlbach (1965) reported gravid females from mid-May through mid-July. Females in the Tularosa Basin (White Sands National Monument) are gravid from mid-June to mid-July with oviposition observed during one year between July 8–16 (Dixon, 1967). The first appearance of hatchlings varies yearly and with geographic location, but can occur from the beginning of July through early September (Dixon, 1967; Gennaro, 1974; Jones and Ballinger, 1987). Hatchlings are about 25 mm SVL, grow slowly prior to hibernation, and do not reach minimum reproductive size until about ten months of age.

Testes of male *H. maculata* are enlarged at spring emergence and remain so through the end of June (Jones and Ballinger, 1987). Enlarged testes were found in lizards as young as seven months of age (34 mm SVL) Males from southern Doña Ana County, New Mexico, possessed mature spermatozoa from May through July (Parker, 1973b).

Maximum ages of four and five years in the field have been reported for females and males (Jones and Ballinger, 1987). Yearling and 2-year-old lizards contributed 31% and 52% of the net replacement rate of this population in western Nebraska over a 7-year period. Gennaro (1974) reported an annual turnover of 80% in an eastern New Mexico population and longevities of three years to be rare.

Food Habits: Burt (1928) reported grasshoppers to be 43% and hemipterans 27% of the diet in Kansas, with butterflies and moths, spiders, and hymenopterans contributing about 10% each. On White Sands National Monument, lizards forage in open spaces between vegetation (Dixon and Medica, 1966); coleopterans and hymenopterans were the most important food items consumed (each 22% by volume), followed by heteropterans (13%) and lepidopterans (7%). Hatchling lizards (*Sceloporus undulatus* and *Cnemidophorus inornatus*) were also eaten (6% by volume). Barbault and Maury (1981) recorded termites (13–43%), ants (1–18%), adult coleopterans (6–40%), lepidopteran larvae (3–52%), orthopterans (29–31%), and neuropteran larvae (11%) as important foods by volume in different seasons over a 2-year period in northeastern Durango, Mexico.

These lizards have been observed drinking water in captivity (Meyer, 1966), although it is assumed that metabolic water combined with behavioral mechanisms allow this species to avoid desiccation under natural circumstances.

Known predators in New Mexico include *Gambelia wislizenii*.

Remarks: The name *maculata* (Latin, "stained") refers to the blotched dorsal pattern of the nominotypical race of this species. Goldberg and Bursey (1992b) found helminth parasites in 13% of the lizards they examined from southern Arizona.

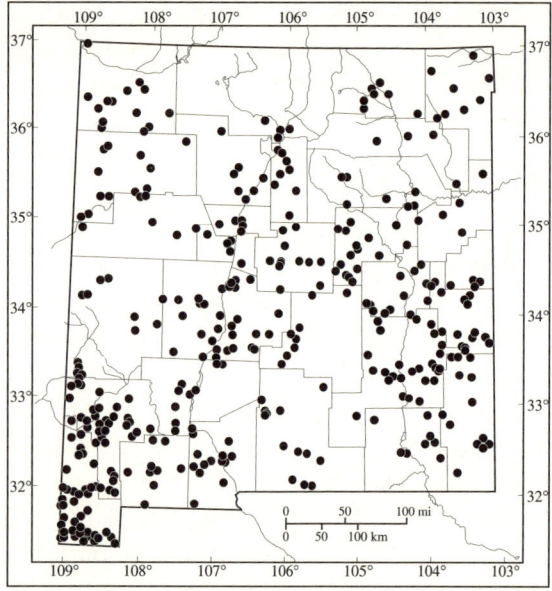

Distribution of Holbrookia maculata in New Mexico.

PHRYNOSOMA CORNUTUM (Harlan, 1825) — *Texas Horned Lizard*

See Plate 42.

Type: No type material was designated in the original description (see Remarks). The type locality was given as the "Plains of Arkansas," and restricted to Fort Riley, Geary County, Kansas, by Smith and Taylor (1950a,b).

Distribution: This species ranges from central Kansas southward through Texas, and from the Missouri/Oklahoma border westward to the southeastern corner of Arizona. It ranges southward in Mexico east of the Sierra Madre Occidental to eastern Durango, and eastward across Zacatecas and San Luis Potosí to the border between Tamaulipas and Veracruz along the coast of the Gulf of Mexico. The natural eastern boundary of the species' range may never be known with certainty; it has been widely introduced throughout most of the United States for at least 115 years, and there are established populations in Florida and South Carolina (Price, 1990).

In New Mexico, the species occurs throughout the southwestern counties south of the Mogollon Rim, and it extends northward in the Rio Grande Valley to the vicinity of Bernardo, Socorro County. East of the Rio Grande, it ranges south of the Sacramento Mountains to the Texas border, and up the Pecos River drainage to the vicinity of Santa Rosa, Guadalupe County. It is also found all along the eastern border. Elevation ranges from 900 m to approximately 2100 m. *Phrynosoma cornutum* is not known to occur naturally in the Albuquerque or Las Vegas areas (see Bragg and Dundee, 1949 [1950]); we believe specimens from these localities represent escaped pets or deliberate introductions.

Description: The morphology of this species has been described in detail by Price (1990). This is a large, sexually dimorphic species. The maximum size reported for males is 94 mm SVL (Munger, 1984a) and 130 mm SVL for females (Brown and Lucchino, 1972). The ventral scales are weakly keeled and non-mucronate. There is a series of enlarged gular scales that project horizontally. The dorsal body scales are highly heterogeneous in size and shape, with the largest variously modified as short, vertical spines. The scales on the anterior surfaces of the limbs are large, pointed, and strongly keeled. There is a single pair of well-developed, widely-spaced, acute occipital spines on the head, along with three pairs of temporal spines that are situated at or above the level of the eye. The occipital spines are usually at least twice the length of the temporal spines. There is a single small interoccipital spine. Each supraorbital ridge terminates posteriorly in a short, thick superciliary spine. A postrictal scale is absent. There are three groups of small spines on each side of the throat. A single row of spinose chinshields enlarging anteroposteriorly occurs along each margin of the lower jaw. There are two complete rows of lateral abdominal fringe scales on each side of the body, and one such row on each side of the tail. The tail is usually at least twice the length of the head. There are 8–9 femoral pores on each side, separated medially by 18–20 preanal scales. Some males have enlarged postanal scales. Hatchlings lack head spines, but they develop rapidly as individuals grow (Blaney and Kimmich, 1973).

The basic dorsal color may be brown, reddish-brown, yellow, tan, or gray, and varies with geographic location and substrate type. There is a distinct middorsal light line. A series of longitudinal dark-brown dorsal blotches occurs on either side of the midline; each is edged with white, cream, or yellow-orange posteriorly. The blotches next to the midline are most distinct. The lateral blotches may fuse to produce a series of undulating dark bars. There is a large brown patch on either side of the neck behind the occipital and temporal spines. Two or three distinct dark bars radiate from the eye, one posteriorly to the temporal spines and the remainder anteriorly and/or vertically to the line of the mouth. The ventral surface is white to dirty yellow. Xanthic pigments may intensify in gravid females.

Similar Species: *Phrynosoma solare* has four long occipital horns, the bases of which are in contact, and a single complete row of lateral abdominal fringe scales. Both *P. douglasii* and *P. modestum* lack well-developed horns and spines on the head. In addition, *P. douglasii* has a single complete row of lateral abdominal fringe scales and *P. modestum* lacks them altogether.

Systematics: No subspecies have been described. Montanucci (1987) regarded this species as a relatively highly derived member of the desert-adapted, "northern radiation" of the genus. He hypothesized the origin of the species early in the diversification of the genus, sometime during the Pliocene.

Habitat: This is a species of open deserts and grasslands up to 1830 m in elevation. It occurs on sandy to gravelly soils. Milstead and Tinkle (1969) found this species in low, gently rolling sand dunes with 20% cover of *Prosopis glandulosa*, *Gutierrezia sarothrae*, and *Yucca angustifolia* in west Texas. In southern New Mexico, it is common in *Yucca-Prosopis-Ephedra* associations around playas (Whitford and Creusere, 1977) as well as in *Larrea-Acacia-Fouquieria* associations of bajadas and mountain foothills. This species has been reported by Gehlbach (1956) to be common in the sandy, mesquite-covered floodplain of the Black River in Eddy County, along with *Sceloporus undulatus*, *Uta stansburiana*, and *Cnemidophorus inornatus*. Clark et al. (1982) recorded these lizards in association with black-tailed prairie dog towns. Whiting et al. (1993) estimated a density of 3 lizards/hectare in central Texas. Worthington (1972) recorded an unusually low figure of 1.64 lizards/hectare during July 1971 in southern Doña Ana County. Whitford and Creusere (1977) estimated densities of up to 30 lizards/hectare during a 5-year period near Las Cruces.

Behavior: New Mexico specimens have been collected between 1 April (Eddy Co.) and 27 September (Luna Co.). Lizards are active in south-central New Mexico from 0800 through 1800 during the summer (Whitford and Bryant, 1979). Activity of individuals after dark usually results from disturbance or other abnormal circumstances (Williams, 1959). We observed a 94 mm SVL male actively *in copulo* with a 96 mm SVL female during a partial solar eclipse at midday on 30 May 1984 in Doña Ana County, when activity of most other diurnal organisms had temporarily ceased. Most thermoregulatory basking takes place in the early morning or late afternoon, and most foraging activity takes place between 0900 and 1100. These lizards move an average of 46.8m/day (9–91 m). They spend the remainder of the day immobile underneath or perched in shrubs such as *Yucca elata* or *Ephedra trifurca*; we have also observed them on top of *Krameria* sp. Ambient temperatures in vegetative canopies at these times range from 35–40°C. Sheffield and Carter (1994) reported arboreal activity to 1 m, presumably in response to heavy rainfall. At night, lizards bury themselves in the soil or take refuge in rodent burrows or similar shelters. Regal (1967) reported that *P. cornutum* selects body temperatures between

10°C and 26°C at night in laboratory heat gradients, well below normal daytime activity temperatures, even when the latter temperatures are available.

Pianka and Parker (1975) reported an average active body temperature of 37.3°C, and Prieto and Whitford (1971) reported critical thermal maximum and minimum mean temperatures of 47.9°C and 9.5°C for lizards in a thermal gradient.

Munger (1984a) estimated average home range sizes of 24,013 m² and 13,775 m² for males and females in southeastern Arizona. Worthington (1972) recorded home range sizes of up to 1.08 hectares for lizards from southern Doña Ana County, New Mexico.

Horned lizards employ a sequence of crypsis, flight to the nearby shelter of vegetation or other objects, and defensive behavior involving their spiny armament to avoid and escape predators (Sherbrooke, 1981, 1987; Sherbrooke and Montanucci, 1988). Sherbrooke (1990) described drinking during rainstorms in this species, a behavior he termed "rain-harvesting." Lizards arch the body and flatten it laterally, lower the head and tail, and water is transported through interscalar capillary channels from the dorsal body surface to the mouth.

Reproduction: The minimum size at maturity for females is 68 mm SVL. Females probably do not reproduce until they are two years old (Ballinger, 1974). Yolked follicles are present from April through the end of June (Ballinger, 1974; Howard, 1974). Oviposition takes place between late May and early August (Ballinger, 1974; Howard, 1974; Pianka and Parker, 1975). We have observed gravid females during June and July in Doña Ana County. Individual females may produce multiple clutches during a season. Average clutch sizes between 23 and 30 have been recorded, with a range of 13–49. Ballinger (1974) found that larger females laid larger clutches, with approximately 1 egg added per 3 mm increment in SVL. Hatchlings (20–32 mm SVL) appear from the beginning of June through September (Howard, 1974). We have observed hatchlings (25–33 mm SVL) in August and September in Doña Ana County. Marked hatchlings recaptured the following May and June have only grown by about 10 mm SVL. Females dig nest chambers at the end of a tunnel 15–20 cm below the surface (Ramsey, 1956).

Males attain sexual maturity at a minimum size of 72 mm SVL, usually during their second spring (Howard, 1974). Males contain mature sperm in decreasing numbers from April through mid-July (Ballinger, 1974; Howard, 1974) and testes and associated reproductive structures reach maximum size in May and early June (Ballinger, 1974; Pianka and Parker, 1975). Mating occurs during this period. Testes regress rapidly in size from mid-July through August but may begin recrudescence prior to hibernation. The fat body cycle inversely follows the sex organs, with minimal values in May and maximum values in August (Ballinger, 1974).

Baur (1986) reported a life span of almost ten years in captivity.

Food Habits: Hotton (1955) reported at least 85% of the diet to be ants. Pianka and Parker (1975) stated that ants comprised 69% by frequency and 61% by volume of the diet of 351 lizards examined. In one study in southern New Mexico, both adult and juvenal lizards were found to prey almost exclusively on harvester ants of the genus *Pogonomyrmex* (Whitford and Bryant, 1979). Ants are taken a few at a time from nest disks, foraging columns and open areas, and individual lizards consume from 30 to more than 100 ants per day. Milstead and Tinkle (1969) reported the stomach contents of 3 specimens from west Texas to consist of 38% beetles, 37% grasshoppers, 21% ants and 4% hemipterans by volume.

Harvester ants respond to the loss of foraging workers by shutting down above-ground nest activity (Munger, 1984b,c), so *P. cornutum* has evolved a "prudent predator" strategy (Whitford and Bryant, 1979), taking a few ants at each feeding opportunity and then moving on to the next nest. The availability and productivity of harvester ant colonies may regulate to some degree the density of lizards of this species in a given area (Munger, 1984b,c).

Cohen and Cohen (1990) reported the ingestion of 11 blister beetles (*Megetra cancellata*), representing 15–30% of its body weight, by a single female lizard in 30 minutes. These beetles are extremely toxic to most vertebrates. Schmidt et al. (1989) demonstrated that an unknown chemical in the blood plasma of this species is able to detoxify the venom of its primary prey, *Pogonomyrmex*, at doses lethal to small mammals and other reptiles.

Known predators of this species in New Mexico include Harris hawks (*Parabuteo unicinctus*; J. N. Stuart, pers. comm.), roadrunners (*Geococcyx californianus*), and *Masticophis flagellum*. Grasshopper mice (*Onychomys* spp.) are apparently effective predators on young lizards of this species (Sherbrooke, 1991). Some mammals, such as coyotes and foxes, may be deterred by the lizards' ability to squirt blood from sinuses behind the eyes, presumably an antipredator defense (Middendorf and Sherbrooke, 1992). Munger (1986) calculated predation probabilities during one season of between 14% and 52% for this species in southeastern Arizona.

Remarks: Harlan (1825) described this species on the basis of 2 stuffed specimens in the collection of the Academy of Natural Sciences in Philadelphia (Price, 1990). These specimens (possibly ANSP 8642–3) are missing and lack catalog annotations. The first specimen of this species known to science was apparently collected by the Lewis and Clark expedition. It was brought back alive and given to Thomas Jefferson, who was an active member of the Philosophical Society of Philadelphia. He in turn presented it to the Philadelphia Academy.

Best et al. (1983) reported melanism in this species from the Pedro Armendariz lava flow in central New Mexico. The species has been reviewed by Price (1990).

This species is protected by state law from commercial and other collecting in many U.S. states, including New Mexico. It has declined or disappeared over a significant portion of its range in Oklahoma and Texas from habitat destruction, the invasion of the introduced red fire ant (*Solenopsis invicta*), continued illegal collection for the pet and curio trades, and other factors (Price, 1990; Donaldson et al., 1994). *Phrynosoma cornutum* is listed as a Category 2, Notice of Review species by the U.S. Fish and Wildlife Service (USDI, 1994).

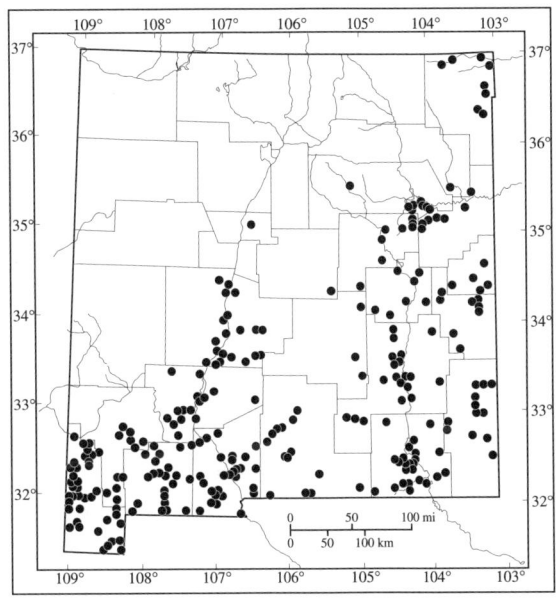

Distribution of Phrynosoma cornutum *in New Mexico.*

PHRYNOSOMA DOUGLASII (Bell, 1828 [1833])   *Short-horned Lizard*

See Plate 43.

Type: The original description did not specify the type-material. The syntypes (BMNH 1946.8.10.52–53), an adult male and a juvenile of unknown sex, were collected by David Douglas "in ora occidentali Americae Borealis ad ripas fluminis Columbiae," (given as "California" by Boulenger, 1885), in 1825 or 1826.

Distribution: *Phrynosoma douglasii* is the most widespread species of horned lizard. It ranges from southern Alberta, Canada, southward throughout the Rocky Mountain region to the Mexican border in Arizona and New Mexico, and down the axis of the Sierra Madre Occidental to southern Durango. It ranges from the western Dakotas and Nebraska westward to the northeastern corner of California. There are disjunct populations in Montana, Nevada, Utah, Arizona, and Trans-Pecos Texas.

Phrynosoma douglasii occurs throughout the western two-thirds of New Mexico, except for the desert regions on either side of the Rio Grande Valley south of Albuquerque. It occupies the isolated mountain ranges of the bootheel, and there are scattered records south of the Mogollon Rim. The range extends southward from the Sandia Mountains along the Manzano-Gallinas-Sacramento mountains axis, and onto Otero Mesa to the

Texas border. The species occurs in the upper Pecos and Canadian river drainages, but not downstream beyond the high country. Elevation ranges from 1200–3424 m.

There is a specimen (ANSP 21119) from Deming, Luna County, although we doubt the provenance of this specimen. There are no other *P. douglasii* known from this location and the area around Deming is not suitable habitat for the species. Deming may have been the shipping point for a specimen actually collected in the Black Range farther to the north. Such misapplication of locality data was a common occurrence during early western herpetological explorations. It is possible this specimen was collected in the nearby Florida Mountains and that the species still exists there; however, no additional specimens are known from that area.

Description: This is a large, sexually dimorphic horned lizard, with a maximum SVL of 125 mm (Conant and Collins, 1991). Gehlbach (1965) recorded body sizes of 56–83 mm SVL for males and 54–98 mm SVL for females in the Zuni Mountains. The ventral scales are smooth and flat, and overlap slightly. The gular scales are all small. The dorsal body scales are highly heterogeneous in size and shape, with the largest variously modified as short, keeled spines. A single row of small, somewhat flattened, laterally projecting chinshields occurs along each margin of the lower jaw. There are one occipital and three temporal spines at the posterior margin of the head shield on each side. All are about the same size, and all are short and blunt. There is a single, small interoccipital spine. The superciliary spine is small. There are two groups of small spines on each side of the throat. A large, cone-shaped postrictal scale is present. There is a single complete row of lateral abdominal fringe scales on each side of the body. There are 15–16 femoral pores on each side, separated medially by 3 preanal scales. Males have enlarged postanal scales.

The dorsal ground color can be dull yellow, brick red, brown, or gray. There may be a suggestion of a broad, faint light middorsal stripe, but usually it is absent. A longitudinal series of irregular dark brown blotches occurs on either side of this area, generally lacking any bright pigment on the margins. Additional lateral series of smaller blotches may be present; if so, they are less well-defined. These blotches occasionally spread laterally across the body to give the appearance of irregular, undulating crossbands. There is a large brown patch on either side of the neck behind the occipital and temporal spines. The head lacks markings. The ventral surface, especially the chin, is often mottled or suffused with gray, and may be bright reddish or orange in males. Xanthic pigments may intensify in gravid females.

Similar Species: *Phrynosoma solare* has four long occipital horns, the bases of which are in contact. *Phrynosoma cornutum* has well-developed head and body spines and two complete rows of lateral abdominal fringe scales. *Phrynosoma modestum* is smaller and lacks abdominal fringe scales.

Systematics: Six subspecies of *P. douglasii* have been recognized (Reeve, 1952; Stebbins, 1954; Collins, 1990), but the infraspecific taxonomy of this species is badly in need of revision. Most authors prior to 1956 recognized two subspecies in New Mexico, *P. d. hernandesi* and *P. d. ornatissimum*. Girard (1858) did not designate the specimens upon which he based the description of either taxon. Stejneger (1890a) designated three specimens (USNM 107 [actually 197], a subadult male and one of undetermined sex exchanged to MNNP and USNM 198, an adult male), collected by John H. Clark from "Sonora" in September 1851 and "Santa Fe," date of collection unknown (catalogued in December, 1857) as the syntypes for *P. d. hernandesi*. Stejneger (1890a) designated two specimens (USNM 204), both adult females collected by S.W. Woodhouse from the "Zuni Mountains, New Mexico" on the Sitgreaves expedition in 1851 as the syntypes of *P. d. ornatissimum*.

Confusion continued over the number and provenance of the designated type specimens, and over the occurrence of the two taxa on the landscape. Stejneger (1890a) considered *P. d. hernandesi* to be characteristic of the wooded Colorado Plateau of Arizona, Utah, Colorado, and New Mexico, and *P. d. ornatissimum* as the form in the desert regions of this area. Cope (1898 [1900]) assigned individual specimens from several identical localities in Arizona and New Mexico, including the type localities, to both subspecies. Van Denburgh (1922a) and Smith (1946) assigned the names *P. d. ornatissimum* and *P. d. hernandesi* to populations in northern and southern New Mexico, whereas Stebbins (1954) applied the latter name only to lizards from Hidalgo County. Smith and Taylor (1950a), possibly following Cope (1898 [1900]), restricted the type locality of *P. d. hernandesi* to Santa Fe, New Mexico, an action followed by Cochran (1961), but specifically refuted by Reeve (1952). Gehlbach (1965) reviewed the taxonomic confusion surrounding these two forms and the reasoning behind it, and placed *P. d. ornatissimum* in the synonymy of *P. d. hernandesi*. The two forms continue to be recognized as distinct by some (e.g. Collins, 1990), but only the latter subspecies occurs in New Mexico (Conant and Collins, 1991).

Habitat: This species occupies a wide variety of habitats in New Mexico, from semi-desert shrublands through shortgrass prairies with various grasses including *Agropyron*, *Bouteloua*, *Festuca*, and *Stipa*, and piñon-juniper woodland to spruce fir–Douglas fir forest (Montanucci, 1981; our obs.). It is most common on relatively open terrain in ponderosa pine and piñon-juniper woodlands, mountain meadows, oak thickets, and sagebrush flats between 2061–2424 m (Bragg and Dundee, 1949 [1950]; Harris, 1963; Gehlbach, 1965; Douglas, 1966). It can be found in abundance in northern New Mexico in association with the towns of Gunnison's prairie dog (Clark et al., 1982). Gehlbach (1965) collected specimens from alpine tundra on Mt. Taylor at 3424 m. Reynolds (1979) reported that the clearing and replanting of native sagebrush habitats in Idaho with *Agropyron* sp. dramatically reduced the density of this species.

Behavior: Mating behavior in this species consists of stereotypical displays by both sexes, involving head-bobbing and olfactory cues in the male's approach to the female (Montanucci and Baur, 1982). The male mounts the female from behind, and usually grasps the nuchal fold behind her head or, occasionally, a temporal horn for purchase.

An average active body temperature of 34.9°C, and critical thermal maximum and minimum temperatures of 43.5°C and 2.7°C have been reported (Prieto and Whitford, 1971; Pianka and Parker, 1975). Lizards engaged in normal activity patterns at temperatures as low as 6°C. New Mexico specimens have been collected between 2 March (Grant Co.) and 15 November (Catron Co.). In southeastern Idaho, individuals are active from mid-April to late September, with peak seasonal activity occurring in June and July (Guyer and Linder, 1985). Adult lizards emerge prior to juveniles in the spring and enter hibernation first in the fall. Daily activity occurs between 0800 and 2100 with no diurnal peaks.

Reproduction: Minimum size at sexual maturity is 63 mm SVL for females and 59 mm SVL for males (Howard, 1974). Individual lizards may be 2 years old before first reproduction. Yolk deposition begins in mid-October prior to hibernation (Howard, 1974). Mating and ovulation occur in March through May depending upon geographic location (Goldberg, 1971b; Howard, 1974; Guyer and Linder, 1985). Gestation has been reported as 63 days in captivity (Montanucci, 1983) or 3 months in wild populations (Goldberg, 1971b). The young are born alive, enclosed in a clear amniotic sac, and will die if not strong enough to free themselves (Tanner, 1954). Females may carry embryos for up to 3 months before giving birth. The young are able to fend for themselves within an hour of birth, and the mother provides no obvious parental care (Milne and Milne, 1950). Parturition has variously been reported from the end of June into September, depending on latitude and elevation (Bragg and Dundee, 1949 [1950]; Woodin, 1953; Gehlbach, 1956; Harris, 1963; Douglas, 1966; Goldberg, 1971b; Pianka and Parker, 1975; Guyer and Linder, 1985). Howard (1974) reported an average litter size of 23.5 (9–48) for 32 lizards from the southwestern United States and adjacent Mexico. Goldberg (1971b) reported an average litter size of 16.7 (9–30) for 11 females from southeastern Arizona, with litter size positively correlated with female body size. One female was observed by Gehlbach (1965) in the Zuni Mountains giving birth to a litter of 10 in a clump of rabbitbush [rabbitbrush]. The young, between 24 and 26 mm SVL, were all born within 10 minutes and immediately buried themselves in the soil after escaping the birth membranes.

Male testes are at maximum size and the epididymides packed with sperm when males emerge from hibernation between March and May in southeastern Arizona (Goldberg, 1971b). Gonadal regression is complete by the end of April and testes remain quiescent throughout the summer. Testicular recrudescence begins in August, and a renewal of germinal epithelial activity is apparent by the time males enter hibernation in late October to mid-November.

Montanucci (1983) reported longevity of 5 ½ years in captivity for lizards that were wild-caught as adults.

Food Habits: Hotton (1955) reported at least 85% of the diet in this species to be ants. Pianka and Parker (1975) reported ants comprising 81% by frequency and 52% by volume of the diet of 50 lizards examined. Powell and Russell (1984) reported ants as comprising 76% by frequency and 36% by weight of the diet in Canada, with beetles (13%, 25%) and grasshoppers (2%, 22%) also important. Montanucci (1981) recorded ants (42%), acridid grasshoppers (18%), and carabid beetles (11%) by volume as important items in the diets of lizards from northern Mexico. He also noted that diets of local populations may vary substantially due to factors like prey population phenology and competition.

Remarks: The species is named for the collector of the type specimens, David Douglas, who traveled extensively throughout the Pacific Northwest collecting plants for the London Horticultural Society, in collaboration with the Hudson's Bay Company. Hammerson and Smith (1991) discussed the nomenclatural history of this species. We follow their revision although we retain the suffix *ii* of the original description by Bell (1828 [1833]), as provided for in the third edition (1985) of the International Code of Zoological Nomenclature, Articles 31 and 33(d).

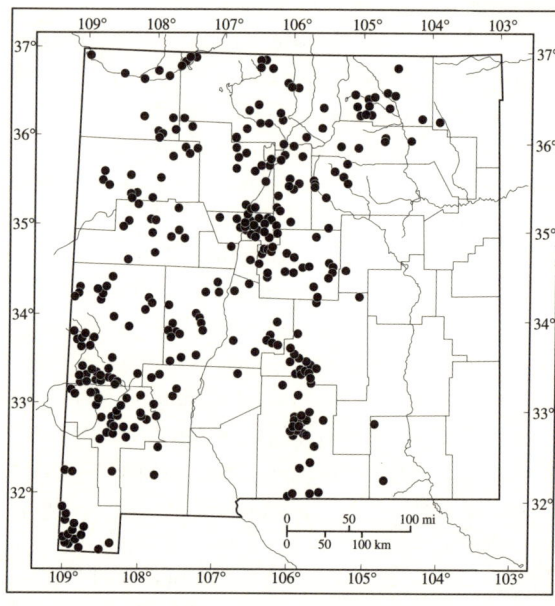

Distribution of Phrynosoma douglasii *in New Mexico.*

PHRYNOSOMA MODESTUM Girard in Baird and Girard, 1852 *Roundtail Horned Lizard*

See Plate 44.

Type: The original description (Baird and Girard, 1852a) states that the type material was "brought from the valley of the Rio Grande west of San Antonio, by Gen. Churchill, and from between San Antonio and El Paso del Norte, by Col. J. D. Graham," but does not specify the nature of that material. Reeve (1952) lists USNM 163 (Gen. Churchill, 1 specimen) and USNM 164 (Col. Graham, 8 specimens) as the syntypes; the former is now lost (R. P. Reynolds, pers. comm.). One syntype of the USNM 164 series was exchanged to the University of Illinois Museum of Natural History in 1956 (Cochran, 1961); however, seven specimens (a subadult male, an adult male, and five adult females) remain in the USNM 164 series and one (an adult male) has been recatalogued as USNM 165660 (R. P. Reynolds, pers. comm.). Smith and Taylor (1950a,b) restricted the type locality to Las Cruces, Doña Ana County, New Mexico, a decision with which Axtell (1988b) disagreed. According to Axtell (1988b), the first specimen found was USNM 163, a juvenal specimen obtained by General Churchill when his army crossed the Rio Grande at Presidio del Norte (= Ojinaga, Chihuahua, Mexico) in 1846 during the Mexican War (see Baird and Girard, 1851 [1852], p. 351, and Girard, 1851 [1852], p. 358). Therefore, the type locality should logically be somewhere in west Texas, the proper restriction deferred until a lectotype is designated. The specimens represented by USNM 164 were taken by J. H. Clark in May or June 1851.

Distribution: This species occurs in west-central Texas from the Devils River in Val Verde County northward along the eastern margin of the Caprock Escarpment to the Salt Fork of the Red River (Axtell, 1988b). There are disjunct populations along several river valleys eastward in central Texas to the vicinity of Abilene, as well as in the Canadian River valley in the Texas Panhandle. Records from elsewhere in Texas apparently represent introductions (Dixon, 1987; Axtell, 1988b). The species ranges westward to the southeastern corner of Arizona and adjacent northeastern corner of Sonora, Mexico, and from northern New Mexico southward throughout the Chihuahuan Desert region (Morafka, 1977) of the Mexican states of Chihuahua, Coahuila, Durango, Nuevo León, San Luis Potosí, and Zacatecas.

This species occurs from the Arizona border in southwestern New Mexico eastward, south of the Mogollon Rim, to the Pecos River valley. It extends up the Rio Grande drainage basin to the vicinity of San Ildefonso Pueblo, and up the Pecos River to the San Miguel County line. There are scattered records between these two basins, particularly along tributary drainages, as well as in the Canadian River system below Conchas Lake and on the Llano Estacado. We agree with Gehlbach (1965) that the McKinley County record represents an escaped or released pet.

Description: This is a small species of *Phrynosoma*. During 1987–92, 119 specimens collected at Antelope Pass in Hidalgo County averaged 45.8 (13–70) mm SVL and 6.0 (0.3–16) gm. Thirty-three adult males averaged 53.4 (42–64) mm SVL and 7.4 (2.2–11.8) gm; 43 adult females averaged 57.1 (43–70) mm SVL and 9.3 (3.7–16) gm (our unpubl. data). There are few, if any, dorsal body spines, and the horns on the head are poorly developed. The dorsal scales are mostly granular, the ventral scales are smooth, and the gular scales are all small. There are two occipital and two temporal spines, one on each side, at the posterior margin of the head shield. A series of five spines, decreasing in size anteriorly, extends forward from the temporal region to below the eye. A postrictal scale is absent. There is no interoccipital spine. The superciliary spine is small. The chinshields are small, increasing in size anteroposteriorly with the penultimate largest. There is no lateral fringe of scales along the side of the body and the tail is round in cross-section. The ear opening may be open or concealed by a scaly integument. There are 7–13 femoral pores on each side, separated medially by up to five scales. Males have enlarged postanal scales.

The ground color in this species may be various shades of white, pink, yellow, gray, red, brown, and even blue (Sherbrooke and Montanucci, 1988), and is closely correlated with substrate coloration on which specific populations live (Bundy and Neess, 1958). Nevertheless, numerous color morphs may occur within a single population (Sherbrooke, 1981). Superimposed on the dorsal surface is a peppering of dark brown or black spots. A pair of distinct dark patches occur on the lateral surface of the neck, and a more diffuse pair extends anteriorly from the groin along each side of the body as far as the axillary region. The tail has several dark bands, and the ventral surface of the body is generally immaculate white to pale yellow.

Similar Species: *Phrynosoma cornutum* and *P. solare* are larger, have well-developed head and body spines, and two and one complete row(s) of lateral abdominal fringe scales, respectively. *Phrynosoma douglasii* is larger and has one complete row of lateral abdominal fringe scales.

Systematics: No subspecies are recognized. Reeve (1952) regarded this species as a primitive member of the genus, not closely related to any other species. However, Montanucci (1987) considered it to be a morphologically specialized species, having evolved relatively recently during the Pleistocene, and being most closely related to the desert horned lizard, *P. platyrhinos*.

Habitat: This species occupies a wide variety of desert-grassland and desert shrubland habitats between elevations of 900–2200 m, characterized by *Acacia* spp., *Larrea tridentata*, *Prosopis glandulosa*, *Yucca* spp., and *Ephedra* spp. (Weese, 1917; Nickerson and Mays, 1969 [1970]; Whitford and Creusere, 1977). It generally avoids sandy soils, and is particularly common on bajada slopes and along the edges of arroyos on gravelly to rocky soils (Little and Keller, 1937; pers. obs.). Shaffer and Whitford (1981) estimated densities of 2 lizards/hectare in central Doña Ana County.

Behavior: These lizards are primarily diurnal, but we have observed enough individuals through the years "active" after dark in west Texas and southern New Mexico to make us question the assertion by Williams (1959) that this behavior is strictly abnormal. Diurnal activity begins by 0830 in Doña Ana County (Shaffer and Whitford, 1981). Pianka and Parker (1975) reported an average active body temperature of 36.5°C for this species. Preferred substrate temperatures for activity are 36–40°C, and lizards seek shelter or burrow when temperatures exceed this range (Weese, 1917, 1919; Shaffer and Whitford, 1981). Lizards may remain active and continue foraging throughout the day in the shelter of vegetation under favorable thermal conditions. New Mexico specimens have been collected between 16 February (Doña Ana County) and 2 November (Bernalillo County). The activity season in southwestern New Mexico extends from March to October (Baltosser and Best, 1990). Munger (1984a) estimated average home range sizes of 4101 m² for males and 1355 m² for females.

Individual *P. modestum* can undergo profound changes in color pattern hue and intensity associated with thermoregulatory and cryptic behaviors (Sherbrooke and Frost, 1989). These lizards may actually mimic small stones in the habitats in which they live, as a primary defense against predators (Sherbrooke and Montanucci, 1988).

Reproduction: Minimum size at sexual maturity for females is 42 mm SVL (Howard, 1974). Female fat bodies decrease in size from spring emergence to minimal values in May, and then increase until hibernation. Yolk deposition may begin as early as April, and oviposition occurs from May through July. Most females are postreproductive by August. Some females may produce 2 clutches per season, and larger females produce larger clutches. Howard (1974) reported an average clutch size of 13.4 (6–18) for 14 females from southeastern Arizona, Vitt (1977) reported an average clutch size of 12.2 (10–19) for 6 females, and Pianka and Parker (1975) reported an average clutch size of 10.6 eggs for 40 lizards examined. Oviductal eggs may represent as much as one-third of an individual female's body weight prior to oviposition (Parker, 1973b). Hatchlings (21–30 mm SVL) first appear in early July and can be found active to the end of October (Howard, 1974; our unpubl. data), at which time they may have grown to minimum reproductive size.

Males larger than 41 mm SVL by the mating season in May are reproductively mature (Howard, 1974). Testes and epididymides are fully developed upon spring emergence, do not decrease significantly in size, and possess spermatozoa through the end of July (Parker, 1973b; Howard, 1974; Pianka and Parker, 1975).

Baur (1986) reported a lifespan of close to five years in captivity.

Food Habits: Hotton (1955) reported at least 85% of the diet as ants. Pianka and Parker (1975) reported ants to comprise 86% by frequency and 66% by volume of the diet of 130 lizards examined throughout the species' range. In Durango, Mexico, the diet consisted of ants (48–86%), termites (17%), lepidopteran larvae (2–19%), and hemipterans (24%) by volume in different seasons over a 2-year period (Barbault and Maury, 1981).

The primary prey in central Doña Ana County is honey pot ants (*Myrmecocystus*); individual lizards take between 40 and 45 of these ants each day, usually under shrubs as foraging ants descend from the foliage (Shaffer and Whitford, 1981). Peak feeding times are between 1000 and 1300. Other ant species (*Pogonomyrmex californicus*, *Pheidole* spp.) are taken when available; when this occurs the consumption of honey pot ants is approximately halved.

Known predators in New Mexico include *Crotaphytus collaris*, loggerhead shrikes (*Lanius ludovicianus*), and roadrunners (*Geococcyx californianus*). Grasshopper mice (*Onychomys* spp.) are apparently effective predators on this species (Sherbrooke, 1991). Munger (1986) calculated annual predation probabilities of between 52% and 86% in southeastern Arizona.

Remarks: Best et al. (1983) reported melanistic individuals from the Pedro Armendariz lava field of central New Mexico. We follow Banta (1971) for the citation of the original description of this species.

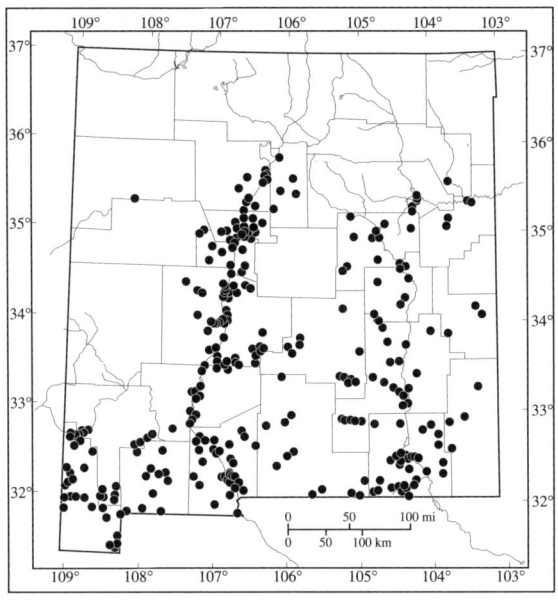

Distribution of
Phrynosoma modestum
in New Mexico.

PHRYNOSOMA SOLARE Gray, 1845 — *Regal Horned Lizard*

See Plate 45.

Type: The holotype (BMNH XXIII.125.d) is an adult male from "California," collector and date of collection unknown. The type locality was restricted to Tucson, [Pima County], Arizona by Schmidt (1953).

Distribution: This species occurs throughout the Sonoran Desert of Arizona southward through the Mexican state of Sonora to northern Sinaloa (Hardy and McDiarmid, 1969). It occurs on Isla Tiburón in the Sea of Cortéz (Murphy and Ottley, 1984). We have suspected its presence in New Mexico for some time; both Reeve (1952) and Parker (1974b) showed the range close to or within Hidalgo County, although specimens were lacking. An adult female (MSB 55920) collected in June 1993 from Guadalupe Canyon confirmed the presence of this species in New Mexico (Painter, 1993b) where it is known only from elevations of 1321–1387 m. The data accompanying UWZM H20145 from Lincoln County are likely in error.

Description: This is a large, sexually dimorphic horned lizard. The maximum SVL reported for males is 95 mm and 117 mm for females (Parker, 1974b). The ventral scales are keeled, sometimes only on the anterior chest region, and non-mucronate. There is a series of enlarged gular scales that project horizontally. The dorsal body scales are highly heterogeneous in size and shape, with the largest variously modified as short, vertical spines. The scales on the anterior surfaces of the limbs are large, pointed, and strongly keeled. There are two pairs of broad, well-developed, acute occipital spines, the bases of which are in contact, forming a continuous row along the rear margin of the head. There are four temporal spines on each side of the head at or above the level of the eye. A small, weakly keeled postrictal scale is present. The supraorbital ridges are prominent, each terminating posteriorly in a broad, pyramidlike scale. There are two groups of small spines on each side of the throat. A single row of spinose chinshields enlarging anteroposteriorly occurs along each margin of the lower jaw. There is a single complete row of lateral abdominal fringe scales on each side of the body. The tail is relatively short, usually less than twice the length of the head. There are 14–26 femoral pores on each side, separated medially by 1–6 scales. Males have enlarged postanal scales.

The dorsal ground color is yellowish-brown to grayish-tan to reddish, darker laterally, and varies with geographic location and substrate type. The head is usually a lighter shade of yellowish or yellow-brown. A middorsal light line, if present, is pale and indistinct. There is a series of paravertebral dark spots, often appearing as undulating, broken lateral bands without light borders. There is a large dark brown or black patch on either side of the neck behind the occipital and temporal spines, often extending posteriorly along the lateral portion of the back. There is a series of narrow, dark bands on the tail. The ventral surface is white with scattered dark spots.

Similar Species: *Phrynosoma cornutum* has a single pair of well-developed occipital horns that are widely separated at the base, and two complete rows of lateral abdominal fringe scales. *Phrynosoma douglasii* and *P. modestum* lack well-developed horns and spines on the head, and the latter species also lacks lateral abdominal fringe scales.

Systematics: No subspecies have been recognized. Montanucci (1987) regarded this species as a relatively highly derived member of the desert-adapted "northern radiation" of the genus. He hypothesized that it arose somewhat later in the diversification of the genus, probably during the middle Pliocene.

Habitat: This species prefers relatively level terrain in open vegetative communities dominated by *Larrea tridentata, Prosopis glandulosa, Franseria* spp., *Cercidium* spp., *Encelia farinosa, Fouquieria splendens, Opuntia* spp., and *Cereus giganteus* on gravelly to sandy soils at elevations up to 1280 m (Lowe, 1954; Nickerson and Mays, 1969 [1970]; Parker, 1971). Other lizards which can be found sympatrically include *Callisaurus draconoides, Sceloporus magister, Urosaurus ornatus,* and *Cnemidophorus tigris.* (See Remarks).

Behavior: Individuals can be found active all year, but most activity takes place between April and December (Huey, 1942; Gates, 1957; Parker, 1971). Active lizards appear to prefer body temperatures between 34–38°C (Heath, 1965). Baharav *in* Pianka and Parker (1975) measured the average body temperature for 151 active lizards as 34.7°C. Lizards burrow at body temperatures below 30°C and begin to pant when body temperatures exceed 40°C.

These lizards appear to be rather sedentary. Lowe (1954) and Parker (1971) have estimated home ranges as 84 m^2 and 133 m^2.

Parker (1971) reported captive individuals remaining in "tonic immobility" on their backs with legs extended and lungs deflated; they retained this posture even when turned right side up.

Reproduction: Females larger than 70 mm SVL are reproductively mature. Vitellogenesis begins in May and ovulation occurs in July and August, with deposition of a single clutch of about 20 (7–33) eggs (Parker, 1971; Van Devender and Howard, 1973; Howard, 1974). Clutch size is strongly correlated with female body size (Howard, 1974). Nests are semicircular tunnels dug in soft sand or silt, 10.5 cm wide and 35 cm deep, dug at a 30° angle and ending in a bulb-shaped cavity (Van Devender and Howard, 1973). Incubation takes about 50 days. Hatchlings are 28–33 mm SVL. Tony A. Snell (pers. comm.) marked and released a hatchling (28 mm SVL, 1.1 gm) in Guadalupe Canyon in extreme southwestern Hidalgo County on 30 September 1993.

Males larger than 70 mm SVL are reproductively mature. Testes are prereproductive upon spring emergence, and spermiogenesis begins in mid-May (Howard, 1974). Maximum testes size is achieved in late June–early July (Parker, 1971; Howard, 1974) and regression occurs prior to hibernation.

Parker (1971) estimated annual turnover as one-third to one-half of the population he studied. Baur (1986) reported a life span of more than eight years in captivity.

Food Habits: This species is an ant specialist. Pianka and Parker (1975) found ants to comprise 90% by frequency and 89% by volume of the prey items eaten. An adult female from Guadalupe Canyon, Hidalgo County, had a stomach full of harvester ants *(Pogonomyrmex barbatus)* (Barney Tomberlin, pers. comm.).

Vorhies (1948) recorded an instance in which a juvenal *Crotalus atrox* died attempting to eat an individual *P. solare*.

Remarks: *Phrynosoma solare* has been reviewed by Parker (1974b). The name *solare* (Latin, "of the sun") may refer to the desert dwelling habits of this species, or it may allude to the fact that the arrangement of the occipital spines is reminiscent of the sun's rays (N. J. Scott, pers. comm.).

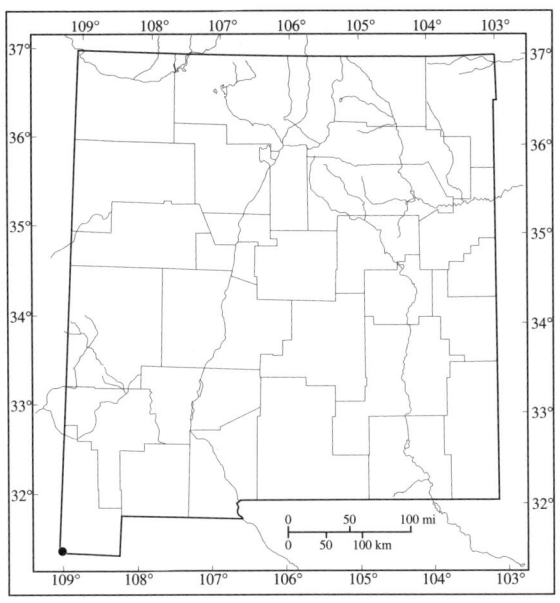

In New Mexico, *P. solare* is known only from a riparian zone of rocky or fine sandy soils grown to *Acacia* spp., *Prosopis* spp., *Platanus wrightii*, *Celtis* spp., *Juglans* spp., and various grasses and forbes. Sympatric lizards include *Cnemidophorus burti*, *C. sonorae*, *Holbrookia maculata*, *Sceloporus clarkii*, and *Urosaurus ornatus*.

Distribution of Phrynosoma solare in New Mexico.

SCELOPORUS ARENICOLUS Degenhardt and Jones, 1972 *Sand Dune Lizard*

See Plate 46.

Type: The holotype, MSB 23621, an adult male, was collected from "Mescalero Sands 3 ½ miles N and 44 miles E Roswell, Chaves Co., New Mexico, [T10S R31E]" by Kirkland L. Jones on 27 April 1968.

Distribution: *Sceloporus arenicolus* was first reported in New Mexico by Sabath (1960), and occurs as disjunct populations in southeastern New Mexico (approximately 1050–1400 m) and adjacent Texas (Andrews, Crane, Ward, and Winkler counties). In southeastern New Mexico, it has been found mainly on the Mescalero Sands, which extend in a broad arc from the vicinity of San Juan Mesa in northeastern Chaves County southward and eastward through eastern Eddy County and southern Lea County (Sena, 1985). Specimens have recently been collected in sandy habitat in southern Roosevelt County near Milnesand (L. A. Fitzgerald, pers. comm.).

Description: This is a small species of *Sceloporus*. Females reach a maximum of 70 mm SVL, and males a maximum of 65 mm SVL. Sena (1985) reported means of 56 mm and 54 mm SVL for males and females. During 1992–95, 1094 *S. arenicolus* collected on the Mescalero Sands in Chaves County averaged 51.1 (22.4–64.9) mm SVL; 507 adult males averaged 54.5 (49–64.9) mm SVL and 5.1 (2.6–8.6) gm; 339 adult females averaged 53.8 (49–62.2) mm SVL and 4.8 (2.8–8.3) gm (L. W. Gorum, unpubl. data). The dorsal scales are keeled and pointed, but do not greatly overlap. There are 41–52 (mean 48) scales around midbody (Degenhardt and Jones, 1972). The supraoculars are small and separated from the superciliaries as well as the median head scales by at least one row of smaller scales. There are 9–16 femoral pores on each leg, the two series usually separated medially by 13 or more (minimum 9) scales. The scales extending backward over the ear openings are small and number 6 or 7. The scales on the posterior surface of the thigh are relatively small and mostly granular.

In lizards from southeastern New Mexico, the dorsum is light brown. They lack a pattern except for a poorly defined grayish-brown band, extending from the upper margin of each ear opening posteriorly onto the tail. The blue coloration of the chin and throat is reduced to scattered flecking or is absent altogether, and that of the ventral body surface is reduced and widely separated. Females develop a lateral yellow-orange suffusion from the throat posteriorly onto the tail during vitellogenesis (Sena, 1985).

Similar Species: *Sceloporus undulatus* lacks the granular scales on the posterior surface of the thigh, has 7 or fewer scales separating the femoral pore series, and usually has paired blue throat spots (faint in females) rather than a mottled or flecked pattern. *Sceloporus scalaris* has the lateral scales arranged in rows parallel to the dorsal scale rows. *Sceloporus graciosus* has a well-defined dorsal pattern, relatively more (48–60) scales around midbody, and is known only from northwestern New Mexico in Cibola, McKinley, Rio Arriba, San Juan, and Sandoval counties.

Systematics: No subspecies of *S. arenicolus* are recognized. This taxon has been treated by most previous workers as a subspecies of *S. graciosus*. We agree with Collins (1991) and Smith et al. (1992), following the arguments of Frost and Hillis (1990), that it merits specific rank.

Habitat: *Sceloporus arenicolus* is restricted to the vicinity of active and semi-stabilized sand dunes within the Mescalero Sands. These dunes occur to an elevation of 1190 m above sea level and support scattered stands of *Quercus havardii* and *Artemisia filifolia* as co-dominant plant species (Sena, 1985). Significant reductions of lizard population sizes are associated with removal of *Q. havardii* (Snell et al., 1993). Sena (1985) found a median sand grain size of 0.201 mm in habitats occupied by *S. arenicolus*.

Other lizard species that can be found in the same or immediately adjacent habitats as *S. arenicolus* include *Holbrookia maculata, Phrynosoma cornutum, P. modestum, S. undulatus, Uta stansburiana, Eumeces obsoletus, Cnemidophorus sexlineatus*, and *C. tigris*. There is a marked negative relationship in relative abundance between *S. arenicolus* and *U. stansburiana*, suggesting that ecological factors that tend to favor one species are detrimental to the other as well as the existence of exploitative competition between the two (Snell et al., 1993). Sena (1985) suggested that these two species occupy different microhabitats in areas of sympatry.

Behavior: New Mexico specimens have been collected between 27 April and 15 September in Chaves County. Lizards are active from 0800 until dusk during May, June, and July (Sena, 1985), but confined their activity during midday (1200–1400) to shaded areas beneath vegetation. Individuals are extremely wary, and are quick to seek shelter in burrows, beneath leaf litter, or by burrowing in loose sand. Sena (1985) recorded an average body temperature of 33.4°C and an average critical thermal maximum of 45.5°C in southeastern New Mexico. Individuals may gape and attempt to bite if handled, but they are too small to inflict any damage on people.

Reproduction: Vitellogenesis begins in late April (Sena, 1985). Degenhardt and Jones (1972) reported that female *S. arenicolus* can reach sexual maturity during the first spring following hatching. The smallest female containing oviductal eggs reported by Sena (1985) was 49 mm SVL. There are two distinct size classes of reproductively active females in the spring, suggesting that some individuals reach at least 2 years of age. Individual females produce one or two clutches a year averaging about 5 (range 3–6) eggs each, with the first clutch laid in late June and the second in late July to early August (Degenhardt and Jones, 1972; Sena, 1985). Clutch size is positively correlated with female body size (Sena, 1985). Hatchlings appear between the end of July and the end of September.

Sexually mature males (at least 49 mm SVL) emerge in April with testes at maximum size (Sena, 1985). Mature sperm are present throughout the reproductive tract through June, although significant testicular regression occurs at this time. Testes reach minimum size in July, with significant testicular recrudescence occurring prior to hibernation in September (Sena, 1985).

Food Habits: *Sceloporus arenicolus* feeds upon ants and their pupae, small beetles (including ladybirds) and their larvae, crickets, grasshoppers, and spiders. Most feeding appears to take place within or immediately adjacent to patches of vegetation.

Remarks: *Sceloporus arenicolus* is listed as a Category 2, Notice of Review species by the U.S. Fish and Wildlife Service (USDI, 1994) and as endangered by the New Mexico Department of Game and Fish (NMGF, 1990). The specific epithet was misspelled "*arenicolous*" in the original description and should be spelled *arenicolus* (Smith et al., 1992).

Axtell (1988a) reviewed this species (as *S. graciosus arenicolous*) in Texas.

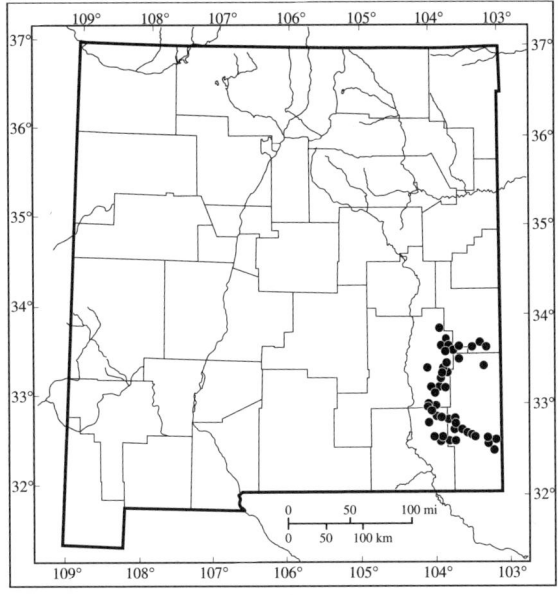

Distribution of Sceloporus arenicolus in New Mexico.

SCELOPORUS CLARKII Baird and Girard, 1852 — *Clark's Spiny Lizard*

See Plate 47.

Type: The original description (Baird and Girard, 1852a) did not specify the type material, which came from the "province of Sonora." Smith (1939) designated an adult male (USNM 2940) as the lectotype, and Smith and Taylor (1950a,b) restricted the type locality to the Santa Rita Mountains [Pima or Santa Cruz County], Arizona. The specimen(s) was collected by John H. Clark in September 1851, during the expeditions of the Scientific Corps of the U.S. and Mexican Boundary Commission under Col. J. D. Graham.

Distribution: The species occurs along the Mogollon Rim of west-central Arizona and the Mogollon Plateau of southwestern New Mexico southward along the Sierra Madre Occidental in Mexico to the mountains of northern Jalisco. It occurs on Tiburón and San Pedro Nolasco islands in the Sea of Cortéz (Murphy and Ottley, 1984), and approaches the coastline on the Pacific versant of the mainland opposite Bahía Kino and extends southward into Nayarit. In New Mexico, *S. clarkii* occurs from Reserve in Catron County southward to the Gila River and in the Peloncillo Mountains to the international border, with seemingly unsuitable habitat intervening. It occurs eastward along the Mogollon Rim from the Arizona border to and throughout the Black Range. Several isolated populations occur in Hidalgo County, including those in the Alamo Hueco Mountains and on the western flank of the Little Hatchet Mountains. Elevation ranges from 1060–1848 m.

Description: This is a large *Sceloporus*, with a maximum reported SVL of 144 mm for males and 120 mm for females. During 1992–95, 310 individuals collected in Guadalupe Canyon averaged 80.7 (27–117) mm SVL and 22.8 (0.5–55.8) gm (our unpubl. data). The dorsal scales are large, strongly keeled, mucronate, and imbricate. There are 10–16 femoral pores on each leg. There are 4–6 enlarged supraocular scales, the posterior two in contact with the median head scales. There is a single parietal scale on either side of the interparietal. The sublabials are usually separated from the mental by a series of small scales. There are three (occasionally four) relatively short and rounded scales extending backward over the ear opening, with the upper one the largest.

There is a distinct black wedge-shaped mark in the shoulder area on each side of the body. The dorsal coloration varies from gray to brown and many specimens, especially adult males, have a distinctly greenish or bluish tint. Young lizards have 5–7 distinct, narrow, and undulating transverse brown or black bands across the body between the limb insertions, and the dorsal surfaces of the limbs and tail are similarly banded. Females retain vestiges of this pattern into adulthood, but males lose it almost entirely, becoming almost uniformly gray or brown in color dorsally. Faint tail rings may be present in some adults, and the dorsal surface of the forearms retains distinct narrow crossbands throughout life in both sexes. Males have blue or green patches of varying hues on the throat, often bordered with black, and a similar pattern along the sides of the belly between the limb insertions; in large specimens the belly markings may fuse along a black border medially. Faint blue markings on the throat and sides may be present in females, but usually there is only a diffuse dark mottling.

Similar Species: *Sceloporus poinsetti* and *S. jarrovii* have complete neck bands. *Sceloporus magister* lacks distinctly barred forearms, has five or more distinctly pointed auricular scales with the median scales the largest, and has the sublabial scales in contact with the mental.

Systematics: Two subspecies are recognized (Hardy and McDiarmid, 1969); only the nominotypical form occurs in New Mexico. Although Stejneger (1893) quite clearly set out the differences between *S. clarkii* and *S. magister*, Cope (1898 [1900]) and Burt (1935 [1936]) continued to regard them as conspecific. Cope (1898 [1900]) had previously listed only two specimens under USNM 2940, but Smith (1939) listed three specimens under USNM 2940 as the original syntypes of *S. clarkii*, and pointed out that two of the three were actually representatives of *S. magister*.

Habitat: This species is abundant in evergreen woodlands (oak and pine-oak) at elevations between 1273–1848 m, and penetrates into more arid habitats along riparian woodland communities of broadleaf deciduous trees including cottonwood, willow, sycamore, ash, walnut, and mesquite, that are commonly confined to major drainages extending from large and usually forested mountains (Nickerson and Mays, 1969 [1970]; Hulse, 1973; Tanner, 1987). It also occurs in rocky habitats at the lower edges of mountain foothills, and oak-grassland communities as low as 1060 m (pers. obs.; Lowe et al., 1967). The strongly arboreal habitat preferences of this species have been noted by others (e.g. Hardy and McDiarmid, 1969; Tinkle and Dunham, 1986). It is a common lizard in the remnant gallery forests along Whitewater Creek in Catron County and along the Gila River below its confluence with Turkey Creek. Tinkle and Dunham (1986) estimated densities of 6–46 lizards/hectare over a 6-year period in central Arizona.

Behavior: This is a wary lizard, and does not seem to bask as readily as other sympatric species of *Sceloporus*. Individuals are quick to seek shelter when approached, and often retreat to the opposite side of tree trunks from observers and then high into the branches if persistently pursued. They also utilize the underside of boulders and crevices. Ballinger (pers. comm.) has observed these lizards from a distance of 40 m basking, sometimes for long periods of time, up to 10 m off the ground in the branches of oak trees in the Chiricahua Mountains of southeastern Arizona. The activity season lasts from April into October in central Arizona (Tinkle and Dunham, 1986). New Mexico specimens have been collected between 3 March (Grant Co.) and 16 October (Catron Co.).

Reproduction: The minimum size at sexual maturity is 89 mm SVL for females (Tinkle and Dunham, 1986), attained at an age of 22 months. During 1992–95, 55 adult females collected in Guadalupe Canyon averaged 96.7 (89–110) mm SVL and 31.6 (19.8–49.3) gm (our unpubl. data). Vitellogenesis begins between March and mid-May in central Arizona, and oviposition occurs in June (Tinkle and Dunham, 1986). Most females produce a single clutch annually, averaging 19.6 (7–28) eggs. A few females may produce a second, much smaller, clutch. Clutch size is significantly correlated with female body size, with approximately one egg added for each 3 mm increment in SVL. Hatchling mortality is about 90%. Hulse (1973) reported an average clutch size of 14.1 (9–19) eggs in June for nine females from east-central Arizona. Vitt (1977) recorded two clutches of 19 and 23 eggs. Clutch weight may represent up to 28% of female body weight prior to oviposition (Parker, 1973b).

Data on male reproductive characteristics are lacking. If we assume that males mature at approximately the same size as females (i.e., 89 mm SVL [Tinkle and Dunham, 1986]) then the following data are useful. During 1992–95, 102 adult males collected in Guadalupe Canyon averaged 98.5 (89–117) mm SVL and 36.1 (22.8–55.8) gm (our unpubl. data). Tinkle and Dunham (1986) estimated between-year survivorship for male and female lizards regardless of age as 28% and 50%.

Food Habits: Very little quantitative dietary information exists for this species. Smith (1939) reported caterpillars, ants, grasshoppers, and beetles as prey items, and Stebbins (1954, 1985) added wasps, crickets, and occasional leaves, buds, and flowers. Painter (1985) reported beetles (69.5%), ants and wasps (52%), hemipterans (48%), butterflies and moths (30%), spiders (22%), and grasshoppers (22%) as important food items by frequency in a sample of 28 lizards from the Gila National Forest, Grant and Catron counties.

Remarks: According to Baird and Girard (1852a), the species name is "dedicated to John H. Clark, to whose skill as a collector, and untiring zeal for science, the world is indebted for the splendid zoological collections sent and brought home by Col. Graham from the survey of the Mexican boundary."

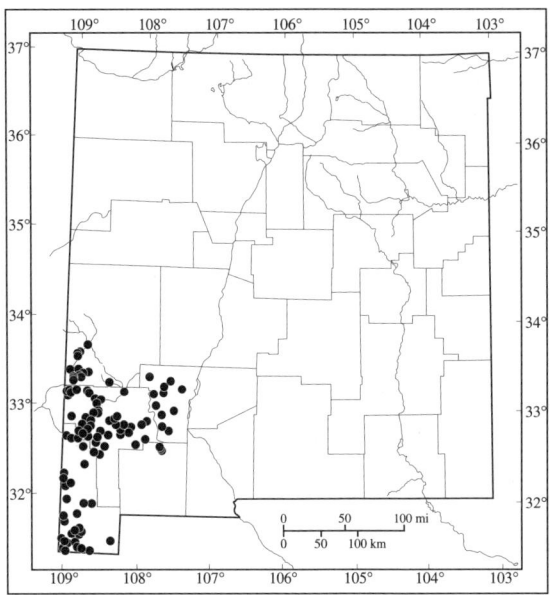

Distribution of Sceloporus clarkii *in New Mexico.*

SCELOPORUS GRACIOSUS Baird and Girard, 1852 *Sagebrush Lizard*

See Plate 48.

Type: The type specimens (USNM 2877; 4 specimens) were collected from the "valley of the Great Salt Lake [Utah]" during the expedition led by Captain Howard Stansbury, U.S. Army Corps of Topographical Engineers, to explore and survey that area for a transcontinental railroad route in 1850. The specific collection date is unknown (Baird and Girard, 1852a).

Distribution: *Sceloporus graciosus* occurs from north-central Washington southward through the Great Basin Desert to southern Nevada, and eastward to western Colorado, northern Arizona, and northwestern New Mexico. In California, it occurs along the Coast Range south to San Francisco and southward in the Sierra Nevada to the southern part of that range (Censky, 1986). A number of disjunct populations occur at the margins of this range.

Sceloporus graciosus occurs in the San Juan Basin and adjacent portions of the Colorado Plateau eastward to the margin of the Rio Grande Rift and southward along the Arizona border to the Zuni River.

Description: This is a small species of *Sceloporus*. Females reach a maximum of 76 mm SVL, and males a maximum of 63 mm SVL. The dorsal scales are keeled and pointed, but do not greatly overlap. The supraoculars are small and separated from the superciliaries as well as the median head scales by at least one row of smaller scales. Scales around midbody number 48–60 (mean 51) (Kerfoot, 1968; Degenhardt and Jones, 1972). There are 9–16 femoral pores on each leg, the two series usually separated medially by 13 or more (minimum 9) scales. The scales extending backward over the ear openings are small and number 6 or 7. The scales on the posterior surface of the thigh are relatively small and mostly granular.

The dorsal ground color is gray to brown. Two longitudinal series of more-or-less irregularly shaped dark brown spots extend posteriorly, one from the parietal region onto the base of the tail, and one from behind the eye to the groin. Each series may be fused to form a continuous, irregular dark stripe. The two dark stripes or series of spots are separated by a distinct light band, and the lateral series is bordered below by another light stripe that extends from the upper lips posteriorly through the anterior limb insertions to the groin. Males have dark blue lateral belly patches between the limb insertions. These belly patches are edged with black and may or may not contact medially. The ventral surface of the throat is usually flecked or mottled with blue. The ventral surface of females is usually plain white, although there may be a slight bluish tint on the throat. Females develop a lateral orange-red suffusion from the throat posteriorly onto the tail during vitellogenesis, which becomes most pronounced while carrying oviductal eggs, and fades and disappears after oviposition. Males may also develop brilliant nuptial colors of orange, light and dark blue, and black, involving the ventral and ventrolateral portions of the head, neck, chest, abdomen, and tail.

Similar Species: *Sceloporus undulatus* lacks the granular scales on the posterior surface of the thigh, has 7 or fewer scales separating the femoral pore series, and usually has paired blue throat spots (faint in females) rather than a mottled or flecked pattern. *Sceloporus scalaris* has the lateral scales arranged in rows parallel to the dorsal scale rows. *Sceloporus arenicolus* has a pale, unblotched dorsum, relatively fewer (41–52) scales around midbody, and occurs on active or semi-stabilized sand dunes in Chaves, Eddy, Lea, and southern Roosevelt counties.

Systematics: Four subspecies (including *S. g. arenicolus* treated here as a distinct species) were recognized by Censky (1986). The nominotypical form occurs in New Mexico.

Habitat: *Sceloporus graciosus* is normally terrestrial, but individuals can be found basking on dead snags one meter or more above ground. Adolph (1990) reported that the use of various habitat structural features such as tree limbs by lizards of this species is strongly constrained by thermoregulatory needs. *Sceloporus g. graciosus* occurs on distinctive soil associations which support piñon-juniper woodlands, sagebrush, and a variety of native grasses (Maker et al., 1978). It can be found sympatrically with *S. undulatus*, but seems to prefer more densely vegetated habitats (Douglas, 1966). It occurs in piñon-juniper woodlands of canyon or mesa slopes and ponderosa pine–Douglas fir forests of canyon bottoms, and boulder fields within oak *(Quercus gambelii)* thickets. In the San Juan Basin, it is often found on sandy soils which support or are near to areas of relatively dense vegetation (Harris, 1963). Gehlbach (1965) found this lizard in saltbush–sage ecotypes dominated by *Atriplex canescens, A. confertifolia,* or *Artemisia tridentata* on relatively level clay or sandy-clay soils between 1758–2182 m in the Zuni Mountains region, McKinley County. This species can be abundant; densities of 63–216 lizards/hectare have been reported (Tinkle, 1973; Burkholder and Tanner, 1974; Congdon and Tinkle, 1982; Deslippe and M'Closkey, 1991; Tinkle et al., 1993). Sagebrush lizards are sometimes found abundantly in association with prairie dog towns (Clark et al., 1982). Reynolds (1979) reported that clearing and replacing of native sagebrush habitats with crested wheatgrass *(Agropyron)* dramatically reduced the abundance of this species at several sites in Idaho.

Other lizard species that can be found in the same or immediately adjacent habitats as *Sceloporus graciosus* include *Gambelia wislizenii* (which probably includes sagebrush lizards in its diet), *Holbrookia maculata, Sceloporus undulatus, Cnemidophorus tigris,* and *C. velox*.

Behavior: Guyer and Linder (1985) reported diurnal activity beginning at 1000 and reaching maximum levels between 1100 and 1500 in Idaho. Adults became active in mid-April and entered hibernation at the end of August; corresponding juvenile seasons were from the end of May to mid-September. Adult activity peaked in May whereas that of juveniles peaked in August. In Utah, adults and juveniles have been observed active from March to August, and hatchlings until October (Burkholder and Tanner, 1974). New Mexico specimens have been collected between 5 May (San Juan County) and 9 October (McKinley County).

Congdon and Tinkle (1982) recorded an average body temperature of 35.1°C for 96 active lizards in southeastern Utah. Burkholder and Tanner (1974) reported that lizards are active at body temperatures between 28–37°C in north-central Utah, and that the critical thermal maximum is 43°C. An average preferred body temperature of 33.9°C for active lizards in Idaho was recorded by Guyer and Linder (1985). In southern California, Adolph (1990) measured average body temperatures of 34.1°C and 34.8°C for active lizards from high and low elevation populations, with an individual lizard's body temperature dependent upon the microhabitat in which it was collected. Individual lizards are able to withstand supercooling to −5.99° C (Lowe et al., 1971).

Both males and females of this species maintain exclusive territories in nature, although female territories are smaller and several may be overlapped by that of a single male (Deslippe and M'Closkey, 1991; Martins, 1991). Territories are formed and maintained both during and outside the breeding season. Average home range sizes of 423 m^2 and 563 m^2 were reported by Burkholder and Tanner (1974) for females and males in north-central Utah. Territory status and quality are visually advertised by individual lizards (Martins, 1991) using stereotyped push-up displays that serve to make an individual lizard appear bigger than it actually is and to expose bright throat and belly colors, the hues of which are enhanced during the breeding season.

These lizards are wary and difficult to catch by hand. Individuals may gape and attempt to bite if handled, but they are too small to inflict any damage on people.

Reproduction: Female *S. g. graciosus* larger than 50 mm SVL are sexually mature (Tinkle, 1973; Deslippe and M'Closkey, 1991; Tinkle et al., 1993); this size is not attained until the spring of their second full year of life in southwestern Utah, at approximately 22 months. Vitellogenesis is underway upon spring emergence in March, and may begin in September prior to hibernation. Mating activity occurs from mid-May through mid-June, with 1–2 clutches of 2–10 eggs laid by the end of July (Gehlbach, 1965; Douglas, 1966; Tinkle, 1973; Burkholder and Tanner, 1974; Deslippe and M'Closkey, 1991; Tinkle et al., 1993). Larger females produce larger clutches. Clutch weight may represent up to 37% of female body weight prior to oviposition (Burkholder and Tanner, 1974). Eggs hatch after an incubation period of 44–52 days. Hatchlings (23–27 mm SVL) first appeared at the end of July or the beginning of August during an 11-year study in southwestern Utah (Tinkle et al., 1993).

Tinkle (1973) and Burkholder and Tanner (1974) estimated hatchling mortality at 73–83%. Tinkle et al. (1993) found considerable annual variation in estimates of egg to yearling survival during an 11-year study, with values of 0.12–0.66. Adult females generally survived better than adult males in all but the oldest age-class. Individual lizards live to be older than 6 years (Tinkle et al., 1993) and may exceed 8 years of age (Stebbins, 1948).

Parker (1973b) and Burkholder and Tanner (1974) reported that male *S. g. graciosus* attain sexual maturity somewhere between 40–51 mm SVL. This size may be attained by juveniles late in the summer of their first full year of life, but these individuals are physiologically incapable of mating until after their second hibernation. Testes increased in size and weight to reach maximum values in May, followed by rapid regression to reach minimal values in July. Mature sperm are present in the reproductive tract from the end of May until the middle of July. Testicular recrudescence begins by August and continues through winter dormancy.

The fat body cycle is essentially similar for both sexes (Burkholder and Tanner, 1974). Fat bodies are depleted in both sexes during reproduction, and are replenished through the remainder of the activity season prior to winter dormancy. Some energy reserves are utilized during hibernation. Derickson (1974) reported that energy reserves stored in fat bodies, throughout the body, and in the tail are utilized for reproduction, as well as for other functions such as tail regeneration and hibernation. These reserves are insufficient for successful reproduction, however, and must be supplemented by feeding in the spring and early summer prior to breeding.

Food Habits: *Sceloporus g. graciosus* is reported to eat a wide variety of arthropods, including grasshoppers, termites, ants, a wide variety of beetles, hemipterans, homopterans, lepidopterans, flies, lycosid spiders, and pseudoscorpions (Douglas, 1966; Knowlton, 1938, 1969 [1971]). Ants and beetles were the most frequently consumed prey items reported by Burkholder and Tanner (1974), representing up to 74% and 12% of the diet. Ants and grasshoppers may be dietary staples for this species

(Hotton, 1955). An incidence of cannibalism was reported by Knowlton (1938). Sexual differences in the diet have been reported (Burkholder and Tanner, 1974), but this phenomenon requires further investigation.

Known predators of *S. graciosus* in New Mexico include *Masticophis taeniatus* and *Crotaphytus collaris*.

Remarks: The species was reviewed by Censky (1986). We follow Banta (1971) for the citation of the original description of this species.

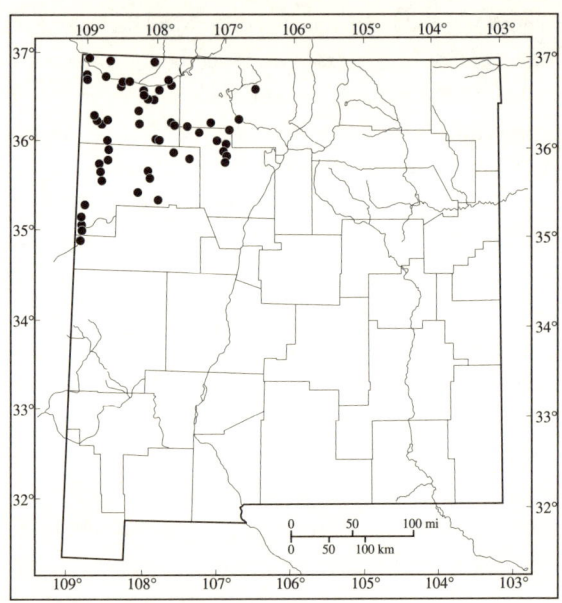

Distribution of Sceloporus graciosus in New Mexico.

SCELOPORUS JARROVII Cope *in* Yarrow, 1875 — *Yarrow's Spiny Lizard*

See Plate 49.

Type: There is much confusion regarding the material upon which the description of this species was based (Webb and Axtell, 1986). Three syntypes were indicated by Yarrow (1875) following the description by Cope, only two of which are currently extant (USNM 8495 and 8611) (Cochran, 1961). They were collected by Henry W. Henshaw on the topographical surveys west of the 100th meridian under command of 1st Lt. George M. Wheeler, Corps of Engineers, U.S. Army, from "southern Arizona" during 1873. Webb and Axtell (1986) designated USNM 8495, an adult male, and USNM 8611, a subadult female, as lectotype and paralectotype of *S. jarrovii*. They restricted the type locality to an area encompassing the southeastern Pinaleño, Dos Cabezas, and northwestern Chiricahua Mountains, Arizona.

Distribution: This species is widely distributed in the cordilleras of northern and central Mexico. It occurs southward in the Sierra Madre Occidental from the international boundary in Sonora and Chihuahua through Zacatecas and Guanajuato to Morelos, and southward in the Sierra Madre Oriental from the vicinity of Monclova and Bustamante in Coahuila and Nuevo Leon, to northern Veracruz, Hidalgo, and Querétaro (Morafka, 1977; Smith and Smith, 1976). In the United States, the species occurs as isolated populations in the Chiricahua, Dragoon, Huachuca, Pinaleño, and Santa Rita mountain ranges in southeastern Arizona. In New Mexico, *S. jarrovii* is found only in Hidalgo County at elevations of 1320–2597 m. It is widespread in the Peloncillo Mountains along the western border of the county south of Rodeo, the Animas Mountains, and the Alamo Hueco Mountains in the southeastern corner of the bootheel. There is an isolated record for the southern foothills of the Big Hatchet Mountains, and the species can also be found on scattered rocky outcrops and boulder fields in the intervening countryside. There is a single juvenal specimen (LACM 4472) from Redrock, Grant County; we consider this locality data to be in error.

Description: This is a medium-sized species of *Sceloporus*, with a maximum SVL of 105 mm for males and 97 mm for females (Bursey and Goldberg, 1994). Males are generally larger than females (Ruby and Dunham, 1984; Smith and Ballinger, 1994b). The dorsal scales on the body are weakly keeled and mucronate and only slightly imbricate, becoming more so on the tail. The supraocular scales are small, sometimes occurring in more than one row, and are separated from the median head scales by a series of small scales. Males are brightly patterned; the ground color is black, with a white spot in the center of each scale. There is a broad black collar around the neck, usually about four scales wide, and edged posteriorly with a narrow white band. The white band is usually interrupted medially, and the black collar may also be; if not, the white band usually has a distinct posterior indentation. The top of the head is usually black and often connects with the black collar, although there usually is a narrow longitudinal light line or scattered light scales on each side of the neck and head. The tail has narrow, indistinct black crossbands and becomes completely black toward the tip. The ventral surface is usually gray to black, and there is no gular fold. Large postanal scales are present in males, which also have distinctive blue patches ventrolaterally and on the throat. Females retain the basic dorsal pattern but are less brightly colored; they are generally irregularly spotted with gray dorsally, and the ventral surfaces are dull white. Blue throat and belly markings similar to those of males are subdued or absent.

Similar Species: *Sceloporus poinsetti* has distinctly imbricate scales, a distinctly crossbanded tail, and broad white borders to the neck collar. *Crotaphytus collaris* has small, nonimbricate scales and a double black collar around the neck.

Systematics: The infraspecific taxonomy of this species is badly in need of revision. Eight subspecies have been recognized (Smith and Smith, 1976); only the nominotypical form occurs in New Mexico.

Habitat: This species is found in rocky habitats including cliffs, boulder fields, and talus slopes in oak and pine-oak and pine woodlands from 1370–3550 m (Ortega et al., 1982; Stebbins, 1985; Middendorf and Simon, 1988). These lizards tend to avoid north-facing slopes or areas with heavy forest canopy (Burns, 1970). Ruby and Dunham (1984) estimated annual densities of 121–145 lizards/hectare over three years in the Pinaleño Mountains of southeastern Arizona, with considerable seasonal variation in density. Ballinger (1979) estimated densities of 32–71 and 148–229 lizards/hectare at two sites in southeastern Arizona, the first at an elevation of 1675 m and the other at 2542 m, over a 4-year period.

Behavior: Activity of individual lizards is strongly influenced by season, geographic location, and climatic conditions. New Mexico specimens have been collected between 5 January and 30 October. Individual adult lizards from a high elevation population (2895 m) in the Pinaleño Mountains of southeastern Arizona were active six out of every seven days during the summer; daily activity of lizards was bimodal on sunny days but not on cloudy ones (Beuchat, 1986, 1989). At lower elevations (ca. 1500 m), larger lizards are active early in the day due to thermoregulatory constraints (Simon, 1976a) and are active only a few days per week. Smaller lizards are active later in the day and on almost every day during the activity season. Body temperature of lizards from high elevations are significantly lower than those from low elevations (Smith and Ballinger, 1994b). Middendorf and Simon (1988) recorded average body temperatures of 33.6°C for adults and 32.7°C for juveniles in southeastern Arizona. Beuchat (1986) recorded average body temperatures of 34.5°C for males and post-parturient females and 32.0°C for pregnant females; Smith and Ballinger (1994b) also measured significantly lower body temperatures for pregnant females (30.7°C) than for non-pregnant ones (31.8). Burns (1970) measured average body temperatures of active lizards in June (33.2°C) and November (27.2°C), and Smith and Ballinger (1994b) found body temperatures to be significantly lower in winter (October–April) than in summer (May–September). Pregnant females were more active on sheltered substrates (cracks, under vegetative cover) than on exposed ones (rocks, logs, sand, tree trunks) compared to non-pregnant females (Smith and Ballinger, 1994b).

Territorial behavior (chasing other lizards, bobbing and pushup displays) is expressed by lizards as young as 13 days old (Simon and Middendorf, 1980). The selection of home ranges by individual lizards is strongly influenced by the presence of rocks with abundant crevices (Ruby, 1986). Both sexes establish territories upon commencement of seasonal activity in the spring, well before the reproductive season begins. Female home ranges average about 300 m^2 and do not vary significantly in size over the course of the activity season (Ruby, 1978). Male home ranges are significantly larger in the fall mating season (about 800 m²) than during the summer (500 m²). Simon (1975) found territory size to be inversely correlated with food abundance in all lizards except yearling females in southeastern Arizona. She estimated average home range sizes of 130 m² and 40 m² for adult males and females whereas home range sizes for yearling male and female lizards were considerably smaller (40 m² and 20 m²). Most lizards in the population studied by Ruby (1978) defended their entire home range. Both sexes were aggressive toward other individuals of either sex. As the breeding season approaches, males shift their territories to maximize overlap with those of females. Territorial behavior is apparently mediated by increasing plasma levels of testosterone in both sexes (Moore, 1986). Ruby and Baird (1994) observed significant social structure and behavioral differences in populations at different elevations in southeastern Arizona.

Ruby (1978) reported that territorial behavior temporarily ceased at the beginning of the rainy season in July. Swarms of ants or termites caused lizards to abandon aggressive interactions and form feeding aggregations until the insect swarms thinned.

These lizards abandon territories maintained during the summer to form winter aggregations of as many as 63 individuals, usually on south-facing rocky cliffs with deep crevices (Ballinger, 1973; Ruby, 1977). These sites are not usually occupied during the summer, and individuals may travel more than 200 m to reach them (Ruby, 1977). Homing behavior is well-developed in this species, but is dependent upon age and distance (Ellis-Quinn and Simon, 1989). Lizards are often active during the winter under favorable conditions, and basking speeds embryonic development in gravid females and growth in yearling animals (Ruby, 1977). The mean body temperature of 44 lizards measured by Congdon et al. (1979) basking during the winter was 32.7°C. Winter activity is facilitated by the ability of lizards to reduce their metabolic requirements by up to 40% compared to summer levels (Congdon et al., 1979; Gatten, 1985). Congdon et al. (1979) reported that lizards are relatively inactive from December through March with metabolic rates averaging only 15% of those during the summer active season. The ability of these lizards to tolerate harsh winter conditions was investigated by Lowe et al. (1971), who found that lizards could be maintained at body temperatures of −3°C for over 30 hours without adverse effects; lizards were able to withstand supercooling to −5.49°C.

Reproduction: This species is live-bearing. There are annual and geographic variations in reproductive parameters among populations (Ballinger, 1979; Ruby and Dunham, 1984). Female *S. jarrovi* are sexually mature at approximately 50–55 mm SVL. Ovulation occurs from mid-November to early December, with fertilization soon thereafter (Goldberg, 1971a). Individual females are able to store sperm over the winter. Embryonic development is delayed until April when warmer weather initiates placental development. The young are born in late June after a gestation period of approximately 6.5 months. Follicular development commences shortly thereafter, with increasingly rapid rates of yolk deposition from September to November (Goldberg, 1970).

Approximately 60% of the females in a population studied by Ballinger (1973) in the Chiricahua Mountains of southeastern Arizona attained sexual maturity at an age of 4–5 months, and the remainder did so the following year. All females greater than 75 mm SVL were found gravid during the reproductive season. Ballinger (1979) discovered that 41–84% of females at low elevations (ca. 1675 m) attained sexual maturity during their first year of life, whereas none did so at high elevations (ca. 2542 m); these differences may reflect a genetic basis (Smith et al., 1994). Litter sizes of 2–14 have been reported and are significantly correlated with body size (Goldberg, 1971a; Ballinger, 1973, 1979; Ruby and Dunham, 1984; Bursey and Goldberg, 1994); approximately one embryo is added for each 3 mm increment in SVL. Prenatal mortality in this population is approximately 4.5%.

Newborn lizards are 25–32 mm SVL and weigh 0.5–1.2 gm. Eighty to 90% of newborn lizards in the low elevation population studied by Ballinger (1973) did not survive their first month of life, and half of the remainder did not live to be one year old. Neonates at higher elevations survive significantly better (Ballinger, 1979), with an average of 31% reaching one year of age. For every 1000 young born in the low-elevation population studied by Ballinger (1973), 146 lizards reach one year of age, 86 reproduce at least once, 60 reproduce at least twice, and 20 reproduce three times. Ruby and Dunham (1984) recorded survivorship of male and female lizards less than 1 year of age marked during one summer to the following summer as 33% and 46%; comparable values for older lizards were 53% and 36%. Simon and Middendorf (1980) observed that only 43% of 56 young survived to 1 year of age and only one-third of those (8 lizards) survived to hibernate. Individual lizards may live to be 8 years old (Bursey and Goldberg, 1994).

Most males attain sexual maturity at 5–6 months prior to their first hibernation (Ballinger and Nietfeldt, 1989). The minimum recorded size at sexual maturity is 46 mm SVL. Male testes are at maximum size in September and October (Goldberg, 1971a). Testicular regression is complete by December, and testes remain small until July. The reproductive tract contains mature sperm from late August until early December depending upon annual and geographic variation.

Food Habits: These lizards eat a wide variety of insects and other arthropods using a sit-and-wait foraging strategy. The kinds and numbers of prey taken are associated with seasonal and yearly variation in primary productivity of the local habitat. Activity levels, growth rates, and survivorship of individual lizards are significantly influenced by food availability (Ballinger, 1980; Smith and Ballinger, 1994d, 1994e). Barbault et al. (1985) found caterpillars (40%), beetles (34%), and grasshoppers (26%) to be important by frequency, and caterpillars (61%) and grasshoppers (44%) by volume, in the diet of a population over a 2-year period in Durango, Mexico. Simon (1975) found ants and wasps (33.6–64.5%), flies (6.4–30.6%), and beetles (7.0–20.9%) important by frequency over a summer season in southeastern Arizona. Beetles (15–27%), grasshoppers (10–19%), ants and wasps (15–34%) and insect larvae (4–16%) were the most important food items volumetrically (Ballinger and Ballinger, 1979) at different elevations during a year of high resource availability in southeastern Arizona. A subsequent year of low resource availability found the same populations consuming mostly beetles (17–56%) and ants and wasps (5–47%) by volume; insect larvae were consumed in low numbers (1–9%) and grasshoppers (2%) were only eaten by the low elevation population. Bursey and Goldberg (1993) found homopterans (81% by frequency, 41% by volume) and ants (68% and 17%) to be important components of the diet of lizards less than 14 days old. Potential intraspecific competition for food is minimized through active prey-size selection by individual lizards (Simon, 1976b; Bursey and Goldberg, 1993). Larger lizards of both sexes eat larger prey than do smaller ones, and females of a given size eat larger prey than do males.

As noted above, lizards of this species are active during the winter. Individuals lose about 22% of their body weight during the winter aggregation period (Congdon et al., 1979). Only about one-half of total energy expenditure during this period can be accounted for by stored body fat; therefore, lizards feed during this time, but at a rate approximately 1/5 of that during the summer. Goldberg and Bursey (1990) recorded some striking differences in prey taken between seasons; ants were much more important during the summer and were replaced by hemipterans during the winter.

Remarks: Mahrt (1989) reported that about 40% of lizards in eight different populations in southern Arizona are naturally infected with the malarial parasite *Plasmodium chiricahuae*. Neonatal *S. jarrovii* exhibit a high prevalence of infection by the nematode, *Spauligodon giganticus*, apparently as a direct consequence of innate substrate-licking behaviors by the lizard (Goldberg and Bursey, 1992c). Chigger mites also infect this species (Smith, 1946; Goldberg and Bursey, 1993).

This species was named for Henry C. Yarrow, who served for a time following the Civil War as honorary curator of amphibians and reptiles at the National Museum of Natural History.

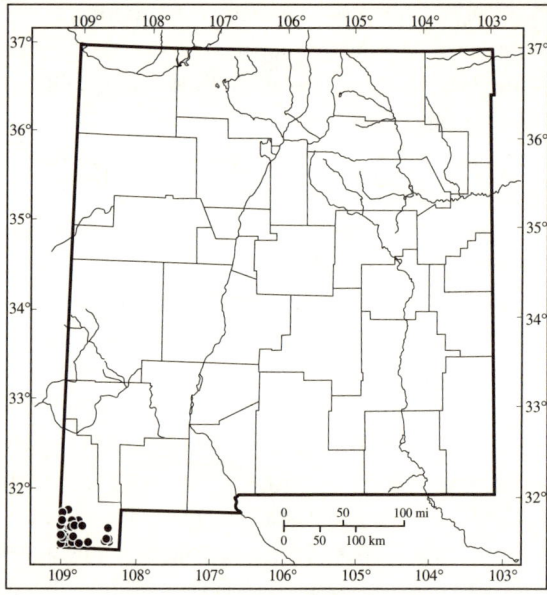

Distribution of Sceloporus jarrovii in New Mexico.

SCELOPORUS MAGISTER Hallowell, 1854   *Desert Spiny Lizard*

See Plate 50.

Type: The type specimen (USNM 2967) is an adult male, collected from "near Fort Yuma, at junction of Colorado and Gila, also near Tucson in Sonora" by A. L. Heermann sometime in 1853. The type locality was restricted to Yuma, Yuma County, Arizona, by Smith and Taylor (1950b).

Distribution: This species occurs in the southern Great Basin Desert from the vicinity of Pyramid Lake in western Nevada, southward through the Mojave and Sonoran deserts of Arizona and California to the southern tip of Baja California, and to northern Sinaloa in mainland Mexico. It also occurs on a number of islands in the Sea of Cortéz (Murphy and Ottley, 1984). It ranges eastward through the upper Colorado River Basin of the Four-Corners region, and southward through Arizona and northern Mexico through the Chihuahuan Desert of southern New Mexico and Trans-Pecos Texas as far south in Mexico as the vicinity of Torreón, Coahuila, and northeastern Durango (Bolsón de Mapimí) (Morafka, 1977; Parker, 1982a).

We are confident that this species is more widespread in New Mexico than the distributional records indicate, and that these latter reflect sampling biases of herpetologists as well as difficulties encountered in collecting this shy and secretive lizard. There are scattered records all along the Arizona border, from the Four-Corners region to the area of San Simon Cienega and Granite Gap in the Peloncillo Mountains, Hidalgo County, as well as in the Pecos River valley south of Carlsbad. The majority of the distribution of this species occurs within the Rio Grande Valley from the Texas border northward to Los Lunas, Valencia County, and is striking in the almost precise congruence with the thermic, light-colored soils defining the northern extension of the Chihuahuan Desert in New Mexico (Morafka, 1977; Maker et al., 1978). Elevation ranges from 900 m to about 1980 m.

Description: This is a large, sexually dimorphic species of *Sceloporus*. Males attain a maximum SVL of 140 mm and females a maximum SVL of 112 mm (Parker, 1982a). The average size recorded for adult males and females by Vitt et al. (1981) were 105 mm SVL and 48 gm, and 92 mm SVL and 32 gm. The body is stocky and the hind legs relatively short. The dorsal scales are large, strongly keeled, mucronate, and imbricate. There are 10–16 femoral pores on each leg. There are five enlarged supraocular scales, the two posterior in contact with the median head scales. There are two parietal scales on either side of the interparietal scale. The sublabial scales are in contact with the mental. The 5–7 scales extending backward over the ear opening are long and pointed with the median ones largest.

There is a distinct black wedge-shaped mark on the shoulder on each side of the body. The dorsal coloration varies from pale yellow to gray or brown. The dorsal surfaces of the legs are not distinctly marked, or may have thin, faint longitudinal dark striping. The tail lacks distinct crossbands. Juvenal lizards possess dorsal and lateral longitudinal rows of dark brown blotches that become more diffuse or merge transversely as the lizards age. Lizards from northwestern New Mexico have yellowish-orange heads as adults, and the juvenal body blotches have merged to form a series of 5–6 chevron-like markings across the body between the head and tail. Lizards from elsewhere in the state lack the yellow head coloration, usually possess a distinct dark stripe behind the eye, and have two parallel rows of diffuse dark dorsal blotches; the lateral juvenal series may be obscured or entirely absent. Adult males have blue, blue-green, or black throats, lateral chests and abdomens; the belly patches are lined with black and may meet ventrally. The ventral coloration of females is creamy white, with the blue-green coloration reduced or absent.

Similar Species: *Sceloporus poinsetti* and *S. jarrovii* have complete neck bands. *Sceloporus clarkii* has distinctly barred forearms, has 3 or fewer auricular scales that are rounded with the uppermost largest, and the sublabial scales are separated from the mental.

Systematics: Nine subspecies have been recognized (Parker, 1982a); however, Murphy and Ottley (1984) returned three of the forms found on islands in the Sea of Cortéz to full species. *Sceloporus m. cephaloflavus* occupies the San Juan Basin in the northwestern corner of the state. This subspecies was originally described by Tanner (1955). The holotype (BYU 11270) was collected from "Bentley's Cabin, approximately 15 miles NW of Hole-in-the-Rock, Kaiparowits Plateau, Kane County, [Utah]" by D. Elden Beck on 16 July 1953. The remaining range of the species in New Mexico is occupied by *S. m. bimaculosus*, described by Phelan and Brattstrom (1955) based on an adult male (CAS 91199) collected from "6.6 mi. E. of San Antonio, Socorro County, New Mexico" by Richard G. Zweifel and Kenneth S. Norris on 24 August 1948. The infraspecific taxonomy of this species is badly in need of revision (Parker, 1982a).

Habitat: This lizard is distinctly arboreal throughout most of its range, but can be saxicolous or terrestrial (Parker and Pianka, 1973). Tinkle (1976) found hatchlings to be primarily terrestrial in southern Utah. The species appears to reach its greatest abundance in riparian habitats (Vitt et al., 1981); densities of 7–50 lizards/hectare have been reported (Tanner and Krogh, 1973; Tinkle, 1976). Tanner and Krogh (1973) reported higher densities in structurally complex habitats. Throughout much of New Mexico, these lizards are mostly terrestrial and closely associated with dense stands of mesquite, creosotebush, or tarbush along arroyos or playa edges, often utilizing packrat middens of the white-throated woodrat (*Neotoma albigula*) for shelter (Parker and Pianka, 1973; Whitford and Creusere, 1977; Baltosser and Best, 1990; our unpubl. data).

Behavior: The activity season for this species has been reported to extend from March through October (Tanner and Krogh, 1973; Baltosser and Best, 1990). New Mexico specimens have been collected between 31 March (Doña Ana Co.) and 14 November (Socorro Co.). Creusere and Whitford (1982) recorded a bimodal activity period from May through October in southern New Mexico, with individuals active until 1130 and again after 1530. Parker and Pianka (1973) recorded an average body temperature of 34.8°C for 92 active lizards throughout the range of the species. In Arizona, individuals maintain daily body temperatures between 35° and 36°C by shuttling back and forth between open sunlight and shelter on the ground, in rock crevices, or under tree bark (Vitt et al., 1981). Bogert (1949) and Tanner and Krogh (1973) recorded average body temperatures of 34.9°C (32–37°C) and 32.8°C (29–35°C) for active lizards in Arizona and southern Nevada. Individuals are able to withstand supercooling to –5.25°C (Lowe et al., 1971).

These lizards appear to be territorial, and very often associate in pairs consisting of an adult male and female (Parker and Pianka, 1973; Tanner and Krogh, 1973).

Reproduction: The minimum size recorded at sexual maturity for females is 80 mm SVL, which is achieved at an age of 21–23 months (Tinkle, 1976, our unpubl. data). Gravid females have been recorded from April through August throughout the range of the species, and individual females may produce two clutches of eggs annually (Parker and Pianka, 1973; Tinkle, 1976). Smaller females produce smaller clutches; Parker and Pianka (1973) recorded an average clutch size of 7.4 (5–10) eggs for females 81–99 mm SVL and 8.7 (3–12) for females 100–112 mm SVL. Tinkle (1976) determined an average clutch size of 5.3 (2–7) eggs for females breeding for the first time and 6.8 (4–9) for older females. Tanner and Krogh (1973) reported one clutch/year in Nevada averaging 7 (4–10) eggs. Hatchlings (30–40 mm SVL) have been recorded from late July to late August in Nevada (Tanner and Krogh, 1973) and from mid-May through September elsewhere (Parker and Pianka, 1973; Tinkle, 1976; our unpubl. data).

Male testes are enlarged from April to mid-June, with mature spermatozoa present in the reproductive tract from early May through July (Parker and Pianka, 1973). Decline in testicular size is rapid in late June and July, and then begins enlarging again following a quiescent period in August. Males do not attain sexual maturity in Nevada until the breeding season following their second hibernation (Tanner and Krogh, 1973). The smallest mature male these authors measured was 83 mm SVL.

Longevity in nature may be as great as 6 years (Tanner and Krogh, 1973).

Food Habits: Hotton (1955) reported the bulk of the diet to consist of ants and caterpillars. Occasionally, lizards of other species may be eaten. In southwestern Colorado, ants are a major component of the diet, with beetles second in importance. Females eat more food as a percentage of body weight in the spring than males, presumably as a proximate boost to reproductive efforts (Johnson, 1966). Tanner and Krogh (1973) also found ants to be the primary component of the diet (62%) in southern Nevada. Parker and Pianka (1973) reported ants as the most frequently eaten prey items (80%), with adult beetles the most important volumetrically (51%).

Known predators of this species in New Mexico include *Gambelia wislizenii* and *Masticophis flagellum*.

Remarks: This species was reviewed by Parker (1982a). The etymology of the Latin name *magister* (meaning "teacher") is obscure; however, many early authors (Stejneger, 1890a; Cope, 1898 [1900]) remarked upon the pugnacious behavior of individuals of this species, and the first collectors of large specimens may well have been "taught" a lesson carelessly handling these lizards.

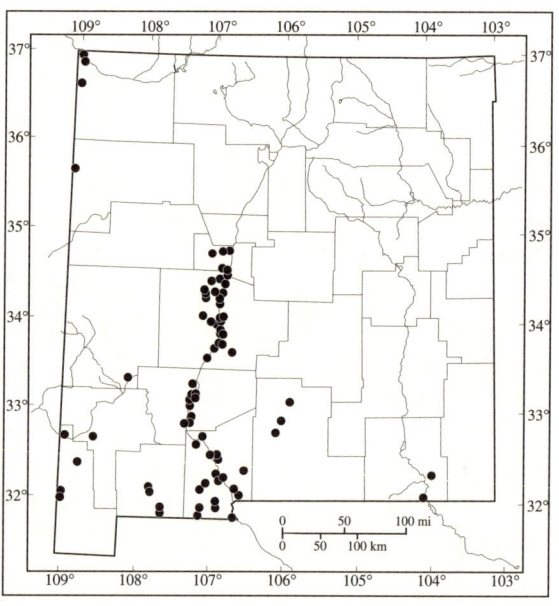

Distribution of Sceloporus magister *in New Mexico.*

SCELOPORUS POINSETTI Baird and Girard, 1852 *Crevice Spiny Lizard*

See Plate 51.

Type: Five syntypes (USNM 2948, 2 specimens; USNM 2952, 2 specimens; and USNM 131668) have come to be associated with the original description by Baird and Girard, 1852a (Cochran, 1961). The specimens were collected from the "Rio San Pedro of the Rio Grande del Norte, and the province of Sonora" by John H. Clark, under Col. J. D. Graham, during the Mexican boundary survey. Webb (1988) designated the adult male and female labeled with USNM 2952 as lectotype and paralectotype. Following an extensive reconstruction of the itinerary of the parties involved, Webb (1988) restricted the type-locality to "either the southern part of the Big Burrow [*sic*] Mountains, or the vicinity of Santa Rita, Grant Co., New Mexico" and the date of collection to late August 1851.

Distribution: This species is widespread in desert and semi-desert regions from southwestern New Mexico eastward to the eastern margin of the Edwards Plateau in central Texas, and southward in central Mexico to the states of Chihuahua, Coahuila, Durango, Sonora, Sinaloa, northern Zacatecas, and eastern Nuevo León (Dixon, 1987; Morafka, 1977; Smith and Smith, 1976; R. G. Webb, pers. comm.). Populations on the eastern margin of the range in Texas are closely associated with major river canyons draining the Edwards Plateau, and do not occur on the intervening uplands (Axtell, 1987). The eastern border of its range in Mexico appears to be circumscribed by the limits of the Chihuahuan Desert as defined by Morafka (1977).

In New Mexico, the range of the species is bisected by the valley of the Rio Grande and the associated northward extension of the Chihuahuan Desert along this corridor. *Sceloporus poinsetti* is abundant throughout the Mogollon Plateau and the Black Range, and occurs as isolated populations in associated mountain ranges such as the Magdalenas and San Mateos. To the east, it occurs in the Sacramento Mountains southward throughout the Diablo Plateau and the Guadalupe Mountains to the Texas border. The two specimens from the Rio Grande Valley proper are curious; they may represent waifs (cf. Axtell, 1987) or they may reflect a former widespread abundance in suitable habitat along this corridor. In the latter event, the apparent absence of *S. poinsetti* from the Organ–San Andres mountain axis is also curious. We question the specimen record from Deming, Luna County (ANSP 21121), collected by H. A. Pilsbry during August 1915. It is possible that the species occurs in the nearby Florida Mountains, but there are no additional records from there. Given that Pilsbry also collected during this time in the Mimbres Valley and Black Range (E. V. Malnate, *in litt.*), we suspect that this specimen may have come from the latter area.

Description: This is a large *Sceloporus*, with a maximum recorded SVL of 137 mm for males, which are the larger sex (Ballinger, 1973). The body is somewhat flattened dorsoventrally, and the dorsal scales are large, strongly keeled, mucronate, and imbricate, especially on the tail. The enlarged supraocular scales are in two rows. The dorsal ground color is usually gray or olive-gray, but may be yellowish to reddish. There is a distinct black collar around the neck, 2–3 scales wide, and not interrupted medially. It is bordered anteriorly and posteriorly by a white band usually two scales wide. Young lizards manifest a distinct series of dark brown or black dorsal crossbands from the head just behind the ear openings to the tip of the tail. This pattern is retained by adult females, but in adult males it is often reduced to a longitudinal series of middorsal blotches on the body or it may be lost altogether except on the tail. The throat is bright blue in adult males, which also manifest a bright blue patch ventrally, bordered with black medially, on either side of the body. The throat region of adult females is usually grey but may be pale blue with some indication of belly patches in some specimens, and the remainder of the ventral surface is usually unmarked.

Similar Species: *Sceloporus jarrovii* is smaller, has smaller, weakly keeled, and imbricate scales, lacks any crossbanding on the body, and lacks an anterior white border to the collar band. *Crotaphytus collaris* has small, non-imbricate scales and a double black collar around the neck.

Systematics: Three subspecies have been recognized

(Smith and Smith, 1976); only the nominotypical form occurs in New Mexico.

Habitat: This is a species of rocky habitats, wooded or otherwise, from mesquite-grasslands and desert shrublands through oak-piñon-juniper habitats to spruce-fir forests. It has been reported from elevations as high as 2818 m in the Guadalupe Mountains of Texas (Mosauer, 1932). It is one of the most common lizard species encountered in the rugged, broken limestone country between the Sacramento and Guadalupe mountains and the Texas border.

Behavior: New Mexico specimens have been collected between 12 February (Eddy County) and 24 November (Socorro County). These lizards are wary and invariably do not stray far from narrow fissures in bedrock in which they take shelter when disturbed (Mosauer, 1932; Ballinger, 1973). Once in a crevice, individual lizards inflate their bodies which, combined with their large and strongly keeled scales, makes them difficult to extract. Bogert (1949) recorded an average body temperature of 34°C (31–38°C) for 19 active lizards at 1100 m elevation in Durango, Mexico. Adult lizards pair during the mating season; otherwise they live in social groups of 3–6 individuals dominated by one adult male.

Reproduction: *Sceloporus poinsetti* mates in the fall. Testes weights increase an order of magnitude from minimal values in April to maximal extent in October, accompanied by corresponding morphological changes in associated structures (Ballinger, 1973, 1978). Sperm formation begins in August and peaks in October.

This species is live-bearing and produces one litter a year in western Texas and southwestern New Mexico (Ballinger, 1973, 1978). Females possess yolked follicles in October and embryos or embryo discs by March or early April. Development proceeds rapidly thereafter to parturition during the last half of June. Female *S. poinsetti* less than 85 mm SVL are sexually immature in western Texas, whereas minimum reproductive size in southwestern New Mexico is 75 mm SVL. Individual lizards attain sexual maturity in their second mating season following birth, at 15–17 months of age. Litter size averages 10.45 and is correlated with female body size; litter size increases by approximately one embryo for each 3 mm increment in SVL (Ballinger, 1973). Greene (1970) reported a litter of 16 born to a single 102 mm SVL female from western Texas. Approximately 19.7% of the eggs of females in New Mexico were unfertilized versus 2.7% of those in western Texas (Ballinger, 1978). The corresponding male/female sex ratios are 1:2 and 1:1. Newborn lizards are 29–35 mm SVL and weigh 1–1.6 grams. Very few hatchlings in New Mexico survive; the ratio of yearlings to adults is 1:9 (Ballinger, 1978). Eighty to 90% of hatchlings in western Texas do not survive their first month of life, and half of the remaining cohort do not reach one year of age. For every 1000 eggs ovulated in the population studied by Ballinger (1973), 155 young reached one year of age, 65 lizards reproduced at least once, 54 reproduced twice, and 43 reproduced three times during their adult lives.

Food Habits: Barbault et al. (1985) reported termites (55%), caterpillars (46%), and plants (23%) by frequency, and caterpillars (52%), plants (40%), and grasshoppers (33%) by volume as important dietary components of this species over a 2-year period in Durango, Mexico. The plant material consisted mainly of flowers of the family Asteraceae. There were marked differences in the rank-order of prey categories in different years. Grasshoppers (42%) and ants (14%) were important components of the diet by volume of a small sample of lizards taken by Smith and Milstead (1971) from Big Bend National Park, Texas. Plant material, including fruit from *Opuntia* and leaves of *Aloysia* and *Parthenium*, composed 26% of the diet in this sample. Ballinger et al. (1977) reported an ontogenetic shift in diet for this species from southwestern New Mexico. Juveniles were primarily insectivorous (94% by volume, with beetles, ants, and insect larvae important) whereas adults were significantly herbivorous (42% by volume). These authors suggest that older adults increase the plant proportion of their diet because it is energetically expensive to chase small insects, and they note that large food items such as beetles and grasshoppers become more important late in the season, after summer rains, with a concomitant decrease in vegetation (22% by volume).

Remarks: According to Baird and Girard (1852a), "at the request of Col. J. D. Graham, we have dedicated this species to the memory of the late Hon. Joel R. Poinsett,

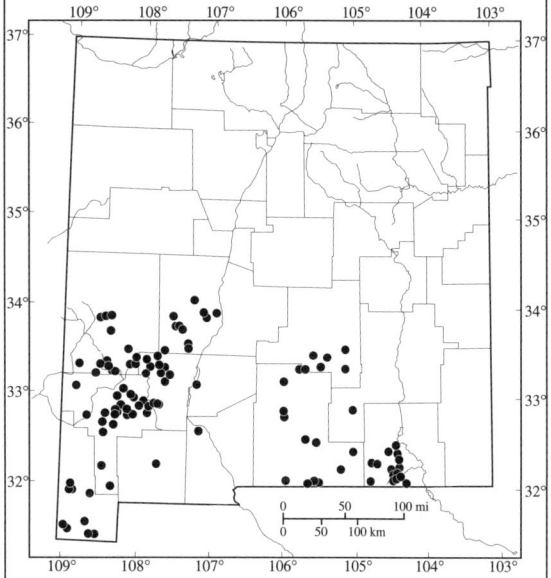

Distribution of Sceloporus poinsetti in New Mexico.

whose name is associated with the progress of science and the useful arts throughout his public career, especially while Secretary of War of the United States."

We retain the use of a single *i* in *poinsetti* following the original description (Baird and Girard, 1852a).

SCELOPORUS SCALARIS Wiegmann, 1828   *Bunch Grass Lizard*

See Plate 52.

Type: No type specimen was designated nor type locality given in the original brief description by Wiegmann (1828), in which he described a number of genera and species from specimens collected by Ferdinand Deppe between 1824 and 1828 in Mexico and sent to the University of Berlin. Smith and Taylor (1950b) restricted the type locality to the Distrito Federal, Mexico, and stated that the type specimen was in the Zoological Museum of Berlin; it was probably destroyed during World War II along with that institution, if indeed it was still extant at that time.

Distribution: The species is distributed along both axes of the Sierra Madre northward. In the east it occurs from the western border of Veracruz and the state of Puebla to central Nuevo Leon and southeastern Coahuila, Mexico, and in the west to southeastern Arizona and adjacent New Mexico. The species apparently occurs as a series of isolated populations in the northern portions of its range, such as the Sierra del Nido in Chihuahua and the Huachuca and Chiricahua mountains in southeastern Arizona. It was first reported from New Mexico by Dixon and Medica (1965) and is currently known from only the southern end of the Animas Valley between the Animas and Peloncillo mountains in Hidalgo County at elevations of 1560–1610 m.

Description: This is a small species of *Sceloporus*, with a maximum SVL of 68 mm. Females are larger than males, averaging 50–56 mm SVL and 46–48 mm SVL (Newlin, 1976; Ballinger and Congdon, 1981; Smith et al., 1993). During 1994–95, 47 adult females collected from the Animas Valley in Hidalgo County averaged 51.9 (41–68) mm SVL and 3.6 (1.8–7.3) gm. Based on maximum size of adults, if we assume that males mature at slightly smaller SVL than females, then 59 adult males averaged 47.9 (35–59) mm SVL and 28 adult males averaged 3.3 (1.4–4.6) gm (our unpubl. data). The dorsal scales are keeled and mucronate, but only slightly overlap. The lateral scales can be keeled or smooth. There are 38–45 dorsal scale rows around midbody, and the lateral scales are arranged in rows parallel to the dorsal scale rows. There are four or five relatively large supraocular scales, usually separated from the superciliaries and the median head scales by at least one complete row of smaller scales. There are only two postrostral scales. There are 12–18 femoral pores on each side, and the two series are in contact medially or separated by one or two scales. There are usually three to four short, broad scales extending backward over the ear opening. The scales on the posterior surface of the thigh are small and nonoverlapping. There is no gular fold.

Some lizards of this species from southeastern Arizona lack a dorsal pattern altogether, and the dorsal ground color in these individuals is gray or yellowish to dark brown. Otherwise, the ground color is light brown and a pair of distinct light lines, one scale wide and separated from each other by 7–8 scale rows, extends from the parietal region of the head posteriorly onto the tail. These lines may join anteriorly in a broad, diffuse light band through the lower portion of the eye. A paired series of 10–13 irregular markings extend backward from behind the head to the base of the tail. These markings are crescent-shaped and concave anteriorly, are dark brown and broadly edged posteriorly with black and then narrowly with white. Another longitudinal light line extends laterally from the lips through the ear opening onto the front of the hind leg. Another series of dark crescent-shaped blotches extends along the sides of the trunk between the two light lines. There is usually a black patch enclosing a blue spot at the base of each front leg. Transverse lateral dark bars, if present, do not extend onto the abdomen which is immaculate except for paired lateral navy blue patches in males. There is a continuous median dark dorsal stripe on the tail.

Similar Species: *Sceloporus virgatus* and *S. undulatus* have blue patches on the throat and broad longitudinal light dorsal lines. They lack granular scales on the posterior surface of the thigh and the longitudinal series of dorsal chevron markings. *Sceloporus graciosus* has two narrow, serrated, lateral light stripes, and the femoral pores are separated by 9 or more scales medially. All other species of *Sceloporus* in the United States, including those just mentioned, have lateral scale rows that are arranged obliquely to the dorsal scale rows, and trend upward posteriorly.

Systematics: Although Thomas and Dixon (1976) regarded the species as monotypic, others (i.e., Collins, 1990) apparently follow Smith and Hall (1974) in recognizing four subspecies within this taxon. In this case, the subspecies *S. s. slevini* Smith, 1937 occurs in New Mexico. The type specimen (CAS 48103) was collected from "Miller Peak, Huachuca Mountains, Cochise County, Arizona" by Joseph R. Slevin on 7 July 1920.

Habitat: As the vernacular name suggests, this species is strongly associated with habitats supporting dense grass cover, generally at elevations between 1500 and 3000 m (Ortega and Barbault, 1986). Van Devender and Lowe (1977), however, found it in forest and woodland habitats in Chihuahua, Mexico, and stated that it did not require "bunch-grass" habitat. Newlin (1974), Ballinger and Congdon (1981), and Smith et al. (1993) found this species to inhabit areas of 80–90% grass cover (*Blepharoneuron tricholepis*, *Muhlenbergia virescens*, and *Bromus frondosus*) in the Chiricahua Mountains of southeastern Arizona. They recorded densities of 80–300 lizards/hectare over several years, with the greatest abundance occurring on east- and southeast-facing slopes. Bock et al. (1990) found this species common in ungrazed grama grasslands of the Sonoita Plains in southeastern Arizona at elevations as low as 1300 m. It was virtually absent in grazed areas within this region, and Bock et al. (1990) believe that its apparent restriction to montane meadows may be an historic artifact associated with chronic and ubiquitous grazing of lower-elevation perennial grasslands. In New Mexico, Dixon and Medica (1965) recorded this species from dense alkali sacaton (*Sporobolus airoides*) at an elevation of 1575 m in the Animas Valley between the Animas and Peloncillo Mountains in southwestern Hidalgo County. Recent trapping efforts in the same area indicate the species occurs in a variety of grassland types that include *Bothriochloa barbinodis*, *Bouteloua gracilis*, *B. hirsuta*, *Buchloe dactyloides*, *Hiluria mutica*, *Panicum hallii*, and *P. obtusum* (Painter, 1995).

Behavior: These small, secretive lizards scurry between bunch grasses in sunlit areas, and may also seek shelter under rocks, logs, and other objects. Individuals rarely climb. This species is apparently not territorial, and aggressive behavior between individuals is rarely observed (Newlin, 1976). Lizards are active throughout the year. Lizards of this species are active thermoregulators, maintaining body temperatures significantly higher than ambient; Smith et al. (1993) measured an average body temperature of 32.6°C for 851 lizards in southeastern Arizona. Body temperatures were higher in the summer than during winter, and vary considerably with location and microhabitat. Newlin (1974) measured the average critical thermal maximum temperature as 44°C, and found the highest body temperature for an active lizard during the summer to be 38°C. Ballinger and Congdon (1981) recorded the average body temperature of 23 lizards active during one February day as 31.1°C when ambient air temperature at the same time averaged 12.0°C. Males had significantly higher body temperatures than females (mean 33.3°C vs. 32.4°C), but average body temperatures of gravid females did not differ significantly (Smith et al., 1993). This species is able to withstand supercooling to −6.68°C (Lowe et al., 1971).

Reproduction: The minimum size at sexual maturity for females is 41 mm SVL (Newlin, 1976; Ballinger and Congdon, 1981), and all females are reproductive in their first spring at approximately 8 or 9 months of age. Mating takes place during April in southeastern Arizona (Ballinger and Congdon, 1981) and has been observed in May and July in Durango, Mexico (Ortega and Barbault, 1986). Ovulation occurs from late May through mid-June, oviposition begins in late June–early July, and hatchlings appear in September. Hatchlings as small as 20 mm SVL and 0.2 gm were collected in the Animas Valley in southwestern Hidalgo County on 5 August (our unpubl. data). Considerable embryonic development occurs in the eggs prior to oviposition (Newlin, 1976; Ballinger and Congdon, 1981; Ortega and Barbault, 1986), oviductal eggs being retained for at least 50 days. Fat bodies increase from August through October in females to reach up to 10% of the total body weight, decrease slightly during the winter, and rapidly decrease at the onset of vitellogenesis until oviposition.

Clutch size apparently does not vary annually in relation to primary productivity at a given locality (Ballinger and Congdon, 1981) but may do so geographically. Two separate studies in southeastern Arizona reported females less than 53 mm SVL averaged 6.22 (Newlin, 1976) and 5.90 eggs/clutch (Ballinger and Congdon, 1981), whereas females larger than 52 mm SVL averaged 10.54 and 9.99 eggs/clutch. A single clutch is produced each year in Arizona, the size of which varies significantly with female body size. Approximately one egg is added to the clutch for each 3 mm and 2 mm increment in SVL for yearling and older females (Newlin, 1976). Clutch weight approaches 50% of female body weight prior to oviposition (Newlin, 1976). About 40% of the females studied in the Durango population produced a second annual clutch. The average clutch size reported there was 8.79 (5–10) eggs, and a clutch represented 37% of an individual female's body weight just prior to oviposition (Ortega and Barbault, 1986).

Testes reach maximum size in April, followed by a marked decrease in testicular activity to minimal levels from August through November (Newlin, 1976; Ortega and Barbault, 1986). There is a sharp increase in testicular size and weight from November to April. Histological examination of the male reproductive tract corresponded with gross morphological changes (Newlin, 1976). Mature sperm were present from mid-May through mid-June, followed by rapid regression and recrudescence beginning in September. Fat bodies were smallest in May and June, representing less than 1% of individual male body weight, and increased to 5% of body weight in February prior to sperm production.

Ballinger and Congdon (1981) estimated hatching failure as 12.7% and yearling mortality as 75%. They found that although longevity may be as long as five years, only 63 of 1000 lizards lived to be two years old or older, and 83% of the reproductive effort of the population under study by these authors was contributed by lizards less than two years old. Females outnumber males by as much as 2 to 1 in several populations in southeastern Arizona.

Food Habits: Not much is known concerning this species. Newlin (1974) found homopterans/hemipterans (53% by frequency, 31% by volume) and ants (36% by frequency, 17% by volume) to be the most important food item in southeastern Arizona. Lizards in this population were selective feeders; termites, the second most abundant arthropod in the area, were not eaten at all. Barbault et al. (1985) reported adult beetles (40%), ants (36%), and hemipterans (26%) by frequency, and grasshoppers (68%)

and beetles (35%) by volume as important food items over a 2-year period in Durango, Mexico. There were marked differences in the rank-order of prey categories in different years of the study.

Remarks: Growth rates of individual lizards are significantly correlated with primary productivity of the environment, and are higher during years of high annual rainfall (and increased insect abundance) than during drought years (Ballinger and Congdon, 1980). Goldberg and Bursey (1992a) found 8% of a sample of lizards from southeastern Arizona to be infected with helminth parasites. This species was thought to give birth to live young by many early workers (e.g. Smith, 1946). The bunch grass lizard is listed as endangered by the New Mexico Department of Game and Fish (NMGF, 1990). Overgrazing has probably reduced habitat for this species in southwestern New Mexico (*fide* Bock et al., 1990).

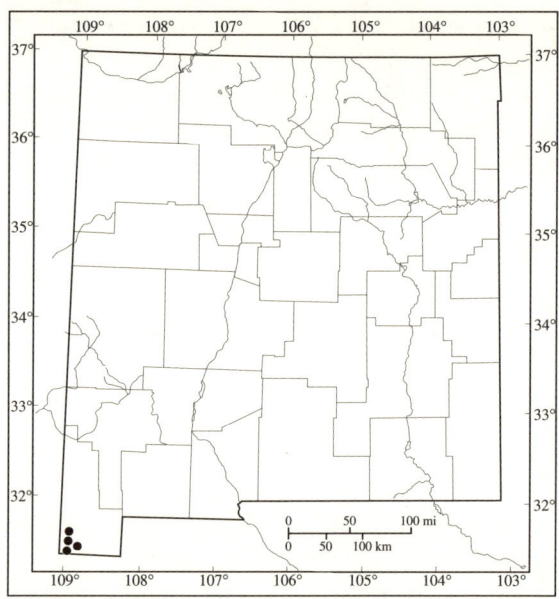

Distribution of Sceloporus scalaris *in New Mexico.*

SCELOPORUS UNDULATUS (Bosc and Daudin, *in* Sonnini and Latreille, 1801)   *Prairie Lizard*

See Plate 53.

Type: No type material was designated in the original description, which was in a manuscript by Louis A. G. Bosc (then the French consul at Charleston, South Carolina) and later sent to Paris after the French Revolution. The type locality was given as "les grands bois de la Caroline," and has been restricted by Smith (1938a) to near Charleston, South Carolina. See Smith (1938a) and Harper (1940) for further discussions on this matter, and Adler (1989) for bibliographic sketches of some of the principals involved.

Distribution: This is an extremely wide-ranging species, occurring in 32 U.S. states, and the Mexican states of Coahuila, Chihuahua, Durango, Nuevo León, San Luis Potosí, Sonora, and Zacatecas. See Conant and Collins (1991, and references therein) for further details on the distribution of this species in the United States. The species occurs throughout New Mexico at elevations of 900 m to approximately 2750 m where suitable habitat exists.

Description: Size, color pattern, and scalation vary with geographic location. Generally, females are larger than males. Gehlbach (1965) reported mean size for males from the Zuni Mountains as 51.8 (40–63) and females as 59.1 (43–70) mm SVL. Douglas (1966) reported that males averaged 63 mm and females 67 mm SVL in the Four-Corners area of Colorado, with the largest individuals 80 mm and 90 mm SVL. Tinkle and Ballinger (1972) reported the mean body size of reproductive females from Colorado and Texas as 57 mm and 70 mm SVL, and Ferner (1976) reported adult males and females from Colorado averaging 62 and 67 mm SVL. During 1987–92, 259 specimens collected at Antelope Pass in Hidalgo County averaged 54.5 (14–76) mm SVL and 6.2 (0.2–16.9) gm. One hundred one adult males averaged 57.9 (42–75) mm SVL and 6.9 (2.8–14) gm; 81 adult females averaged 64.1 (53–76) mm SVL and 8.9 (4.4–16.9) gm (our unpubl. data). The dorsal scales are keeled and mucronate, but only slightly overlap. There are 37–55 dorsal scale rows around midbody. There are five or six relatively large supraocular scales, separated from the superciliaries and the median head scales by a complete row of smaller scales. There are 10–23 femoral pores on each leg, and the two series are separated medially by 3–11 scales.

The ground color of selected populations may closely match that of the prevailing substrate (Gillis, 1989). The dorsal ground color varies from grayish-white for populations living on sandy substrates such as the White Sands National Monument through dark brown. Dorsolateral light stripes, one to three scales wide, extending posteriorly from behind each eye onto the tail may be present or not; if they are, then they extend through and interrupt a series of dark brown or black, irregularly shaped blotches. These blotches may spread and fuse sufficiently to form broad dark longitudinal bands on either side of the white stripes. In this case, the dorsolateral bands are separated by a lighter middorsal band, tan to light blue in color, which extends without blemishes from the head onto the tail. If the white stripes are not present, then these blotches fuse into a series of broad or narrow transverse crossbands, which may meet medially or be offset slightly. The throats of males usually possess markings of some kind (except for some populations along the northeastern border): a pair of blue spots, one on each side, which may meet at the ventral midline; the throat may be entirely suffused with black; in some northern populations, the anterior throat and lips may be orange or red. Males also have bright blue belly patches edged with black between the limbs, which are usually widely separated at the ventral midline. These colors may be present in females but are usually reduced or absent altogether. Gravid females show increasingly intense yellow and orange dorsal and dorsolateral hues during ovulation. These hues are particularly evident in lizards from populations with lighter dorsal coloration (e.g., Dixon, 1967).

Similar Species: *Sceloporus arenicolus* and *S. graciosus* have granular scales on the posterior surface of the thigh, 9 or more scales separating the femoral pore series medially, and have a mottled or flecked throat pattern or none at all. *Sceloporus scalaris* lacks throat coloration altogether, has granular scales on the posterior surface of the thigh, and has lateral scale rows that are parallel to the dorsal scale rows. *Sceloporus virgatus* is smaller, has a relatively shorter tail, lacks ventral markings of any kind, and usually has notched femoral pore scales.

Systematics: Eleven subspecies are currently recognized (Collins, 1990; Smith et al., 1995b). Seven occur in New Mexico, and four of these (*consobrinus*, *elongatus*, *garmani*, and *tristichus*) were originally described as distinct species early in the biological explorations of the continent before sufficient comparative material was available. *Sceloporus u. consobrinus* was originally described by Baird and Girard (1853b) from a specimen collected on 6 June 1852, during Captain Randolph B. Marcy's exploration of the Red River. Neither a type specimen nor type locality were designated; the type was USNM 2855 according to Cope (1898 [1900]), but has been destroyed (Jones, 1926). Webb (1970b), tracing Marcy's itinerary, restricted the type locality to "about four miles east-southeast of Sayre, near or at the confluence of Timber Creek and the North Fork of the Red River, Beckham County, Oklahoma."

Sceloporus u. cowlesi was described by Lowe and Norris (1956); the type (UAZ 682) is an adult male, collected by the authors from "White Sands, 3 miles northwest of the Monument headquarters, Otero County, New Mexico" on 22 August 1949. *Sceloporus u. elongatus* was described by Stejneger (1890a); the type (USNM 15858) is an adult male, collected by C. Hart Merriam from "Moa Ave, Painted Desert, Arizona" on 23 September 1889. *Sceloporus u. erythrocheilus* was described by Maslin (1956); the type (USNM 137833) is an adult male, collected by T. Paul Maslin and H. Adair Fehlmann from "Purgatoire River, 19 mi. E. Model, Las Animas Co., Colorado" on 17 September 1949.

Sceloporus u. garmani was described by Boulenger (1882) from 5 specimens sent to him by Samuel Garman from "Dacota" (restricted by Smith [1938a] to Pine Ridge, South Dakota), collector and date of collection unknown. *Sceloporus u. tristichus* was originally described by Cope (*in* Yarrow, 1875); the type (USNM 4137, recatalogued to USNM 8613) is a subadult female collected by W. G. Shedd from "Taos, [Taos County,] New Mexico" in August 1874.

Smith et al. (1992) have recently described *S. u. tedbrowni* from the Mescalero Sands of eastern Chaves County. The type (MSB 33859) is an adult male from "Chaves Co., New Mexico, 6 mi. W. Caprock, Lea Co., large dune, Waldrop Peak, 0.5 mi. S. Hwy 380," collected by A. and H. Sena on 19 June 1978.

Lowe and Norris (1956) and Dixon (1967) indicated that meristic characters in *S. u. cowlesi* and *S. u. consobrinus* overlap, and that the only differences between the two forms is color. Dixon (1967) further suggested that substantial gene flow occurs between the two populations, but since lizards on the white sands (White Sands National Monument) spend most of their time on the ground or under vegetation, selection pressure for substrate matching maintains the color differences between them. Similarly, Smith (1946) and Maslin (1956) indicated substantial levels of intergradation between the forms discussed above. Smith et al. (1992) provided the most recent discussion of the relationships among the populations of this species. We follow Cole (1983) in considering *Sceloporus undulatus* as a single polytypic species.

Habitat: Lizards of this species can be arboreal, saxicolous, or terrestrial in New Mexico. In mountainous parts of the state, they can be found in piñon-juniper woodlands, on bedrock slopes, in boulder fields and drainages with shrubs, broadleaf (*Quercus, Populus*) and evergreen trees, on sandstone cliffs and mesas, and along riparian corridors within xeric environments (Douglas, 1966; our obs.). When living in sagebrush or desert-grassland habitats, lizards are invariably found in clumps of debris, rock piles, or on yuccas (Harris, 1963; Vinegar, 1975b; Jones and Droge, 1980; our obs.). Ballinger and Jones (1985) found these lizards in open sandy areas with less than 10% vegetative cover, or associated with *Yucca glauca*, in western Nebraska. Spring and summer densities in the Nebraska population averaged 19.4 and 16.5 lizards/hectare over a 7-year period (Jones and Ballinger, 1987). Tinkle and Ballinger (1972) estimated population densities of 1.98 and 13.59 lizards/hectare in Texas and Colorado. Ferner (1974) reported a density of 25–35 lizards/hectare in central Colorado. Vinegar (1975b,c) estimated densities of 8.2 and 10.4 lizards/hectare over two years for a desert-grassland population (1219 m) in Hidalgo County, New Mexico, whereas densities for a nearby montane population (2057 m) in Grant County were estimated as 26 and 42 lizards/hectare over the same time period. This species can be found sympatric with *Sceloporus graciosus, Urosaurus ornatus,* and *Uta stansburiana* in extreme southwestern Colorado (Douglas, 1966). It is common in the depressions between active gypsum dunes on White Sands National Monument wherever *Chrysothamnus pulchellus, Yucca glauca, Ephedra torreyana* and *Oryzopsis hymenoides* are present (Dixon, 1967). It is very abundant on lava flows of the Tularosa Malpais (Lewis, 1949). Gehlbach (1956) has reported this lizard to be common in the sandy, mesquite-covered floodplain of the Black River in Eddy County along with *Phrynosoma cornutum, Uta stansburiana,* and *Cnemidophorus inornatus.*

Behavior: These lizards use fallen tree trunks and other elevated sites for sunning and as vantage points, but are distinctly terrestrial in foraging (Gehlbach, 1965). On White Sands National Monument, they are reluctant to leave the cover of vegetation (Dixon, 1967). There is a major activity period between 0900 and 1200 during the summer when most lizards are actively thermoregulating at the base of vegetation. Lizards usually remained inactive between 1200 and 1600, when body temperatures recorded were between 33.0°C and 34.7°C. Another activity peak occurred after 1600. Douglas (1966) found that lizards in the Four-Corners region didn't appear in the spring until the snow was completely gone and the ground was warm. In this study, emergence in all three years followed the first major insect bloom. Hibernation begins by mid-September and the last lizards were seen in late October. New Mexico specimens have been collected between 11 February (Grant County) and 9 November (Grant County).

Body temperatures of active lizards vary significantly over the active season and with geographic location. Average body temperatures of lizards in a population at 1750 m elevation in Bernalillo County, New Mexico, varied between 33.5°C in May and 36.8°C in August, whereas similar values for a population at 2400 m in southern Colorado varied between 27.9°C in May and 35.3°C in July (Crowley, 1985). Gehlbach (1965) recorded an average body temperature of 33.1 (29–37)°C for 10 actively foraging individuals at 2242 m in the Zuni Mountains. The mean preferred body temperature of 46 lizards measured from Los Alamos County was 35.9°C (Bowker et al., 1986). Lizards were shown to be precise thermoregulators by passive means (i.e., shifting positions), and spent 54% of their time inactive. Dixon (1967) measured body temperatures between 32.3°C and 37.2°C for 140 lizards on White Sands National Monument. The critical thermal maximum and minimum for this species has been recorded as 42.7°C and 6.3°C (Crowley, 1985). Individual lizards are able to withstand supercooling to −5.45°C (Lowe et al., 1971).

Both sexes appear to be territorial, with males engaging in significantly more social behavior during the breeding season (mid-May to mid-July) than later in the summer (Ferner, 1976; Gillis and Ballinger, 1992). Ferner (1974) reported a significant degree (52%) of overlap among territories of individual males in central Colorado, and that home range size decreased significantly after the breeding season. He reported an average home range size of 826.3 m^2 (309.5–2506.7 m^2) for males and 362.6 m^2 (163.9–593.3 m^2) for females. Jones and Droge (1980) have estimated home range sizes of 717.2 m^2 and 851.5 m^2 for males and females in western Nebraska.

Reproduction: The timing of the reproductive cycle and other reproductive parameters varies greatly with latitude and altitude, and is somewhat dependent on annual primary productivity (e.g. Ferguson and Brockman, 1980). The female reproductive season has been reported to extend from late March through mid-July (Tinkle and Ballinger, 1972; Gillis and Ballinger, 1992). Females from southwestern New Mexico laid eggs as early as mid-May and as late as mid-August (Vinegar, 1975b). Females from Texas and Colorado reach minimum reproductive sizes of 47 and 58 mm SVL at 3 months and 12–13 months of age but don't reproduce until the following spring (Tinkle and Ballinger, 1972; Ferner, 1976; Gillis and Ballinger, 1992). Vinegar (1975b) reported that females from southwestern New Mexico attain reproductive maturity at a minimum size of 53 mm SVL at an age of 11–14 months. The fat body cycle closely approximates the reproductive cycle. Fat bodies are an important energy source for production of the first clutch in Colorado lizards (Gillis and Ballinger, 1992), and may represent up to 7% of individual body mass as reproduction commences.

Clutch size and clutch frequency vary with female age and geographic location (Tinkle and Ballinger, 1972; Vinegar, 1975b; Ferner, 1976; Jones and Ballinger, 1987; Gillis and Ballinger, 1992). Older females produce larger clutches and may produce more clutches during a reproductive season than younger females. More clutches are produced in populations subject to longer seasons of primary productivity. Average clutch sizes have been reported from Colorado (7.9, 9.4, 10.94), Nebraska (4.26, 4.5, 6.57, 9.0), New Mexico (7.2, 9.9), and Texas (9.5), and yearly clutch frequencies vary between one and four in these populations. Clutch size and female body size are positively correlated; approximately one egg was added to the clutch for each 3 mm (Colorado) or 4.5 mm (Nebraska) increment in SVL (Jones and Ballinger, 1987; Gillis and Ballinger, 1992). Late season eggs and hatchlings are larger than those in the early part of the season, and larger hatchlings at this time have a significantly greater probability of surviving than smaller ones (Ferguson et al., 1982; Ferguson and Snell, 1986).

Emergence of hatchlings occurs from mid-June through late September, and varies annually and with latitude and altitude (Gehlbach, 1965; Douglas, 1966; Dixon, 1967; Tinkle and Ballinger, 1972; Vinegar, 1975b; Gillis and Ballinger, 1992). Some hatchlings that emerge early grow rapidly, achieving minimum reproductive size (42 mm SVL) prior to hibernation. Some predepositional embryonic development occurs in this species (Gehlbach, 1965), perhaps as an adaptation to short reproductive seasons in some localities. Christian et al. (1986) have shown that exposure to cold temperatures for a week or more significantly extends the duration of incubation and increases egg mortality. Eggs can represent up to a third of the total body weight of individual females prior to oviposition (Tinkle and Ballinger, 1972; Vinegar, 1975b; Gillis and Ballinger, 1992).

Testes of male *S. undulatus* were enlarged at spring emergence in western Nebraska and remained so through late May (Jones and Ballinger, 1987). Regression began in early June, but then testes enlarged once again prior to hibernation. In Colorado, Ferner (1976) reported maximum and minimum testicular sizes in early April and late July, with recrudescence in September. Spermatogenesis is well underway upon spring emergence.

Jones and Ballinger (1987) reported maximum field longevity for both males and females as four years. Yearling and 2-year-old females contributed 53% and 40% of the net replacement rate for this population over a 7-year period in western Nebraska. Tinkle and Ballinger (1972) estimated that 70% and 85% of eggs survived to hatch during two years in Texas, but that hatchling survival to the following season was quite low (5% in Texas, 12.5% in Colorado). Most Texas lizards do not survive to be three years old, whereas 35% of Colorado lizards do. Jones et al. (1987) estimated overwinter mortality in hatchling lizards in western Nebraska at 78%. Vinegar (1975b) estimated that less than 5% of the eggs laid in two southwestern New Mexico populations produce young that survive to one year of age, but that annual survivorship thereafter improved (14–50%).

Food Habits: This species is a sit-and-wait forager. Ferner (1976) reported flies, grasshoppers, beetles, insect larvae, and arachnids as common food items in central Colorado, and Douglas (1966) recorded ants, wasps, lepidopterans, and a wide variety of beetles and flies from the Four-Corners area. On White Sands National Monument, these lizards forage in and immediately beneath vegetation (Dixon and Medica, 1966); hymenopterans (38%) and coleopterans (32%) were important by volume, and hatchling *Cnemidophorus inornatus* were also eaten. In Arizona, Toliver and Jennings (1975) reported this species to eat termites (winged adults, 22% by volume, 46% by frequency), ants (9%, 28%), beetles (15%, 12%), lepidopteran larvae (10%, 4%), and grasshopper nymphs (8%, 1%). Barbault and Maury (1981) recorded (by percent volume) coleopteran adults (6–77%), lepidopteran larvae (19–69%), and hemipterans (4–36%) during June and July over a 3-year period in Durango, Mexico.

DeMarco et al. (1985) showed that prey size influences the types of prey taken by lizards of different size classes; smaller lizards eat smaller prey items. Johnson (1966) reported sex and age differences in the diet of this species from southwestern Colorado. Adult lizards ate 18% by weight grasshoppers and 3% hemiptera, whereas these percentages were reversed for juvenal lizards. Females ate more food as a percentage of body weight in the spring than males, presumably as a proximate boost to reproductive activities.

Known predators in New Mexico include *Crotaphytus collaris*, *Gambelia wislizenii*, *Holbrookia maculata*, *Cnemidophorus inornatus*, and *Masticophis taeniatus*.

Remarks: Lewis (1949) and Best et al. (1983) have reported melanistic individuals from lava fields in south-central New Mexico. Dixon (1967) reported homing movements of 600 m for three lizards. We follow Harper (1940) in the citation of the description of this species.

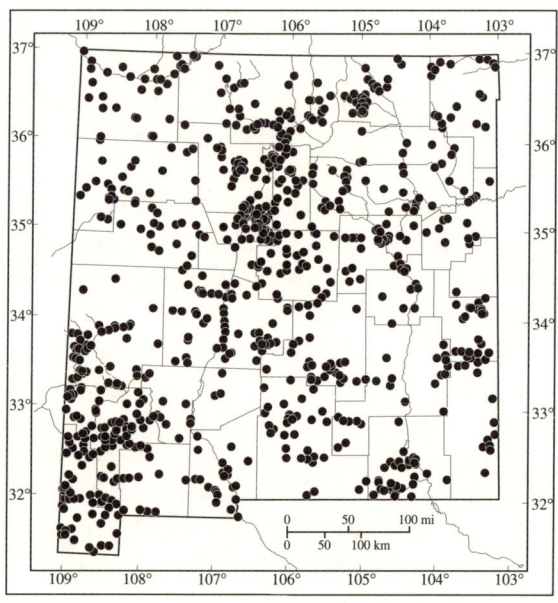

Distribution of Sceloporus undulatus *in New Mexico.*

SCELOPORUS VIRGATUS Smith, 1938 — *Striped Plateau Lizard*

See Plate 54.

Type: The type specimen (UMMZ 81912) is an adult male collected from "above Santa Maria Mine, El Tigre Mountains, Sonora, Mexico" by Berry Campbell on 27 June 1935 (Smith, 1938a).

Distribution: This species occurs in the Chiricahua Mountains of southeastern Arizona and the Animas, Peloncillo, and San Luis mountains in Hidalgo County, New Mexico, and southward in the Sierra Madre Occidental of Chihuahua and Sonora, Mexico. The southern limit of its distribution is not well known (Cole, 1968; Morafka, 1977). It occurs as a number of small, isolated populations within this geographic region because of the restricted types of habitat it occupies.

Description: This is a small member of the genus, with a maximum SVL of 61 mm for males and 70 mm for females (Rose, 1981; Smith, 1985). The ground color is brown. There is a light gray vertebral band, about four scales wide, from the head to the base of the tail (Cole, 1963). This is bordered on either side by a light brown stripe; the two fuse at the base of the tail. These stripes each contain a longitudinal series of irregularly shaped, dark brown-and-white, or dark brown, spots. A distinct yellowish-white stripe extends dorsolaterally from behind each ear opening to the base of the tail, and less distinctly onto the tail itself. Another distinct white line extends laterally from the supralabials to the anterior surface of the thigh. A uniform dark brown band extends posteriorly between the two light stripes from the eye through the ear opening onto the base of the tail. Another extends posteriorly from the ear below the lateral white stripe onto the anterior surface of the thigh. The limbs are striped or banded. Individuals of both sexes have a blue patch posteriorly on either side of the throat; during the breeding season, this patch in mature females becomes surrounded by, or replaced with, orange (Cole, 1963; Vinegar, 1972). This change begins when eggs are well-developed, but prior to ovulation. It becomes most intense following ovulation, and fades after oviposition. The ventral surface of the body, with the exception of occasional scattered and minute black flecks on the chest and throat, is immaculate white. Sexual dimorphism is evident in the color pattern; males are usually more boldly striped than females and more frequently exhibit reduction in or absence of dorsal spotting (Cole, 1963).

Similar Species: The only other species likely to be confused with *Sceloporus virgatus* is *S. undulatus*. The former species has a relatively shorter tail than the latter, and also averages about 10 mm less in adult body size. The ventral surface of the body is immaculate in both sexes of *S. virgatus*, whereas male *S. undulatus* have pronounced lateral blue belly patches. In addition, the scales on the underside of the leg, which contain the femoral pores in *S. virgatus*, are almost always notched on their posterior edges; the corresponding scales in *S. undulatus* are not.

Systematics: This lizard was originally described as a subspecies of the widespread and polytypic *Sceloporus undulatus*. Cole (1963) demonstrated that it was a distinct species.

Habitat: The species is found only in pine-oak and pine woodlands and savannas to an elevation of 2515 m (McCranie and Wilson, 1987; Tanner, 1987) and may follow canyons with riparian oak woodlands downward to elevations of 1515 m (Cole, 1963). This species usually inhabits moderately rocky areas within these habitats, with boulders, logs, leaf litter, and scattered grasses providing required cover. It is most abundant in New Mexico at elevations between 1600–2100 m in wooded canyons and ravines, although it occurs as high as 2597 m on Animas Peak. Smith (1985) estimated densities ranging between 42 and 130 lizards/hectare over a 3-year period in the Chiricahua Mountains of southeastern Arizona, whereas Vinegar (1975a, c) estimated densities of 61–97 lizards/hectare in the same area.

Behavior: Individual lizards of both sexes are territorial (Rose, 1982; Smith, 1985) and spend most of their time on the ground, although they use trees and boulders for lookout posts. Males are more active during the reproductive season than at other times, suggesting that male territoriality facilitates breeding access to females (Rose, 1981). Males move more and are active on more days during the breeding season. They also spend little time feeding and consequently lose body mass during this time (Rose, 1981; Merker and Nagy, 1984), as field metabolic rates are positively correlated with daily activity. Merker and Nagy (1984) reported that males and females were active for an average of 7.6 and 3.6 hrs/day in May; comparable values for August were 4.5 and 6.5 hrs/day. Males generally cease activity by the beginning of June (Rose, 1981). Female activity levels do not differ between the breeding and non-breeding seasons. Females ate twice as much as males during the breeding season despite being less active, and used the energy acquired beyond that necessary to sustain basic metabolism for reproduction (Merker and Nagy, 1984). Females, however, remain active through mid-October, presumably to replenish energy stores depleted during reproduction. Hatchlings may remain active throughout the winter on warm days. The 151 New Mexico specimens available to us were collected between 11 April and 18 September.

Rose (1982) estimated average home range sizes of males and females during the breeding season as 411 m^2 and 102 m^2 and during the nonbreeding season as 287 m^2 and 233 m^2. Average and median home range size were reported by Smith (1985) for females as 527 m^2 and 351 m^2, and there was almost no overlap between neighbors. In contrast, male home ranges had average and median values of 1570 m^2 and 1376 m^2 and overlapped extensively, although a core area was defended. Consequently, females interact very little, whereas a single female can be courted by as many as six different males within her territory. Female home ranges overlap more with those of males during the breeding season than at other times (Rose, 1982). In aggressive encounters between males for access to females, larger and older males (3 or 4 years old) almost always win, unless the smaller male is closer to the center of his home range than his adversary (Smith, 1985).

Smith and Ballinger (1994a) measured an average body temperature of 33°C for 383 active lizards in southeastern Arizona. Males and females had similar body temperatures, and individual lizards appeared to actively thermoregulate, maintaining relatively constant body temperatures throughout the year. The average body temperature of gravid females was significantly lower (28°C) than that of non-gravid females. Andrews and Rose (1994) also found the average body temperature of active gravid females (34°C) to be significantly lower than that of non-gravid females (35°C). Individual lizards are able to withstand supercooling to –4.9°C (Lowe et al., 1971).

Reproduction: The minimum size at sexual maturity is 47 mm SVL for females and 43 mm SVL for males (Vinegar, 1975a). Courtship and mating occurs from the end of April through the end of May (Rose, 1981). Ovulation begins by late May and egg laying occurs from late June through late July (Vinegar, 1975a). Nests are constructed in areas of open sun (Andrews and Rose, 1994). Individual lizards may retain oviductal eggs for up to one month until favorable conditions prevail (Rose, 1981). Females produce one clutch annually averaging 9.45 eggs in southeastern Arizona and southwestern New Mexico. Hatchlings appear from the end of August through September. Females may either reproduce during their first year at an age of approximately 10 months or delay reproduction until their second year, at an age of approximately 22 months (Vinegar, 1975a). More eggs are produced during wet years than during dry ones, and more eggs hatch and hatchlings survive under these conditions as well (Vinegar, 1975a). Hatchling survivorship is low; over 70% of the reproductive replacement for the population studied by Vinegar (1975a) was contributed by females two years old or older. Vinegar (1975a) estimated survivorship from the egg to one year as 19% for males and 21% for females; comparable values to two years were 57% and 44%. Occasional lizards of both sexes may live to be four years old or older.

All males two years old or older (49–61 mm SVL) were found to be reproductively mature in a population studied in the Chiricahua Mountains of southeastern Arizona (Ballinger and Ketels, 1983). Some yearling males (40–49 mm SVL) may attain sexual maturity late in the reproductive season, but it is unclear whether they participate in mating activity. All males less than 40 mm SVL are immature. Males emerge from hibernation in March, at which time testes are already about 3.5% of body mass (Rose, 1981). Testes reach maximum size in April and decrease to minimum values (ca. 0.5% of body mass; Rose,

1981) by the end of June. Testicular recrudescence commences in September and October prior to hibernation. Spermatogenesis is evidently a protracted process in this species; division in germinal reproductive cells can be demonstrated histologically in lizards collected from June through October and again in March. Mature sperm were only found in lizards in May (Ballinger and Ketels, 1983).

The fat body cycle of both sexes mirrors reproductive activity (Rose, 1981; Ballinger and Ketels, 1983). In males, fat bodies represent 2% of body weight in March, decline to less than 0.5% in May and June, increase to 2% in July and 3.5% in September prior to hibernation. Female fat bodies increase to as much as 5.5% of body weight following oviposition.

Food Habits: The species is probably an opportunistic insectivore, but specific dietary information is lacking. Phelan and Niessen (1989) reported individuals taking grasshoppers.

Remarks: The species has been reviewed by Cole (1968). The specific epithet is derived from the Latin word meaning "striped," and presumably refers to that distinctive feature of the dorsal color pattern.

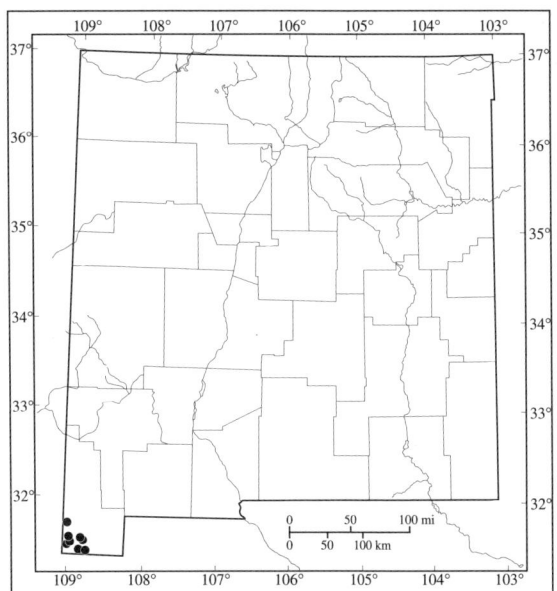

Distribution of Sceloporus virgatus *in New Mexico.*

UROSAURUS ORNATUS (Baird and Girard, 1852)   *Tree Lizard*

See Plate 55.

Type: The original description (Baird and Girard, 1852a) did not specify the type material, which was collected "on the Rio San Pedro (Texas) and provinces of Sonora." The specimens were collected by John H. Clark in May or June 1851 during expeditions of the Scientific Corps of the U.S. and Mexican Boundary Commission under Col. J. D. Graham. Yarrow (1882) designated the syntypes, an adult male and female (USNM 2750), and restricted the type locality to the Rio San Pedro (now the Devils River, Val Verde County), Texas.

Distribution: The tree lizard ranges from the eastern edge of the Edwards Plateau in central Texas westward, and from Rio Grande City up the Rio Grande on both sides of the border, through Trans-Pecos Texas, northern Chihuahua and Sonora to the Sea of Cortéz. It ranges from the border area common to Colorado, Utah, and Wyoming south throughout Arizona and into Mexico along the coast to Nayarit, and on Isla Tiburón in the Sea of Cortéz (Murphy and Ottley, 1984). In New Mexico, *U. ornatus* occurs throughout the state west of the Rio Grande and in many of the mountain ranges bordering the eastern margin of the Rio Grande Valley northward from the Texas border to Santa Fe. It also occurs in the Guadalupe Mountains of Eddy and Otero counties, and isolated records suggest that it may occur in the rugged, relatively poorly explored Otero Mesa area along the southern margins of the Sacramento Mountains.

Description: *Urosaurus ornatus* is a small lizard, with a maximum SVL of 69 mm. Gehlbach (1965) measured adult males between 39–52 mm SVL (mean 48.3 mm) and adult females between 42–53 mm SVL (mean 47.5 mm) in northern New Mexico, and Vitt et al. (1981) found adult males averaged 52.5 mm and adult females

51.2 mm SVL in central Arizona. During 1987–92, 462 specimens collected at Antelope Pass in Hidalgo County averaged 39.9 (17–59) mm SVL and 1.9 (0.4–5.5) gm. One hundred forty adult males averaged 45.2 (40–59) mm SVL and 2.6 (1–5.5) gm; 85 adult females averaged 45.9 (41–53) mm SVL and 2.5 (1–5) gm (our unpubl. data). The body is slim, the tail relatively long, and the limbs slender. The dorsal scales are somewhat heterogeneous in size. There is a single, sometimes incomplete, row of small granular scales arranged longitudinally along either side of the midline; lateral to these are two rows of enlarged, keeled scales extending from the neck onto the tail. The scales lateral to these become small again, and grade into even smaller scales on the sides of the body. There is a distinct dorsolateral fold containing numerous enlarged scales or tubercles that may be arranged in diagonal rows. There may be a second, less distinct fold laterally. The subcaudal scales are keeled to within 5 mm of the cloaca, and the keels of the caudal scales run the entire length of each scale at the same height, producing a series of continuous parallel longitudinal ridges surrounding the tail (Ballinger and Tinkle, 1972). The scales on the ventral surface are smooth, flat, and imbricate. The head scales are uneven in size but relatively large. There are no supranasal scales between the nasals and the internasals. There are 4–5 enlarged supraocular scales. The interparietal scale is relatively small and flanked laterally by supratemporal scales, which are not appreciably larger than adjoining lateral head scales. There is a single pair of enlarged occipital scales. A distinct gular fold is present. There are 10–11 femoral pores on each thigh, and males have enlarged postanal scales.

The dorsal ground color can be tan, gray, brown or black, depending on local substrate conditions. The dorsal pattern consists of a series of paired dark brown or black blotches or transverse bands, edged posteriorly with blue or white. These markings begin at the shoulder and continue posteriorly onto the tail. They may be arranged in irregular longitudinal lines, or they may occur as transverse bands that may fuse or be interrupted medially. The medial edges of the transverse bands are often flared and the pairs offset. The head is usually marked with an intricate pattern of fine, dark lines, and a distinct dark line passes anteriorly from the first dorsal blotch through the eye to the tip of the snout. The forelimbs are distinctly barred and the hindlimbs less so. Both male and female lizards possess colored throat patches (see Behavior); the lateral ventral surfaces of males are also blue.

Similar Species: Spiny lizards of the genus *Sceloporus* lack a gular fold, have larger and/or stouter bodies, and have large, keeled, imbricate dorsal body scales that are all approximately equal in size. *Uta stansburiana* has a distinct black spot on the side of the body behind the front limbs, possesses supranasal scales, lacks subcaudal keels except on the distal portion of the tail, lacks keels extending the entire length of each caudal scale, and the dorsal scales of the body are either not heterogeneous in size or grade smoothly from the largest along the midline to the smallest laterally.

Systematics: The phylogenetic relationships of populations within this species are badly in need of study. Wiens (1993) considered this species monotypic. Nine subspecies have been recognized (Smith and Smith, 1976; Collins, 1990), and four of these occur in New Mexico. *Urosaurus o. levis* was originally described as a distinct species by Stejneger (1890a); the holotype (USNM 11474) is an adult male, collected by E. D. Cope from "Tierra Amarilla [Rio Arriba County], New Mexico," date of collection unknown. *Urosaurus o. linearis* was described by Baird (1859), who designated USNM 2759 as the holotype, collected by Dr. Kennerly from "Los Nogales, Sonora [Mexico]" during the U.S. and Mexican Boundary Survey, precise date unknown. This specimen was apparently lost; Mittleman (1942) designated an adult female (USNM 62077) from the same locality as the neotype, collected by F. J. Dyer, date of collection unknown (catalogued 26 July 1919). *Urosaurus. o. schmidti* was described by Mittleman (1940); the holotype (USNM 32929) is an adult male, collected by Vernon Bailey from "Fort Davis, Jeff Davis Co., Texas" on 3 January 1896. *Urosaurus o. wrighti* was originally described as a distinct species by Schmidt (1921); the holotype (AMNH 18097) is an adult male, collected by B. T. B. Hyde from "Grand Gulch, San Juan County, Utah" on 9 November 1920.

Habitat: Despite its common name, this lizard is more often found on rocks in New Mexico (Gehlbach, 1965;

our obs.). It is typically found at low to moderate elevations in relatively xeric environments, but can be found from elevations of 950 m to about 2500 m. It is abundant on lava fields in southern New Mexico (Lewis, 1951; Ballinger, 1977; our obs.). Zucker (1989) found this lizard common on boulders associated with *Rhus trilobata*, *Garrya wrightii*, *Haplopappus laricifolius*, *Morus microphylla*, and *Quercus arizonica* in the Organ Mountains, Doña Ana County. The species can be arboreal, particularly in riparian zones (Vitt et al., 1981; Thompson and Moore, 1991b) or where abundant rocky habitat is sparse (Baltosser and Best, 1990).

Dunham (1981) estimated densities of 30.5–58.5 adult lizards/hectare over a 4-year period in Big Bend National Park; densities were highest in years of above-average rainfall and lowest during drought years. Smith (1981) estimated a density of 143 lizards/hectare in pine-oak woodland at 1745 m elevation in southeastern Arizona, and Tinkle and Dunham (1983) estimated densities of 42–171 lizards/hectare in a 7-year interval along a riparian corridor at 1077 m in south-central Arizona. Dunham (1982) estimated a density of 721 lizards/hectare for a population occupying a restricted rocky habitat west of Animas in Hidalgo County, and Ballinger (1984) reported that the density of this population decreased by 62% following a severe drought and remained low for three years. Smith and Ballinger (1994c) reported an inverse relationship between density and individual growth rates at this same site. Zucker (1989) recorded 19 and 20 individuals over two years within a three-dimensional rocky habitat (39 x 6 x 3 m) at Aguirre Springs in the Organ Mountains, Doña Ana County.

Behavior: New Mexico specimens have been collected between 10 January (Catron Co.) and 10 November (Catron Co.). The activity season in southwestern New Mexico extends from March through November (Baltosser and Best, 1990). These lizards are active thermoregulators, shuttling back and forth between sun and shade, and maintaining body temperatures during activity between 34–37°C (Vitt et al., 1981). Most daily activity takes place between 0900 and 1200 at lower elevations during the summer. Zucker (1987) reported that males emerge earlier during the active season, at cooler temperatures and lower light intensities, than do females. Tinkle and Dunham (1983) and Boykin and Zucker (1993) reported aggregation during winter hibernation in this species. Individuals are able to withstand supercooling to -5.56°C (Lowe et al., 1971).

Males of this species are highly territorial and possess brightly colored throats that can be extended in a display, much as that seen in males of many New World *Anolis*. Throat colors may be green, yellow, orange, blue, or a bicolor combination of these, and they reliably signal the status of individual lizards (Hover, 1985; Thompson and Moore, 1991a,b). Territories vary in quality and are actively defended by males to guarantee exclusive mating opportunities with females resident within them (M'Closkey et al., 1987). Some populations are composed of males with a single dewlap color whereas other populations have males showing up to five different dewlap colors (Thompson and Moore, 1991b); these geographic differences may be related to differential selection operating in structurally dissimilar habitats. Zucker (1989) reported a population in which dominant males assume an almost uniformly black dorsal coloration, and maintain exclusive territories which overlap those of subordinate, normally colored males and females. Dark males had larger home ranges (87 m^2) than did light males (53 m^2), and females had the smallest home ranges (39 m^2).

Female tree lizards also have colored throat patches that may be uniformly orange or vary from greenish-yellow through reddish-orange (Zucker and Boecklen, 1990). Throats of gravid females tend to change to a deep orange color as oogenesis progresses. Females are relatively sedentary within individual home ranges, but significantly increase activity levels prior to oviposition associated with finding suitable nesting sites (Deslippe et al., 1990).

Reproduction: This lizard has been extensively studied (Asplund and Lowe, 1964; Ballinger, 1977; Dunham, 1981; Tinkle and Dunham, 1983; Van Loben Sels and Vitt, 1984; Thompson and Moore, 1991b). Yearly variations in primary productivity and availability of resources have profound effects upon reproduction in this species. Females reproduce at a minimum size of 41 mm SVL and at an age of 9–12 months. Individual females may produce up to 6 clutches (usually 2–3) annually, and a single clutch may represent up to 33% of a female's body weight (Tinkle and Dunham, 1983). Larger females produce more eggs per clutch than do smaller ones; in southwestern New Mexico approximately one egg is added for each 2 mm increment in SVL (Michel, 1976). Vitellogenesis begins in May or June with oviductal eggs present from June through August. Michel (1976) found that approximately 90% of females produced two annual clutches with a clutch size of 5–16 eggs. The first clutch averaged 10.9 eggs and the second 7.5 eggs. In northwestern New Mexico, females were gravid from mid-June through late July, and produced clutches of 2–5 eggs (Gehlbach, 1965; Douglas, 1966). Hatchlings may appear from July through September.

The fat bodies of females in southwestern New Mexico are at maximum size in mid-June, representing up to 5.5% of an individual lizard's body weight (Michel, 1976). They are rapidly depleted as vitellogenesis progresses, reaching minimum values (.05% of body weight) by the beginning of August. Fat bodies are substantially replenished prior to hibernation.

Minimum reported size at sexual maturity for males is 40 mm SVL, achieved during the first reproductive season following hatching. Males emerge from hibernation in west Texas with testes near maximum size (Dunham, 1981); however, Parker (1973a) reported males from southern New Mexico and west Texas emerging in March with testes at minimum size. Both authors agree that testes regress from maximum size in May to minimal size by September. All males in southwestern New Mexico contain mature sperm in their reproductive tracts and are reproductively active between early June and mid-August (Michel, 1976). Testes represent 3% of lizard body weight in mid-June and decline to 0.47% of body weight by late August and September. Courtship behavior is synchronized with the testicular cycle in males. Fat bodies are smallest during the spring and summer reproductive season (Van Loben Sels and Vitt, 1984) and begin to be replenished following the cessation of spermiogenesis in August. They may not be used appreciably for basic metabolic activities during the winter, but rather serve as energy stores for spermiogenesis and territory maintenance the following spring.

Annual turnover in populations is substantial (ca. 50–90%), and very few animals live to be three years old or older (Dunham, 1981; Smith, 1981; Ballinger, 1976, 1984). Mortality of both sexes increased substantially during the breeding season (Tinkle and Dunham, 1983).

Food Habits: These lizards eat a wide variety of arthropods including aphids, thrips, plant lice, beetles, bugs, flies, ants, bees and wasps, termites, lepidopteran adults and larvae, grasshoppers, chrysopid larvae, mites, and lycosid spiders (Knowlton, 1938; Hotton, 1955; Asplund, 1964; Douglas, 1966). Asplund (1964) demonstrated significant seasonal variation in the diet of a population near Tucson, Arizona, and attributed this to seasonal variation in prey abundance and microhabitats utilized by lizards. Lizards grow slower and weigh less for a given body size during dry years when prey abundance is reduced than they do in wet years when prey is more plentiful (Ballinger, 1977; Dunham, 1981).

These lizards are potential prey to a wide variety of avian, mammalian, and reptilian predators (Dunham, 1981).

Remarks: Gehlbach (1965) reported melanistic individuals from lava fields in the Grants vicinity, Valencia County. Lewis (1951) reported melanism in this species from the Afton lava field in southern Doña Ana County; we have collected melanistic individuals from nearby Aden Crater. Goldberg et al. (1993) found infection rates of parasitic helminths to be 4% and 38% in lizards from two different populations in southern New Mexico.

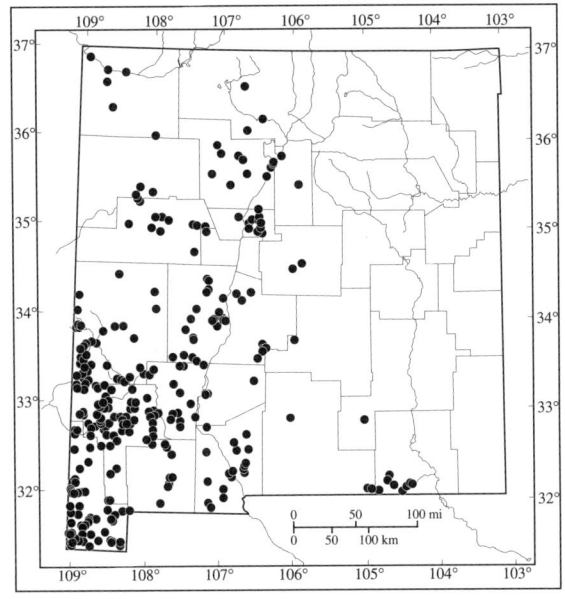

Distribution of Urosaurus ornatus in New Mexico.

UTA STANSBURIANA Baird and Girard, 1852   *Side-blotched Lizard*

See Plate 56A, 56B.

Type: The species was described on the basis of four specimens (USNM 2753) obtained by Captain Howard Stansbury from the "Valley of the Great Salt Lake, Utah" during the railroad survey of 1850 (Baird and Girard, 1852a).

Distribution: This polytypic species is widely distributed throughout the western United States and northern Mexico, and it occurs on many islands in the Sea of Cortéz (Murphy and Ottley, 1984; Stebbins, 1985). The eastern and western portions of the range are disjunct, divided by unsuitable habitat along the mountains of the Sierra Madre Occidental in Mexico and northward extensions (Chiricahua and Graham Mts.) in southeastern Arizona.

This species is common in New Mexico across the southern tier of counties, and it extends northward along the Rio Grande and Pecos River basins and associated drainages to San Ildefonso, Santa Fe County, and Storrie Lake, San Miguel County. It occurs in the San Juan Basin of northwestern New Mexico but is absent from the intervening high country to the southeast. It occurs on the Llano Estacado along the eastern border of the state northward to Clovis in Curry County, and also along the Canadian River between Logan and Conchas Lake. We expect the species to turn up in the Plains-Mesa Grasslands and Juniper Savannas (Dick-Peddie, 1993) of Torrance, Lincoln, and southwestern Chaves counties, east of the Sacramento Mountains, where specimens are currently lacking. There are several specimens in the Museum of Southwestern Biology with locality data indicating they were collected from the Gila National Forest in Catron County. We doubt the reliability of this information. We would expect this species, should it occur in Catron County, to be to the east in the Plains of San Augustin.

Description: This is a small lizard exhibiting slight sexual dimorphism in size, with a maximum SVL of 64 mm for males and 58 mm for females (Parker and Pianka, 1975). During 1992–95, 2,440 *U. stansburiana* collected at Mescalero Sands in Chaves County averaged 42.5 (19.2–58.7) mm SVL and 2.7 (0.1–5.7) gm; 1137 adult males averaged 48.6 (42.1–58.7) mm SVL and 3.7 (1.6–5.7) gm; 656 adult females averaged 45.9 (42–52.4) mm SVL and 3.0 (1.3–4.7) gm (L.W. Gorum, unpubl. data). During 1987–92, 734 specimens collected at Antelope Pass in Hidalgo County averaged 40.9 (11–57) mm SVL and 2.5 (0.1–9.8) gm. One hundred ninety-five adult males averaged 49.3 (42–57) mm SVL and 3.7 (1.8–9.8) gm; 233 adult females averaged 47.2 (42–54) mm SVL and 3.3 (1.5–9) gm (our unpubl. data). The dorsal scales are small and weakly keeled (Ballinger and Tinkle, 1972), and may either be equal in size (*U. s. uniformis*) or grade smoothly from the largest along the midline to the smallest laterally (*U. s. stejnegeri*). The body and limbs are relatively stout, and the tail relatively short. The subcaudal scales are keeled only on the distal third of the tail. The scales on the dorsal surface of the tail are larger and more distinctly keeled than those on the body, but the keels extend approximately half the length of each caudal scale and do not produce continuous ridges the length of the tail. The scales on the ventral surface are larger, smooth, and flat. The head scales are uneven in size but relatively large. There is at least one pair of supranasal scales between the nasals and the internasals. There are 3–6 enlarged supraocular scales. The interparietal scale is relatively large and flanked laterally by a complete row of enlarged supratemporal scales. There are usually several pairs of enlarged occipital scales. A distinct gular fold is present. There are 11–19 femoral pores on each thigh, and males have enlarged postanal scales.

The dorsal ground color is usually brown or gray but can be yellowish or occasionally black. The color pattern of this species is characterized by a large black patch on the lateral flanks behind the axillary region on each side of the body. In *U. s. stejnegeri*, juveniles of both sexes possess a white dorsolateral stripe on each side of the body, beginning just below the nasal opening and extending posteriorly through the eye, over the shoulder and onto the tail. These stripes may be intersected laterally by a longitudinal series of short black lines, which extend diagonally downward from the stripes, or a longitudinal series of chevrons, which actually break the stripes. Both lines and chevrons are edged with white, extend posteriorly from behind the head onto the tail, and are most pronounced on the trunk. Females retain the striped juvenal pattern into adulthood, but this pattern becomes progressively obscure as males age, remaining distinct only on the head and neck. Adult males also lose the chevron markings, and the dorsal pattern transforms into a salt-and-pepper pattern of light and dark spots. Occasionally individual lizards can be found that lack a dorsal pattern altogether (Ballinger and McKinney, 1967). Lizards from northwestern New Mexico (*U. s. uniformis*) are characterized by an almost complete absence of the dorsal pattern. Whatever pattern exists consists of rows or irregularly scattered small dark brown spots on a dorsal color of gray or reddish-brown. Lizards from the sand dune areas of southeastern New Mexico are often patternless (see photo).

Similar Species: Spiny lizards of the genus *Sceloporus* lack a gular fold, have larger and/or stouter bodies, and have large, keeled, imbricate dorsal body scales that are all approximately equal in size. *Urosaurus ornatus* lacks a distinct black spot on the side of the body behind the front limbs, lacks supranasal scales, has subcaudal keels virtually the entire length of the tail and complete keels on the caudal scales that produce continuous parallel longitudinal ridges, and has enlarged vertebral and paravertebral scales.

Systematics: Six subspecies of *U. stansburiana* are currently recognized (Ballinger and Tinkle, 1972; Collins, 1990), two of which occur in New Mexico. *Uta s. uniformis* occurs in the San Juan basin and vicinity, San Juan and McKinley counties. It was described by Pack and Tanner (1970), based on an adult male (BYU 10035) collected from "Split Mountain, Uintah County, Utah" by Wilmer W. Tanner on 21 May 1950. *Uta s. stejnegeri* occurs throughout the remainder of the state. This subspecies was described by Schmidt (1921) based on an adult male (AMNH 348) collected from the "mouth of Dry Cañon, Alamogordo, [Otero County,] New Mexico" by Alexander G. Ruthven on 23 July 1906. Collins (1991) has suggested that *U. s. stejnegeri* be recognized as a distinct species.

Habitat: Side-blotched lizards seem not to prefer any given substrate or soil type; they may be found from boulder fields to sand dunes (Tinkle, 1967). Areas of sparse vegetative cover are preferred. The species is not commonly found in grasslands, woodlands, or forests. This species may be expanding its range in the desert and semi-arid areas of the southwest where these habitats have been converted to desert grasslands or desert scrub. Peterson and Whitford (1987), however, reported that reduction of shrub cover and increase of grass cover favored this species. Side-blotched lizards are not common in the San Juan Basin of northwestern New Mexico, where they are found in canyons, arroyos, and rocky outwash alluvial fans or bajada slopes (Harris, 1963). The species is reasonably common in Chaco Canyon National Monument, and occurs on boulders, ruins, and other rocky substrates (Jones, 1970). In western Colorado, Waldschmidt (1979) studied this species inhabiting south-facing rocky hillsides at 1658 m with 20% vegetative cover, primarily *Juniperus osteosperma* and *Bromus tectorum*. Clark et al. (1982) recorded this species in association with Gunnison's prairie dog towns in this region, but did not find these lizards associated with the towns of black-tailed prairie dogs in the southeastern half of the state.

This species is common in western Texas in sandy habitats with sparse (20%) vegetative cover of *Prosopis glandulosa, Yucca* spp., *Quercus havardii, Gutierrezia sarothrae*, and *Artemisia filifolia* (Tinkle, 1967; Milstead and Tinkle, 1969). This is one of the most common lizards as recorded from the following suitable habitats in southern New Mexico: the sandy, mesquite-covered floodplain of the Black River in Eddy County (Gehlbach, 1956); the creosotebush-alkali sacaton-grama grass flats around the periphery of the white sands (White Sands National Monument) in Otero County (Dixon and Medica, 1966); rocky to sandy soils where the dominant vegetation is *Larrea tridentata* and *Gutierrezia sarothrae* on the large bajada of the Peloncillo Mountains southwest of Granite Gap in Hidalgo County (Baltosser and Best, 1990). Tinkle et al. (1962) suggested a definite association of this species with packrat (*Neotoma* spp.) nests; lizards appear to be absent where these structures are also absent from otherwise suitable habitat in west Texas.

Densities of this species have been reported from 5–62 lizards/hectare, and depend markedly upon local primary productivity and geographic location (Tinkle, 1967; Tinkle and Woodward, 1967; Worthington and Arvizo, 1973; Parker, 1974a; Whitford and Creusere, 1977). Other lizards that can be found sympatrically with this species in New Mexico include *Crotaphytus collaris, Gambelia wislizenii, Phrynosoma cornutum, Sceloporus magister, Sceloporus undulatus, Eumeces obsoletus, Cnemidophorus inornatus, Cnemidophorus neomexicanus*, and *Cnemidophorus tigris*.

Behavior: These lizards are generally active at the base of or beneath plants that form dense ground cover, such as *Rhus trilobata, Poliomintha incana, Scleropogon* sp., *Bouteloua* sp., *Yucca glauca, Flourensia cernua*, and *Larrea tridentata*. They are active throughout the day during the summer in western Texas, but activity is strongly bimodal with most occurring between 0800–1300 and after 1700 (Irwin, 1965). Dixon (1967) found lizards in Otero County active between 0700 and 1200 when exposed soil temperatures exceeded about 45°C. Lizards in western Texas usually bury in the soil to pass the night, but may occasionally sleep in low vegetation (Tinkle, 1967). Individual lizards are active on sufficiently warm days throughout all months of the year in southern New Mexico (Little and Keller, 1937; Baltosser and Best, 1990; our obs.); New Mexico specimens have been collected between 2 January (Doña Ana Co.) and 4 December (Luna Co.). Alexander and Whitford (1968) recorded winter activity on clear days when substrate temperatures exceeded 20°C. The preferred body temperature is typically 33–39°C, and varies with geography, weather conditions, and time of day (Dixon, 1967; Medica [pers. comm. *in* Alexander and Whitford, 1968]); Roberts, 1968; Parker and Pianka, 1975; Waldschmidt and Tracy, 1983). The critical thermal maximum is about 42°C. This species is able to withstand supercooling to –6.19°C (Lowe et al., 1971); however, Tinkle (1967) reported that lizards are often killed by sudden cold snaps.

Lizards in western Texas appear to form monogamous pairs during the mating season (Irwin, 1965; Tinkle, 1965). Lizards of southern populations (*U. s. stejnegeri*) are highly aggressive toward members of the same sex and are strongly territorial, whereas lizards belonging to northern populations (*U. s. stansburiana*) are less aggressive and maintain social hierarchies (Tinkle, 1969). Territories are maintained through ritualized displays involving head-bobbing, lateral body expansion, push-ups and, rarely, overt fighting (Tinkle, 1967). These behaviors are also utilized by males in courtship behavior (Ferguson, 1970), and are expressed by hatchlings (Tinkle, 1967).

Tinkle (1967) found annual average home range sizes of both adult male and female lizards to vary widely over successive years. Average home ranges of 1030 m² (190–2880 m²) for males and 400 m² (60–660 m²) for females in southern Doña Ana County, New Mexico, were reported by Worthington and Arvizo (1973). Waldschmidt (1983) reported average home range sizes of 284.3 m² and 153.6 m² for males and females, from a sand-dune habitat 50 km east of Carlsbad. Other studies have been done in Arizona (Parker, 1974a) and Colorado (Tinkle and Woodward, 1967; Waldschmidt, 1979; Waldschmidt and Tracy, 1983). Males patrol the borders of their home ranges. Territoriality breaks down during the winter, and as many as six adult lizards have been found hibernating together (Tinkle, 1967). Ted Brown (pers. comm.) collected five males, 13 females, and 13 juveniles from two horizontal holes in a roadcut in Albuquerque on 18 January 1975. Hatchling male and female lizards moved an average of 25.5 m and 16 m from hatching points to adult home ranges in Texas (Tinkle, 1967); however, Spoecker (1967a) reported dispersal movements of up to 545 m by juveniles in California. Lizards from Texas rarely (<0.02%) move more than 50 m from resident home ranges, whereas lizards from Colorado frequently (>10%) do so (Tinkle, 1969).

Reproduction: This has been one of the most intensively studied of any lizard species; we refer interested readers to Tinkle (1961, 1967, 1969) and Parker and Pianka (1975) for details. Minimum reproductive size is 42 mm SVL for both sexes. Early season hatchlings may attain these sizes at an age of three months (Colorado) or four months (western Texas) prior to hibernation, but neither will reproduce until the following spring. The extent of the breeding season varies geographically. Mature females may lay as many as three clutches annually of 2–5 eggs each but females breeding for the first time may produce only two clutches. Communal nesting may occur (Mautz, 1982). Incubation takes 60–80 days depending upon temperature and moisture regimes. Hatchlings (18–23 mm SVL) first appear in mid-June and continue to appear throughout the remainder of the season. Individual females may be able to store sperm for up to three months following insemination and use them to fertilize subsequent clutches (Cuellar, 1966).

Dixon (1967) observed gravid females in mid-June and mid-July in Otero County, with the subsequent appearance of hatchlings at the end of July and August, suggesting multiple annual clutches there. Fitch (1985) reported an average clutch size of 2.96 eggs (1–5) from a sample of 53 females from the Rio Grande valley of New Mexico.

Larger females produce larger clutches (Tinkle, 1961, 1967). Larger hatchlings enjoy a survival advantage over smaller ones under a variety of environmental circumstances (Fox, 1978; Ferguson et al., 1982; Ferguson and Fox, 1984). Worthington (1982) and Ferguson et al. (1990) discovered that annual rainfall patterns significantly influenced body sizes of adult lizards and correlated reproductive parameters in western Texas.

Population turnover in this species is essentially annual in western Texas, although individual lizards occasionally live to be three years old (Tinkle, 1967). Average life expectancy at hatching is about 18.5 weeks. Less than 1% of the lizards studied over a six-year period lived 60 weeks. In southwestern Colorado, annual turnover involves only two-thirds of the population, and about 20% of juveniles survive to reproduce (Tinkle, 1967).

Maximum testicular size is achieved by males during December and January, followed by decline to minimal values by August and September (Asplund and Lowe, 1964; Hahn, 1964; Parker and Pianka, 1975). Spermatids do not mature until the spring and summer. The fat body cycles of male and female lizards are similar and inversely proportional to cycles of gonadal activity (Hahn and Tinkle, 1965). Female fat bodies may represent as much as 4% of individual body weights in December. They are rapidly mobilized beginning in February and completely depleted by June.

Food Habits: This species is an opportunistic, generalized insectivore, taking prey of a suitable size utilizing a "sit-and-wait" foraging strategy (Tinkle, 1967; Parker and Pianka, 1975). Diet is dependent on seasonal variation in prey availability and abundance, geographic location, and local environmental features (i.e., plant phenology). Dixon and Medica (1966) reported hymenopterans (53%)

and beetles (18%) by volume as important food items for a population from White Sands National Monument of Otero County. Best and Gennaro (1984) reported hymenopterans (80%, mostly ants), beetles (67%), hemipterans (46%), grasshoppers (38%, mostly acridids) and spiders (32%) as important food items by frequency of occurrence for this species in southeastern New Mexico. Other dietary studies include Knowlton (1938), Hotton (1955), Milstead and Tinkle (1969), and Barbault and Maury (1981). Individual lizards may occasionally consume plant material, perhaps as a supplement to metabolic water (Hoff and Kay, 1970).

Uta stansburiana is preyed upon by a wide variety of vertebrate and invertebrate predators, and predation is a major factor influencing the demographics of populations throughout the range of the species (Wilson, 1991). Known predators of this species in New Mexico include *Cnemidophorus tigris* and *Masticophis flagellum*.

Remarks: Chiggers have been reported infesting this species (Spoecker, 1967b). Tail regeneration takes between nine and 76 days, depending upon position of the break on the tail, age of the lizard, and time of year (Tinkle, 1967). Lewis (1949) and Best et al. (1983) reported melanistic individuals from lava flows in south-central New Mexico, but Lewis (1951) noted the absence of melanism in specimens from the Afton lava flows in southern Doña Ana County. We follow Banta (1971) in the citation of the original description of this species.

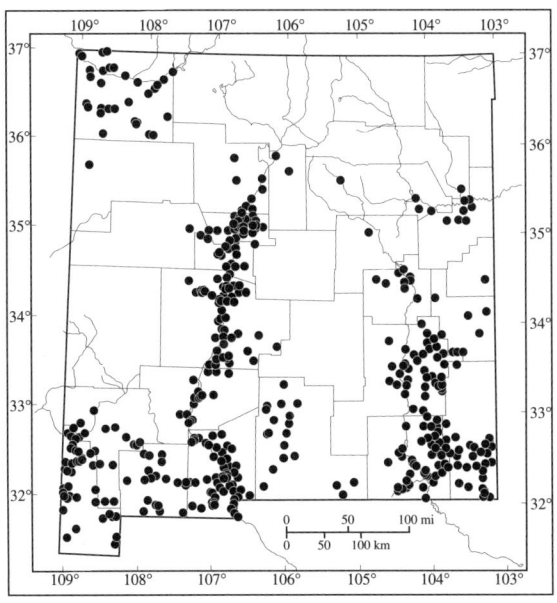

Distribution of Uta stansburiana *in New Mexico.*

Geckos **FAMILY GEKKONIDAE**

This is a large family of about 78 genera and 710 species, representing about 25% of worldwide lizard biodiversity (Kluge, 1987). It is found throughout all the major continents except Antarctica and on many oceanic islands, with limited distributions in North America, north of the Mediterranean Sea in Europe, and north of central Asia. As might be expected, geckos come in a variety of shapes and sizes, ranging from 16 mm SVL to over 250 mm SVL (Zug, 1993). They are terrestrial to arboreal in habitat. Most geckos are nocturnal, and have a transparent covering ("spectacle") formed by fusion of the eyelids over the eyes, with vertically elliptical pupils. Their skins are relatively thin and in some cases translucent, smooth or with a variety of tubercles. Many arboreal species have expanded toepads on the feet. Most geckos have a fixed clutch size of two eggs, although some species lay only one. Some species, especially island forms, are parthenogenetic (Darevsky et al., 1985), and appear to be derived from hybridizations between bisexual species. Two genera and three species occur in New Mexico. Many authorities have arranged gecko diversity into two or more separate families (Kluge, 1987; Grismer, 1988; Zug, 1993), in which case our native species would belong to the family Eublepharidae.

COLEONYX BREVIS Stejneger, 1893 — *Texas Banded Gecko*

See Plate 57A, 57B.

Type: An adult male (USNM 13627), collected from "Helotes, Bexar County, Texas" by G.W. Marnock on 30 November 1883.

Distribution: The species ranges in Texas along the southern edge of the Balcones Escarpment west of San Antonio, the western face of the Bordas Escarpment in southern Texas, and along the Rio Grande from McAllen northward throughout much of the Stockton Plateau and Trans-Pecos Texas (Axtell, 1986; Dixon, 1987). In Mexico, it ranges throughout the northern two-thirds of the Chihuahuan Desert as defined by Morafka (1977), eastward through Nuevo Leon and the northwestern panhandle of Tamaulipas. In New Mexico, it occurs northward along the Pecos River valley to Lake McMillan, westward across Otero Mesa to the Tularosa Basin and then northward to the Oscura Mountains (see Remarks). The western range appears to be limited by the eastern edge of the Hueco Bolson (Axtell, 1986), perhaps related to edaphic constraints.

Description: This is a small lizard, typically weighing about two grams as adults (Dial and Fitzpatrick, 1982). Females are slightly larger than males; the maximum recorded SVL for each sex is 63 mm and 56 mm. The legs are small, and the unregenerated tail is about as long as the body. The digits are short and narrow. The skin is soft, and unattached to the bones of the skull. The dorsal scales are finely granular without scattered, enlarged tubercles. The ventral surface of the body is unpigmented, thin, and somewhat translucent. There are fewer than six (usually four) preanal pores in males, and the scales which carry them (or the corresponding scales in females) are separated at the ventral midline by one or more small scales. Males possess cloacal bones, which are small, flattened, broad-tipped, and project outward and forward from either side of the base of the tail. Females possess a similarly shaped flap of skin in the analogous position. The eyes are large, the pupils vertically elliptical, and functional eyelids are present.

The ground color is light cream to tan. The dorsal surface of the limbs is unmarked. The edges of the eyelids are white. There are usually four dark brown transverse bars across the dorsal surface of the body between the limbs, and another anterior to the front limbs on the neck. These bars are usually 2–3 times as wide as the light spaces between them and their edges are darker than the centers, resulting in a superficial "double-barred" effect. Below each eye there is a narrow light line passing posteriorly above the ear and joining to form a loop on the anterior part of the neck, just anterior to the dark neck crossbar. The dorsal surface of the head is conspicuously mottled or spotted with brown markings, and these markings occur in the light fields of the dorsum between the transverse body bars as well. Dorsal spotting increases as lizards age to produce a leopard-like effect; sometimes the original transverse bars are obliterated entirely, or they expand to leave the dorsal interspaces as thin light lines. There are 6–11 dark tail rings, which are usually narrower than the light interspaces and which become increasingly indistinct and incomplete distally; regenerated tails usually lack the above pattern and are spotted or speckled instead. Hatchlings and juveniles lack the dorsal spotting and mottling, and the transverse body bars and tail rings are well marked and lack the differential intensity of pigmentation within them.

Similar Species: *Coleonyx variegatus* usually has seven or more preanal pores that are continuous across the ventral midline, usually lacks the degree of dark dorsal mottling and speckling, retaining more of a banded pattern, and the cloacal bones of males are narrow and pointed. *Hemidactylus turcicus* has conspicuously enlarged dorsal tubercles, expanded toepads, and lacks functional eyelids.

Systematics: This species is monotypic. It was originally confused with *Coleonyx variegatus*, which was described by Baird (1858 [1859]) with specimens representing both species. Even though Stejneger (1893) clearly recognized the differences between the two species, they continued to be confused by some workers for over 100 years (Dixon, 1970a).

Habitat: This gecko is a common inhabitant of rocky, limestone foothills with desert scrub vegetation such as

Larrea tridentata, *Acacia* spp., and *Juniperus* spp. It has been found to elevations of 1520 m.

Behavior: This species is nocturnal. The 134 New Mexico specimens available to us were collected between 7 April and 16 September. Dial (1978b) reported the mean body temperature of 52 active lizards as 28.6°C, when ambient air temperatures ranged between 21–33°C (mean 26.9°C). Individuals actively thermoregulate when they are inactive under shelter by pressing their backs against the underside of rocks, thus maintaining body temperatures significantly higher than that of the substrate. This behavior is particularly important on cool evenings or sunny but cool winter and spring days, allowing lizards to increase metabolic efficiency and expand the species' range into otherwise unsuitable regions (Dial, 1978b). Dial and Fitzpatrick (1982) reported that lizards kept at a body temperature of 32°C lose approximately 3% of their body mass per day.

These geckos frequently emit a series of high-pitched vocalizations when disturbed; "when handled, [this animal] chirrups and squeals feebly like a singing mouse" (Cope, 1880). Individuals often employ tail autotomy and associated stereotyped behaviors as a first line of defense against predators (Dial, 1978a). The percentage of regenerated tails is as high as 80% in natural populations and increases with age. Individuals eat their shed skins (Cope, 1880; Axtell, pers. comm.), a habit that perhaps serves as a means to ensure that potential predators will not use skins to find the areas where lizards are active.

Reproduction: Very little information is available for this species. Dial and Fitzpatrick (1981) reported that 2–3 clutches per year are produced in Texas. Hatchlings are 20–25 mm SVL. Snider and Bowler (1992) reported that a wild-caught adult lived for almost three years in captivity.

Food Habits: This species is an opportunistic insectivore, and employs visual and chemosensory cues in stalking its prey (Punzo, 1974b; Dial, 1978a). Dial (1978a) reported termites (42%), moths (adults and larvae, 14%), homopterans (cicadelids, 11%), spiders and solpugids (7% each), and crickets (5%) as important components by volume of the diet in 59 lizards examined from the Black Gap Wildlife Management Area, Brewster County, Texas. Punzo (1974b) recorded beetles (18%), hymenopterans (10%), lepidopterans (6%), spiders (6%), and orthopterans (4%) as important items by frequency of 56 lizards examined from the Chisos Mountains, Brewster County, Texas.

This species is probably preyed upon by a variety of vertebrate predators. We observed a *Sonora semiannulata* eating a dead gecko on a paved road in Big Bend National Park, Texas.

Remarks: An unverified sight record exists for Sierra County in the San Andres Mountains ca 18 km west of the Otero County line. The species has been reviewed by Dixon (1970a).

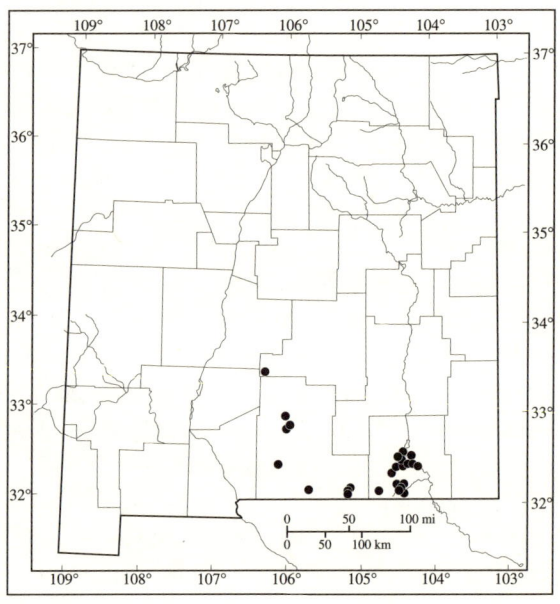

Distribution of Coleonyx brevis *in New Mexico.*

COLEONYX VARIEGATUS (Baird, 1858 [1859]) — *Western Banded Gecko*

See Plate 58.

Type: A subadult male (USNM 3217), collected from the "Rio Grande and Gila Valleys" (restricted to Winterhaven, Imperial County, California by Klauber, 1945) by Arthur Schott in 1852 during the course of the Mexican Boundary survey.

Distribution: This species occurs throughout the Sonoran and Mojave deserts, southward from a line between Bishop, California, and St. George, Utah, through southern California and southwestern Arizona to the tip of the Baja California Peninsula and along the coast of Sonora (see Systematics). It also occurs on a number of islands in the Sea of Cortéz (Murphy and Ottley, 1984). *Coleonyx variegatus* was first reported from New Mexico by Lowe (1955a), although Klauber (1945) had earlier anticipated its occurrence in the state. It is a Sonoran species that has penetrated the Cochise Filter Barrier (Morafka, 1977), extending up the Gila River Valley to Redrock and across the Continental Divide to the pass between the Little Hatchet and Big Hatchet mountains in Hidalgo County.

Description: This is a small lizard, although slightly larger than its eastern congener; the maximum SVL is 72 mm for females and 66 mm for males. During 1987–92, 131 individuals collected at Antelope Pass in Hidalgo County averaged 50.0 (22–67) mm SVL and 2.2 (0.3–5) gm; 40 adult males averaged 56.2 (52–64) mm SVL and 2.7 (1.5–4.5) gm; 24 adult females averaged 58.5 (52–67) SVL and 3.2 (2–5) gm (our unpubl. data). The legs are small, and the unregenerated tail is about as long as the body. The digits are short and narrow. The skin is soft, and unattached to the bones of the skull. The dorsal scales are finely granular without tubercles. The ventral surface of the body is unpigmented, its skin thin and somewhat translucent. There are usually 8–10 preanal pores in males, and the scales carrying them (or the corresponding scales in females) are continuous across the ventral midline. Males possess cloacal bones, which are small and pointed, and project outward and forward from either side of the base of the tail. Females possess a similarly shaped flap of skin in the analogous position. The eyes are large, the pupils vertically elliptical, and functional eyelids are present.

The ground color is light cream to tan. The dorsal surface of the limbs is unmarked. The edges of the eyelids are white. There are usually four dark brown transverse bars across the dorsal surface of the body between the limbs; these bars are usually equal to or narrower than the spaces between them and their edges are darker than the centers, resulting in a superficial "double-barred" effect. Below each eye there is a narrow light line passing posteriorly above the ear opening and joining to form a loop on the anterior part of the neck, posterior to which there is a broad crossband with forward pointing extensions carried antero-laterally toward the ears (this pattern fades and becomes inconspicuous with age). The dorsal surface of the head is conspicuously mottled or spotted with brown, and these markings occur in the light fields of the dorsum between the transverse body bars as well. There are 9–14 dark tail rings, which become increasingly indistinct and incomplete distally; regenerated tails usually lack this characteristic pattern and are spotted or speckled. Hatchlings and juveniles lack the dorsal spotting and mottling and have the transverse body bars and tail rings well marked, without the differential intensity of pigmentation within them.

Similar Species: *Coleonyx brevis* usually has fewer than six preanal pores interrupted across the ventral midline by one or more small scales, is usually less distinctly banded as an adult, and the cloacal bones of males are broad and flat. *Hemidactylus turcicus* has conspicuously enlarged dorsal tubercles, expanded toepads, and lacks functional eyelids.

Systematics: Eight subspecies of *C. variegatus* have been recognized (Dixon, 1970b), although one of these, *C. v. fasciatus*, has subsequently been recognized once again as a distinct species (Grismer, 1988). *Coleonyx v. bogerti* occurs in New Mexico. It was described by Lawrence M. Klauber in 1945; the holotype (SDSNH 32486), an adult male, was collected from "Xavier, Pima County, Arizona" by Lee W. Arnold on 17 July 1939.

Habitat: This species occurs in desert vegetation communities on a wide variety of soil types, from sandy arroyos to bedrock, between sea level and 1485 m. Nickerson and Mays (1969 [1970]) found these lizards from 1212–1485 m in *Prosopis-Acacia* habitat in southeastern Arizona. Ortenburger and Ortenburger (1926) found an individual about 45 cm below the ground surface and 2.4 m from the entrance of the mammal burrow it was using. Parker (1972a) reported densities of 12–25 lizards/hectare in southern Arizona.

Behavior: Adults are active above ground from April through October in southern Arizona (Parker, 1972a). Immature lizards can be found year-round, but only intermittently from November through March. The 73 New Mexico specimens have been collected between 12 April and 28 August. This gecko is nocturnal, although some diurnal activity may take place in daytime retreats (Cooper et al., 1985a). Parker and Pianka (1974) recorded an average body temperature for 35 active lizards in nature as 28.35°C (22.7–33.7°C). Body temperatures were positively correlated with air temperatures. The average preferred body temperature selected by lizards in a thermal gradient was 28.6°C; no lizard voluntarily tolerated temperatures below 24°C or above 33°C (Vance, 1973). Critical thermal minimum and maximum temperatures were 7.7°C and 44.0°C. Males may be more active than females (Kingsbury, 1989) partly because of searching for mates. Individual geckos do not move much; linear distances greater than 50 m are exceptional (Parker, 1972a).

Captive specimens eat their own shed skins (Kauffeld, 1943a); whether this behavior is of nutritional value, removes physical or olfactory cues to predators, or serves some other function is unknown. Geckos readily lose their tails, and employ specific behaviors in the presence of predators that draw their attention to the wiggling tail instead of to the body (Congdon et al., 1974; Johnson and Brodie, 1974). Parker and Pianka (1974) have suggested that these lizards also mimic the gait and carriage of venomous scorpions (*Hadrurus*) found in the same habitats. Geckos are able to detect and identify potential snake predators by chemical means (Dial et al., 1989). Lizards of this species have been reported in laboratory experiments to aggregate in diurnal shelters and to choose sites already occupied by other geckos (Cooper et al., 1985b); the significance of this behavior for wild geckos is unknown, however, as these lizards display well-developed aggressive behavior toward conspecifics. Burke (1994) recorded free-ranging individuals sharing shelters more frequently than expected by chance.

Reproduction: The reproductive season in southern Arizona reportedly extends from April through September (Parker, 1972a) but may be shorter depending on seasonal environmental conditions. Peak egg-laying seems to occur during May and June. The minimum size at maturity for females was reported by Parker (1972a) to be 52 mm SVL, with the mean size of 54 gravid females as 62.4 mm SVL (52–70). The clutch size is almost invariably two eggs (Klauber, 1945; Nickerson and Mays, 1969 [1970]; Parker, 1972a; Parker and Pianka, 1974; Vitt, 1977). Many individuals produce at least two clutches annually and the incubation time is about six weeks. Females may store sperm from early mating to produce clutches later in the year (Parker, 1972a). Hatchlings are 25–30 mm SVL and reach adult size by the following spring.

Parker (1972a) reported minimum adult size for males as 52 mm SVL. Adult males emerge in the spring with testes maximally enlarged, and these decline steadily in size throughout the active season.

Snider and Bowler (1992) reported a living wild-caught adult female *C. v. bogerti* that lived for more than 15 years in captivity.

Food Habits: Little quantitative data exist for this species. Beetles, grasshoppers, and sowbugs are eaten (Knowlton, 1938; Klauber, 1945). Parker and Pianka (1974) reported termites (75% by number, 24% by volume), beetles (7%, 18%), orthopterans (1%, 14%), and insect larvae (3%, 12%) as major components of the diet in a sample of 185 geckos from Arizona. Insect larvae, beetles, and grasshoppers are important in the spring, whereas termites, spiders, and solpugids predominate by late summer.

Remarks: The species has been reviewed by Dixon (1970b).

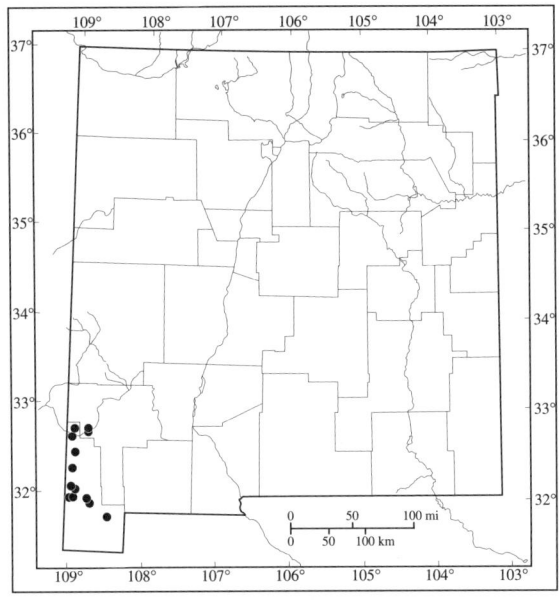

Distribution of
Coleonyx variegatus
in New Mexico.

HEMIDACTYLUS TURCICUS (Linnaeus, 1758)   *Mediterranean Gecko*

See Plate 59.

Type: The holotype of *Hemidactylus turcicus* is unknown. It was collected from "Oriente" (restricted to somewhere in Turkey by Mertens and Müller, 1940). The collector and date of collection are unknown.

Distribution: *Hemidactylus turcicus* is native to coastlines of the Mediterranean and Red seas, and the Arabian Peninsula to northwestern India. It was first reported in the United States from Key West, Florida (Fowler, 1915 [1916]; Stejneger, 1922), and has been widely introduced in the Caribbean (McCoy, 1970; Conant and Collins, 1991). Conant (1955b) first reported *H. turcicus* from Texas and Dixon (1987) recorded it from 44 Texas counties. It was first reported from metropolitan areas in Arizona by Robinson and Romack (1973) and from El Paso by Price (1980b), who anticipated its eventual occurrence in southern New Mexico. The first verified occurrence in New Mexico was reported by Painter et al. (1992) in Las Cruces at 1200 m, and the species appears to be spreading throughout that city.

Description: This is a small lizard; maximum SVL in males is 60 mm. The dorsal scales are granular; interspersed with these are numerous conspicuously enlarged and keeled tubercles. The tubercles are arranged in up to 18 irregular longitudinal rows on the back, scattered over the surfaces of the limbs, and up to eight longitudinal rows on each side of the tail. The ventral trunk scales are smooth and cycloid. The eyes are large, lack eyelids, and have a vertically elliptical pupil. The digits are expanded, except for the terminal phalanx, with two complete rows of subdigital lamellae. The terminal phalanx possesses a prominent claw. Males have 3–10 preanal pores.

The dorsal ground color is highly variable; nearly white, sandy yellow or gray, pinkish-brown, or grayish-brown. The skin tubercles are white. Irregular dark brown spots are scattered over the dorsal surface, and may form a banded pattern on the tail. A vague, dark crossbanded pattern may also be present on the dorsal surface of the limbs. The ventral surface is usually white or cream, with faint dark speckling throughout, especially on the tail.

Similar Species: *Coleonyx brevis* and *C. variegatus* lack enlarged dorsal tubercles on the body and expanded toepads, and both species possess functional eyelids.

Systematics: Three subspecies have been recognized,

but all introduced populations in the Western Hemisphere have been referred to the nominotypical subspecies (McCoy, 1970).

Habitat: *Hemidactylus turcicus* has long been a commensal with humans, and introduced populations are found on buildings and natural habitats which have been disturbed or altered through human activities. Selcer (1986) estimated densities of 544–2210 lizards/hectare in south Texas.

Behavior: These lizards are primarily nocturnal. They can often be found congregating around lights that attract insects on which they feed. The expanded toepads and long claws on their feet allow these lizards to climb vertical walls and traverse ceilings upside down. They are vocal, capable of emitting faint squeaks and a series of clicking noises when alarmed or in social situations. Males are aggressive toward one another during the breeding season (Selcer, 1987).

Reproduction: The minimum size at maturity in the United States is 49 mm SVL, attained at approximately 9 months of age (Selcer, 1986). The clutch size of two eggs is invariant (Rose and Barbour, 1968; Robinson and Romack, 1973), and an individual clutch may represent up to 27% of an individual female's body weight (Selcer, 1986). The reproductive season extends from April through early August (Louisiana) or early September (Texas), and individual females produce 1–3 clutches a year (Rose and Barbour, 1968; Selcer, 1986). Fat body mass is inversely proportional to clutch mass, and is essentially depleted during formation of the first annual clutch (Selcer, 1987). The reproductive season in Arizona may extend throughout the year (Robinson and Romack, 1973). Considerable variation exists, although not related to female body size, in the size and weight of eggs produced by individual females (Selcer, 1990); larger eggs produce larger hatchlings. Hatchlings are 20–30 mm SVL and have been reported from Texas in July, and from Arizona in February, August, and September (Robinson and Romack, 1973; Trauth, 1985).

Females will construct nests if appropriate material (i.e. debris) is available, and individuals may use the same nest sites repeatedly over several years (Rose and Barbour, 1968; Trauth, 1985; Selcer, 1986). Different females may use the same nest (Selcer, 1986). The eggs have hard shells which are resistant to desiccation.

Males reach sexual maturity at a minimum size of 44 mm SVL (Selcer, 1986). Male testes are of minimum size by late fall or early winter (Rose and Barbour, 1968; Selcer, 1986). Spermatogenesis is underway in December, and by April the epididymides are packed with sperm. Rapid regression of the reproductive tract ensues by August, although reproductively competent males can be found throughout the year (Selcer, 1986). Selcer (1987) found that fat bodies increased in size during testicular regression and decreased during testicular recrudescence.

Mortality appears to be highest for juveniles immediately after hatching, with annual survivorship of about 20% (Selcer, 1986). Individual lizards may live to be three years old.

Food Habits: No quantitative data exist for United States populations. Rose and Barbour (1968) reported *H. turcicus* in New Orleans to be an opportunistic insectivore. Caterpillars, ants, beetles, hemipterans, homopterans, and grasshoppers were major components of the diet.

Remarks: Many people throughout the world have the mistaken belief that these lizards are venomous; they are actually quite harmless.

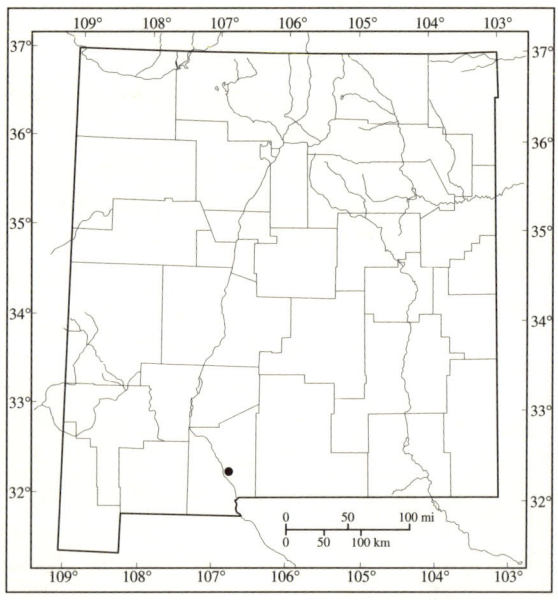

Distribution of Hemidactylus turcicus *in New Mexico.*

FAMILY TEIIDAE *Whiptails*

This is a small New World family of 39 genera and about 130 species (MacLean, 1974; Harris, 1985) that occurs from the northern United States to Argentina and throughout the Caribbean. These lizards are diurnal and occupy a variety of habitats from deserts to rain forests. The microteiids are small (<60 mm SVL) lizards with reduced limbs and elongated bodies, and are generally secretive and fossorial. The macroteiids are large (70–500 mm SVL) terrestrial lizards with well-developed limbs and relatively compact bodies, and are alert, mobile predators. Microteiid body scales are heterogeneous while macroteiids possess granular dorsal scales and plate-like scales on the head and belly. Some authorities (e.g., Zug, 1993) have recognized these two groups as different families. Both groups are oviparous; microteiids produce two eggs/clutch (MacLean, 1974), macroteiids usually more. A number of species within the family, particularly in the genus *Cnemidophorus*, are parthenogenetic; all of the individuals in a species are female, and offspring are virtually genetically identical to mothers generation after generation. This phenomenon was first noted in the genus by Minton (1958 [1959]), Tinkle (1959), and Duellman and Zweifel (1962). Maslin (1962) was the first to examine large series of lizards from several nominal taxa within the genus and explicitly suggest that true parthenogenesis occurred. Considerable research on this phenomenon has ensued (see Wright and Vitt, 1993, for the latest synopsis) and has clearly shown that the parthenogenetic species were derived from hybridization events between bisexual congeners, followed in some cases by further hybridizations between diploid parthenogenetic lineages and bisexual ones to produce triploid parthenogenetic species. Cuellar (1971) clearly demonstrated the genetic mechanism of parthenogenesis in the triploid species *Cnemidophorus uniparens*. The somatic number of chromosomes is doubled early in oogenesis presumably by a premeiotic endoduplication, and the 3N level is restored by two subsequent maturation divisions. Elucidating the phylogenetic relationships and systematics of members of the genus *Cnemidophorus* continues to be one of the more vexing problems in herpetology (Wright, 1993, 1994). This genus with 13 species, eight of which are parthenogenetic, occurs in New Mexico.

	MALES	LONGITUDINAL DORSAL STRIPES	LIGHT SPOTS IN DORSAL FIELDS
BURTI	YES	USUALLY 6 IN JUVENILES; INDISTINCT OR ABSENT IN ADULTS	NONE IN JUVENILES; PRESENT IN ADULTS
DIXONI	NO	10–14 IN JUVENILES; ABSENT IN ADULTS	PRESENT IN JUVENILES; ABSENT IN ADULTS
EXSANGUIS	NO	6–7	FEW IN JUVENILES; MANY IN ADULTS; SPOTTING MAY OVERLAY OR IMPINGE ON LIGHT STRIPES
FLAGELLICAUDUS	NO	6	ABSENT IN JUVENILES; FEW IN ADULTS; USUALLY NO SPOTS BETWEEN PARAVERTEBRAL STRIPES
GRAHAMII	NO	6–10 IN JUVENILES; ABSENT IN ADULTS	FEW, IF ANY IN JUVENILES; ABSENT IN ADULTS
GULARIS	YES	USUALLY 8	FEW; DISTINCT IN LATERAL FIELDS, FAINT OR ABSENT IN PARAVERTEBRAL FIELDS
INORNATUS	YES	6–8	ABSENT
NEOMEXICANUS	NO	7	FEW; DISTINCT IN DORSOLATERAL AND UPPER LATERAL FIELDS; ABSENT LOWER LATERAL AND PARAVERTEBRAL FIELDS
SEXLINEATUS	YES	USUALLY 7	ABSENT
SONORAE	NO	6	ABSENT IN JUVENILES; MANY IN ADULTS
TIGRIS	YES	4–6 IN JUVENILES; USUALLY ABSENT IN ADULTS	PRESENT IN JUVENILES; ABSENT IN ADULTS
UNIPARENS	NO	USUALLY 6	ABSENT
VELOX	NO	USUALLY 6	USUALLY ABSENT

	MESOPTYCHIAL SCALES	POSTANTEBRACHIAL SCALES	GAB*	GBPS**
BURTI	ABRUPTLY ENLARGED	ABRUPTLY ENLARGED	98–115	5–11
DIXONI	ABRUPTLY ENLARGED	GRANULAR	94–112	NONE
EXSANGUIS	ENLARGED	GREATLY ENLARGED (4X)	62–86	2–8
FLAGELLICAUDUS	ENLARGED	ENLARGED	77–84	3–5
GRAHAMII	ABRUPTLY ENLARGED	GRANULAR	75–112	NONE
GULARIS	ENLARGED	ENLARGED	75–90	13–18
INORNATUS	GRANULAR TO SLIGHTLY ENLARGED	GRANULAR TO SLIGHTLY ENLARGED	52–78	3–15
NEOMEXICANUS	GRANULAR	GRANULAR	71–80	7–13
SEXLINEATUS	ABRUPTLY ENLARGED	GRANULAR	62–91	8–13
SONORAE	ENLARGED	ENLARGED	74–80	5–8
TIGRIS	GRANULAR	GRANULAR	68–114	NONE
UNIPARENS	ABRUPTLY ENLARGED	ABRUPTLY ENLARGED	59–78	4–8
VELOX	ABRUPTLY ENLARGED	ABRUPTLY ENLARGED	63–85	3–11

Table 2 Characteristics of Whiptail Lizards in New Mexico.
*GAB = granular scales around the body, measured at midbody
**GBPS = granular scales between the paravertebral stripes

	VERTEBRAL STRIPE	TAIL COLOR	SUPRAORBITAL SEMICIRCLES EXTEND TO:
BURTI	IF PRESENT, 1 DISTINCT	REDDISH OR ORANGE AS JUVENILES; FADING TO GRAY-BROWN AS ADULT	THIRD SUPRAOCULAR
DIXONI	ABSENT	ORANGE-BROWN	SECOND SUPRAOCULAR; INCOMPLETELY SEPARATES IT FROM MEDIAN HEAD SHIELDS
EXSANGUIS	INCOMPLETE AND INDISTINCT, IF PRESENT	GREENISH-GRAY	THIRD SUPRAOCULAR
FLAGELLICAUDUS	ABSENT	OLIVE-GREEN WITH BLUE TINGE	THIRD SUPRAOCULAR
GRAHAMII	ABSENT	LIGHT BROWN W/ LIGHT AND DARK SPOTTING; OCCASIONALLY YELLOWISH	SECOND SUPRAOCULAR; INCOMPLETELY SEPARATES IT FROM MEDIAN HEAD SHIELDS
GULARIS	1–2 COMPLETE BUT INDISTINCT	PINKISH TO REDDISH BROWN	THIRD SUPRAOCULAR
INORNATUS	1–2 COMPLETE, SOMETIMES INDISTINCT	BRIGHT BLUE TO BLUISH WHITE	FOURTH SUPRAOCULAR
NEOMEXICANUS	DISTINCT AND WAVY	BRIGHT BLUE AS HATCHLINGS; BLUISH TIP AND GREENISH BLUE TO BROWN IN ADULTS	SECOND SUPRAOCULAR; COMPLETELY SEPARATES IT FROM MEDIAN HEAD SHIELDS
SEXLINEATUS	DISTINCT; OFTEN DIVIDED ANTERIORLY, OCCASIONALLY 2 SEPARATE STRIPES	BRIGHT BLUE AS HATCHLINGS; BROWN AS ADULTS	THIRD SUPRAOCULAR
SONORAE	ABSENT	ORANGE-TAN	THIRD SUPRAOCULAR
TIGRIS	ABSENT	BROWN	USUALLY THIRD SUPRAOCULAR
UNIPARENS	INCOMPLETE POSTERIORLY IF PRESENT	BRIGHT BLUE AS HATCHLINGS; BLUISH GREEN TO OLIVE-GREEN AS ADULTS	THIRD SUPRAOCULAR
VELOX	COMPLETE, INDISTINCT, AND UNDULATING, IF PRESENT	BRIGHT BLUE IN HATCHLINGS, LIGHT BLUE AS ADULTS	THIRD SUPRAOCULAR

*Table 2 (**continued**) Characteristics of Whiptail Lizards in New Mexico.*

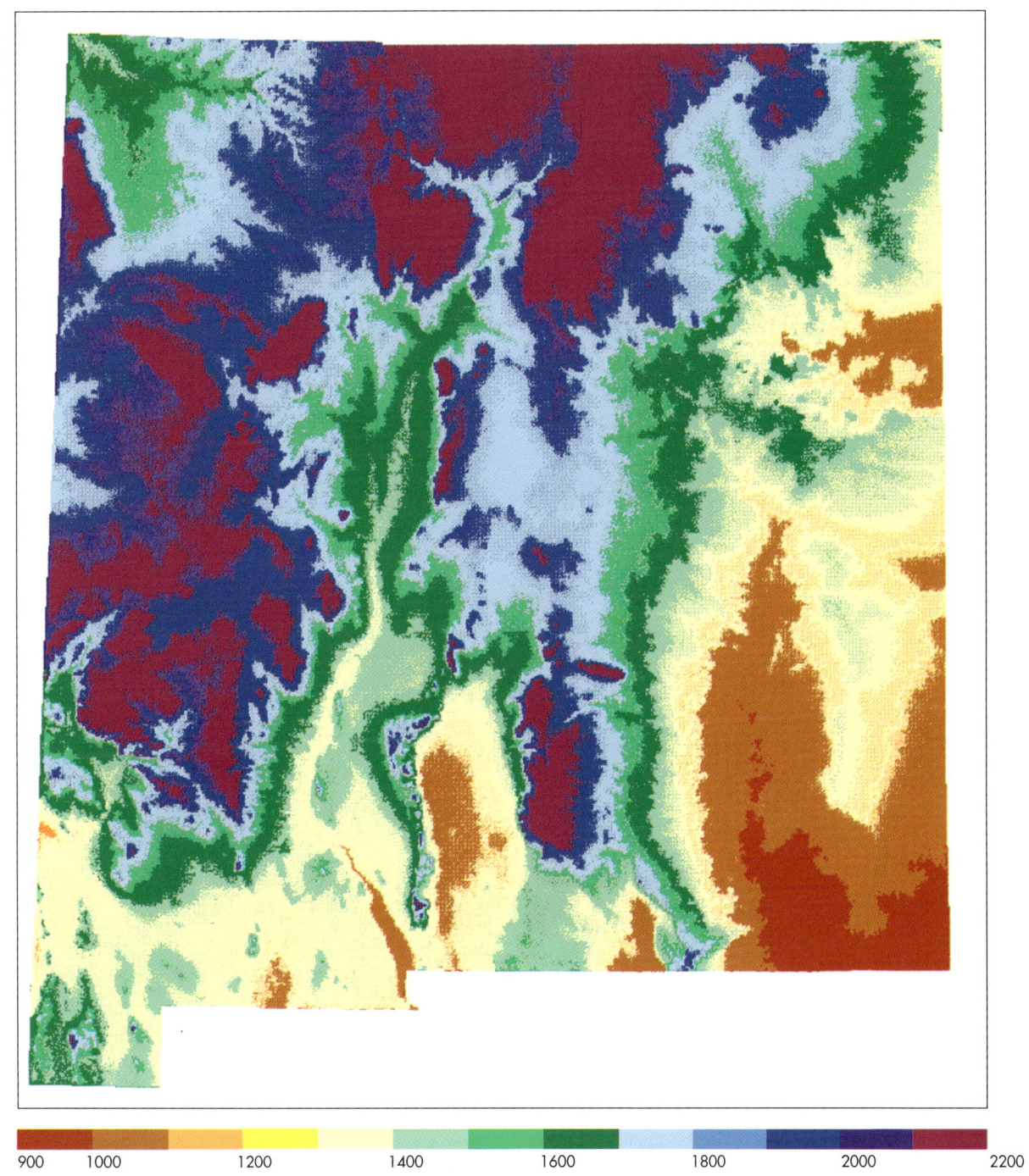

Figure 2 Topographic Map of New Mexico showing 100 meter contour intervals.

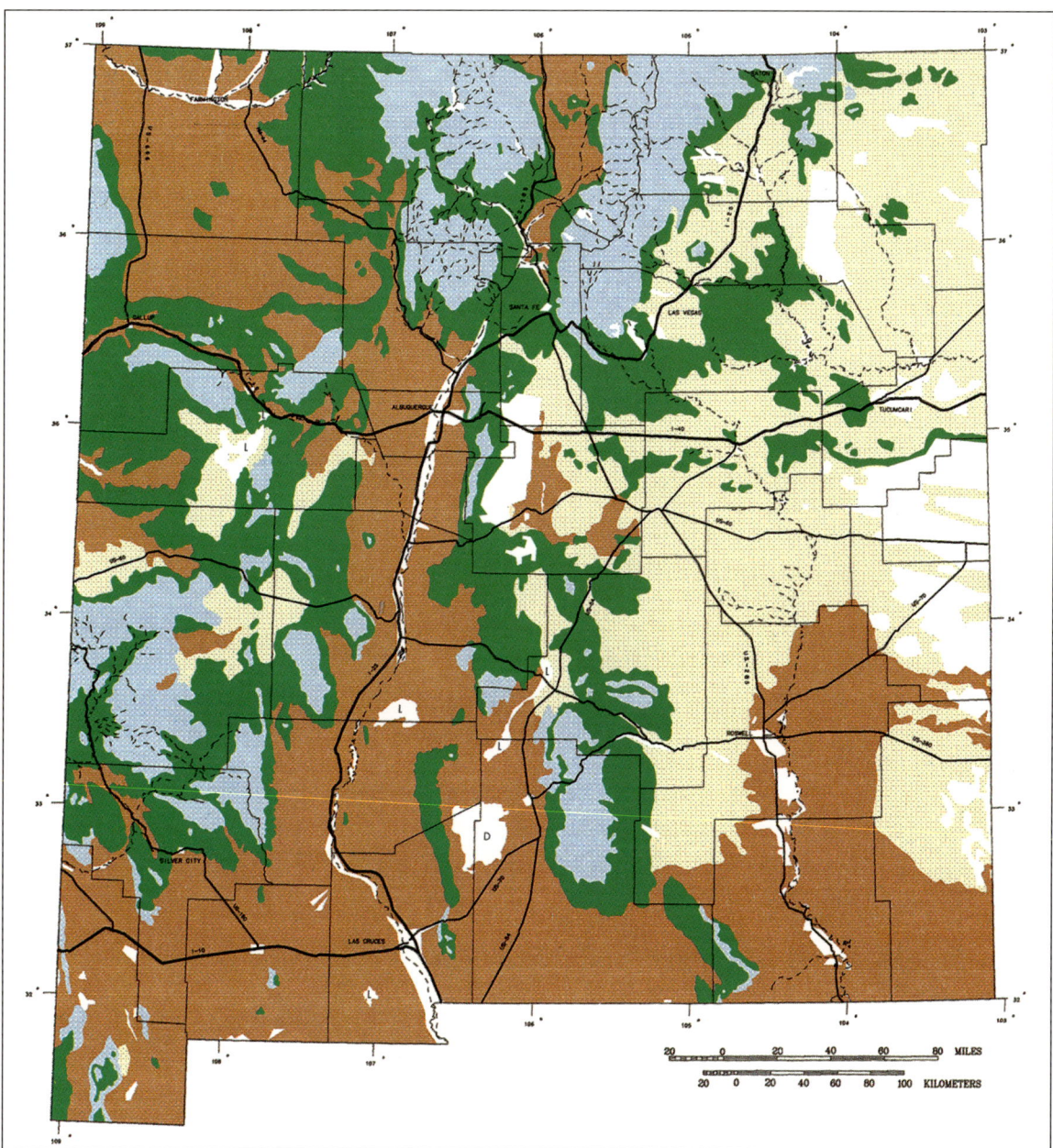

Figure 3 Vegetation Map of New Mexico adapted from Dick-Peddle (1993), showing major zones of vegetation.

Montane associations include the Alpine tundra, Subalpine coniferous forest, Montane coniferous forest, Montane grassland, and Montane scrub. These are mostly coniferous forest habitat types with some grassland and tundra at high elevations and scrub vegetation at low elevations.

Woodland and savannah associations include the Coniferous and mixed woodland and Juniper savannah. These consist largely of piñon, junipers, oaks, and grasses and are found at intermediate elevations and moisture conditions.

Plains-mesa grassland. Vast areas of eastern New Mexico and other scattered locations in the state where adequate moisture conditions prevail are grass-covered.

Desert associations include Desert grassland, Chihuahuan desert scrub, Great Basin desert scrub, Plains-mesa sand scrub, and Closed basin scrub. These associations of drought adapted plants are found mostly where the evaporotranspiration rate exceeds precipitation.

Urban and cultivated areas.
These areas may contain small isolated patches of native vegetation.

L Lava beds.

D Dune areas with little vegetation.

Figure 4
Chihuahuan Desert Scrub
This area around Carlsbad is home to a diverse array of amphibians and reptiles, including Bufo speciosus, Eleutherodactylus augusti, Coleonyx brevis, Cnemidophorus gularis, Lampropeltis alterna, *and* Crotalus l. lepidus. *Photograph by Don L. MacCarter.*

Figure 5
Creosotebush Desert
This widespread, somewhat homogeneous habitat is widespread at low elevations in southern and central New Mexico. The herpetofauna in this region is composed of mainly drought tolerant species with Scaphiopus couchii, Bufo debilis, Gambelia wislizenii, Cnemidophorus tigris, Arizona elegans, Masticophis flagellum, Rhinocheilus lecontei, *and* Crotalus scutulatus. *Photograph by Charles W. Painter.*

Figure 6
White Sands National Monument
Pale forms of amphibians and reptiles that inhabit the gypsum sands in this unique area include Scaphiopus couchii, Holbrookia maculata ruthveni, Cnemidophorus inornatus gypsi, *and* Sceloporus undulatus cowlesi.
Photograph by Don L. MacCarter.

Figure 7
Riparian Association
The San Francisco River at S. Dugway Canyon in Catron County provides habitat for Bufo microscaphus, Hyla arenicolor, Rana chiricahuensis, Kinosternon sonoriense, Thamnophis cyrtopsis, *and* T. rufipunctatus.
Photograph by Don L. MacCarter.

Figure 8
Riparian Association
Sitting Bull Falls in the Guadalupe Mountains of Eddy County is a permanent water source that attracts many amphibians and reptiles including Rana berlandieri, Eumeces multivirgatus, Nerodia erythrogaster, *and* Thamnophis cyrtopsis.
Photograph by Don L. MacCarter.

Figure 9
Juniper Savannah
This spacious and widespread habitat supports a wide variety of amphibians and reptiles. Species collected near this site in Cibola County include Spea bombifrons, S. multiplicata, Holbrookia maculata, Cnemidophorus uniparens, C. velox, Heterodon nasicus, Masticophis flagellum, Pituophis melanoleucus, *and* Crotalus viridis.
Photograph by William G. Degenhardt.

Figure 10
Piñon-Juniper Woodland
This rocky, foothill habitat is widespread at mid-elevations in New Mexico. It contains relatively few species of amphibians and reptiles. Conspicuous species from this area in Santa Fe County include Phrynosoma douglasii, Sceloporus undulatus, Cnemidophorus exsanguis, C. velox, Lampropeltis triangulum, Thamnophis elegans, *and* Crotalus viridis. *Photograph by William G. Degenhardt.*

Figure 11
Great Basin Desert Scrub
This high elevation, scrub-dominated habitat in San Juan County typifies many areas in northwestern New Mexico. Species present include Spea bombifrons, Holbrookia maculata, Sceloporus graciosus, Uta stansburiana, Cnemidophorus velox, Pituophis melanoleucus, *and* Crotalus viridis. *Photograph by William G. Degenhardt.*

Figure 12
Montane Grassland
This ancient caldera in the Jemez Mountains is a beautiful example of montane grasslands found throughout the higher elevations in New Mexico. Long cold winters limit the distribution of amphibians and reptiles here, although Ambystoma tigrinum, Pseudacris triseriata, Rana pipiens, Liochlorophis vernalis, *and* Thamnophis elegans *are found in the area.* Plethodon neomexicanus *occurs on the rocky mountain slopes surrounding this area. Photograph by Don L. MacCarter.*

Figure 13
High Elevation Montane Forest
Numerous areas in New Mexico reach elevations exceeding 3000 m. This area in the Pecos Wilderness Area of northern New Mexico contains mostly spruce and fir and is almost devoid of amphibians and reptiles. Phrynosoma douglasii *and* Thamnophis elegans *may endure the severe winters in this habitat where their activity season is limited to brief periods of the summer. Photograph by Don L. MacCarter.*

Figure 14
High Elevation Montane Forest capped by Alpine Tundra
This cold, wind-swept environment on Santa Fe Baldy in the Pecos Wilderness Area in northern New Mexico is completely devoid of amphibians and reptiles. Photograph by Don L. MacCarter.

Plate 1
Adult tiger salamander (Ambystoma tigrinum) *from Santa Fe County.*

Plate 2
Adult Sacramento mountain salamander (Aneides hardii) *from Lincoln County.*

Plate 3
Adult Jemez Mountains salamander (Plethodon neomexicanus) *from Sandoval County.*

Plate 4
Adult Couch's spadefoot
(Scaphiopus couchii)
from Roosevelt County.

Plate 5
Adult plains spadefoot
(Spea bombifrons)
from Chaves County.

Plate 6
Adult New Mexico spadefoot
(Spea multiplicata)
from Torrance County.

Plate 7(A)
Adult barking frog
(Eleutherodactylus augusti)
from Chaves County.

Plate 7(B))
Juvenal barking frog
(Eleutherodactylus augusti)
from Eddy County.

Plate 8
Adult Colorado River toad
(Bufo alvarius)
from Hidalgo County.

Plate 9
*Adult female boreal toad
(Bufo boreas) from
Larmier County, Colorado.
Photograph by
P. Steven Corn.*

Plate 10
Adult Great Plains toad
(Bufo cognatus)
from Hidalgo County.

Plate 11
Adult male green toad
(Bufo debilis)
from Luna County.

Plate 12
Adult southwestern toad
(Bufo microscaphus)
from Grant County.

Plate 13
Adult red-spotted toad (Bufo punctatus) from Eddy County.

Plate 14
Adult Texas toad (Bufo speciosus) from Eddy County.

Plate 15
Adult Woodhouse's toad
(Bufo woodhousii)
from Hidalgo County.

Plate 16
Adult cricket frogs
(Acris crepitans)
from Eddy County.

Plate 17
Adult canyon treefrog (Hyla arenicolor) from Hidalgo County.

Plate 18
Adult mountain treefrog (Hyla eximia) from Sierra County. Photograph by William G. Degenhardt.

Plate 19
Adult western chorus frog (Pseudacris triseriata) from Colfax County.

Plate 20
Adult male narrowmouth toad (Gastrophryne olivacea) *from Luna County.*

Plate 21
Adult Rio Grande leopard frog (Rana berlandieri) *from Eddy County.*

Plate 22
Adult plains leopard frog (Rana blairi) *from San Miguel County.*

Plate 23
Adult bullfrog
(Rana catesbeiana)
from Socorro County.

Plate 24
Adult Chiricahua leopard frog
(Rana chiricahuensis)
from Socorro County.

Plate 25
Adult northern leopard frog
(Rana pipiens)
from Mora County.

Plate 26
Adult lowland leopard frog (Rana yavapaiensis) *from Graham County, Arizona.*

Plate 27
Adult snapping turtle (Chelydra serpentina) *from Quay County.*

Plate 28
Adult painted turtle (Chrysemys picta) *from Socorro County.*

Plate 29
Adult western river cooter (Pseudemys gorzugi) from Eddy County.

Plate 30
Adult ornate box turtle (Terrapene ornata) from Hidalgo County.

Plate 31
Adult Big Bend slider (Trachemys gaigeae) from Sierra County.

Plate 32
Adult slider
(Trachemys scripta)
from Eddy County.

Plate 33
Adult yellow mud turtle
(Kinosternon flavescens)
from Eddy County.

Plate 34
Adult Sonoran mud turtle
(Kinosternon sonoriense)
from Sonora, Mexico.

Plate 35
Adult smooth softshell
(Trionyx muticus)
from Quay County.

Plate 36
Adult spiny softshell
(Trionyx spiniferus)
from Quay County.

Plate 37
Adult male collared lizard
(Crotaphytus collaris)
from Chaves County.

Plate 38
Adult leopard lizard
(Gambelia wislizenii)
from Hidalgo County.

Plate 39
Adult zebratail lizard
(Callisaurus draconoides)
from Hidalgo County.

Plate 40
Adult greater earless lizard
(Cophosaurus texanus)
from Hidalgo County.

Plate 41
Adult male (top) and female (bottom) lesser earless lizard
(Holbrookia maculata)
from Chaves County.

Plate 42
Adult Texas horned lizard
(Phrynosoma cornutum)
from Socorro County.

Plate 43
Adult short-horned lizard
(Phrynosoma douglasii)
from Sandoval County.

Plate 44
Adult roundtail horned lizard
(Phrynosoma modestum)
from Socorro County.

Plate 45
Adult regal horned lizard (Phrynosoma solare) from Hidalgo County.

Plate 46
Adult sand dune lizard (Sceloporus arenicolus) from Chaves County.

Plate 47
Adult Clark's spiny lizard (Sceloporus clarkii) from Hidalgo County.

Plate 48
Adult sagebrush lizard (Sceloporus graciosus) *from Sandoval County.*

Plate 49
Adult Yarrow's spiny lizard (Sceloporus jarrovii) *from Hidalgo County.*

Plate 50
Adult desert spiny lizard (Sceloporus magister) *from Hidalgo County.*

Plate 51
*Adult crevice spiny lizard
(Sceloporus poinsetti)
from Hidalgo County.*

Plate 52
*Adult bunch grass lizard
(Sceloporus scalaris)
from Hidalgo County.*

Plate 53
*Adult prairie lizard
(Sceloporus undulatus)
from Harding County.*

Plate 54
Adult striped plateau lizard
(Sceloporus virgatus)
from Hidalgo County.

Plate 55
Adult tree lizard
(Urosaurus ornatus)
from Grant County.

Plate 56 (A)
Adult male side-blotched lizard
(Uta stansburiana)
from Hidalgo County.

Plate 56 (B)
Patternless adult morph of side-blotched lizard (Uta stansburiana) *from Chaves County.*

Plate 57 (A)
Adult Texas banded gecko (Coleonyx brevis) *from Eddy County.*

Plate 57 (B)
Juvenal Texas banded gecko (Coleonyx brevis) *from Eddy County.*

Plate 58
Juvenal western banded gecko
(Coleonyx variegatus)
from Hidalgo County.

Plate 59
Adult Mediterranean gecko
(Hemidactylus turcicus)
from Doña Ana County.

Plate 60 (A)
Adult canyon spotted whiptail
(Cnemidophorus burti)
from Hidalgo County.
Photograph by
Clay M. Garrett.

Plate 60 (B)
Juvenal canyon spotted whiptail
(Cnemidophorus burti)
from Hidalgo County.

Plate 61
Adult gray-checkered whiptail
(Cnemidophorus dixoni)
from Hidalgo County.

Plate 62
Adult Chihuahuan
spotted whiptail
(Cnemidophorus exsanguis)
from Socorro County.

Plate 63
Adult Gila spotted whiptail (Cnemidophorus flagellicaudus) *from Catron County.*

Plate 64
Adult checkered whiptail (Cnemidophorus grahamii) *from Doña Ana County.*

Plate 65
Adult Texas spotted whiptail (Cnemidophorus gularis) *from Eddy County.*

Plate 66
Adult little striped whiptail (Cnemidophorus inornatus) *from Eddy County.*

Plate 67
Adult New Mexico whiptail (Cnemidophorus neomexicanus) *from Socorro County.*

Plate 68
Adult male six-lined racerunner (Cnemidophorus sexlineatus) *from Chaves County.*

Plate 69
Adult Sonoran spotted whiptail (Cnemidophorus sonorae) *from Hidalgo County.*

Plate 70 (A)
Adult western whiptail (Cnemidophorus tigris) *from Chaves County.*

Plate 70 (B)
Adult western whiptail (Cnemidophorus tigris) *from Hidalgo County.*

Plate 71
Adult desert grassland whiptail (Cnemidophorus uniparens) from Hidalgo County.

Plate 72
Adult plateau striped whiptail (Cnemidophorus velox) from Santa Fe County.

Plate 73 (A)
Adult many-lined skink (Eumeces multivirgatus) from Socorro County.

Plate 73 (B)
Adult many-lined skink
(Eumeces multivirgatus)
from Santa Fe County.

Plate 73 (C)
Adult many-lined skink
(Eumeces multivirgatus)
from Eddy County.

Plate 73 (D)
Juvenal many-lined skink
(Eumeces multivirgatus)
from Eddy County.
Photograph by
L. William Gorum.

Plate 74 (A)
Adult Great Plains skink
(Eumeces obsoletus)
from Union County.

Plate 74 (B)
Juvenal Great Plains skink
(Eumeces obsoletus)
from Eddy County.

Plate 75
Adult four-lined skink (Eumeces tetragrammus) *from Hidalgo County.*

Plate 76
Adult Madrean alligator lizard (Elgaria kingii) *from Hidalgo County.*

Plate 77
Gila monster (Heloderma suspectum) *from Hidalgo County.*

Plate 78
Adult Texas blind snake (Leptotyphlops dulcis) from Rio Arriba County.

Plate 79
Adult western blind snake (Leptotyphlops humilis) from Hidalgo County. Photograph by Don Sias.

Plate 80
Adult glossy snake (Arizona elegans) from Santa Fe County.

Plate 81
Adult Trans-Pecos rat snake
(Bogertophis subocularis)
from Otero County.

Plate 82
Adult racer
(Coluber constrictor)
from Socorro County.

Plate 83
Adult ringneck snake
(Diadophis punctatus)
from Mora County.

Plate 84
*Adult corn snake
(Elaphe guttata)
from Eddy County.*

Plate 85
*Adult western hooknose snake
(Gyalopion canum)
from Sierra County.*

Plate 86
*Adult western hognose snake
(Heterodon nasicus)
from DeBaca County.*

Plate 87
Adult night snake
(Hypsiglena torquata)
from Doña Ana County.

Plate 88
Adult gray-banded kingsnake
(Lampropeltis alterna)
from Eddy County.

Plate 89
Adult common kingsnake
(Lampropeltis getula)
from Hidalgo County.

Plate 90
Adult Sonoran mountain kingsnake (Lampropeltis pyromelana) *from Hidalgo County.*

Plate 91
Adult milk snake (Lampropeltis triangulum) *from Chaves County.*

Plate 92
Adult smooth green snake (Liochlorophis vernalis) *from Lincoln County.*

Plate 93
Adult Sonoran whipsnake
(Masticophis bilineatus)
from Hidalgo County.

Plate 94 (A)
Adult coachwhip
(Masticophis flagellum)
from Santa Fe County.

Plate 94 (B)
Juvenal coachwhip
(Masticophis flagellum)
from Socorro County.

Plate 95
Adult striped whipsnake
(Masticophis taeniatus)
from Grant County.

Plate 96
Adult plainbelly water snake
(Nerodia erythrogaster)
from Eddy County.

Plate 97
Adult gopher snake
(Pituophis melanoleucus)
from Santa Fe County.

Plate 98
Adult longnose snake (Rhinocheilus lecontei) *from Socorro County.*

Plate 99
Adult Big Bend patchnose snake (Salvadora deserticola) *from Sierra County.*

Plate 100
Adult mountain patchnose snake (Salvadora grahamiae) *from Catron County.*

Plate 101
Adult green rat snake (Senticolis triaspis) from Hidalgo County.

Plate 102 (A)
Adult ground snake (Sonora semiannulata) from San Miguel County.

Plate 102 (B)
Adult ground snake (Sonora semiannulata) from Sierra County.

Plate 102 (C)
Adult ground snake (Sonora semiannulata) *from Hidalgo County.*

Plate 103
Adult southwestern black-headed snake (Tantilla hobartsmithi) *from Hidalgo County.*

Plate 104
Adult plains black-headed snake (Tantilla nigriceps) *from Socorro County.*

Plate 105
Adult Yaqui black-headed snake (Tantilla yaquia) from Hidalgo County.

Plate 106
Adult blackneck garter snake (Thamnophis cyrtopsis) from Sierra County.

Plate 107
Adult western terrestrial garter snake (Thamnophis elegans) from Sandoval County.

Plate 108
Adult Mexican garter snake
(Thamnophis eques)
from Grant County.

Plate 109
Adult checkered garter snake
(Thamnophis marcianus)
from Eddy County.

Plate 110
Adult western ribbon snake
(Thamnophis proximus)
from Harding County.

Plate 111
Adult plains garter snake (Thamnophis radix) from Harding County.

Plate 112
Adult narrowhead garter snake (Thamnophis rufipunctatus) from Catron County.

Plate 113
Adult common garter snake (Thamnophis sirtalis) from Socorro County.

Plate 114
Adult lyre snake
(Trimorphodon biscutatus)
from Doña Ana County.

Plate 115
Adult lined snake
(Tropidoclonion lineatum)
from San Miguel County.

Plate 116
Adult western coral snake
(Micruroides euryxanthus)
from Hidalgo County.

Plate 117
Adult western diamondback rattlesnake (Crotalus atrox) *from Eddy County.*

Plate 118 (A)
Adult mottled rock rattlesnake (Crotalus l. lepidus) *from Eddy County.*

Plate 118 (B)
Adult banded rock rattlesnake (Crotalus lepidus klauberi) *from Hidalgo County. Photograph by Clay M. Garrett.*

Plate 119
Adult blacktail rattlesnake
(Crotalus molossus)
from Hidalgo County.
Photograph by
Clay M. Garrett.

Plate 120
Adult Mojave rattlesnake
(Crotalus scutulatus)
from Hidalgo County.

Plate 121
Adult western rattlesnake
(Crotalus viridis)
from Chaves County.

Plate 122
Adult New Mexico ridgenose rattlesnake (Crotalus willardi obscurus) *from Hidalgo County.*

Plate 123
Adult massasauga (Sistrurus catenatus) *from Chaves County.*

CNEMIDOPHORUS BURTI Taylor, 1936 [1938]   *Canyon Spotted Whiptail*

See Plate 60A, 60B.

Type: Edward H. Taylor (1936 [1938]) described *Cnemidophorus burti* and designated the type locality as "near La Posa, 10 miles northwest of Guaymas, Sonora, Mexico." The holotype is CNHM 100004 collected on 4 July 1934 by E. H. Taylor.

Distribution: *Cnemidophorus burti* ranges from southern and southeastern Arizona and extreme southwestern New Mexico southward into Sonora, Mexico. It occurs from near sea level to around 1375 m elevation (Stebbins, 1985). In New Mexico, *C. burti* is confined to the southwestern corner of Hidalgo County where it is known only from Guadalupe Canyon at elevations of 1321–1387 m. Reports from adjacent localities in the Peloncillo and Alamo Hueco mountains of New Mexico need verification. Records from New Mexico evidently represent an extension northward from Sonora rather than eastward from south-central Arizona (Duellman and Zweifel, 1962). There is a single record (UTACV-R 17552) from the Sierra San Luis in Sonora, Mexico. This locality is just south of the international border and only a few kilometers west of Antelope Wells. The Whitewater Mountains of south-central Hidalgo County should be investigated for the presence of this species.

Description: This large spotted and striped *Cnemidophorus* may reach 140 mm SVL and 510 mm TL. During 1992–95, 206 individuals examined from Guadalupe Canyon averaged 87.4 (33–124) mm SVL and 21.4 (0.6–61.9) gm (our unpubl. data). There are more than 95 granules around the midbody. The adults are spotted; the hatchlings and juveniles are unspotted and have a dark background color with six (rarely seven) light lines that disappear or become obscure with age. The ventral surfaces are pale and immaculate in both sexes. The tail is orangish tan or reddish.

Cnemidophorus burti undergoes a distinct ontogenetic change in color pattern as it matures. The longitudinal stripes on the body become progressively less distinct and may completely disappear on the upper surfaces of the body. The pattern shifts from one dominated by light longitudinal stripes to one completely or predominately characterized by numerous large and conspicuous light spots. The ground color gradually changes in the neck and head region to reddish.

Similar Species: In New Mexico, the only described member of the genus *Cnemidophorus* that coexists with *C. burti* is *C. sonorae*. Young individuals of these species are very difficult to distinguish, although the tail and feet of *C. burti* are often orangish tan or reddish. The best character for distinguishing these species is the number of granules around midbody. The published range of variation for *C. sonorae* is 74–80, and for *C. burti stictogrammus* 98–115. Thus, specimens with fewer than 90 granules around midbody are assignable to *C. sonorae*, those with more than 90 are *C. burti*.

Systematics: Four subspecies of *C. burti* are recognized (Wright, 1993). Only the giant spotted whiptail, *C. b. stictogrammus* occurs in New Mexico. This is the form described as *C. sacki stictogrammus* by Burger (1950). The holotype, USNM 132456, is a female collected at Yank's Spring, 6 miles southeast of Ruby, Santa Cruz County, Arizona on 17 August 1948 by M. M. Hensley and W. L. Burger.

As with most *Cnemidophorus*, *C. burti* has a rather complex taxonomic history. Burger (1950) included two distinct and sympatric forms in his type series of *C. sacki stictogrammus*. Lowe (1956) referred to these as *C. stictogrammus* and *C. sacki exsanguis*, although later, Duellman and Zweifel (1962) included Lowe's *C. stictogrammus* under the current name, *C. b. stictogrammus*.

Habitat: In New Mexico, *C. burti* occurs in riparian habitat dominated by sycamore, cottonwood, ash, and various grasses and forbs. It is often found in heavily shaded areas among rocks, logs, and leaf litter in the vicinity of streams. Open areas of bunch grass within these riparian habitats are also occupied.

Behavior: These large lizards are extremely wary and difficult to approach. Approximately 28% (159 of 564) of the *Cnemidophorus* collected during 1992–94 in Guadalupe Canyon were *C. burti*; the remainder were *C. sonorae*. Activity occurred between 28 April and

14 October, although 60% of the *C. burti* were collected during May and June (our unpubl. data).

Reproduction: Very little has been published on the reproduction of *C. burti*. Mating of captives held in a large, semi-natural enclosure was observed on 24 May 1993. A 98 mm SVL female laid seven eggs on 17 June. These eggs were 8 mm x 17 mm and weighed 0.6 gm each. A wild-caught gravid female, 110 mm SVL, laid 10 eggs during mid-June (B. R. Tomberlin, pers. comm.). Goldberg (1987) collected 56 *C. burti* from Pima County, Arizona, during May–August 1966. The following information is based on his findings. Lizards emerged from hibernation during April or early May depending upon yearly climatic conditions. The earliest individual to emerge was a juvenal male collected on 27 April. Spermiogenesis occurred in all mature males from May–July. Five of 10 males collected in August had regressed testes indicating the testicular cycle was ending. The smallest mature adult male was 87 mm SVL. During 1992–95, 72 adult males collected in Guadalupe Canyon averaged 107 (88–124) mm SVL and 34.7 (14.2–61.9) gm (our unpubl. data).

Yolk deposition was first observed in a female collected 6 May. Another female, collected on 26 May contained five oviductal eggs. Each of five females collected in June and five collected in July had evidence of reproductive activity. Clutch size averaged 4.33 eggs (3–5). A female collected on 5 July contained corpora lutea from a previous clutch and vitellogenic follicles, suggesting that more than one clutch can be produced in a reproductive season. The smallest female with vitellogenic follicles measured 90 mm SVL. During 1992–95, 45 adult females collected in Guadalupe Canyon averaged 102.2 (90–115) mm SVL and 27.5 (15.6–49.1) gm (our unpubl. data).

The smallest *C. burti* examined during a study in Guadalupe Canyon, Hidalgo County, was 33 mm SVL and 0.9 gm. Hatchlings were observed as early as 7 August. An obviously gravid female 110 mm SVL and 35 gm was collected on 27 June (our unpubl. data).

Food Habits: Food habits of *C. burti* have not been studied in New Mexico, although they likely feed on a large variety of insects, spiders, and other terrestrial invertebrates and their larvae. Prey is captured with a sudden dash while the lizard is actively foraging between cover objects and bushes.

Remarks: This is the largest species of *Cnemidophorus* occurring in the United States. Price (1983) provided an annotated bibliography of all species of *Cnemidophorus* occurring in New Mexico. *Cnemidophorus burti* is listed as a Category 2, Notice of Review species by the U.S. Fish and Wildlife Service (USDI, 1994) and as endangered by the New Mexico Department of Game and Fish (NMGF, 1990).

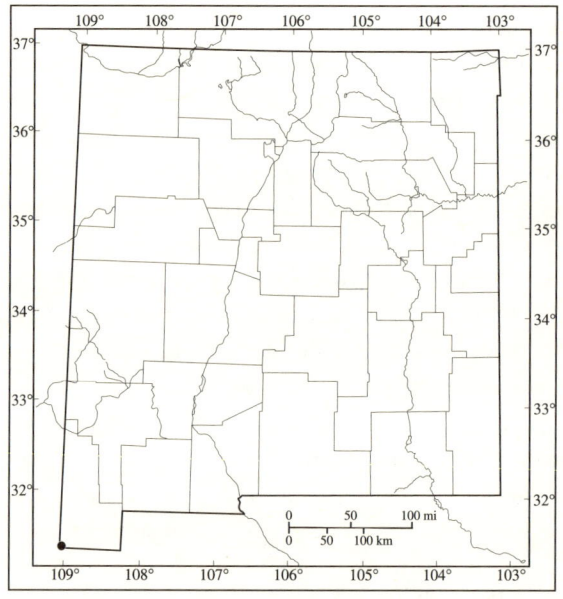

Distribution of Cnemidophorus burti *in New Mexico.*

CNEMIDOPHORUS DIXONI Scudday, 1973 — *Gray-checkered Whiptail*

See Plate 61.

Type: An adult female (TCWC 40691) collected "24.5 mi NW Presidio (16.9 mi from Jct US 67 and FM 170, then 7.6 mi NE) on Ireneo Gonzales Ranch, [Presidio Co., Texas]" by Doug Stine on 4 July 1970.

Distribution: This species occurs in two disjunct areas, the southwestern slopes of the Chinati Mountains in Presidio County, Texas, and the vicinity of Antelope Pass in the Peloncillo Mountains, Hidalgo County, New Mexico. Elevation in New Mexico ranges from approximately 1300–1450 m.

Description: This is a large, all-female species with a maximum SVL of 108 mm. During 1987–93, 402 specimens collected at Antelope Pass in Hidalgo County averaged 81.0 (34–108) mm SVL and 15.5 (0.8–36.8) gm. If we assume that *C. dixoni* from this region mature at 67 mm SVL as those studied in southwestern Texas by Walker et al. (1991), then the following data are pertinent; 320 adults averaged 87.7 (67–108) mm SVL and 18.30 (5.5–36.8) gm (our unpubl. data). The dorsal scales are granular, and there are 94–112 scales around midbody. There are 39–44 femoral pores. The mesoptychial scales are abruptly enlarged, whereas the scales of the anterior gular fold and those bordering the mesoptychium are minute and granular. The supraorbital semicircles usually do not completely separate the second supraocular scale from the frontal. The postantebrachial scales are normal or slightly enlarged.

Cnemidophorus dixoni is a boldly marked lizard, but the pattern changes with age and geographic location. Hatchlings have 10–14 dorsal longitudinal light stripes, cream to yellow, on a dark brown or black ground color. There are three well-defined lateral light stripes on each side of the body. The remaining dorsolateral and paravertebral light stripes are well-defined anteriorly, but begin to lose definition at midbody and disintegrate completely at the base of the tail. This pattern is modified ontogenetically to a greater or lesser degree, depending on geographic location, by (1) appearance and/or spreading of the light spots in the dark fields, sometimes fusing with one or both longitudinal light stripes bordering the field, and (2) spreading and fusion of segments of the dark fields, disrupting the light stripes. The result is a finely vermiculated dorsal pattern of small squarish blotches obscuring the original lined pattern. Many individuals have a conspicuous orange-brown coloration on the posterior half of the body dorsally that extends onto the tail.

Similar Species: The lizards most likely to be confused with *C. dixoni* in New Mexico are *C. tigris* and *C. grahamii*. The former differs in having mesoptychial scales that are not abruptly enlarged, and in having considerably more dark pigmentation ventrally on the chin, chest, and tail as adults. The latter has a much more boldly reticulated dorsal pattern of 6–10 longitudinal stripes broken by frequent fusions between adjacent dark fields, and it is not known to occur in Hidalgo County.

Systematics: This is a diploid parthenogenetic species of hybrid origin belonging to the *Cnemidophorus tesselatus* species group. The parental species are *C. septemvittatus* and *C. tigris marmoratus* (Brown and Wright, 1979; Wright, 1993). *Cnemidophorus dixoni* and *C. grahamii*, along with *C. tesselatus*, were considered to be the same species for many years (Maslin and Secoy, 1986; Price, 1986a). Recent work involving mitochondrial DNA restriction site analyses have demonstrated that *C. dixoni* and *C. grahamii* arose from separate hybridization events involving the same two parental species (Wright, 1993). We follow Wright (1993) in regarding these two taxa as separate species under the criteria established by Frost and Wright (1988) and Frost and Hillis (1990). Walker et al. (1994) have suggested that the New Mexico population of *C. dixoni* is distinct from those in Texas.

Habitat: In west Texas, *C. dixoni* occurs on generally rocky soils in desert shrublands and degraded grasslands on alluvial benches, canyon bottoms, and the lower southwestern slopes of the Chinati Mountains between 909 and 1515 m (Scudday, 1973). Characteristic vegetation includes *Larrea tridentata, Acacia* spp., *Prosopis glandulosa, Lycium berlandieri, Condalia* sp., *Jatropha dioica, Fouquieria splendens, Opuntia leptocaulis, Erioneuron pulchellum, Aristida ternipes,* and *Setaria leucopila.*

In New Mexico, *C. dixoni* is found on creosotebush flats with little or no shrubby undergrowth on sandy to gravelly soils at 1200–1400 m (Painter, 1991b, 1992). Characteristic vegetation includes *Flourensia cernua, Prosopis glandulosa, Koeberlinia spinosa, Ephedra* spp., *Opuntia* spp., *Fallugia paradoxa, Chilopsis linearis, Rhus microphylla, Atriplex canescens,* and *Celtis pallida.* It has not been found in the sandy arroyo bottoms of Antelope Pass nor in surrounding desert grasslands. It occurs sympatrically with at least 17 other species of lizards in Antelope Pass, including four species of *Cnemidophorus* (*C. exsanguis, C. sonorae, C. tigris,* and *C. uniparens*).

Behavior: Painter (1992) reported this species active in New Mexico from the beginning of April through the beginning of October, with peak seasonal activity occurring in June. Adults cease surface activity by mid-August. Walker et al. (1994) found lizards active throughout the day in west Texas from May through July, with most lizards active before 1445 and again after 1645. Walker et al. (1994) reported an average body temperature of 40°C for 210 active *C. dixoni* from west Texas.

Reproduction: Walker et al. (1991) reported the smallest reproductive female in west Texas as 67 mm SVL. The reproductive season lasts from May through July. The average clutch size for west Texas lizards is about 3 eggs (range 1–6, N = 148), whereas 12 lizards from New Mexico averaged 4.66 eggs per clutch with a maximum of eight (Walker et al., 1994). Larger females produced larger clutches. Eggs are parchment-shelled, creamy-white, and elliptical (typical size 10 x 18 mm). Neonate lizards are 32–36 mm SVL in west Texas. Painter (1992) reported two clutches of 3 and 5 eggs from New Mexico, with hatchlings between 33 and 38 mm SVL. Lizards from the Texas population may live to be three years of age, whereas those from New Mexico may reach an age of four years (Walker et al., 1994).

Food Habits: Twenty-seven *C. dixoni* (78.4–102.6 mm SVL) collected at Antelope Pass in early July 1993 were stomach-flushed following the technique described by Legler and Sullivan (1979). Termites were found in 82% of the 22 lizards that contained food, ants in 41%, beetles in 32%, and spiders in 23%. Other items, including solpugids, desert roaches, beetle larvae, robber flies, lizard skin, and plant debris were found in less than 10% of the samples examined (our unpubl. data). *Cnemidophorus dixoni* may be assumed to be a deliberate forager and a generalized, opportunistic insectivore similar to its close relative, *C. grahamii.*

Remarks: McAllister et al. (1991) reported on parasitic helminths of this species. The scientific name of this species honors Dr. James R. ("Bwana") Dixon, a gentleman, scholar, and longtime student of the herpetofauna of the deserts of North America.

Cnemidophorus dixoni is listed as a Category 2, Notice of Review species by the U.S. Fish and Wildlife Service (USDI, 1994) and as endangered by the New Mexico Department of Game and Fish (NMGF, 1990).

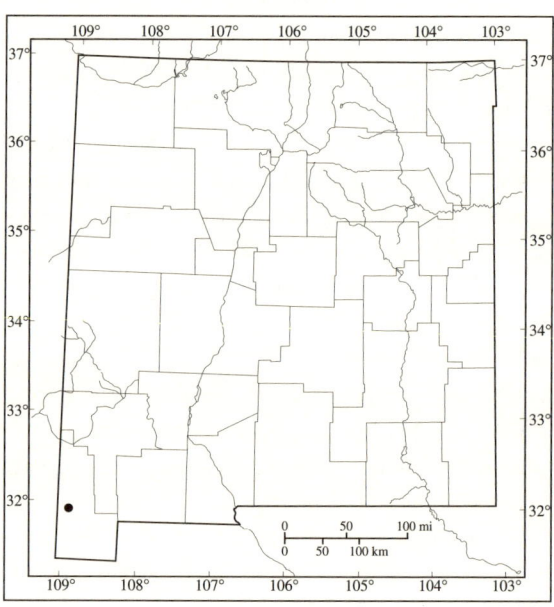

Distribution of Cnemidophorus dixoni *in New Mexico.*

CNEMIDOPHORUS EXSANGUIS Lowe, 1956 — *Chihuahuan Spotted Whiptail*

See Plate 62.

Type: An adult female (UAZ 16188) collected from "Socorro, Socorro County, New Mexico" by Richard G. Zweifel and Kenneth S. Norris on 10 August 1948.

Distribution: *Cnemidophorus exsanguis* occurs from northern New Mexico southward through the western half of Trans-Pecos Texas into Chihuahua, Mexico, east of the Sierra Madre Occidental as far south as Meoqui. It ranges from the Glass Mountains northeast of Marathon, Brewster County, Texas, westward to extreme northeastern Sonora, Mexico, and the eastern border of Arizona as far north as the Gila and San Francisco river drainages, extending up these drainages in New Mexico well onto the Mogollon Plateau. It extends eastward to Carlsbad, and northward in the Rio Grande and Pecos River valleys to White Rock and Pecos, and occurs in the Canadian River drainage as far north as Mills, Harding County. The range in southeastern New Mexico appears to be circumscribed by the escarpment of the Llano Estacado (Tanner, 1975).

Description: *Cnemidophorus exsanguis* is a large, all-female species with a maximum SVL of 100 mm (Stuart, 1991b). The dorsal scales are granular; there are 62–86 scales around midbody and 2–8 scales anteriorly between the paravertebral light stripes. There are 33–39 femoral pores. Both the mesoptychial and postantebrachial scales are enlarged, the latter usually four or more times the size of neighboring scales. The supraorbital semicircles do not extend anteriorly beyond the third supraocular scale.

This is a striped and spotted whiptail. The dorsal ground color in hatchlings and juveniles is black or dark brown. There are six longitudinal light stripes; a 7th vertebral stripe may be present anteriorly but, if so, it is incomplete and indistinct. There are many indistinct light spots arranged in multiple rows within the dark dorsal fields. The dorsal surfaces of the limbs have a bold vermiculate pattern. The tail is greenish-gray. Hatchlings have very few, if any, distinct spots. As individuals grow, the dark dorsal fields may lighten to tan or reddish-brown, the light spots increase in number, intensify in hue and expand, some onto the longitudinal light stripes, and the light stripes and vermiculate limb pattern fade in intensity. The ventral surface of the body is milky-white.

Similar Species: *Cnemidophorus exsanguis* is easily confused with *C. flagellicaudus* and *C. sonorae*. It differs from them (where they occur together) in having many more light spots (light spots on the light stripes and between the paravertebral stripes), a greater tendency for spots to occur on the neck and head, on the upper surface of the hind legs and across the hip region, and a marked fading of the light stripes in the adult pattern. *Cnemidophorus gularis* has eight longitudinal light stripes, 13–18 scales between the paravertebral light stripes, lacks spots in the paravertebral fields, and has a pinkish tail. *Cnemidophorus burti* is much larger, has more than 95 granular scales around the midbody, and stripes which tend to disappear with age.

Systematics: Duellman and Zweifel (1962) first recognized this taxon, originally described as a subspecies of *Cnemidophorus sacki*, as a distinct species and noted the absence of males. This is an allotriploid, parthenogenetic species of the *Cnemidophorus sexlineatus* species group. It was derived through one or more hybridization events between males of the gonochoristic species *C. inornatus* and female *C. costatus barrancorum* or *C. burti stictogrammus* to produce an intermediate (presumably parthenogenetic and extinct) lineage, followed by backcross hybridization with male *C. septemvittatus* (Good and Wright, 1984; Moritz et al., 1989; Wright, 1993). Dessauer and Cole (1989) have proposed a slightly different scenario involving hybridizations between *C. burti*, *C. inornatus*, and *C. scalaris*. There are probably several independent clonal lineages in New Mexico, as lizards from different localities exhibit distinct color and pattern variations.

Habitat: *Cnemidophorus exsanguis* is typically found in relatively mesic habitats, such as juniper-grassland and piñon-juniper woodland. It extends into Ponderosa pine forest to 2121 m and rarely as high as 2424 m in the northern end of its range (Lowe and Zweifel, 1952; our obs.). Medica (1967) found this species in a severely degraded riparian habitat within the Rio Grande floodplain west of Las Cruces. It occurs in piñon-juniper habitat in the Organ and San Andres mountains east and north of Las Cruces, and follows major drainages out of these mountains into desert vegetation at lower elevations. It is sympatric with *C. flagellicaudus* along the San Francisco River and in the Big Burro Mountains in Grant and Catron counties, with *C. sonorae* in the Peloncillo Mountains in southwestern Hidalgo County, and with both species along the Gila River in the vicinity of Redrock.

Behavior: Lizards become active when soil temperatures reach 26–30°C and seek shelter when soil temperatures exceed 50°C (Medica, 1967). Lizards were reported to be active in the morning until about 1300 and then again in the evening between 1600 and 1800 along a riparian corridor in south-central New Mexico. Preferred body temperatures of active lizards were reported by Medica (1967) to fall within a narrow range of 39.0–39.8°C. New Mexico specimens have been collected between 6 February (Grant County) and 16 October (Socorro County). In west Texas, Schall (1977) reported a mean body temperature for 203 active lizards as 39.9°C, with cooler temperatures for basking lizards (39°C) than those actively moving or sheltering in the shade of vegetation (41°C).

Schall and Pianka (1980) reported that this species is remarkably unwary, and may depend upon crypsis to escape predators. Lizards of this species are able to incur a significant lactic acid debt through anaerobic metabolism without overt detrimental effects upon their daily activities (Pough and Andrews, 1985).

Reproduction: The minimum size at reproductive maturity is 63 mm SVL (Schall, 1978). Clutch size averaged 2.96 (1–6) eggs in one study in Trans-Pecos Texas (Schall, 1978). Larger females produce larger clutches.

Medica (1967) reported a short breeding season for this species in south-central New Mexico, with oviductal eggs present only from late June through late July. Clutch size averaged 2.7 eggs. Parker (1973b) recorded an average clutch size of 4.4 (3–5) eggs for lizards from southern Doña Ana County, along with evidence that individual females may produce two clutches annually. Individual eggs weigh between 0.25 and 1.10 gm.

Cnemidophorus exsanguis has been successfully raised in the laboratory through at least 7 generations, further demonstrating its parthenogenetic mode of reproduction; offspring were genetically identical to their mothers (Dessauer and Cole, 1986).

Food Habits: This species is a deliberate and methodical forager, digging under objects in its environment in search of food. Milstead (1957) and Scudday and Dixon (1973) recorded the following food items by volume in Trans-Pecos Texas: termites (76%, 24%), grasshoppers (8%, 34%), and beetles (7%, 14%). In addition, the latter authors recorded arachnids (14%) and lepidopteran larvae (10%) in their sample. Smith (1989) reported grasshoppers (42–70%) and termites (21–42%) by volume in the diet of this species in Trans-Pecos Texas over a 3-month study period. Lepidopteran larvae constituted 24% of the diet by volume during the last month of the study, following a month of abnormally heavy rainfall. Termites (58%), grasshoppers (29%), and spiders (10%) were important food items by volume for hatchlings.

Medica (1967) reported a wide diversity of insect prey over a 2-year period for this species in south-central New Mexico, with lepidopteran larvae (38%), beetles (20%), ants (10%), spiders (9%), and grasshoppers (8%) constituting the bulk of the diet by volume. The diet changed radically from the first year, the driest on record, to the second, one of normal rainfall. Lepidopteran larvae increased from 17% to 43% while ants and beetles fell from 33% and 31% to 4% and 17%.

Remarks: This species was reviewed by Stuart (1991b). The name *exsanguis* (Latin, "without blood") refers to the difference between this species and *C. burti stictogrammus*, with which it was formerly included (Lowe, 1956).

Taylor et al. (1967) reported a male lizard referable to this species from 2 miles west and 1 mile south of Mesilla, Doña Ana County. Unlike occasional males similarly reported in other parthenogenetic lineages of the genus, this specimen could not be demonstrated (on the basis of erythrocyte size) as the product of backcross hybridization with a sympatric gonochoristic species (Taylor et al., 1989).

FAMILY TEIIDAE

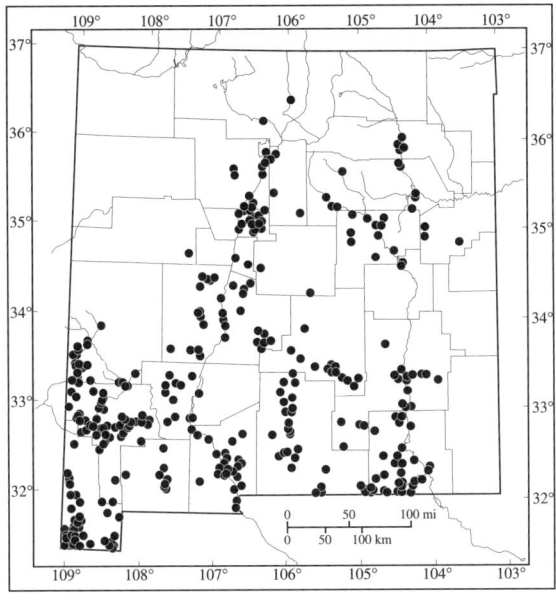

Distribution of Cnemidophorus exsanguis *in New Mexico.*

CNEMIDOPHORUS FLAGELLICAUDUS Lowe and Wright, 1964   *Gila Spotted Whiptail*

See Plate 63.

Type: An adult female (UAZ 11775), collected "at San Francisco Hot Springs (Frisco Hot Springs), 4800 ft. elev., Catron County, New Mexico" by John W. Wright on 16 July 1964.

Distribution: This species ranges from the Gila and San Francisco river drainages and the Big Burro Mountains in Catron and Grant counties, New Mexico, northwestward throughout the Mogollon Plateau to the Colorado River in northwestern Arizona. It also occurs south of the Mogollon Rim to the Gila River, and as isolated populations in the Chiricahua and Santa Catalina mountains in southeastern Arizona.

Description: *Cnemidophorus flagellicaudus* is a large, all-female species with a maximum SVL of 99 mm (Stebbins, 1985). The dorsal scales are granular; there are 77–84 scales around midbody and 3–5 scales anteriorly between the paravertebral light stripes. There are 35–41 femoral pores. Both the mesoptychial and postantebrachial scales are enlarged. The supraorbital semicircles do not extend anteriorly beyond the third supraocular scale. There are usually two enlarged preanal scales.

Adults are striped and spotted, but juveniles lack spots. The dorsal ground color is dark brown to black. There are six complete longitudinal dorsal light stripes that remain distinct on the neck. As lizards age, they develop relatively few and indistinct beige to golden-yellow light spots in the dark dorsal fields. These spots are arranged in one or two irregular longitudinal rows, and may touch or occur within the light longitudinal stripes, especially the anterior paravertebrals. The dorsal surfaces of the hind limbs are indistinctly mottled. The original tail is olive-green with a slight bluish tint. The ventral surface of the body is creamy white.

Similar Species: *Cnemidophorus flagellicaudus* is easily confused with both C. *exsanguis* and C. *sonorae*. It differs from both these species, where they occur together, in usually having only two enlarged preanal scales, and in having an indistinct vermiculate patterning of the dorsal surface of the thigh. It differs from C. *exsanguis* in having fewer and less distinct light spots in the dark dorsal fields, and by retaining its boldly striped pattern throughout life. It differs from C. *sonorae* in having the light spots in the dark dorsal fields that touch or lie within the longitudinal light stripes. *Cnemidophorus burti* is much larger, has more than 95 granular scales around the midbody, and stripes that tend to disappear altogether with age.

Systematics: This is an allotriploid, parthenogenetic species of hybrid origin (Wright and Lowe, 1968). Recent mtDNA analyses suggest that this species arose from one or more hybridization events between male *C. burti* or *C. costatus* and female *C. inornatus arizonae* to create a diploid (presumably parthenogenetic and now extinct) intermediate lineage, followed by backcrossing hybridization(s) with one or both of the paternal species (Densmore et al., 1989b; Wright, 1993). There may be more than one independently derived lineage within this species (Wright, 1993. 1994).

Habitat: This species occurs in oak-piñon-juniper woodlands from 1220–1910 m and riparian woodland extensions into desert-grassland ecotones (Lowe and Wright, 1964). Hulse (1973) reported this lizard (as *C. exsanguis*) as common in east-central Arizona on riparian terraces, piñon-juniper, and deciduous riparian woodlands between 1060–1515 m. It is sympatric with both *C. exsanguis* and *C. sonorae* in the vicinity of Redrock along the Gila River in Grant County, and with the former species in the Big Burro Mountains and along the San Francisco River in Catron County.

Behavior: *Cnemidophorus flagellicaudus* are active in central Arizona from May until August, when they enter hibernation. The 99 New Mexico specimens available to us were collected between 31 March (Grant County) and 24 September (Grant County). Stevens (1980) recorded an average body temperature of 39.9°C for 27 active lizards.

Reproduction: Vitellogenesis begins in May in central Arizona (Stevens, 1980) and oviposition occurs in June or July. Most females produce a single clutch annually, but older animals may produce a second. Limited data (n = 13) suggest an average clutch size of 4.3 (2–6) eggs, with clutch size positively correlated with female body size. Fat bodies increase from as low as 0.1% of female body weight during reproduction to as much as 6% of female body weight prior to winter dormancy.

Food Habits: No quantitative data are available for this species. Like many congeners, these lizards are probably actively foraging, opportunistic insectivores, taking a wide variety of palatable arthropods.

Remarks: McAllister (1992) studied the helminths parasitic in this species. The name *flagellicaudus* is a Latin combination and literally means "whip-tail."

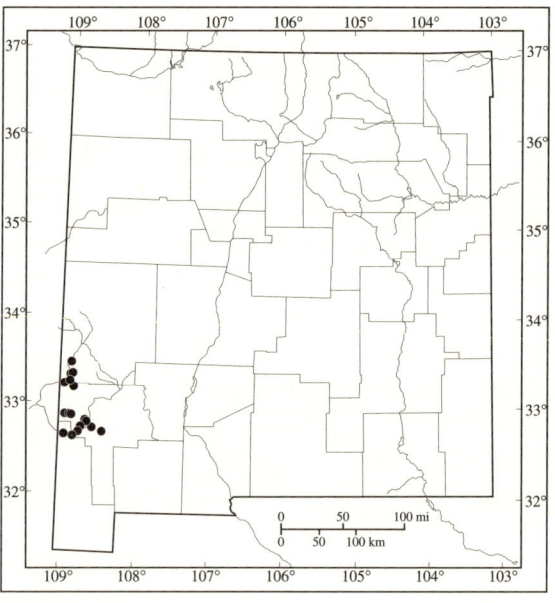

Distribution of Cnemidophorus flagellicaudus *in New Mexico.*

CNEMIDOPHORUS GRAHAMII Baird and Girard, 1852   *Checkered Whiptail*

See Plate 64.

Type: The original description (Baird and Girard, 1852a) was based on two specimens (USNM 3046) which came from "between San Antonio and El Paso del Norte" (given as "El Paso" in the USNM catalogue; Cochran, 1961). Smith and Taylor (1950a,b) restricted the type locality to Fort Davis, Jeff Davis Co., Texas. Smith and Burger (1949) designated USNM 3046a, an adult female, as the lectotype.

Distribution: *Cnemidophorus grahamii* ranges from

the vicinity of the Arkansas River near Rocky Ford, Otero County, Colorado, southward through the tip of the Oklahoma Panhandle to the high plains of New Mexico, with an eastward extension into the Texas Panhandle along the Canadian River. It ranges southward within the Pecos and Rio Grande river basins to their confluence, throughout Trans-Pecos Texas, and upstream along the Río Conchos in Chihuahua, Mexico, to the vicinity of La Cruz. The southern and eastern range margins are not well known (Price, 1986a). This species tends to occur as isolated local populations even where suitable habitat is continuous, especially where congeneric sexual species are abundant (Knopf, 1966; Wright and Lowe, 1968; our obs.).

In New Mexico, the bulk of the species' distribution occurs southward in the Pecos River and Rio Grande valleys from Santa Rosa and San Ildefonso. It occurs on the Otero Mesa and in the Tularosa Basin in Otero County, and occurs as far west as the Cook's Range in northern Luna County. The southeastern margin of the range in New Mexico appears to be circumscribed by the escarpment of the Llano Estacado (Tanner, 1975). There are several records within the Cimarron River drainage in northeastern Union County that are isolated from the nearest records around Conchas Lake in the Canadian River system. We suspect that additional collecting will reveal the presence of this species in the intervening area.

Description: The morphology of this species has been described in detail by Zweifel (1965). This is a large, all-female species with a maximum SVL of 106 mm. The dorsal scales are granular, and there are 75–112 scales around midbody. There are 36–48 femoral pores. The mesoptychial scales are abruptly enlarged, whereas the scales of the anterior gular fold and those bordering the mesoptychium are minute and granular. The supraorbital semicircles usually do not completely separate the second supraocular scale from the frontal. The postantebrachial scales are granular or slightly enlarged.

Cnemidophorus grahamii is a boldly marked lizard, but the pattern changes with age and geographic location. The basic color pattern of hatchlings consists of usually six primary to as many as 10 longitudinal light stripes, cream to yellow, on a dark brown or black ground color. The tail is light brown with light and dark spots scattered throughout. The light portions of the dorsal color pattern on the posterior half of the body are often suffused with yellow, especially in gravid females, and the yellow often extends as lateral stripes on the tail. The six primary stripes occur in pairs, with the paravertebrals extending posteriorly from the parietal scales, the dorsolaterals arising at the posterior corner of each eye, and the laterals passing through the ear opening. Occasionally light spots occur in the dark fields between stripes and the secondary stripes, especially the vertebral (if present), and may be represented by a series of light dashes posterior to the neck region. This pattern is modified ontogenetically to a greater or lesser degree, depending on geographic location, by (1) appearance and/or spreading of the light spots in the dark fields, sometimes fusing with one or both longitudinal light stripes bordering the field, and (2) spreading and fusion of segments of the dark fields, disrupting the light stripes. Thus, lizards from northeastern New Mexico retain a distinctly striped appearance as adults, whereas lizards from the lower Rio Grande Valley are nearly completely tesselated as adults, with vertical lateral dark rectangles irregularly bordered with light bars. The ventral surface is milky white with the exception of occasional scattered black flecks.

Similar Species: The lizards most likely to be confused with *C. grahamii* in New Mexico are *C. tigris* and *C. dixoni*. The former species differs in having mesoptychial scales that are not abruptly enlarged, and in having considerably more dark pigmentation ventrally on the chin, chest, and tail as adults. The latter species has a much more finely reticulated dorsal pattern of 10–14 longitudinal stripes broken by frequent fusions between adjacent dark fields, and it is known in New Mexico only from the vicinity of Antelope Pass in the Peloncillo Mountains, Hidalgo County.

Systematics: *Cnemidophorus grahamii* belongs to the *C. tesselatus* species group and has a long history of confusion with *C. tigris*. Smith and Burger (1949) first pointed out that the name *tesselatus* (or "*tessellatus*"), as applied to this species (*sensu lato*) in the literature up to that time, referred to *C. tigris* and that the name *tesselatus* correctly belonged to the species Baird and Girard (1852a) had described as *Cnemidophorus grahamii*. *Cnemi-*

dophorus tesselatus has a complex nomenclatural history (Price, 1986a) due, in part, to the recent discovery of a number of closely related, hybrid-derived parthenogenetic lineages within the genus.

Zweifel (1965) described six pattern classes (A–F) to encompass the discordant morphological variation seen in the *C. tesselatus* complex. Subsequent work has demonstrated that this complex consists of both diploid and triploid lineages of hybrid origin (Lowe and Wright, 1966a,b; Wright and Lowe, 1967a; Neaves, 1969; Neaves and Gerald, 1968, 1969; Brown and Wright, 1979). The parental species of diploid lineages (pattern classes C–F of Zweifel) are *C. septemvittatus* and *C. tigris marmoratus*. Triploid lineages (pattern classes A–B of Zweifel, and some C (Parker, 1979a)) are derived from hybridization between diploid *C. tesselatus* complex lizards and *C. sexlineatus*. Extensive electrophoretic and morphological analyses indicate that multiple hybridizations between the same suite of parental species may have occurred within both diploid and triploid lineages of the *Cnemidophorus tesselatus* complex (Parker and Selander, 1976; Parker, 1979a,b; but see Frost and Wright, 1988 and Wright, 1993 for an alternative viewpoint). We follow Wright (1993) and consider *grahamii* as the proper name for all of the diploid lineages of the complex except for pattern class F *(C. dixoni)*, with the name *tesselatus* restricted to some of the triploid lineages. We note, however, that Zweifel (1965) stated that a triploid lineage (pattern class A) has apparently replaced a diploid lineage (pattern class D) at the type-locality of *C. tesselatus* (*sensu lato*) since the original description by Say (*in* James, 1823). If this is true, then *grahamii* would remain a junior synonym of *C. tesselatus*, and another name would have to be found for the unnamed diploid lineages of the *C. tesselatus* complex.

Habitat: *Cnemidophorus grahamii* occurs in a variety of situations, including piñon-juniper, *Yucca*-grassland, mesquite creosotebush, and cottonwood-tamarisk-willow associations on rocky, gravelly, or sandy soils from 250 m to 1829 m (Zweifel, 1965; Price, 1986a). In the Guadalupe Mountains, Eddy County, it occurs sympatrically with *C. inornatus, C. tigris, C. exsanguis, C. gularis,* and *Cophosaurus texanus* (Zweifel, 1965). Knopf (1966) reported 25–100 lizards/hectare in southern Colorado. Price et al. (1993) recorded densities of 2–10 lizards/hectare over a 5-year period in southern Doña Ana County.

Behavior: Juveniles become active first on an annual basis, emerging from hibernation in early April in southern New Mexico. Adults emerge in the beginning of May, and remain surface active until mid-August. Hatchlings are active until October. New Mexico specimens have been collected between 7 April (Chaves County) and 25 September (De Baca County). Schall (1977) recorded a mean body temperature for 154 active lizards in west Texas as 40.1°C, with cooler temperatures for basking lizards (38°C) than those actively moving (40°C) or in the shade of vegetation (41°C). Knopf (1966) recorded an average body temperature of 39.3°C (34.0–42.6°C) for active lizards in Colorado. Individual lizards are not active every day (Knopf, 1966; our obs.), and may spend several days in shelter depending on hunger, disturbance by predators, or other motivations.

This species is remarkably unwary (Holland, 1965; Price, 1992; our obs.) and individuals can often be approached within arm's length during the course of normal activities. A lizard being pursued at Conchas Lake attempted escape by diving underwater (J. N. Stuart, pers. comm.).

Knopf (1966) reported home ranges of 0.06–0.10 hectare in southern Colorado. Individual burrows may or may not be defended (Knopf, 1966; Leuck, 1982, 1985) but home ranges overlap. Gravid females become quite sedentary and do not exhibit much surface activity for a week or two prior to oviposition (Saxon, 1970; Schall, 1978; our obs.). This species may nest communally (Knopf, 1966), and individual nest burrows are defended prior to and immediately after oviposition.

Cnemidophorus grahamii sometimes behave as males in captivity, engaging in pseudocopulatory behavior (Crews and Fitzgerald, 1980); see the *C. uniparens* account for further discussion of this phenomenon.

Reproduction: *Cnemidophorus grahamii* reproduces parthenogenetically. Females from southern New Mexico and adjacent west Texas attain reproductive maturity at a minimum size of 66 mm SVL (Schall, 1978). In Colorado, this occurs at an approximate age of 22 months (Knopf, 1966) and in Texas at 13–14 months (Saxon, 1970). Females are gravid from May through July in the south (Parker, 1973b; Schall, 1978), and in June and July in Colorado (Knopf, 1966). Individual females lay an average of 3–4 (1–8) eggs per clutch (Parker, 1973b; Schall, 1978). Individual eggs usually weigh less than a gram. Larger females produce larger clutches (Saxon, 1970; Schall, 1978), and larger females may produce two clutches annually. Clutch weight is about 12% of female body weight prior to oviposition. Eggs hatch in 60–74 days, and some females may retain oviductal eggs for a week or more (Knopf, 1966). The first hatchlings appear from late July in southern New Mexico to mid-September in Colorado (Knopf, 1966; our obs.). Neonates are 39–48 mm SVL in southern Colorado (Knopf, 1966). Fat bodies approximately double in size in July and August following the reproductive season (Parker, 1973b).

The age structure of a Colorado population was reported by Knopf (1966) to consist of 19.5% juveniles, 19.5% 2-year-olds, 34.5% 3-year-olds, and 26.4% lizards four years old or older. In contrast, Saxon (1970) suggested that lizards from west Texas rarely live to be three years old.

Food Habits: These lizards are deliberate foragers, spending considerable time in one place digging for food in the soil or rooting through debris. Milstead (1957) recorded termites as 33–54% by volume of the diet of three geographically dispersed populations in Trans-Pecos Texas; other prey taken included beetles (6–15%), grasshoppers (6–26%), and butterflies and moths (9–22%). Scudday and Dixon (1973) recorded termites (28%), larval and adult beetles (22%), and grasshoppers (14%) as the most important foods by volume in another study in Trans-Pecos Texas; a wide variety of insect prey were taken. They suggested that this species selects relatively larger prey items than other sympatric congeners. Lepidopteran larvae were the most important food item recorded by Saxon (1970) for a population of lizards near Presidio, Texas; grasshoppers and spiders were also significant components of the diet. Paulissen et al. (1993) found seasonal and ontogenetic differences in the diets of populations from southern Colorado. Grasshoppers (4–46%), larval and adult butterflies (3–40%), ants (1–24%), termites (4–21%), larval and adult beetles (14–19%), and spiders (7–17%) were the most frequently consumed prey by adults, whereas juveniles ate primarily mites (21%), spiders (21–36%), termites (3–20%), and homopterans (8–13%).

This species is probably prey for a number of snakes and birds of prey; Knopf (1966) recorded heavy predation by *Masticophis flagellum*.

Remarks: The name *grahamii* is a patronym honoring Col. J. D. Graham, who commanded the Boundary Survey expedition on which the type material was collected.

Taylor et al. (1967) reported two male lizards referable to this species from Sierra County. On the basis of erythrocyte size, both specimens appear to be products of backcross hybridization between this species and a sympatric gonochoristic species (Taylor et al., 1989). McAllister (1990a) reported on the parasitic helminths of this species.

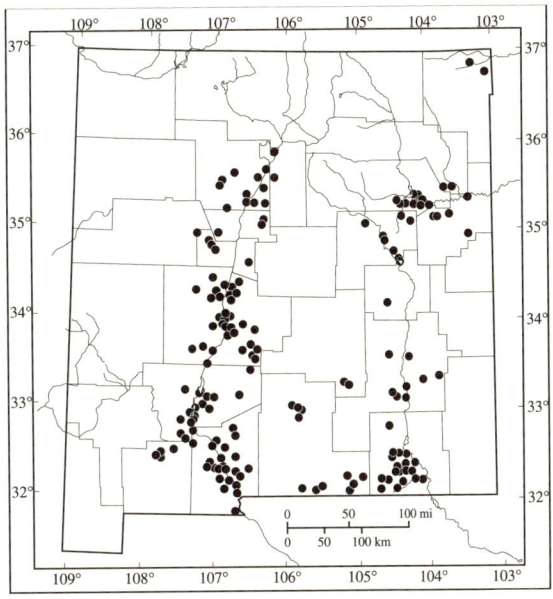

Distribution of Cnemidophorus grahamii *in New Mexico.*

CNEMIDOPHORUS GULARIS Baird and Girard, 1852 — *Texas Spotted Whiptail*

See Plate 65.

Type: The original description (Baird and Girard, 1852a) did not specify the type material, which is "from Indianola, [Texas] and the valley of the Rio San Pedro, a tributary of the Rio Grande del Norte." The 13 specimens (Cochran, 1961) were collected by John H. Clark during the expeditions led by Col. J. D. Graham of the Scientific Corps of the U.S. and Mexican Boundary Commission. The lectotype (USNM 3022a) is an adult male. The type locality has been restricted to the mouth of the Devils River, Val Verde County, Texas by Smith and Taylor (1950a,b).

Distribution: *Cnemidophorus gularis* ranges from the Red River Valley along the Oklahoma-Texas border southward throughout most of Texas and into Mexico on both sides of the Sierra Madre Oriental to the states of Aguascalientes, Querétaro, and Veracruz. In New Mexico, it has only been found in Eddy and Lea counties from 900 m to about 1300 m. It was first reported from New Mexico by Wright (1963).

Description: *Cnemidophorus gularis* is a moderate-sized whiptail, with a maximum SVL of 105 mm (Dundee, 1995). The dorsal scales are granular and there are 75–97 scale rows at midbody. There are 28–33 femoral pores. The paravertebral light stripes are separated by 13–18 scales. Both the mesoptychial and postantebrachial scales are enlarged. The supraocular semicircles do not extend forward beyond the anterior margin of the third supraocular scale.

This is a striped and spotted whiptail. The dorsal ground color is dark brown to black. There are seven (vertebral stripe entire) or eight (vertebral stripe divided) longitudinal light stripes, with the vertebral stripe(s) less distinct than the others. In New Mexico this species usually has eight stripes (Christiansen and Degenhardt, 1969). The dorsolateral stripes continue onto the tail. Both juveniles and adults have creamy-yellow spots in the dark dorsal fields, usually in a single row within each field. These spots are distinct in the lateral fields but faint or absent in the paravertebral fields. They become more distinct and may expand with age, but never involve the longitudinal stripes. Many specimens have a greenish-yellow tint on the dorsal surface of the body. The tail is pinkish to reddish-brown in color. The ventral surface of females is creamy-white, but adult males have pink, red, or orange throats and blue-black chests and abdomens, often with a large black patch.

Similar Species: *Cnemidophorus exsanguis* has only six longitudinal light stripes, spots in the space between the paravertebral stripes, only 2–8 scales between the paravertebral stripes, an unmarked ventral surface at all ages, and a greenish-gray tail. *Cnemidophorus sexlineatus viridis* lacks spots on the dorsal surface and enlarged postantebrachial scales, and the anterior body color is bright green in males.

Systematics: Seven subspecies have been recognized (Maslin and Secoy, 1986), but several of these are considered by many workers to belong to a different species (Frost and Wright, 1988; Moritz et al., 1989). Wright (1993) recognizes only two subspecies. The nominotypical form (*C. g. gularis*) occurs in New Mexico.

Habitat: Typically, this lizard is found in shortgrass prairies, shrublands, riparian areas, and rocky hillsides or plateaus. Christiansen and Degenhardt (1969) collected this species in dense mesquite within the city limits of Carlsbad. Dixon and Medica (1965) collected a specimen from grama grasslands in Lea County. This species is abundant along the Delaware River in southern Eddy County in habitats dominated by *Sporobolus airoides* and *Tamarix*.

Behavior: The 133 New Mexico specimens available to us were collected between 14 April (Lea Co.) and 7 September (Eddy Co.). *Cnemidophorus gularis* is wary and depends on speed to evade predators (Schall and Pianka, 1980). In west Texas, Schall (1977) recorded an average body temperature of 40.2°C for 149 active lizards, with cooler temperatures for basking lizards (38°C) than those actively moving (41°C) or sheltering in the shade of vegetation (42°C).

Reproduction: Minimum size at sexual maturity has been reported by Ballinger and Schrank (1972) and Schall (1978) as 59 mm SVL in west-central and Trans-

Pecos Texas. Ballinger and Schrank (1972) considered females less than 70 mm SVL to be yearlings, maturing in late June or July at an age of 10–11 months. These females produce a single clutch averaging 3.22 eggs. Clutch size for older females averaged 4.98 eggs, and individuals in this age-class produce two and perhaps three annual clutches (Ballinger and Schrank, 1972). The peak of egg-laying activity occurs from late May through mid-July in this population, but reproduction probably begins in early May and extends into August. Clutch size is significantly correlated with female body size; approximately one egg is added to the clutch for every seven mm increment in SVL. Clutch weight averages 18% (12–27%) of female body weight prior to oviposition. Schall (1978) reported an average clutch size of 3.13 (1–5) eggs for this species in Trans-Pecos Texas. Females dig nesting chambers at the ends of burrows as much as 30 cm below the surface and cover the eggs with dirt (Trauth, 1987). Hatchlings first appear in August.

The fat body cycle is inversely proportional to the reproductive cycle in females (Ballinger and Schrank, 1972; Schall, 1978). Fat bodies are smallest during follicular development and begin increasing in size subsequent to oviposition to as much as 7% of female body weight at maximum size.

Testes reached maximum size in mid-May following April emergence of lizards in a population in west-central Texas (Schrank and Ballinger, 1973). Fat bodies were essentially depleted by the beginning of June, and began to be replenished in July after spermiogenesis ceased. Testicular regression is complete by August. Males emerge from hibernation in late May in Trans-Pecos Texas with testes at maximum size (Schall, 1978). Testes decrease to minimal values by early August. The fat body cycle is inversely proportional to the testicular cycle in this population.

Food Habits: Milstead (1957) and Scudday and Dixon (1973) recorded the following food items by volume in Trans-Pecos Texas: grasshoppers (31%, 24%), lepidopteran larvae (26%, 19%), beetles (15%, 5%), termites (15%, 19%). In addition, the latter authors recorded ants (12%) and spiders (8%) in their sample. Lizards shifted their diets following mid-season rains from ants and termites to lepidopteran larvae. Smith (1989) reported grasshoppers (46–67%) and termites (14–42%) as major components by volume of the diet during a 3-month study period in Trans-Pecos Texas. Spiders (24%) and lepidopteran larvae (21%) by volume became important items in the last month of the study following a month of abnormally heavy rainfall. Hatchlings concentrated on termites (56%) and grasshoppers (36%) as food in the same study.

Remarks: One agent implicated in natural egg mortality is the larva of sarcophagid flies (Trauth, 1987). McAllister (1990c) reported on the parasitic helminths of this species.

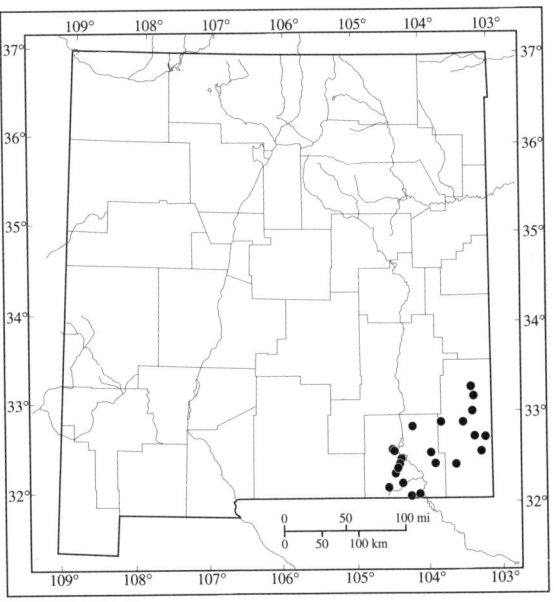

Distribution of Cnemidophorus gularis *in New Mexico.*

CNEMIDOPHORUS INORNATUS Baird, 1858 [1859]   *Little Striped Whiptail*

See Plate 66.

Type: The original description was based on two specimens (USNM 3032; Cochran, 1961), collected from "New Leon" by Lt. Darius N. Couch. The type locality has been restricted to 25° 49' N, 100° 35' W Pesquería Grande (= Villa de Garcia), Nuevo Leon, Mexico by Axtell (1961b), who also designated USNM 3032A, an adult female (but see Wright and Lowe, 1993 and discussion in Axtell, 1994) collected by Couch in March or April 1853, as the lectotype.

Distribution: *Cnemidophorus inornatus* is widespread throughout the Chihuahuan Desert of the southwestern United States and Mexico as defined by Morafka (1977), with disjunct populations in the northern, southern, and western portions of its range. It occurs from the Mexican states of Chihuahua and Coahuila northward into Texas eastward to the headwaters of the Concho River, and southward along the Rio Grande to the vicinity of Del Rio (Dixon, 1987; Wright and Lowe, 1993; Axtell, pers. comm.). In Arizona, there are disjunct populations in the southeastern corner of the state, along the central portion of the Mogollon Rim, and on the plateaus on either side of the Colorado River in northwestern Arizona.

The range occupied in the San Juan Basin in northwestern New Mexico is disjunct. The species extends historically northward in the remainder of the state up the Pecos River and Rio Grande valleys to the foothills of the Jemez and Sangre de Cristo mountains and westward south of the Mogollon Rim to the Arizona border. It occurs on broad elevated mesas and plains between the river valleys in central New Mexico, and penetrates through the central mountain chain of the Basin and Range along several areas of low relief (e.g., Tijeras Canyon in Bernalillo County, and Chupadera Mesa in Socorro County). This species appears to be declining or has disappeared over a considerable portion of its range in New Mexico, especially in the southwestern corner (Wright and Lowe, 1993), due to habitat degradation or destruction from anthropogenic sources, notably overgrazing and urbanization. These changes favor the spread of several parthenogenetic lineages such as *C. neomexicanus* and *C. uniparens* (Wright and Lowe, 1968). The southeastern margin of the range in New Mexico appears to be circumscribed by the western escarpment of the Llano Estacado (Tanner, 1975).

Description: The morphology of this species has been described in detail by Wright (1966, 1968). This is a small whiptail, with a maximum SVL for males of 85 mm and for females of 78 mm (Stevens, 1983). Christiansen (1971) found males and females (excluding hatchlings) ranging between 49–71 mm and 51–66 mm SVL in Albuquerque. The ground color of the body is black in hatchlings and may lighten to dark gray with age. The dorsal surface of the head is usually greenish-olive to brown. The sides and ventral surface of the head, the ventral surface of the body, and the tail are bright sky-blue in adult males and light blue to bluish-white in adult females. Tails of juveniles are blue distally. There are six, seven, or occasionally eight complete longitudinal light stripes on lizards from New Mexico, and there are no spots in the dark dorsal fields. The vertebral light stripe, usually present, can be markedly less distinct than the rest. The paravertebral light stripes often extend anteriorly onto the head to above the eye. There are 52–78 dorsal granular scales around midbody and 3–15 between the paravertebral light stripes anteriorly (Lowe and Zweifel, 1952; Wright and Lowe, 1993). There are usually only two or three enlarged preanal scales. The mesoptychial and postantebrachial scales are only slightly enlarged, if at all, and the supraorbital semicircles are normal. There are 6–12 interlabial scales on each side, often arranged in double rows. The ground color of lizards living on the white sands (White Sands National Monument) in Otero County is pale yellowish-gray to pale bluish-gray with the longitudinal light stripes obscured or absent (Dixon, 1967; Wright and Lowe, 1993).

Similar species: *Cnemidophorus exsanguis*, *C. flagellicaudus*, *C. sonorae*, and *C. gularis* all are distinctly spotted in the dark dorsal fields. *Cnemidophorus neomexicanus* lacks the vivid blue suffusion of body color, usually has some spotting in the dark dorsal fields, and has a distinctly wavy vertebral stripe. *Cnemidophorus sex-*

lineatus is bright green over the anterior portion of the body and lacks the bright blue suffusion elsewhere, and has conspicuously enlarged mesoptychial scales. *Cnemidophorus velox* lacks the bright blue suffusion of body color and blue tail, occasionally has spots in the dark dorsal fields, usually has more than three enlarged preanal scales, and usually has conspicuously enlarged mesoptychial and postantebrachial scales. *Cnemidophorus uniparens* usually has six longitudinal light stripes, lacks the bright blue body color and blue tail, and has abruptly enlarged mesoptychial and postantebrachial scales.

Systematics: Baird (1858 [1859]) described two species on the same page of the Proceedings of the Academy of Natural Sciences of Philadelphia based on specimens collected from the same place by Couch in 1853: *C. inornatus* (a stripeless form) and *C. octolineatus* (a distinctly striped form). Both names subsequently slipped into the confusing synonymy of the genus. Axtell (1961b) revisited the locality in question and determined that both pattern morphs represent the same species; the stripeless morph is found only at the type locality. Since the name *inornatus* precedes the name *octolineatus* in the paper by Baird, Axtell (1961b) followed the recommendation contained in Article 24 of the rules of the International Code of Zoological Nomenclature and selected *inornatus* as the formal name to apply to this species. Wright and Lowe (1993) raised the possibility that more than one species may occur at the type locality, in which case the name *octolineatus* would be the appropriate name for this strikingly patterned lizard.

Cnemidophorus inornatus belongs to the *C. sexlineatus* species group. Ten subspecies are recognized (Wright and Lowe, 1993), four occur in New Mexico. The holotype of *C. i. gypsi*, described by Wright and Lowe (1993), is an adult female (UAZ 16187) collected by Charles H. Lowe from "White Sands National Monument, at 3 miles (by road) NW Monument Headquarters, 4020 ft, Otero County, New Mexico" on 22 August 1949. The holotype of *C. i. heptagrammus*, described by Axtell (1961b), an adult male collected by and in the personal collection of Ralph W. Axtell (RWA 1758), is from "30° 11' 30" N, 103° 09' W, 5 mi. ESE Marathon, Brewster County, Texas" and was collected on 16 May 1959. The holotypes of *C. i. juniperus* and *C. i. llanuras*, described by Wright and Lowe (1993), are both adult males (MSB 5026 and MSB 10357) collected by John W. Wright on 22 June 1961 and 19 June 1962. The respective type localities given are "San Pedro Creek, 3 miles north and 2 miles east of San Antonio, 4550 ft, Bernalillo County, New Mexico" and "Carthage, 4990 ft, Socorro County, New Mexico" (see Remarks).

Habitat: This species is most common where grasses and/or low-growing shrubs, including disclimax invaders like *Salsola* and *Gutierrezia*, are the major component of the ground cover (Wright, 1966). Lizards occupy dense grassland habitats in the Albuquerque area dominated by *Hilaria jamesii*, *Muhlenbergia torreyi*, and *Aristida fendleriana* (Christiansen et al., 1971). It is a common species of yucca-grasslands throughout the state, and also occurs in juniper-grassland and piñon-juniper woodland (Lowe and Zweifel, 1952) to elevations of 2272 m. In the San Juan Basin, it occurs in grasslands with shrubs (i.e. *Ephedra*) on markedly sandy soils (Harris, 1963). This species was reported common in the sandy, mesquite-covered floodplain of the Black River in Eddy County by Gehlbach (1956), along with *Sceloporus undulatus consobrinus*, *Uta stansburiana* and *Phrynosoma cornutum*. Medica (1967) reported this species common in the sandy floodplain dominated by *Distichlis* and *Salsola* of the Rio Grande in Doña Ana County. The species has been reported to be abundant in prairie dog towns throughout New Mexico (Clark et al., 1982).

Behavior: Individuals hibernate in narrow burrows, the entrances of which are plugged with dirt, about 30 cm beneath the surface of the ground (Christiansen, 1969). Body temperatures during hibernation are less than 10°C. Lizards first appear during April (Christiansen, 1971), with juveniles and adult males more active than adult females for the first 6 weeks of the activity season. New Mexico specimens have been collected between 26 March (Bernalillo Co.) and 29 October (Doña Ana Co.). Dixon (1967) found that lizards on White Sands National Monument became active during the summer when soil temperatures reached 33°C and became inactive after soil temperatures reached 49°C; no afternoon activity period was observed. He recorded body temperatures between 37–39°C for 84 active lizards. Lizards seek temporary shelter in burrows 10–15 cm below the surface of the ground, usually at the base of vegetation such as *Oryzopsis hymenoides*, *Yucca glauca*, and *Chrysothamnus pulchellus*. Medica (1967) reported that lizard activity occurs when soil temperatures range between 26–50°C, resulting in a bimodal peak of activity along a riparian corridor in south-central New Mexico. Body temperatures ranged between 39.04–39.8°C. In west Texas, Schall (1977) recorded a mean body temperature for 318 active lizards of 40.2°C, with cooler temperature for basking lizards (39°C) than those actively moving (40°C) or sheltering in the shade of vegetation (42°C).

Male lizards apparently do not defend exclusive territories (Milstead, 1957), although they can be highly aggressive toward conspecifics of both sexes (Scudday, 1971). This species often escapes potential predators by running into the nearest clump of vegetation until the danger has passed (Christiansen, 1971; Christiansen et al., 1971; Schall and Pianka, 1980).

Reproduction: The smallest reproductively active male and female lizards reported by Christiansen (1971) from Albuquerque were 49 mm and 51 mm SVL. Most females do not reproduce until their third year at approximately 22 months of age. Follicular development in females begins by late May, and most eggs are laid in June. Eggs are smooth, leathery-shelled, creamy white, and about twice as long as they are wide. Incubation times were estimated to be 43–59 days (Medica, 1967; Christiansen, 1971). Average clutch size is two eggs, and as many as one-third of the females may produce a second clutch. Hatchlings (32–36 mm SVL) appeared from mid-July through September, depending upon weather conditions. Gravid females were observed during late June and July on White Sands National Monument (Dixon, 1967) and hatchlings were first observed at the end of August. Medica (1967) observed oviductal eggs from late May through late July in south-central New Mexico. He reported an average clutch size of 2.15 eggs and suggested that multiple annual clutches are produced. Fat body cycles in females from Albuquerque were inversely proportional to the oogenic cycle, reaching maximum size in August and September and minimum size in June (Christiansen, 1971). Other reproductive studies include Schall (1978) in Texas and Stevens (1983) in Arizona.

In the high-elevation population studied by Stevens (1983), spermiogenesis in males appeared to already be underway as they emerged from hibernation in April. Sperm production reached a peak in May, presumably concomitant with mating activity, and testicular regression was evident as early as the first week of June. Most males were completely post-reproductive by July. Christiansen (1971) reported a similar reproductive cycle for males from two populations in northern New Mexico. Testicular regression was maximal in August and began enlargement prior to hibernation. Spermatogenesis proceeded during the winter months, with all stages of primary sex cells present except mature spermatozoa. Testes achieved maximum size and weight in April, and were an order of magnitude larger in both parameters than in August. The fat body cycle of males inversely parallels the testicular cycle.

Food Habits: In one study on the white sands (White Sands National Monument) in Otero County, lizards were found to forage in organic debris beneath vegetation where lepidopterans and coleopteran larvae (41%), adult beetles (12%), heteropterans (7%), and an occasional lizard (hatchling *Sceloporus undulatus*, 2%) constituted the majority of the prey items eaten (Dixon and Medica, 1966). Medica (1967) reported a varied diet for this species over a 2-year period along a riparian corridor in south-central New Mexico. Larval and adult butterflies and moths (16%), ants (15%), other insect larvae (14%), homopterans (14%), beetles (12%), and tarantulas (8%) by volume constituted the bulk of the diet. Lepidopterans were not eaten at all during the first year of the study, which was the driest on record. Ants decreased in importance from 61% during the first year to 6% during the second, a year of normal rainfall for the area. Studies elsewhere have demonstrated a similar dietary profile for this species, with termites relatively more important (Milstead, 1957; Mitchell, 1979; Barbault and Maury, 1981).

Individuals of this species undoubtedly fall prey to a variety of vertebrate and invertebrate predators. Known predators in New Mexico include *Holbrookia maculata* and *Sceloporus undulatus*.

FAMILY TEIIDAE

Remarks: *Cnemidophorus inornatus* has been reported to be sympatric with *C. exsanguis* in Otero County (Alamogordo), with *C. neomexicanus* in Bernalillo County (Albuquerque), Cibola County (near Correo), Doña Ana County (near Mesilla and 8.4 mi. W. Hatch), Sandoval County (San Pedro Creek and Tanque Arroyo), and Socorro County (9 mi. E. La Joya and 3 mi. S. San Antonio), with *C. tigris* in Socorro County (vicinity Bosque del Apache National Wildlife Refuge), with *C. uniparens* in Grant County (jct. Hwys. 61 & 180), Luna County (vicinity of Nutt and along highway 180 northwest of Deming to the county line), and Socorro County (3 mi. S. San Antonio and northwest of Socorro) and with *C. velox* in Bernalillo County (San Pedro Creek), McKinley County (14 mi. N. Gallup), San Juan County (near Aztec), Santa Fe County (near Santa Fe and near San Ildefonso), and Socorro County (northwest of Socorro) (Taylor, 1965; Axtell, 1966; Taylor and Medica, 1966; Wright, 1966, 1968; Wright and Lowe, 1967b; Christiansen et al., 1971; Neaves, 1971; Cuellar and McKinney, 1976; Cuellar, 1979). Hybrid individuals between *C. inornatus* and the above-named species have been reported from most of these sites.

Cnemidophorus inornatus has been implicated in the generation of at least 13 parthenogenetic lineages in the genus (Wright, 1993; Wright and Lowe, 1993).

The holotype of *C. i. juniperus* (MSB 5026) actually comes from 3 miles north and 2 miles east of San Antonito, and not "San Antonio" as stated by Wright and Lowe (1993). San Antonito is 8 airline kilometers northeast of San Antonio in Bernalillo County.

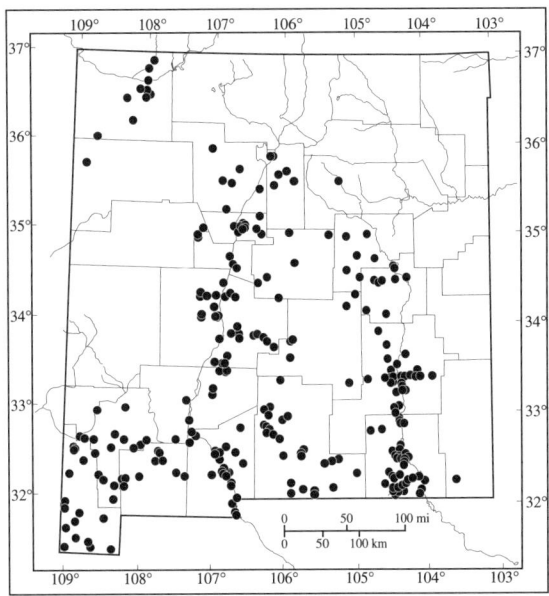

Distribution of Cnemidophorus inornatus *in New Mexico.*

CNEMIDOPHORUS NEOMEXICANUS Lowe and Zweifel, 1952   *New Mexico Whiptail*

See Plate 67.

Type: An adult female (MVZ 55807), collected at "McDonald Ranch Headquarters, 4800 feet elevation, 8.7 miles west and 22.8 miles south of New Bingham Post Office, Socorro County, New Mexico" by Charles H. Lowe, Jr. on 2 August 1947.

Distribution: Most of the range lies within the valley of the Rio Grande in New Mexico from Española south to the Texas border, where periodic flooding historically maintained perpetually disturbed habitats favored by this species (Wright and Lowe, 1968). It occurs to the east in the Tularosa Basin and extends westward from the Rio Grande Valley to the Arizona border south of the Gila River Valley. The species was first reported in the vicinity of Lordsburg by Pough (1961), and may be extending its range in the southwestern corner of the state (Cole et al., 1988). It extends downstream along the Rio Grande in Texas to the vicinity of Candelaria, Presidio County, and eastward to the Hueco Mountains in Hudspeth County (Axtell, 1966; Dixon, 1987; Cordes et al., 1989). The population in the vicinity of Conchas Lake, San Miguel County, represents an introduction (Leuck et al., 1981; Cole et al., 1988).

Description: *Cnemidophorus neomexicanus* is a small, all-female species, with adults averaging 70 mm SVL and reaching a maximum SVL of 82 mm (Christiansen,

1971). The dorsal scales are granular and there are 71–80 around the midbody. There are 34–43 femoral pores. The paravertebral light stripes are separated anteriorly by 7–13 scales. Neither the mesoptychial nor the postantebrachial scales are abruptly enlarged. The supraocular semicircles extend far forward on the head, separating the third and often the second supraocular scale from the frontal.

Cnemidophorus neomexicanus is a striped and spotted whiptail. The dorsal ground color is brown or black. There are seven well-defined longitudinal light stripes with the middorsal stripe, and often the flanking paravertebral stripes, serrate and distinctly wavy or zigzagged. A single longitudinal row of small, diffuse, light-colored spots is present in each of the two dorsolateral and upper lateral dark fields but absent in the lower lateral and paravertebral fields. The ventral surface is immaculate, occasionally with a light blue suffusion throughout. The tail is bright blue in hatchlings, fading to blue only at the tip and greenish-blue to brown anteriorly in adults. What little ontogenetic change occurs in this species involves fading of color and dulling of pattern, both of which are sharp and distinct in juveniles (Lowe and Zweifel, 1952).

Similar Species: No other species of *Cnemidophorus* in New Mexico exhibits the combination of normal-sized mesoptychial and postantebrachial scales, supraorbital semicircles penetrating far forward on the head, and a zigzag vertebral stripe.

Systematics: *Cnemidophorus neomexicanus* is an allodiploid, parthenogenetic species belonging to the *C. tesselatus* species group. It was created through hybridization between the gonochoristic species *C. inornatus* (as the paternal parent) and *C. tigris marmoratus* (as the maternal parent) (Lowe and Wright, 1966a; Brown and Wright, 1979; Cole et al., 1988; Densmore et al., 1989a). The name *perplexus* has often been applied to this species (Wright, 1971). Wright and Lowe (1967b) demonstrated that the specimen to which this name was originally assigned was an unusual hybrid individual *C. neomexicanus* x *C. inornatus*, and demonstrated that such hybridization still occurs at several sites within the range of these two species.

Individuals from central Socorro County north to south-central Rio Arriba County may represent a single lineage derived from one or a very few females resulting from the same hybridization event (Cuellar, 1977). Further electrophoretic evidence suggests that the entire range within the Rio Grande Valley, as well as in the vicinity of Lordsburg, Hidalgo County, is inhabited by lizards from a single parthenogenetic lineage (Parker and Selander, 1984; Cole et al., 1988). Cordes et al. (1990) reported that skin allografts between single lizards from the introduced Conchas Lake population in New Mexico and a population from Presidio County, Texas, healed without any signs of rejection.

Habitat: *Cnemidophorus neomexicanus* is characteristic of perpetually disturbed, disclimax habitats within the Rio Grande drainage, and desert-grassland ecotones in areas west of the Rio Grande (Cole et al., 1988). It occupies areas of subdued relief associated with internal basins on sandy and/or fine detrital soils, and can be found in open grasslands or shrublands and into piñon-juniper woodlands at elevations between 1000–1900 m (Axtell, 1966; Parker and Selander, 1984). Lowe and Zweifel (1952) reported this species to be the most abundant of five species of *Cnemidophorus* (also including *exsanguis, grahamii, inornatus,* and *tigris*) at the type locality on the Jornada del Muerto near the western foothills of the San Andres Mountains. This site borders an *Atriplex canescens–Sporobolus airoides* playa to the west, where the species was common, and a *Yucca elata–Aristida adscensionis–Bouteloua barbata* sandy grassland to the east, where it was absent. Medica (1967) reported this species to inhabit the mesquite-creosotebush community at the western margin of the Rio Grande floodplain west of Las Cruces, where it is also common in vacant lots, around buildings, and in xeric landscapes. Disturbed urban areas with sparse shrubby vegetation were also preferred by this species in Albuquerque (Christiansen et al., 1971). We have collected individuals in drainages midway between the Organ Mountains and the Rio Grande floodplain in Doña Ana County.

This species can occur sympatrically with *C. exsanguis, C. grahamii, C. inornatus, C. tigris, C. uniparens,* and *C. velox* (Axtell, 1966).

Behavior: Individuals of this species hibernate about 30 cm beneath the surface in narrow burrows, the entrances of which are plugged with dirt (Christiansen, 1969). Body temperatures during hibernation may drop to 10°C. Lizards emerge from hibernation in April (Christiansen, 1971), with adults remaining active through mid-August, and hatchlings until early October. New Mexico specimens have been collected between 21 March (Bernalillo and Doña Ana counties) and 18 October (Doña Ana County). Medica (1967) reported a bimodal period of daily activity for this species along a riparian corridor in south-central New Mexico. *Cnemidophorus neomexicanus* became active when soil temperatures reached 26–30°C and sought shelter when it reached 50°C, resulting in lizard activity prior to 1300 and again between 1600 and 1800. Body temperatures of active lizards were reported to be between 39.04 and 39.8°C.

Reproduction: Follicular enlargement begins by late April, and most eggs are laid in June in northern New Mexico (Christiansen, 1971), extending through July in southern New Mexico (Medica, 1967). Eggs are smooth, leathery-shelled, and creamy white in color, and are about twice as long as wide. Incubation takes from 40 to 60 days. Approximately 2 eggs per clutch are laid, and as many as one-third of the females may lay a second clutch each season (Medica, 1967; Christiansen, 1971). Hatchlings appear in late July through early August, depending on weather conditions. Fat body cycles are inversely proportional to the reproductive cycle in northern New Mexico. The fat bodies reach maximum size in August and September and minimum size in June (Christiansen, 1971).

Food Habits: Medica (1967) reported larval and adult lepidopterans (29%), beetles (22%), unidentified insect larvae (12%), grasshoppers (9%), and ants (8%) to constitute by volume the major components of the diet of this species during a 2-year period in south-central New Mexico. Insect larvae were not eaten at all during the first year of study, the driest year on record to that date. Lepidopterans increased (24% to 30%) whereas ants and beetles decreased (21% to 3% and 37% to 16%) from the drought year to the second year, one of normal rainfall.

Known predators of this species in New Mexico include *Gambelia wislizenii*.

Remarks: Walker et al. (1990) described a hybrid individual from a cross with *Cnemidophorus sexlineatus* at Conchas Lake in San Miguel County. McAllister (1990b) reported on the parasitic helminths of this species.

Cole et al. (1988) provided a comprehensive review of the biology of this species.

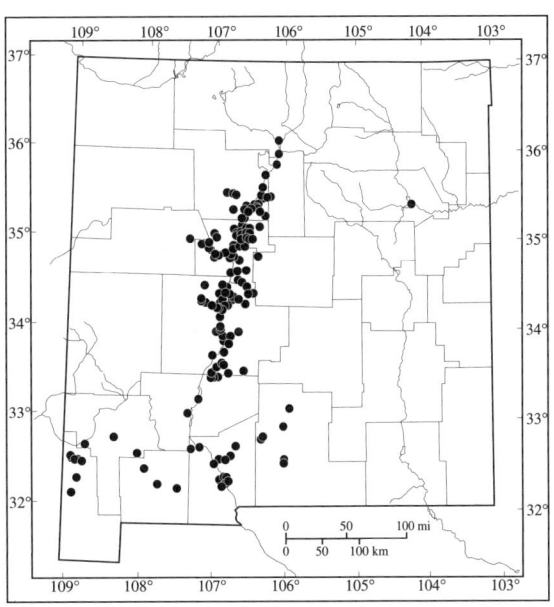

Distribution of Cnemidophorus neomexicanus *in New Mexico.*

CNEMIDOPHORUS SEXLINEATUS (Linnaeus, 1766)   *Six-lined Racerunner*

See Plate 68.

Type: The type specimen no longer exists (Maslin and Secoy, 1986). It was collected by Dr. Alexander Garden from "Carolina," date of collection unknown. The type locality was restricted by Smith and Taylor (1950a) to Charleston, Charleston County, South Carolina.

Distribution: This species occurs along the eastern seaboard from the north end of Chesapeake Bay south to Key West, and westward through Virginia, Tennessee, and western Kentucky to the shortgrass prairies extending from southern South Dakota to eastern New Mexico. It ranges south through most of Texas, except the Trans-Pecos, to Brownsville. It extends narrowly up the Mississippi and Illinois river valleys to Minneapolis–St. Paul and Lake Michigan. It is absent from the Florida Everglades and the delta region of the Mississippi River. In New Mexico, *C. sexlineatus* occurs along the eastern border east of the Pecos River, and within the Canadian and Cimarron river basins. It ranges from about 950–2000 m elevation. The species was first reported from New Mexico by Lowe (1955c).

Description: *Cnemidophorus sexlineatus* is a relatively small whiptail, with adults typically 55–75 mm SVL (Paulissen, 1987a) but reaching a maximum size of 85 mm SVL (Fitch, 1958). Females are slightly larger than males. During 1992–95, 390 *C. sexlineatus* collected on Mescalero Sands in Chaves County averaged 59.6 (28.8–70.9) mm SVL; 182 adult males averaged 61.7 (55.3–70.0) mm SVL and 5.3 (2.4–12.9) gm; 146 adult females averaged 62.5 (55.2–70.9) mm SVL and 5.3 (2.9–8.9) gm (L.W. Gorum, unpubl. data). The dorsal scales are granular; there are 62–91 scales around the midbody and 8–13 between the paravertebral light stripes. There are 30–34 femoral pores. The mesoptychial scales are conspicuously enlarged, but the postantebrachial scales are not. The supraorbital semicircles do not extend forward beyond the anterior margin of the third supraocular scale.

This is a striped whiptail without spots. The dorsal ground color is greenish-brown to black. The dorsal and lateral surfaces of the anterior body, neck, and head are distinctively bright yellow-green, especially in males. There are seven longitudinal light stripes; the vertebral light stripe is often partially divided anteriorly and occasionally separated into a pair of distinct but thin vertebral stripes. The dorsal surface between the paravertebral light stripes is often a shade lighter than the remainder of the dark dorsal fields. The tail is bright blue in hatchlings, fading to brown in adults. The dorsal surface of the hind limbs are unpatterned. In males, the lateral stripes are often obscured by a suffusion of bright green anteriorly and the belly takes on a bluish tint.

Similar Species: All other species of whiptails in New Mexico lack the bright green anterior dorsal body color. *Cnemidophorus inornatus* lacks conspicuously enlarged mesoptychial scales. *Cnemidophorus gularis* has light spots in the dark dorsal fields and enlarged postantebrachial scales.

Systematics: Three subspecies have been recognized (Maslin and Secoy, 1986; Trauth, 1992). *Cnemidophorus s. viridis* occurs in New Mexico. It was described by Lowe (1966) based on an adult female (UAZ 14800) he collected from "7.6 mi. S. Tucumcari, along state rd. 18, Quay County, New Mexico" on 13 August 1949.

Habitat: Throughout its range this species characteristically occupies open, relatively xeric habitats with patchy vegetative cover and well-drained, usually sandy, soils. Paulissen (1988b) found this lizard on sandy or clay soils with open vegetation such as *Ambrosia psilotachya*, *Cenchrus pauciflorus*, *Monarda fistulosa*, *Trifolium* sp., composites, and grasses in southern Oklahoma. Juveniles preferred more open habitats than adults. In western Nebraska, Ballinger and Jones (1985) found this species to occupy areas of relatively dense (>50%) grass cover. Fitch (1958) recorded densities of 98.8–177.8 lizards/hectare in Kansas.

Behavior: *Cnemidophorus sexlineatus* emerge from hibernation in April (Fitch, 1958) and reach peak seasonal activity during June. Hatchlings appear by early to mid-August in Kansas and southern Oklahoma (Fitch, 1958; Hardy, 1962; Paulissen, 1987a) by which time adults have largely ceased surface activity. New Mexico specimens have been collected between 18 April (Roosevelt County) and 14 November (Roosevelt County).

Fitch (1958) recorded body temperatures of 100 active lizards between 30°C and 45°C, with 75% of the readings between 38°C and 42°C. Paulissen (1988a) found the mean body temperature selected in a thermal gradient to be 36.8°C and 37.1°C and the critical thermal maximum to be 51.0°C and 49.8°C for adult and juvenal lizards. Lizards seek temporary shelter for thermoregulatory purposes or to spend the night under rocks, logs, buildings, trash piles, and a variety of similar objects. They also dig their own burrows or utilize those made by other animals (Edgren, 1955; Fitch, 1958; Hardy,

1962). As with other species of whiptails, *C. sexlineatus* depends upon speed as a first line of defense against predators (Ballinger et al., 1979).

Reproduction: Female lizards reach reproductive maturity during their second season at approximately 68 mm SVL (Fitch, 1958); less than 10% lived to be four years old in this Kansas study. Larger and older females lay larger clutches with a smaller interval between annual multiple clutches (Carpenter, 1960; Hardy, 1962). Females are gravid from May through August in Oklahoma with a peak in July (Carpenter, 1960). The average clutch size is 2.46 (1–6), incubation lasts 46–63 days, and hatchlings first appeared by mid-July. In Kansas, oviposition takes place between June and August, and the incubation period is about 50 days (Hardy, 1962). Brown (1956) reported an average clutch size of 2.9 (1–5) for 67 natural nests excavated in North Carolina. Hatchlings are 31–35 mm SVL and attain a size larger than 50 mm SVL in Oklahoma prior to hibernation (Paulissen, 1987a).

Food Habits: *Cnemidophorus sexlineatus* are active foragers, finding prey opportunistically by sight and olfaction. They forage primarily on the ground, often digging up hidden prey and occasionally climbing into low vegetation after insects and other arthropods (Fitch, 1958; Hardy, 1962). Paulissen (1987b) reported that adults forage at a more rapid rate than juveniles and cover a much wider area in search of relatively rare but large and mobile prey like grasshoppers, whereas juveniles methodically search for cryptic arthropods on plants and in detritus. Burt (1928) reported grasshoppers (41%), spiders (24%), butterflies and moths of all life stages (20%), and land snails (7%) to be important components of the diet in Kansas. Paulissen (1987a) reported adults and juveniles eating different prey in southern Oklahoma. Adults preferred large jumping planthoppers (63%) and grasshoppers (mostly acridids, 40%) whereas juveniles preferred leafhoppers (62%), salticid spiders (40%), microlepidopterans (23%), caterpillars (13%), and juvenal hemipterans (13%) by frequency. Chrysomelid beetles (11% and 10%), beetle larvae (29% and 39%), and ants (26% and 31%) were also eaten by adults and juveniles.

These lizards are eaten by a wide variety of vertebrate predators in other parts of the species' range (Fitch, 1958; Hardy, 1962). Snakes (*Masticophis, Salvadora*), hawks, owls, and roadrunners (*Geococcyx californianus*) are probably important predators of this species in New Mexico.

Remarks: Walker et al. (1990) reported a hybrid male from a cross between this species and *C. neomexicanus* at Conchas Lake in San Miguel County. These species are normally allopatric; this is further evidence of the propensity of species within the *C. sexlineatus* species group to hybridize.

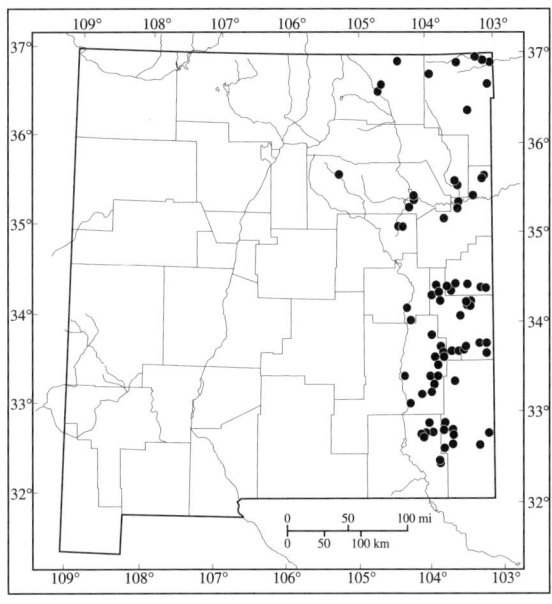

Distribution of Cnemidophorus sexlineatus *in New Mexico.*

CNEMIDOPHORUS SONORAE Lowe and Wright, 1964 — *Sonoran Spotted Whiptail*

See Plate 69.

Type: An adult female (UAZ 11777), collected "at 2 miles southwest of Oracle (in vicinity of old C.C.C. Camp), 4500 ft. elev., near the north base of the Santa Catalina Mountains, Pinal County, Arizona" by John W. Wright on 20 July 1964.

Distribution: *Cnemidophorus sonorae* ranges throughout the Río Yaqui and Río Sonora basins in northern Sonora (north of 29°30' N.) northward and westward to the Santa Catalina and Baboquivari mountains, Pima County, Arizona, and eastward to the Peloncillo Mountains and the Gila River in New Mexico. It extends up the Gila River to the vicinity of Redrock, and there are isolated records from the San Francisco River in Catron County (UAZ 9248) and from the Animas Mountains, Hidalgo County (MSB 39268 and 54651). We have recently taken specimens from the vicinity of Antelope Pass in the Peloncillo Mountains. It does not occur in the southeastern corner of New Mexico as indicated by Maslin and Secoy (1986).

Description: This is a moderate-sized, all-female *Cnemidophorus*, with a maximum SVL of 89 mm. During 1992–95, 504 individuals collected in Guadalupe Canyon in Hidalgo County averaged 59.8 (22–89) mm SVL and 6 (0.4–15.7) gm; 261 adults averaged 73.7 (62–89) mm SVL and 9.5 (3.8–15.7) gm (our unpubl. data). The dorsal scales are granular; there are 74–80 scales around midbody and 5–8 scales anteriorly between the paravertebral light stripes. There are 32–39 femoral pores. Both the mesoptychial and postantebrachial scales are enlarged. The supraorbital semicircles do not extend anteriorly beyond the third supraocular scale.

Cnemidophorus sonorae is a striped and spotted whiptail except that juveniles lack spots. The dorsal ground color is dark brown to black. There are six complete dorsal longitudinal light stripes which remain distinct on the neck. As lizards age they develop many indistinct white to dull yellow light spots within the dark dorsal fields. These spots do not involve the longitudinal light stripes. The dorsal surface of the hind limbs have a bold vermiculate pattern in juveniles, fading to a uniform tan in large adults. The original tail is orange-tan in color. The ventral surface of the body is creamy-white.

Similar species: This species is easily confused with both *C. exsanguis* and *C. flagellicaudus*. It differs from both in lacking distinct spots on the hind limbs. It differs further from *C. exsanguis* by retaining distinct stripes and indistinct spots throughout life, and the spots not involving the longitudinal light stripes. It differs further from *C. flagellicaudus* by usually possessing three or more enlarged preanal scales, a higher count of light spots in the dorsal fields that do not involve the longitudinal lights stripes, and by retaining the bold vermiculate pattern on the dorsal surface of the hind limbs throughout most of life. *Cnemidophorus burti* is much larger, has more than 95 granular scales around midbody, and tends to lose its stripes altogether with age.

Systematics: *Cnemidophorus sonorae* is an allotriploid, parthenogenetic species of hybrid origin belonging to the *C. sexlineatus* species group (Wright and Lowe, 1968; Wright, 1993). Recent mtDNA analyses have identified the probable paternal and maternal parent species of an intermediate (presumably parthenogenetic and extinct) lineage as either *C. burti* or *C. costatus* and *C. inornatus arizonae* (Densmore et al., 1989b), followed by back-crossing hybridization(s) with one or both of the paternal species. There may be more than one independently derived lineage within this species (Wright, 1993, 1994).

Habitat: *Cnemidophorus sonorae* occurs in oak-woodlands and oak-grasslands in New Mexico at elevations from 1065–2130 m, and ranges into pine-oak woodlands at higher elevations in Arizona and lower elevations and more xeric habitats, usually along riparian corridors, in both Arizona and Mexico (Lowe and Wright, 1964; Wright and Lowe, 1968). It is sympatric with *C. burti* and *C. exsanguis* in the Peloncillo Mountains, Hidalgo County, with both *C. exsanguis* and *C. flagellicaudus* along the Gila River in the vicinity of Redrock, Grant County, and with *C. dixoni*, *C. exsanguis*, *C. tigris*, and *C. uniparens* in the creosotebush flats at Antelope Pass in Hidalgo County.

Behavior: Daily activity in Arizona is inhibited by hot weather, with lizards becoming inactive on the surface when soil temperatures exceed 50°C. Adults enter hibernation by mid-August although hatchlings remain active through the end of September (Routman and Hulse, 1984). The 82 New Mexico specimens available to us were collected between 27 April (Hidalgo County) and 26 August (Grant County). We have recorded this species active from 13 April to 14 October in Guadalupe Canyon. Lizards of this species are able to incur a significant lactic acid debt through anaerobic metabolism without overt detrimental effects upon their daily activities (Pough and Andrews, 1985).

Reproduction: Vitellogenesis begins in mid-May and the reproductive season extends through the end of July in southeastern Arizona (Routman and Hulse, 1984). Average clutch size of 58 specimens examined was 3.7 (1–7) eggs. Individual lizards produce 2 or 3 clutches per year. Clutch size is significantly correlated with body size, and approximately one egg is added per 7 mm increment in SVL. Hatchlings first appear by the end of July. Echternacht (1967) reported the reproductive season extending from late June through late August in southeastern Arizona, with a clutch size of 3 (1–5). Congdon et al. (1978) reported an average clutch size of 4.1 (1–7) eggs.

Food Habits: Termites (96%), spiders (44%), beetles (40%), ants and grasshoppers (36% each), and homopterans (17%) were important foods by frequency for this species in southeastern Arizona (Echternacht, 1967). Mitchell (1979) reported termites as 96% of the diet by frequency during one season in southeastern Arizona.

Remarks: This species has been confused with *Cnemidophorus exsanguis* in some of the historical literature (Wright, 1993). McAllister (1992) studied the helminths which parasitize this species.

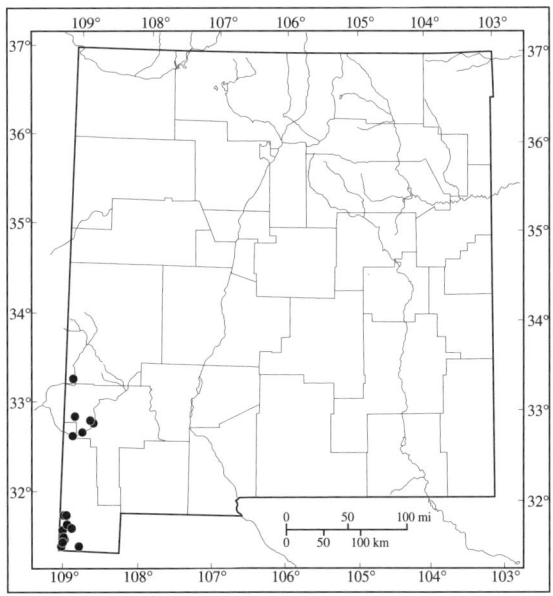

Distribution of Cnemidophorus sonorae *in New Mexico.*

CNEMIDOPHORUS TIGRIS Baird and Girard, 1852   *Western Whiptail*

See Plate 70A, 70B.

Type: No type material was designated in the original description (Baird and Girard, 1852a). Burt (1931) designated an adult male (USNM 4103) as the lectotype. It was collected from "the Valley of the great Salt Lake" by Captain Howard Stansbury, U.S. Army Corps of Topographical Engineers, sometime during the spring of 1850.

Distribution: *Cnemidophorus tigris* occurs widely in arid and semi-arid portions of western North America, including all or parts of Oregon, Idaho, California, Nevada, Utah, Colorado, Arizona, New Mexico, and Texas, and the Mexican states of Baja California del Norte, Sonora, Sinaloa, Chihuahua, Durango, and Coahuila (Pianka, 1970; Hendricks and Dixon, 1986). The species also occurs on many islands in the Sea of Cortéz (Murphy and Ottley, 1984).

Cnemidophorus tigris occurs in the San Juan Basin of northwestern New Mexico. It also occurs across the southern tier of counties in the state (south of the Mogollon Rim) and in the foothills of the Sacramento Mountains. It extends up the Rio Grande Valley to the vicinity of Albuquerque, throughout the Tularosa Basin (apparently absent in White Sands National Monument), and up the Pecos River Valley to the vicinity of the Chaves/DeBaca county line. This distribution coincides almost precisely with the distribution of desert soils within the state as defined and mapped by Maker et al. (1978). Elevation ranges from approximately 900–1675 m.

Description: *Cnemidophorus tigris* is a large species, with a maximum SVL of 107 mm in males and 96 mm in females (Hendricks and Dixon, 1986). During 1987–92, 2,798 specimens collected at Antelope Pass in Hidalgo County averaged 73.2 (22–110) mm SVL and 12.0 (0.5–34) gm. Nine hundred eighty-six adult males averaged 86.6 (70–106) mm SVL and 17.9 (6.2–31.3) gm; 810 adult females averaged 78.5 (60–100) mm SVL and 13.1 (4–34) gm (our unpubl. data). The dorsal scales are granular with 68–114 around midbody. There are 33–48 femoral pores. Neither the mesoptychial nor the postantebrachial scales are abruptly differentiated in size from neighboring scales. The supraocular semicircles may occasionally extend beyond the anterior margin of the third supraocular scale, but never completely separate the second supraocular from the frontal.

The dorsal ground color of hatchlings is dark brown to black and the pattern is striped and spotted. This pattern changes quickly as lizards grow, and adults are brown to tan to gray with more of a reticulated pattern, often losing the stripes altogether. In *C. t. marmoratus*, there are many small, light round or oval spots arranged in longitudinal rows between the stripes. Adults can retain a striped appearance despite losing the distinctness of the spots, or develop a marbled pattern with no indication of stripes or crossbars, or with crossbars connected in a linear pattern across the dorsum and the sides barred or mottled. A patternless morph with a uniformly brownish-gray dorsum has been reported from Crane County, Texas (Ballinger and McKinney, 1968). In *C. t. gracilis*, hatchlings have four or six distinct longitudinal light stripes with spots in the dark fields. The stripes fade as lizards grow and the spots expand, forming a reticulated dorsal pattern in adults. Usually a striped pattern persists although laterally adults are distinctly marbled. The ontogenetic change in dorsal color pattern of *C. t. septentrionalis* is very similar, except that it is less distinct. The ventral surface of hatchlings is immaculate, and adult females usually have some scattered black flecking. Adult males of *C. t. marmoratus* and *C. t. septentrionalis* develop more spotting, and usually have sooty gray throats. Adult male *C. t. gracilis* typically have almost completely black chins and chests and darkening over the majority of the rest of the ventral surface.

Similar Species: *Cnemidophorus dixoni* and *C. grahamii* are the species most likely to be confused with *C. tigris* in New Mexico. Both differ in having abruptly enlarged mesoptychial scales, an immaculate ventral surface marred by at most a few black flecks, and a boldly reticulate dorsal pattern.

Systematics: The nomenclatural history of this polytypic species, belonging to the *C. tigris* species group, is as byzantine as any in lizard systematics. Twelve subspecies are currently recognized (Wright, 1993; but see Maslin and Secoy, 1986, and Dessauer and Cole, 1991), and four occur in New Mexico. Both *C. t. gracilis* and *C. t. marmoratus* were originally described by Baird and Girard (1852a) as distinct species without designating any type material. The type localities were given as "Desert of Colorado" and "between San Antonia [sic] (Texas) and El Paso del Norte." The lectotype of *C. t. gracilis*, designated by Cope (1898 [1900]), is a subadult female (USNM 3034), collected by Dr. John L. LeConte, specific date unknown. The lectotype of *C. t. marmoratus*, designated by Cochran (1961), is an adult female (USNM 3024a), collected by John H. Clark in May or June 1851. Both were collected during the U.S. and Mexican Boundary Survey. The type localities were restricted by Smith and Taylor (1950a) to Yuma, Yuma County, Arizona, and El Paso, El Paso County, Texas.

Cnemidophorus t. septentrionalis was described by Burger (1950), who designated a subadult female (FMNH 38217) as the holotype, collected from "Una, Garfield County, Colorado" by E.V. Prostov on 25 June 1941. *Cnemidophorus t. reticuloriens* was first named by Vance (1978; see Vance et al., 1991) and subsequently thoroughly described by Hendricks and Dixon (1986) based on a series of 24 lizards at Texas A&M University (holotype, TCWC 39509, an adult male) collected from "24.7 mi. NW Fort Stockton, Pecos Co., Texas; 31° 06' N latitude, 103° 13' W longitude; south side of U.S. Highway 285 near roadside park; elevation 2850 feet" by Fred S. Hendricks on 1 June 1972.

The subspecies *C. t. gracilis* and *C. t. marmoratus*, which are widespread in desert scrub habitats in the Sonoran and Chihuahuan deserts, respectively, hybridize within a narrow zone in southwestern New Mexico where suitable habitat transcends low passes in the Peloncillo Mountains. These two forms are morphologically and genetically distinctive, and have been the subject of intensive study in this zone of contact for over 30 years (Zweifel, 1962b; Dessauer et al., 1962; Dessauer and Cole, 1991). Hendricks and Dixon (1986), based on an extensive morphological analysis of *Cnemidophorus tigris* east of the Continental Divide, re-elevated *marmoratus* to full species status and described two new subspecies within this form for a total of five subspecific taxa. Two subspecies of *C. marmoratus* would then occur in New Mexico; the nominotypical form west of the Guadalupe Mountains escarpment, and *C. m. reticuloriens* from the Pecos River valley eastward. This taxonomic arrangement has been followed in some recent checklists (e.g., Collins, 1990) and field guides (e.g., Conant and Collins, 1991). Dessauer and Cole (1991, pers. comm., and unpubl. data) have argued that speciation between *gracilis* and *marmoratus* has not yet been completed because panmixia occurs within the hybrid zone and there appears to be no evidence of selection against hybrids. We regard the taxonomic question as unresolved, recognizing that this situation is the sort that evolutionary biologists and others interested in biodiversity dream of discovering and observing, and we prefer to retain *marmoratus* as a subspecies of the widespread polytypic species *Cnemidophorus tigris* (see also Wright, 1993).

Habitat: This species is characteristic of open desert shrublands on a variety of soils, from sandy to rocky alluvium (Tinkle, 1959; Zweifel, 1962b; Peterson and Whitford, 1987). In the San Juan Basin, it occupies relatively open habitats of piñon-juniper, grassland, or riparian situations on ground with meager soils, rocky, or exposed bedrock (Harris, 1963). In west Texas and southern New Mexico, it can be found on low, gently rolling sand dunes with 20% cover of *Prosopis*, *Gutierrezia sarothrae*, and *Yucca glauca* (Milstead and Tinkle, 1969;

our obs.). Other characteristic shrub species of occupied habitat include *Larrea* and *Ephedra trifurca* (Whitford and Creusere, 1977; Cole et al., 1988; Baltosser and Best, 1990). Clark et al. (1982) recorded *C. tigris* in association with Gunnison's prairie dog towns in northwestern New Mexico, but not with black-tailed prairie dog towns in the southeastern part of the state.

Milstead (1965) estimated a population size of 44 lizards per hectare during an extended drought in west Texas; ten years later after the drought the population on the same site had increased to 183 lizards per hectare. Whitford and Creusere (1977) and Price et al. (1993) recorded densities of 40–70 and 10–40 lizards/hectare over different 5-year periods in Doña Ana County, New Mexico. Parker (1972b) estimated spring densities of 13 lizards/hectare and late summer densities of 36 lizards/hectare in central Arizona.

Behavior: Barbault (1977) recorded daily activity between 0800–1900 with a sharp peak at 1300 and 70% of activity occurring between 1000–1500 in Mexico. Medica (1967) reported a bimodal activity period; lizards were active at soil temperatures of 26–30°C and sought shelter at a soil temperature of 50°C in southern New Mexico. Creusere and Whitford (1982) found lizards active between 0700–2000 at a nearby site from May through October. The activity season in southwestern New Mexico extends from April through September (Baltosser and Best, 1990), with occasional individuals active on warm days in March and October. New Mexico specimens have been collected between 17 March (Eddy County) and 2 November (Doña Ana County).

In west Texas, Schall (1977) reported the mean body temperature for 289 active lizards as 40.4°C, with cooler temperatures for basking lizards (39°C) than those actively moving (41°C) and those in the shade (42°C). Anderson and Karasov (1981) reported that lizards of this species spend up to 91% of a 10-hour daily activity period during the summer in movement, with active body temperatures of 41.0°C, in southern California. Pianka (1970) and Barbault (1977) have recorded average body temperatures of 37–40°C. Individual lizards are able to withstand supercooling to –4.81°C (Lowe et al., 1971).

Milstead (1957) regarded this species as non-territorial in Trans-Pecos Texas because male home ranges widely overlapped. Minton (1958 [1959]) noted that these lizards appear to have recognizable home ranges and behave aggressively toward other trespassing lizards, including conspecifics; we have noted the same in lizards from southern New Mexico. Unpublished data from the removal experiments conducted by Price et al. (1993) suggest that adults of both sexes maintain overlapping home ranges, whereas subadults and juveniles do not. The latter class behave as a surplus, "floating" population, with individuals moving into and establishing permanent home ranges upon the demise of a vested adult. Home range sizes in Nevada have been estimated as 0.22 hectares for juveniles, 0.52 hectares for females, and 0.29 hectares for males (Jorgensen and Tanner, 1963).

Cnemidophorus tigris is a very wary lizard that depends upon speed to escape its enemies (Schall and Pianka, 1980; Price, 1992). James F. Scudday (pers. comm.) calls this species the "roadrunner of the lizard world" and considers it one of the most difficult to collect by hand.

Reproduction: Males and females achieve sexual maturity at minimum body sizes of 70 mm and 60 mm SVL (Pianka, 1970; Parker, 1973b; Schall, 1978; Hendricks and Dixon, 1984). Pianka (1970) observed oviductal eggs from mid-April to late July in southern populations and from late May to early June in northern populations. Clutch size ranged from 1–5, with an overall average of 2.63 eggs; clutch size of a given population was strongly correlated with prior short-term annual primary productivity. Taylor et al. (1992, 1994) found that differences between mean clutch size in northern and western populations was significantly influenced by female body size and elevation. McCoy and Hoddenbach (1966) reported that female *C. tigris* from Colorado lay one clutch averaging 3.4 eggs whereas females from west Texas produce two annual clutches averaging 2.2 eggs each, a latitudinal trend corroborated by Parker's (1973b) data from Utah, Arizona, and New Mexico. Females in their first reproductive season produce smaller clutches in both populations than do older females (2.9 vs. 3.9 in Colorado, 2.1 vs. 2.8 in Texas). Schall (1978) recorded an average clutch size of 2.02 (1–4) eggs in Trans-Pecos Texas. Medica (1967) reported 1–2 clutches of two eggs each for females in south-central New Mexico. Clutch weight prior to oviposition may vary between 10% (New Mexico) and 20% (Utah) of an individual female's body weight (Parker, 1973b).

Upon emergence from winter dormancy in March or April, spermatogenesis begins, and male testes and seminiferous epithelial height and tubule diameter increase through the mating season to reach maximum size in June and July (Goldberg and Lowe, 1966; Parker, 1973b). These structures undergo rapid regression in August to reach minimum sizes by September and October; a gradual recrudescence takes place during hibernation.

Our data for southern New Mexico corroborate those of Medica (1967) and Hendricks and Dixon (1984) concerning seasonal activity. Juveniles are the first to appear in the spring, as early as mid-March. Adult lizards appear in abundance in early May and generally disappear by mid to late August. Hatchlings are first observed in late July and are present until early October. Our observations in southern New Mexico support those of Pianka (1970) in other portions of the species' range in that males are generally active earlier in the season than females, and females often remain active longer in the late summer period of activity than males. Our observations also support those of Goldberg and Lowe (1966), Schall (1978), and Hendricks and Dixon (1988) that gravid females may retire underground for up to three weeks prior to oviposition, and then reappear to mate again or forage to supply energy stores for hibernation.

Individual lizards generally live 3–4 years (Pianka, 1970), with a maximum lifespan of eight years (our unpubl. data).

Food Habits: *Cnemidophorus tigris* are active foragers with a keen sense of olfaction, "moving through the environment, constantly grubbing, digging, and frequently extending their tongues" (Pianka, 1970). Echternacht (1967) reported lizards of this species occasionally foraging in shrubs off the ground. They are opportunistic insectivores, and the recorded diet of this species varies widely with geographic location, year, and season. Pianka (1970) reported that insect larvae constitute an important food source early in the spring, and that beetles are a staple food source. Termites are a staple food item for many populations in the Sonoran and Mojave deserts (Pianka, 1970; Echternacht, 1967; Mitchell, 1979). Additional food habits studies have been conducted in Arizona (Bickham and MacMahon, 1972; Vitt and Ohmart, 1977b), Colorado (Johnson, 1966), Mexico (Barbault and Maury, 1981), Texas (Milstead, 1965; Scudday and Dixon, 1973), and Utah (Pack, 1923b: Knowlton, 1938).

Grasshoppers (84% by frequency, mostly acridids); beetles (45%), termites (32%), and butterflies and moths (28%) were reported by Best and Gennaro (1985) as significant items in the diet of this species over a 4-year period from an area 40 km E of Carlsbad. There was little seasonal or annual variation in staple food items eaten. Medica (1967) reported that butterfly and moth larvae (38%) and beetles (19%) were the most important items by volume during two years in the diet of a population along the Rio Grande west of Las Cruces. Beetles and termites were more important during the first year, 1964, the driest on record to that date. Insect larvae, particularly lepidopterans, became more important during 1965, a year of normal rainfall, whereas the proportion of beetles eaten was halved and termites disappeared altogether.

A wide variety of other lizards, snakes, and birds are known to include *C. tigris* in their diets (Pianka, 1970). Jim Stuart (pers. comm.) has observed *Parabuteo unicinctus* (Harris's hawk) taking these lizards in southeastern New Mexico. Known reptilian predators of this species in New Mexico include *Gambelia wislizenii* and *Masticophis flagellum*.

Remarks: We follow Banta (1971) in the citation for the original description of this species.

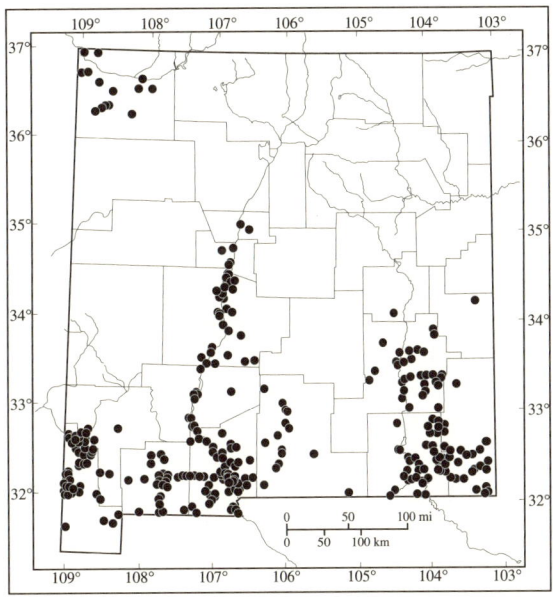

Distribution of Cnemidophorus tigris *in New Mexico.*

CNEMIDOPHORUS UNIPARENS Wright and Lowe, 1965   *Desert Grassland Whiptail*

See Plate 71.

Type: The holotype is an adult female (UAZ 5278), collected by John W. Wright, Robert L. Bezy, and Charles H. Lowe at "Fairbank, Cochise County, Arizona" on 15 July 1962.

Distribution: *Cnemidophorus uniparens* ranges from the vicinity of Ciudad Chihuahua, Chihuahua, Mexico, northward in the Chihuahuan Desert into New Mexico, and westward into extreme northeastern Sonora, Mexico, and adjacent southeastern Arizona. It extends northward along the eastern border of Arizona to the vicinity of the Gila River, and northwestward below the Mogollon Rim through Phoenix to the vicinity of the Hualapai Mountains. It occurs throughout southwestern New Mexico south of the Mogollon Rim, although it extends into the Mogollon Highlands up a number of drainages along the southern margin. It extends throughout the Rio Grande drainage from the Texas state line northward to La Joya, Socorro County, with a northwestward extension up the Rio Salado to the Cibola/Valencia county line. The Bernalillo County records probably represent introductions (Stuart and Degenhardt, 1986). Elevation ranges from 1120 m to about 2100 m.

Description: The morphology of *C. uniparens* has been described in detail by Wright (1968). This is a small, all-female *Cnemidophorus*, with a maximum SVL of 86 mm. During 1987–92, 1,117 specimens collected at Antelope Pass in Hidalgo County averaged 52.8 (21–86) mm SVL and 4.1 (0.3–15.5) gm. Four hundred thirty-one adults averaged 66.6 (60–86) mm SVL and 6.8 (3.5–15.5) gm (our unpubl. data). The dorsal scales are granular and there are 59–78 rows around midbody. There are 28–41 femoral pores. The paravertebral light stripes are separated anteriorly by 4–8 scales. Both the mesoptychial and postantebrachial scales are abruptly enlarged. The supraocular semicircles do not extend beyond the midpoint of the third supraocular. There are usually three enlarged preanal scales.

Cnemidophorus uniparens is a striped whiptail without spots. There is little ontogenetic color pattern change in this species. The ground color is olive-brown to black. There are six cream to white dorsal longitudinal stripes extending from the head to the tail. There may be a suggestion of a seventh middorsal stripe, but it extends at most one-quarter of the way from the occipital region to the tail. The tail is bright blue in hatchlings, fading to bluish-green to olive-green in adults. The ventral surface is fleshy white; some adults have a faint bluish tint on the chin and throat.

Similar Species: *Cnemidophorus inornatus* is smaller and more gracile, has considerable blue pigmentation throughout the ventral surface and the tail at all ages, usually has seven complete longitudinal light stripes, has more than six interlabial scales often in two rows, and lacks enlarged mesoptychial and postantebrachial scales. *Cnemidophorus velox* is more robust in body form, has seven complete longitudinal light stripes in areas of sympatry, and often has a few spots in the lateral dark fields.

Systematics: *Cnemidophorus uniparens* is an allotriploid, parthenogenetic species of hybrid origin belonging to the *C. sexlineatus* species group (Wright and Lowe, 1965, 1968; Wright, 1968, 1993). Recent mtDNA analyses suggest that *C. uniparens* arose from one or more hybridization events between male *C. burti* or *C. costatus* and female *C. inornatus arizonae* to create a diploid (presumably parthenogenetic and extinct) intermediate lineage, followed by backcrossing hybridization(s) with *C. inornatus* (Densmore et al., 1989b; Wright, 1993). Populations from Bosque del Apache National Wildlife Refuge and from the Fra Cristobal Mountains near the north end of Elephant Butte Lake appear to be genetically identical (Cuellar, 1976).

Habitat: *Cnemidophorus uniparens* is common in desert-grassland habitats and in grasslands that have been degraded, allowing the expansion of shrubby species at the expense of perennial grasses (Wright and Lowe, 1968). Hulse (1981) reported density estimates of 103 and 78 adult lizards per hectare during a 2-year period in southeastern Arizona. Cuellar (1993) estimated a maximum density of 89 lizards/hectare in the floodplain of the Rio Grande in Socorro County. It is sympatric with both *C. inornatus* and *C. velox* over a wide area north and west of Socorro, east of the Magdalena Mountains and west of the Ladrone Mountains, north to the Cibola/Valencia county line (Wright, 1968; our unpubl. data).

Behavior: The daily activity pattern of a population of *C. uniparens* studied for two years near Portal, Cochise County, Arizona, was distinctly bimodal, with 75% of all animals collected between sunrise and 1130 (Hulse, 1981). A smaller peak of activity occurred between 1730 and 1930. Lizards were active throughout the day on overcast days; activity ceased during rainfall. New Mexico specimens have been collected between 11 March (Grant County) and 29 October (Grant County).

Hulse (1981) reported home range sizes of 120 m^2 to 2386 m^2 (average 728 m^2), for 47 lizards recaptured 5 or more times. Home range size was independent of body size and home ranges of different individuals overlapped widely. Average home range size was twice as large during a year of low primary productivity (and low insect prey availability) than during a year of high primary productivity; presumably individual lizards had to forage more widely under the former conditions.

Cnemidophorus uniparens frequently behave as males in captivity, engaging in pseudocopulatory behavior (Crews and Fitzgerald, 1980; Moore et al., 1985). The same individual can express both male and female-like behavior, but these behaviors are only expressed by reproductive individuals. Female-like behavior occurs most frequently during the period when a clutch of eggs is being produced, and male-like behavior occurs most frequently just prior to or subsequent to oviposition and prior to the beginning of another vitellogenic cycle. The significance of this behavior in natural populations is unknown. Crews et al. (1986) have suggested that pseudocopulatory behavior serves to stimulate ovulation and reduce the latency between ovulations in reproductive individuals, although the evolutionary advantage to this behavior isn't immediately obvious if all individuals are genetically identical. See Crews and Moore (1993) and Cuellar (1993) for further discussions of this issue.

Reproduction: Females attain sexual maturity at a minimum size of 60 mm SVL during their first reproductive season after hatching (Hulse, 1981; Cuellar, 1993). The reproductive season extends from May through July in southeastern Arizona, and individuals produce 2–3 clutches per year with intervals between clutches of 21–28 days. Cuellar (1971) reported this interval to be 23 days. Clutch size ranged from 1–4 in 105 individuals (average 2.77) and is significantly correlated with body size; lizards add one egg to the clutch with approximately each seven mm increment in SVL (Hulse, 1981). Cuellar (1993) reported an average clutch size of 3.6 eggs for a population from Socorro County. Females smaller than 65 mm SVL averaged two eggs/clutch, those 65–70 mm SVL averaged three, and those larger than 70 mm SVL averaged four. Cuellar (1971) reported an average clutch size of 3.32 eggs from a widely dispersed geographic sample throughout Arizona and New Mexico, and Cuellar (1981) reported an average clutch size of four and two clutches per year from a population in central New Mexico. Freshly laid eggs are cream-colored with a parchment-like shell, oval in shape (8 x 16 mm), and weigh about 0.4 gram.

Hulse (1981) reported almost a complete annual turnover in the population he studied, with only 12 of 135 lizards marked during the first year recaptured the second.

Food Habits: Little quantitative information on the food habits of *C. uniparens* exists. Mitchell (1979) reported termites as 94% of the diet by frequency during one season at a study site in southeastern Arizona.

Known predators of this species include *Gambelia wislizenii*, *Geococcyx californianus* (roadrunner), *Speotyto cunicularia* (burrowing owl), and *Lanius ludovicianus* (loggerhead shrike) (Hulse, 1981).

Remarks: The range of *C. uniparens* is apparently expanding in New Mexico with the advent of overgrazing and increasing desertification of desert grasslands and riparian areas.

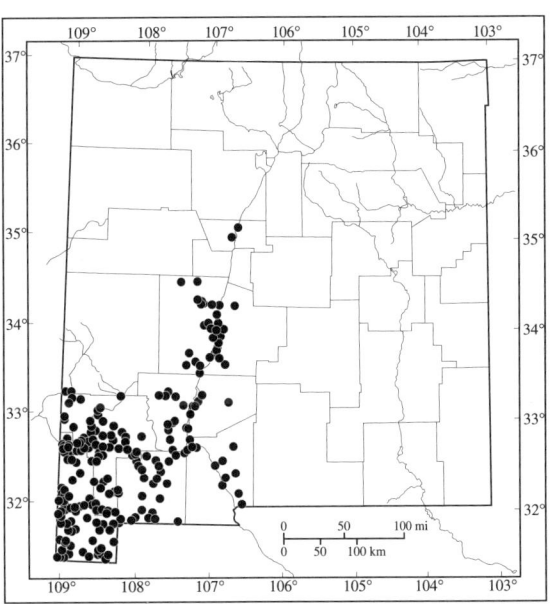

Distribution of Cnemidophorus uniparens *in New Mexico.*

CNEMIDOPHORUS VELOX Springer, 1928

Plateau Striped Whiptail

See Plate 72.

Type: An adult female (MCZ 37208), one of a series of four specimens used in the original description by Springer (1928) from "Oraibi, Arizona, and . . . Pueblo Bonito, New Mexico," restricted to Oraibi, Navajo County, Arizona by Lowe (1955b). This series was collected by Stewart Springer and his party sometime during August 1928, and was received by the MCZ as a gift from J. Piatt on 11 January 1934.

Distribution: *Cnemidophorus velox* occurs primarily on the Colorado Plateau of Arizona, Utah, Colorado, and New Mexico. In Arizona, it ranges southward to the edge of the Mogollon Rim and into several isolated mountain ranges in the central highlands, and westward to the margins of the Coconino and Kaibab plateaus. It ranges northeastward along the Colorado River basin through southern Utah to the vicinity of Grand Junction, Colorado, remaining west of the Rocky Mountains. There is an introduced population in Jefferson County, Oregon (Stebbins, 1985; Moritz et al., 1989), and another in Phoenix, Arizona (L. J. Vitt, pers. comm.).

In New Mexico, *C. velox* occurs throughout the northern half of the state, eastward to Ute Creek, Union County, and southward from there through the upper Canadian and Pecos river drainages to Sumner Lake. West of the Rio Grande, it occurs southward to the high plains between the Ladrone and Magdalena mountains in Socorro County. There are isolated populations near Quemado and in the Plains of San Augustin (Old Horse Springs westward to Apache Creek) in Catron County, and an isolated record from near Mule Creek, Grant County. Elevational range is approximately 1200–2400 m.

Description: The morphology of this species has been described in detail by Wright (1966, 1968). This is a moderately-sized, all-female *Cnemidophorus*, with adults 73–80 mm SVL and reaching a maximum SVL of 85 mm (Lowe, 1955b; Gehlbach, 1965). The dorsal scales are granular and there are 63–85 scales around midbody. There are 27–39 femoral pores. The paravertebral light stripes are separated anteriorly by 3–11 scales. Both the mesoptychial and postantebrachial scales are abruptly enlarged. The supraocular semicircles do not extend beyond the midpoint of the third supraocular scale. There are usually four or more enlarged preanal scales. An interlabial scale series is usually lacking; if present, it usually consists of four or fewer scales on each side which are never arranged in double rows.

Cnemidophorus velox is a striped whiptail, occasionally with lateral and/or dorsolateral "fade-out" spots in the dark dorsal fields (Wright and Lowe, 1965). There is little ontogenetic color pattern change. The dorsal ground color is dark brown to black. There are six distinct dorsal longitudinal light stripes extending from the head to the tail; southern populations possess a seventh, complete, less distinct and somewhat undulating middorsal stripe. The distal portion of the tail is bright blue in hatchlings, fading to light blue in adults. The ventral surface is milky-white, with an occasional faint trace of a bluish tint on the chin.

Similar Species: *Cnemidophorus inornatus* is smaller and more gracile, exhibits considerable blue pigmentation throughout the ventral surface of the body and the tail at all ages, has six or more interlabial scales on each side often arranged in double rows, and lacks enlarged mesoptychial and postantebrachial scales. *Cnemidophorus uniparens* has six complete longitudinal light stripes in areas of sympatry and lacks any spotting in the dorsal dark fields. *Cnemidophorus sexlineatus* is smaller and more gracile, has a suffusion of bright green pigmentation throughout the dorsal and lateral anterior surfaces of the body, often has a partially or completely divided middorsal stripe, and lacks enlarged postantebrachial scales.

Systematics: This is an allotriploid, parthenogenetic species of hybrid origin belonging to the *C. sexlineatus* species group (Wright, 1966, 1968, 1993; Wright and Lowe, 1968). Recent mtDNA analyses (Moritz et al., 1989) have identified *C. inornatus* as the paternal parent and either *C. burti stictogrammus* or *C. costatus barrancorum* as the maternal parent involved in the production of a diploid (and presumably parthenogenetic) intermediate lineage, followed by backcrossing with male *C. inornatus* to form *C. velox*.

Cnemidophorus velox from Ojo Caliente, Taos County, and Española, Rio Arriba County, appear to be genetically identical, but differ substantially in major histocompatibility genes from lizards in Colorado and Utah, as do those populations from each other (Cuellar, 1977). Cuellar and Wright (1992) suggested that some northern New Mexico populations may consist of diploid individuals, raising the possibility that *C. velox* is paraphyletic.

Habitat: *Cnemidophorus velox* occurs in oak-mountain mahogany shrublands at lower elevations, and also higher into piñon-juniper woodlands with considerable shrubby undergrowth of oaks, sagebrush, sumac (*Rhus*), rabbitbrush (*Chrysothamnus*), and similar vegetation (Wright, 1966). Harris (1963) recorded the species in open sagebrush and piñon-juniper, often along riparian corridors or on gently sloping bajadas in the San Juan Basin. In the Zuni Mountains, it is most abundant in open, rocky, savannah-like piñon-juniper ecotypes between 2061–2424 m (Gehlbach, 1965). Elsewhere it occurs at elevations as low as 1370 m (Stebbins, 1985). It is sympatric with both *C. inornatus* and *C. uniparens* over a wide area north and west of Socorro, east of the Magdalena Mountains and west of the Ladrone Mountains, north to the Cibola/Valencia county line (Wright, 1968; our unpubl. data).

Behavior: Bowker et al. (1986) recorded a mean preferred body temperature of 37.2°C for lizards from Los Alamos County. These lizards achieve precise thermoregulation by actively shuttling between thermal microenvironments; lizards spent 84% of their time in motion. New Mexico specimens have been collected between 10 April (Socorro County) and 17 November (San Juan County).

Cnemidophorus velox sometimes behave as males in captivity, engaging in pseudocopulatory behavior (Crews and Fitzgerald, 1980); see the *C. uniparens* account for further discussion of this phenomenon.

Reproduction: Little data are available on reproduction in *C. velox*. Gravid females have been reported from the end of May to the end of June, with clutch sizes of 3–5 eggs (Gehlbach, 1965; Douglas, 1966). Hatchlings, 30–40 mm SVL, appear during the last half of August.

Food Habits: No empirical data are available on the food habits of *C. velox*. Like other whiptails, this species is probably an opportunistic insectivore.

Remarks: McAllister (1992) studied the parasitic helminths of *C. velox*. This species is in need of a detailed life history study in New Mexico.

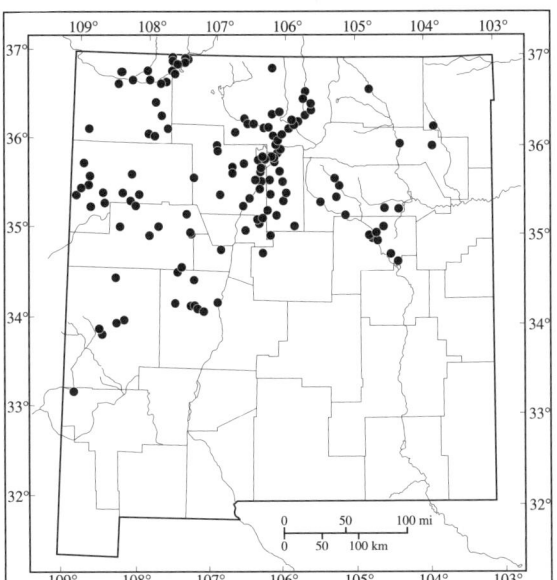

Distribution of Cnemidophorus velox *in New Mexico.*

Skinks — FAMILY SCINCIDAE

This is a large family of more than 100 genera and 1000 species (Zug, 1993). It is found throughout all the major continents except Antarctica and on many oceanic islands, with limited distributions north of the Mediterranean Sea in Europe and in northern Asia. Skinks range in size from 30–350 mm SVL (Zug, 1993), and most have cylindrical bodies and tails with smooth to moderately keeled, shiny, cycloid scales. Legs are moderately well-developed to vestigial in this group. Skinks are semiaquatic to arboreal and occupy a variety of habitats from near sea level to treeline on mountain peaks. Skinks are generally active during the day but are secretive in habits and seldom seen. Reproductive modes vary from egg-laying to live-bearing, and clutch/litter size is generally less than ten. One genus and three species occur in New Mexico.

EUMECES MULTIVIRGATUS (Hallowell, 1857 [1858])   *Many-lined Skink*

See Plate 73A, 73B, 73C, 73D.

Type: The type specimen, ANSP 9371, was collected from "Posa Creek, 460 miles west of Fort Riley, [Geary County], Kansas" (amended to "Cow Creek, Larimer County, Colorado" by Taylor, 1935) by Dr. William A. Hammond, United States Army, date of collection unknown.

Distribution: This species occurs southward from the extreme southern edge of South Dakota, west of the Missouri River, through the southeastern corner of Wyoming, much of northwestern Nebraska, through central Colorado east of the Rocky Mountains (Hammerson, 1982), and through much of New Mexico to the Guadalupe Mountains in Culberson and Hudspeth counties, Texas. There are scattered records eastward on the High Plains into Texas as far as the Caprock Escarpment east of Lubbock, and there are several other isolated records farther south in western Texas (Dixon, 1987). It ranges westward on the Colorado Plateau throughout much of New Mexico to the Mogollon Rim in north-central Arizona and as far as the Colorado River in southern Colorado and southeastern Utah. The existence of an apparently isolated population near Galeana, Chihuahua, Mexico, was questioned by Mecham (1980).

The distribution of individual records from New Mexico accurately reflects our current knowledge about the habitats occupied by this species, although it is undoubtedly more common and widely distributed than the records indicate. The species appears to be absent from the Chihuahuan Desert except along the Pecos River and in the nearby Guadalupe Mountains where it is quite common. It occurs on the Llano Estacado east of the Pecos River, and extends northward along the Pecos and the broken chain of mountains formed by the Guadalupe, Sacramento, and Sandia mountains to the high country of north-central New Mexico. It may be absent from smaller intervening ranges such as the Gallinas Mountains for vicariant reasons, although records from the Carrizozo malpais and the Magdalena Mountains in Socorro County would suggest otherwise. It occurs in isolated ranges such as the Zuni Mountains in the west, and is to be expected throughout the Mogollon-Datil highlands.

Description: This is a medium-sized *Eumeces*, with a maximum SVL of 73 mm (Mecham, 1980). Females are larger; Gehlbach (1965) found males and females from northwestern New Mexico averaging 55.6 mm (42–68) and 65.5 mm (49–70) SVL. There are 23–26 scale rows around midbody, and the dorsal rows are aligned parallel to those laterally. Postnasal scales are usually present on both sides of the head, and there are usually 7 upper labials. The unregenerated tail is usually 1.5–2.0 times the length of the body.

There are two basic color patterns, a striped and a stripeless phase. The latter morph appears to be more prevalent in drier habitats as well as in the southern portion of the range in New Mexico (Mecham, 1957; Gehlbach, 1965), but both forms occur in the same populations. The dorsal ground color of the stripeless (= "*E. taylori*") phase is a uniform light tan to olive brown, and there is usually an indistinct dorsolateral dark streak on each side of the head and body. Sometimes the dorsal scales are darker where they overlap with adjacent scales, giving individuals the appearance of indistinct, fine spotting throughout. Occasionally, small specimens have a lighter, ill-defined streak over the eye. The throat and lips of all pattern morphs are pale, and the ventral surfaces are cream-colored to gray.

Juveniles of the striped morph are dark brown or black dorsally with four light longitudinal stripes. The pair closest to the midline extends onto the parietal region, and sometimes fuses into a single vertebral stripe behind the head. Another light stripe extends dorsolaterally through the eye posteriorly on the third scale row below the midline to the tail. An additional lateral light stripe extends posteriorly from the ear opening above the anterior limb insertions to the groin. A series of dark longitudinal stripes occupies the areas between the light stripes but are obscured by the dark ground color. The tail may be striped or not but, if not, the pattern is less distinct and the stripes are not continuous with those of the body. The tail is bright blue in hatchlings, and this coloration may be retained as a faint wash or disappear altogether in adults.

The ontogenetic changes in color pattern of the striped phase (= *"E. gaigei"*) are lightening of the dark ground color to olive-gray, increased contrast of the dark longitudinal stripes with a reduction in width of the dorsal-most one, gradual to complete loss of the middorsal light stripe, and a rapid lightening of the broad lateral band and its replacement with three narrow dark stripes (Mecham, 1957).

Similar Species: *Eumeces obsoletus* is larger, lacks stripes, is black as a juvenile or light yellowish-brown to tan as an adult, and has the lateral scale rows oblique to the dorsal scale rows. *Eumeces tetragrammus* lacks a light middorsal stripe, has the dorsolateral light stripe confined to the 4th scale row from the midline, often has a light Y-shaped mark on the head with its base located on the neck, has an undivided postmental scale or, if divided, has lateral light stripes passing above the ear opening, and in New Mexico is found only in the Peloncillo Mountains of extreme southwestern Hidalgo County.

Systematics: Lizards of this species were originally described as three separate species *(E. gaigei, E. multivirgatus,* and *E. taylori)*, until Mecham (1957) pointed out the existence of both pattern morphs in a single egg clutch, and a considerable amount of ontogenetic pattern change (see above). Two subspecies are currently recognized, but infraspecific variation in this taxon is badly in need of study (Mecham, 1980). The form currently recognized in New Mexico is *E. m. epipleurotus*. It was originally described as a distinct species by Cope (1880) based on three specimens from the "northern Boundary of Texas and from Nebraska, at Fort Kearney." Cope (1898 [1900]) subsequently designated an adult female specimen (USNM 5263) as the type, collected by John H. Clark from the "northern boundary of Texas." As pointed out by Axtell (1961a), Texas has three northern boundaries, and he restricted the type-locality to the "northern boundary of Trans-Pecos Texas, in the vicinity of the Guadalupe Mountains, Culberson County."

Habitat: This species can be found in mesic situations along streams in mountain canyons or enclosed basins, along river valleys, within shortgrass prairies and in similar submesic oases in otherwise xeric situations such as dense grassland edges of playas, at elevations between 910 and 2575 meters. These lizards also occupy talus slopes or otherwise rocky areas in open pine woodland and forests in the northern and western portions of the state. Gehlbach (1965) recorded them from canyon sides with heavy cover of *Quercus gambeli* and *Cercocarpus montanus*, and with *Ribes* sp., *Rhus trilobata*, and *Yucca baccata* between 2061–2424 m in the Zuni Mountains.

Behavior: The 183 New Mexico specimens have been collected between 15 February (Chaves County) and 20 October (Santa Fe County). A hibernaculum containing individuals of this species along with *Liochlorophis vernalis* and *Thamnophis elegans* was excavated from beneath a large boulder in Santa Fe County (Stuart and Painter, 1993b).

Reproduction: Little is known about the reproductive characteristics of this species. Oviposition takes place from the end of June through the beginning of August (Gehlbach, 1956); between three and five eggs per clutch have been recorded. Like many skinks, this species apparently broods its eggs and young (Mecham, 1957; Gehlbach, 1965); nests can be in shallow depressions excavated under rocks or in similar situations. An adult female and eight hatchlings averaging 26.3 (25.1–27.1) mm SVL were found in a nest under a rock on the west shore of Avalon Lake in Eddy County on 16 June 1993 (M. D. Doles, pers. comm.). Hatchlings are 20–30 mm SVL.

Food Habits: No quantitative data are available. Taylor (1935) recorded lizards feeding on ant larvae, and being preyed on by American kestrels (*Falco sparverius*).

Remarks: The species has been reviewed by Mecham (1980).

FAMILY SCINCIDAE

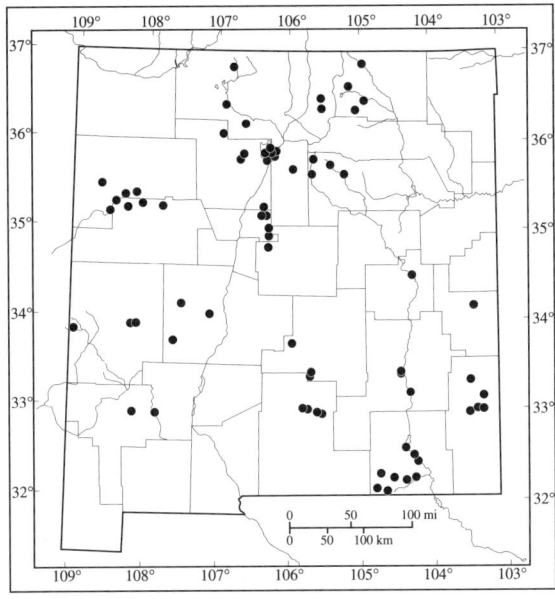

Distribution of Eumeces multivirgatus *in New Mexico.*

EUMECES OBSOLETUS (Baird and Girard, 1852) *Great Plains Skink*

See Plate 74A, 74B.

Type: The original description (Baird and Girard, 1852a) did not specify the type material; the type locality was given as the "valley of the Rio San Pedro [= Devils River] of the Rio Grande del Norte [Val Verde County, Texas]." The holotype was collected by John H. Clark during the expeditions led by Col. J. D. Graham of the Scientific Corps of the U.S. and Mexican Boundary Commission in May or June of 1851. Cope (1898 [1900]), Taylor (1935), and Cochran (1961) listed the holotype as USNM 3133.

Distribution: This species is widespread throughout the southern Central Plains of the United States, from the extreme southwestern corner of Iowa, western Missouri, and southern Nebraska southward throughout Texas west of the Balcones Escarpment, and westward onto the Mogollon Plateau of central Arizona. In Mexico, it occurs from northeastern Sonora southeastward east of the Sierra Madre Occidental to central Durango, Nuevo Leon, and northern Tamaulipas.

Specimen records for this species are scattered widely throughout New Mexico except for the Mogollon and Colorado plateaus. In New Mexico, this species is often associated with major rivers and their tributaries. We suspect that it also occurs more widely than current records indicate in remnant grasslands in the southern part of the state, in shortgrass prairies along the eastern border, and in the central Basin and Range.

Description: This is a large lizard, with adults between 77–115 mm SVL and a maximum size of 130 and 142 mm SVL for females and males (Hall, 1976). Belfit and Belfit (1985) recorded a mean of 97 mm SVL for 16 adults from Sierra County. During 1987–92, 77 individuals collected at Antelope Pass in Hidalgo County averaged 82.5 (32–125) mm SVL (our unpubl. data). The limbs are short and the body is stout. The unregenerated tail is about 1.5 times the length of the body. The scales are large, smooth, and cycloid. There are 25 to 30 scale rows around midbody. The lateral scale rows are aligned obliquely to, and trend upward posteriorly toward the dorsal scale rows. There are four supraoculars and seven upper labials; the number and condition of the remaining head scales are variable. The postnasal scales are usually present and the postmental scale is usually divided. The parietal scales usually do not enclose the interparietal posteriorly. Enlarged pyramidal scales on the soles of the forefeet are evidently an adaptation for digging (Fitch, 1955).

Newly hatched young are jet black, with a bright blue tail and white spots on the labial scales and occasionally elsewhere on the head. These colors fade as the lizards grow, and generally are lost by spring emergence following first hibernation (Fitch, 1955). This occurs at a body size of 55–60 mm SVL in lizards from southern New Mexico (our obs.). Adults are uniformly yellow to greenish-gray to grayish-brown in color with irregular black flecks scattered on the free edges of scales throughout the dorsal surface, sometimes producing indistinct dorsolateral lines.

Similar Species: Both *E. multivirgatus* and *E. tetragrammus* are smaller, are usually darker as adults and/or manifest striped patterns, and have dorsal scale rows aligned parallel to the lateral rows.

Systematics: No subspecies are recognized. The ontogenetic pattern change in this species is so marked and abrupt, however, that adult and juvenal specimens were originally described as two distinct species in two separate genera (*Plestiodon obsoletum* and *Lamprosaurus guttulatus*) during one year in the same journal (Baird and Girard, 1852a; Hallowell, 1852 [1853]). This distinction remained more or less universal until the monographic treatment of the genus by Taylor (1935), who discussed the history of the problem in some detail.

Habitat: Lizards of this species can be found in a wide variety of habitats in New Mexico: creosotebush desert, desert-grasslands, riparian corridors with permanent or intermittent streams, piñon-juniper woodlands, and pine-oak woodlands. They occur on sandy to gravelly soils, and to elevations of 2300 m. They are always found associated with mesic microclimates (see Distribution), and can be found under rocks, logs, in rodent burrows, and among leaf litter and organic debris along floodplains and elsewhere. The species may occur as localized populations; Ballinger (pers. comm.) found it abundant in mesquite sand hills near Loco Hills, Eddy County, yet very rare in the same habitat near Maljamar, Lea County. Belfit and Belfit (1985) found these lizards in artificial, rocky, mesic areas near Elephant Butte Dam. This is a common lizard in places like the malpais at Grants, the lower slopes of the Sandia Mountains, along irrigation ditches in the valley south of Albuquerque, and around Sitting Bull Falls in the Guadalupe Mountains. We have found this species around the edges of playas in the Jornada del Muerto, and along the mainly dry drainages between the Organ Mountains and the Rio Grande in Doña Ana County. It is also a common lizard in the Kiowa National Grasslands in Union County. New Mexico specimens have been collected between 15 February (Grant County) and 11 December (Roosevelt County). Individual lizards appear quite sedentary (our unpubl. data); Hall (1971) reported home ranges to be 800 m^2 or less.

Behavior: These lizards are secretive and primarily fossorial, often sheltering in rodent burrows. They do not tolerate desiccating conditions very well (Hall and Fitch, 1971; our obs.) and are surface active in shaded microclimates, on cool mornings in desert and semiarid habitats often following rainstorms, and on cool days elsewhere. Lizards are active at a preferred activity range between 31 and 34°C in Kansas (Fitch, 1955). Dawson (1960) recorded the critical thermal maximum temperature as 40°C.

These lizards are aggressive and readily bite when molested, and do not usually lose these traits in captivity, although individuals may readily take hand-held food items. The bite of a large adult can be quite painful.

Livo (1989) reported individuals escaping by entering water.

Reproduction: Females reach sexual maturity in about 3 years and produce a single clutch of eggs annually (Hall, 1971; Fitch, 1985). Females guard their clutches during the entire one- to two-month incubation period (Hall and Fitch, 1971). Individual females may not breed every year following sexual maturity (Fitch, 1955). Clutch size varied from seven to 24 eggs, with an average of 12.3 eggs in 37 Kansas clutches. Larger (and therefore older) females produce larger clutches; Hall and Fitch (1971) reported average clutch sizes of 9.00, 10.59, and 13.25 eggs for first-time, second-time, and older females. Belfit and Belfit (1985) recorded the first appearance of hatchlings in the first week of July near Elephant Butte Reservoir. We have recorded the first appearance of hatchlings (33–43 mm SVL) during the last two weeks of July in Doña Ana County.

Very little is known about the breeding cycle of males. Males collected by Parker (1973b) from west Texas and southern New Mexico in July and August were post-reproductive. Hall (1971) found males to be sexually mature following emergence from their third hibernation at an age of approximately 32 months in Kansas.

Food Habits: Caterpillars and grasshoppers comprised 57% and 42% of the diet of this species in Kansas (Burt, 1928). Lizards locate prey by sight or olfaction (Fitch, 1955). He reported grasshoppers (33%), spiders (24%), and beetles (19%) as important food items by frequency in the diet of Kansas lizards. Little quantitative data exist for other populations of this species.

These lizards are probably preyed upon by several birds of prey, snakes, and small mammals such as skunks and badgers (Hall, 1971). James N. Stuart (pers. comm.) observed *Parabuteo unicinctus* taking these lizards in southeastern New Mexico. Domestic cats may be serious predators in urban areas like Albuquerque and Las Cruces (J. Wirt Atmar, pers. comm.; our obs.).

Remarks: The name *obsoletus* is derived from the Latin term with the same meaning, and refers to the fact that dorsal lines, characteristic of the pattern of most North American skinks, are absent in this species (Hall, 1976).

Snider and Bowler (1992) reported a life span of greater than six years in captivity for a wild-caught adult. Hall and Fitch (1971) reported a natural life span of at least eight years in Kansas, a figure supported by our very limited data from southern New Mexico.

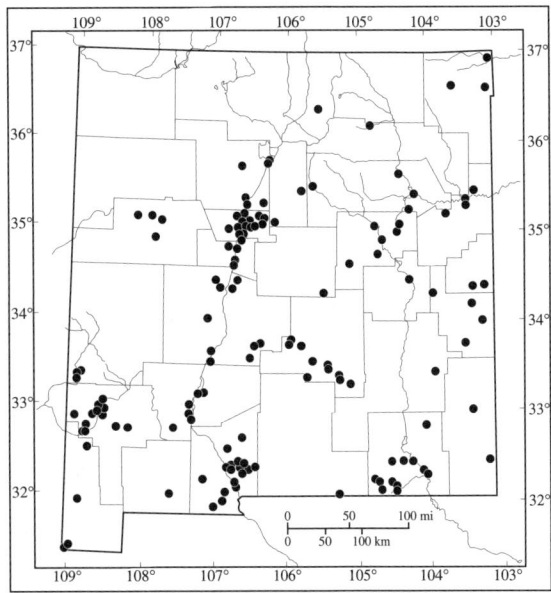

Distribution of Eumeces obsoletus *in New Mexico.*

EUMECES TETRAGRAMMUS (Baird, 1858 [1859])   *Four-lined Skink*

See Plate 75.

Type: The type specimen is USNM 165662, an adult of unknown sex, collected from "Lower Rio Grande" (restricted to Matamoros, Tamaulipas, Mexico, by Taylor, 1935) by either John Louis Berlandier or Darius Nash Couch, date of collection unknown (see Remarks).

Distribution: The range of *E. tetragrammus* is disjunct. The eastern part of its range occupies north-central Texas westward through the southern Trans-Pecos, and southward through the Edwards Plateau and southern Texas along the northern and eastern margins of the Sierra Madre Oriental through Tamaulipas to northern Veracruz and eastern San Luis Potosí. In the west, it occurs from southeastern Arizona (Baboquivari, Huachuca, Pajarito, and Santa Rita mountains) southward along the Pacific versant of the Sierra Madre Occidental to coastal Sinaloa, Nayarit, and northern Jalisco. There are disjunct populations in the Sierra del Nido of central Chihuahua and the Cuatro Ciénegas basin of central Coahuila (Lieb, 1990). In New Mexico, this species is only known from Guadalupe Canyon in the Peloncillo Mountains of extreme southwestern Hidalgo County. There is also a single specimen (UTA R–17891) from Geronimo Trail in Clanton Canyon, Peloncillo Mountains, near the New Mexico/Arizona border in Hidalgo County. In New Mexico, *E. tetragrammus* has been collected from

1327–1387 m in Guadalupe Canyon and has been observed as high as approximately 1950 m along Geronimo Trail (B. R. Tomberlin, pers. comm.).

Description: This is a medium-sized species of *Eumeces*, with a maximum SVL of 76 mm (Lieb, 1990). During 1992–95, 87 individuals collected in Guadalupe Canyon averaged 52.9 (21–69) mm SVL and 3.1 (0.2–6.3) gm (our unpubl. data). The scales are smooth and cycloid, and there are 26 or 28 parallel rows around midbody. There are 4 supraoculars and 7 upper labials; the number and condition of the other head scales are variable. The postnasals are usually absent and the postmental is usually undivided. The legs are small, and the unregenerated tail is approximately twice as long as the body.

The ground color is olive gray, and usually the center of each scale is slightly darker than the periphery. Juveniles are dark brown overall and have bright blue tails. Adult males have bright red lips. A distinct dark lateral band, 2–3 scales wide, extends from the neck to the groin. It is bordered above on each side by a narrow light line that extends posteriorly from above the eye to about midway between the limb insertions on the 4th scale row from the midline. A pale lateral light line extends posteriorly from the nostril through the ear opening to the shoulder on each side. A light median line is usually present, extending anteriorly from the shoulders onto the head and bifurcating above the eyes before continuing to the nose. In large adults the lined pattern may fade considerably or be absent altogether.

Similar Species: *Eumeces obsoletus* is larger, lacks distinct stripes, is black as a juvenile or light yellowish-brown to tan as an adult, and has lateral scale rows that are oblique to the dorsal scale rows. *Eumeces multivirgatus* either lacks stripes altogether or has a light middorsal band and four light longitudinal stripes, with the dorsolateral pair confined to the 3rd scale row from the midline, lacks a Y-shaped mark on the head, and has a divided postmental scale.

Systematics: Three subspecies are recognized (Lieb, 1990); *E. t. callicephalus* Bocourt, 1879 occurs in New Mexico. The nomenclatural history of this taxon is complex, and we refer interested readers to Lieb (1985, 1990) for a full consideration. We follow Lieb (op. cit.) in our taxonomy, but other workers consider *E. callicephalus* a distinct species (e.g. Tanner, 1987; Collins, 1990, and see Frost and Hillis, 1990, and Frost et al., 1992). Further investigation may validate this viewpoint.

Habitat: This species occupies a wide variety of habitats, from semiarid grasslands to dense thornscrub to pine-oak woodlands, at elevations between sea level and 2300 m. Individuals are usually found in sheltered, mesic situations such as leaf litter, detritus, and under rocks and logs. The riparian habitat in New Mexico where specimens have been collected is characterized by loose rocky soils with numerous tree species including Arizona sycamore, Arizona walnut, mesquite, and various oaks.

Behavior: No detailed information on the behavior of *E. tetragrammus* exists. These lizards are active during the day, but generally avoid open habitats and desiccating conditions. They are secretive, spend most of their time underneath objects, and seldom venture far from shelter. Most of the specimens observed in New Mexico have been taken from pitfall traps set amongst leaf litter in the riparian zone along Guadalupe Canyon. Specimens have been collected in New Mexico from April through October, with 56% (50 of 89) encountered during April and May (our unpubl. data).

Reproduction: Little is known regarding reproduction in *E. tetragrammus*. We have found hatchlings as early as 23 July in Guadalupe Canyon. Tanner (1987) found hatchlings in northern Chihuahua still being brooded by females in nests under rocks in mid-July. Campbell and Simmons (1961) reported similar brooding behavior of a single female around a clutch of three eggs in Jalisco, Mexico. Two additional clutches of six eggs each and brooding behavior were recorded by Hardy and McDiarmid (1969). Zweifel (1962a) recorded a clutch of six eggs for one individual female, and indicated that considerable embryonic development may occur prior to egg deposition. Taylor (1985) reported an individual from southern Arizona giving birth to live young. Hatchlings are 20–25 mm SVL.

Food Habits: No quantitative data are available. Taylor (1935) recorded small beetles as predominant stomach contents in three specimens from Arizona, along with dipterans and blattids. Insects and spiders have been

FAMILY SCINCIDAE

reported as dietary items (Stebbins, 1954, 1985; Behler and King, 1979).

Remarks: Although Cochran (1961) listed Couch as the collector of the lectotype, it is unclear whether he or Berlandier actually collected the specimen (Taylor, 1935; Lieb, 1990). The specimen was collected during February or March, 1853, if the collector was Couch; if not, then it was collected sometime during the preceding 20 years when Berlandier was a resident near the type locality (Conant, 1968).

The name *callicephalus* is a Greek word meaning "beautiful head," and probably refers to the distinctive lyre-shaped light line on the top of the head.

Distribution of Eumeces tetragrammus *in New Mexico.*

Alligator Lizards **FAMILY ANGUIDAE**

This is a small family of about 12 genera and about 70 species (Zug, 1993). Members can be found throughout the Americas, the Caribbean, Europe, the Middle East, and in China. They range in size between 55–520 mm SVL and all are covered with heavy scales reinforced with bony plates, or osteoderms. A ventrolateral fold running the length of the body characteristically separates the dorsal from the ventral body scales, allowing the body to expand during breathing activities, when distended with prey, or when females become gravid. Many species have small legs and are incapable of running fast and some are limbless. Some species are arboreal, but most are terrestrial and some partly fossorial. Some species lay eggs and some give birth to living young. Oviparous species often brood their eggs. One genus with one species occurs in New Mexico.

ELGARIA KINGII Gray, 1838 *Madrean Alligator Lizard*

See Plate 76.

Type: An adult of unknown sex (BMNH 1946.8.29.46, originally V.25a), presented by T. Bell, date of collection unknown. The original type locality of "Mexico" was subsequently twice restricted, first to Chihuahua and again to Arizona (Webb, 1970a). The collector and date of collection are unknown.

Distribution: This species ranges from the central Arizona highlands near Prescott and the San Mateo Mountains in New Mexico southward in a narrow band along the axis of the Sierra Madre Occidental to the Mexican state of Jalisco; the southern boundary of the range is imprecise (Webb, 1970a). In New Mexico, *E. kingii* occurs throughout the Mogollon Plateau, and in several isolated mountain ranges to the south between the Arizona border and the Rio Grande. The species has recently been discovered in relatively xeric, low-lying mountains and foothills in Doña Ana and Hidalgo counties (Knight and Duerre, 1987; our obs.).

Description: This is a large, relatively slender lizard with a maximum SVL of 133 mm (Bowker, 1994). The legs are short, only about 25% of the length of the body. Males have stouter bodies than females and broader heads, especially in the temporal region. There is a well-defined lateral fold, containing granular scales, along both sides of the body extending backward from the ear. Both dorsal and ventral scales are relatively large and the posterior edges overlap slightly with the adjacent scales. Most of the dorsal body scales are smooth; scales in the 6–8th middorsal rows are weakly keeled. There are 48–60 transverse rows from the interparietal scale to the base of the tail, and 14–16 (rarely 18) longitudinal dorsal scale rows at midbody. The ventral scales are larger than the dorsals and more platelike. The unregenerated tail is more than twice the length of the body, and there are 121–150 whorls of scales from base to tip.

Juveniles and adult males are boldly patterned dorsally with 8–15 dark brown transverse bands about 3 scales wide, edged posteriorly with black, on a pale grayish-brown ground color. These bands extend onto the tail. The dorsal surfaces of the limbs are unpatterned except for scattered dark flecks. There are 3–5 white spots along the upper lip. The ventral surface is yellowish-white, often clouded with gray, with numerous indistinct black spots which tend to be arranged in longitudinal rows. The dorsal pattern in adult females is less distinct; the crossbands are narrower, pale and lack a distinct black posterior border (Knight and Duerre, 1987).

Similar Species: The only lizards in New Mexico likely to be confused with *E. kingii* are skinks belonging to the genus *Eumeces*. Skinks have smooth, shiny, cycloid scales, and lack the lateral body fold.

Systematics: Three subspecies have been recognized (Webb, 1970a); only *E. k. nobilis* occurs in New Mexico. It was originally described by Baird and Girard (1852a) as a distinct species, from "Fort Webster, Copper mines of the Gila (Santa Rita del Cobre) [Grant County], New Mexico" (see Webb, 1970a, for a discussion of changes in the Fort Webster location). No type material was specified. Two syntypes (USNM 3076; Cochran, 1961), an adult male and female, were collected by John H. Clark during the expeditions led by Col. J. D. Graham of the Scientific Corps of the U.S.-Mexican Boundary Commission in 1851.

Habitat: This species is often characterized as an inhabitant of pine-oak woodlands and riparian habitats of canyon floors in mountain ranges at moderate elevation and up to 2675 m (Medica, 1965; Webb, 1970a; McCranie and Wilson, 1987). Lizards have also been found in more xeric habitats including juniper-grassland, desert shrubland, and creosotebush flats at elevations below 1200 m (Lowe, 1964; Knight and Duerre, 1987; our obs.). Sheltered microhabitats such as dense leaf litter, rodent burrows, and packrat middens were preferred in these latter situations. It is common on talus slopes in the Animas Mountains in Hidalgo County, where it reaches the highest known elevations within its range in New Mexico. Specimens have been collected from San Simon Cienega at 1180 m to about 2500 m near Animas Peak.

Behavior: This is a diurnal and largely terrestrial species. These lizards are deliberate and methodical in their movements. They are often heard rustling through leaf litter or other organic detritus before they are seen, as their movements and color pattern render them quite cryptic in their natural environment. The 79 New Mexico specimens available to us were collected between 9 April (Hidalgo County) and 17 November (Grant County).

Bowker (1987) reported that an individual seized its own tail in its mouth upon being seized by a *Masticophis taeniatus*, and maintained this "loop" until released by the snake. This posture apparently makes it impossible for the lizard to be swallowed.

Reproduction: Little quantitative data are available. *Elgaria kingii* apparently breeds in the fall (Goldberg, 1975). Male testes are regressed from May through July, and reach maximum size in September and October. Sperm formation begins in late August, and females evidently store sperm and embryonic development is arrested during the winter. Females contain enlarged follicles during October and March (Goldberg, 1975), and possess oviductal eggs in April and May. Cunningham (1958) listed a clutch of 12 eggs laid in June, and another of 9 eggs laid in July. Stebbins (1958) listed a clutch of 15 eggs laid in June. We collected a single female with 12 eggs in the process of hatching from a horizontal crevice in August in the Animas Mountains, suggesting the possibility of parental care in this species. Snider and Bowler (1992) reported a maximum lifespan in captivity of more than 7 years. Fred Gehlbach (pers. comm.) reported a maximum lifespan in captivity of more than 15 years.

Food Habits: No quantitative data are available for this species. Stebbins (1954) mentioned praying mantids, grasshoppers, caterpillars, moths, and scorpions as food.

Remarks: The scientific name for this species in much of the older literature is *Gerrhonotus kingi*. The species has been reviewed by Webb (1970a).

Distribution of Elgaria kingii *in New Mexico.*

FAMILY HELODERMATIDAE *Venomous Lizards*

The family of beaded lizards, Helodermatidae, contains only two living species: the Mexican beaded lizard, *Heloderma horridum,* which occurs along the Pacific coast of Mexico and Guatemala, and the Gila monster, *H. suspectum,* found from the southwestern United States to northwestern Mexico (Bauer, 1992). Both are recognized by their small bead-like scales that do not overlap, stocky bodies, and broad heads with black snouts. They are large, slow-moving lizards with black or yellowish to reddish background coloration, and varying amounts of yellow or pinkish spotting or reticulations. These are the only venomous lizards known. The bite of the Gila monster results in localized swelling and severe pain but is not known to be fatal to humans. Gila monsters in New Mexico are most active during April and May, with egg laying occurring in July and August. The reticulated Gila monster occurs in southwestern New Mexico.

HELODERMA SUSPECTUM Cope, 1869 *Gila Monster*

See Plate 77.

Type: During the U.S. and Mexican boundary survey of 1855, three specimens of *Heloderma suspectum* were collected by Arthur Scott at Sierra de la Unión in Sonora, Mexico. These specimens were deposited at the Smithsonian Institution, and Baird (1859) misidentified them as *Heloderma horridum* following an earlier description by Wiegmann (1829). He illustrated one of these specimens for the Boundary Survey that actually became the first illustration of *Heloderma suspectum*. Cope (1869) noticed differences between these specimens and *Heloderma horridum*, and designated these three specimens, USNM 2971, as syntypes. In 1869, Cope formally described *Heloderma suspectum* in a publication of the Academy of Natural Sciences of Philadelphia. Bogert and Martín del Campo (1956) considered USNM 2971a, the adult female illustrated by Baird (1859) as the lectotype.

Distribution: *Heloderma suspectum* ranges from extreme southwestern Utah to northern Sinaloa, and from southwestern New Mexico west to the Colorado River in San Bernardino County of extreme southeastern California. It is found from sea level to about 1525 m elevation (Stebbins, 1985). In New Mexico, *H. suspectum* is known from Hidalgo, Grant, Luna, and perhaps Doña Ana counties in southwestern New Mexico at elevations of 1180–1950 m. It is commonly encountered in and near the Redrock Wildlife Area in Grant County and at Granite Gap in Hidalgo County. Records of occurrence at Kilbourne Hole in Doña Ana County (NMSU 2857) and most areas east of a line drawn from Silver City southward to Animas may represent displaced, released, or escaped captive individuals. To understand the distribution of *H. suspectum* in New Mexico, additional areas of suitable habitat east of the known range need to be investigated.

Although Cope (1898 [1900]), Bailey (1913), and Van Denburgh (1924) believed *H. suspectum* occurred in New Mexico, its presence was not confirmed until Shaw (1950) and Koster (1951) independently documented specimens from the state.

Description: *Heloderma suspectum* is the largest lizard native to the United States and the only venomous one. It is characterized by a large body with a robust tail that comprises approximately 30% of the body length. Adults may reach approximately 570 mm TL although most are under 500 mm. Nineteen adult and subadult specimens from Redrock averaged 447.7 (325–503) mm TL; 15 adults averaged 623.6 (418–1012) g. The large flattened head terminates in a broad black snout. The jaws are powerful and the recurved, slightly lance-shaped teeth have sharp cutting edges that are grooved for venom delivery. Unlike those of snakes, the venom glands are in the lower jaw rather than the upper jaw. The dorsal surface is covered by rounded tuberculate bead-like scales; the belly scales are squarish and plate-like. Overall coloration is a pattern of black or brown blotches that overlay a background color of orange or orangish-red. There are 4–5 uniquely-shaped alternating black bands around the tail. The legs are short and stout, and the third toe of the hind foot is nearly equal in length to the fourth toe.

The juvenile is banded, not blotched, and the tail is distinctly ringed. In the New Mexico form, this juvenal pattern gradually changes to the blotched adult markings as the individual matures. Bogert and Martín del Campo (1956) illustrated this ontogenetic color change.

Similar Species: In New Mexico, *H. suspectum* should not be confused with any other lizard. It is easily recognized by its large size, distinctive coloration, and characteristic bead-like scales.

Systematics: Two subspecies are recognized. Only the nominotypical form occurs in New Mexico.

Dr. Charles H. Lowe, Jr., and his students at the University of Arizona have extensively studied *H. suspectum* in Arizona. In 1992, a 5-year study of *H. suspectum* was initiated at the Redrock Wildlife Area in Grant County by Daniel D. Beck and Charles W. Painter. The following information on *H. suspectum* in New Mexico, unless otherwise credited, is taken from preliminary results of that study.

Habitat: In New Mexico, *Heloderma suspectum* occurs in desert scrub and, more rarely, woodland and grassland habitats most commonly associated with rocky regions of mountain foothills and canyons. Dominant vegetation often includes creosotebush, mesquite, acacia, ocotillo, and snakeweed. *Heloderma suspectum* is also encountered along the lower fringes of piñon-juniper and oak woodlands. They have been rarely observed in agricultural areas near Silver City and Cotton City. Near Redrock, *H. suspectum* uses shelters in rock crevices, burrows normally excavated by other animals, and packrat mounds. They generally prefer southeast facing slopes during the spring, and southwest facing slopes during the fall and winter.

Behavior: The seasonal activity period extends from March to November, although *H. suspectum* can be encountered basking at shelter entrances during the winter and early spring. There is a peak in activity during April or May when *H. suspectum* forages for vertebrate eggs and nestlings, and searches for mates. Another activity peak occurs in July and August with the onset of the summer rains. Throughout their range, *H. suspectum* is primarily diurnal (Lardner, 1969; Porzer, 1981; Lowe et al., 1986; Beck, 1990). In New Mexico activity can occur at any time depending upon ambient temperature and other factors, although at Redrock there is a slight peak in activity in the morning between 0730 and 1200 hrs., with an apparent smaller activity peak in the afternoon between 1600 and 2000 hrs.

Heloderma suspectum spends over 96% of their time in sub-surface refugia. They return year after year to the same foraging and overwintering areas, sometimes reusing the same shelters. Estimates of approximately five *H. suspectum* per square km have been recorded in a study area near Redrock. Home range sizes vary from 10 to over 50 hectares, but are usually less than one km across. There is considerable overlap in home ranges of individual lizards; some shelters are reused by two or more *H. suspectum*. Above ground travels range from a few meters around shelter entrances, to forays over 1.5 km. Cross and Rand (1979) discussed the climbing behavior of free-ranging *H. suspectum* and reported one individual that climbed to a height of 3 m in a desert willow.

During activity, *H. suspectum* prefers a body temperature around 30°C (22–35°C), which is fairly low for a diurnal desert lizard (Porzer, 1981; Lowe et al., 1986; Beck, 1990). At Redrock, body temperatures of winter-dormant animals drop below 10°C.

Heloderma suspectum pairs up in April, May, and June, with males and females often sharing the same shelter. During this time males perform spectacular fighting behaviors that bear a striking resemblance to varanid lizard combat and to the entwining postures exhibited by viperid snakes. The objective of these male-male agonistic behaviors is apparently to maintain a superior position by forcing the subordinate lizard to the ground. These tests of dominance may play an important role in determining which males gain access to females (Lowe et al., 1986; Beck, 1990).

Reproduction: Little is known of the reproductive biology of free-ranging *H. suspectum*, including nest sites, nest conditions, and exact incubation periods. Two individual females studied at Redrock laid eggs during late July or early August. In Arizona, *H. suspectum* lays from 2–12 eggs, averaging five. Egg laying coincides with the early portion of the southwestern summer monsoon season and usually occurs during mid-July to mid-August (Lowe et al., 1986). Recently hatched *H. suspectum* begins to emerge in April and May, suggesting that the eggs may over-winter after an incubation period of 8–10 months. The eggs average 59.9 mm in length by 30.6 mm in diameter and weigh an average of 35.6 gm. Newborn *H. suspectum* average 165.3 mm total length and 32.7 gm. Rundquist (1992) described the reproduction of captive *H. suspectum*.

Food Habits: *Heloderma suspectum* employs a "search-and-dig" foraging strategy (Lowe et al., 1986) to encounter a food resource upon which few other reptiles specialize. They feed on the eggs of ground-nesting birds and other reptiles, and on juvenal birds and mammals taken from the nest. During April–May, and in August–September hairs of juvenal cottontail rabbits are frequently found in fecal samples from *H. suspectum* near Redrock. Eggs from ground nesting birds and reptiles also form an important part of their diet in New Mexico. Preliminary evidence suggests that Gila monsters, especially adult males, seem to feed most heavily when they first come out of winter dormancy, eat less in May and June, then feed more often again later in the summer and early fall. In Arizona and Utah, *H. suspectum* feeds on eggs of Gambel's quail, mourning doves, and desert tortoises, juvenal mammals (including cottontail rabbits, ground squirrels, and rock squirrels), and carrion (Arnberger, 1948; Hensley, 1949; Stahnke, 1950, 1952; Kauffeld, 1949, 1954; Tinkham, 1971; Jones, 1983; reviews in Lowe et al., 1986; Beck, 1990). Young *H. suspectum* can consume quantities equivalent to over 50% of their body weight in a single feeding; adults can consume about 35%. With their very low metabolic rates (Beck and Lowe, 1994), healthy adult *H. suspectum* can store enough fat from only 3–4 large meals to sustain themselves for an entire year.

Remarks: No other species of our herpetofauna has been the source of more superstitions, the subject of as many myths, or the object of more exaggerated claims than *H. suspectum*. Brown and Carmony (1991) have written a delightful and informative book on Gila monsters that presents many historical accounts and anecdotes of this unique and interesting lizard. Russell and Bogert (1980) provided a review of the venom and bite. They discussed clinical symptoms of the bite and presented the case history of several bites. *Heloderma suspectum* is listed as endangered by the New Mexico Department of Game and Fish (NMGF, 1990).

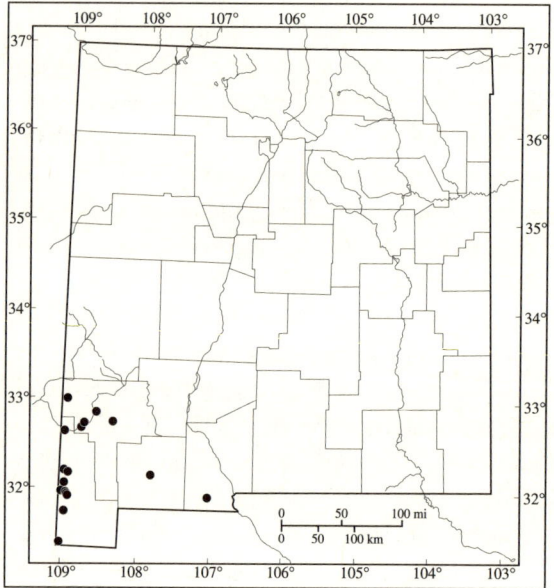

Distribution of Heloderma suspectum *in New Mexico.*

A KEY TO THE SNAKES OF NEW MEXICO

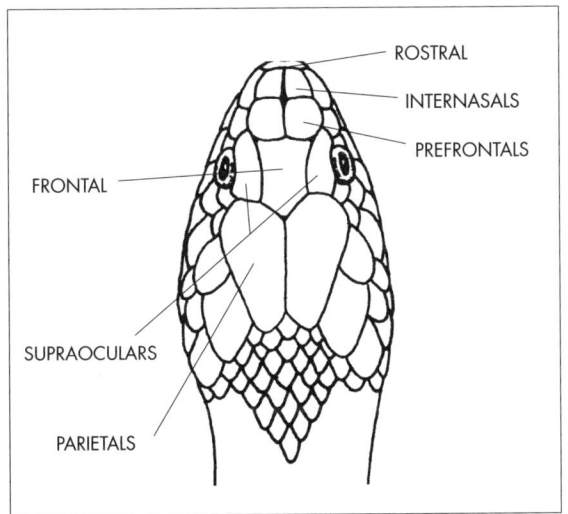

Dorsal view of typical snake head showing scales.

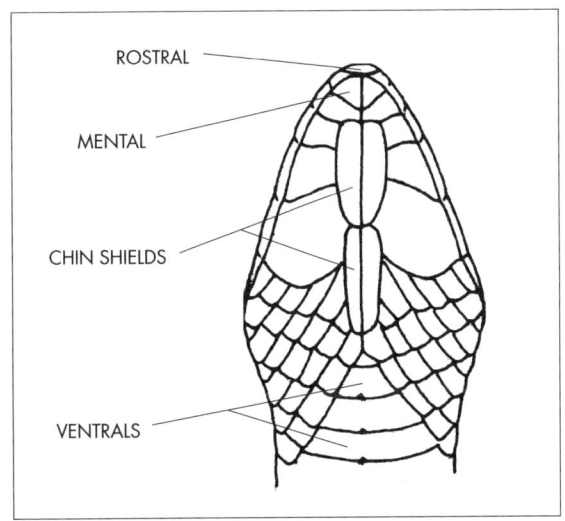

Ventral view of typical snake head showing scales.

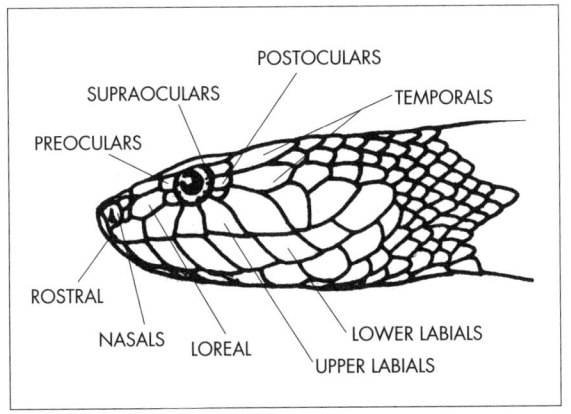

Lateral view of typical snake head showing scales.

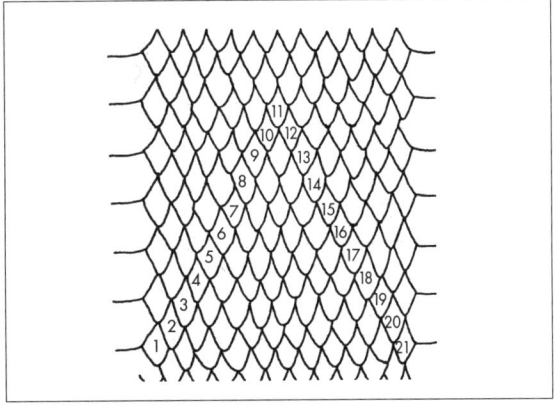

Method of counting dorsal scale rows.

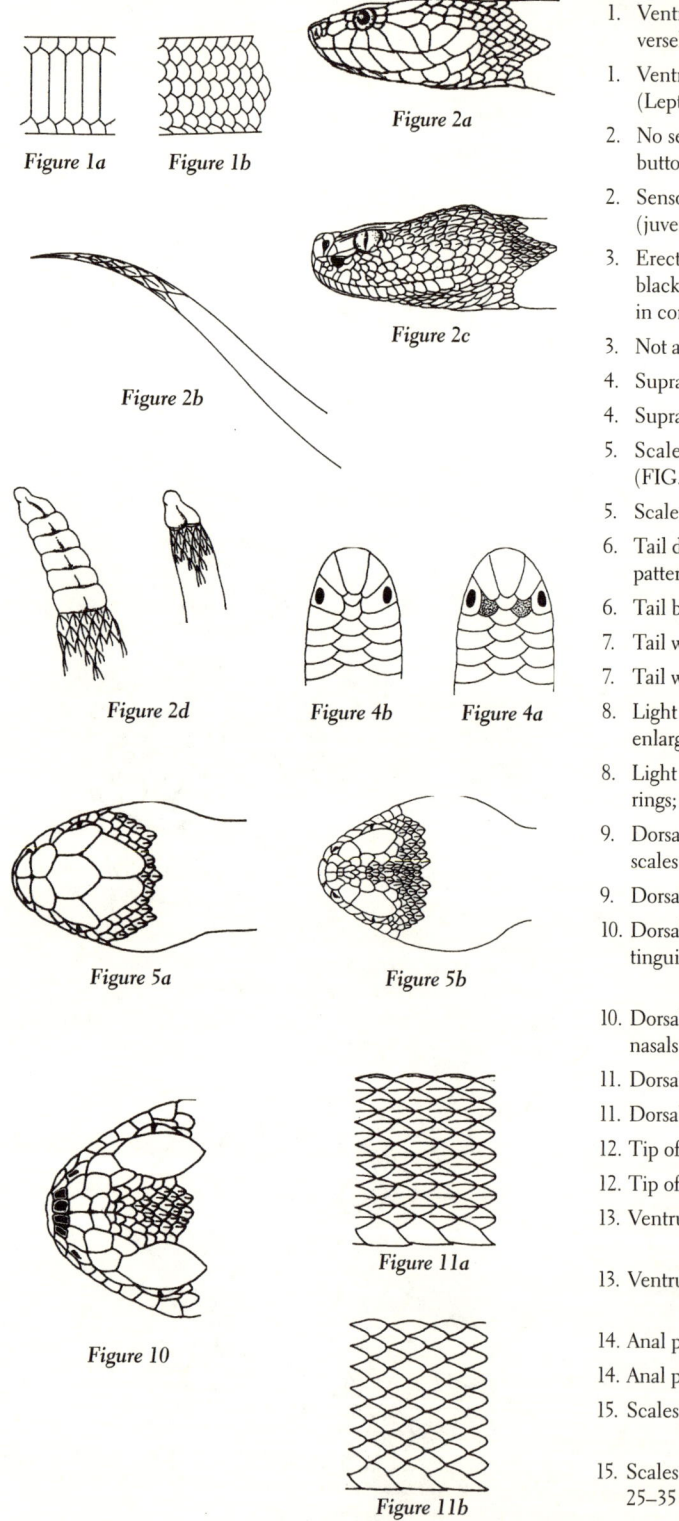

Figure 1a
Figure 1b
Figure 2a
Figure 2c
Figure 2b
Figure 2d
Figure 4b
Figure 4a
Figure 5a
Figure 5b
Figure 10
Figure 11a
Figure 11b

1. Ventral scales longer than dorsal scales and elongated transversely (FIG. 1a) 2
1. Ventral scales like dorsals, not transversely elongated (FIG. 1b) (Leptotyphlopidae) 4
2. No sensory pit between eye and nostril (FIG. 2a); no rattles or button on tail (FIG. 2b) 3
2. Sensory pit between eye and nostril (FIG. 2c); rattles or button (juveniles) on tail (FIG. 2d) (Viperidae) 5
3. Erect grooved fangs on anterior part of the upper jaw; rings red, black, and yellow (or whitish) on body with red and yellow rings in contact (Elapidae) 6
3. Not as above (Colubridae) 11
4. Supraoculars present (FIG. 4a) .... *Leptotyphlops dulcis*
4. Supraoculars absent (FIG. 4b) .... *Leptotyphlops humilis*
5. Scales on top of head enlarged, platelike and symmetrical (FIG. 5a) .... *Sistrurus catenatus*
5. Scales on top of head small and unsymmetrical (FIG. 5b) 6
6. Tail dark and unicolored, in strong contrast with body (juvenal pattern with suggestion of bands) .... *Crotalus molossus*
6. Tail banded, or not in strong contrast with body 7
7. Tail with distinct light and dark rings 8
7. Tail without distinct light and dark rings 9
8. Light rings on tail distinctly broader than dark rings; 2 (rarely 3) enlarged interocular scales .... *Crotalus scutulatus*
8. Light rings on tail usually about the same width as the dark rings; 3 or more smaller interocular scales .... *Crotalus atrox*
9. Dorsal pattern of 19–23 short transverse bands of white; dorsal scales of snout forming a ridge .... *Crotalus willardi*
9. Dorsal pattern not as above 10
10. Dorsal pattern of 16–18 transverse blackish bands usually distinguishable; 2 internasals in contact with rostral .... *Crotalus lepidus*
10. Dorsal pattern with dark blotches on back; more than 2 internasals in contact with rostral (Fig. 10) .... *Crotalus viridis*
11. Dorsal scales keeled (FIG. 11a) 12
11. Dorsal scales smooth (FIG. 11b) 27
12. Tip of snout (rostral) turned upward or pointed 13
12. Tip of snout not turned upward 14
13. Ventrum, including tail, black; snout strongly upturned .... *Heterodon nasicus*
13. Ventrum not black, with lighter tail; snout not strongly upturned .... *Heterodon platirhinos*
14. Anal plate divided (FIG. 14a) 15
14. Anal plate single (FIG. 14b) 18
15. Scales strongly keeled; scale rows 21–25 (rarely 27) at midbody .... *Nerodia erythrogaster*
15. Scales weakly keeled middorsally to smooth on sides; scale rows 25–35 at midbody 16

KEY TO THE SNAKES OF NEW MEXICO

16. Body color light (yellow to yellow orange or tan) with brown to black "H-like" blotches on dorsum; 31–35 dorsal scales
Bogertophis subocularis 16
16. Body color not as above; with or without blotches 17
17. Dorsal color olive, green, or greenish-gray (young may be blotched); ventrum unmarked whitish or cream
Senticolis triaspis
17. Dorsal color light gray with dark-edged darker blotches; ventrum with squarish black markings (checkerboard-like)
Elaphe guttata
18. Blotched pattern; 27 or more dorsal scale rows; 4 prefrontals
Pituophis melanoleucus
18. Striped or spotted pattern; 21 or fewer dorsal scale rows; two prefrontals 19
19. Less than 8 lower labials; 17 (rarely 19) dorsal scales; 2 rows of black spots on light ventrum (FIG. 19)
Tropidoclonion lineatum
19. 8 or more lower labials; 19–21 dorsal scales; ventrum not as above 20
20. No stripes; dorsum olive or brown with dark spots
Thamnophis rufipunctatus
20. Stripes present with or without spots 21
21. Lateral stripe restricted to 3rd scale row on neck, on rows 2 and 3 farther back; distinct checkered pattern may obscure some stripes
Thamnophis marcianus
21. Lateral stripe involves at least two scale rows on anterior part of body 22
22. Lateral stripe involves 4th scale row, usually rows 3–4 (occasionally 2–4) (FIG. 22a) 23
22. Lateral stripe does not involve 4th scale row, usually on rows 2–3 (occasionally 1–3) (FIG. 22b) 25
23. Paired black blotches behind head; dorsum brown to olive with yellowish stripes; sides checkered with dark spots
Thamnophis eques
23. No paired black blotches immediately behind head 24
24. Upper labials with bold dark bars on edges; dorsum dark with black spots between stripes *Thamnophis radix*
24. Upper labials immaculate; dorsum boldly striped without spots
Thamnophis proximus
25. Upper labials always 7; 19 scale rows at midbody; no dark blotches on neck behind head *Thamnophis sirtalis*
25. Upper labials usually 8 (rarely 7); 19 or 21 scale rows at mid-body (if 19 has dark blotches on neck); dark blotches present or absent (if absent has 21 scale rows) 26
26. 19 scale rows (usually); distinct large black blotches on neck behind head; usually 7 upper labials *Thamnophis cyrtopsis*
26. 21 scale rows; no distinct black blotches behind head (may be some darkening); 8 upper labials *Thamnophis elegans*
27. Anal plate not divided 28
27. Anal plate divided 34

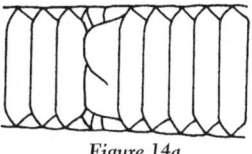
Figure 14a

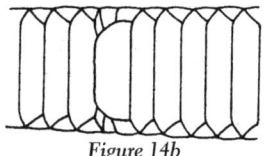
Figure 14b

Figure 19

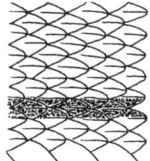
Figure 22a

Figure 29a

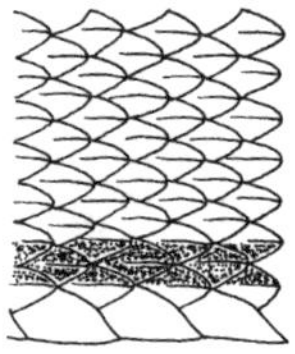

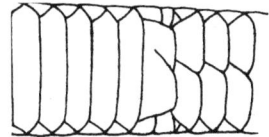
Figure 22b

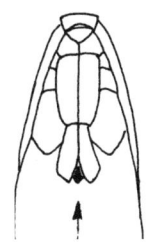
Figure 29b

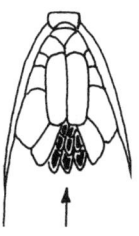
Figure 36a

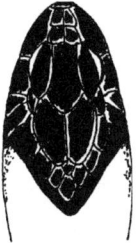
Figure 36b

Figure 40b

Figure 40a

28. Pupils elliptical　　　　　　　　*Trimorphodon biscutatus*
28. Pupils round　　　　　　　　　　29
29. Caudal scales mostly single (FIG. 29a)　*Rhinocheilus lecontei*
29. Caudal scales all divided (FIG. 29b)　　30
30. Ventrum light without markings; 29–31 dorsals; lower labials 12–15 (usually 13 or 14)　　*Arizona elegans*
30. Ventrum with at least some dark markings; 21–25 dorsals; lower labials 7–10 (rarely 11 or 1)　　31
31. Body bands of black, red, and yellow or white　　32
31. Body without bands or without yellow bands　　33
32. Body bands of black, red, and white; black bands may become narrow or disappear on the sides; snout white, yellowish, or black flecked with white; over 210 ventrals
　　Lampropeltis pyromelana
32. Body bands of black, red, and yellow or white; white bands tend to become wider on lowermost scales; snout dark but may be flecked with white; fewer than 210 ventrals
　　Lampropeltis triangulum
33. Banded pattern of white-bordered gray and black bands that may be bisected with red bands of varying widths
　　Lampropeltis alterna
33. Dorsum with white or yellow spots on scales; spotted scales may form a chain-like pattern around solid dark middorsal blotches or tend toward a "salt and pepper" appearance
　　Lampropeltis getula
34. Rostral turned up at tip and pointed　*Gyalopion canum*
34. Rostral not as above　　35
35. Rostral enlarged and leaf-like with free lateral edges; body striped　　36
35. Rostral not as above; striped or unstriped　　37
36. Upper stripe broad and distinct; lower stripe if present is on third scale row at midbody; 8 upper labials; chin shields in contact or separated by a single scale (FIG. 36a)　*Salvadora grahamiae*
36. Distinctly 4-striped; lower stripe is on fourth scale row at midbody; mid-dorsal band darker than ground color on the sides; 9 upper labials; chin shields never in contact and separated by 2–3 small scales (FIG. 36b)　*Salvadora deserticola*
37. Pupil elliptical; dark body blotches on lighter ground color; scale rows usually 21 at midbody　*Hypsiglena torquata*
37. Pupil round; no body blotches (except young *Coluber*); scale rows 17 or less at midbody　　38
38. Head markedly darker than body; 15 scale rows at midbody; no loreal scale　　39
38. Head not markedly darker than body; 15–17 scale rows at midbody; loreal present　　41
39. Distinct white ring present on neck and on posterior upper labials　*Tantilla yaquia*
39. Distinct white ring absent　　40
40. Rear of black cap convex or pointed at rear and extends back 2–5 scale lengths from parietals (FIG. 40a); first lower labials usually meet beneath chin　*Tantilla nigriceps*
40. Rear of black cap straight or slightly convex and extends back 1–3 scale lengths from the parietals (FIG.40b); first lower labials usually fail to meet beneath chin　*Tantilla hobartsmithi*
41. Dark blotches (not saddles or rings) on lighter ground color
　　Coluber constrictor
41. No dark blotches (may have saddles or rings) on lighter ground color　　42
42. Longitudinal stripes present　　43
42. No longitudinal stripes present　　44
43. 15 dorsal scale rows at midbody; lateral stripes continue onto tail; venter of tail orange or reddish　*Masticophis taeniatus*
43. 17 dorsal scale rows at midbody; lateral stripes do not continue onto tail; venter of tail cream colored
　　Masticophis bilineatus
44. Dorsum plain green, greenish, or bluish with unmarked cream to yellow ventrum　　45
44. Dorsum not as above　　46
45. Small size; 1 anterior temporal; 15 dorsal scale rows at midbody; each nostril within single scale　*Liochlorophis vernalis*
45. Medium size; more than 1 anterior temporal; usually 17 (rarely 15) dorsal scale rows; nostril involves more than 1 scale
　　Coluber constrictor
46. Solid color dorsum with dark markings most prominent on neck; medium to large size　*Masticophis flagellum*
46. Dorsum not as above but may be solid colored, with dark saddles, dark rings or neck rings　　47
47. Gray to slate or almost black dorsum usually with yellow or orange neck ring; ventrum yellow or orange to reddish with black spots　*Diadophis punctatus*
47. Tannish or reddish dorsum plain colored or with various combinations of saddles, rings, or middorsal red longitudinal band; ventrum plain or with rings continued from dorsum
　　Sonora semiannulata

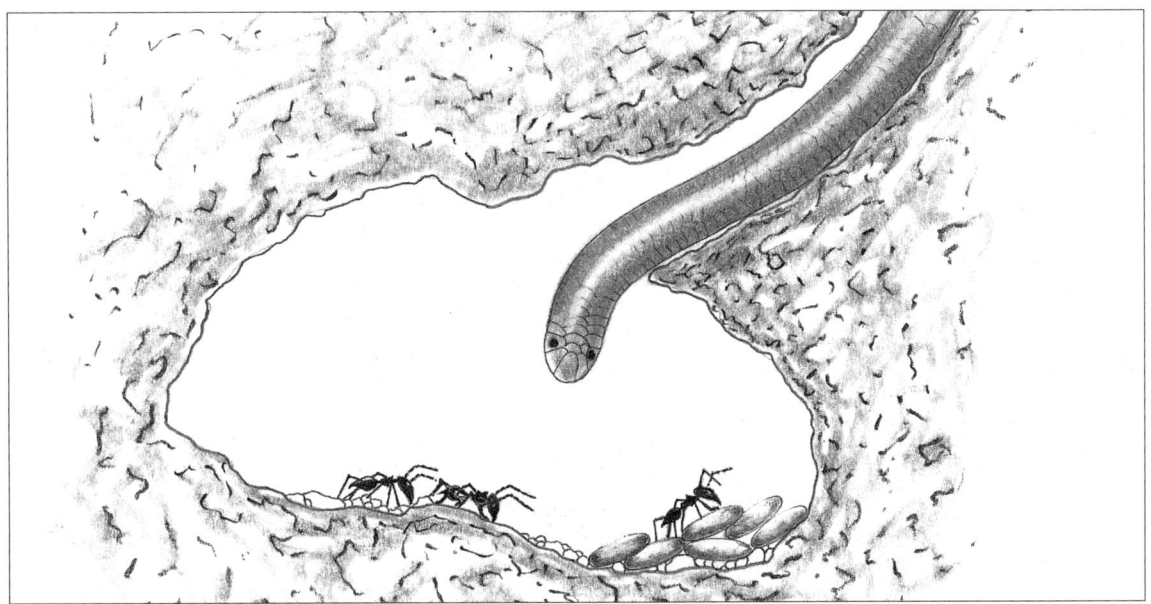

FAMILY LEPTOTYPHLOPIDAE *Blind Snakes*

This family of small, slender, wormlike, burrowing snakes includes about 64 species found in the southern United States, the West Indies, Central America, Africa, Arabia, and Pakistan (Shine, 1992). There are two species in a single genus found in New Mexico. They are often referred to as wormsnakes in reference to their cylindrical bodies, blunt heads, rudimentary eyes, and their superficial resemblance to large earthworms. The tail is very short, with a distinct apical spine. Their eyes are reduced to small, darkly pigmented spots. All species possess a vestigial pelvic girdle although it is not visible externally. Blind snakes in New Mexico reproduce by egg-laying. Females of *Leptotyphlops dulcis* may curl around the eggs and stay with them until hatching. Food consists primarily of soft-bodied invertebrates and the eggs and larvae of ants and termites.

LEPTOTYPHLOPS DULCIS (Baird and Girard, 1853)   *Texas Blind Snake*

See Plate 78.

Type: The holotype, USNM 7296, is an adult of unspecified sex collected by J. H. Clark (Baird and Girard, 1853a). The specimen was collected in 1851, but the exact date is not known. The type locality is "between San Pedro and Camanche [sic] Springs (= Fort Stockton), Texas." (See Remarks.)

Distribution: *Leptotyphlops dulcis* is found from central Oklahoma and southwestern Kansas to southeastern Arizona and south into northeastern Mexico. In New Mexico, it is most often found along the major drainages, but it is not restricted to these.

Description: *Leptotyphlops dulcis* has a wormlike body less than 30 cm TL. Dorsal color varies from pink to reddish-brown and the belly is pinkish. The head is blunt, not enlarged, without a neck constriction, and has small dark spots representing the location of vestigial eyes. The tail is short and blunt, ending in a down-curved spine. The scales are smooth, hard, glossy, and uniform in size, including those on the venter. There are 14 scale rows around the body, middorsal scales (from the rostral to tip of tail) number from 199–255 (224–246 in New Mexico), and there are 10 scale rows around the tail. There are one or two upper labials anterior to the ocular (two in New Mexico). There are usually three scales (prefrontal with a supraocular on each side) between the enlarged scales (oculars) covering the eyespots.

Similar Species: *Leptotyphlops humilis* has only one scale (prefrontal) between the oculars rather than three. Earthworms are scaleless and slimy.

Systematics: Three subspecies are recognized by Hahn (1979b); only *L. d. dissectus* (Cope, 1896) occurs in the state. The holotype of *L. d. dissectus* (ANSP 10752) is an adult collected by E. D. Cope with no sex or date specified. (See Remarks.)

Habitat: The principal requirement is light soil suitable for burrowing, but most collection sites also include abundant surface rocks. In New Mexico, limestone or sandstone flags on slopes or canyon sides often support sizeable populations of these snakes. *Leptotyphlops dulcis* ranges from desert into the piñon-juniper woodland of mountain foothills and from the lowest elevations at 900 m to near 2135 m. Deep burrows allow escape from even the driest surface conditions. Painter and Brown (1991) reported a specimen discovered during excavations approximately 15 m below the surface. *Leptotyphlops dulcis* may occur in association with *Sonora semiannulata*, *Tantilla* spp., and *L. humilis*, but requires more soil moisture than those snakes.

Behavior: *Leptotyphlops* is secretive and obviously specialized for burrowing. When first touched this snake will writhe, coil, and often assume an almost ball-like posture with moving coils. These actions apparently help to distribute the concurrently voided anal discharge over the body surface. At the same time, the scales may be tilted giving the body a silvery appearance overall. The "excrete and smear" technique seems to afford protection against some potential predators and discourages attacks by ants.

Activity is mainly nocturnal, but some forays on the surface may occur during wet and/or cool periods. These snakes thermoregulate under warm rocks and will rapidly attempt to escape down their burrows after the sheltering rock is turned. They are adept at following pheromone trails of ants and termites as well as trails of conspecifics. Gehlbach et al. (1971) detailed this trailing behavior and referenced important previous work on the subject. Annual activity based on collection records extends from 1 April to October in New Mexico. Hibernacula lie deep below ground; there is a tendency for aggregation, which may be enhanced by their trailing ability.

Reproduction: Information on reproduction in *L. dulcis* is scarce. Force (1936) found gravid females as small as 193 mm, but there appears to be no other information on sexual maturity in this species. Mates are located primarily by olfaction. Gehlbach et al. (1971) found that blind snakes follow trails of individuals of the opposite sex farther than their own trails or those of consexuals, and they are more proficient at trail-following than the small colubrids tested. Tennant (1984), reporting on *L. d. dulcis* in Texas, says that "groups of 3 to 12 copulating plains blind snakes are not uncommonly found twisted together in clusters of up to golf ball size." Males typically outnumber females 3–1 in these clusters.

Klauber (1940a) reported that most adult females collected in early summer contained large well-developed eggs suggesting laying in late summer. Nest eggs or fully formed eggs in the single right oviduct have numbered from 2–7. Hibbard (1964) discovered nests 56 cm deep in a joint crack as he was excavating fossils in Kansas on 10 July. Females were coiled around the clutches, and there were old, dry, open eggshells, seemingly hatched in previous years. The previous year on 5 July he found four gravid females in a bank 46 cm below the surface and 61 cm back from the exposed soil of the cut bank. These females contained four, three, three, and two eggs and were within 92 cm of each other. Eggs are long and thin and as large as 4.5 x 15 mm in blind snakes, a size which seems very large for a snake this slender. The smallest size reported for young is 6.5 cm TL (Conant and Collins, 1991).

Food Habits: Punzo (1974a) analyzed and compared the diets of *L. dulcis* and *L. humilis* where they are sympatric in southeastern Arizona. Ants and termites made up 54–64% of the total diet, but a large variety of other arthropods were included. *Leptotyphlops dulcis* appeared to be more fossorial in its feeding activity than *L. humilis* as fewer surface-dwelling arthropods were included. There is a definite preference for soft-bodied prey, and commensals of ant and termite nests are readily consumed along with their hosts. Punzo's study suggests a more opportunistic feeding strategy than previously supposed.

Ant pheromone trails are used by this species to locate prey sources; larvae and pupae are preferred over workers (Watkins et al., 1967). Blind snakes have been observed to feed on the soft parts of adult termites by extraction of the body contents (Smith, 1957) or by breaking off the head as the softer abdomen was ingested (Reid and Lott, 1963).

Baldridge and Wivagg (1992) showed that *L. dulcis* was able to overcome the defenses of the imported fire ant and feed on ant brood in the field. However, their data suggested that blind snakes have only a minor effect on these ant populations.

Remarks: The holotype was part of the collection obtained during the Mexican Boundary Survey from San Antonio to El Paso, Texas, in 1851. Specimens were collected by John H. Clark, under Col. J. D. Graham (in charge of the survey), and sent to the USNM by Graham in 1852. See the *Salvadora grahamiae* account and Webb (1988) for more detail.

Smith and Brodie (1982) considered New Mexico *L. dulcis* to be a different species, *L. myopicus*. Webb (1970b) and Smith and Chiszar (1993) have argued against this arrangement because of apparent intergradation between the two forms. We are following the review by Hahn (1979b) in recognizing *L. myopicus* as a Mexican subspecies of *L. dulcis* and referring all New Mexico populations to *L. dulcis dissectus*. The genus was reviewed by Hahn (1979a).

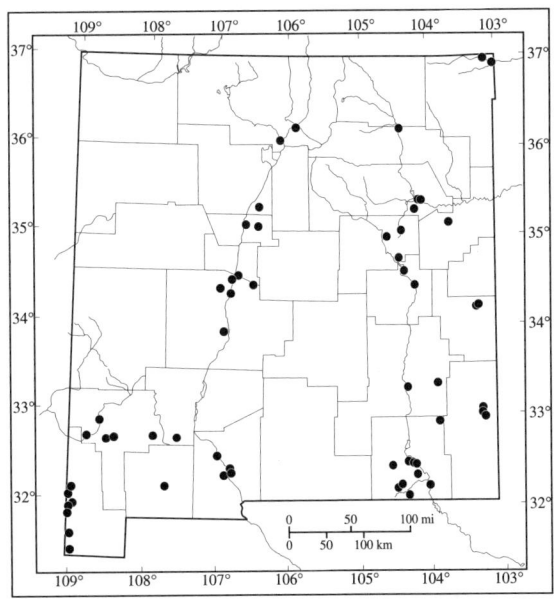

Distribution of Leptotyphlops dulcis *in New Mexico.*

LEPTOTYPHLOPS HUMILIS (Baird and Girard, 1853)   *Western Blind Snake*

See Plate 79.

Type: The holotype, USNM 2101, was collected by J. L. LeConte in 1850 (Baird and Girard, 1853a). No sex was specified, but the total length was given as six inches. The type locality was given as "Valliecitas, Cal." This type locality was restricted by Klauber (1931) to the vicinity of Vallecito, eastern San Diego County, California, and later by Brattstrom (1953) to the Upper Sonoran Life Zone of the Vallecito area.

Distribution: *Leptotyphlops humilis* is found in the lowland southwestern United States and Mexico. In New Mexico, it occurs primarily at low elevations from 900–1425 m near the Pecos, Rio Grande, and Gila river drainages. It may occur distant from these river valleys in Doña Ana and Hidalgo counties.

Description: *Leptotyphlops humilis* has a slender wormlike body with a maximum size of 339 mm TL (Punzo, 1974a) in the New Mexico form. The dorsal color ranges from pink to reddish-brown or purplish with a lighter venter. The head is blunt, not enlarged, and without a neck constriction. There are small, dark spots representing the location of vestigial eyes. The tail is short and blunt, ending in a down-curved spine.

The similar dorsal and ventral scales are smooth, hard, glossy, and uniform in size. There are 14 scale rows around the body, the middorsal scales (from the rostral to tip of tail) number 210–308. There are 12–21 subcaudals and 10 or 12 scale rows around the tail. There is one prefrontal scale between the enlarged ocular scales.

In New Mexico, dorsal scales number 261–275, subcaudals 12–16, and there are 10 scale rows around the tail.

Similar Species: *Leptotyphlops dulcis* has more than one scale (usually three) between the oculars, unlike *L. humilis*, which has only one. Earthworms are scaleless and slimy and lack vestigial eyes.

Systematics: Nine subspecies are recognized by Hahn (1979c) but only one of these, *L. h. segregus* Klauber, 1939, occurs in New Mexico. The holotype, USNM 103670, is an adult of undetermined sex collected by T. F. Smith on 11 August 1936. The type locality is "Chalk Draw, Brewster County, Texas, 3000 ft."

Habitat: *Leptotyphlops humilis* is found in desert or grassland where the soil is suitable for burrowing. In New Mexico, most collection sites have been along major drainages of tributary arroyos where the combination of slopes and surface rocks is present. *Sonora semiannulata*, *Tantilla* spp., and *Coleonyx* spp. often share this habitat in desert or grassland at elevations below 1220 m. The range overlaps that of *L. dulcis* in New Mexico, and both species have been taken on the same date in pitfall traps at Antelope Pass, Hidalgo County, and from adjacent collection sites near Carlsbad, Eddy County, and Las Cruces, Doña Ana County.

Behavior: These secretive and active little snakes are sometimes difficult to grab when first discovered and then may be difficult to hold as they attempt to force their way through your fingers. When more than one are found beneath a rock, as is often the case, quickness is important in order to collect all before they escape into the nearby burrow. This species is mostly nocturnal with some crepuscular activity if conditions permit. Road collecting can be successful for those with very sharp eyes who drive slowly, for the slender body of the blind snake is difficult to see and they move rapidly. We have had more success finding them under warm rocks during daytime while they are thermoregulating, or in pitfall traps.

Annual activity has been reported for all months of the year in southern populations of the species (i.e., southern California). In New Mexico, our records, based on 22 dates, extend from April to September with one specimen collected from the southern Rio Grande Valley in February. No information is available on hibernacula or the temperature at which this species becomes inactive, but since their burrows extend deep into the soil, a hibernation site should not be a problem. It is possible that this species does not hibernate but remains active throughout the year in most portions of its range.

Reproduction: Information on reproduction in *L. humilis* is scarce. Fox (1965) studied the spermatogenic cycle of *L. humilis* from southern California. He found that spermatogenesis was well under way in late March although no active sperm were present until April, suggesting spring breeding.

Klauber (1940a) found that most of the females he captured in early summer were carrying well-developed eggs, and he suggested that most eggs are laid in late summer. The smallest specimen with eggs in his series was 248 mm TL and contained 3 eggs. Egg number was correlated with body size in the series and ranged from 2–6. The eggs were long and thin and averaged about 4.5 x 15 mm in size. He indicated that the young at birth are about one third of the maximum adult size, which is 389 mm TL for *L. h. coahuilae* and 319 mm TL for *L. h. segregus*. The smallest specimens Klauber (1931) examined were 90 mm TL.

Fox and Dessauer (1962) reported that females in this genus possess only the right oviduct, a condition that is rare in snakes, and as far as was known, shared only with *Typhlops*, a genus in a different family. Clark (1970) has since found that most of the species of *Tantilla* that he examined have a single functional right oviduct. This condition is probably an advanced adaptation to the burrowing habit, which selects for a slender body shape in reptiles.

Food Habits: Punzo (1974a) analyzed and compared the diets of *L. dulcis* and *L. humilis* where they are sympatric in southeastern Arizona. Ants and termites made up 54–64% of the total diet, but a large variety of other arthropods were included. *Leptotyphlops humilis* seems to feed on more surface-dwelling arthropods than *L. dulcis*. There is a definite preference for soft-bodied prey types and commensals of ant and termite nests are readily consumed along with their hosts. This study suggests a more opportunistic feeding strategy than previously supposed.

Leptotyphlops dulcis uses pheromone trails to locate ant nests (Watkins et al., 1967) and *L. humilis* probably also has this ability. See the *L. dulcis* account.

Remarks: The specimen from Socorro County was taken inside a building and was possibly an introduction (G. M. Stolz, pers. comm.).

Hulse (1971) discovered that *L. humilis* will fluoresce under ultraviolet light. The rostral scale reflects U-V light and appears bright blue while the rest of the body surface glows with a pale green fluorescence.

This genus was reviewed by Hahn (1979a) and the species reviewed by Hahn (1979c).

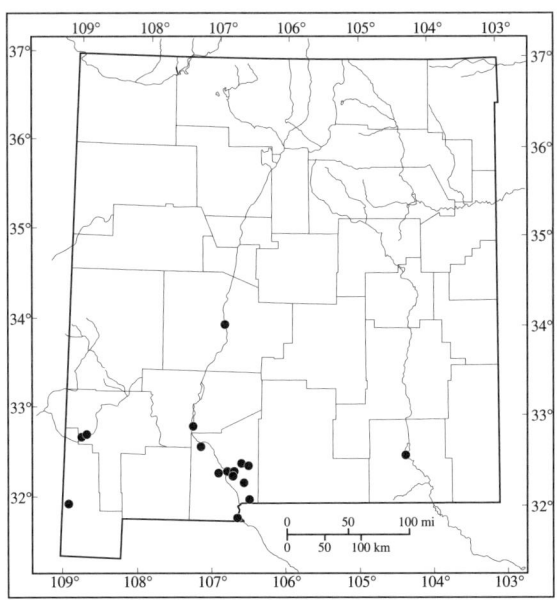

Distribution of Leptotyphlops humilis *in New Mexico.*

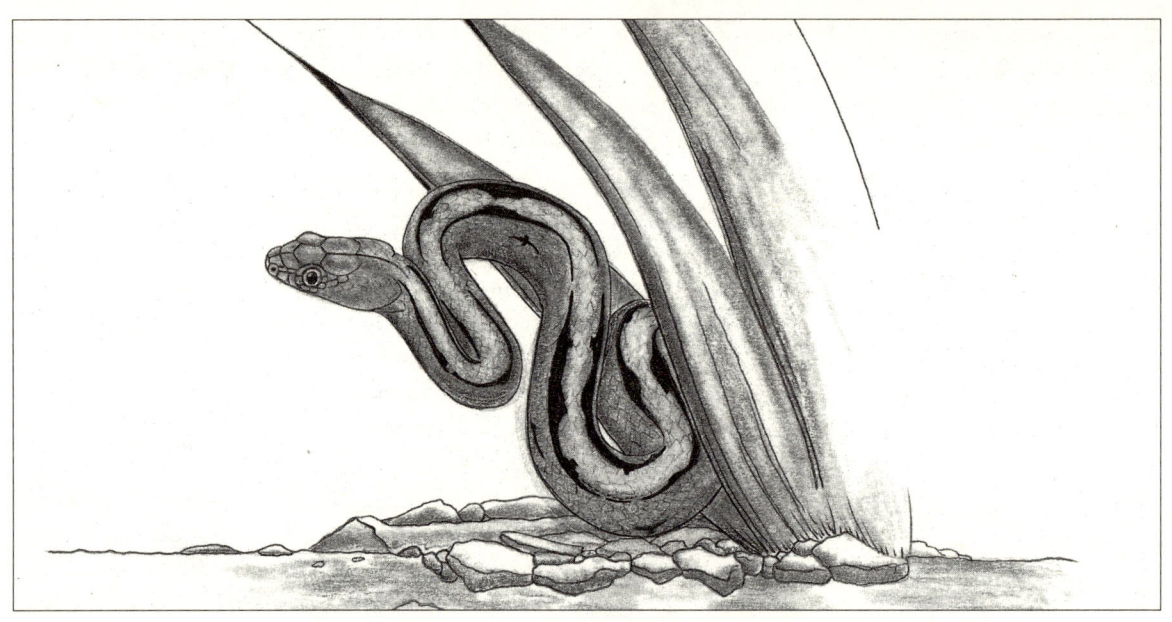

Colubrids # FAMILY COLUBRIDAE

Most living snakes belong to a single family, Colubridae, with about 1,600 species occurring on all continents except Antarctica (Shine, 1992). Colubrids are generally the most abundant snakes where snakes occur, except in Australia where the venomous elapid snakes are most abundant. In New Mexico, there are 36 species in 21 genera, which make up about 78% of the entire snake fauna of the state. These are often referred to as the "harmless snakes," and although a few are rear-fanged and have toxic saliva, none are harmful to humans. Rear-fanged colubrids in New Mexico include *Diadophis*, *Heterodon*, *Hypsiglena*, *Tantilla*, *Trimorphodon*, and *Sonora*.

Modes of reproduction in the Colubridae are varied, with some genera laying large clutches of eggs, others, including the water snakes and garter snakes, produce living young. In many colubrid snakes, adult females reproduce only every second or third year. These snakes live in a variety of habitats and take a wide assortment of prey.

ARIZONA ELEGANS Kennicott *in* Baird, 1859 — *Glossy Snake*

See Plate 80.

Type: The syntypes, USNM 1722, a male, from the "Rio Grande" and USNM 4266, a female, from "between Arkansas and Cimarron" were used by Kennicott (*in* Baird, 1859) in his original description. The collection date was not specified. The locality, "Lower Rio Grande," was used in Yarrow's (1882) checklist for USNM 1722 and has been interpreted as a restriction of type locality, possibly because he listed it first in his "reserve series." The Stejneger and Barbour checklists (1917, 1923, 1933, 1939, 1943) did not recognize the apparent restriction, even though Blanchard (1924) did so in his paper naming *A. e. occidentalis*. However, Schmidt's (1953) revision of the checklist recognized this restricted type locality. Dixon and Fleet (1976) examined USNM 1722 and proposed that "Lower Rio Grande" was in error since the "salient features" of this specimen best agree with Trans-Pecos Texas specimens. They suggested that Schott, the collector, probably used Eagle Pass, on the "lower Rio Grande," as a shipping point. They recognized USNM 1722 as the lectotype, and did not change the type locality restriction.

Distribution: *Arizona elegans* is found mainly in the southwestern United States, from the southern half of California to southern Nebraska, and as far south as Aguascalientes in central Mexico. In New Mexico, *A. elegans* is widespread and abundant between 900 and 2200 m.

Description: This is a medium sized snake; adults measure 24–142 cm SVL. It has the general appearance of a slender *Pituophis* with a washed out or faded pattern, and has been called the "faded snake" in the past. The dorsal color is light tan, cream, or pinkish with darker brown, tan, or grayish-pink blotches that are usually edged with a color darker than the blotches. A dark tan, brown, or brownish-black line is usually present from the eye to the posterior edge of the jaw. The venter is cream to white.

The snout is narrow and pointed and the tail is approximately one ninth of the total length. The scales are smooth and range from 25–31 rows at midbody. There are 185–241 ventrals, 39–63 subcaudals, and the anal plate is undivided. There are two prefrontals, one loreal, one preocular, and two postocular scales.

Similar Species: *Pituophis* has keeled scales and four prefrontals. *Hypsiglena* has a divided anal plate, a vertically elliptical pupil, and bold elongated dark blotches on each side of the neck.

Systematics: Klauber (1946) and Dixon and Fleet (1976) recognized nine subspecies, although using a slightly different arrangement. More recently, seven subspecies have been recognized in the United States (Smith and Brodie, 1982; Stebbins, 1985). Two of these, *A. e. elegans*, the nominotypical form, to the east and *A. e. philipi* Klauber, 1946 to the west, occur in New Mexico. The expected zone of intergradation, roughly along a broad corridor on both sides of the Pecos River, has not been defined due to the paucity of specimens from this area. Recently collected material from this area has not been examined.

The holotype of *A. e. philipi* (SDSNH 34456), an adolescent male, was collected by C. E. Shaw and C. Engler 29 July 1941 from 10 miles east of Winslow, Navajo County, Arizona.

Habitat: In general, *A. elegans* seems to prefer sandy grasslands and shrublands rather than woodlands. They are excellent burrowers but will often use rodent burrows or excavations under rocks for retreats. Although the species ranges to about 2200 m, glossy snakes are more abundant at lower elevations. *Pituophis*, *Rhinocheilus*, *Hypsiglena*, *Crotalus viridis*, and *C. atrox* are frequent cohabitants.

Behavior: *Arizona elegans* rarely attempts to bite but will often vibrate the tail and thrash vigorously when first picked up. In New Mexico, they are primarily nocturnal. During the day they make use of rodent burrows or other natural entrances to their subterranean retreats. A number of authors mention diurnal or crepuscular activity occurring in some populations (Klauber, 1946). We suspect that differences in humidity and temperature cause variation in the diel activity pattern. Warm and not excessively dry nights with air temperatures over 22°C (27°C near optimum) may result in surface activity.

Under laboratory conditions, Cowles and Bogert (1944) found the critical thermal maximum to average about 41.8°C. Most annual activity extends from April to September and varies with latitude and elevation, resulting in a shorter period at higher elevations in the north. Collection records are available from February–November in southern New Mexico. Like other species of snakes in this habitat, *A. elegans* uses rodent burrows as entrances to hibernacula. In Albuquerque, one was found 26 December about 60 cm below the surface in a rodent burrow (T. L. Brown, pers. comm.).

Reproduction: Fitch (1970) summarized clutch sizes from previous authors, and these have ranged from 3–23 with a mean of 8.5. The clutch of 23 eggs was laid on 8 July by a snake from southern California. These eggs hatched after 68 days. Anderson (1992) reported that a captive pair from southern California were observed copulating four times from 23–29 May, at which time they were separated. On 28 June the female laid eight eggs measuring 38 x 16 mm. The eggs were incubated at 27–28°C, and shells were slit 8 September with all emerging within two days (72 days incubation). The young were 22.5–25.4 cm long and able to eat pink mice. A female from Bernalillo County collected 13 June 1982 measured 60 cm SVL, and contained eight shelled eggs with no discernible embryos (J. N. Stuart, pers. comm.). Aldridge (1979a, 1979b) studied reproduction in a central New Mexico population and reported activity from April–September based on 243 capture records. Ovulation occurs in late June to early July and oviposition in early to mid-July. His data suggest that some females do not reproduce every year, that multiple annual clutches are impossible in the population he studied, and highly unlikely in any species of north temperate snake. Aldridge (1979a) also found that temperature, but not photoperiod, is a major factor in the control of spermatogenesis, and that both sperm and egg production are initiated in the year prior to fertilization.

Food Habits: Lizards, small rodents, and sometimes small snakes and birds may be eaten by *A. elegans* (Wright and Wright, 1957). Constriction or pressing of the prey against a firm surface, such as the side of a burrow, may be used to kill prey. Small easily subdued prey may be swallowed alive. Feeding is probably mostly nocturnal.

Remarks: *Arizona elegans* was reviewed by Klauber (1946) and Dixon and Fleet (1976).

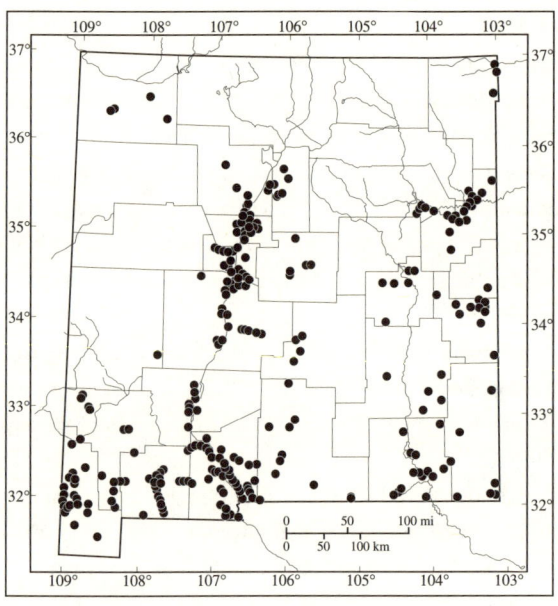

Distribution of Arizona elegans *in New Mexico.*

BOGERTOPHIS SUBOCULARIS (Brown, 1901) *Trans-Pecos Rat Snake*

See Plate 81.

Type: The holotype, ANSP 13733, described by Brown (1901a) is an adult male (159 cm TL) collected by E. Meyenberg in the spring of 1901. The type locality is 50 miles southwest of Pecos, near the head of Toyah Creek, in the Davis Mountains, Jeff Davis County, Texas.

Distribution: *Bogertophis subocularis* is found from southern New Mexico and southwestern Texas as far south as Durango and Nuevo León in Mexico. In New Mexico, it occurs in the south-central and southeastern portions of the state where suitable habitat is available from about 1000–1600 m elevation.

Description: Adult *B. subocularis* measure 85–168 cm TL and are relatively slender. Color and pattern vary, but there are usually 21–28 distinctive black or brown H-shaped dorsal blotches on a tan, yellow, or olive-yellow background. The sides of the H's may join to form stripes down each side, especially toward the head. There are 7–10 blotches on the tail. The head is patternless, as is the whitish to buff venter. The young are similar to the adults in general pattern but are paler. Dorsal scales are weakly keeled along the middle of the back and are in 31–35 rows at midbody. There are 260–277 ventrals, 69–79 subcaudals, and the anal plate is divided. Upper labials range from 9–12 and lower labials from 13–17. There is a single loreal, a single preocular, and the nasal is divided. The eyes are large, and there is a row of small scales (suboculars or lorilabials) between the eye and the labials.

Similar Species: No other snakes in New Mexico have the characteristic H-markings and few have subocular scales.

Systematics: Two subspecies of *B. subocularis* have been named but only the nominotypical form occurs in New Mexico (see Remarks).

Habitat: The habitat of *B. subocularis* was described in detail by Degenhardt and Degenhardt (1965). This ratsnake needs rocky terrain, which allows access to deep retreats below the soil surface. The type of rock may vary from limestone to granite or lava so long as sufficient depth is available. Surface rocks or debris are seldom used for cover. In our experience with over 60 snakes, none have been found in a daytime retreat; all were active on the surface when encountered. One published instance of a snake taken "in an outbuilding near the eastern slope of the Organ Mountains" (Lewis, 1948) gives no other details of capture. The highest elevation record for natural occurrence was about 1600 m in the Tularosa Basin, and this record is also the northernmost. Soil and plant types are unimportant so long as they afford sufficient cover and food for the prey on which these snakes feed. The statement by Price (1990), that specimens have been collected, "often in association with permanent water," suggests that surface water is an important habitat parameter, which it is not. Price was apparently unaware of the many published statements on habitat that were listed in Worthington (1980), for he cited only Kauffeld (1969) in his discussion of habitat. Published distribution maps often show plotted localities that seem to follow rivers (Degenhardt and Degenhardt, 1965). Most of these collection sites are on roads that provide access to habitat which happens to be near rivers. Cohabitants with nearly similar range and habits include *Lampropeltis alterna* and *Trimorphodon biscutatus*.

Behavior: *Bogertophis subocularis* is gentle and seldom attempts to bite when first collected, although occasional individuals may do so as captives. The large eyes of this snake suggest nocturnal activity. We know of very few individuals observed in daylight, and these were seen at dusk. This snake never basks in the sun and seems to shun it. Since this habit conserves moisture, it may allow these snakes to live in arid areas far from water. On two consecutive nights in Big Bend National Park, a snake was observed crossing the road at about the same time and place (Degenhardt and Degenhardt, 1965). This suggests a home range, or at least use of the same (nearby?) daytime retreat. Little early or late annual activity data are available, and to our knowledge no hibernacula or deep retreats have been discovered. The unique snake-tick association of this species supports the assumed drought avoidance habits of strictly nocturnal activity and use of deep retreats (see Remarks).

Reproduction: Mature Trans-Pecos ratsnakes range from 850–1676 mm TL. Captive-bred litter-mates have bred in the third year at just under two years of age; however, their development may have been accelerated by the prolonged season of activity and feeding that captivity affords (Tryon, 1976). Courtship includes chasing and biting the female by the male, and this activity was observed to continue for eight hours (Tryon, 1976). These snakes mate in the spring and eggs are laid June–August. Eggs number 3–7, are non-adhesive, and have measured from 19–30 x 48–73. Incubation time has ranged from 73–89 days and is temperature dependent (Campbell, 1972; Degenhardt and Degenhardt, 1965; Tryon, 1976). Hatchlings measure from 280–357 mm TL.

Food Habits: In order of importance, small mammals, lizards, and birds are eaten by these snakes. Captives have eaten *Peromyscus* sp., *Dipodomys merriami, Perognathus pennicillatus, Neotoma mexicana, Sigmodon hispidus, Pipistrellus hesperus, Uta stansburiana, Cophosaurus texanus, Urosaurus ornatus, Crotaphytus collaris, Cnemidophorus tigris, C. septemvittatus, Coleonyx brevis, Anolis carolinensis, Lanius ludovicianus,* laboratory rats, mice, and rabbits. There is a marked preference for nestling rodents. Constriction is used for larger live prey but not for smaller or dead prey items. Jameson (1956) reported this snake's ability to capture and constrict a second live mouse while a first is held in its coils.

Remarks: We are using the newly proposed genus of *Bogertophis* (Dowling and Price, 1988) with some reservation. Worthington's (1980) review of the species uses *Elaphe subocularis*. Webb (1990b) in naming a new subspecies from Mexico used *Bogertophis* "on the basis of precedent only." Van Devender and Bradley (1994) "find the arguments to elevate *E. triaspis* to the genus *Senticollis* (Dowling and Fries, 1987) and *E. rosaliae* and *E. subocularis* to *Bogertophis* (Dowling and Price, 1988) unsatisfying and prefer to keep them in *Elaphe*." Van Devender and Bradley (1994) gave reasons for not recognizing the new generic names.

Degenhardt and Degenhardt (1965) discussed the unique host-parasite relationship between this snake and the tick, *Aponomma elaphense*. The tick is apparently host-specific on this snake. There is seemingly little harm to the host regardless of the density of the tick, which may approach 100 on a single snake. The authors proposed that this relationship is due primarily to the habits of the snake in relation to the sun, which it avoids, and which might kill the ticks. Also, in a hot and arid habitat, vulnerable tick life history stages, such as eggs and larvae, would be best protected deep underground, where we assume snake retreats to be. Webb (1990b), in naming a Mexican subspecies, found these ticks on most of the Durango specimens. This is a range extension for *Aponomma elaphense*, since the single Durango snake (DOR) examined by Degenhardt and Degenhardt (1965) was free of ticks.

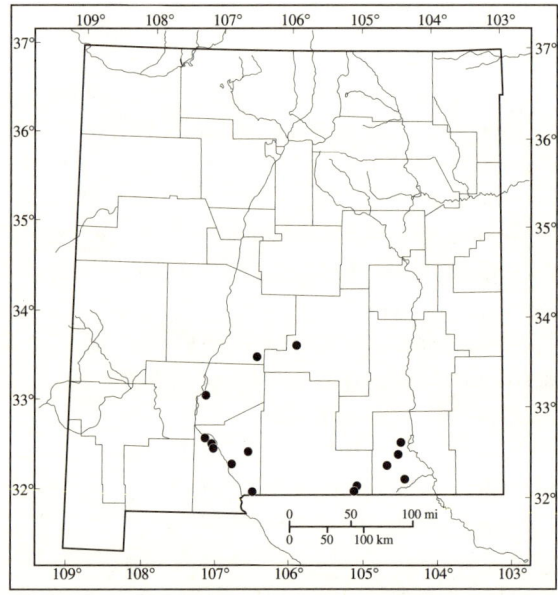

Distribution of Bogertophis subocularis *in New Mexico.*

COLUBER CONSTRICTOR Linnaeus, 1758 — *Racer*

See Plate 82.

Type: The type specimen, collected by Peter Kalm, was not so designated but was probably in Linnaeus' personal collection (Klauber, 1948). The type locality, "America septentrionale," was restricted to "Canada" by Schmidt (1953), but Dunn and Wood (1939) indicated that the type locality is probably in the vicinity of Philadelphia, Pennsylvania.

Distribution: The racer is found from the Canadian border region south through Mexico to Guatemala, but is apparently absent or rare in most of the southwestern United States and western Mexico. In New Mexico, most records are from the northeastern corner of the state and the Rio Grande Valley. Nowhere in New Mexico is it common, and it is rare or absent from most portions of the state. Most localities on the map represent single specimens.

Description: The racer is a medium to large snake; adults range from 60–190 cm SVL. There are many color phases of the adult over the very large range of the species. In the West, dorsal color varies mainly from greenish to bluish to brownish, and ventral color varies from pale cream to yellow. The young are very different from the adults. Most young are strongly patterned with a middorsal row of dark blotches or crossbands on a lighter ground color; some southwestern populations have scattered small dark spots anteriorly, and sometimes transverse lines on a greenish ground color. The scales are smooth and in 15–17 rows at midbody. Ventrals range from 158–193, subcaudals from 66–119, and the anal plate and subcaudal scales are divided. The eyes are large with round pupils. There are two preoculars, with the lowermost wedged between upper labials, and two anterior temporals. Each nostril is located between two scales.

Similar Species: *Liochlorophis vernalis* is always green dorsally (blue or gray in preservative), with one anterior temporal, and each nostril is centrally located in a single scale. Young *Pituophis melanoleucus* have keeled scales, *Lampropeltis* spp. and *Arizona elegans* have a single anal plate, *Hypsiglena torquata* has elliptical pupils, and *Elaphe* spp. have weakly keeled scales.

Systematics: Wilson (1978) recognized eleven subspecies of the racer. *Coluber c. flaviventris* Say *in* James, 1823 is the principal race found in New Mexico, but western and southern material from this state may be referable to *C. c. mormon* Baird and Girard, 1852[b]. Arguments to elevate *C. c. mormon* to species status were presented by Fitch et al. (1981), but others (Greene, 1984; Corn and Bury, 1986) have concluded that the form should continue to be considered a subspecies.

Three syntypes, whereabouts unknown, of *C. c. flaviventris* were collected by Say 12 December 1819. The type locality was restated by Rossman (1963) as "approximately 3 miles ENE Fort Calhoun, Washington County, Nebraska."

The holotype of *C. c. mormon*, USNM 2012, is a juvenile (male ?) collected by H. Stansbury on an unknown date. The type locality is "Valley of the Great Salt Lake, Utah."

Habitat: The racer has been found in a variety of grasslands, brushlands, and woodlands in New Mexico from 1160–2000 m. When found in the more arid portions of the state, water (including rivers, lakes, springs, and irrigation canals) is usually available. Rocks, deadfall, old foundations, and lumber from old houses or corrals are frequently used for cover, so soil type is less important than it would be to a burrowing species. Common cohabitants are *Crotalus viridis*, *Diadophis punctatus*, *Lampropeltis getula*, *Tantilla nigriceps*, *Thamnophis elegans* (in montane grassland), *T. sirtalis* (the Rio Grande Valley), *T. radix*, and *Tropidoclonion lineatum*.

Behavior: As the name indicates, the racer is one of the swiftest of our snakes and depends on speed to escape enemies and capture prey. Though most often terrestrial, it may climb into shrubs and small trees where branches are available for support. When cornered or captured, the racer usually bites and chews aggressively and may discharge musk and feces. Scars are still visible on Degenhardt more than fifty years after he was bitten by a racer from Long Island, New York. They generally make poor captives; they are nervous and excitable, and seldom live more than a few years even when they can be induced to feed. Most individuals will readily cross bodies of water; they swim at the surface with the head well elevated.

Activity is diurnal, and body temperature controls basking and hunting activities. Fitch (1963) recorded body temperatures of newly captured racers in Kansas and found the modal temperature to lie between 34–35°C. This is a high temperature range for active snakes and is indicative of the high heat tolerance of this species.

These snakes hibernate underground during the winter, and numerous accounts of aggregations with other species as well as conspecifics are available (Brown and Parker, 1976; Rosen, 1991). Entrances to underground hibernacula may be natural or made by other burrowing animals. The tendency to return year after year to the same hibernacula was not strong in Kansas (Fitch, 1963). However, Brown and Parker (1976) studied a population in Utah and reported the tendency for most surviving snakes to return to the same communal dens used the previous winter, suggesting homing ability. In New Mexico, dates for 38 collecting records indicate annual activity from 27 April to 21 October.

Reproduction: Most of the basic reproductive information that may be applied to New Mexico populations has been gathered by Fitch (1963). Although sexual maturity may occur earlier, at 13–14 months of age, mating will not occur until the third season or about 20 months. Cottam (1937) published a photograph and observations on copulation of a pair of racers in Utah. The main breeding season is probably similar to that observed in Kansas for most of our populations. Breeding likely occurs during the last three weeks in May, with egg laying in late June or early July. A Chaves County female (MSB 55926) laid six eggs 29 June measuring 16–17 x 36–47 mm; the eggs did not hatch (G. E. Knadle, pers. comm.).

The eggs are elongate and non-adhesive with tough, leathery, and granular surfaced shells. Egg number varies from 2–26 and is proportional to female size with an average near 12 for *C. c. flaviventris* and near six for *C. c. mormon*, the smallest form. In our dry climate, they are probably laid most often in rodent burrows. Incubation time averages about 50 days but is inversely correlated with temperature. Swain and Smith (1978) summarized earlier reports on communal nests in North American snakes, and reported an apparently communal nest in Colorado containing 89 eggs and 29 empty shells from previous years. They suggested that communal nesting may be an environmental necessity due to the paucity of suitable nest sites in a given area.

Food Habits: Racers depend primarily on eyesight and speed to locate and capture prey during their daytime hunting forays. They feed somewhat opportunistically, resulting in the use of a large variety of prey items. Fitch (1963) summarized published reports, and presented his own exhaustive work, which included data from scats, stomach contents, palping, and direct feeding observations. He found the most important foods were small mammals, lizards, snakes (including racers), frogs, and soft-bodied orthopterans.

In spite of the specific epithet, *constrictor*, the racer does not constrict, but swallows prey "alive and kicking."

Remarks: Banta (1971) argued convincingly that the original description for *C. c. mormon* was first published in Baird and Girard (1852a) and not in Baird and Girard (1851 [1852]) or (1852b). Wilson (1978) reviewed this species and included a spot in Catron County on his map. We cannot locate a specimen for this locality and Wilson (pers. comm.) said that the spot was probably an error.

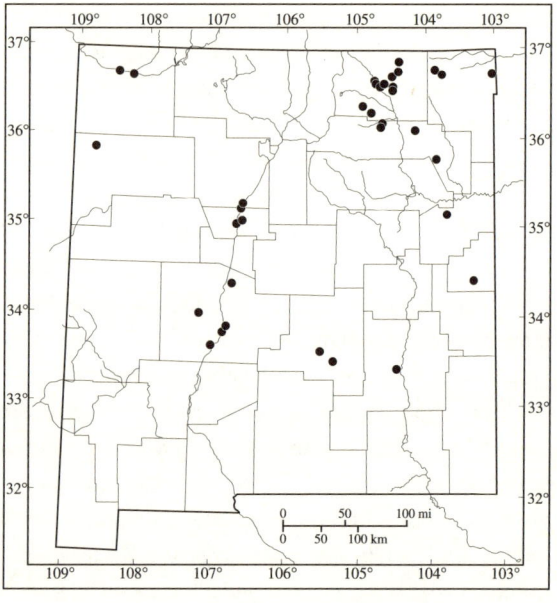

Distribution of Coluber constrictor *in New Mexico.*

DIADOPHIS PUNCTATUS (Linnaeus, 1766)   *Ringneck Snake*

See Plate 83.

Type: The type was in the Linnaeus private collection; its present location is unknown. The original material was sent to Linnaeus by Dr. Alexander Garden, who lived in Charleston, South Carolina at the time. The shipment consisted of other animals besides reptiles and was said to be from "Carolina." Schmidt (1953) restricted the type locality to Charleston, S.C., for Dr. Garden was known to have collected in that vicinity.

Distribution: *Diadophis punctatus* is widespread in the United States and northern Mexico except in the northern plains, northern Rocky Mountains, and most of the driest areas of the western deserts. It has been found throughout New Mexico to near 2200 m except in the northwestern and north-central portions of the state.

Description: This small slender species usually measures 25–45 cm TL, but records of over 80 cm are available for New Mexico. Most snakes are some shade of gray or olive dorsally, orange to red with black spots ventrally, and have a yellow or orange ring around the neck. In some individuals, the neck ring may be broken or absent and the ventral spotting reduced. The dorsal scales are smooth and in 15 or 17 rows at midbody. There are 111–239 ventrals, 30–76 subcaudals, and the anal plate is divided. There are 2–3 preoculars, two postoculars, a single anterior temporal, seven upper labials, and a single loreal.

Similar Species: The usually present light neck ring and smooth scales are unique to *D. punctatus* in New Mexico. In the absence of the ring, the plain dark dorsal color, plus the orange to red venter with black spots, distinguishes the ringneck from all other snakes in New Mexico.

Systematics: We consider all *Diadophis* to be conspecific based on Gehlbach (1974) and most recent authors. Twelve subspecies of *D. punctatus* are currently listed in Collins (1990). Two of these, *D. p. arnyi* Kennicott, 1859 and *D. p. regalis* Baird and Girard, 1853, are found in New Mexico.

The holotype of *D. p. arnyi*, USNM 1968, was collected by S. Arny; the sex and collection date were unspecified. The type locality is Hyatt, Anderson County, Kansas.

The holotype of *D. p. regalis*, USNM 2062, is an adult; the sex and collection date were unspecified. The collector was probably J. H. Clark (see Remarks). The type locality is Sonora, Mexico.

Habitat: In New Mexico, *D. punctatus* is found in a wide variety of habitats from woodland as high as 2200 m in piñon-juniper-ponderosa ecotone, to desert grassland as low as 980 m. Moisture at the surface or below is required. The most dense population known in the state occurs at about 1820 m in Lincoln County along a leaky irrigation ditch that keeps the soil on the downhill side moist. Turning rocks or other surface debris along this ditch has produced dozens of snakes, often more than one under the same rock. Shelter is important in all habitats, and snakes may use flat rocks, boards or other building debris, trash, or dead vegetation and "cow-pies" for cover.

Behavior: In New Mexico, the ringneck snake is rarely seen abroad in daylight. When disturbed, this gentle snake often coils its tail in a spiral and exposes the brightly colored underside to the potential danger. The name "thimble snake" is often applied because of this habit of coiling the tail. Handling usually causes voiding of musk, semi-liquid feces, and uric acid from the cloaca; this voiding probably deters predators. Some individuals may bite. In some populations, rough handling may elicit a sequence of writhing, head-hiding, head-cocking and eye rotation, and finally a sudden, limp, upside-down position exposing the brightly marked underparts (Gehlbach, 1970). Shaw and Campbell (1974) have discussed the possible value and evolutionary history of this habit. This death-feigning display has not been observed in New Mexico ringneck snakes.

Spaces under warmed rocks and debris, rather than sunning in the open, are used for thermoregulation, a habit also serving to protect snakes against moisture loss and predation. Fitch (1975) found that his Kansas population has a relatively long season of activity, averaging about 213 days, and excluding only that part of the year in which freezing regularly occurs. He attributes this long season to small size and efficient thermoregulatory behavior permitting rapid temperature adjustment. Body temperatures averaged 26°C in 129 snakes found beneath surface objects. The highest body temperature recorded was 34.4°C, but the experimental critical maximum is approximately 41°C. The critical minimum is approximately freezing, and Fitch (1975) found that the temperature in hibernacula 30–80 cm beneath the surface stayed within a 0–10°C range. In New Mexico, our 131 collection dates range from March 10 to October 4, an annual activity period of 208 days, similar to the 213 days given above for Kansas. Deep crevices in rock outcrops, rodent burrows, or man-made structures such as old rock walls, foundations, abandoned wells, and cisterns may serve as hibernacula in New Mexico.

Reproduction: Maturity in this oviparous species occurs before the second hibernation when the snakes are approximately 13–14 months old, although they do not mate until the third year (Fitch, 1960). Males were at least 168 mm SVL and females 210 mm SVL at maturity in Fitch's (1975) Kansas study, but the *D. p. regalis* populations in New Mexico probably mature at a larger size. Active sperm is found in males throughout the activity period. Females are receptive primarily in April and May; ovulation occurs in June, and egg-laying from mid-June to late July. The timing of the cycle, however, varies with weather conditions in the particular area. Egg number is related to size of the female, but usually ranges from 1–10 and averages about four. Gehlbach (1965) reported a large *D. p. regalis* measuring 650 mm TL containing 18 eggs. Stebbins (1954) mentioned a large female from Utah with five eggs, and Hammerson (1982) noted that a captive Colorado snake laid 5 eggs in Mid-July. Gehlbach (1965) reported a captive bred pair (Arizona *regalis* x Texas *arnyi*) that produced six fertile eggs. Where snakes are abundant or nests scarce, communal sites may be utilized, and as many as 48 eggs have been found in one rotting log in Michigan (Blanchard, 1936 [1937]). The leathery-shelled eggs average about 25 mm in length, are often sausage-shaped, but may vary in both size and shape. We are not aware of nests being found in New Mexico, but they are probably located in soil where permanent moisture is available. Hatching has occurred under laboratory conditions in as little as 65 days. Under natural conditions in New Mexico, the incubation time is probably longer, since the requirement of permanent moisture is usually found deeper where temperatures are lower.

Food Habits: Fitch (1975) summarized the known food items of *D. punctatus*. They include a wide variety of small invertebrates and vertebrates. Partial constriction is used in subduing prey. Earthworms and salamanders often serve as the principal diet in areas where they are abundant. In the dryer conditions of the Southwest, snakes and lizards are very important foods. Anton (1994) reported *Urosaurus ornatus* taken by a captive specimen from Hidalgo County. Gehlbach (1974) observed that small snakes grasped by *D. punctatus* from the Guadalupe Mountains in Eddy County were apparently immobilized by toxic salivary secretions. The production of a weak venom is suspected, since there are enlarged rear maxillary teeth which in other snakes are accompanied by venom. Shaw and Campbell (1974) reported that an observer felt a burning sensation after being bitten by a ringneck snake.

Remarks: The large western type (*"regalis"*) and the small eastern type (*"punctatus"*) are both present in New Mexico and should be studied further. The possibility that they are distinct species exists, although Gehlbach (1974) reported a captive bred pair that produced viable eggs, and Mecham (1956) considered the two forms conspecific based on scale counts.

In the description of *D. p. regalis* (Baird and Girard, 1853[a]), the name of Col. J. D. Graham is given at the end and appears to be the name of the collector. The specimens from that boundary survey collection, however, were collected by Mr. J. H. Clark. Baird (1859) cited Clark as the collector, as did Cochran (1961). Cope (1898 [1900]) and Yarrow (1882) listed Graham in their "from whom received" column; this is misleading since the name of the collector usually appears there. A somewhat similar case occurred with *Salvadora grahamiae*. Cochran (1961), in error, listed Graham as the collector of that type. Details concerning this boundary survey collection are found in Webb (1988) and in the *Salvadora grahamiae* account.

This species was reviewed by Blanchard (1942) and in part by Mecham (1956) and Gehlbach (1974).

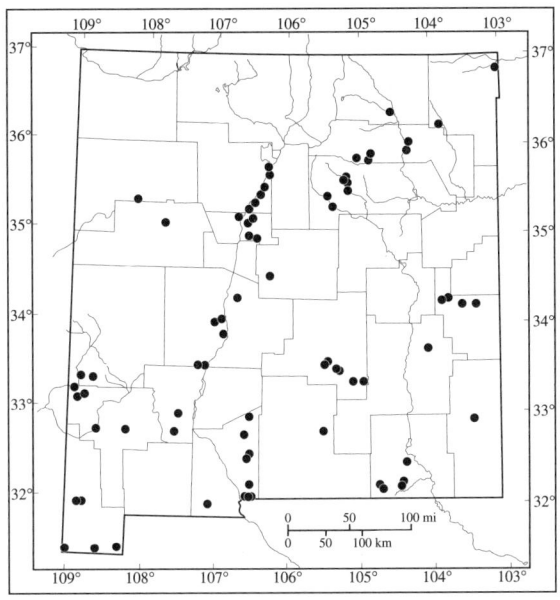

Distribution of
Diadophis punctatus
in New Mexico.

ELAPHE GUTTATA (Linnaeus, 1766) — *Corn Snake*

See Plate 84.

Type: The holotype, a specimen in the Zool. Mus. Royal U. Upsala, Sweden, was collected by Dr. Alexander Garden. The type locality, "Carolina," was restricted to the vicinity of Charleston, S.C., by Dowling (1952).

Distribution: *Elaphe guttata* is found throughout the United States from southern New Jersey to western Utah and south into Mexico to Veracruz. In New Mexico, this species has been found largely in the eastern half of the state from about 900–2450 m elevation.

Description: Adults are mostly 60–120 cm TL in size, but may reach 180 cm. *Elaphe guttata* is a blotched snake with the blotches outlined in black. In the east, the species has red to reddish-brown blotches on a gray to orange background. In New Mexico, it has grayish or brownish blotches on a light gray background, but there is variation in the intensity of colors. The venter is checkered with black on white. The first blotch on the neck divides into two branches anteriorly to form lines that converge on the head, forming a spearpoint between the eyes, but in older individuals the head markings become less distinct. The scales are weakly keeled, and are in 25–31 rows at midbody. There are 207–245 ventrals, 60–97 subcaudals, and the anal plate is divided. There are usually eight, occasionally nine, upper labials, and lower labials range from 12–14. A loreal is present, and there is one preocular and two postoculars.

Similar Species: *Pituophis melanoleucus* and *Nerodia erythrogaster* have strongly keeled scales. *Lampropeltis* spp. and *Arizona elegans* have a single anal plate and smooth scales. Young of *Coluber constrictor* and other *Elaphe* spp., if blotched, do not have the characteristic spearpoint between the eyes.

Systematics: Three subspecies are recognized; two of these, *E. g. emoryi* (Baird and Girard, 1853[a]) and *E. g. meahllmorum* Smith, Chiszar, Staley, and Tepedelen, 1994 occur in New Mexico (see Remarks). Older references refer to these subspecies as *Elaphe laeta*, the name used from 1917–1951.

The type of *E. g. emoryi* was apparently lost from the United States National Museum. It was collected at Howard Springs, about 20 miles southwest of Ozona, Crockett County, Texas, by J. H. Clark. The sex and date of collection are unknown.

The type of *E. g. meahllmorum* is a young adult female (UCM 46009) from El Salto, San Luis Potosí, Mexico collected 13 June 1964 by T. Paul Maslin et al.

Habitat: *Elaphe guttata* occurs in a variety of habitats in the southwestern portion of its range. Two important habitat constituents are a permanent source of water and good cover for a daytime retreat or hibernation. In northern New Mexico, *E. guttata* is found mainly along rocky arroyos or mountain streams to about 2200 m in grassland, shrubland, or woodland. One location that supported a population, before it was developed into a national forest picnic area and trailhead, was the site of an old fish hatchery with spring-fed, rock-bordered pools at the edge of montane woodland. In the more open country of eastern New Mexico, they are found primarily along watercourses or near ponds, lakes, and springs. In the hot and arid southeastern portion of the state, cracks and fissures of subsurface gypsum deposits may serve as cover. This gypsum habitat supports other drought-avoiding species such as *Eleutherodactylus augusti* and *Eumeces multivirgatus*. Man-made habitat and rodent-rich locations such as barns, deserted houses, old wells, cisterns, and foundations may also be used. We have found individuals on roads at night and others under rocks and surface debris in the daytime. One snake was found lying at the side of a rock-bordered shallow pool; another was about 1.5 m off the ground in a small tree near a seep that was fed from a leaking irrigation canal.

Behavior: In New Mexico, *E. guttata* is primarily nocturnal; remaining under cover of rocks, logs, surface debris or in crevices, rodent burrows, and caves during daytime. When first captured it may vibrate the tail, bite, void feces, release contents of anal scent glands, or any combination of these, but usually quiets with handling. Collection records indicate activity from mid-April through October, but activity may begin in March if temperatures permit. We know of no discoveries of hibernacula in New Mexico, but Webb (1970b) records an Oklahoma hibernating aggregation of six *E. guttata* and 27 *Coluber constrictor* that was discovered during removal of the cement foundation of a house on 17 December.

Reproduction: The youngest breeding age recorded is 18 months in captive bred *E. g. guttata*. The smallest of the four females which bred measured 550 mm TL and the largest 750 mm. The authors, Bechtel and Bechtel (1958), suggested the early age and small size was due to the long annual activity period and the regular feeding that their captive snakes enjoyed. Under natural conditions, breeding, at least in females, probably does not occur until the third year. This potential for "forcing" breeding stock to mature early is well known among reptile breeders. Courtship and breeding normally occur in the spring after emergence from hibernation, but a delay of 3–4 weeks may occur between these events (Ford and Cobb, 1992). Eggs are usually laid 30–40 days after fertilization. The long period between first-observed mating and egg-laying in the Bechtel and Bechtel (1958) females (54–68 days), suggests sperm retention until ovulation occurred. The eggs are white, smooth-shelled, usually adhesive, and may number from 3–30. Seigel and Ford (1991) showed that clutch size, but not size of hatchlings, is related to food intake. Data for *E. g. emoryi* are scarce. Perkins (1943) reported a clutch of 14 eggs laid 8–9 July by a Texas snake. Two of these hatched 18 September and the hatchlings measured 142 mm TL. Werler (1951) reported a 1136 mm TL female laid 15 eggs 14–15 June. Seven of these hatched 7–12 August, and the hatchlings measured 370–397 mm TL. Ted Brown (pers. comm.) collected a 920 mm TL female and a 1150 mm TL male in Oklahoma near the New Mexico border. The pair mated 29 May, and the female laid four eggs measuring 18–19 x 46–58 mm 7 July, but the eggs did not hatch.

Food Habits: *Elaphe guttata* is a powerful constrictor, preying on small mammals, lizards, birds, and bird eggs. In New Mexico, foraging is mostly nocturnal during hot and dry weather, but may be diurnal if more favorable temperature and moisture conditions are available. These snakes are excellent climbers and may forage in trees and shrubs as well as on the ground. Throughout the western Great Plains, they are regular cave inhabitants and are important bat predators and adept at climbing cave walls (McCoy, 1975).

Remarks: Smith et al. (1994) described a new subspecies, *Elaphe g. meahllmorum*, which ranges into New Mexico only in the Las Cruces area of Doña Ana County.

FAMILY COLUBRIDAE

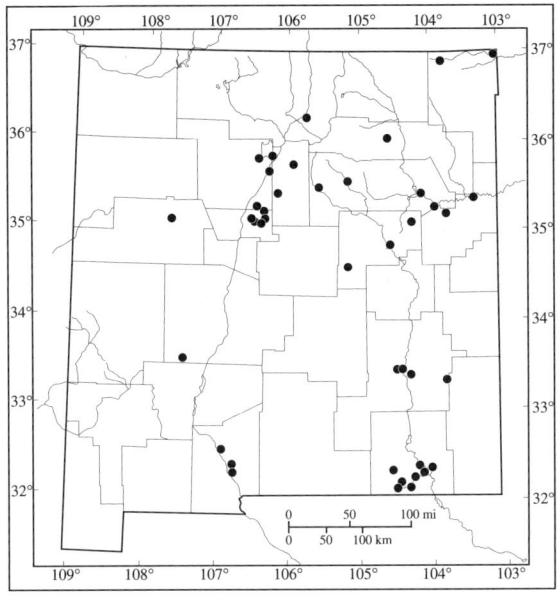

Elaphe guttata was reviewed by Dowling (1952) and Smith et al. (1994).

Distribution of Elaphe guttata in New Mexico.

GYALOPION CANUM Cope, 1860 — *Western Hooknose Snake*

See Plate 85.

Type: The holotype, USNM 5284 (formerly USNM 4675), was collected by Dr. B. J. D. Irwin; the sex and date of collection were unspecified. The type locality was near Fort Buchanan, Arizona. Webb (1966) placed this locality "near Sonoita Creek above Patagonia, 45 mi. SE Tucson, Santa Cruz County, Arizona."

Distribution: *Gyalopion canum* is found from west Texas to southeastern Arizona and south in Mexico to Zacatecas and San Luis Potosí. In New Mexico, it is most common in the southern half of the state but has been taken as far north as Sandoval and Guadalupe counties.

Description: These are small, heavy-bodied snakes seldom attaining more than 28 cm TL. The ground color is tan with 25–48 dark-edged brown crossbars becoming most pronounced nearer the head. The venter is whitish but may have a salmon tinge near midbody. As the common name implies, the snout is upturned and pointed. There are 122–140 ventrals, 26–37 subcaudals, and the anal plate may be single or divided. Dorsal scales are smooth and in 17 rows at midbody. There are seven upper and seven lower labials. Internasals and a loreal are present.

Similar Species: *Gyalopion canum* may be confused with *Heterodon* spp. that also have an upturned snout, although in *Heterodon* the scales are keeled.

Systematics: *Gyalopion canum* is considered monotypic by Hardy (1976).

Habitat: This snake has been described as largely a desert species, often found in areas where mesquite, creosotebush, and agave are dominant plants, although it may ascend to piñon-juniper in the mountains (Conant and Collins, 1991). In New Mexico and elsewhere, we have found it most often in grass-covered foothills at intermediate elevations, ascending and descending to woodland and desert habitats in small numbers. Most of the range coincides with the Chihuahuan Desert, which is rich in these intermediate elevation grassland and foothill habitats. Northward extension may be prevented by the use of shallow hibernacula and retreats since these are inadequate protection against severe winter temperatures. Collection sites in the state range from about 950 m in Eddy County to 1675 m in Sandoval County. Elsewhere, both lower and higher elevations are recorded. We found one at 2100 m on Lost Mine Ridge in the Chisos

Mountains of Texas. This snake probably survives in desert habitats by using efficient water conservation habits such as inactivity on the surface during dry conditions. Surface rock or loose soils for burrowing are used for cover. Frequent cohabitants are *Crotalus viridis, Diadophis punctatus, Salvadora grahamiae, Sonora semiannulata*, and *Tantilla* spp.

Behavior: Probably the most memorable habit of *G. canum* is the "anal popping" that many individuals exhibit. When first touched, the tail is flipped toward the place of contact and the cloacal lining is everted causing a bubbling or popping sound. This is usually accompanied by writhing movements of the body, and the action is repeated a number of times. As in the "play dead" behavior of *Heterodon*, individuals vary in the extent to which they use this defensive ploy. These snakes may also strike with closed mouth, but we have never seen one bite. The upturned rostral suggests burrowing, and Kauffeld (1948) reported a captive doing this in the sand of its cage. Surface activity is mainly nocturnal, and most specimens have been collected on roads at night, although daylight activity has been reported. Curtis (1950) found one near Alpine, Texas, crawling among rocks on a hillside at 1815 in August. Kauffeld (1948) discussed a snake found on the east slope of the Franklin Mountains, Texas, at dusk. Many have been collected on cold nights, after rains, and even on wet pavement (Stebbins, 1954) or ground (Taylor, 1931). One recently killed snake was taken in Durango, Mexico (Webb, 1960) in a light rain at 0230. Two DOR specimens were found by Degenhardt and J. W. Wright in Grant County on 3 September, after a cold and rainy night that was assumed to be too cold for road cruising. Hardy (1975) stated that most specimens have been found from April to September, usually after a light rain. In 51 New Mexico collection records, dates range from April to September, with one snake taken on 3 November, although these particular records probably define the activity of the collector rather than that of the snake, since road cruising before April or after September is usually unproductive in New Mexico. We suspect annual activity is longer than records indicate since one was taken in November and these snakes can be active at relatively low temperatures. No information on hibernation or hibernacula is available.

Reproduction: *Gyalopion canum* is oviparous, although little is known about reproduction in this species. Hardy (1975) reported on a Texas snake that "laid one white egg measuring 6 x 29 mm on July 1, 1970," which did not hatch. A single specimen (275 mm SVL) collected in Hidalgo County laid four eggs on 10 June 1994. These eggs averaged 25 mm (23–27) in length and 2.1 g (1.9–2.3) (our unpubl. data).

Food Habits: Arthropods, mostly spiders, have been reported as the principal food of *G. canum*. Kauffeld (1948) also mentioned that a captive ate spiders and a centipede but refused millipedes, lizards, small snakes, frogs, and newborn mice. Webb (1960) mentioned parts of a scorpion voided by a freshly killed specimen. A very large *G. canum*, collected by Ernest Tanzer in Chihuahua, readily ate newborn lab mice; another snake from Texas ate lizard legs, rat legs, rat meat, and a piece of beef. Stebbins (1985) added snakes as a food item. Vaeth (1980) reported a captive *G. canum* that ate a dead *Diadophis punctatus* almost as long as itself. However, captive animals often will eat unnatural foods. In Big Bend National Park, Texas, a *G. canum* and a *D. punctatus* were found with each gripping the body of the other in its jaws. At the time, it was presumed that the *Diadophis*, since it was larger and is known to eat snakes, was the aggressor.

Remarks: In much of the literature, *G. canum* was known as *Ficimia cana*. Hardy (1976) reviewed the species.

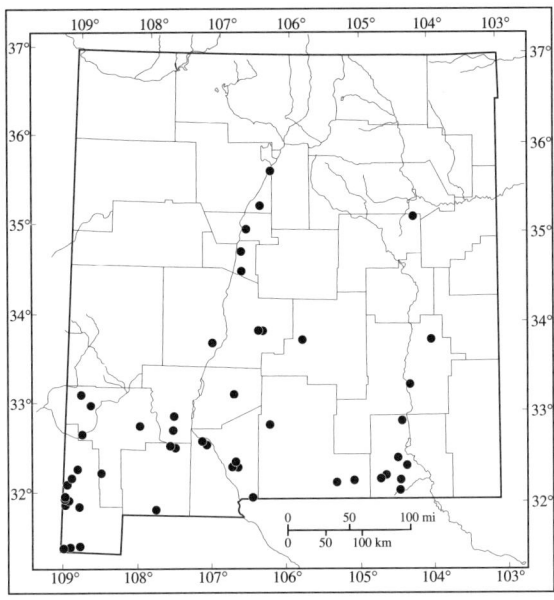

Distribution of Gyalopion canum in New Mexico.

HETERODON NASICUS Baird and Girard, 1852 — *Western Hognose Snake*

See Plate 86.

Type: The holotype, USNM 1272, is a juvenile received from Gen. S. Churchill on 16 February 1838 (Baird and Girard, 1852a). The type locality, "Valley of the Rio Grande," is often given in error as "Texas" (see Remarks).

Distribution: The western hognose snake is found in the Great Plains from southern Canada to central Mexico, with disjunct populations found both east and west of the general range. In New Mexico, it occurs statewide.

Description: These small heavy-bodied snakes are usually 38–64 cm TL when adult, but larger individuals of over 90 cm are occasionally found (See Remarks). Dark, well-defined vertebral blotches and smaller blotches or spots on the sides extend from behind the head onto the tail. The venter is heavily pigmented with black on both body and tail. The snout is sharply upturned and pointed (plow-shaped). The dorsal scales are heavily keeled and in 21–23 rows at midbody. There are 125–152 ventrals, 27–47 subcaudals, and the anal plate is divided. There are 2–28 small scales behind the rostrum (azygous area), one or two loreals, and the eight upper labials are much higher than broad. The eyes are small, with round pupils, and separated from the labials by subocular scales.

Similar Species: *Gyalopion canum* has smooth scales and a white venter. *Heterodon platirhinos*, which may occur in extreme east-central New Mexico, has a light venter which may have dark markings, although the tail is always light colored.

Systematics: Three subspecies are currently recognized (Platt, 1969). *Heterodon n. nasicus* Baird and Girard, 1851 [1852], occurs throughout most of the species range in New Mexico. *Heterodon n. kennerlyi* Kennicott, 1860 occurs in the southwestern portion of the state at least as far north as southern Sierra County and as far east as southern Eddy County based on low azygous scale counts. Intergradation with *H. n. gloydi* Edgren, 1952 may occur in the extreme southeastern corner of New Mexico based on dorsal blotch counts of the specimens we have examined from Eddy and Lea counties, but the status of *H. n. gloydi* as a subspecies was questioned by Platt (1969).

Cochran (1961) listed three syntypes for *H. n. kennerlyi*. She stated that two of these, USNM 1282, from "Sonora" (restricted to Lower Rio Grande, Texas and Sonora, Mexico by Smith and Taylor, 1945), were collected by D. N. Couch; the third, USNM 7290, from "Lower Rio

Grande," was collected by A. Schott. The original description (Kennicott, 1860) recorded Dr. Kennerly as the collector. Other authors (Yarrow, 1882; Cope, 1898 [1900]; Smith and Taylor, 1945, 1950b) listed only USNM 1282 as syntypes. Smith and Taylor (1950b) restricted the type locality to Brownsville, Texas, however, Cochran (1961) was unaware of or chose to ignore this further restriction.

The holotype of *H. n. gloydi* is USNM 5083a, a female, from Wheelock, Robertson County, Texas, collected by F. Kellogg on an unspecified date. Two paratypes, USNM 5083b and USNM 5083c, are males with the same collection data as the holotype. Platt (1969) questioned the validity of this subspecies, attributing the few differences between it and *H. n. nasicus* to clinal variation.

Habitat: Based on studies throughout the range of *H. nasicus*, grasslands with sandy soils serve as typical habitat. In New Mexico, habitat is found in many places from creosotebush desert to open montane woodland where sandy or gravelly soils suitable for burrowing are available. Most collections are from open plains areas, but open valleys and bajadas also support populations. Elevations range between the lowest point in the state, about 920 m, to about 2200 m. Common snake cohabitants are *Arizona elegans*, *Crotalus atrox*, *C. viridis*, *Gyalopion canum*, *Masticophis flagellum*, *Pituophis melanoleucus*, and *Rhinocheilus lecontei*.

Behavior: Hognose snakes are best known for their behavior when threatened. At first the snake may bluff with neck spreading, body inflating, hissing, and striking, but will seldom actually bite. If the threat remains, the snake may turn belly up with body writhing, mouth open, tongue out, and may even regurgitate food and void feces. After these actions it will "play dead" by remaining belly up and will return to this position if turned over. Sexton (1979) described this defensive behavior in detail and explored the possible adaptive value of the various components. There is variation in individual behavior, and some individuals may hide the head beneath the body and will feint at the molester with the coiled tail if touched.

Heterodon nasicus avoids the midday heat, and daily activity occurs mostly in the morning or evening. In Kansas, Platt (1969) measured cloacal temperatures of 21.4–36.2°C in normally active snakes at ground surface temperatures of 17.6–36.0°C. He reported that snakes with a cloacal temperature of 13.7°C were sluggish, but those with temperatures over 15.4°C were able to writhe, feign death, and crawl normally. In cooling experiments, Platt found that hognose snakes have a high normal activity range when compared with other snakes.

Western hognose snakes use below ground retreats when inactive and only rarely use rocks, logs, or other surface objects as hiding places. They use temporary shallow burrows just below the surface or enlarged burrows of other animals. Platt (1969) found that the greatest activity occurred from May to early August in Kansas. Based on New Mexico collection records for May–October, *H. nasicus* appears to be active later in the season here. This is expected in more southern populations. Platt (1969) found *H. nasicus* in his Kansas study first emerged from hibernation from 24 April to 23 May, and the latest captured snakes were taken from 11–31 October. New Mexico data closely approach this annual activity period. In 213 collection records listing date of capture, 18% were taken in May and 10% in October. There is a single record for March and none in November. No hibernacula have been described, but this species probably hibernates singly below the frost line, using self-dug burrows or abandoned burrows of other animals.

Reproduction: Platt (1969) and Fitch (1970, 1985) have published most of the available information on reproduction in the western hognose snake. Males are mature in their third year at an average size of 309 mm TL. Females may mature in either the third or fourth year at 350–400 mm. Most mating is in May, but occurs as late as August. Sperm is produced after mating and stored during hibernation. Most females lay eggs every other year, and the nesting period may extend from June into August, but July is most common. A female from Valencia County laid eggs in early June and two of these hatched 12 September (T. L. Brown, pers. comm.). The eggs are laid in nests a few inches below the surface in loose or sandy soils. They average about 17.9 x 32.5 mm in size and have white leathery shells. The clutch size of 4–23 eggs, averaging 9.4, shows little geographic variation but is correlated with female body size. Incubation time varies with temperature, and hatching normally occurs in August and September, after 52–64 days. Hatchlings average about 180 mm TL, and the sex ratio nears 1:1. The two hatchlings from Valencia County referred to above, measured 150 and 165 mm TL (T. L. Brown, pers. comm.).

Food Habits: *Heterodon nasicus* has a much more varied diet than its eastern congener, *H. platirhinos*, which feeds primarily on anurans. Anurans, lizards, small snakes, rodents, and reptile eggs are foods of choice, but birds may also be eaten by larger individuals. Captives often feed and do well on mice. Prey items are located by sight or smell, grasped by the jaws, and swallowed. Constriction is not used, but prey may be restricted to some extent with a body coil. Buried prey may be dug up when located by smell (Platt, 1969). Iverson (1990) reported this species as an important predator on *Kinosternon flavescens* eggs in his Nebraska study area. There are enlarged teeth at the rear of the upper jaw that may function in holding prey firmly during swallowing, or introducing the slightly toxic salivary secretions into the prey. The ability to neutralize certain anuran secretions (mainly the digitaloids) is shared with other toad-eating snakes (Smith and White, 1955).

Remarks: Banta (1971) argued convincingly that the original description for *Heterodon nasicus* was first published in Baird and Girard (1852a) and not in Baird and Girard (1851 [1852]) or (1852b).

Designation of the holotype is uncertain and confusing. Baird and Girard (1853a) gave a detailed description of the type, a juvenile collected by Gen. S. Churchill from the "Valley of the Rio Grande." Yarrow (1882) does not list a USNM specimen from the Rio Grande nor one collected by Churchill, but lists two "types," USNM 1285 from Red River, Arkansas, and USNM 4863 from Santa Fe, New Mexico. Cope (1898 [1900]) also listed USNM 4863 as the type but placed USNM 1285 further down on his list of USNM specimens with no type designation. Cochran (1961) listed no type specimens for *H. nasicus*. Robert P. Reynolds, of the USNM, informed us that USNM 1272 is the holotype, but this specimen was found to be missing in 1957. This may be why Cochran (1961) did not list the type.

We doubt that the type locality, often given as "Texas," is correct. The description of the type (Baird and Girard, 1853a) best fits *H. n. nasicus*, and this subspecies, as presently defined, does not occur in the Rio Grande Valley of Texas, with the possible exception of the El Paso area. Schmidt (1953) recorded "Texas; restr. to Amarillo by R. A. Edgren (p.c.)." Edgren (1952), in his synopsis of the species, said nothing about the type locality of *H. nasicus*, and any other information that Edgren may have had was never published. Based on the original description, if the missing type was indeed from the Rio Grande Valley, it was probably collected in New Mexico.

During September 1995, an adult female *H. n. kennerlyi* was collected DOR in the southern Animas Valley by B. R. Tomberlin. The specimen measured 760 mm TL, exceeding the 656 mm record length of Werler (1951).

This species was reviewed by Edgren (1952), Smith and Smith (1962), and Platt (1969). The genus was reviewed by Platt (1983).

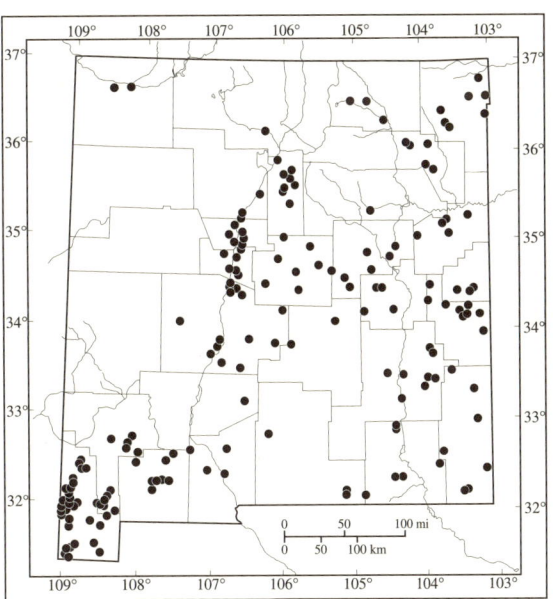

Distribution of Heterodon nasicus *in New Mexico.*

HYPSIGLENA TORQUATA (Günther, 1860) *Night Snake*

See Plate 87.

Type: The existing type is BMNH 61–12–30,97, one of two "cotypes" (= syntypes) described by Günther in the original description. The type locality, Laguna Island, Nicaragua, is probably in error (see Remarks).

Distribution: *Hypsiglena torquata* is widespread throughout the southwestern United States from California to Kansas and in Mexico as far south as Guerrero. It is widespread in New Mexico, but most common in the desert lowlands.

Description: This is a small slender-bodied snake with elliptical pupils. Adults measure 300–642 mm TL (Tanner, 1944). The ground color is tan or gray with brown or dark gray blotches on the back and sides, producing a spotted appearance. There are usually large elongated dark blotches on either side of the neck, but these vary considerably. The top of the head is patternless, but there is a dark band above the light-colored labials on the sides of the head. The venter is white or yellowish with no markings. There are 162–204 ventrals, 38–66 subcaudals, and the anal plate is divided. The dorsal scales are smooth and in 19–21 rows at midbody. There are two preoculars and two postoculars; the lower preocular is wedged between two upper labial scales. The loreal is single. There are usually eight upper and 10 lower labials. Ungrooved, fanglike teeth are present on the rear upper jaw.

Similar Species: *Trimorphodon biscutatus* also has elliptical pupils but has light-centered crossbands and usually a V- or lyre-shaped marking on the head. *Arizona elegans* has a single anal plate. Young *Coluber constrictor* has round pupils. *Pituophis melanoleucus* has keeled scales. Young *Crotalus* spp. and *Sistrurus catenatus* have keeled scales and a horny button on the tip of the tail.

Systematics: Statistical data from Dixon and Dean (1986) suggest that one subspecies, *H. t. jani* (Dugès, 1865), occurs throughout most of New Mexico. *H. t. loreala* Tanner, 1944, may be present in the extreme northwestern corner of the state.

The type of *H. t. jani* is unknown; the type locality is Guanajuato, Mexico (see Remarks).

The holotype of *H. t. loreala*, BYU 2829, is an adult female collected at the west edge of Castle Dale, Emery County, Utah, 17 June 1939 by V. M. Tanner.

Habitat: This widely distributed species is found in a variety of habitats from temperate into tropical regions. In New Mexico, *H. torquata* reaches its highest densities in desert and semi-desert, is well represented in foothill grasslands, and is found in montane woodland from 900 m to at least 2200 m. Rocks are preferred as cover, but other types of surface debris, either natural or man-produced, are used for daytime retreats. Cracks in rocky outcrops may be used, and these are also inviting to lizards that may serve as the next meal. Commonly found snake cohabitants are *Sonora semiannulata*, *Tantilla* spp., *Leptotyphlops* spp., and *Rhinocheilus lecontei*.

Behavior: The elliptical pupils of *H. torquata* suggest its nocturnal habit, and it is most often seen while crossing the road at night. Night snakes do not bite defensively but do have enlarged rear teeth and a mild venom used for subduing prey. Price (1987) described an interesting defensive behavior of some individuals, which form the body into a tight coil with the head in the center. This somewhat ball-like position is held even if the snake is picked up or turned over. Stuart (1988) described another defensive behavior in which body rigidity and "body-snapping" actions were elicited when the snake was startled and touched.

Annual activity normally extends from April–October, but there are records as early as January, February, and March (one each) from southern New Mexico. Little information is available on hibernacula. One snake emerged from a rabbit burrow near Los Lunas, Valencia County, after an attempt was made to smoke out a rabbit (Koster, 1940). The snake was probably hibernating in the burrow, since it was a cold day in February and the ground was partially covered with the previous night's snow.

Reproduction: Little is known of the reproductive activities of *H. torquata*. Female sizes that have been reported at parturition range from 307–425 mm SVL (Clark and Lieb, 1973). Spring mating is assumed, since eggs have been laid or found in gravid females from 25 April (4 laid) to as late as 1 September (3 laid). Vitt (1975) suggested the possibility of two clutches per year, in reporting that a September clutch from an Arizona snake may have been the second clutch. Fitch (1970) assumed an extended breeding season in the southwestern United States and Mexico to explain these late clutches. Data from Texas night snakes (Clark and Lieb, 1973) suggest a north-south cline in timing of egg production (later farther north). Clark and Lieb (1973) also suggested that the timing of the reproductive cycle in arid environments may be controlled by such factors as rainfall. Eggs usually number from 3–6, vary in size from 9–12 mm x 15–32 mm, and hatch in about two months. Wright and Wright (1957) reported the smallest snake as 102 mm TL. Werler (1951) recorded the size of two hatchlings from Texas at 169 and 175 mm TL, and Clark and Lieb (1973) recorded southern Arizona hatchlings at 192 mm (♂), 157 mm (♀), and 133 mm TL (♂). Ted L. Brown (pers. comm.) found a 195 mm TL juvenile on 4 July in Socorro County.

Food Habits: *Hypsiglena torquata* feeds mainly on lizards, small snakes, and amphibians. Kauffeld (1943a) recorded one feeding on a DOR *Scaphiopus couchi* in Arizona. A captive Texas snake fed on small frogs such as *Pseudacris* spp., but upon grasping a *Gastrophryne carolinensis*, went into violent body contortions and apparently tried to rub off the toad secretions from its mouth on the cage bottom. After flushing the snake's mouth with water, the snake recovered, apparently with little harm done. Webb (1970b) recorded a night snake eating a *Leptotyphlops dulcis* and T. L. Brown (pers. comm.) removed a *Leptotyphlops dulcis* (MSB 13403) from the stomach of a small *Hypsiglena* and fed a snake in captivity on "*Uta*, small whiptails, small *Sceloporus*, and small *Scaphiopus*." Other small snakes often found in the same microhabitat (*Tantilla, Sonora*) are probably frequent prey items. Cannibalism has occurred in captivity. Two night snakes were kept in the same cage for weeks before the smaller snake served as a meal for the larger. Most feeding is done at night and the venom apparatus, though primitive, is useful in subduing prey. Lizards grasped by night snakes have been observed to become limp before they were swallowed. Inactive diurnal lizards in retreats or flattened on the surface of the ground, as well as nocturnally active prey, may serve as food.

Remarks: Dixon (1965) obtained information on the type from the late Mr. J. C. Battersby of the British Museum. Dixon concluded that the identity seems to be correct, but the locality appears to be in error based on two important facts. The location, Lake Nicaragua, lies 1850 km from the nearest verified locality for the species, and Günther never travelled to Central America. Tanner (1944) suggested that USNM 9889 and 11369, collected by Dugès from Guanajuato, Mexico, may represent type material of *H. t. jani*.

Three captives from New Mexico and southeastern Arizona lived for 92, 98, and 106 months (T. L. Brown, pers. comm.).

This species has been reviewed by Taylor (1938a), Tanner (1944), and Dixon and Dean (1986).

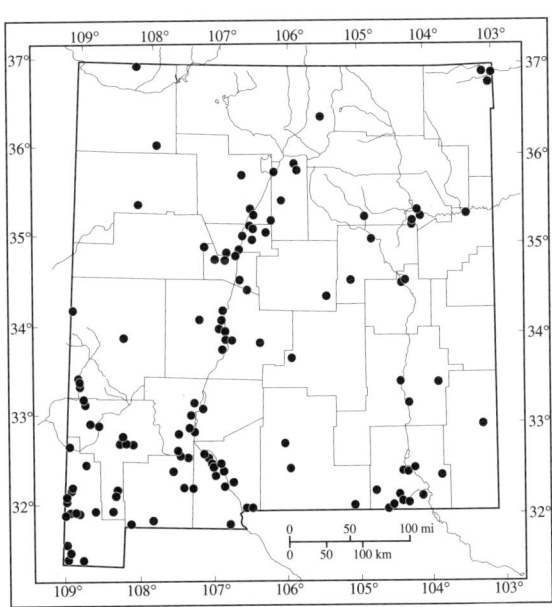

Distribution of Hypsiglena torquata *in New Mexico.*

LAMPROPELTIS ALTERNA (Brown, 1901) — *Gray-banded Kingsnake*

See Plate 88.

Type: *Lampropeltis alterna* was first described as *Ophibolus alternus* by Brown (1901b). The type specimen, ANSP 14977, is a female collected by E. Meyenberg. The type locality is "Davis Mountains, Jeff Davis County, Texas." The collection date is unknown (Malnate, 1971).

Distribution: *Lampropeltis alterna* ranges from Trans-Pecos Texas and southeastern New Mexico, east to Edwards County, Texas, then south to northeast Durango and extreme western Nuevo León (Conant and Collins, 1991; Painter et al., 1992). In New Mexico, *L. alterna* is known from extreme south-central Eddy County in the Guadalupe Mountains at an elevation of 1160 m. We expect this species to occur on Otero Mesa and also in the western foothills of the Guadalupe Mountains in extreme southeastern Otero County, at elevations from approximately 1070–2040 m. Hakkila (1994) provided a detailed map of the expected range of *L. alterna* in New Mexico.

Description: The most striking feature of *L. alterna* is the brilliant color pattern, particularly in the *"blairi"* morph. It is extremely variable, with dark gray to black crossbands either narrow and unicolor or expanded with varying amounts of red or orange. The large eyes are slightly protuberant, and the broad head is noticeably wider than the neck. The scales are smooth and glossy. The iris is a distinctive silver-gray color. Currently only the "alterna" color phase is known from New Mexico. This color phase is characterized by 15–23 thin, occasionally white-edged, black bands that are sometimes split with red. These black bands are separated by broad areas of light to medium gray. There is a strong tendency toward melanism in some populations. Several illustrations of the wide diversity of pattern types found in this species are in Miller (1979) and Conant and Collins (1991). The single known museum specimen from New Mexico is a male with a slate gray background and 15 solid black crossbands that do not encircle the body. Maximum length reported for adults is 147.1 cm (Conant and Collins, 1991) although most are 51–90 cm TL. This species has 23–27 scale rows at midbody, 208–229 ventrals, and 55–67 subcaudals. The anal plate is entire. There are seven upper labials, one preocular, 2–3 postoculars, and usually two anterior temporals (Smith and Brodie, 1982).

Similar Species: Although the gray-and-black banded *"alterna"* morph somewhat resembles the sympatric mottled rock rattlesnake, the lack of a rattle, slender form, and the smooth scales will immediately distinguish *L. alterna*. This species has 19 or more posterior scale rows while most other *Lampropeltis* have only 17.

Systematics: No subspecies of *L. alterna* are currently recognized, although considerable taxonomic confusion has existed in the past. Information in the literature may be found under the names *L. mexicana*, *L. blairi*, and *L. alterna* and various combinations thereof (Gehlbach and Baker, 1962; Gehlbach, 1967b; Garstka, 1982). Flury (1950) described the orange or Blair's color phase of *L. alterna* as a separate species, *L. blairi*, from a specimen collected "8.8 miles west of Dryden, Terrell County, Texas." Later studies of the clinal variation in scutellation and coloration by Gehlbach and Baker (1962) suggested that these two "species" were actually subspecies of the widespread *L. mexicana*. Reproduction of *L. alterna* was studied by Tanzer (1970), who observed five siblings that exhibited the color and pattern types of both *L. alterna* and *L. blairi*. Based on this color variation, Tanzer (1970) suggested that these two presumed subspecies were actually polymorphs of the same taxon. His view is accepted today.

Habitat: *Lampropeltis alterna* is a secretive inhabitant of the Chihuahuan Desert, where it is usually found in rocky igneous or limestone areas associated with various species of desert vegetation including sotol, cacti, acacia, mesquite, ocotillo, or creosotebush. The single individual collected in New Mexico was found on a blacktop road that follows a large rocky canyon with steep canyon walls and abundant Chihuahuan desert vegetation.

Behavior: *Lampropeltis alterna* appears to be primarily nocturnal. Most specimens have been found while crossing blacktop roads at night or in rocky situations where there are abundant large cracks between rock layers such as in highway road cuts or canyon walls. In Texas, *L. alterna* becomes active in mid- to late May with a peak in activity in late June, tapering off in July.

Activity may begin during mid- to late April with warm evening temperatures that permit nocturnal activity early in the year. Late season activity occurs into mid- to late October (Miller, 1979).

Reproduction: Reproduction of *L. alterna* has not been studied in New Mexico. Murphy et al. (1978) reported three clutches of eggs laid by captive females on 7, 9, and 23 July. One clutch of nine eggs was infertile whereas two clutches of six and eight eggs were incubated and successfully hatched. Hatching occurred from 9–12 September after incubation for approximately 63 days. The hatchlings were between 257 and 323 mm total length and averaged 284 mm. Gehlbach and McCoy (1965) described five eggs taken from a gravid female collected in Brewster County, Texas, on 16 May that were, "apparently ready for oviposition." These eggs averaged only 12 x 30 mm. Miller (1979) provided measurements for two clutches of nine eggs laid on 6 and 18 June that averaged 34.2 x 23.4 mm, much larger than those reported by Gehlbach and McCoy (1965).

Food Habits: Although *L. alterna* forages mostly at night, diurnal lizards constitute the chief prey. Those reported by Gehlbach and Baker (1962) include *Eumeces* sp., *Sceloporus undulatus*, *S. poinsetti*, and *Cnemidophorus* sp. We found the plateau spotted whiptail, *Cnemidophorus septemvittatus*, in the stomach of a specimen collected DOR in the Chisos Mountains. Damon Salceies (pers. comm.) found an adult male from Brewster County, Texas, that regurgitated an adult *Cophosaurus texanus* shortly after capture. Rodents are occasionally eaten also. Miller (1979) reported a specimen (SRSU 1968) that contained *Hyla arenicolor*. Captive specimens are known to eat domestic white mice, *Mus musculus*, and a variety of lizards. Cannibalism is rare in captives and ophiophagy is unknown.

Remarks: Several workers (Gehlbach and Baker, 1962; Gehlbach, 1967b; Garstka, 1982) have suggested that this species occurs in New Mexico, although no specimen was available from the state until 1991. Painter et al. (1992) provided information on a single specimen (MSB 52000) collected from Carlsbad Caverns National Park in southern Eddy County. There are recent photographs of this species from southeastern Otero County.

As a result of widespread successful captive breeding, these kingsnakes are plentiful in private collections, and the small population thought to occur in New Mexico should be preserved in the wild.

Gehlbach (1967b) reviewed this species (as *Lampropeltis mexicana*). Hakkila (1994) reviewed the status and distribution of *L. alterna* in New Mexico.

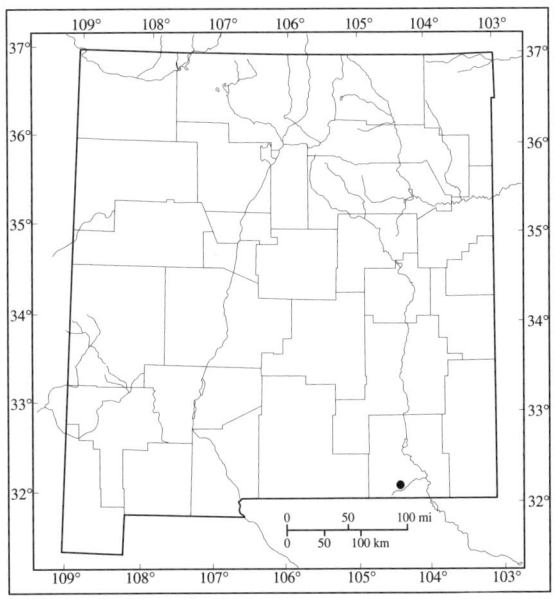

Distribution of Lampropeltis alterna *in New Mexico.*

LAMPROPELTIS GETULA (Linnaeus, 1766)   *Common Kingsnake*

See Plate 89.

Type: The holotype was in the private collection of Linnaeus and the present whereabouts is unknown. It was probably a female, based on the low subcaudal count (44), and was collected by Dr. Garden. The original description gives "Carolina" as the locality. Klauber (1948) restricted the type locality to Charleston, South Carolina.

Distribution: *Lampropeltis getula* is widely distributed throughout the southern two-thirds of the United States and northern Mexico to Zacatecas and San Luis Potosí. In New Mexico, it has been found statewide except at higher elevations and in the northwest corner (see Remarks).

Description: *Lampropeltis getula* is of medium size and girth, adult size ranging from 78–210 cm TL. New Mexico individuals rarely exceed 150 cm TL and usually are smaller than 120 cm TL. The basic color pattern is black or dark brown with white or yellow markings. These markings may form a chain-like pattern on the dorsum, or a white spot occurs on each dorsal scale, or the pattern is a combination of the two. A black form occurs in southern Arizona, and a striped form is found in southern California. In New Mexico, the most frequent pattern is white speckled with dark dorsal blotches. Speckling is less pronounced in snakes from the southwest part of the state. The venter is dark gray to black. In the young, the white is more pronounced. The head is only slightly distinct from the neck, and the pupils are round. The scales are smooth and glossy and in 19–25 rows at midbody, usually 21 in New Mexico. There are 199–254 ventrals, 37–56 divided subcaudals, and the anal plate is single. There are seven upper and 9–10 lower labials, a single preocular, usually two postoculars, and a single loreal. The eye contacts the third and fourth upper labials.

Similar Species: Considering the distinctive coloration, *L. getula* should not be confused with any other snake in New Mexico.

Systematics: The most recent taxonomic study by Blaney (1977) indicated that one subspecies, *L. g. splendida* (Baird and Girard, 1853[a]), occurs in New Mexico, although eastern populations show influence from *L. g. holbrooki* Stejneger, 1902, and *L. g. californiae* (Blainville, 1835) may occur in the northwestern corner of the state (see Remarks).

The holotype of *L. g. splendida*, USNM 1726, is an adult male collected by J. H. Clark (see Remarks) in September 1851. The type locality is "Sonora," Mexico, restricted by Smith and Taylor (1950b) to the Santa Rita Mountains, Arizona.

Neither the type nor the type locality of *L. g. holbrooki* was designated in the original description. Stejneger and Barbour (1917) gave the type locality as "valley of the Mississippi."

A type of *L. g. californiae* is not designated in the original description. The type locality, "California," was restricted to San Diego by Smith and Taylor (1950b) and to the vicinity of Fresno by Schmidt (1953).

Habitat: *Lampropeltis getula* occupies a wide diversity of habitats from desert to coniferous forest. In New Mexico, preferred habitat is riparian or grassland, but some are found in piñon-juniper, and low elevation desert areas. The majority of specimens have been taken along our principal river systems. The elevational limits are 900 m to about 1800 m. *Lampropeltis getula* uses all kinds of surface cover, but in desert areas, it uses rock outcrops or rodent burrows that allow it to penetrate to sufficient depth to escape hot dry conditions. It often uses old houses, barns, foundations, wells, and cisterns for retreats.

Behavior: When first captured or disturbed, the common kingsnake may coil, hiss, vibrate the tail, strike, and bite. With handling, most individuals become gentle and make good pets. These snakes may be active either at night or during morning or evening, depending on climatic conditions. Most New Mexico specimens are taken at night while crossing the road, or in the daytime under rocks or surface debris. Snakes active in the daytime have often been found near water. Annual activity in New Mexico extends from April to October with a peak in July. Gibbons and Semlitsch (1987) reported that annual activity in *L. getula* was unimodal and peaked in June. Price and LaPointe (1990) studied snake activity by road cruising in the lower Rio Grande Valley of New Mexico over four years. Based on 46 records, their data show a unimodal activity pattern with a peak in July for *L. getula*. By tabulating all 151 records with collecting dates available to us (excluding those from Price and LaPointe, 1990), we find a similar pattern with a peak in July. June is our hottest and driest month; the shift of maximum activity to July coincides with the beginning of the summer monsoon season. Common kingsnakes probably hibernate below the frostline using rodent burrows, fractured rocky outcrops, or man-made structures such as foundations, wells, and cisterns, but no hibernacula have been discovered in New Mexico.

Reproduction: Zweifel (1980) published most of the information on reproduction in *L. getula* based on a laboratory population at the American Museum of Natural History. This population was somewhat mixed genetically, but it did include a good representation of southwestern races. The earliest age that any of the males or females bred was two years and eight months after hatching, but breeding at nearly four years was much more common. Since growth was not "forced" by abundant feeding, these data may approximate natural conditions. Fitch (1970) reported mating from April to June in six non–New Mexico populations, and Zweifel (1980) reported February to May in his mixed population. The mating period may actually be extended, for multiple clutching has been reported. There was a high correlation between mating and egg laying; unmated females rarely laid eggs whereas most mated females did. Zweifel (1980) also noted that unreceptive females often attacked a prospective mate. Successful copulations averaged 4.3 hours and males tended to alternate hemipenes in repeated copulations.

The leathery-shelled, adhesive eggs measure 18–30 mm x 35–69 mm. Fitch (1970) summarized clutch sizes and found they ranged from 5–17 with a mean of 10.1. Zweifel (1980) reported a range of 2–7 with a mean of 4.4 in the hybrids and 2–9 with a mean of 6.3 in pure *L. g. californiae* stocks. Egg numbers have ranged from 2–24 in literature reports. Clutch size is probably related to body size in this species, as it is in most snakes. Therefore the smaller-sized races of *L. getula*, such as *L. g. splendida* in New Mexico, probably conform most closely to the smaller range and mean. A female from Valencia County (100 cm TL) laid seven eggs on 4 July and four of these hatched 12–19 September, after 70–77 days of incubation (T. L. Brown, pers. comm.). Brown had another female from Sierra County that laid five eggs on 9 July; these all hatched on 18 September after 71 days of incubation. Hammerson (1982) recorded a late July clutch of eight eggs laid by a 940 mm female from southeastern Colorado, and Webb (1970b) reported a snake from central Oklahoma that laid 12 eggs. Humphrey (1956) reported a pair found mating in southern Arizona on 17 April. The female laid six eggs on 17 June, and a single egg hatched on 11 August. This was an apparent gestation of 60 days and incubation period of 54 days. Meade (1932) reported a gestation of 73 days and an incubation time of 69 days. Zweifel (1980) found that gestation time for 58 clutches ranged from 37–73 days at variable room temperatures and found the expected inverse relationship between mean room temperature and gestation time. In 51 clutches, incubation times ranged from 51–78 days at room temperatures, and showed the same inverse relationship between temperature and time. In small samples of eggs incubated at different temperatures, Zweifel (1980) found temperatures below about 31.2°C to be best for normal development, and that a constant temperature above 32.8°C approached the limit of embryonic tolerance. Snout-vent length of hatchlings ranged from 230–315 mm TL.

Food Habits: Kingsnakes are well known for their snake-eating habits, and we often hear of their success in subduing and eating venomous species. However, *L. getula* also feeds on a wide variety of both endothermic and ectothermic vertebrates and their eggs. They are relatively immune to snake venom, as are many other snakes, and include both pit vipers and coral snakes in their diet. Venomous snakes occurring in sympatry with kingsnakes often exhibit defensive behavior when confronting a kingsnake or its odor. Rather than showing a preference for venomous snakes, the opposite seems to be true if the kingsnake is given a choice. Otherwise, the feeding pattern is generally opportunistic with a preference for familiar foods. One newly acquired captive Florida kingsnake steadfastly refused to feed on laboratory mice until the fast was broken with a northern watersnake, and thereafter it fed ravenously on mice. *Lampropeltis getula* is a powerful constrictor and has been known to subdue and feed on snakes longer than itself. Kingsnakes will feed anytime they are active, and activity may be nocturnal or diurnal depending on climatic conditions.

Remarks: *Lampropeltis g. californiae* is expected to occur in the San Juan River valley of northwestern New Mexico since this race has been found as near as 2 km north of the New Mexico border in southwestern Colorado (Hammerson et al., 1991).

This is another case where Col. J. D. Graham is falsely listed as the collector, rather than J. H. Clark. Specimens in the Graham collection, reported on by Baird and Girard (1853a), were collected by Clark and sent to the USNM by Graham since he was in charge of the expedition. See the discussion in the *Salvadora grahamiae* account for additional information.

The genus was reviewed by Blaney (1973) and the species by Blaney (1977).

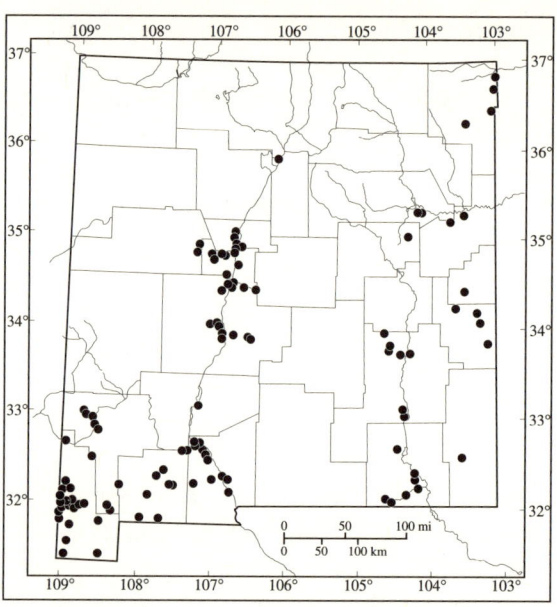

Distribution of Lampropeltis getula *in New Mexico.*

LAMPROPELTIS PYROMELANA (Cope, 1866 [1867])   *Sonoran Mountain Kingsnake*

See Plate 90.

Type: The lectotype, USNM 11421 (see Remarks), was collected by Dr. E. Coues. Cope (1866 [1867]) did not specify a type locality, but Coues (1875) gave "Fort Whipple [Yavapai County, Arizona], August, 1864" as collection data. USNM 7845 and USNM 11421 were syntypes.

Distribution: *Lampropeltis pyromelana* has a broken and spotty distribution mainly west of the Rocky Mountains from eastern Nevada and western Utah, south through Arizona and southwestern New Mexico, and into eastern Sonora and western Chihuahua of northern Mexico. In New Mexico, it is confined to the mountains of the southwest where it has been taken only in Catron, Grant, Hidalgo, and Sierra counties.

Description: *Lampropeltis pyromelana* is a medium sized snake, with adults ranging from 45–109 cm TL. The color consists of bands of red, black, and white. The red bands are bordered by black bands and form wedges which split the black bands. There is much variation in both these dorsal bands and those that extend onto the venter. There may be fewer than 43 to over 80 white bands in the species, but the New Mexico population usually has 45–75. The snout is white or yellowish and may have black or red flecking. The scales are smooth and in 23–25 rows at midbody. There are 213–238 ventrals, 61–79 divided subcaudals, and the anal plate is single. There are usually 7 (occasionally 8) upper labials, usually 10 lower labials in New Mexico individuals, a single preocular, and a loreal which is longer than high.

Similar Species: *Lampropeltis triangulum* has a black (not white) snout, fewer than 45 white bands, and fewer than 210 ventrals. *Micruroides euryxanthus* has a much more regular and distinct banding pattern with red bands bordered by yellow. *Rhinocheilus lecontei* has white spots in the black bands, although this may be hard to discern in young, and subcaudals undivided anteriorly with some divided posteriorly.

Systematics: Three subspecies are recognized by Tanner (1983), although only the nominotypical form, *L. p. pyromelana*, occurs in New Mexico.

Habitat: The Sonoran mountain kingsnake is primarily a montane species, as the common name suggests, but it does descend to lower elevations in mesic canyons and valleys. It has been taken as low as 1380 m in oak-juniper to about 2500 m in coniferous forest. Talus or rock piles in stream beds, allowing fairly deep penetration, are preferred, although individuals regularly make use of other types of surface cover. Snakes have been collected after they emerged from an old concrete slab in oak-juniper woodland near a small permanent stream (W. Joyce, pers. comm.). Many have been taken from a relatively wet, north facing canyon with an abundance of loose talus on the sides. Occasionally they are encountered on highways that traverse such areas. Frequent snake cohabitants are *Crotalus lepidus, C. willardi,* and *Masticophis bilineatus. Sceloporus jarrovi* is usually abundant in the southern portions of the habitat in New Mexico.

Behavior: *Lampropeltis pyromelana* is often found in the daytime along trails or on forested slopes. Thermoregulation may involve basking in the open, or probably, more often, remaining hidden under warmed surface rocks. There are 24 capture dates for New Mexico and these indicate annual activity from April to October. No information is available on hibernation.

Reproduction: Little information is published on reproduction in *L. pyromelana*. Painter (1985) observed a copulating pair 5 May in southeastern Arizona. Stebbins (1985) reported that 3–6 adhesive eggs are laid in June or July. Zweifel (1980) reported on nine clutches from a female collected in southeastern Arizona. Three to six eggs were laid from 23 April to 27 May; there was no indication of clutch size increasing with age. Incubation times ranged from 58–78 days and were inversely related to temperature. Hatching success was 70%. Tanner and Cox (1981) reported that a snake from Utah laid five eggs on 23 June which measured about 44 x 18 mm. Incubation time was 57–58 days, and the hatchlings measured from 205–294 mm TL. The smallest specimen seen by Wright and Wright (1957) was 203 mm TL.

Food Habits: Lizards are the principal food items of *L. pyromelana*, although small snakes and mammals are also eaten. Constriction is used for killing. Tanner and Cox (1981) reported hatchlings from Utah that showed a preference for baby lizards. Small laboratory mice are usually accepted by captives.

Remarks: The syntypes, USNM 7845 (collected by Palmer) and USNM 11421 (collected by Coues), were both listed by Cope in his 1898 [1900] work but only USNM 11421 was designated as "type." Cochran (1961) mentioned only USNM 7845 as a syntype. In his review, Tanner (1983) cited Cochran (1961) and reported USNM 7845 as a "syntype" and seemingly ignored Cope's (1898 [1900]) designation of USNM 11421 as the type. We believe Cochran (1961) was in error and USNM 11421 is the lectotype and USNM 7845 is a paralectotype.

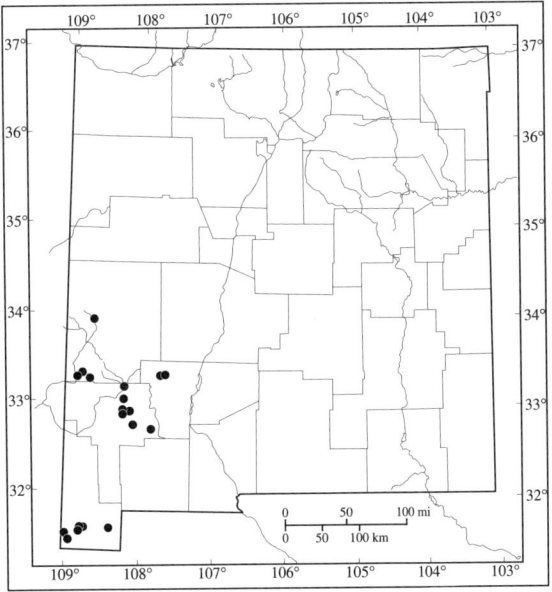

Distribution of Lampropeltis pyromelana *in New Mexico.*

LAMPROPELTIS TRIANGULUM (Lacepède, 1789)   *Milk Snake*

See Plate 91.

Type: The type is unknown. The type locality "America," was restricted to the vicinity of New York City by Schmidt (1953).

Distribution: *Lampropeltis triangulum* is widespread in North America from southeastern Canada through the United States, except west of the Rocky Mountains, south through Mexico and Central America into northwestern South America. The milk snake probably occurs throughout New Mexico, but has yet to be found in most counties west of the Rio Grande Valley, except San Juan, Rio Arriba, and Sandoval in the northwest. Holycross and Schwalbe (1995) reported this species from southwestern New Mexico in Hidalgo County. An unverified sight record exists for El Morro National Monument in Cibola County (Gary Stolz, pers. comm.).

Description: *Lampropeltis triangulum* may attain 132 cm TL in the northern United States (Conant 1975), but most individuals in the Southwest measure from 60–90 cm. The color and pattern vary markedly within the large range of the species. Northern and some eastern *L. triangulum* have bordered darker blotches on a light background; others have a banded pattern with combinations of red, black, yellow, or white bands, often resembling sympatric coral snakes. New Mexico individuals have triads consisting of red bands bordered with black, and these triads are separated by the white ground color forming annuli. There are fewer than 45 white annuli in New Mexico milk snakes. The snout is black but may be mottled with white. The scales are smooth and in 21 rows at midbody. The anal plate is single, subcaudals divided, and ventrals range from 170–194.

Similar Species: *Lampropeltis pyromelana* has a white snout, usually more white annuli (43–80+), and more than 210 ventrals. *Rhinocheilus lecontei* usually has white spots in the black bands and subcaudals are undivided (a few may be divided). *Micruroides euryxanthus* has a more regular and distinct banding pattern with the red bands, rather than the black bands, bordered by white or yellow.

Systematics: One subspecies, *L. t. celaenops* Stejneger, 1902 [1903], is known to occur in New Mexico and another, *L. t. gentilis*, (Baird and Girard, 1853a) may occur (see Remarks).

The holotype of *L. t. celaenops* is USNM 22375, an adult male, collected by H. B. Lane on an unknown date, from the Mesilla Valley, Doña Ana County, New Mexico.

The lectotype of *L. t. gentilis*, USNM 1853, was designated by Blanchard (1921) from one of the syntypes. It is an adult male collected by Capt. R. B. Marcy 14 June 1852 from the north fork of the Red River, near Sweetwater Creek, Wheeler County, Texas.

Habitat: That this species occupies a wide variety of habitats is not surprising considering its extremely wide range. Even in New Mexico where all specimens have been referred to one subspecies, *L. t. celaenops*, diverse habitats are occupied. Most specimens were collected in high foothill grassland and coniferous forest habitats to near 2400 m, but others are from river valley, piñon-juniper, sagebrush, short grass prairie, desert scrub, and sandhill-shinnery oak habitats. Cover for retreats may be rocks, logs, any other kind of surface debris, or animal burrows. In Bernalillo County, a snake was active on a pine-needle-covered, moderately rocky hillside at 2230 m in piñon-juniper-ponderosa woodland. Another was taken from under rotting boards in the remains of an old collapsed house in piñon-juniper woodland. Where snakes have been collected in sand hill habitats, the only available cover is in rodent burrows. Other reptiles often found with milk snakes and with approximately similar habitat preferences are *Pituophis melanoleucus*, *Sceloporus undulatus*, and *Eumeces* spp.

Behavior: *Lampropeltis triangulum* is shy and retiring but may void feces or deliberately bite when captured. They may be active day or night depending on weather conditions and the type of activity. Many low elevation records are of snakes taken on roads at night or of those taken under rocks, boards, logs, and other surface objects during the day. Rather than basking in the open, sun heated retreats are usually used for thermoregulation. Annual activity extends from at least March–October, even in the northern portions of the state. Nothing is known concerning hibernacula in New Mexico. In Utah, apparently hibernating snakes have been found in December during the digging of holes for power line poles and gravel from a pit (Wright and Wright, 1957), suggesting the use of animal burrows for access to retreats below the frost line. Woodbury and Hanson (1950) reported *L. triangulum* entering hibernation with other species in a "long horizontal tunnel" on 16 October. Evidently any access to subterranean retreats below the frost line may be used.

Reproduction: Little information is available for New Mexico populations of *L. triangulum*. Mating probably occurs shortly after emergence from hibernation, as it does in other southwestern forms, and as observed in captives. The male holds the female during copulation by grasping her neck in his jaws. Eggs are normally laid in May and June. Fitch (1985) reported the clutch size for all *L. triangulum* to range from 3–24, but the larger-sized northern populations normally lay about twice as many eggs as the southern populations. Ted L. Brown (pers. comm.) captured a 60 cm female in Roosevelt County which laid three eggs measuring 13–14 x 37–44 mm on 4 July. Two of these eggs hatched 24–25 August; the hatchlings measured 190 and 195 mm TL. The mother died after 13 years in captivity. Eggs are adhesive and reported sizes for all subspecies range from 13–22 x 24–52 mm. Another specimen in the MSB collection, from Bernalillo County, laid eggs on 1 July which hatched on 25 August. Fitch (1970) recorded other laying to hatching dates of 5 June to 24 and 25 July for one clutch, and 21 July to 26 and 28 August for a second. He gave two other laying-only dates of 21 June and 3 July.

Food Habits: Milk snakes are so called because of the fallacious old belief that they suckled milk from cows. Their presence in rodent-rich barns and the sometimes poor milk production of a cow seemed proof enough to many persons. Lizards, small snakes, and rodents make up the largest part of the diet. Constriction may be used to kill prey, but small prey is often swallowed directly or while a loop or coil is used for restraint. A southern Colorado snake disgorged seven reptile eggs (Hahn, 1968), and such foods as earthworms, birds, and insects have been reported. Captives can usually be induced to feed on mice, but some individuals are erratic feeders or will not feed at all.

Remarks: Lacepède did not sign his signature with the acute accent on the first "e" (Adler, 1989). Based on the proposed distribution of subspecies in the reviews by Williams (1988, 1994) and the reports by Webb (1970b) and Blair et al. (1994), *L. t. gentilis* may occur in the extreme northeastern corner of the state. The single juvenile specimen known from Hidalgo County is unassignable to a particular subspecies.

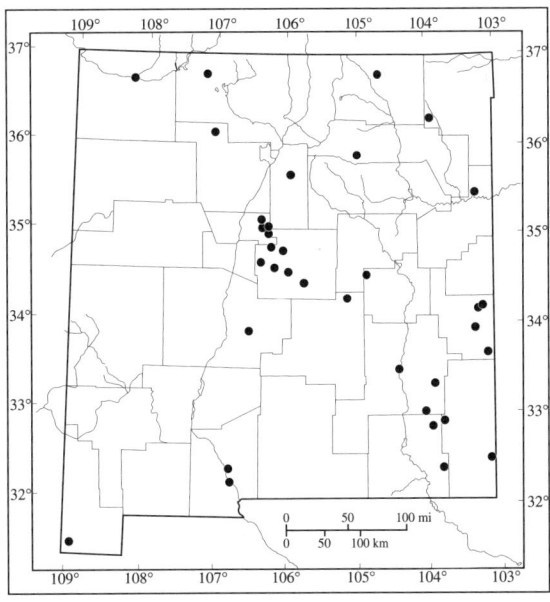

Distribution of Lampropeltis triangulum *in New Mexico.*

LIOCHLOROPHIS VERNALIS (Harlan, 1827) — *Smooth Green Snake*

See Plate 92.

Type: The type specimen is lost, although it was originally in the collection of the Philadelphia Academy of Natural Sciences. No type locality was given in the original description, but Pennsylvania and New Jersey were mentioned as habitat. Schmidt (1953) restricted the type locality to "vicinity of Philadelphia."

Distribution: *Liochlorophis vernalis* is widespread and the range continuous in the northeastern United States and southeastern Canada including offshore islands (e.g., Long Island, Martha's Vineyard). Toward the southwest, in Utah, Colorado, New Mexico, and Chihuahua, the range becomes discontinuous, and the isolated populations are relictual where suitable habitat exists.

In New Mexico, *L. vernalis* has been verified in the Sangre de Cristo, Jemez, Manzano, and Sacramento mountains (see Remarks).

Description: This small slender species ranges in adult size from 30–80 cm TL, but individuals over 50 cm TL are rare. As the generic name implies, *L. vernalis* is bright green dorsally. It has a white or yellow venter; the yellow is most pronounced toward the tail. Young individuals are much darker and may be olive, bluish-gray, or even brownish. Preserved specimens are usually blue.

The scales are smooth and in 15 rows at midbody. The ventrals range from 116–154, subcaudals 59–98, and the anal plate is divided. There is a single anterior temporal, usually one preocular, two postoculars, and a single loreal. There are seven upper and eight lower labials. Each nostril lies within a single scale.

Similar Species: Adult *Coluber constrictor*, if greenish, are always larger than the largest smooth green snake, have two anterior temporals, and each nostril is located between two scales. *Senticolis triaspis* has 25 rows of scales and the middorsal rows are usually keeled. Unringed *Diadophis punctatus* have an orange to red ventrum, 15 or 17 dorsal scale rows, and are olive, blue-gray, or dark slate.

Systematics: Three subspecies are recognized by Grobman (1992) and one of these, *L. v. blanchardi* (Grobman, 1941), occurs in New Mexico. The holotype (UMMZ 62439) of *L. v. blanchardi* is an adult male collected by H. T. Gaige on 25 July 1925. The type locality is Spanish Peaks, 8000 feet, Colorado. See Remarks for our use of *Liochlorophis*.

Habitat: *Liochlorophis vernalis* prefers grassy or herbaceous areas with permanent moisture. Grobman (1941) suggested that essential requirements are "a high altitude or latitude and a relatively moist grassy situation." In the Southwest where moisture is at a premium, these snakes occur only in mountains, foothills, and associated valleys. In New Mexico, they are found mostly in open grassy areas along watercourses or on adjoining rocky slopes from 1525–2450 m. Surface objects such as rocks, bark, logs, and old boards are used for cover when available. Snakes have been taken at locations devoid of obvious surface cover, and here their protective coloration serves them well in the usually dense herbaceous vegetation. The most typical reptile cohabitants are *Diadophis punctatus*, *Thamnophis elegans*, and *Eumeces multivirgatus*.

Behavior: This very gentle species is most often seen while active in the open. The cool mountain and valley habitats where *L. vernalis* is found in New Mexico dictate that activity be confined largely to the warmest parts of the day. Individuals may climb into low vegetation for sunning or foraging, especially where ground vegetation is dense.

The annual activity period is apparently short. Hammerson (1982) stated that most activity occurs from May to September in Colorado. In New Mexico, the collection dates of 41 specimens range from 13 May to 13 September, lending support to a short annual activity period. A small hibernaculum was discovered on 20 February 1992 at 2250 m on a 45° slope above the Santa Fe River (Stuart and Painter, 1993b). Large granite slabs with narrow horizontal spaces between rock and soil served as the "den." In addition to *L. vernalis*, *Eumeces multivirgatus* and *Thamnophis elegans* were present. The site was buried after discovery and redug for further examination on 16 March. At that later time, two single *L. vernalis* and four others in a "ball" were discovered. The six skinks and three garter snakes originally observed had evidently moved off or penetrated more deeply into the den site.

In other parts of the range, L. vernalis has been found to share communal hibernacula with other species. Criddle (1937) reported the use of an ant hill in Manitoba that was shared with *Storeria occipitomaculata* and *Thamnophis radix*. Lang (1969) reported another anthill hibernaculum in Minnesota that was shared with *Storeria occipitomaculata* and *Thamnophis sirtalis*. Johnson (1987) reported the possible use of small mammal burrows in Missouri.

Reproduction: Minimum size at maturity has been given in most recent accounts as approximately 300 mm TL for males and 280 mm TL for females. Grobman (1989) made reference to a female, 245 mm SVL, that laid two eggs. In his sample of 49 gravid individuals, he demonstrated that larger females contain more eggs than smaller ones. Spring mating has been suggested as the norm (Shaw and Campbell, 1974; Stebbins, 1954), but copulating pairs have been observed in August. Also, individuals seem to be more frequently encountered in August, suggesting mating activity in a population. Sperm retention by the overwintering female is a possibility.

Eggs are laid mostly in July and August and number from 2–18, although six or seven eggs are most commonly produced in a single clutch. Western populations probably produce fewer eggs due to their smaller body size. One *L. vernalis* from New Mexico laid four eggs (T. L. Brown, pers. comm.). Nesting may be communal where green snakes are abundant (Cook, 1964). The eggs are white, thin-shelled, vary in size and shape, and measure 8–18 x 19–34 mm. Embryos are at an advanced state of development at laying time and may hatch within a few days in northern parts of the range; data are lacking for southwestern populations. Five hatchlings from a 370 m TL South Dakota snake measured 125–130 mm (T. L. Brown, pers. comm.).

Food Habits: Mostly terrestrial small-bodied insects and spiders are consumed, these being very abundant in green snake habitat. One instance of a small crayfish being eaten (Hammerson, 1982) suggested the possibility that aquatic arthropods may occasionally be used as food.

Remarks: The old locality record (USNM 22377) from the Mesilla Valley, Doña Ana County, is probably in error. There is no suitable habitat in this hot and dry valley.

We are following Oldham and Smith (1991) who presented convincing morphological and physiological evidence in their review that *L. vernalis* is generically different from *Opheodrys aestivus*, the rough green snake, and should be placed in the new genus, *Liochorophis*. This species was reviewed by Grobman (1941, 1992).

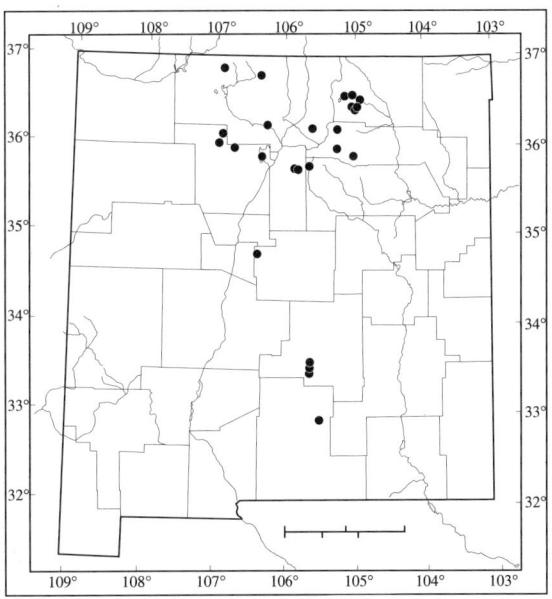

Distribution of Liochlorophis vernalis *in New Mexico.*

MASTICOPHIS BILINEATUS Jan, 1863　　*Sonoran Whipsnake*

See Plate 93.

Type: The only information available concerning the holotype is that it was a specimen in the Leipzig Museum. The vague type locality of "Western Mexico," was designated as Casas Grandes, Chihuahua, by Schmidt (1953), but was restricted to Guaymas, Sonora, by Smith and Taylor (1950b).

Distribution: *Masticophis bilineatus* is found from southeastern Arizona and southwestern New Mexico southward through western Mexico to Oaxaca. In New Mexico, it is restricted to southwestern Hidalgo County.

Description: This slender snake has large eyes, a distinct neck, and a long tail. It ranges in length from 60–170 cm TL. The dorsal color may be olive, bluish gray, or brownish gray, becoming lighter toward the rear. There are usually two or three light-colored stripes on each side that fade toward the tail. The venter is cream, becoming pale yellow toward the tail.

The dorsal scales are smooth and in 17 rows at midbody. The anal plate is divided. There are usually eight upper and nine lower labial scales. The lower preocular is smaller than the upper preocular and is wedged between two upper labials. The loreal is longer than high.

Similar Species: *Masticophis taeniatus* is dark colored dorsally, pink toward the tail ventrally, has 15 rows of dorsal scales at midbody, and has four light stripes on each side. Other sympatric snakes with lateral stripes have keeled scales *(Thamnophis)* or an enlarged rostral with free edges *(Salvadora)*.

Systematics: Two subspecies are recognized (Hensley, 1950), but only the nominotypical subspecies is found in New Mexico.

Habitat: Sonoran whipsnakes are most abundant in canyons and on mountain slopes where rocks and dense vegetation are present. Rocky canyon bottoms with water and abundant vegetation have produced the most snakes of this species in New Mexico. The mountain habitats are often steep-sided, rocky, and covered with low woody vegetation such as scrub oaks or manzanita. A snake was taken, along with scrapes and bruises to the collector, on an oak-covered and almost impenetrable slope on Animas Peak at about 2200 m. Some snakes have also been found in the desert lowlands where they make use of rocky outcrops or rocky shrub-filled arroyos. Individuals have been collected on barren creosotebush flats as low as 1200 m, suggesting that rodent burrows are used for cover. Snake cohabitants at particular locations in the mountains may be *Crotalus willardi*, *C. lepidus*, *C. molossus*, *Senticolis triaspis*, and *Lampropeltis pyromelana*. *Masticophis taeniatus*, although syntopic in Arizona, has not been found within the range of *M. bilineatus* in New Mexico.

Behavior: This agile and fast-moving species is active in the daytime and difficult to capture in brushy habitat. It is equally at home in shrubs, low trees, or on the ground. Nickerson and Mays (1969 [1970]) reported that, in Arizona, five of approximately 20 Sonoran whipsnakes were in trees when observed.

Annual activity has been recorded from March–October but details of hibernation or periods of inactivity are unknown.

Reproduction: Little reproductive information has been published for this oviparous species. No data are available from New Mexico, but three clutches of eggs have been reported from Arizona. Individual eggs in a clutch of seven laid on 7 June measured 19 X 40 mm and had leathery granular-surfaced shells (Van Denburgh, 1922b). Vitt (1975) recorded an Arizona female collected between 10 and 15 April, measuring 792 mm SVL, that contained 10 large yolked follicles. He listed another individual 890 mm SVL that laid six eggs measuring 19.2 X 54.0 mm on 12 July. The eggs of this species are non-adhesive.

Food Habits: Young birds, lizards, and frogs are regular food items of this species. Nickerson and Mays (1969 [1970]) reported *M. bilineatus* eating dead *Peromyscus* that they were using for carnivore bait in Arizona. Since *M. bilineatus* is not a constrictor, food selection may be limited by the size, strength, or ferocity of the potential prey items.

Remarks: There is a specimen of *M. bilineatus* (LSUMZ 14644) with locality data for Doña Ana County. Based on our knowledge of the range and habitat in New Mexico, these data are likely in error.

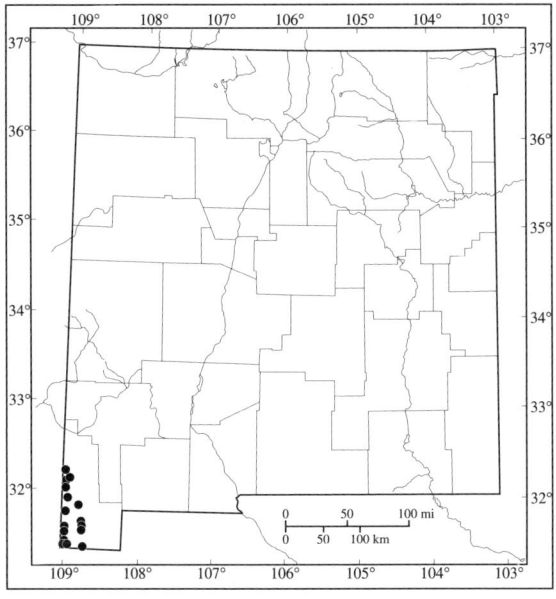

Distribution of Masticophis bilineatus *in New Mexico.*

The genus was reviewed by Wilson (1973a), but the species is in need of review.

MASTICOPHIS FLAGELLUM (Shaw, 1802) — *Coachwhip*

See Plate 94A, 94B.

Type: No holotype was designated. The type locality, "Carolina and Virginia," was restricted to Charleston, South Carolina, by Schmidt (1953).

Distribution: *Masticophis flagellum* is widely distributed throughout the southern half of the United States and the northern half of Mexico. It is generally absent from mountainous terrain and the Mississippi Valley and Delta. In New Mexico, *M. flagellum* is widely distributed except in the northwestern quadrant and mountainous areas above about 2200 m.

Description: This slender snake varies from 90–255 cm TL and is one of our longest snakes. Dorsal coloration may be tan, gray, pink, reddish brown, or almost black. Dark crossbars that are most prominent on the neck may be present. Gehlbach (1965) described a specimen from Cibola County that has both anterior crossbars and longitudinal dark lines on the dorsal scales. Some populations in the Southwest may be bicolored with a dark dorsum and reddish venter. Juveniles are strongly marked with dark crossbars which become less prominent with age.

The dorsal scales are smooth, and in 17 rows at midbody. There are 173–212 ventrals, 91–129 subcaudals, and the anal plate is divided. Two preoculars are present with the lowermost very small. There are eight upper and 10 or more lower labial scales.

Similar Species: All other *Masticophis* spp. are lined. *Coluber constrictor* and *Senticolis triaspis* never have dark crossbars on the dorsum, are bluish or greenish, never yellowish or reddish, and may be blotched (juveniles). *Liochlorophis vernalis* is green and has 15 dorsal scale rows at midbody.

Systematics: Seven subspecies based on color and pattern differences are recognized by Wilson (1973b). The western coachwhip, *M. f. testaceus* (Say *in* James, 1823), is the principal form in New Mexico, but it intergrades with the red coachwhip, *M. f. piceus* (Cope, 1892) in the southwestern portion of the state (Wilson, 1973b). (See Remarks).

The holotype of *M. f. testaceus* is lost. The type locality was given as the headwaters of the Arkansas River near the Rocky Mountains. This locality was defined more precisely by Wilson (1973b) as "the junction of Turkey Creek with the Arkansas River, 12 miles W Pueblo, Pueblo County, Colorado."

The holotype of *Masticophis f. piceus* is USNM 7891,

a skin of an adult female(?), collected by E. Palmer on an unknown date. The type locality is Camp Grant (= Fort Grant), Graham County, Arizona.

Habitat: *Masticophis flagellum* occupies a variety of habitats from the lowlands at 900 m into the mountain foothills to about 2200 m. Within New Mexico, creosotebush desert, short-grass prairie, and shrub-covered flats and hills at our lowest elevations are preferred, although sagebrush desert and piñon-juniper woodland to near 2200 m are also occupied. The substrate may be sandy to rocky, or surfaced by desert pavement. Common snake species associated with the coachwhip are *Crotalus atrox*, *Arizona elegans*, *Heterodon nasicus*, *Hypsiglena torquata*, and *Rhinocheilus lecontei*.

Behavior: This diurnal and fast-moving snake is the most frequently observed snake in the Southwest, as it is often seen crossing highways. It is one of the speediest snakes and was clocked at 3.6 mi/hr by Mosauer (1935), but when chasing one through the brush it seemingly travels much faster. Few live coachwhips seen from moving automobiles are collected. By the time the automobile can be stopped, and the potential collector returns to the sight location, the snake is long gone. Besides eluding an enemy by flight, the snake may take refuge in brush, rodent burrows, or rocky outcrops. When cornered *M. flagellum* will often defend itself by actually advancing on the enemy. When grasped, it will thrash and twist the body, will usually bite repeatedly, and produce bleeding lacerations. Unlike most snakes, it does not release at once when biting defensively, but chews at length. Like other whipsnakes, *M. flagellum* may climb into bushes and heavily branched trees for foraging, predator escape, or to reach cooler temperatures above the hot desert floor. Cowles and Bogert (1944) observed that the mean normal activity temperature was 33.0°C and the maximum observed voluntary temperature was 37.0°C for coachwhips in Arizona. These temperatures were the highest in the group of desert snakes they observed, but significantly lower than those for the diurnal lizards in their study area. Most diurnal activity in lowland deserts is restricted to morning and evening in order to avoid the heat and dry air at mid-day. Brattstrom (1965), in his study of 57 snake and 95 lizard temperatures, obtained similar results for *M. flagellum* (37°C). Only three snake species, *Coluber constrictor* (37.4°C), *Elaphe obsoleta* (38°C), and *Salvadora hexalepis* (37.5°C), had a higher voluntary maximum temperature than *M. flagellum*.

In New Mexico, *M. flagellum* has one of the longest annual active periods of any of our snakes. Our records indicate activity from mid-March to mid-November. No information on hibernation in New Mexico is available, but elsewhere, *M. flagellum* makes use of rock crevices and rodent burrows. Both of these possible entrances to below-ground hibernacula are available in New Mexico.

Reproduction: Published information on reproduction in this widespread and abundant species is sparse. Fitch (1970) suggested that this genus conforms to the breeding schedule of snakes living in the temperate zone.

In New Mexico, *M. flagellum* probably mates in April and May with egg-laying from May to July. A copulating pair was collected 30 April in Trans-Pecos Texas (Minton, 1958 [1959]). Copulation was observed in southern Eddy County on 29 April (our unpubl. data) and in Guadalupe County in late May (J. N. Stuart, pers. comm.).

A snake collected 20 June in Bernalillo County contained seven eggs. A gravid female taken by T. L. Brown in Sierra County on 24 May laid 11 eggs on 14 June. Five of these eggs hatched 9–10 September. Two of the hatchlings were preserved and measured 326 and 342 mm TL (J. N. Stuart, pers. comm.). Ted L. Brown (pers. comm.) took a large female, 190 cm TL, from Tucumcari, Quay County, that laid 18 infertile eggs 1 July measuring 23–27 x 34–41 mm. Brown also found a DOR snake from Sandoval County, measuring 170 cm TL, on 4 June with six eggs. A 128 cm TL snake taken by Brown in Grant County laid seven eggs on 22 June. One of these hatched on 9 September and the hatchling measured 29 cm TL. Another snake collected by Brown in Sierra County measured 132 cm TL, laid 11 eggs on 13–14 June, and 7 of them hatched on 9 September. In other areas, eggs have numbered from 4–20 and averaged about 10 (Fitch, 1970). Incubation time has ranged from 44 days (Fitch, 1970) to 88 days (Brown, pers. comm.). Incubation time appears to be negatively correlated with temperature. Eggs vary in size from 15–27 x 25–57 mm, are granular-surfaced, and non-adhesive.

Food Habits: *Masticophis flagellum* preys mostly on vertebrates, and the list of prey items is long and diverse. Literature references mention lizards, snakes, small mammals, birds, frogs, young turtles, eggs, carrion, lubber grasshoppers, and cicadas (Whiting et al., 1992). Cannibalism has been observed in captivity. Foraging is done in the daytime and *M. flagellum* uses its keen eyesight and speed for prey capture or employs a "sit-and-wait" strategy. Small mammals or reptiles using burrows may be captured below ground, suggesting the use of chemosensory cues to locate hidden prey (Whiting et al., 1992). Feeding on road-kills has been observed (Cowles, 1946; Small et al.,1994). These snakes do not constrict but may hold larger prey items down with their body as they swallow it. Snake prey may be killed before being eaten. Smith (1956) described how a coachwhip grasps the snake's head in its jaws, holds the body down with its own, and then rasps its teeth through the flesh behind the victim's head.

Remarks: This species was reviewed by Wilson (1970, 1973b) and the genus by Wilson (1973a). Wilson (1970, 1973b) incorrectly used Cope (1875) rather than Cope (1892) for the original description for *M. f. piceus*. The source of this error is discussed by Beltz and Smith (1990).

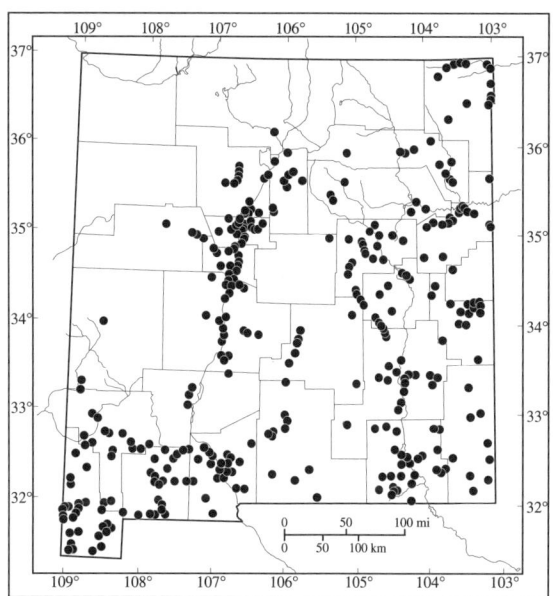

Distribution of Masticophis flagellum *in New Mexico.*

MASTICOPHIS TAENIATUS (Hallowell, 1852)   *Striped Whipsnake*

See Plate 95.

Type: The holotype, USNM 2110, is a male collected by S. W. Woodhouse on an unknown date. Hallowell (1852) gave the type locality as "New Mexico, west of the Rio Grande," but this was later restricted to Shiprock, San Juan County, New Mexico, by Smith and Taylor (1950b).

Distribution: *Masticophis taeniatus* is found within a broad band of the western United States from southeastern Washington southeastward as far as Hidalgo and Michoacán, Mexico. This species has been found throughout most of New Mexico except along the extreme eastern edge and in the higher mountains.

Description: The slender *M. taeniatus* has large eyes, a distinct neck, and a long tail. It varies in length from 90–183 cm TL. The ground color may be black, dark brown, or dark gray dorsally with four light lines on each side. These lines start at the labial scales, are continuous and uniform throughout their length, and reach the anal region. The venter is white to yellow or cream, grading to pink toward the tail.

The scales are smooth and in 15 rows at midbody. There are 188–214 ventrals, 122–160 subcaudals, and the anal plate is divided. There are eight upper and nine lower labials. Two preoculars are present; the lower is smaller and wedged between two upper labials. The loreal scale is longer than high.

Similar Species: *Masticophis bilineatus* is light olive, bluish gray, or brownish gray dorsally, the light lines do not touch the labial scales, there is no pink ventrally toward the tail, and there are 17 dorsal scale rows. *Salvadora* spp. have an enlarged rostral scale with free edges. All other snakes in New Mexico with lateral stripes, including *Thamnophis* spp., have keeled scales.

Systematics: Parker (1982b) recognized five subspecies, but only the nominotypical subspecies, *M. t. taeniatus*, is found in New Mexico.

Habitat: *Masticophis taeniatus* occupies many of the diverse habitats associated with the mountains and basins of western North America. Within this snake's large range, low mountains, foothills, rocky escarpments, and brushy basins are used. In New Mexico, it is primarily a foothills species and occupies the usually rocky and brushy habitats found there. It may range in elevation from about 950 m in desert scrub to 2200 m in piñon-juniper or oak-piñon-juniper woodland. Lowland habitats are often dissected by arroyos. Desert flatlands, especially those with reduced shrub cover, are avoided. Common species of snakes found associated with *M. taeniatus* include *Crotalus atrox*, *C. molossus*, *C. viridis*, *Gyalopion canum*, and *Salvadora grahamiae*. No sympatry with *Masticophis bilineatus* has been found in New Mexico.

Behavior: The well-known *Masticophis* characteristics of alertness, speed, and heroic defense behavior are present in *M. taeniatus*. It is an excellent climber, and its tree-climbing ability may be ranked above that of *M. flagellum* and similar to *M. bilineatus*. It is strictly diurnal and individuals may bask, sometimes in shrubs, upon emergence from their nighttime retreats. They are most active in morning and late afternoon.

Based on our collection dates in New Mexico, annual activity of *M. taeniatus* extends from late March to early November. The length of the active period varies, and is determined by seasonal temperatures and latitude. In Utah, *M. taeniatus* uses multispecies dens for hibernation, which are accessed through rock crevices, or in one case, an old volcanic vent (Woodbury and Parker, 1956; Parker, 1976). Communal denning is more prevalent at higher latitudes and/or where suitable dens are scarcer than in New Mexico. Few hibernation data for our populations are available, however it often hibernates with *C. atrox* in rocky densites in southeastern New Mexico.

Reproduction: There is little published information on age and size of *M. taeniatus* at maturity. A Utah female measuring 760 mm SVL seems close to the minimal size, and her clutch was small (3 eggs). Studies in Utah by Parker and Brown (1980) found the minimum SVL of mature males and females to be 530 mm and 740 mm, with females maturing when three years old. Details of courtship and mating are described by Bennion and Parker (1976). They found that *M. taeniatus* remains close to the hibernation site after emergence in the spring and that males actively hunt for the more sedentary females. Males were aggressive toward each other only when one individual's mating territory appeared to be violated by another male. Soon after mating, *M. taeniatus* dispersed to their summer foraging areas.

Goldberg and Parker (1975) described the male reproductive cycle of *M. taeniatus*. As is common in snakes, spermiogenesis occurs in late summer to early autumn, the sperm overwinter, and breeding occurs the following spring or early summer.

Egg clutches in *M. taeniatus* range from 3–12, and eggs are laid in June and July. Parker and Brown (1972) gave an account of two gravid females apparently hunting for and finally locating a satisfactory nest site. After investigating numerous rodent burrow systems, those chosen were in horizontal portions of the burrow measuring 360 mm and 410 mm deep. Egg size may vary from 14–19.8 x 40–65.4 mm. The non-adherent eggs have leathery shells with a rough sandpaper-like surface. Parker and Brown (1972) reported that incubation time under natural conditions ranged from 44–58 days. Eggs in the laboratory have incubated from 62–81 days prior to hatching. Hatchlings have measured from 254–355 mm TL. A female from Albuquerque, measuring 1350 mm TL, laid five eggs on 30 June that averaged 15 x 62 mm (T. L. Brown, pers. comm.). Brown reported that another female from Grant County, measuring 1170 mm TL, laid six eggs on 12 July that measured 15–18 x 36–52 mm. Neither clutch hatched.

Food Habits: *Masticophis taeniatus* feeds on lizards, snakes, small rodents, birds, frogs, and insects. This is a widely foraging species and often uses speed to capture moving prey such as lizards. As these snakes are not constrictors, larger prey items are held down with the body

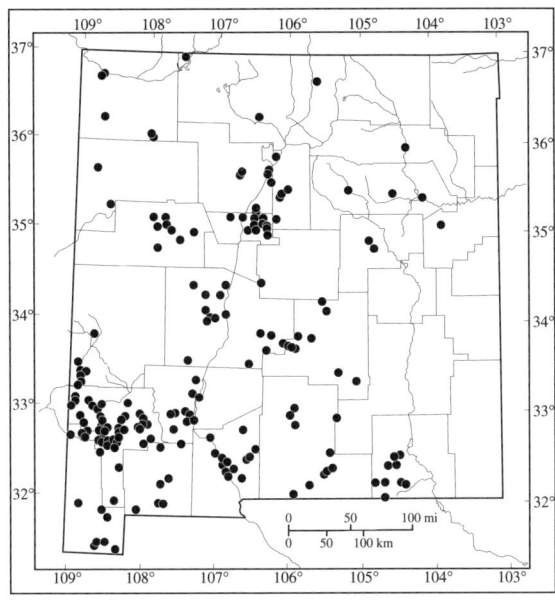

Distribution of Masticophis taeniatus *in New Mexico.*

while swallowing. They may use their climbing ability to forage for birds in shrubs and trees.

Remarks: Parker (1982b) reviewed this species.

NERODIA ERYTHROGASTER (Forster, 1771) — *Plainbelly Water Snake*

See Plate 96.

Type: This species was originally described as *Coluber erythrogaster* by Forster (1771) with neither a type specimen nor type locality designated. Conant (1949) selected a neotype, USNM 126890, collected 19 March 1948 by T. M. Beckett. It was obtained "near Parker's Ferry, Edisto River Swamp, Charleston County, South Carolina," which Conant (1949) designated as the type locality.

Distribution: *Nerodia erythrogaster* ranges from southern Michigan and southern Delaware to the Gulf Coast and Nuevo Léon with isolated populations in east-central Durango and Zacatecas, Mexico, and from the Atlantic Coast to extreme western Oklahoma and southeastern New Mexico (Conant, 1975). In New Mexico, *N. erythrogaster* is confined to Eddy County in the lower Pecos River drainage, including the Black and Delaware rivers and Rocky Arroyo from 900 m to approximately 1100 m. Records from Lea County (Hubbard et al., 1978) are based on misidentified specimens. Milstead (1960) considered it to be a relict species of the Chihuahuan Desert and pointed out that it exists there as disjunct populations considerably removed from the main portions of the species' range.

Description: This large semiaquatic snake is characterized by a dorsal coloration of various shades of gray or brown and an unmarked and uniformly pale yellow venter. There is often dark pigment on the base or on the lateral edges of the ventral scales. The dorsal scales are heavily keeled with prominent apical pits. The head is broad and distinctly wider than the neck, especially in adults. There are dark bars on the upper labials. The young are strongly marked with alternating dorsal and lateral blotches that may be joined anteriorly to form transverse bars. Adults may retain the juvenal pattern in the form of dark-bordered, light crossbars in the middorsal region, although larger adults are sometimes uniformly dark. Average size is 76–122 cm; record size is 157.5 cm. TL.

Scutellation in the New Mexico subspecies is as follows: 23–27 (usually 25) rows of keeled scales at midbody; 132–159 ventrals; 75–87 subcaudals in males, 61–79 in females. There are 7–10 (usually 8) upper labials, 1–2 preoculars, 1–4 postoculars, 2–3 posterior temporals, and a single loreal. The fourth and fifth upper labials enter the orbit. The anal plate is usually divided, but it may be single (Conant, 1949). Roger Conant (pers. comm.) advised us that the frequency of an undivided anal plate is about 11% throughout the *N. erythrogaster* complex.

Similar Species: No other snake in New Mexico has a plain yellow belly and heavily keeled scales. The young, with the prominent alternating dorsal and lateral blotches and keeled scales, are not easily confused with any other species.

Systematics: There are six subspecies (McCranie, 1990); only *Nerodia erythrogaster transversa*, the blotched water snake, is found in New Mexico. This form was named by Hallowell (1852) as *Tropidonotus transversus*. The type locality was given as "Creek boundary, found near the banks of the Arkansas and its tributaries." The holotype is ANSP 5044, a young adult male collected sometime in 1849 or 1850 by Dr. Samuel Washington Woodhouse (Conant, 1969; Webb, 1970b). Conant (1969) revised the type locality to "Arkansas River between Keystone and Tulsa, Tulsa County, Oklahoma."

Habitat: In New Mexico, *N. erythrogaster* is confined to rivers, main irrigation diversion drains, or rocky intermittent streams where large deep pools with abundant fish and frogs remain. Emergent aquatic vegetation may or may not be present. It has not been collected in stock tanks, nor has it been collected in the headwaters of the Black River where the deep spring-fed pools may be too cool to provide suitable habitat. Conant (1969) reported finding specimens in marshes, streams, and springs in northeastern Mexico.

Behavior: Very little is known regarding the behavior of *N. erythrogaster* in New Mexico. It forages along rocky streams both at night and during the day and has been discovered in dense streamside vegetation. It basks on branches overhanging the water and is quick to flee into the water when approached. These water snakes are pugnacious and will flatten their bodies, strike, and bite savagely when captured. Cloacal glands at the base of the tail discharge a foul-smelling musk when individuals are handled roughly. In New Mexico, specimens of the blotched water snake have been collected from April through September, although approximately 70% of these were found in May.

Reproduction: Sexual size dimorphism is pronounced in this species, the females being significantly larger than the males. *Nerodia erythrogaster* is viviparous, with the young born during the late summer. Of 10 broods from Oklahoma, Texas, and Louisiana reported by Fitch (1985), brood size averaged 12 and ranged from 2–22. Conant (1969) reported on mensural characteristics, female size, and birth date of seven broods of *N. erythrogaster transversa* from northeastern Mexico. Brood size averaged 12.3 (2–32) young. Among 82 captive-born juveniles, the measurements varied from 191–283 mm TL, and the mean length, calculated separately for each of six broods, varied from 227.8–263.3 mm. Female *N. erythrogaster* may store active sperm for almost two years after mating (Conant, 1965). Little is known about the reproduction of *N. erythrogaster* in New Mexico. On 27 June 1992, we collected a 361 mm TL juvenile along the Delaware River south of Loving. The umbilical scar was evident on this specimen, and it likely represented a yearling from the previous summer's brood.

Food Habits: No studies of the food habits of *N. erythrogaster* have been conducted in New Mexico, although it is thought to feed to a large degree on fish and frogs, including *Acris crepitans*. A single specimen collected along the Delaware River contained a sunfish, *Lepomis* sp. Mushinsky and Lotz (1980) and Mushinsky et al. (1982) found *N. erythrogaster* to undergo distinct ontogenetic changes in prey preference. At approximately 50 cm SVL these snakes show an abrupt and virtually complete shift from fish to anurans. Conant (1969) suggested that this species probably feeds on most kinds of fish and frogs and some species of toads. He described a feeding incident where a captive blotched water snake consumed a dried-out leopard frog, which suggested that some dead and desiccated food may occasionally be accepted.

Remarks: The presence of this species in New Mexico was not confirmed until Conant (1955a) published data on a specimen (MSB 101) collected during 1949 southeast of Carlsbad at Six-Mile Dam. Previously, the single record of *N. erythrogaster* from New Mexico was based on a badly mutilated specimen collected 25 July 1901 (USNM 32758).

Habitat destruction and/or human contact have caused the extirpation of *N. erythrogaster* from parts of its former range (Scudday, 1977 [1978] and references therein). Conant (1969) discussed the distribution of *N. erythrogaster* in the Pecos River drainage and mentioned that it was becoming confined to relictual areas as were also *Acris crepitans* and *Thamnophis proximus*.

Nerodia erythrogaster is listed as endangered by the New Mexico Department of Game and Fish (NMGF, 1990). These snakes are often shot by uninformed fishermen for sport or because of the mistaken beliefs that they are venomous or that they eat large quantities of sport fish.

Nerodia erythrogaster was reviewed by McCranie (1990) who states that a thorough taxonomic study similar to that done for the Mexican portion of the range (Conant, 1969) is needed for the U.S. populations of this wide-ranging species. Roger Conant (pers. comm.), long contemplating a detailed review of the *N. erythrogaster* complex throughout its range, assembled a large data bank on the species. Because of his now advanced age, it is unlikely that he will ever complete this project.

Much of the earlier literature on this species is found under the name *Natrix erythrogaster*.

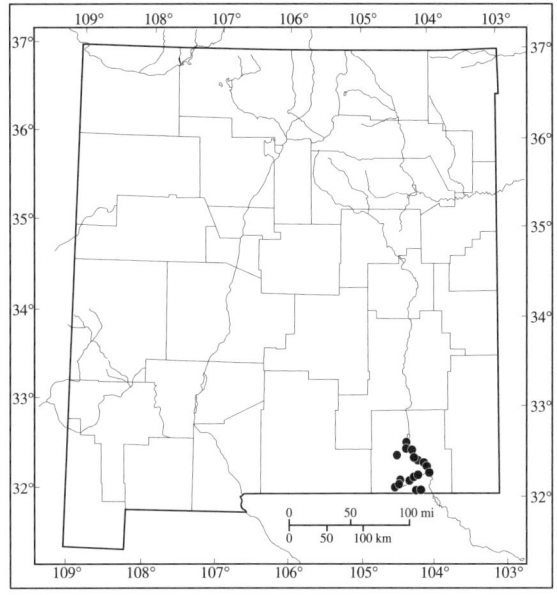

Distribution of Nerodia erythrogaster *in New Mexico.*

PITUOPHIS MELANOLEUCUS (Daudin, 1803) *Bullsnake, Gopher Snake*

See Plate 97.

Type: The type locality, "Florida," was restricted to Charleston, South Carolina, by Schmidt (1953). No type has been designated.

Distribution: *Pituophis melanoleucus* is widespread from southwestern Canada to southern New Jersey and south into Mexico to Guatemala and the southern tip of Baja California (Conant, 1975). In New Mexico, it is found everywhere from 900 m in the lowest desert to about 2800 m in the mountains.

Description: *Pituophis melanoleucus* may exceed 250 cm TL and is the longest snake reported from New Mexico. Most individuals are between 90–180 cm TL. There are 33–66 large, light brown to dark brown, middorsal body blotches or saddles on a ground color of tan, cream, or yellow. Color is variable but the pattern tends to be more contrasting toward both the head and tail. The blotches become saddle-like posteriorly. There is usually a dark band extending from the eye to the angle of the jaw. The venter is white, cream, or yellow with bold dark spots. The scales are strongly keeled dorsally, smooth on the lower sides, and in 27–37 rows at midbody. There are 205–259 ventrals, 46–89 subcaudals, and the anal plate is single. There are usually four prefrontals.

Similar Species: *Arizona elegans* is most similar in pattern but has smooth scales and 2 prefrontals and does not attain the large size of *P. melanoleucus*. Young *Coluber constrictor* are blotched but have smooth scales. Young *Masticophis flagellum* may have dark crossbands but have smooth scales. *Elaphe* spp. have a divided anal plate and feebly keeled dorsal scales. *Trimorphodon biscutatus* is very slender, has elliptical pupils, and light-centered saddles. *Crotalus* spp. have very short tails ending in a rattle or button and have elliptical pupils. *Heterodon* spp. have a sharply pointed rostral.

Systematics: Sweet and Parker (1990) recognized 15 subspecies of *P. melanoleucus*, which include all of the pine, bull, and gopher snakes (see Remarks). In New Mexico, *P. m. affinis* Hallowell, 1852 is most widespread, *P. m. sayi* (Schlegel, 1837) occurs east of the Pecos River, and *P. m. deserticola* Stejneger, 1893 is found in the northwestern corner of the state.

The holotype, ANSP 3792, of *P. m. affinis* is a subadult male collected by S. W. Woodhouse in September, 1851. The type locality, "New Mexico," was restricted to "Zuñi River, New Mexico" (which included Arizona) by Hallowell (*in* Sitgreaves, 1853), and to Zuni, McKinley County, New Mexico, by Smith and Taylor (1950b).

The holotype of *P. m. sayi* was collected by T. P. Say on an unknown date, but is not known presently to exist. The type locality, "Missouri," was restricted to Carthage, Jasper County, Missouri by Smith and Taylor (1950b), and to St. Louis, Missouri by Schmidt (1953).

The lectotype (see Klauber, 1947) of *P. m. deserticola* (USNM 18070) is an adult male collected by C. H. Merriam 11 May 1891. The type locality is the "east slope of the Beaverdam Mts., Utah."

Habitat: *Pituophis melanoleucus* is likely the most widespread and abundant snake in New Mexico, largely resulting from the wide variety of habitats occupied. It is found from the driest desert elevations at 900 m to mixed coniferous forest at 2800 m. Within these extremes, areas of deep sedimentary soils or rocky terrain may be occupied. Open ponderosa pine forest, piñon-juniper woodland, grasslands, shrublands, sandhills, marshes, and cultivated fields have been mentioned as habitats in the voluminous literature on this species. Bullsnakes are most abundant where small mammal activity is the greatest. These serve as food, and their excavations are used for shelter. Buildings on ranches and farms, as well as those found in suburbia, may also provide shelter. The most common snake cohabitants in New Mexico are *Crotalus atrox, C. viridis, Heterodon nasicus,* and *Salvadora grahamiae.*

Behavior: Large and slow-moving, *P. melanoleucus* can demonstrate an impressive display of defensive actions when threatened. Flattening the head to a triangular shape, coiling and puffing up the body, vibrating the tail, hissing loudly, and striking repeatedly, are actions that can cause a potential predator to wonder if the meal is worth the effort. These actions, along with the pattern and color, tend to make this species a convincing rattlesnake mimic. Furthermore, when in dry leaves or grass, the vibration of the tail makes a sound much like a rattle. Hissing is amplified by an enlarged epiglottis characteristic of this species. It is unfortunate that this same defense behavior may lead to the animal's death by the hand of fearless humans wishing to demonstrate their bravery, often before an admiring audience. Some individual bullsnakes are quite passive and allow handling even when initially approached. Practically all will become tame in captivity. Bullsnakes are usually most active during the day but become more active at night under hot and/or dry conditions. They are generally active from April to October although some March and November specimen records are available. Hammerson (1982) cited volcanic rock crevices and lime concretions formed by spring water as hibernacula in Colorado. These were communal dens shared with *Crotalus viridis, Thamnophis elegans,* and *Lampropeltis triangulum.* Parker and Brown (1973) found aggregations of *Coluber constrictor, Masticophis taeniatus,* and *Crotalus viridis* hibernating with *P. melanoleucus* in rock accumulations in Utah. Collins (1982) stated that deep crevices on rocky hillsides and burrows of small mammals are used as hibernacula in Kansas. No hibernacula have been described in New Mexico, but crevices in rock accumulations or small mammal burrows in rockless habitats are probably important winter retreats for *Pituophis* in this state.

Reproduction: *Pituophis melanoleucus* matures at 760 mm TL or more in the third or fourth year. Courtship and mating occur after emergence from hibernation in the spring, and males often compete for a female or territory during this period. Males may engage in a combat ritual similar to that often seen in rattlesnakes, where each combatant tries to maintain the uppermost position and keep his opponent's head depressed. There is seldom any biting or injury inflicted, and the loser moves off. Sperm is produced in late summer and early autumn and stored in the epididymis and ductus deferens until the spring. There is variation in the breeding time of *P. melanoleucus* throughout its range, partially due to differences in latitude, habitat, and annual local weather cycles. Most often, mating occurs in early spring, egg-laying in late spring to early summer, and hatching in late summer to early fall. The eggs are large, about 30 x 90 mm. Clutches are relatively small, numbering 3–19 eggs, and clutch size is correlated with parental body size. There is normally one clutch produced per year. Depending on the availability of food, occasional individuals may skip a year or, if well-fed, may produce a second clutch within a few weeks of the first one. The eggs usually hatch in 2–2.5 months. Three clutches of eggs from wild-caught Illinois *P. m. sayi* were incubated at a constant 24°C. The 37 eggs that hatched took 53–54 days with no apparent defects in the hatchlings (Snow, 1993). Newly hatched young measure between 300 and 550 mm TL. Ted L. Brown (pers. comm.) bred a male with two females, all from Bernalillo County. The male measured 1475 mm TL and the females were both 1420 mm TL. The first female mated 16 May, laid eight eggs averaging 33 x 62 mm on 3 July, but they did not hatch. The second female mated on 18 May, laid 11 eggs on 2 July measuring 20–29 x 48–59, and these hatched 65 days later on 5 September. Another female from Torrance County that Brown captured on 9 May measured 1245 mm SVL. This snake laid 10 eggs on 1 July. After 62 days, 9 of them hatched on 1 September. The hatchlings measured 355–380 mm TL.

Food Habits: *Pituophis melanoleucus* feeds mostly on small mammals, although birds and their eggs are readily used when available. Lizards are also eaten, but mostly by smaller and newly hatched individuals. Prey is killed by constriction or by being forced against a firm surface by a loop of the body. This latter method is especially useful in a confined space such as a mammal burrow where coiling is impossible. Also, multiple prey may be killed at the same time by this method.

Remarks: Bullsnakes do not eat or drive away rattlesnakes as is commonly believed, although they may compete for the same prey. Both of these snakes are important for rodent control, and in turn, are also important in controlling plague, a disease transmitted to humans by way of rodents and their fleas. Ted L. Brown of the New Mexico Environmental Health Department (pers. comm.) says, "In my presentations on plague I emphasize these snakes' capabilities as rodent-catchers in reducing risk of plague to people." Recently, *Peromyscus maniculatus*, the deermouse, has been identified as the most important vector in transmitting the Hantavirus, a life-threatening virus with flu-like symptoms. Bullsnakes and rattlesnakes are both important predators in deermouse habitat.

This species has often been considered to consist of multiple species. The split most often used places the pine snakes into *P. melanoleucus* and the bull and gopher snakes into *P. catenifer* (i.e., Collins, 1990). We are following the review by Sweet and Parker (1990) in placing all pine, bull, and gopher snakes in *P. melanoleucus*.

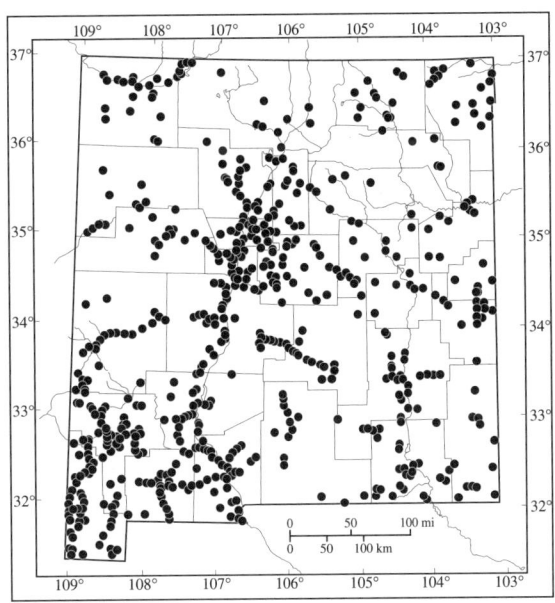

Distribution of Pituophis melanoleucus *in New Mexico.*

RHINOCHEILUS LECONTEI Baird and Girard, 1853   *Longnose Snake*

See Plate 98.

Type: The holotype, MCZ 137, is a young adult male collected by J. L. LeConte on an unspecified date (Baird and Girard, 1853a). The type locality is "San Diego," California.

Distribution: *Rhinocheilus lecontei* is found in the southwestern United States from northern California to southwestern Kansas and south to Jalisco, San Luis Potosí, Tamaulipas, and Zacatecas, Mexico. The higher mountains within this range are excluded. In New Mexico, it is widely distributed throughout the southern two-thirds of the state from 900 m to near 1900 m.

Description: The slender *R. lecontei* usually ranges from 50–80 cm TL, although occasional individuals may grow to over 90 cm TL. Typically, there are black saddles, outlined with light spots on the sides. The interspaces are usually pink, rose, or red and flecked with dark spots on the sides. The venter is white, cream, or yellow and usually has dark spots toward the sides. Young snakes differ in having less flecking on the scales, and the reddish coloring is paler, giving them a more black-and-white appearance.

The scales are smooth and usually in 23 rows (occasionally 25) at midbody. There are 181–213 ventrals and 41–61 subcaudals (both highest in males). The anal plate is single, as are most of the subcaudal scales. There are eight upper and nine lower labial scales. The loreal is single.

Another color form, the "*clarus* morph," may be found in southwestern Hidalgo County (see Remarks). This form has longer and fewer black saddles, usually lacks red in the interspaces, and has less black spotting on the sides.

Similar Species: *Micruroides euryxanthus* has no flecking along the sides, the white or yellow rings are broad, and all rings completely encircle the body. Ringed *Lampropeltis* spp. and *Sonora semiannulata* have divided, rather than single, caudal scales.

Systematics: Two subspecies occur in New Mexico (Medica, 1975). *Rhinocheilus l. tessellatus* Garman, 1883 occurs throughout the species' range in the state, except in southwestern New Mexico, where the nominotypical subspecies, *R. l. lecontei*, is found.

The holotype, MCZ 4577, of *R. l. tessellatus* is a young female collected by Edward Palmer in 1880. The type locality is Monclova, Coahuila, Mexico.

The subspecies, *R. l. clarus*, was named by L. M. Klauber (1941). The holotype, SDSNH 31440, is an adult male collected by Richard Neil on 7 May 1939. The type locality is "in the Borego Valley, 2 miles north of The Narrows, San Diego County, California." This form is no longer recognized as a subspecies but is considered a morph of *R. l. lecontei* (see Remarks).

Habitat: *Rhinocheilus lecontei* generally avoids the mountains, residing primarily in the valleys and plains with grassy or shrubby vegetation. It is most often found in areas with sandy soils, with or without surface rock for cover. Common snakes often found associated with *R. lecontei* in New Mexico are *Crotalus atrox*, *C. viridis*, *Heterodon nasicus*, *Hypsiglena torquata*, and *Masticophis flagellum*.

Behavior: Like many other snakes that are confronted by a potential enemy, *R. lecontei* may exhibit a repertoire of defensive behavioral displays. These consist of coiling, striking, twisting the body, and defecating. Biting is uncommon, except in the case of juveniles, which usually exhibit a fierce defense display. Of special interest is the peculiar anal and nasal bleeding that may occur in females. McCoy and Gehlbach (1967) described this behavior in detail. It accompanies a side-to-side movement of the posterior body, tail vibration, and anal distention. The question of whether this hemorrhaging has survival value, or is simply a side effect of the other defensive actions, is not answered. It was noted that cloacal hemorrhaging has also been observed in female *Lampropeltis getula*. *Rhinocheilus lecontei* are nocturnal and seldom venture abroad in daylight. According to Klauber (1941), in southern California they seem to be less affected than other snakes by low temperatures or wind. He found them most active in early evening at air temperatures ranging from 15.5–30.5°C. Shaw and Campbell (1974) also noted *R. lecontei* active at temperatures as low as 15°C. These snakes are excellent burrowers in loose sandy soils but prefer to use rocks, surface debris, or rodent burrows for daytime retreats when these are available.

Annual activity in New Mexico normally extends from April to September, but warm seasonal temperatures may extend this period. Our records indicate collection as early as 26 March and as late as 24 October. Klauber (1941) recorded May as the most active month in southern California. In New Mexico, most *R. lecontei* have been collected in May and August. Rodent burrows or rock crevices probably provide access to underground hibernacula.

Reproduction: There are few published data on the age and size at maturity for *R. lecontei*. Klauber (1941) gave the total length of four gravid females as 612, 615, 672, and 678 mm. Fitch (1970) suggested that *R. lecontei* conforms to the usual breeding schedule for snakes in the temperate zone. If this is so, then mating should occur after emergence from hibernation in April and May.

Available egg-laying records are mostly from June and July, but a snake captured in the first week of July laid eggs on 15 and 16 August (Woodin, 1953), suggesting double-clutching as a possibility in the warmer parts of the range. Clutch size varied from three to nine, with white leathery-shelled eggs about 36 x 16 mm. Incubation varied from about six weeks (Woodin, 1953) to as long as 88 days (Shannon and Humphrey, 1963). Hatchlings are usually just over 200 mm TL, although Klauber (1941) recorded an individual from New Mexico 186 mm TL.

Food Habits: Lizards and small mammals serve as the principal foods, but *R. lecontei* may also eat snakes, reptile eggs, and large insects. Large prey are constricted. The serum of this species has shown some marginal neutralization capacity for several crotaline venoms (Weinstein et al., 1992).

Remarks: *Rhinocheilus l. clarus* was once considered to be a distinct form sympatric with *R. l. lecontei*. Shannon and Humphrey (1963) obtained both *clarus* and *lecontei* morphs from the same clutch of eggs, resulting in the suppression of *R. l. clarus*. The ratio of the two morphs in different populations varies from pure *clarus* to pure *lecontei* (Klauber, 1941; Medica, 1975).

This species was reviewed by Medica (1975).

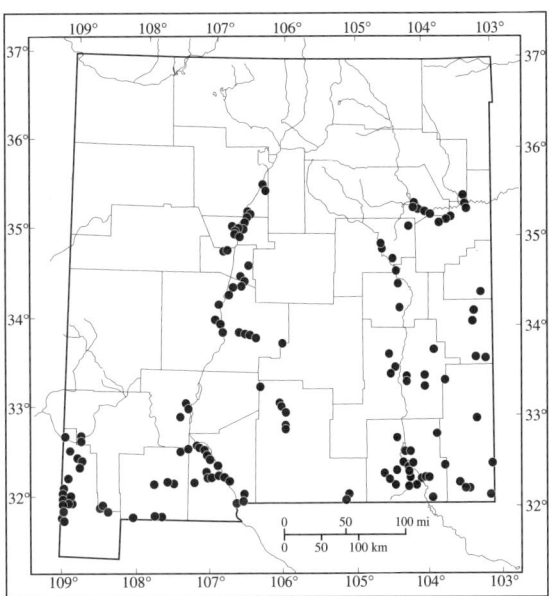

Distribution of Rhinocheilus lecontei *in New Mexico.*

SALVADORA DESERTICOLA Schmidt, 1940 — *Big Bend Patchnose Snake*

See Plate 99.

Type: The holotype, FMNH 26615, was collected by Tom Carney in 1935. The type locality is Government Spring, near Chisos Mountains, Brewster County, Texas.

Distribution: *Salvadora deserticola* ranges from southeastern Arizona, southwestern New Mexico and Trans-Pecos Texas, south through Sonora and Chihuahua into Sinaloa, Mexico. In New Mexico, it occurs mostly in the southwestern quadrant, although there is a single record from Carlsbad Caverns National Park in southern Eddy County.

Description: This slender species usually measures from 600–800 mm TL but there is a record of 1143 mm TL (Tennant, 1984). The pattern is four-lined with the uppermost black or dark brown stripe on each side being the most prominent and widest. The dorsal area between the uppermost stripes varies from tan to brownish-orange and forms a distinct longitudinal band darker than the paler grayish ground color of the sides. The lowermost stripe is on the fourth scale row at midbody. The venter tends to be peach-colored. Dorsal scales are smooth and in 17 rows at midbody. There are 180–196 ventrals (higher in Mexico), 66–87 subcaudals, and the anal plate is divided. There are usually nine upper labials and 10–12 lower labials. The posterior chin shields are separated by two or three small scales. The large triangular rostral has a loosely attached appearance that is characteristic of the genus.

Similar Species: *Salvadora grahamiae* has a wider and very distinct upper stripe. The lower stripe is on the third scale row or is absent; there are eight upper labials, and the posterior chinshields make contact or are separated by a single scale. No other striped species has the characteristic patch-like rostral.

Systematics: We treat *S. deserticola* as a monotypic species. This form was originally described as a subspecies of *S. hexalepis* (Schmidt, 1940), but unpublished data and the conclusions of the late C. M. Bogert deny conspecificity. Other authors have used Bogert's arrangement (Smith and Brodie, 1982; Dixon, 1987; Conant and Collins, 1991; Flores-Villela, 1993).

Habitat: *Salvadora deserticola* occupies desert habitats with sandy or gravelly soils. It is found mostly within the range of creosotebush, both geographically and altitudinally, but may be found at elevations from 950 m to about 1600 m. Some common perennial plants often found in the habitat are tarbush, catclaw, mesquite, snakeweed (broomweed), ocotillo, plus various cacti and grasses. Rodent burrows are usually present, and in many of the desert flats, are the only cover available for nighttime retreats. In New Mexico, some common snakes associated with *S. deserticola* are *Arizona elegans*, *Crotalus atrox*, *Heterodon nasicus*, *Hypsiglena torquata*, *Masticophis flagellum*, *Pituophis melanoleucus*, and *Rhinocheilus lecontei*.

Behavior: *Salvadora deserticola* is an active diurnal species with general behavior much like that of the whipsnakes and racers. It crawls rapidly and is chiefly ground-dwelling, but may crawl into low vegetation on occasion. Burrowing in loose sand has been observed in this species by Minton (1958 [1959]) and the closely related *S. hexalepis* by Van Denburgh (1922b). When captured, these snakes typically react violently by thrashing and striking as they attempt to escape. Diel activity varies from crepuscular during the hottest weather, to midday when temperatures moderate. They may bask in direct sun or partially or wholly buried in warm sand.

In New Mexico, records indicate a long annual activity period extending from early April to early November. Farther south, at lower elevations in Trans-Pecos Texas and northern Mexico, *S. deserticola* may be active on warm and sunny days year-round. There is no information on hibernation, but during this snake's temperature-determined winter dormancy, they may use shallow retreats such as rodent burrows, rock outcroppings, or other surface debris accumulations. In a laboratory study of mixed *S. deserticola* and *S. h. hexalepis*, Jacobson and Whitford (1971) found that these snakes had a wide thermal tolerance. The low critical thermal minimum allows emergence from a cool retreat in order to bask and raise the body temperature to a level allowing maximal activity in search of prey. The high thermal maximum protects them from thermal damage if they are forced to remain exposed to high temperatures for an extended period. This wide thermal tolerance may also explain the long annual activity period observed in these snakes.

Reproduction: There is little published information on age or size at maturity for *S. deserticola*. Wright and Wright (1957) suggested 684 and 709 mm TL as minimal lengths for sexual maturity in male and female snakes, whereas Conant and Collins (1991) gave 610 mm TL as the minimal adult length. Recently, working with 35 male *S. hexalepis* from Arizona (representing both *S. hexalepis* and *S. deserticola*), Goldberg (1995) found that the smallest spermiogenic male measured 468 mm SVL. He also found a female with three oviducal eggs measuring 584 mm SVL.

The reproductive cycle is probably similar to that of *S. hexalepis*, with mating from April–June and egg-laying from May–August. Goldberg (1995) gave data on the male and female reproductive cycles of *S. hexalepis* from Arizona. A female *S. deserticola* from Cochise County, Arizona, laid 8 eggs on 5 June (T. L. Brown, pers. comm.).

Fitch (1970) noted that records of egg-laying in August suggest that more than one clutch per season may be produced by some females. Goldberg's (1995) data suggest that not all S. *hexalepis* females breed each year. Eggs of captive S. *hexalepis* have numbered from 4–10. These eggs measured 9–12 x 27–40 mm and were white, smooth-shelled, and adhesive. In the laboratory, 85 days have been required for incubation (Stebbins, 1954). The newly hatched young are patterned like the adults and measure 21–28 cm TL. A Sierra County female (MSB 54363), measuring about 690 mm SVL, laid eight eggs within two days of capture on 3 June (J. N. Stuart, pers. comm.).

Food Habits: *Salvadora deserticola* eats lizards, reptile eggs, and small mammals. A 900 mm SVL female from Texas ate *Cnemidophorus tigris*, *Sceloporus undulatus*, *Anolis carolinensis*, and a pocket mouse, *Perognathus penicillatus*. Painter (1985) reported that lizards and grasshoppers were used as prey in southwestern New Mexico. Foraging coincides with the morning and evening activity of lizards, the most often consumed prey. Live prey may be pursued on the surface or captured in burrows.

Remarks: Shaw and Campbell (1974) suggested that the enlarged rostral scale is an adaptation for the excavation of eggs. They also noted that patchnose snakes regularly feed on reptile eggs, with the danger that eggs of their own species might be eaten. However, they point out that a female patchnose snake lays a pheromone scent with her eggs that "warns off" hungry snakes of the same species.

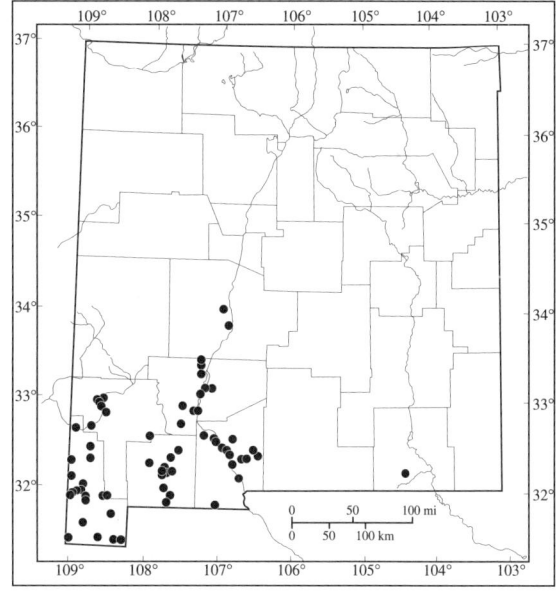

Distribution of Salvadora deserticola *in New Mexico.*

SALVADORA GRAHAMIAE Baird and Girard, 1853   *Mountain Patchnose Snake*

See Plate 100.

Type: The holotype, USNM 2081, is a male collected by J. H. Clark in September 1851 (Baird and Girard, 1853a; see Remarks). The type locality was originally given as "Sonora, Mexico" (= southern Arizona), but was later restricted by Schmidt (1953) to the Huachuca Mountains, Cochise County, Arizona.

Distribution: *Salvadora grahamiae* is found in Arizona, across New Mexico to Trans-Pecos and central Texas, and south into northern Mexico. In New Mexico, it is widely distributed except in the northeastern and northwestern corners and along the eastern edge of the state.

Description: This slender snake usually measures 55–75 cm TL but occasionally may exceed 90 cm TL. Ted L. Brown (pers. comm.) measured a male 90 cm TL and a female 94 cm TL. The pattern is strongly two-lined. The wide, dark, brown to black dorsolateral stripes stand out sharply against the pale gray to slightly olive or brownish ground color. The middorsal color matches, or may be a shade brighter than, the color on the sides. The venter is plain white or cream. Some individuals may have another, very faint, dark stripe on the third row of scales.

The large triangular rostral has a loosely attached appearance that is characteristic of the genus. Dorsal scales are smooth and in 17 rows at midbody. There are 178–197 ventrals, 85–112 subcaudals, and the anal plate is divided. There are eight (occasionally nine) upper labials, the second and third reaching the single loreal. There are nine (occasionally 10) lower labials. The posterior chin shields touch or are separated by one scale width.

Similar Species: *Salvadora deserticola* has four distinct longitudinal lines on the dorsum with the outer pair of lines on the fourth scale row at midbody; the middorsal color is tan to brownish-orange and is darker than the paler ground color of the sides; there are nine upper labials, and the chinshields are always separated by 2–3 scale widths. No other lined species has the patch-like rostral.

Systematics: Only the nominotypical subspecies, *S. g. grahamiae*, occurs in New Mexico.

Habitat: As the common name implies, the mountain patchnose snake is found most often in mountains and foothills, reaching elevations of about 2200 m. When found at elevations as low as 950 m in New Mexico, it occupies rough or broken terrain at the base of mountains, along arroyos, in canyons, or on rocky flats. Trees, shrubs, or other types of plant cover are usually abundant in the habitat. Common snake cohabitants when their ranges overlap are *Crotalus atrox*, *C. molossus*, *C. lepidus*, *Bogertophis subocularis*, *Masticophis taeniatus*, *M. bilineatus*, *Pituophis melanoleucus*, and *Tantilla* spp. In New Mexico, *S. deserticola* is rarely found in the same habitat.

Behavior: *Salvadora grahamiae* is very active and fast-moving with general behavior much like that of the whipsnakes. It is chiefly ground-dwelling, but may ascend into low vegetation to forage, escape predators, or bask. This snake will often attempt to escape from potential enemies by taking refuge in thick brush or grasses. When the bases of such plants as *Agave*, *Nolina*, *Dasylirion*, or various cacti are used for cover, it is very difficult for the collector to locate and extricate the potential specimen. *Salvadora grahamiae* is active in the daytime, especially during the morning hours, when it is often seen basking or foraging.

The annual activity period is long. New Mexico collection records range from March–November, with one taken near Carlsbad on 8 December. It would seem that, like *S. deserticola*, this species has a low critical thermal minimum temperature, and will remain active as long as daytime temperatures permit warming above that point. It should prove rewarding to determine the thermal attributes of this species compared with *S. deserticola*, as Jacobson and Whitford (1971) did for *S. h. hexalepis* and *S. deserticola*. The results of such a study might help to explain the usually exclusive range differences between these two otherwise similar species. More than simple habitat differences may be involved in their distributions.

Reproduction: Published information on reproduction in *S. grahamiae* is scarce. Wright and Wright (1957) mentioned an adult male measuring 67 cm TL. Conant and Collins (1991) and Stebbins (1985) gave 56 cm TL as the minimum mature size. Goldberg (1995) found spermiogenic activity in Arizona males as small as 483 mm SVL and eight oviducal eggs in a female measuring 568 mm SVL.

Courtship and mating probably occur shortly after the snakes become active in the spring, since eggs have been laid as early as 1 April in Texas (Fitch, 1970). Clutch dates and size of snakes from Texas are: 1 April (10), the first week in May (5,7), 29 May (6), and sometime in June (9). The eggs have smooth shells, are adhesive, and take 2–3 months to hatch. Minton (1958 [1959]) gave sizes for three hatchlings from Big Bend as 263, 264, and 267 mm TL. A female, 94 cm TL, from Santa Fe County, laid six eggs 3 July measuring 13–15 x 31–35 mm. These hatched 10–12 October and the hatchlings measured 275–290 mm TL (T. L. Brown, pers. comm.). Another female from Bernalillo County collected by Brown laid 3 eggs measuring 15 x 40 mm, although these did not hatch.

Food Habits: Lizards, reptile eggs, and small mammals serve as food. Live prey may be chased down on the surface or taken in burrows. The specialized rostral scale of *S. grahamiae* is likely helpful in excavating reptile eggs.

Remarks: In the description of *S. grahamiae* (Baird and Girard, 1853a), the name Col. J. D. Graham is given at the end and appears to be the name of the collector. However, it is stated in the preface of that same publication that the specimens from that U.S. and Mexican Boundary Survey (directed by Graham) were collected by Mr. J. H. Clark. Details of routes and dates of this survey are discussed in Webb (1988). Baird (1859) also listed

J. H. Clark as the collector of the type. Cope (1898 [1900]) listed Graham in his "from whom received" column. This is misleading since the name of the collector usually appears there. Cochran (1961) listed Graham as the collector. A similar problem exists for other snakes in the Graham survey collection that were treated in Baird and Girard (1853a).

Reviews of the genus that included this species are Smith (1938b), Schmidt (1940), and Hartweg (1940). The late C. M. Bogert, considered an authority on this group, was actively preparing a monograph on the genus just prior to his death.

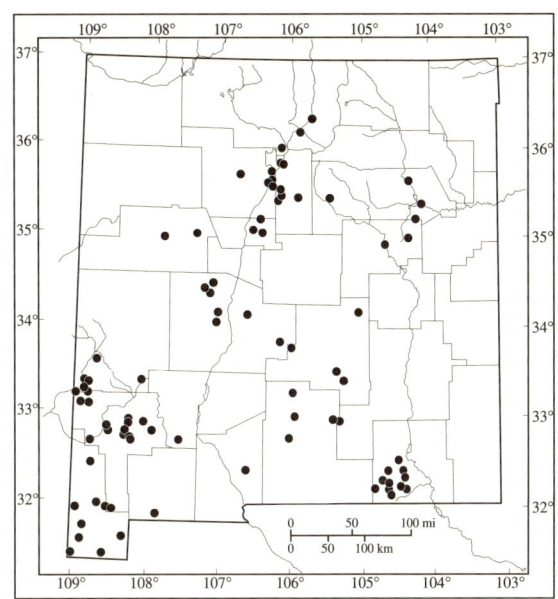

Distribution of Salvadora grahamiae in New Mexico.

SENTICOLIS TRIASPIS (Cope, 1866) — *Green Rat Snake*

See Plate 101.

Type: This species was described by Cope (1866a) as *Coluber triaspis*. The holotype, USNM 24903, is a juvenal male collected by D. B. Parsons. The date of collection is unknown (Price, 1991). The type locality was given as "Belize" (formerly British Honduras), although it was later restricted to the city of that name by Smith and Taylor (1950b).

Distribution: *Senticolis triaspis* has a broad geographic distribution, ranging from southwestern New Mexico and southeastern Arizona and southern Tamaulipas southward through Mexico and much of central Mexico to Costa Rica (Price, 1991; Garrett and Painter, 1992). The species was not recorded from the United States until reported by Stone (1911). The elevational range is from near sea level to over 2200 m. In New Mexico, *S. triaspis* is restricted to Hidalgo County where its occurrence in the Guadalupe Mountains has only recently been documented by a museum specimen (Garrett & Painter, 1992). Photo documentation exists for specimens from Post Office Canyon in the Peloncillo Mountains. Sight records from the Animas Mountains in Hidalgo County and the Mule Mountains in Grant County have not been verified.

In New Mexico, *S. triaspis* probably occurs from 1260 m near Rodeo to at least 2286 m in the mountains; it has been collected from 1330–1387 m in Guadalupe Canyon.

Description: Adult *S. triaspis* are slim-bodied and uniformly greenish in color. The venter is uniformly colored, usually whitish or cream colored. The scale rows along the back are weekly keeled and have paired indistinct apical pits. The narrow head is distinct from the neck. There are three elongate temporals, only two of which may touch the postoculars. The juveniles are blotched dorsally, although these blotches tend to fade in approximately two years. Adults may reach 1.6 m in length and weigh 600 grams. Sexually dimorphic characters seen in adults include tail length and total length, with males having a longer tail, and females longer in overall length (Dowling, 1960). The tail is equal to 30–35% of SVL in adult males and 23–26% in adult females.

There are 30–39 lightly keeled scale rows at midbody. Ventral scales range from 241–264 in males and 256–282 in females; subcaudals range from 95–126 in males and 87–110 in females (Price, 1991). There are usually eight upper labials. There is one loreal, one preocular, two postoculars, and three elongate temporal scales. The distinctive hemipenis is illustrated in Dowling and Fries (1987) and Price (1991).

Similar Species: The unique uniform greenish coloration and long narrow head will distinguish adult *S. triaspis* within its restricted range in New Mexico. The juveniles resemble young *Masticophis flagellum* or *Arizona elegans*, although they may be separated by three elongate temporal scales and weakly keeled scales along the back. *Liochlorophis vernalis* has only 15 rows of scales at midbody, seven upper labials, and a short head that is not distinct from the neck.

Systematics: Three subspecies are recognized (Price, 1991). Only *S. triaspis intermedius* (Boettger, 1883) occurs in New Mexico. The holotype is SMF 34575, a juvenal male from the collection of Dr. Pagenstecher. The type locality "Mexico" was later restricted to Hacienda El Sabino, Michoacán, Mexico, by Dowling (1960). *Senticolis triaspis* was formerly known as *Elaphe chlorosoma* or *E. triaspis* until a reclassification by Lawson and Dessauer (1981) and Dowling and Fries (1987), who provided biochemical and morphological characters to support the erection of a separate genus, *Senticolis*. *Senticolis* may be distinguished from closely related forms by its unique hemipenial morphology (Price, 1991).

Habitat: Most specimens have been taken in montane mesophytic forests along the slopes of the Mexican highlands and in Central America (Price, 1991). In New Mexico, *Senticolis* is known only from the foothills and lower elevations of the Peloncillo and Guadalupe mountains where it has been encountered in rocky riparian areas grown to cottonwood, sycamore, various oaks, ash, and mesquite. These snakes are secretive but not uncommon along Cave Creek in the Chiricahua Mountains in adjacent Cochise County, Arizona.

Behavior: Although others (Wright and Wright, 1957; Shaw and Campbell, 1974; Stebbins, 1985) believed *S. triaspis* to be primarily arboreal, Cranston's (1989a) observations of over a dozen specimens on the ground in the Chiricahua Mountains may indicate otherwise. Middendorf and Lawler (*in* Cranston, 1989a) have noted *S. triaspis* climbing into low shrubs, perhaps in search of potential prey. All specimens Cranston (1989a) observed were on the ground, and none climbed into low shrubbery.

Reproduction: Virtually nothing is known about the reproduction or other aspects of the natural history of *S. triaspis* in New Mexico. An individual 300 mm SVL and 9.3 grams was captured in Post Office Canyon near Rodeo on 4 June. This specimen likely represents a hatchling, based on the size at hatching reported by Cranston (1989b). Cranston (1989a, 1989b) discussed captive reproduction in four individuals taken from southeastern Arizona. In four observations of reproductive behavior that resulted in egg clutches, courtship was initiated as early as 16 March and continued to 11 May. After copulation, egg formation took approximately 70 days with an average of only 29% of the eggs being fertile. Incubation ranged from 77–88 days at 26°C. Ten neonates averaged 18.5 gm in weight (11–22.5 g) and 340 mm (320–360 mm) TL. Females invested approximately 25–30% of their biomass in egg weight.

Food Habits: Practically nothing is known of the food habits of *S. triaspis* in the wild. Duellman (1958) reported a single specimen from Colima, Mexico, that contained a house mouse (*Mus*). Cranston (1989a) reported a white-footed mouse (*Peromyscus*) from a specimen collected in the Chiricahua Mountains of southeastern Arizona, and C. M. Bogert (*in* Stebbins, 1954) reported a specimen that had eaten a large woodrat (*Neotoma*). Observations reported in Cranston (1989a) indicate that birds might also be eaten. In captivity, this species is known to feed on laboratory mice.

Remarks: The young were not described until Smith (1941a) discussed the ontogenetic changes that occur in this species. The name *Senticolis* is from the Latin *sentis*, a thorn or bramble, and *colis*, a penis, in reference to the large spines on the hemipenis. McAllister et al. (1993) presented detailed information on four species of parasites found in this species. *Senticolis triaspis* is listed as endangered by the New Mexico Department of Game and Fish (NMGF, 1990). Price (1991) reviewed this species.

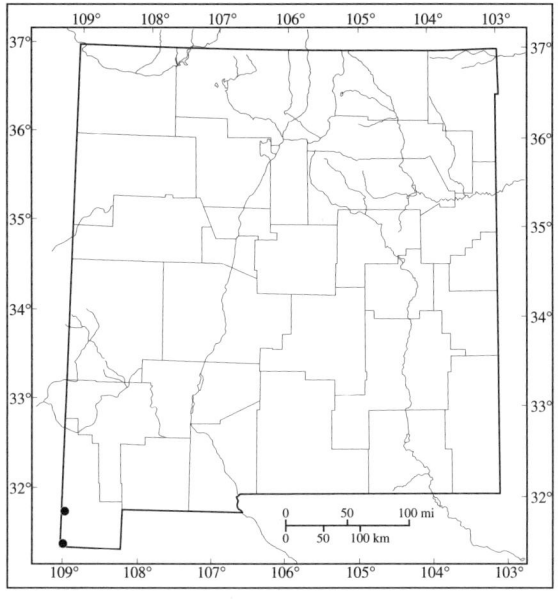

*Distribution of
Senticolis triaspis
in New Mexico.*

SONORA SEMIANNULATA Baird and Girard, 1853   *Ground Snake*

See Plate 102A, 102B, 102C.

Type: The holotype, USNM 2109, is a male collected by J. H. Clark, probably in September 1851 (Baird and Girard, 1853a; see Remarks). The type locality was originally designated as "Sonora, Mexico," but was restricted by Stickel (1943) to the "vicinity of the Santa Rita Mountains, Arizona" in Pima and Santa Cruz counties.

Distribution: *Sonora semiannulata* is found in the grasslands of the central United States from southeastern Colorado to southwestern Missouri and thence south and west to northern Mexico. It ranges across the deserts of west Texas, New Mexico, Mexico, Arizona, Nevada, and California. Apparently isolated populations occur in eastern Oregon and western Idaho, Baja California, northern Utah, and northern Kansas (Frost, 1983b). In New Mexico, *S. semiannulata* occurs mainly southeastern of a diagonal line extending from the northeastern corner to the southwestern corner of the state, at elevations from 900 m to approximately 1600 m.

Description: This small species may reach about 45 cm TL. Individuals of this highly variable snake may be plain-colored, crossbanded, longitudinally banded, or have a combination of these characters. The dark crossbands may encircle the body, form saddles, be reduced to a single neck band, or be absent. If a longitudinal band is present, it is red or orange. The ground color may be tan or reddish. Some plain-colored individuals may have a slightly darkened head. The venter is whitish or cream colored and may or may not have dark crossbands.

Dorsal scales are smooth and in 13–15 (always 15 in New Mexico) rows at midbody. There are 126–186 ventrals, 31–59 subcaudals, with much variation geographically. Males have lower ventral but higher subcaudal counts. The anal plate is divided. A loreal scale is present. There are usually seven upper and seven lower labials.

Similar Species: *Tantilla* spp. have distinctly dark heads, lack crossbands on the body, and have no loreal scale. *Leptotyphlops* spp. have ventral and dorsal scales of the same size and short, stubby, non-tapering tails.

Systematics: Frost (1983b) considered this species to be monotypic (see Remarks).

Habitat: *Sonora semiannulata* occupies plains, valley, and foothill habitats. Higher mountain slopes are avoided, as are areas with heavy, poorly drained soils. Subsurface moisture is required by the snakes as well as their prey. In New Mexico, the most dense populations are on rocky slopes of low hills or on valley sides. Rocks subject to flooding in arroyo bottoms are seldom used. Flat shallow rocks which warm quickly in the sun are preferred, and a single rock may shelter more than one ground snake. The cavities under these rocks often connect to deeper retreats through holes or cracks. Common snake cohabitants in New Mexico are *Leptotyphlops* spp., *Tantilla* spp., *Hypsiglena torquata,* and *Rhinocheilus lecontei.*

Behavior: This gentle species has little in the way of defensive behavior except escape. Most individuals are discovered by rock-turning, and if the potential collector hesitates, they may disappear into a nearby hole or crevice or burrow into loose soil. Ground snakes are primarily active on the surface shortly after dark, but we have observed them at sunset or just before. Most thermoregulation is done while concealed under warm rocks, but some individuals may sun on the surface near entrances to their retreats. One snake from Big Bend, Texas, was observed in July at mid-day in full afternoon sun, but crawled rapidly into a crevice and was lost (Minton, 1958 [1959]). These snakes like moisture and have been found on wet pavement after rains.

Annual activity extends at least from March to October in New Mexico. One specimen was found inside a Socorro residence on 15 December. In an Oklahoma study (Kassing, 1961), *S. semiannulata* was collected during every month of the year except January, and precipitation seemed to be more important than temperature for collecting success. There is no information available on hibernation, but the burrow and crevice systems used during the active portion of the year usually penetrate to below the frost line, and should serve adequately as hibernacula.

Reproduction: Kassing's (1961) study on Oklahoma and Texas populations provides most of the available information on *S. semiannulata.* She suggested that sexual maturity comes at 230–240 mm TL, when the snakes are 1.5–2.5 years old. During courtship, the bodies and tails of the snakes are entwined, while the male rubs the head and neck of the female with his head, sometimes biting her neck. Kroll (1971) described combat between males competing for a female. In these combats, lasting from 1–2 minutes to over 15 minutes, males entwined their bodies and bit their opponent. In all instances, the displaced male was the same size or smaller. Coitus was involved in all cases, before, between, or after male combat episodes. Most mating occurs in April, May, and June, and perhaps again in late summer or fall.

Eggs are laid a little more than a month after copulation and number from 3–6. They measure 5–9 x 13–28 mm and normally hatch in 50–70 days. A female from Big Bend, Texas, laid three eggs on 2 July. James N. Stuart (pers. comm.) collected a gravid female about 340 mm TL near Santa Rosa, New Mexico. It laid five white smooth-shelled eggs, 8 x 23 mm, on 9 June. Four of these hatched 19–20 August. The hatchlings measured 118–127 mm TL and resembled the parent. Hatchlings have been reported as small as 70 mm TL (Kassing, 1961).

Food Habits: *Sonora semiannulata* feeds opportunistically on a wide variety of small arthropods. Scorpions, centipedes, and spiders, including the black widow, apparently present no problem to these snakes, since these food items are common in stomach contents and fecal samples. A captive from Big Bend, Texas, ate a wolf spider, a small solpugid, an unidentified black spider, and baby crickets, but refused thysanurans.

Sonora forages on the surface at night and probably also uses prey associated with the below-ground tunnel and crevice systems that these snakes inhabit. Many of the prey species are the same ones the collector will find when turning rocks during snake-hunting activity. Degenhardt, while road-cruising in Big Bend National Park, found *S. semiannulata* attempting to swallow a DOR *Coleonyx brevis.*

Remarks: The specimens sent to the USNM from the U.S. and Mexican Boundary Survey led by Col. J. D. Graham were actually collected by John H. Clark. However, Cochran (1961) listed Graham as the collector for many of these, and other authors have followed her lead. This problem is discussed in more detail in the *Salvadora grahamiae* account and by Webb (1988).

The specific epithet, *semiannulata,* meaning half-ringed, describes only one of many color morphs of this polymorphic species in which the crossbands fail to cross the venter. The polymorphism in *S. semiannulata* is re-

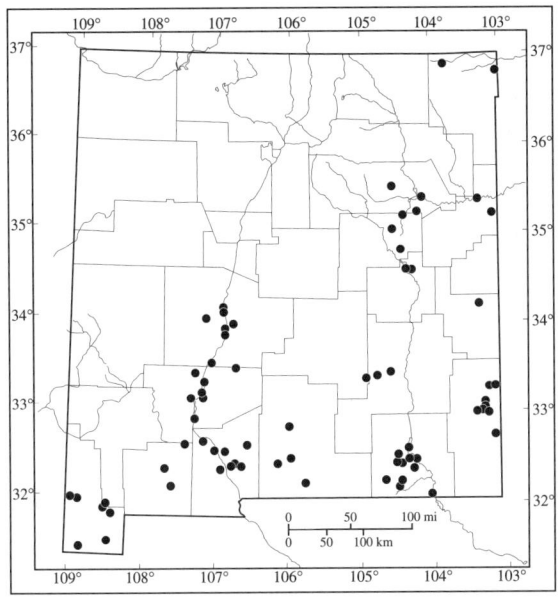

Distribution of
Sonora semiannulata
in New Mexico.

sponsible for the numerous synonyms previously applied to this snake. Much of the literature of this species refers to it as *S. episcopa* (Kennicott *in* Baird, 1859). Frost (1983b) reviewed this species.

TANTILLA HOBARTSMITHI Taylor, 1936 [1937] — *Southwestern Black-headed Snake*

See Plate 103.

Type: The holotype, UIMNH 25066, is an adult male collected by E. H. Taylor on 3 July 1934. The type locality is "near La Posa, 10 mi. northwest of Guaymas," Sonora, Mexico (Taylor, 1936 [1937]).

Distribution: Locality records for *T. hobartsmithi* suggest a disjunct distribution in the six southwestern states, southwest Texas, and northern Mexico. In New Mexico, this species occurs at low elevations from 900 m to about 1600 m within the southern one-quarter of the state.

Description: This diminutive species seldom exceeds 310 mm TL (Cole and Hardy, 1981). During 1992–95, 65 individuals collected in pit-fall traps in Guadalupe Canyon averaged 162.7 (120–204) mm SVL and 2.0 (1.1–2.9) gm (our unpubl. data). The dorsum is tan to light brown and in sharp contrast with a dark brown or black cap. This cap extends 1–3 scales behind the parietals, is straight or slightly convex at the rear edge, and is bordered posteriorly by a somewhat indistinct white to cream collar 0.5–2 scales wide. There is a midventral red or coral stripe that does not extend to the outer edges of the ventrals.

Dorsal scales are smooth and in 15 rows at midbody. There are 124–166 ventrals in males and 130–169 in females. Subcaudals number 48–74 in males and 47–67 in females. The anal plate is divided. The mental scale usually touches the front pair of chinshields, preventing the first lower labials from meeting at the midline. The loreal scale is absent and there are seven upper labials.

Similar Species: *Tantilla yaquia* has a distinct white ring on the neck bordered with the dark hood color both anteriorly and posteriorly. *Tantilla nigriceps* has the black cap extending 3–5 (rather than 1–3) scale lengths behind the parietals, with the posterior margin pointed or strongly convex (rather than straight or weakly convex), and the first lower labials meet at midline. *Sonora semiannulata* has a loreal scale and never a distinct black cap, although the head may be slightly darkened.

Systematics: We follow Cole and Hardy (1983) in considering *T. hobartsmithi* as a monotypic species. Earlier accounts have referred this species to *T. atriceps* or *T. planiceps*, but neither of these species occurs in New Mexico. See Cole and Hardy (1981) for a discussion of the complex nomenclatural history of *T. hobartsmithi*.

Habitat: *Tantilla hobartsmithi* is found in many different southwestern habitats. Stebbins (1985) listed brushland, grassland, sagebrush-greasewood, mesquite-yucca and creosotebush, open chaparral, thornscrub, piñon–juniper woodland, open coniferous forests, persimmon-shin oak, mesquite-creosotebush, and cedar-savannah as habitat. Plants are necessary to support arthropod populations on which these snakes feed, although light well-drained soils and access to subsurface retreats may be more important factors than the actual type of plant cover. These retreats allow protection from drought, heat, cold, and predation. Black-headed snakes are most often collected by turning rocks or surface debris in the daytime or by road cruising at night. Old dumping areas have been productive; they have also been found in rodent burrows. Moisture is important, and during dry conditions they will retreat more deeply into the ground and spend less time on the surface.

In New Mexico, the most productive collecting has been in rocky places at the lowest elevations. The limestone covered hills and valley sides in the lower Pecos drainage support especially large populations of *T. hobartsmithi*. This area also supports the largest assemblage of Chihuahuan Desert reptiles in the state.

Common snake cohabitants are *Leptotyphlops* spp., *Sonora semiannulata*, *Hypsiglena torquata*, and *Rhinocheilus lecontei*. The more widespread *T. nigriceps* may be found sympatric with *T. hobartsmithi*, although where one of these species is abundant, the other occurs in smaller numbers, suggesting microhabitat differences and/or competition. This same situation may occur with *T. yaquia* and *T. hobartsmithi*, for in southwestern New Mexico, where *T. hobartsmithi* is abundant, *T. yaquia* is rare.

Behavior: *Tantilla hobartsmithi* is nocturnal, and by day it will take shelter under many types of surface objects. When discovered, it moves rapidly in its efforts to escape, and may elude capture if the collector is not "on his or her toes." While collecting in the Chisos Mountains, Texas, Murray (1939) found one swimming in a small stream, and Little (1940) collected another in a runoff tank in central Arizona, but these are the only instances of diurnal surface activity we are aware of. Shaw and Campbell (1974) stated that this snake "seldom appears above ground and then only on warm evenings." We regularly found *T. hobartsmithi* on roads in the Chisos Mountains of Texas. Temperatures were usually lower there than in the surrounding deserts, and ranged from 15–25°C. Taylor (1936 [1937]) also referred to surface activity when he described collecting the type specimen at night while it was "running with most surprising rapidity over rough, gravelly terrain, under low shrubs." These very small and slender snakes are hard to see, and most road cruising is done at speeds too high for their detection.

Cole and Hardy (1981), in examining many specimens from throughout the range, found collection dates in all 12 months of the year. Our records from New Mexico range from March into September, but probably define more the activity of the collectors than of the snakes. Little (1940) found one hibernating in gravel along a central Arizona road bank on 24 December. Access to potential hibernacula below the frost line is easily attained by these fossorial snakes, but we have no information on winter dormancy in New Mexico.

Reproduction: Information on reproduction is scarce for *T. hobartsmithi*. Most often one egg is laid, but we recorded a clutch of three laid 19 June by a *T. hobartsmithi* from the Chisos Mountains, Texas. The eggs are elongate, measuring 6–7 x 23–28 mm when laid (Minton, 1958 [1959]; Easterla, 1975). Cole and Hardy (1981), citing Stebbins (1954), gave egg size in two California snakes as "approximately 4 by 17 mm"; however, these were from *Tantilla utahensis*. In Big Bend National Park, Texas, laying dates have extended from 19 June to 28 July. Minton (1958 [1959]) mentioned a snake collected 1 June that contained, "one large egg ready for deposition." The large range in laying dates suggests the possibility of more than one clutch per year.

Clark (1970) noted that in 10 species of *Tantilla*, seven, including *T. hobartsmithi*, had only a single right functional oviduct. Generally, the size of the vestigial left oviduct varied directly with the size of the species. Clark pointed out that the structure of reptilian ovaries and oviducts is such that when an egg is ovulated, nothing guarantees it must move into its respective oviduct. The shorter oviduct, with its ostium farther from the ovary, would be less likely to receive the (usually single) ovum, and might be functionally superfluous. Therefore, fossorial tendencies, which select for an elongated, cylindrical body form, may favor mutations reducing the size of the left oviduct.

Food Habits: *Tantilla hobartsmithi* feeds on a variety of arthropods. Beetle larvae, caterpillars, centipedes, and millipedes have been found in stomachs; mealworms have been eaten in captivity. Grooved teeth on the rear of the upper jaw may deliver venom which is possibly used in subduing prey.

Remarks: We are deviating from Collins (1990) by using the common name of black-headed snake rather than blackhead snake. This species was last reviewed by Cole and Hardy (1983) and the genus by Wilson (1982).

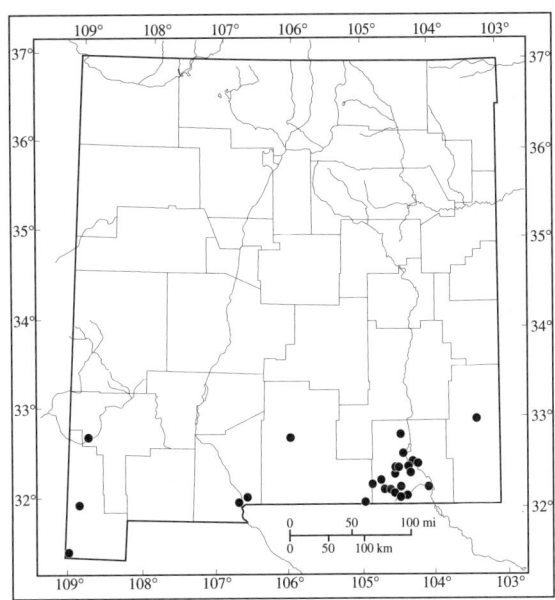

Distribution of Tantilla hobartsmithi *in New Mexico.*

TANTILLA NIGRICEPS Kennicott, 1860 — *Plains Black-headed Snake*

See Plate 104.

Type: Two syntypes were used in the original description; USNM 2040, from "Indianola to Nueces, Texas," was collected by Capt. Pope and USNM 4491, from "Fort Bliss, New Mexico" (Texas) was collected by Dr. Crawford.

Distribution: *Tantilla nigriceps* is found in the central United States from Colorado, Nebraska, and Kansas south into southeastern Arizona and northern Mexico. It is widespread in New Mexico except the northwestern corner and the higher mountains of the north and west. Elevational range is 900 m to about 1900 m.

Description: *Tantilla nigriceps* usually ranges in adult size from 18–25 cm TL, but it may reach over 38 cm TL (Conant and Collins, 1991; our obs.). The dorsum is yellowish or grayish brown with a distinct dark head color extending 3–5 scale rows behind the parietals. The rear margin of the head color pattern is pointed or sharply convex, and is never bordered by a light ring. The venter is white with a broad midventral pink streak.

Dorsal scales are smooth and in 15 rows at midbody. There are 130–161 ventrals, 33–62 subcaudals, and the anal plate is divided. The mental scale does not touch the front pair of chinshields because the first lower labials meet at the midline. The loreal scale is absent, and there are seven upper labials.

Similar Species: *Tantilla yaquia* has a distinct white ring on the neck. *Tantilla hobartsmithi* is smaller, the black cap extends 1–3 (rather than 3–5) scale lengths behind the parietals with the posterior margin straight or slightly convex (rather than pointed or strongly convex), and the first lower labials fail to meet at the midline. *Sonora semiannulata* has a loreal scale and never a distinct black cap, although the head may be slightly darkened.

Systematics: *Tantilla nigriceps* is monotypic. Cole and Hardy (1981) described the hemipenis and concluded that *T. nigriceps* differs from other members of the genus. They also suggested that since subspecies determinations (*T. n. nigriceps* and *T. n. fumiceps* [Cope, 1861]) are based only on variation in numbers of ventral and subcaudal scales, the species should be regarded as monotypic.

Habitat: *Tantilla nigriceps* is less selective in habitat preference than other species of *Tantilla* in New Mexico. It is adapted to the central plains grassland, but has moved into the valleys and foothills of the southwest as conditions allow. Besides the flats, rolling sandhills and rocky slopes may support populations. It is particularly adept at using a variety of surface materials of both natural and man-made origin. Even the lowly "cow-chip" serves as cover when other retreats are unavailable. Snake cohabitants are largely other small species that can make use of similar microhabitat. *Sonora semiannulata* and *Leptotyphlops dulcis* are often found under adjacent rocks. When sympatric with *Tantilla hobartsmithi* in New Mexico, the numbers of one species or the other are reduced. Subtle microhabitat differences are probably present between different species of this genus.

Behavior: This small and gentle species is active on the surface at night, and spends the daylight hours hidden below the ground or under surface debris. Flat rocks, metal sheets, old roofing, boards, or other objects warmed by the sun, are used for thermoregulation as well as protection from predators. *Tantilla nigriceps* is regularly collected by road-cruising at night. Price and LaPointe (1990) found it to be one of the common species in their four-year study of snake activity on a 25 km section of U.S. Hwy 85 between Radium Springs and Hatch, Doña Ana County. Their study also showed that a direct, positive correlation exists between snake activity and temperature and rainfall.

Most annual activity extends from April to September but a few records for both March and October are available. No hibernacula have been discovered in New Mexico, but this small fossorial species can easily penetrate the soil to depths below the frost line when necessary. Tihen (1937 [1938]) reported two specimens found eight feet below the surface of the ground in Kansas on 13 January 1934.

Reproduction: Nothing has been written concerning reproduction in *T. nigriceps*. The pattern probably follows roughly that of *T. gracilis*, with mating in early spring, egg-laying in June and July, and hatching 2–3 months later. Clutch size in other species of *Tantilla* numbers 1–4, and hatchlings measure about 80 mm TL. This species also has a single functional right oviduct. See the discussion in the *T. hobartsmithi* account for additional information.

A female (MSB 50375) from Sabinal, Socorro County, measuring 350 mm TL, laid three eggs on 22 June 1975. The eggs measured 7 × 22–25 mm, and two of them hatched on 26 August. The hatchlings measured 123 and 128 mm TL, and the dead snake in the unhatched egg measured 110 mm TL (T. L. Brown, pers. comm.).

Food Habits: The most commonly recorded food items of *T. nigriceps* are insects (including their larvae and pupae), centipedes, millipedes, and spiders. These snakes probably feed on below-ground food items at any time and on surface active prey at night. Grooved teeth on the rear of the upper jaw can deliver venom that may be used in subduing active prey. Ted L. Brown observed that, "The female who laid the eggs fed well on large centipedes (*Scolopendra*); it grabbed them behind the head, hung on for 3–5 minutes, then devoured them with relish."

Remarks: We prefer the common name black-headed snake rather than blackhead snake of Collins (1990). Taylor (1936 [1937]) reviewed "Certain American and Mexican snakes of the genus *Tantilla*," which included *T. nigriceps*. Cole and Hardy (1981) described the hemipenis, and suggested that hemipenes and head coloration are the most useful characters for distinguishing species of *Tantilla*.

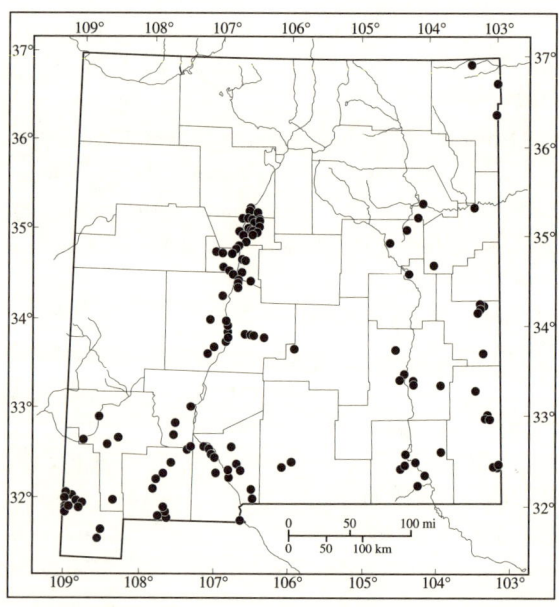

Distribution of Tantilla nigriceps *in New Mexico.*

TANTILLA YAQUIA Smith, 1942 — *Yaqui Black-headed Snake*

See Plate 105.

Type: The type specimen, MCZ 43274, a female collected by H. S. Gentry in August 1936 from "Guasaremos, Río Mayo, Chihuahua, [Mexico.]" remains the only specimen known from Chihuahua (Tanner, 1985).

Distribution: *Tantilla yaquia* occurs from extreme southeastern Arizona southward through Sonora and Sinaloa to the Río Santiago Valley in Nayarit (Stebbins, 1985). In New Mexico, this species is known only from the lower slopes of the Guadalupe and Peloncillo mountains in southwestern Hidalgo County, where specimens have been collected from 1325 m in Guadalupe Canyon to approximately 1586 m in Skeleton Canyon. *Tantilla yaquia* was unknown from New Mexico until Painter et al. (1992) reported a single specimen from Skeleton Canyon.

Description: *Tantilla yaquia* is small, reaching only 325 mm total length. The dorsal coloration is uniformly brown to brownish tan, while the ventral coloration is pinkish-orange, brightest on the posterior third and fading to creamy white on the anterior half of the venter. There is a brown to black cap on the head that extends onto the nape from two to slightly more than four scales on the midline. This dark head cap is bordered posteriorly by a narrow, light, nuchal collar, 1–1.5 scales wide. The black cap extends laterally 0.5–3 scales below the angle of the mouth. Most of upper labials 1,4,5,6, and the lower one-third to one-half of the anterior temporal are white and contrast sharply with the dark cap (McDiarmid, 1977). This species has 46–75 subcaudals, 134–157 ventrals in males, and 145–165 ventrals in females. There are 15 scale rows at midbody, one preocular, and one anterior temporal. The loreal is absent (Smith and Brodie, 1982). Two adult males from New Mexico (MSB 54920, 54921) have seven upper labials, six lower labials, and two postoculars. The anal plate is divided.

The detailed color notes and photograph of *"Tantilla atriceps"* from Arizona presented by Wright and Wright (1957) are actually based on a specimen of *T. yaquia* (Cole and Hardy, 1981).

Similar Species: The three species of *Tantilla* that occur in New Mexico are easily confused, although *T. yaquia* may be distinguished from the others by a white collar 0.5–1.5 scales wide that borders the posterior edge of the black cap and by 100% of the area of upper labials 5–6 being light in color (Cole and Hardy, 1981). This species is sympatric with *T. hobartsmithi* in Guadalupe Canyon in extreme southwestern New Mexico.

Systematics: Two subspecies of this primarily Mexican species have been recognized, although currently the species is considered to be monotypic (Cole and Hardy, 1981; McDiarmid, 1977). Zweifel and Norris (1955) considered *T. yaquia* and *T. bogerti* Hartweg, 1944 as subspecies, although McCoy (1964) recognized them as conspecific after examination of additional specimens.

Habitat: McDiarmid (1968) reported that most specimens from the northern part of the range have been found beneath rocks and surface litter, especially in March, April, August, and September, when the soil is damp after winter and summer rains. One specimen (MVZ 59778) was found in the stomach of *Rana tarahumarae*, a frog rarely seen away from water (McDiarmid, 1968). The riparian habitat in New Mexico where specimens have been collected is characterized by loose rocky soils with numerous tree species including Arizona sycamore, Arizona walnut, mesquite, and various oaks.

Behavior: Little is known of the habits of this nocturnal, fossorial snake. Three specimens collected in New Mexico were captured in pit-fall traps, and another was found under a flat rock. A small amount of musk and feces may be released from the vent when one is captured. This species is occasionally seen during summer nights on the paved roadways near Portal, Arizona (B. R. Tomberlin, pers. comm.).

Reproduction: *Tantilla yaquia* is oviparous. Very few data on reproduction are available. McDiarmid (1968) reported two specimens, 104 mm and 140 mm total length, that retained a visible umbilical scar indicating they had recently hatched. The hemipenes were described and illustrated by Cole and Hardy (1981).

Food Habits: Detailed food habits of *T. yaquia* have not been described, although based on size and distribution of this species Cole and Hardy (1981) speculated that warthogs are not included in its diet. It is expected that small invertebrates such as millipedes, centipedes, larval and adult soft-bodied insects, and spiders comprise the bulk of its diet.

Remarks: We prefer to use the common name blackheaded snake in place of the name blackhead snake used by Collins (1990). The first specimen from the United States (UCM 875) was collected at Bisbee, Arizona, in 1907 by Dr. C.L. Edmonson, although the specimen was not reported until McCoy (1964) added this species to the herpetofauna of the United States.

McDiarmid (1977) reviewed this species.

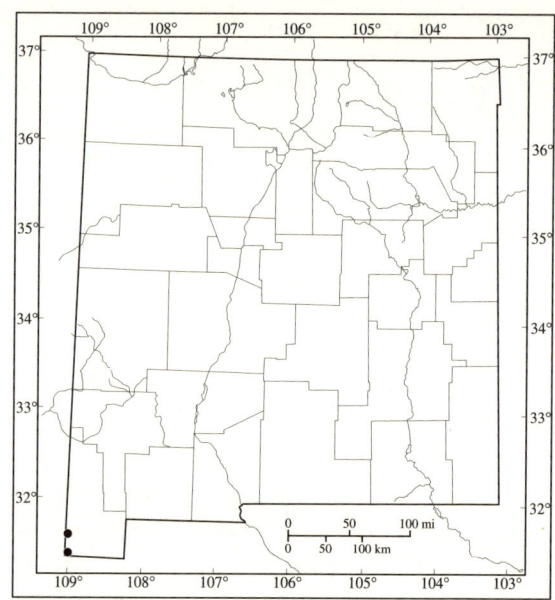

Distribution of Tantilla yaquia *in New Mexico.*

THAMNOPHIS CYRTOPSIS (Kennicott, 1860) — *Blackneck Garter Snake*

See Plate 106.

Type: The holotype, USNM 8067 (formerly USNM 930), is an adult male collected by Lieut. D. N. Couch in 1853. The type locality is "Rinconada," Coahuila, Mexico. Conant (1968) determined that this site is "approximately 20 miles northeast of Ramos Arizpe" (Coahuila), just across the state line in Nuevo León.

Distribution: *Thamnophis cyrtopsis* is found from southern Colorado and Utah south through Arizona, New Mexico, and Texas to Guatemala. In New Mexico, it is widely distributed statewide from 1125 m to about 2400 m in elevation.

Description: *Thamnophis cyrtopsis* normally ranges from 40–70 cm TL, although Conant and Collins (1991) reported a record size of 107 cm TL. Dorsal color varies from olive gray to olive brown, or sometimes a darker brown. The head is gray. The white or yellow midvertebral stripe becomes orange anteriorly and divides two black neck blotches behind the head. The white lateral stripes are on the second and third scale rows. There are two rows of alternating, and somewhat checkerboard-like, dark spots, most distinct anteriorly, between the stripes. Another row of spots, sometimes divided, lies between the lateral stripe and the ventral scales. The venter is unmarked and white to green. The dorsal scales are keeled and in 19 rows at midbody. There are 130–184 ventrals, 64–109 subcaudals, and the anal plate is single. There are eight upper labials, one preocular, and three postoculars.

Similar Species: *Salvadora* spp. and *Masticophis* spp. have smooth scales, no spots, and a divided anal plate. *Thamnophis marcianus* has 21 dorsal scale rows at midbody, and the narrow lateral stripe is only on the third scale row. *Thamnophis elegans* has a paler (and never yelloworange) vertebral stripe and a venter with dark markings; the neck blotches, if present, are indistinct. Other New Mexico *Thamnophis* spp. have no wide, well defined neck blotches.

Systematics: Two subspecies occur in the southwestern United States, but only the nominotypical subspecies, *T. c. cyrtopsis*, occurs in New Mexico. (See Remarks.)

Habitat: *Thamnophis cyrtopsis* is usually associated with water, and has been called an "aquatic habitat specialist" by Jones (1990). Permanent or intermittent streams in valleys and canyons supply choice habitat. Quiet and shallow rocky pools along these watercourses are preferred by the snakes, and also by the tadpoles that serve as favorite prey items. However, aquatic stream habitats change throughout the year from times of more runoff, when there are more stream runs and riffles, to times of less runoff, when there are only scattered pools. Earthen cattle tanks or other man-made containments may be used, and only rarely are individuals found with no obvious water reservoirs nearby. Overnight cover sites include exposed roots along stream banks, rodent holes in adjacent desert, crevices in streambanks and rock, and vegetative debris piles created when streams flood (Jones, 1990).

In New Mexico, common snake associates are usually other species of *Thamnophis*. At higher elevations, and in the mountains and foothills of the north, *T. elegans* is the most common associate. Fleharty (1967) found that both *T. elegans* and *T. rufipunctatus* were associates in the Mogollon Mountains, but the individual species showed at least some ecological separation. Habitat may be occasionally shared with *T. radix* where they are sympatric, and in the Rio Grande Valley both *T. sirtalis* and *T. elegans* may share habitat with *T. cyrtopsis*. At lower elevations, usually below 1500 m, *T. marcianus* may be present in the more open and permanent habitat such as earthen tanks and marshes.

Behavior: Blackneck garter snakes are most often seen in or near water in the daytime, but may also be active at night. When approached, they will attempt to escape into surrounding cover or, if in water, may swim on the surface to the opposite bank. Fleharty (1967) noted that most of these snakes using the water as an escape route, swam on the surface rather than below. Captured individuals will release excrement and pungent anal gland secretions. The teeth are used less often, and then not as vigorously as certain other snakes. Basking on streambanks, especially in the spring, is common behavior, and is more frequent in adults and subadults than in newborn. Mosauer (1932) mentioned three snakes that he observed in Dark Canyon, Guadalupe Mountains. They had a particular basking place on the stream bank, and could be seen there daily, coiled on the moist grass. Rosen (1991) found that active season body temperatures in southeastern Arizona ranged from 22.5–35.0°C and averaged 27.5°C. The annual activity period is dependent on seasonal temperatures, and in New Mexico collecting dates have ranged from March–October. There is no information available on hibernacula.

Reproduction: *Thamnophis cyrtopsis* is viviparous. The young are born alive, and materials are exchanged between the embryonic and maternal membranes. Wright and Wright (1957) stated that males mature at 429–470 mm TL and females at 504–764. Other authors have stated that maturity is reached at 40–107 cm TL (Stebbins, 1985; Conant and Collins, 1991). Sabath and Worthington (1959) recorded a specimen from Texas 395 mm TL that gave birth to seven young. Other garter snakes mature in the second or third year, and we assume that this species does also. Courtship and mating probably occur shortly after emergence from hibernation in the spring, since most births occur in July and August. Ted L. Brown (pers. comm.) collected a snake in Santa Fe County on 26 July that measured 241 mm TL and was probably born sometime in July. Minton (1958 [1959]) suggested the possibility of autumn or winter breeding in Big Bend, Texas, based on very small snakes, "about 8 inches in total length and probably only a few days old," collected on 29 June. Autumn breeding in other *Thamnophis* spp. has been documented (Fitch, 1965, 1970). Young are usually born in or near water, and number from 3–22. The birth of 25 young was mentioned by Fitch (1970) who mistakenly cited Woodin (1953) for this large litter. However, this record was actually based on a litter of *T. eques*, a name that was used earlier for *T. cyrtopsis*, but is now applied to the Mexican garter snake (see Remarks). Within a single Arizona litter of six live and one stillborn young, the live snakes measured 180–230 mm TL.

Food Habits: Adult and larval anurans are the principal food items of *T. cyrtopsis*, although small fishes are readily eaten when available, and such prey as skinks, crustaceans (*Triops* sp.), and earthworms may occasionally be used. The specific diet is determined largely by which prey is most available.

Jones (1990), studying Arizona populations of *T. cyrtopsis*, found that feeding strategy and prey varied both seasonally and by the size of the snake. Active pursuit was more common in the spring, and was replaced by a sit-and-wait strategy later. When tadpoles and small fishes are abundant in contained pools, snakes may swim partly submerged and try to grasp the prey as they make contact with it. Adult and subadult snakes may hunt terrestrially by moving along banks in search of adult frogs and toads. Newborn snakes were observed by Jones (1990) to remain motionless, while floating and semi-submerged, until a fish or tadpole moved. They then struck at the prey and were successful almost half of the time. Mosauer (1932) recorded a snake swallowing a decaying fish. He also observed another snake crawling slowly through a drying isolated pool, "swallowing as he went large numbers of tadpoles of *Rana pipiens* that were lying helpless on the wet mud." Most foraging is done in the daytime, but Minton (1958 [1959]) found *T. cyrtopsis* active at night.

Remarks: At various times in the past, other specific epithets including *T. eques* and *T. dorsalis* have been applied to *T. cyrtopsis*. Since *T. eques* (Reuss, 1834) is now applied to the Mexican garter snake, there has been some confusion in the literature concerning this name. This is a situation where the use of the common name will indicate which species the author is discussing. Webb (1966, 1980) gave historical accounts of these changes. The species was reviewed by Webb (1980).

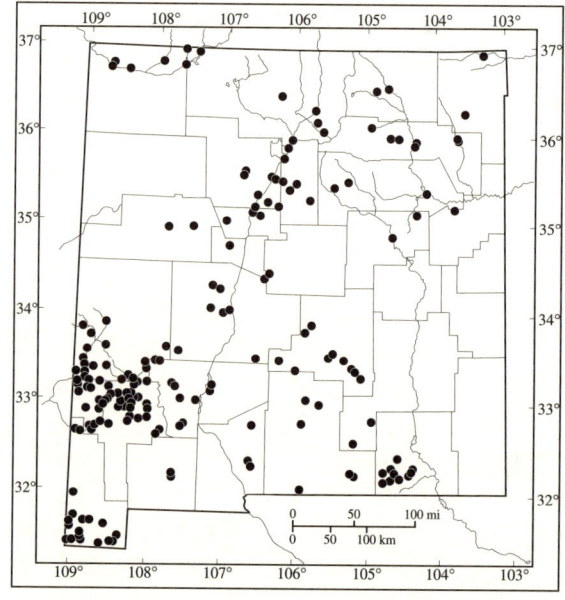

Distribution of Thamnophis cyrtopsis *in New Mexico.*

THAMNOPHIS ELEGANS (Baird and Girard, 1853) *Western Terrestrial Garter Snake*

See Plate 107.

Type: The holotype, USNM 882, is an adult male collected by C. C. Boyle, and sent to the U.S. National Museum by H. W. Henshaw (Baird and Girard, 1853a). The type locality is "El Dorado County, California." The date of collection is unknown.

Distribution: *Thamnophis elegans* is widely distributed in western North America from southern Canada to northern Mexico. The range includes a number of disjunct populations. In New Mexico, *T. elegans* is one of the most widespread and frequently encountered snakes in the mountainous regions and higher river valleys where suitable mesic habitat exists. It has been taken from about 1200–3300 m.

Description: Adult *T. elegans* ranges from 45–75 cm TL, although large females may reach more than 90 cm TL. The dorsal coloration varies from dark brown through various shades of greenish brown and tan to gray. The vertebral stripe is always well defined, narrow, and white to yellow. The lateral stripes are dull yellow to tan and on the 3rd and 4th scale rows. Two rows of alternating rounded spots lie between the stripes. These spots are most distinct anteriorly, and the uppermost spots invade the vertebral stripe slightly in most subspecies. Some individuals have poorly defined dark areas behind the head. The venter usually has irregular dark markings on a pale gray ground color. The dorsal scales are keeled and in 19–21 rows at midbody. There are 146–185 ventrals, 67–101 subcaudals, and the anal plate is single. There are usually eight (occasionally seven) upper labials with numbers six and seven greatly enlarged and higher than wide. There are one or two preoculars and three or four postoculars.

Similar Species: *Salvadora* spp. and *Masticophis* spp. have smooth scales, a divided anal plate, and no spots. *Thamnophis marcianus* has neck blotches, more distinct squarish spots, and the narrow lateral stripe is restricted to the 3rd scale row. *Thamnophis cyrtopsis* has well defined neck blotches and a yellow to orange vertebral stripe. *Thamnophis radix* usually has 19 scale rows, seven upper labials, and the lateral stripe on the third and fourth scale rows. *Thamnophis sirtalis* has seven upper labials, a wide, yellow, vertebral stripe, and often has red coloring between the lateral scales.

Systematics: Since Fitch (1983) recognized six subspecies, Tanner and Lowe (1989) have described two others. Some authors (Tanner, 1959; Webb, 1976; Tanner and Lowe, 1989) recognized a population in Chihuahua as *T. elegans errans* Baird and Girard, 1853. In New Mexico, two subspecies may occur, *T. e. vagrans* (Baird and Girard, 1853a) and *T. e. arizonae* Tanner and Lowe, 1989. The geographic distribution of these two forms has not been determined. (See Remarks.)

Thamnophis e. vagrans was originally based on two "cotypes" (= syntypes), USNM 907 and USNM 908. One of these, USNM 907, is referable to *T. e. elegans*. The other, USNM 908, an adult female collected by Dr. William Gambel, is now the lectotype. The original type locality was given as "California," and later restricted to southwestern Utah by Schmidt (1953); however, Maslin et al. (1958) argued convincingly that the type was from the vicinity of Santa Fe, New Mexico.

The holotype of *Thamnophis e. arizonae*, BYU 13358, is an adult female collected by W. W. Tanner on 20 April 1956. The type locality is "a marsh approximately 2 miles east of Joseph City, Navajo County, Arizona."

Habitat: *Thamnophis elegans* is widespread in a variety of habitats in the northern portion of its range, although it is restricted to mountains or wet valleys in the south. It is most abundant near surface water, and rivers, streams, springs, lakes, earthen stock tanks, or marshes serve equally well in supporting populations. However, individuals are often found great distances from water, especially in the mountains. In New Mexico, we find *T. elegans* at high elevations, often over 3200 m, where it is frequently seen on mountain trails by hikers. Hahn (1968) reported occurrence at approximately 3355 m in Colorado. Rocks, logs, heavy vegetation, or man-made debris may serve as cover.

Other snakes most often associated with *T. elegans* are other members of the genus. Fleharty (1967) found *T. cyrtopsis* and *T. rufipunctatus* sympatric with *T. elegans* at Wall Lake in the Black Range, but there was segregation in microhabitat; *T. elegans* was always found along portions of the lake with more shoreline cover (vegetation and logs). These same three species are found together in the Mogollon Mountains. The northern Rio Grande bottomland supports *T. elegans*, *T. sirtalis*, and *T. cyrtopsis*. In northeastern New Mexico, *T. radix* as well as *T. cyrtopsis* are found with *T. elegans*. *Tropidoclonion lineatum*, *Diadophis punctatus*, and *Liochlorophis vernalis* may share the same habitat with *T. elegans* where their ranges overlap.

Behavior: *Thamnophis elegans* is the most terrestrial of the New Mexico garter snakes. Proximity to surface water is preferred, but it is commonly found many miles away, and even when near water, it will often try to escape to cover on land. When captured, it typically voids excrement and anal gland secretions. This garter snake may attempt to bite, but its usually small size makes this defense ineffective against large predators and humans.

Thamnophis elegans is mainly active in the daytime, but a shift to more nocturnal activity is determined by weather conditions or food availability. The annual activity period extends from March through October at lower elevations or more southerly localities. This period may be shortened at higher elevations or more northerly localities. Deviations from these generalities are common in New Mexico, because of unpredictable spring and fall temperatures. There are collection records for February and November in this state. In Wyoming, Baxter and Stone (1980) reported that *T. elegans* hibernates in the same dens as *Crotalus viridis*, but enters later and leaves earlier. Hibernacula may vary; in Wyoming, *T. e. vagrans*, along with *C. viridis*, gained entrance to dens through openings in the bedrock (Graves and Duvall, 1990). In Utah, Brown et al. (1974) found that *T. elegans* and *Coluber constrictor* were using an old well that had been filled with soil and human debris (i.e., bottles and rusty cans).

In Colorado, Hammerson (1982) reported the use of fissures in volcanic rock and small mammal burrows beneath rocks near ponds. These were often shared with bullsnakes, milk snakes, and rattlesnakes. In Santa Fe County, New Mexico, an excavation on 20 February 1992 uncovered hibernating *T. elegans, Liochlorophis vernalis*, and *Eumeces multivirgatus*. The site was on a steep west-facing slope of a canyon above a river. Spaces between large granitic boulders, embedded in damp, clayey loam soil, served as hibernacula (Stuart and Painter, 1993b).

Reproduction: Male *T. elegans* mature at a smaller size, and probably an earlier age, than females. Wright and Wright (1957) recorded minimum sizes of 510 and 588 mm TL for males and females. Conant and Collins (1991) and Stebbins (1985) suggested that 45 cm TL is the minimum mature size for the species. In the Wyoming den, Baxter and Stone (1980) considered male and female snakes mature at 50 and 56 cm SVL. Of three gravid females from the Zuni Mountains, the smallest was 744 mm TL (Gehlbach, 1965).

Courtship and mating usually take place in the spring. Males leave hibernacula earlier, and courtship occurs as the females emerge. However, autumn mating and sperm storage in the female through the winter may also occur. Brown et al. (1974) found active sperm in one of three females they removed from a Utah hibernaculum on 27 January 1973. They also found active sperm in all six of the males at that time, showing readiness for spring breeding. Mating under natural conditions has been observed on 16 and 21 March and 2 May (Fitch, 1970).

Young are born from July through September and number from 3–21. Some Zuni Mountains dates and litter sizes recorded by Gehlbach (1965) are: 2 July (15), 11 July (13), and 26 August (20). James N. Stuart and R. D. Jennings (pers. comm.) collected a large gravid female (620 mm SVL, 185 mm tail) from a stock pond near Datil, Catron County, that gave birth to 20 young on 7 August. Of nine males and eleven females, the males were slightly larger (175–188 mm SVL) than the females (171–183 mm SVL). The smallest snake in the litter was a 227 mm TL female. Ted L. Brown gave us information on four litters. A 635 mm TL Bernalillo County snake gave birth to 10 young measuring 229–241 mm TL on 22 July. The second, from Sandoval County, 711 mm TL, gave birth to 12 young, 203–216 mm TL, on 5 August. The third, from Catron County, 845 mm TL, gave birth to 26 young, 200–225 mm TL, on 27 July. The fourth, from Catron County, 851 mm TL, gave birth to 27 young, 200–240 mm TL, on 30 July.

It has been shown that some populations of garter snakes do not reproduce annually (Gregory, 1977), and studies by Graves and Duvall (1990) show that certain Wyoming populations follow this pattern. Data are lacking for New Mexico populations, although being farther south, they may not follow this pattern.

Food Habits: *Thamnophis elegans* is carnivorous, with fishes, anurans and tadpoles, salamanders and larvae, lizards, birds, mice, shrews, chipmunks, earthworms, slugs, snails, leeches, and insects being recorded as prey (Tanner, 1949; Wright and Wright, 1957; Stebbins, 1985; pers. obs.). The diet varies with the particular population or locality and may be specific for the area. Fleharty (1967) found that *T. elegans* had the most generalized feeding habits of three sympatric garter snakes he studied at Wall Lake in Catron County. Furthermore, this correlated well with their generalized habitat preference.

Feeding is mostly diurnal and prey are actively hunted. *Thamnophis elegans* have been observed to enter tadpole-filled pools of muddy water, submerge completely, and then emerge again swallowing tadpoles they have captured. A "bump and grab" method of capture is probably used in water where visibility is poor and prey abundant.

Remarks: In describing *T. e. arizonae*, Tanner and Lowe (1989) made the point that "between the San Juan River in southeastern Utah and the Little Colorado River in northeastern Arizona is an important arid and semi-arid area with few aquatic habitats; it has been an effective barrier that continues to isolate populations to the north and south of it." This point was made in order to explain the isolation of the population in the Little Colorado drainage that they named *T. e. arizonae*. In their description, they used six specimens from five localities within New Mexico along the Arizona state line as paratypes (UAZ 36370, 36372, 43141–42, 43145, and 43196). The southernmost three localities are in the

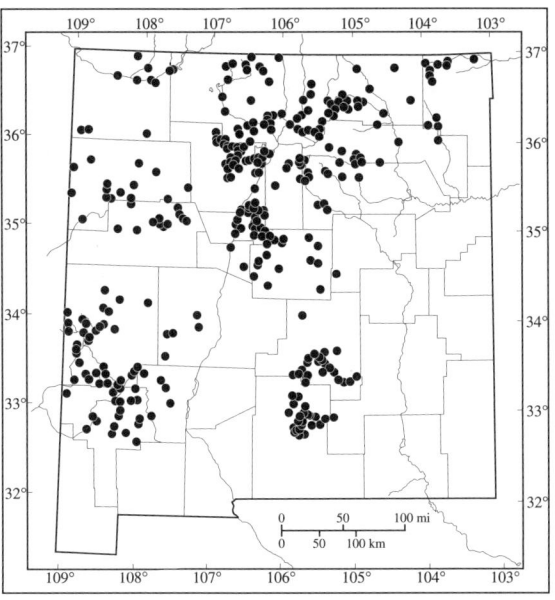

Little Colorado River Basin, to which *T. e. arizonae* is supposedly restricted. However, the two northernmost localities, "ca 10 mi S Newcomb," are actually within the Upper Colorado River Basin in the San Juan drainage, and theoretically should be referable to *T. e. vagrans*, if isolation by drainage systems was the important mechanism of speciation.

Distribution of Thamnophis elegans in New Mexico.

THAMNOPHIS EQUES (Reuss, 1834) — *Mexican Garter Snake*

See Plate 108.

Type: *Thamnophis eques* was first described as *Coluber eques* by Reuss (1834); early misapplication of the name was rectified by Smith (1951). The type locality was given as "Mexico," although it was later restricted to "El Limon Totalco, Veracruz" by Smith and Taylor (1950b). The type specimen (SMF 7209a, now SMF 17179), a female, bears the data "Gesch., 1832 von F. A. Dillenburger" (Boettger, 1898), and was designated as a lectotype by Mertens (1967).

Distribution: *Thamnophis eques* ranges from southeastern Arizona and extreme southwestern New Mexico southward in the highlands of western and southern Mexico to Oaxaca (Conant, 1963; Morafka, 1977; Stebbins, 1985). In New Mexico, *T. eques* occurs from 1125–1650 m in western Grant and Hidalgo counties, where it is known from single localities at and near Mule Creek and along the Gila River near Virden. There is also a single, century-old record (ANSP 10688) from Duck Creek near Cliff. Price (1980c) suggested that the species probably occurred throughout the Gila River and its tributaries wherever permanent water remained. Conant (1977 [1978]) suggested that the vagility of *T. eques*, especially in wet weather, may enable it to reach a number of localities where it survives as a relict, as it may have in southwestern New Mexico. On the basis of known localities, *T. eques* occurs in only two drainage systems, those of the Gila and San Francisco rivers. There is an old record from "Ocate R." (USNM 706) that was catalogued in 1858. Inasmuch as this would have been in Mora County, in northeastern New Mexico, this locality is doubtless in error.

Description: Adult *T. eques* average 45.7–101.6 cm in TL (Stebbins, 1985). There are three bright longitudinal stripes confined to scale rows three and four and set off from a spotted olive or brown background color. There is a white crescent behind each corner of the mouth. There are 8–9 barred upper labials, the dorsal scales are keeled, and the anal plate is single. This species has 21–23 distinctly keeled scale rows at midbody, 149–172 ventrals, 68–69 subcaudals, 8–9 upper labials, one preocular, and 3–4 postoculars (Smith and Brodie, 1982).

Similar Species: *Thamnophis eques* may be confused with other striped species of *Thamnophis* in New Mexico, although it can be separated as follows: *T. sirtalis*, *T. marcianus*, *T. cyrtopsis*, and *T. elegans* have the lateral stripe on scale rows two and three; *T. radix* usually has seven upper labials; the upper labials of *T. proximus* are unmarked. *Masticophis taeniatus* has only 15 scale rows at midbody, the upper stripe is confined to the 4th scale row and the edges of adjacent rows, and its anal plate is divided.

Systematics: Three subspecies are recognized (Smith and Smith, 1993); only *T. e. megalops* (Kennicott, 1860) occurs in New Mexico. The type locality was given as "Tucson, Arizona, or Santa Magdalena, Sonora," although it was later restricted to Tucson, Pima County, Arizona (Smith and Taylor, 1950b; Schmidt, 1953). The syntype is USNM 965, collected by Major Emory and A. Schott, date of collection unknown (Cochran, 1961). Much of the earlier literature on this species is found under the names *Eutaenia megalops*, *T. subcarinatus megalops*, or *T. macrostemma megalops*.

Habitat: All *T. eques* collected in New Mexico, with two notable exceptions, have been encountered around a small series of shallow stock tanks with abundant shoreside vegetation, often including a lush growth of cattails. The two exceptions include a single juvenile collected in a broad, shallow, sandy section of the lower Gila River near Virden, and an adult specimen collected by E. D. Cope in Grant County at Duck Creek, formerly a small, meandering, and well-shaded tributary of the Gila River.

Rosen and Schwalbe (1988) studied this species in Arizona and found *T. eques* most abundant in the densely vegetated habitat surrounding cienegas, cienega-streams, and stock tanks. Javier Manjarrez et al. (unpubl. data) found a dense population of this species on the University Campus of the Escuela de Ciencias, Univ. Autónoma del Estado de México, 23 km NE of Toluca City. Habitat at this site was a flat grassy meadow covered with dense grass and abundant flat rocks. The meadow and nearby ephemeral body of water were surrounded by willow trees.

Behavior: *Thamnophis eques* is active during the warmer months of the year. Rosen (1991) measured the active season body temperature of 18 specimens from Arizona and found it to average 27.26°C (22–33). *Thamnophis eques* may be observed foraging along watercourses where it is quick to seek shelter in the thick streamside vegetation or in the stream. When threatened, it will flatten the head and body and strike repeatedly. As with most garter snakes, when roughly handled these snakes attempt to bite and smear a foul-smelling musk from glands located at the base of the tail.

Reproduction: Males appear to mature at two years of age, females at 2–3 years. Females are larger than males, and begin reproducing at 53–70 cm TL. Reproduction in wild populations of *T. eques* has not been studied in New Mexico, although Ted L. Brown (pers. comm.) provided the following information on two captive-born litters. A 78.7 cm TL female collected at Mule Creek on 30 May gave birth to nine young (23–26.5 cm TL) on 16 July. A 73.6 cm TL male collected at Mule Creek on 23 May mated with a 53.3 cm TL female from Toluca, Mexico, on 29 October. The female gave birth to 10 dead young on 28 September. The same pair mated again on 7 April and the female gave birth to 10 young on 3–4 August.

In Arizona, Rosen and Schwalbe (1988) found that clutch size of eight individuals varied from 7–26 and averaged 13.6. Two clutches of neonates averaged 137 and 194 mm SVL. Gravid females generally move a few meters from the water to warm microenvironments where the young are born from early June to early July. Only half of the females in a population bear young in any one year. Javier Manjarrez et al. (unpubl. data) measured three newborn snakes between 150–203 mm SVL and 42 juveniles ranged from 210–440 mm SVL.

Food Habits: Food habits of *T. eques* have not been studied in New Mexico, although specimens from Chihuahua are known to include minnows, larval and adult leopard frogs, toads, and spadefoots in their diet (Van Devender and Lowe, 1977). Rosen and Schwalbe (1988) found frogs, tadpoles, and native fish to be the dominant items in the diet, although they reported lizards and mice as prey items as well. Native prey species seem to play a larger role in the ecology of *T. eques* than other southwestern striped garter snakes. Rosen and Schwalbe (1988) observed that substantial populations of *T. eques* occurred in Arizona only where anurans and fish are abundant. In areas where such prey was scarce, due to introduced predators or other causes, few *T. eques* were observed.

Remarks: Cope (1898 [1900]) was the first to report *T. eques* (as *Eutaenia megalops*) from New Mexico. He wrote, "I took a specimen on Duck Creek, which is a tributary of the Gila in southwestern New Mexico. It was in swampy ground near the water." The year of collection was almost certainly 1883, inasmuch as Osborn (1931) devoted the end papers of his book on "Cope: Master Naturalist" to maps showing Cope's travels, and he was at Silver City, not far from Duck Creek, in 1883. The specimen (ANSP 10688) has survived, but the habitat has not. Ted E. Brown, Roger Conant, and others found that Duck Creek had been severely eroded, probably a great many years ago, concurrently destroying Cope's "swampy ground."

The present status of T. eques in New Mexico is unknown. Where moist conditions or even semi-permanent water remain near the Gila River and the few small creeks of the general area, this snake may still survive. It is known to persist in very small numbers at one locality along Mule Creek. Although most of the small ponds and overgrown stock tanks in the area are behind tight fences and "No Trespassing" signs, and are thus off limits to casual collectors and observers. Roger Conant, James S. Jacob, and a group of students were fortunate on 30 April 1977. They met a rancher as he unlocked a gate in a large high-fenced area southwest of the village of Mule Creek, asked him if they could look for snakes, and he invited them to hunt all they wanted. In three small ponds within the fenced area and roughly 1.6 km south of the settlement they saw an estimated 100 T. eques, most of which escaped into the water as they were approached. A study series of 26 was collected. Conditions were similar to those observed by Conant (pers. comm.) in many parts of Mexico—large populations around lakes, ponds, and stock tanks.

Attrition of the species continues, however, as habitats are disturbed or destroyed, or new predators are introduced. Rosen and Schwalbe (1988) and Schwalbe and Rosen (1988) implicated the introduced *Rana catesbeiana* in southeastern Arizona as causing the decline of populations of native wetland reptiles and amphibians, including T. eques. At San Bernardino National Wildlife Refuge, they documented a disproportionate number of large T. eques, probably caused by R. catesbeiana consuming all of the small snakes.

Thamnophis eques is listed as a Category 2, Notice of Review species by the U.S. Fish and Wildlife Service (USDI, 1994) and as endangered by the New Mexico Department of Game and Fish (NMGF, 1990).

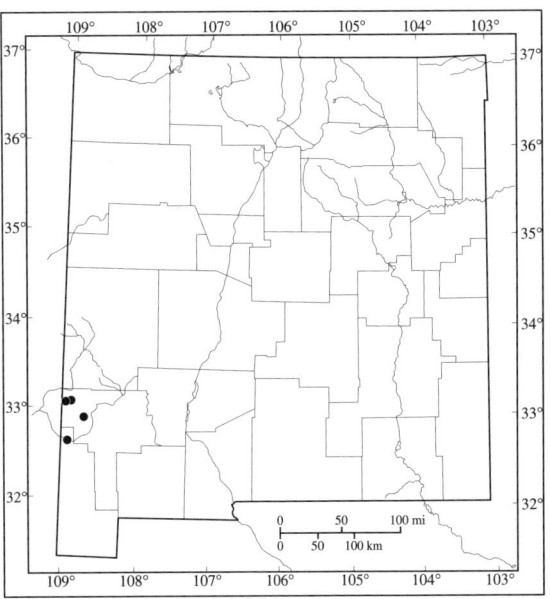

Distribution of Thamnophis eques *in New Mexico.*

THAMNOPHIS MARCIANUS (Baird and Girard, 1853) *Checkered Garter Snake*

See Plate 109.

Type: The holotype, USNM 884, was collected by Capt. R. B. Marcy (and Capt. G. B. McClellan ?, see Remarks) on 20 May 1852 (Baird and Girard, 1853a). The sex of the holotype was unspecified. The type locality, originally given as "Red River, Ark.," was determined by Mittleman (1949) to be in the vicinity of Slough Creek, east of Hollister, Tillman County, Oklahoma.

Distribution: *Thamnophis marcianus* is found in the southwestern United States from southern Kansas, western Oklahoma, and central Texas to southeastern California, and southward through Mexico into Central America. In New Mexico, it is widespread through the eastern, central, and southwestern portions of the state largely excluding the mountains.

Description: *Thamnophis marcianus* normally ranges from 460–600 mm TL, although Vial (1957) reported an adult female 1088 mm TL (978 mm SVL) and T. L. Brown (pers. comm.) collected a female in San Miguel County measuring 1092 mm TL. The dorsal coloration is a pale brown, brownish yellow, or olive with a distinct checkered pattern of large squarish spots. The cream or yellow dorsal stripe is distinct and partially invaded by the spots. The lateral stripes are white, narrow, and on the third scale row anteriorly, but may move to the second scale row posteriorly. On each side of the head behind the mouth there is a distinct light crescent followed by a large dark blotch. The venter is white, tan, yellow, or greenish gray and may have small spots near the ends of the scutes. There are distinct black vertical bars on the labial sutures. Some of these bars extend above the level of the top of the eye. There are 139–173 ventrals, 61–83 subcaudals, and the anal plate is single. The dorsal scales are keeled and in 21 rows at midbody. There are usually eight upper labials, one preocular, and three postoculars.

Similar Species: *Salvadora* spp. and *Masticophis* spp. have smooth scales, a divided anal plate, and no spots. *Thamnophis cyrtopsis* has 19 scale rows at midbody and wider lateral stripes on the second and third scale rows. *Thamnophis radix* has 19 scale rows at midbody, no neck blotches, seven upper labials, and the lateral stripe is on the third and fourth scale rows. *Thamnophis sirtalis* has seven upper labials, a wide yellow vertebral stripe, no neck blotches, and often has red between the lateral scales. *Thamnophis eques* has no neck blotches and the light lateral stripe is on the third and fourth scale rows.

Systematics: Rossman (1971) recognized three geographic races, and of these only the nominotypical form, *T. m. marcianus*, is found in New Mexico.

Habitat: Although widespread from 900–2200 m in New Mexico, *T. marcianus* prefers lowland habitats to those in the mountains. It is basically a Mexican and southwestern United States species that does not range much farther north than New Mexico. It is often found some distance from surface water, but has a decided predilection for wet habitats near rivers, streams, springs, ponds, marshes, and irrigation ditches.

Shaw and Campbell (1974) made the point that, in sharp contrast to the normal trend among snakes, *T. marcianus* has sometimes benefitted from the works of man. They discussed how irrigation activities in the Southwest have produced habitat. Mendelson and Jennings (1992) found the distribution of *T. marcianus* roughly followed the distribution of cattle tanks along NM-AZ Hwy 80 in southwestern New Mexico and southeastern Arizona. Furthermore, they found that over the last 20 years, the population of *T. marcianus* has increased there, in spite of the invasion of semidesert grassland by Chihuahuan desertscrub. Cover may be rocks, logs, boards, trash, heavy vegetation, or rodent burrows. Snake cohabitants in the wetter habitats are *Thamnophis* spp., *Tropidoclonion lineatum*, *Lampropeltis getula*, and *Nerodia erythrogaster* where their ranges overlap. *Elaphe guttata*, *Heterodon nasicus*, and *Tantilla* spp. often share the dryer habitats.

Behavior: *Thamnophis marcianus* is most often found in or near surface water where it may be active at all hours. It is easily captured, but once in hand will exhibit the typical garter snake defenses of defecation and biting. Most individuals become gentle in captivity.

Rosen (1991) found that *T. marcianus* spent a large proportion of its time moving, rather than basking or hidden under superficial cover. Since it is found in warm environments, preferred body temperature is easy to maintain. There is a shift from diurnal to nocturnal activity under arid or high temperature conditions, and these snakes are then often collected on roads at night. Annual activity may extend from March through October if temperatures permit. We have no information on hibernacula but they probably den singly or in small groups.

Reproduction: The minimum size at maturity of *T. marcianus* recorded by Wright and Wright (1957) is 325 mm and 350 mm TL for males and females, respectively. Ford and Karges (1987) combined data from nearly 500 preserved and living snakes from southern Texas and northern Mexico. The smallest female in their sample with corpora lutea was 33 cm SVL. Based on these sizes, maturity is probably attained in the second or third year after birth. The time of courtship and mating in New Mexico populations has not been determined. Ford and Cobb (1992) found that snakes from a southern Arizona population bred immediately on leaving hibernation, while those from southern Texas delayed courtship a minimum of 24 days. They also pointed out that, because of sperm storage capability, the timing of courtship for snakes need not be strongly tied to ovulation, and may evolve to fit parameters important to the survival of the particular species or population. The courtship process is initiated in *T. marcianus* when the male locates the female by pheromone trailing. Apparently, trail pheromones are primarily involved in reproductive activity in *T. marcianus*, as only conspecific males follow the trails (Ford and O'Bleness, 1986). Cloacal alignment, preliminary to intromission, results mainly from alignment of the tail tips, although this method of alignment may result in some difficulty for stub-tailed individuals. However, stub-tailed *T. marcianus* have given birth, so tail-tip alignment may be important but not essential (Perry-Richardson et al., 1990). Ford and O'Bleness (1986) and Perry-Richardson et al. (1990) recorded the courtship process in detail.

All *Thamnophis* are viviparous, and *T. marcianus* young are born from June into October. Ford and Karges (1987) believed that females may be capable of producing two clutches in the same year, differing from most other snakes studied. Larger females produce larger young, as well as larger clutch sizes. Ford and Karges (1987) pointed out that more studies of southern species are needed, since most previously examined North American snakes are species from higher latitudes, where the active season is shorter than that of southern species. Neonate TL has varied from 123–245 mm TL and clutch sizes from 5–31. Ford and Seigel (1989) found that both number of offspring and clutch mass were significantly affected by prey availability. Ford and Karges (1987) recorded neonate size ranging from 131–175 mm SVL and number of young 3–16 in a south Texas population. Ted L. Brown (pers. comm.) supplied notes on four gravid females. One from Curry County was 813 mm TL and had 15 young on 13 August that measured 185–205 mm TL. An unmeasured DOR from Curry County contained 16 young on 29 July. A 711 mm TL female from Guadalupe County produced 15 young, 210–230 mm TL, on 4 August. A large female from San Miguel County gave birth to 25 young, 225–245 mm TL, on 15 August. Brown (pers. comm.) collected a juvenile in Guadalupe County on 30 April measuring 279 mm TL, and another from Eddy County collected on 13 May measured 248 mm TL. The Eddy County juvenile may be an example of a female overwintering with advanced stage follicles (Ford and Karges, 1987), resulting in a shortened development time and early birth the next year.

Food Habits: Prey items of *T. marcianus* include fishes, anurans, tadpoles, salamanders, lizards, mice, slugs, earthworms, and a variety of other invertebrates (Wright and Wright, 1957; Stebbins, 1985; pers. obs.). The individual snake's menu is determined by its size, and the type of prey available in the habitat. Members of this genus seem to have no difficulty with the toxic secretions of anurans such as *Bufo*, *Scaphiopus*, *Spea*, and *Gastrophryne*.

Remarks: Capt. R. B. Marcy is usually listed as the collector of the type specimen, however, in the original description (Baird and Girard, 1853a) both Capts. Marcy and McClellan were given as collectors. Yarrow (1882) and Cope (1898 [1900]) listed only Marcy as the source in their "from whom received" column. This species was reviewed by Mittleman (1949) and Rossman (1971).

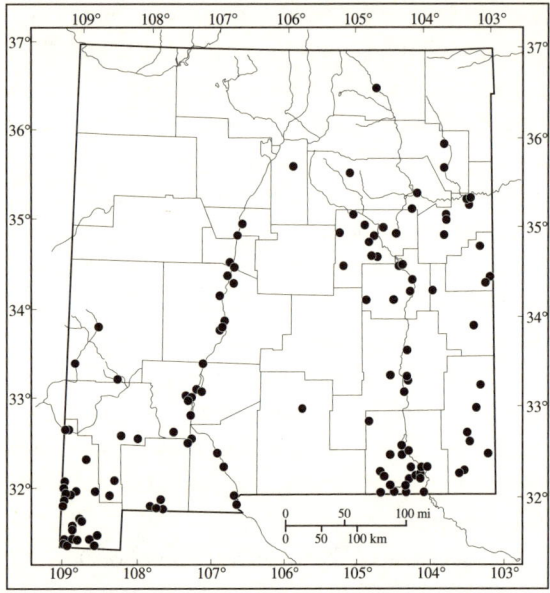

Distribution of
Thamnophis marcianus
in New Mexico.

THAMNOPHIS PROXIMUS (Say *in* James, 1823)   *Western Ribbon Snake*

See Plate 110.

Type: Thomas Say originally described *Thamnophis proximus* as *Coluber proximus* in 1823. He gave the type locality as "Stone quarry on west side of Missouri River, 3 miles above the mouth of Boyers River." This was later restated as "approximately 3 miles ENE Fort Calhoun, Washington County, Nebraska" by Rossman (1963). The collector and date of collection are unknown (Rossman, 1970). According to Smith and Taylor (1945), the holotype was lost, although it was probably deposited in the Academy of Natural Sciences of Philadelphia (H. M. Smith *in* Rossman, 1970).

Distribution: *Thamnophis proximus* ranges from southern Wisconsin to the Gulf Coast, through northeastern Mexico to Costa Rica, and from eastern New Mexico to the Mississippi River. It ranges from sea level to around 2440 m elevation (Stebbins, 1985). In New Mexico, *T. proximus* is known from 900–1500 m at scattered localities in the eastern third of the state where permanent water is present, such as along the Canadian River at Mills Canyon and the lower Pecos River drainage near Artesia, Carlsbad, and Roswell.

Description: Adult *T. proximus* average 50.8–86.4 cm TL, although the New Mexico form may reach slightly over 1.2 m (Conant and Collins, 1991). There are three bright longitudinal stripes that contrast with the dark or olive-brown background; the lateral stripes are confined to scale rows three and four. The tail is long, only slightly less than 1/3 of total length. There are eight unmarked white upper labials, with the 4th and 5th entering the orbit. The dorsal scales are keeled and the anal plate is single. This species has 19 keeled scale rows at midbody, 141–181 ventrals, 82–131 subcaudals, and typically 8 upper labials (Rossman, 1970).

The New Mexico form is characterized by an olive-gray to olive-brown dorsum, an orange vertebral stripe, and a narrow dark ventrolateral stripe. There is often a reduction in the width of the lateral stripe on the posterior portion of the body in the southern part of its range (Rossman, 1970).

Similar Species: *Thamnophis proximus* may be confused with other striped species of *Thamnophis* in New Mexico, although it can be separated as follows: *T. sirtalis*, *T. marcianus*, *T. cyrtopsis*, and *T. elegans* have the lateral stripe on scale rows two and three; *T. eques* and *T. radix* have barred upper labials. *Masticophis taeniatus* has only

15 scale rows at midbody, the anal plate is divided, and the upper stripe is confined to the fourth scale row and the edges of adjacent rows.

Systematics: There are six recognized subspecies (Rossman, 1970); only *Thamnophis proximus diabolicus* Rossman, 1963 occurs in New Mexico. The type locality is "Rio Nadadores, 8 miles W Nadadores, Coahuila, Mexico." The type specimen is University of Florida (UF 12210), an adult female collected by Roger Conant on 26 June 1960. Intergradation between two or three subspecies is common over a wide area within the range of the western ribbon snake.

Habitat: *Thamnophis proximus* is semiaquatic and is rarely found away from permanent water sources in New Mexico. Its ability to subsist around small bodies of water has allowed this westernmost race to colonize the xeric areas of eastern New Mexico and western Texas. *Thamnophis proximus* inhabits rivers and streams, irrigation canals, stock tanks, and rocky intermittent creeks where large deep pools with abundant frogs and fish remain. It may be found among the dense streamside vegetation, often basking on overhanging branches.

Behavior: Slender and agile, *T. proximus* is most often encountered near water where it is quick to seek shelter if pursued. When captured, it usually voids the foul-smelling cloacal contents and smears its captor with musk and feces. It is most active during the day when temperatures are mild, and it may be encountered foraging or basking in the vegetation along the water's edge. During hot weather it may become nocturnal. Rosen (1991) measured the body temperature of 10 *T. proximus* from Tabasco, Mexico, and found the average body temperature to be 29°C (25.5–32.3).

Reproduction: Females are slightly larger than males. *Thamnophis proximus* has been reported to reach sexual maturity at approximately 1–2 years of age and at a minimum of 485 mm SVL for females and 410 mm for males (Tinkle, 1957) or 515 mm for females and 368 mm for males (Clark, 1974). Reproduction of *T. proximus* in New Mexico needs to be studied to support data presented by Tinkle (1957) and Clark (1974). Although males may mate during their first year, females probably first give birth at two years of age. Mating occurs in the spring and parturition occurs in July and August with some broods produced as late as October.

Thamnophis proximus is live-bearing. Brood size averaged 11.6 (4–24) in 41 broods from Oklahoma, Texas, and Louisiana reported by Fitch (1985). Conant (1965) reported on five broods produced from a captive female collected in Nuevo León, Mexico. This female produced a total of 50 offspring in two broods a year during 1956 (June and September) and 1957 (May and August) and a single brood during 1958 (May). The broods ranged from 8–13 young that averaged approximately 235 mm TL. Ted L. Brown (pers. comm.) reported a female 69.8 cm TL collected near Roswell in Chaves County that gave birth to eight young (190–220 mm TL) on 20 September. Conant (1965) suggested that the production of two litters a year may be commonplace in parts of Mexico where the growing season is long and the winters are mild and of short duration. Reproduction has not been studied in New Mexico, although it is likely that only a single brood per year is produced.

Food Habits: Little is known about the food habits of *T. proximus* in New Mexico. Clark (1974) studied the food habits of a population in Brazos County in east-central Texas. Among 24 specimens containing food, 67% had eaten frogs and toads of various species, 4% smallmouth salamanders, 25% leopard frog tadpoles, 8% ground skinks, and 4% the fish, *Gambusia affinis*. Fouquette (1954), also working in Texas, reported that 82% of the diet consisted of amphibians. Although discussing a different subspecies, Dundee and Rossmann (1989) state that *T. proximus* will not eat earthworms or adult toads.

Remarks: Shockey (1992) reported slight localized swelling, discoloration, and reduced mobility of the ring finger on his right hand after a bite by *T. proximus*.

Thamnophis proximus diabolicus is listed as endangered by the New Mexico Department of Game and Fish (NMGF, 1990). Rossman (1970) reviewed this species.

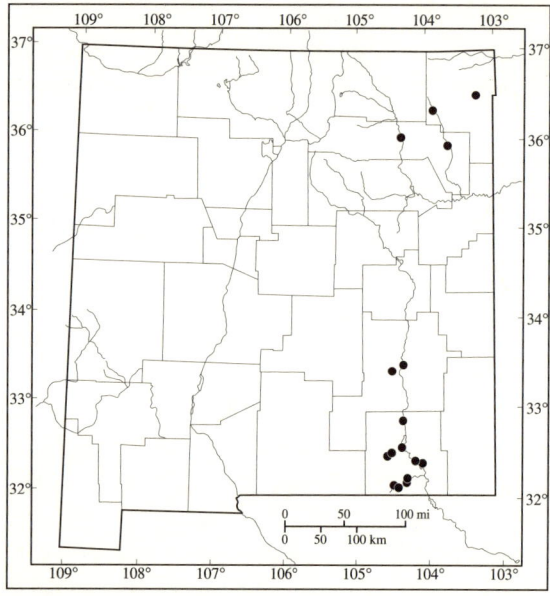

Distribution of Thamnophis proximus *in New Mexico.*

THAMNOPHIS RADIX (Baird and Girard, 1853) — *Plains Garter Snake*

See Plate 111.

Type: The holotype, USNM 719, from "Racine, Wisconsin," was collected by Dr. P. R. Hoy on an unspecified date (Baird and Girard, 1853a). According to Schmidt (1953), the type locality was "Probably *in error*," and he restricted it to the "vicinity of Chicago, Illinois" (see Remarks).

Distribution: *Thamnophis radix* is found primarily in the Great Plains from southern Canada to northern New Mexico, Texas, and Oklahoma. Disjunct populations occur in Ohio, Missouri, and Illinois. In New Mexico, it has been found in only five counties of the northeastern quadrant (Union, Colfax, Mora, Harding, and San Miguel) from about 1300–2200 m in elevation.

Description: The adult size of *T. radix* ranges from 38–70 cm TL, with an individual record of 109 cm TL. The dorsal ground color varies from greenish gray or olive to brown or sometimes red. The vertebral stripe is distinct and yellow to orange but is paler posteriorly. The lateral stripes are greenish or bluish cream and are on scale rows three and four. The venter varies from white or yellow to a darker bluish green or gray, and the undersurface of the tail is usually mottled with black. Other markings consist of black vertical bars on the upper labial sutures, alternating black spots between the vertebral and lateral stripe, a row of black spots below the lateral stripe, and spots down each side of the venter. Some of these markings may be obscured in some individuals. The dorsal scales are keeled and in 19–21 rows (usually 19) at midbody. The anal plate is single. There are 139–176 ventrals, usually seven (sometimes eight) upper labials, and one preocular.

Similar Species: *Salvadora* spp. and *Masticophis* spp. have smooth scales, a divided anal plate, and no spots. *Thamnophis cyrtopsis*, *T. eques*, and *T. marcianus* have prominent dark neck blotches. *Thamnophis elegans* and *T. sirtalis* have the lateral stripe on scale rows two and three.

Systematics: Two subspecies are recognized; only one, *T. r. haydenii* (Kennicott *in* Cooper, 1860), occurs in New Mexico. The holotype, USNM 707, was collected by J. Evans from "Fort Pierre in Nebraska" (= Fort Pierre, Stanley County, South Dakota). No sex or date of collection was specified.

The type specimen was in such poor condition that Smith (1949) designated CA 14498 as a "lectotype" (error for neotype), and CA 14499, 14500, and 13858 as "paralectotypes" (error; no ancillary types are recognized by the Code with neotypes). These specimens were all from near the type locality. Two of these (CA 14498, 14499) were sent to the United States National Museum where they were renumbered USNM 128137 and 128138. The neotype, USNM 128137, was collected at Whitlock Crossing, Davey County, South Dakota, by A. M. Jackley et al. on 6 October 1947 (Cochran, 1961). No sex was indicated.

Habitat: *Thamnophis radix* is a species of Great Plains habitats that extend into the northeastern quadrant of New Mexico. This part of the state is mostly contained within the High Plains, and is largely grass-covered. There is piñon-juniper woodland at higher elevations and ponderosa pine–Douglas fir forest in the more moist or sheltered places. Here, *T. radix* is usually associated with surface waters of streams, lakes, farm ponds, and marshes, but may stray some distance from these. It has been collected at 2170 m in Mora County, but probably occurs higher. It ranges to 2286 m in Colorado (Hammerson, 1982). The plains garter snake was formerly common in vacant lots and parks in many large cities, but these populations have been greatly reduced because of building activities and pesticide use (Conant and Collins, 1991). Although *T. radix* is the most widespread garter snake in the High Plains and prefers the prairie, habitat may be shared with other congeners, especially in prairie-forest ecotone. In a Minnesota study, Jordan (1967) observed that *T. radix* preferred the prairie, and *T. sirtalis* the forest, with little overlap. *Thamnophis elegans* encroaches from the west and has the most similar habitat preferences. The range of *T. cyrtopsis* overlaps from the southwest, although it is most often found along rocky canyon streams, where *T. radix* is less evident. *Thamnophis proximus*, rare at this western edge of its range, may be found along the weedy shorelines of rivers and lakes.

Behavior: Like most of its congeners, *T. radix* may use the double-ended defense of biting and anal discharge to deter potential predators, but most captive individuals quickly become gentle. This species has often been described as being one of the less aquatic garter snakes (Wright and Wright, 1957), since it is often found some distance from water if sufficient moisture or cover is available. This does not seem to be the case in New Mexico, for most of our collections were made near water. *Thamnophis radix* is wary, and if it is able to take refuge in deep or muddy water, its escape is probable, since it may remain submerged for long periods of time.

We have found *T. radix* to be active and often sunning at all hours of the day, but nocturnal activity has been reported elsewhere. Heckrotte (1962) found that *T. r. radix* adjusted its diel activity pattern in response to temperature. He found that at low temperatures snakes were diurnal, at intermediate temperatures they became crepuscular, and they were nocturnal at the highest temperatures. In Colorado, these snakes are diurnal during mild spring and fall weather but may be active day or night during hot summer months (Hammerson, 1982). Annual activity may extend from March to November. In many parts of its extensive range, *T. radix* is one of the first snakes observed in the spring and among the last to be seen in the fall. It has also been observed during warm spells in the coldest winter months (Collins, 1982). We have no information on hibernacula in New Mexico, but in more northern populations snakes often aggregate, frequently with other species. In Manitoba, *T. radix* were in below-ground galleries of an ant hill, along with *Liochlorophis vernalis* and *Storeria occipitomaculata* (Criddle, 1937).

In an interesting study by Lawson and Secoy (1991), *T. radix* used solar cues as orientation guides during fall return to a communal hibernaculum in southern Canada. Their data suggested that polarized light was the particular cue.

Reproduction: Most studies of reproduction in *T. radix* have been on northern populations, and Fitch (1970) summarized most of these data. Sexual maturity is attained in the second or third year at a minimum size of approximately 360 mm TL. Since maturity in the second year occurs late in the active season, breeding in the spring after hibernation is suggested. Documented matings have occurred in April, May, and June. In Colorado, *T. radix* were observed mating in May and June, and pregnant females can be seen as early as mid-May, suggesting that mating may also occur in April, or possibly in the fall (Hammerson, 1982). Other instances of possible fall mating are given by Wright and Wright (1957).

Large litters may be produced by *T. radix*, and numbers of young vary from 5–92 (Fitch, 1970). As in many snakes, females may not reproduce every year. Seigel and Ford (1987) gave 74% as the proportion of females breeding in a given year in Missouri. Litter size in Colorado averages 16; neonates usually appear in late July or early August, but different females continue to give birth through September (Hammerson, 1982). Ted L. Brown (pers. comm.) noted that a Colfax County female measuring 737 mm TL gave birth to 14 young on 9 September. The neonates ranged from 229–241 mm TL. Brown also collected a hatchling measuring 222 mm TL in Colfax County on 18 August. Other New Mexico young were born on 15 August to MSB 40930 from San Miguel County.

Food Habits: The diet is varied and Halloy and Burghardt (1990) considered *T. radix* a generalist. Food items include earthworms, toads, frogs, salamanders, fish, leeches, insects, rodents, and many types of carrion, including dead birds and mice. A snake taken from a small pond in Harding County disgorged several earthworms when captured. This was surprising as the pond contained adult *Pseudacris* and larval *Ambystoma* and we suspected that this snake, and the others that got away, were there for a meal of amphibians. This same snake, a few days later, fed on live mosquito fish, which it was adept at capturing in a large water dish.

Remarks: Douglas A. Rossman (pers. comm.) pointed out that *T. radix* does occur at Racine to this day, and Schmidt was probably in error with his restriction.

This species was found to serve as a reservoir host for western equine encephalitis, a disease transmitted to horses by mosquitos (Gebhardt et al., 1966).

The last review of this species was by Smith (1949).

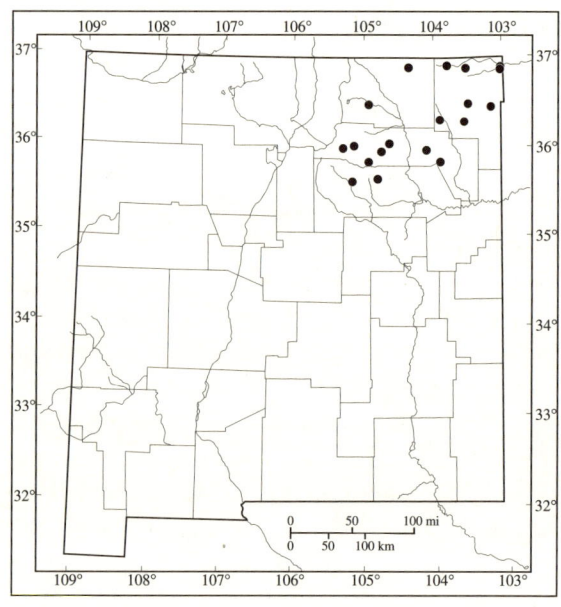

Distribution of Thamnophis radix *in New Mexico.*

THAMNOPHIS RUFIPUNCTATUS (Cope *in* Yarrow, 1875) *Narrowhead Garter Snake*

See Plate 112.

Type: *Thamnophis rufipunctatus* was first described as *Chilopoma rufipunctatum* by E. D. Cope (*in* Yarrow, 1875). The type specimen, USNM 8600, was collected by W. H. Henshaw during October 1874. The type locality was given as "Southern Arizona" although was later restricted to "the vicinity of Fort Apache, Arizona" by Webb and Axtell (1986). Interestingly, Cope (1883) later described the same species from New Mexico as *Atomarchus multimaculatus* from the "San Francisco River, [Catron County] New Mexico, on ranch of Mr. H. C. Wilson, which is near the boundary line of Arizona." This specimen was collected by E. D. Cope; the date of collection and disposition of the specimen is not available (Tanner, 1990).

Distribution: *Thamnophis rufipunctatus* occurs from central Arizona eastward into western New Mexico, then southward into central and western Chihuahua and northern and western Durango, Mexico (Tanner, 1990). In New Mexico, *T. rufipunctatus* is confined to warm water reaches of the Gila and San Francisco river drainages in western Grant, Catron, and Hidalgo counties at elevations between 1125–2100 m.

Description: *Thamnophis rufipunctatus* has an unstriped dorsum with an olive or brown background color and a dark-spotted pattern. The belly is patterned with irregular dark spots or a series of dark brown crossbars. On the anterior edge of the ventrals there is often a row of black, wedge-shaped marks on each side of the belly. The eyes are placed high on the laterally compressed head. Maximum size reported in Stebbins (1985) is 86 cm TL. Females are larger than males. During August and September, 1995 we collected 113 *T. rufipunctatus* along the San Francisco River, Catron County, using standard wire mesh, inverted-cone minnow traps. Forty-six females averaged 483.4 (260–880) mm SVL, 604.5 (338–1115) mm TL, and 74.6 (10.1–285) g; 67 males averaged 425.8 (285–635) mm SVL, 556.5 (373–836) mm TL, and 41.2 (10.8–107.7) g. Mean percent of tail length to total length is 22.1 in females, 23.8 in males. The young resemble the adults, although the dorsal spots tend to be reddish brown. This species has 21 keeled scale rows at midbody, 151–179 ventrals, 62–88 subcaudals, 8–9 upper labials, 2–3 preoculars, 2–4 postoculars, and a single loreal. Ventrals average 159 (151–171) in males, and 165 (155–179) in females; subcaudals average 77 (68–88) in males, 69 (62–79) in females (Tanner, 1990). Smith (1955) described in detail the pattern and scalation of two specimens collected in Durango, Mexico. Van Denburgh (1922b) stated "their general appearance is very different from that of most garter-snakes, the absence of lines, the heavy spotting, and the long narrow head are not suggestive of *Thamnophis*."

Similar Species: Because of the distinctive coloration and morphology and the aquatic habitat, *T. rufipunctatus* is not likely to be confused with any other snake in New Mexico.

Systematics: The name *T. rufipunctatus* was placed in the synonymy of *T. angustirostris* (Kennicott, 1860) by Ruthven (1908) and the species was known by that latter name for a long period of time. Thompson (1957) suggested that the original type material was referable to *T. marcianus*. He placed *T. angustrostris* in synonymy with *T. marcianus*, and applied *T. rufipunctatus* (Cope *in* Yarrow, 1875) to this species. Lowe (1955d) proposed the transfer of *T. rufipunctatus* as *T. angustirostris* [= *T. marcianus*, Thompson, 1957] to the genus *Natrix* (= *Nerodia*), based primarily on the divided anal plate in some specimens from the Arizona population and on its habits and color pattern. Chiasson and Lowe (1989) compared the ultrastructural scale patterns of *Nerodia* and *Thamnophis* and presented further evidence for this transfer. They found the microdermatoglyphic pattern of *T. rufipunctatus* was more similar to that of *Nerodia* than of other *Thamnophis* studied. Thompson (1957), Tanner (1959), and Conant (1963) chose to retain *rufipunctatus* in the genus *Thamnophis* based primarily on the small number of specimens having a divided anal plate. We follow this arrangement pending further study. Tanner (1985, 1990) recognized three subspecies of *T. rufipunctatus*, and assigned populations in the southwestern United States to the nominotypical *T. r. rufipunctatus*.

Habitat: *Thamnophis rufipunctatus* is highly aquatic for a garter snake. In New Mexico, it is closely associated with clear rocky streams with abundant streamside vegetation that is often used for basking or for escape from predators. The greatest number of *T. rufipunctatus* are found in association with riffles and deep pools with abundant large boulders. They often live in the interstitial spaces of partially submerged complexes of rocks and boulders (Rosen and Schwalbe, 1988). River reaches characterized by broad expanses of small rock and sand, and streams that cut through meadows, do not appear to be suitable habitat for *T. rufipunctatus* (Fitzgerald, 1986a, 1986b).

Behavior: *Thamnophis rufipunctatus* is a highly aquatic species that often uses rocky riffles as forage areas. They may use their prehensile tail to anchor themselves among rocks while pursuing prey. Streamside rocks and overhanging vegetation serve as basking sites from which snakes, at the slightest sign of disturbance, take refuge under boulders in the stream. *Thamnophis rufipunctatus* remains active at lower temperatures than other garter snakes in the same habitat (Fleharty, 1967), although it will seek out warm microclimates such as may be found near thermal spring seeps. Rosen (1991) measured the active season body temperature of 18 *T. rufipunctatus* from Arizona and found the average body temperature to be 25°C (19.2–31.4), an average lower than other *Thamnophis* measured. When handled, this species may attempt to bite and will void the foul-smelling contents of its anal scent glands. On two occasions, we have found *T. rufipunctatus* dead in the hot water of Turkey Creek Hot Springs in the Gila Wilderness. It appeared they entered the hot water and were unable to retreat before they succumbed to the lethal high temperatures.

Reproduction: Males mature at approximately 2.5 years of age, females at two years. Essentially all adult females reproduce each year, as indicated by the large percentage of gravid females examined by Rosen and Schwalbe (1988) during the breeding season. Young narrowhead garter snakes are born from late July into early August at an average size of 160–205 mm SVL. Fleharty (1967) reported a near-term gravid female taken during the second week in July, and Fitzgerald (1986a) found a neonate at Turkey Creek in Grant County on 15 July that measured 220 mm SVL. Additionally, Fitzgerald (pers. comm.) measured seven juveniles ranging from 203–220 mm TL near San Francisco Hot Springs in southwestern Catron County on 15 July. A female collected in northern Chihuahua, Mexico, and held in captivity since 15 March 1993, gave birth to eight young on 17 July 1993. Average birth weight was 2.4 (1.9–3.1) grams, and average length was 165.9 (150–193) mm SVL (Doug Burkett, pers. comm.). In Arizona, three gravid females examined by Rosen and Schwalbe (1988) contained 8–17 developing ova.

Food Habits: With its elongated snout and narrow head, *T. rufipunctatus* seems to have become adapted for feeding on small aquatic vertebrates, especially fish. Fleharty (1967) found they preferred fish. Longfin dace, fathead minnow, desert sucker, Sonoran sucker, green sunfish, and rainbow trout have been recorded in the diet of narrowhead garter snakes in the Gila and San Francisco rivers. Adults have also been known to include toads, frogs, tadpoles, and larval tiger salamanders in their diet.

Remarks: The name *rufipunctatus* is derived from the Latin *rufus*, meaning "red" or "ruddy," referring to the red or rust-colored spots on some individuals, and the Latin *punctatus*, meaning "spotted" or "having spots" in reference to the numerous dark body spots.

Thamnophis rufipunctatus is listed as a Category 2, Notice of Review species by the U.S. Fish and Wildlife Service (USDI, 1994) and as endangered by the New Mexico Department of Game and Fish (NMGF, 1990).

Tanner (1990) reviewed this species.

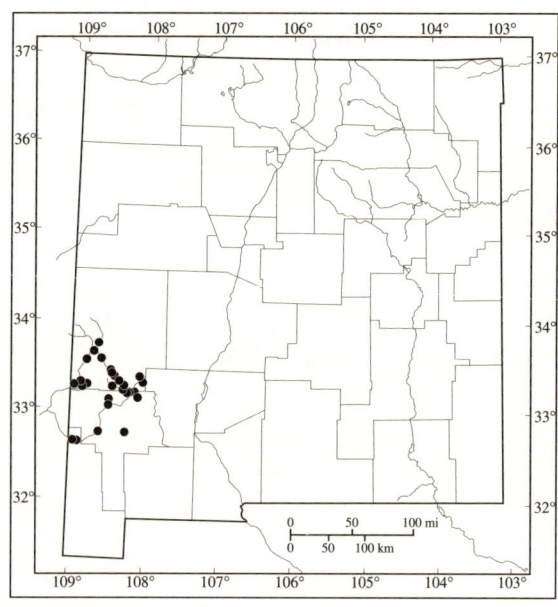

Distribution of Thamnophis rufipunctatus *in New Mexico.*

THAMNOPHIS SIRTALIS (Linnaeus, 1758) — *Common Garter Snake*

See Plate 113.

Type: The type specimen is lost. Linnaeus (1758, 1766) provided little information on the specimen other than "Canada," as the type locality, and P. Kalm as the collector. Harlan (1827) later inferred (but did not designate) a more specific locality, "inhabits Pennsylvania."

Distribution: The many races of *T. sirtalis* range widely through North America from southeastern Alaska and the southern half of Canada southward into north-central Mexico, where there are isolated populations east of the Continental Divide. Within this range, the species is found from sea level to about 2500 m, but it is excluded from much of the arid southwestern United States. In New Mexico, it is restricted to the Rio Grande Valley, occurring from southern Rio Arriba County (1700 m) south to central Doña Ana County (1175 m). Records of *T. sirtalis* from Mora County, northern Taos County, and the Pecos River Valley near Artesia are questionable (see Remarks).

Description: *Thamnophis sirtalis* is largest garter snake found in New Mexico. It seldom exceeds 90 cm TL in New Mexico although has been measured at 1311 mm TL elsewhere in its extensive range (Conant, 1975; Conant and Collins, 1991). *Thamnophis sirtalis* is especially variable in color and pattern over its extensive range. Stripes, when present, may vary in color or width. The vertebral stripe may be tan, gray, yellow, orange, green, or blue. Round or squarish spots may be present between the stripes. A wide array of colors may be present on the dorsal scales and/or between them.

In New Mexico, the dorsal color between the stripes usually consists of black spots on a background of red or dark orange. This reddish color may be reduced in some individuals and is largely confined to the skin between the scales. The color below the lateral stripes is continuous with that of the venter and is tan, yellow, green, or blue, with small indistinct black markings that are most prominent anteriorly. The upper row of dark spots tends to fuse longitudinally and form a continuous black border to the wide, prominent, yellow dorsal stripe. In some pale individuals, the red is reduced and the overall appearance is greenish or brownish with prominent stripes. The pale lateral stripe is on scale rows two and three. The dorsal scales are keeled and in 19 rows at midbody. The anal plate is single. There are 137–178 ventrals, seven upper labials, and one preocular.

Thamnophis sirtalis is our largest garter snake. It seldom exceeds 90 cm TL in New Mexico but has been measured at 1311 mm TL (Conant, 1975) elsewhere in its large range.

Similar Species: *Salvadora* spp. and *Masticophis* spp. have smooth scales, a divided anal plate, and no spots. *Thamnophis cyrtopsis*, *T. eques*, and *T. marcianus* have dark neck blotches. *Thamnophis radix* has the lateral stripe on scale rows three and four. *Thamnophis elegans* usually has eight upper labials, never any red between scales, and the paravertebral spots are distinct and not fused longitudinally.

Systematics: Fitch (1980) recognized 12 subspecies. One of these, *T. s. dorsalis* (Baird and Girard, 1853[a]), occurs in the state. The holotype, an adult male collected by General S. Churchill, was lost from the United States National Museum. The original number is unknown and the type locality uncertain (see Remarks).

Habitat: In much of its extensive range, *T. sirtalis* may be found almost anywhere it is assured of a food supply and a place to hide. It is often abundant in the parks, cemeteries, and vacant lots of our largest and most populous cities. In New York City it is often called "garden snake," a name that is also applied to the northern brown snake, *Storeria dekayi*. When available, water is a favored component of the habitat, and this is especially true in the western populations of this species. In his detailed Kansas study, Fitch (1965) found *T. sirtalis* in many different habitats within his study sites. He ranked these habitats by apparent preference based on numbers of snakes caught in each. The more moist habitats ranked higher than the more arid ones. He was surprised, however, that in dry weather the snakes remained dispersed, and showed little tendency to move to and concentrate in the wettest habitats.

In New Mexico, *T. sirtalis* is seldom found distant from moist river valley habitats. Ponds, marshes, and irrigation canals are favored places, and heavy plant cover is usually present or nearby. The most common snake associates are *T. elegans*, *T. cyrtopsis*, and *Lampropeltis getula* where their ranges overlap.

Behavior: *Thamnophis sirtalis* can move rapidly and usually tries to escape when confronted by a potential enemy. It is quite pugnacious, however, and may "stand and fight" if a quick escape is not possible. Its threat display consists of flattening of the body and head, coiling in a strike position, and actually striking with closed or open mouth if the molester comes within range. In the process, the red skin between the scales is stretched and prominently displayed. If grasped, the snake will void the contents of its anal musk glands and attempt to smear the secretion on the predator as well as itself. If grasped by the tail, the snake will often twist with such force that the tail tip will be lost. Fitch (1965) found that 17.9% of the 940 snakes he examined had incomplete tails. As captives, some individuals may be handled without their performing defensive behavior. Female garter snakes generally reach greater lengths than the male, and large females seem especially pugnacious and tame less readily.

Thamnophis sirtalis is largely diurnal and is often seen basking, but it may become nocturnal in hot or dry weather. It is one of the most cold-tolerant snakes, which may partly explain its occurrence farther north than any other snake in North America. Fitch (1965) experimented with a group of *T. s. parietalis* and found that on many occasions these snakes survived body temperatures between −2° and −3°C. He suggested that the critical minimum was near that lower temperature. On the Kansas Natural History Reservation, snakes were seen from February through November, but most annual activity extended from March or April to October. Fitch (1965) also noted that *T. sirtalis* enters hibernation later than other species using the same dens. In Colorado, it emerges from hibernation in March or April and generally ceases activity in September or October (Hammerson, 1982). In Oklahoma, Force (1930) found annual activity from February–October. Our records show collections from March–October in the Rio Grande Valley, but actual emergence and retreat is determined by temperatures for the particular year. Gibbons and Semlitsch (1987) showed that where some species are seldom active from late autumn until spring, others, such as *T. sirtalis*, become active during any month when environmental temperatures are warm enough. Hibernacula may be anywhere below the frost line. Gregory (1982) listed animal burrows or tunnels in the ground, abandoned ant mounds, rock crevices, and abandoned human structures as hibernacula. In some cases, *T. sirtalis* has been found partly immersed if the water table is high. In Wisconsin, they may hibernate while completely submerged in water for up to 5.5 months during winter. During this time their body processes are slowed and cutaneous diffusion of oxygen is adequate for aerobic metabolism (Costanzo, 1989). *Thamnophis sirtalis* may hibernate singly or aggregate by the hundreds or thousands in some places (Aleksiuk, 1976), and often with other species. In New Mexico, the nature of the available hibernacula should preclude large aggregations.

Reproduction: We have little reproductive information for *T. s. dorsalis*, but the literature abounds with data on other races. Fitch's (1965) report is especially useful and is used here for general information. Sexual maturity in this viviparous species is attained when males are over 400 mm SVL and females over 500 mm SVL. These sizes may occur by the end of the second year, since females normally grow faster than males. Variation in the year when snakes become mature may be caused by factors such as the birth month (early vs. late in the birth year) and food supply. Spring mating, as snakes emerge from hibernation in March and April, is normal. This results in most young being born in July and August. Latitude, elevation, and weather also affect this generalization. Fall mating may occur, but Blanchard and Blanchard (1942) found that females inseminated in the fall retain viable sperm until the following spring when ovulation takes place. They suggested that other males mated with these same females in the spring, resulting in the phenotypic variation in offspring that cannot be explained by simple Mendelian inheritance. In populations where large numbers of this species aggregate for hibernation, multiple inseminations would seem unavoidable. A copulatory plug formed by the male, however, effectively blocks intromission for at least a few days until the plug is expelled by the female (Devine, 1975). Seigel and Ford (1987) showed that females often skip years, with 65–78% of females gravid each year. Larger and older females produce larger litters and are more likely to breed in a given year.

Litter size averages near 16 based on 413 litters from throughout the range of *T. sirtalis*, but larger litters are common in large females, and records as high as 85 young are known (Fitch, 1985). Young range in size from 152–246 mm TL and males are larger than females. A large female from Socorro County, roughly 600 mm SVL, gave birth to 24 young on 28 July with a 12–12 sex ratio. Eight randomly chosen neonates from this litter measured from 180–214 mm TL (J. N. Stuart, pers. comm.). Ted L. Brown (pers. comm.) supplied notes on three wild-captured gravid females. A Bernalillo County snake gave birth to three young, 229–241 mm TL on 26 August. A female from Sandoval County that measured 787 mm TL gave birth to 21 young on 22 August; they ranged in size from 195–225 mm TL. A very large female measuring 1118 mm TL gave birth to 47 young as follows: 2 August (2), 4 August (4), 6 August (7), 7 August (16), 12 August (12), 13 August (2), 14 August (2). These hatchlings measured 200–225 mm TL.

Dunlap and Lang (1990) found that the sex ratio within litters apparently varies with female size; larger females usually produce more male offspring.

Food Habits: *Thamnophis sirtalis* is a dietary generalist (Halloy and Burghardt, 1990) and feeds opportunistically. Prey items in the wild include earthworms, tadpoles, fish, frogs, toads, salamanders, slugs, leeches, small mammals, birds, bird eggs, snakes, and lizards. Insects are often found in the stomach contents. Fitch (1965) suggested that, at least in the western forms of this snake, these insect remains are most often secondary items that were previously eaten by the primary prey. The actual prey used is determined by a number of factors involving relative size of predator and prey, availability of the prey in the habitat, imprinting, and evolutionary history of the population. Frogs and toads always score high on food preference lists of snakes large enough to eat them, while earthworms make up a large proportion of the juvenal diet.

Thamnophis sirtalis shows high resistance to the toxins of *Bufo, Scaphiopus, Spea,* and *Gastrophryne*, which are common foods. Brodie and Brodie (1990) have shown that certain populations of *T. sirtalis* have evolved resistance to the very potent toxin (tetrodotoxin) produced by the roughskin newt, *Taricha granulosa*, while other populations where the newt is absent show less resistance to this toxin.

Earthworms, *Rana catesbeiana*, and several species of *Bufo, Spea,* and *Scaphiopus* are abundant in the Rio Grande Valley, and probably make up a large proportion of the diet of this snake in New Mexico.

Remarks: Bundy (1951) reported *T. sirtalis* from Wade's Swamp, near Artesia, but the specimen cannot be located, and it may have been misidentified. Other specimens (OUVC 2640–41) from the Pecos River near Artesia were correctly identified, but the locality data are questionable. These old records should be kept in mind when working in the lower Pecos Valley, for it is possible that a population of *T. sirtalis* may exist there. Fitch and Maslin (1961) mapped a locality in the Rio Grande Valley near the Colorado border. They cite Little and Keller (1937) for their New Mexico records, but this reference only listed one locality, "Ropes Spring," for this species, and they in turn cite Cockerell (1896). Cockerell collected in Doña Ana County and "Ropes Spring" probably refers to the spring in Ropes Draw northeast of Las Cruces. We suspect that the Ropes Spring record represented a misidentification. The record of *T. sirtalis* from southern Colorado (Yarrow, 1875) would seem to give credibility to the northern New Mexico record; however, the Colorado specimen was later identified as *T. elegans* (Hammerson, 1982). The Mora County record is in error since the specimen it was based on (USNM 706) is *T. eques*. Unfortunately, these old records have been used by Conant and Collins (1991), Smith and Brodie (1982), and Stebbins (1985) in their range maps.

The type locality, "between Monclova, Coahuila, Mexico, and the Rio Grande," is seemingly correct as part of the Churchill itinerary, but the holotype was incorrectly associated with this locality (Fitch, 1980; Webb, 1966). The confusing nomenclatural history of this species is discussed by Webb (1966) and in the review by Fitch (1980).

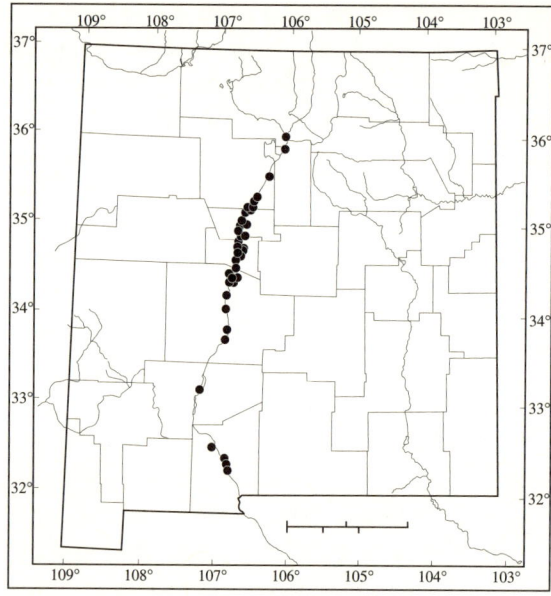

Distribution of
Thamnophis sirtalis
in New Mexico.

TRIMORPHODON BISCUTATUS (Duméril, Bibron, and Duméril, 1854) *Lyre Snake*

See Plate 114.

Type: The holotype is MNHN 5900. The sex, collector, and date of collection are unknown. The type locality was given as "Mexique," but was later restricted to Tehuantepec, Oaxaca, Mexico, by Smith and Taylor (1950b).

Distribution: *Trimorphodon biscutatus* is found from southern California, Nevada, and southwestern Utah southeastward into Trans-Pecos Texas, Chihuahua, Baja California, Sonora, and mostly western Mexico; southernmost populations are in Costa Rica. In New Mexico, it occurs in the southern deserts and foothills, ranging north in the Rio Grande Valley to Truth or Consequences in Sierra County.

Description: In New Mexico, the slender *T. biscutatus* usually measures from 45–75 cm TL, although it may grow to over 100 cm or longer in some Mexican populations. The dorsum is light brown or gray with darker brown saddles. These saddles are widest at the midline, and in western New Mexico populations they may have light centers which vary in width. The species gets its name from a lyre-shaped head marking, but in the more eastern populations (*T. b. vilkinsonii*) the lyre is usually reduced to a short "V" or a few spots. The venter is white, gray, or yellowish tan with dark spots that are most prominent toward the sides. The scales are smooth and in 21–24 rows. The pupils are elliptical. There are 220–244 ventrals, 58–86 subcaudals, and the anal plate is divided. There are usually three preoculars, 2–5 loreals, 8–10 upper labials, and two anterior temporals.

Similar Species: *Lampropeltis alterna* has round pupils and the anal plate is undivided. *Hypsiglena torquata*, another species with elliptical pupils, is usually less than 60 cm TL, has blotches or spots rather than saddles, and has an unmarked ventrum. Other snakes with blotches or saddles have round pupils, or if they have elliptical pupils, have rattles, or at least a horny button, on a short stout tail.

Systematics: Scott and McDiarmid (1984b) recognized six subspecies of *T. biscutatus*. Two of these, *T. b. vilkinsonii* Cope, 1885 [1886] to the east, and *T. b. lambda* Cope, 1885 [1886] to the west, occur in New Mexico. These subspecies apparently intergrade where their ranges contact in south-central New Mexico and Mexico.

The holotype of *T. b. vilkinsonii* is USNM 14268. It was collected by E. Wilkinson; date and sex were not specified. The type locality "near the City of Chihuahua,

Mexico," was later restricted to "vicinity of the city of Chihuahua," by Schmidt (1953), and later expressed as "vicinity of Ciudad Chihuahua," by Gehlbach (1971).

The holotype of *T. b. lambda* is USNM 13487. It is a juvenile collected by H. F. Emerich in 1883. The type locality is "Guaymas, Sonora," Mexico.

Habitat: Within its range, *T. biscutatus* may occupy habitats from the montane woodlands and forests to desert grassland and desert scrub at low elevations. It mainly inhabits the rocky terrain of mountains, mesas, hills, canyons, and arroyos, but is occasionally found on flats. Banicki and Webb (1982) described the rocky Chihuahuan Desert habitats of the Franklin Mountains, north of El Paso, Texas, where the species is commonly collected. They found these snakes from the basal foothills near 1372 m to 1600 m at the crest. In Big Bend National Park, Texas, lyre snakes have been found below 950 m in the Grapevine Hills to over 1800 m in the Chisos Mountains (Degenhardt and Steele, 1957; our unpubl. data). In Arizona, they can be found at elevations over 2200 m in ponderosa pine forest. In New Mexico, this snake has been taken from 1200 m in the lower Rio Grande Valley to 1725 m in the Mogollon Mountains. Natural rock formations such as outcroppings with deep fissures, talus slopes, or boulder fields provide excellent cover, but man-made structures such as culverts and old buildings may be used. Individual snakes in New Mexico have been taken in mid-morning beneath a loose slab of granite, in the rafters of an old mining cabin inhabited by bats, on bare ground beside an old adobe house, and on the ground near a snap trap with a captured pocket mouse. In the last case, the snake had been trying to swallow the dead mouse (Harris, 1959). We have collected specimens along the riparian zone in Guadalupe Canyon, Hidalgo County. Vegetation here includes sycamore, mesquite, and various grasses and forbs.

Trimorphodon b. vilkinsonii often shares Chihuahuan Desert habitats of south central New Mexico with *Bogertophis subocularis*, and *Crotalus lepidus*. The more western *T. b. lambda* may be found with *Lampropeltis pyromelana*, *Crotalus lepidus*, and *Masticophis bilineatus* where their ranges overlap.

Behavior: *Trimorphodon biscutatus* is nocturnal, shy, and secretive. Most individuals have been collected from roads which pass through their habitat. Although largely terrestrial, these snakes are excellent climbers and can ascend rock faces and trees easily. When confronting a potential enemy, they assume a defensive coil and vibrate the tail, but seldom bite.

New Mexico collection records show annual activity from March–October. Mark Doles and Douglas Burkett (pers. comm.) found a single *T. biscutatus* hibernating in a deep rock crevice occupied by *Crotalus atrox*, *C. lepidus*, and *C. molossus*. No other hibernacula have been found, but the usual rocky habitat of *T. biscutatus* should allow access to hibernacula below the frost line.

Reproduction: Little is known of reproduction in this oviparous species. Data from Fitch (1970) suggested that they do not conform strictly to the pattern typical for temperate zone snakes, with mating in spring, egg-laying in early summer, and hatching in late summer. He cited records of November courtship in a zoo pair, which produced 20 eggs between 29 December and 7 January. A female from Colima, Mexico, laid 20 eggs on 3 March, a snake from California laid 12 eggs in September, and newly hatched young were found in the field on 1 October (Fitch, 1970). Stebbins (1985) mentioned a clutch of seven eggs.

Food Habits: Lizards are the principal food of *T. biscutatus*, but small mammals, birds, and other snakes serve as prey. We have fed captives various iguanids, whiptails, skinks, and geckos, which the snakes preferred dead. The same snakes readily ate dead small rodents and bats, including *Perognathus*, *Sigmodon*, *Mus*, and *Tadarida*, but refused dead earth snakes, *Virginia striatula*.

Live prey is usually grasped and held until immobilized by mild venom introduced by the enlarged rear teeth on the upper jaw. Constriction may also be used. Cowles and Bogert (1935) found that chewing movements were apparently necessary in order for the small rear fangs to be effective. They also found that lizards were affected by the venom to a much greater degree than mammals. Foraging is nocturnal, as expected in snakes with elliptical eye pupils.

Remarks: This species was reviewed by Taylor (1938a), Klauber (1940b), Smith (1941b), Gehlbach (1971), and

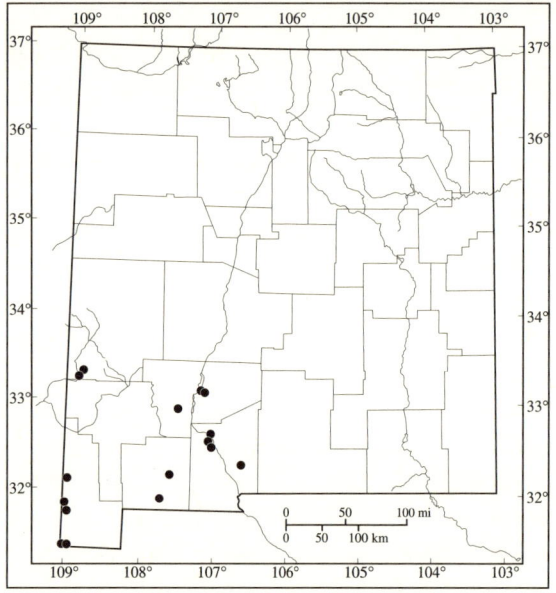

Distribution of *Trimorphodon biscutatus* in New Mexico.

Scott and McDiarmid (1984b). Scott and McDiarmid (1984b) summarized the very complex nomenclatural history of this species. Scott and McDiarmid (1984a) reviewed the genus.

TROPIDOCLONION LINEATUM (Hallowell, 1856 [1857]) *Lined Snake*

See Plate 115.

Type: The holotype is ANSP 5922, a female, collected by Dr. Hammond, date unknown. The type locality, originally given as "Kansas," was restricted by Ramsey (1953) to Fort Leavenworth, Leavenworth County, Kansas.

Distribution: *Tropidoclonion lineatum* is found from Illinois west through the Great Plains to the western half of Colorado and New Mexico and as far south as southern Texas. In New Mexico, it has been found from elevations of 1200–2000 m. Records are primarily in the northeastern and eastcentral parts of the state, but the species ranges south at higher elevations where habitat is available. There is an isolated record for the lower Rio Grande Valley at Las Cruces, Doña Ana County (see Remarks).

Description: Most adult *T. lineatum* vary between 20–38 cm TL, but some individuals may grow to over 57 cm TL (Tennant, 1984). The head is narrow and not wider than the neck. The dorsum is olive-gray or brownish with well-defined middorsal and side stripes. The middorsal stripe varies from whitish or pale gray to yellow or orange. The side stripes are on the second and third scale rows, and the dark area between stripes is without darker spots. The venter is whitish to yellow and is marked with a double row of bold black spots or half-moons which may or may not be connected along the midline. The dorsal scales are keeled and usually in 19 rows at midbody. There are 132–156 ventrals, subcaudals usually number less than 45, and the anal plate is single. There is one preocular, one anterior temporal, and five or six upper labials.

Similar Species: *Thamnophis* spp. have heads that are wider than their necks, a divided anal plate, more than 49 subcaudals, more than six upper labials, and any ventral markings that might be present are not as clearly defined. Other lined New Mexico snakes have smooth scales.

Systematics: We treat *T. lineatum* as monotypic. Smith and Brodie (1982) have referred New Mexico populations to *T. l. lineatum*, the northern lined snake, in the north and *T. l. mertensi* Smith, 1965, the New Mexico lined snake, in the south. More recently, Smith and Chiszar (1994) considered *T. l. mertensi* a junior synonym of *T. l. lineatum*. (See Remarks.)

Habitat: *Tropidoclonion lineatum* is a ground-dwelling species of prairie or woodland habitats. It is often locally abundant in populated areas where it takes advantage of parks, cemeteries, gardens, vacant lots, and city dumps. In the western portions of its range, its distribution is limited by the drier environment. Here it is found only in habitats that will satisfy its high moisture requirement, resulting in a discontinuous distribution pattern. River valleys, canyons, ponds, marshes, and seeps in the eastern half of New Mexico may support local populations. At higher elevations where there is more rainfall and cooler temperatures, these snakes may be more widespread and occupy grassland habitats more like those in the Midwest. An adequate habitat is one that can support earthworms, their principal food. Habitat may be further enhanced by an abundance of surface rocks or debris for cover. Snake cohabitants include *Leptotyphlops dulcis*, *Diadophis punctatus*, *Thamnophis elegans*, and *T. radix* where the ranges overlap.

Behavior: This shy snake is usually active at night and spends the daytime hidden under rocks, logs, leaves, or man-made material. It is a strong burrower and will use loose soil, such as that found in gardens or disturbed areas, for cover. There may be some crepuscular activity, and sunning in the morning is common. We find that it is very active at night after rain showers. When threatened by a potential enemy, it may flatten the body and/or hide the head under a coil. When captured, it will produce an anal discharge while trying to escape, but will never bite. Annual activity in New Mexico extends from April–October, but our knowledge is based on very few early and late records. Ramsey (1953) studied a population in north-central Texas and suggested a hibernation period extending from November until March. He found that snakes hibernated under whatever cover was available, or burrowed in loose or soft soil. However, he collected them during warm spells in every month of the year, suggesting temperature controlled dormancy. Somewhat conflicting evidence comes from Hamilton (1947), also working in north-central Texas. He discovered seven hibernating *T. lineatum* while digging in his garden during a warm spell in January. He reported that they were in clay soil, 15.2–20.3 cm deep, and four of them were together, "similar to preserved specimens coiled to fit into a small jar." He exposed them to bright sunlight for an hour, at an air temperature of 24.4°C, but they showed little movement, retaining their hibernating pose. Citing this occurrence, Gregory (1982) suggested that dormancy may be controlled independently of temperature, and that *T. lineatum* is unlikely to be active in winter. Wright and Wright (1957) noted that in San Antonio, Texas, these snakes were found hibernating in numbers under boards and similar cover during February and March.

Reproduction: This viviparous species grows rapidly, and both sexes are mature by their second year. Krohmer and Aldridge (1985a, 1985b) made detailed studies of the male and female reproductive cycles of a population in St. Louis, Missouri. Males have sperm available in the ductus deferens at about 200 mm SVL in size and eight months of age. This is the earliest known age for potential mating of any snake studied. Females may attain the minimal size for maturity, 221 mm SVL, in their second year, but vitellogenesis does not begin until the following spring (at 20 months of age). Mating occurs in late summer and fall, and sperm overwinter in the oviducts until ovulation in the spring.

Most females produce litters within a short time period in mid-August, after about a 70-day gestation period. Young usually number from 4–10, but extremes of 2 and 17 are recorded (Force, 1931; our unpubl. data). Litter size is correlated with female size, but the size of neonates is not (Krohmer and Aldridge, 1985b). Ninety-eight percent of 45 females studied by Krohmer and Aldridge (1985b) were reproductive. The one that was not contained six large atretic follicles, indicating that vitellogenesis was initiated. Also, the coelomic fat mass of this snake was considerably less (0.39 g) than the mean of other snakes in July (0.62 g). Coelomic fat storage is important for the reproductive cycles in both sexes. Young usually fall within a range of 70–130 mm SVL and the males are longer. Force (1931) described the birth process in detail.

A large female from Bernalillo County, collected on 22 May 1992 by John Woodworth, gave birth to a record litter of 17 young on 13 August. The female measured 343 mm SVL (395 TL). When measured alive two days later, the young ranged from 115–130 mm TL. A more accurate measurement was done 18 days after the first. The results were 118–132 mm TL, averaging 123 mm TL. This larger measurement may be due to greater accuracy or actual growth in this rapidly growing species.

Food Habits: *Tropidoclonion lineatum* feeds almost exclusively on earthworms. Insects and sowbugs are mentioned in the literature as food items (Wright and Wright, 1957), but these foods are exceptional. A single instance of cannibalism was documented by Force (1931) in a captive female that ate her recently born young. The large New Mexico female from Bernalillo County showed little interest in her young, though they were kept together for almost three weeks. Gravid females usually fast during gestation, and especially during the second half of this period. They begin feeding again following parturition, and must accumulate sufficient fat stores in order to reproduce the following year. These snakes are especially active after rains, coinciding somewhat with earthworm surface activity.

Remarks: We have chosen 1857 as the publication date for the original description in spite of the fact that all previous authors (Schmidt, 1953; Ramsey, 1953; Dixon, 1987; Collins, 1990) have used 1856 when citing the publication date. Roughly half of the papers presented to the Philadelphia Academy in 1856 and published in the *Proceedings* were received by subscribers in 1857 (Nolan, 1913). Since Hallowell's paper appeared on pages 238–253, and there were only 322 pages published in the *Proceedings* that year, it seems logical to assume that this paper must have been received by the subscribers in 1857.

The isolated record (UNM 54706) of *T. lineatum* from the Rio Grande Valley at Las Cruces is interesting; no other specimens have been collected there despite years of active collecting in the vicinity by NMSU staff and students. We doubt that this record denotes a breeding population for this snake could have been transported from elsewhere.

Some recent authors (Dixon, 1987; Hammerson, 1982; Stebbins, 1985; Smith and Chiszar, 1994) considered this species polytypic, however, others (Conant and Collins, 1991) considered it monotypic. We have collected many more specimens from New Mexico since Smith (1965) named *T. l. mertensi* from three localities in the state. These specimens, as well as those in other collections, should be reevaluated. The last complete review of the species complex was by Ramsey (1953), and this predated *T. l. mertensi* Smith, 1965. Smith and Chiszar (1994), working with Texas populations, presented evidence in support of recognizing three subspecies, *T. l. lineatum* Hallowell, 1856 [1857], *T. l. annectens* Ramsey, 1953, and *T. l. texanum* Ramsey, 1953. Their key to these subspecies is based on ventrals and subcaudals.

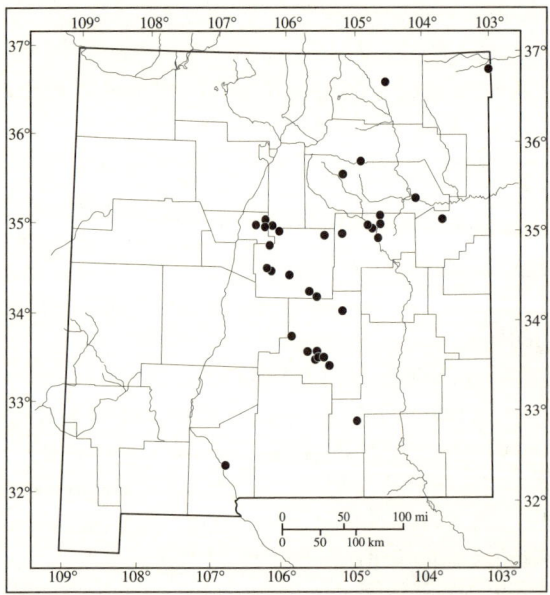

Distibution of Tropidoclonion lineatum *in New Mexico.*

FAMILY ELAPIDAE *Elapids*

This family is represented by about 180 terrestrial species and 56 aquatic species, inhabiting parts of Africa, Asia, Central and South America, North America, and Australia, including the Indian and Pacific oceans. Five subfamilies include the Australian elapids, mambas, cobras and their allies, coral snakes, and the sea snakes, many dangerous and all with highly toxic venom. The fangs of the elapids do not have the ability to rotate out of the way as in the pit vipers. They are attached to the upper jawbone like normal teeth, and therefore must be small enough for the snake to be able to close its mouth without the downward-projecting fangs piercing the lower jaw (Shine, 1992). Most feed on ectothermic prey, including fish, frogs, lizards, and other snakes.

There are about 50 species of coral snakes, subfamily Micrurinae, in two genera, with only a single representative in New Mexico. The brightly colored and semifossorial Arizona coral snake, *Micruroides euryxanthus,* is found in southwestern New Mexico where it is rarely encountered except on the roadways during the periods of summer rainfall.

MICRUROIDES EURYXANTHUS (Kennicott, 1860)   *Western Coral Snake*

See Plate 116.

Type: Robert Kennicott (1860) described *Micruroides euryxanthus* as *Elaps euryxanthus* based on a specimen received from T. H. Webb. The holotype, USNM 1122, is an adult male. Although the type locality was not given in the original description, USNM records show its provenance as "Sonora, Mexico" (Cochran, 1961). Smith and Taylor (1950b) restricted the type locality to "Guaymas, Sonora, Mexico," without substantiating their action. Roze (1974) rejected their restriction of the type locality and left it as presented for the type specimen listed by Cochran (1961).

Distribution: *Micruroides euryxanthus* is found in the southwestern United States from central Arizona and southwestern New Mexico southward into Mexico (see Remarks). The Mexican range extends from western Chihuahua and Sonora (including Isla Tiburón) southward to Mazatlán in southern Sinaloa. The altitudinal range is from sea level to 1900 m (Roze, 1974; Ernst, 1992a). In New Mexico, *M. euryxanthus* has been found only in western Hidalgo, Grant, and Catron counties from approximately 1125–1750 m.

Description: The distinctively marked *M. euryxanthus* has a body circled with complete, wide, red and black bands that are separated by a narrow yellow or white band. Some individuals have black flecking on the red rings. The entire upper part of the head is black and the blunt snout and small head are barely distinct from the body. Gates (1956) cited 551 mm TL as a record length for the species, although Lowe et al. (1986) cited an adult female 619 mm TL.

There are 205–245 ventrals and 19–31 subcaudals. Males average 223 (212–230) ventrals and 27–28 subcaudals, females 233 (219–230) ventrals and 24 subcaudals. There are 15 rows of smooth scales at midbody. There is one preocular, two postoculars, 1–2 temporals, seven upper labials, and 6–7 lower labials. The loreal and suboculars are absent. The anal plate is divided (Roze, 1974; Ernst, 1992a).

Similar Species: No other snake in New Mexico is similarly colored, with red and black rings that completely encircle the body and that are separated by white or yellow rings and with a completely black snout. The old adage "Red touch yellow, kill a fellow; red touch black, friend of Jack" will allow for the quick separation of the venomous coral snake from the harmless *Lampropeltis pyromelana* or *L. triangulum* in the United States.

Systematics: Three subspecies of *M. euryxanthus* are currently recognized (Roze, 1974) on the basis of pattern and scale characteristics. Two occur in Mexico while only the nominotypical form, *Micruroides e. euryxanthus*, occurs in New Mexico and the United States. Zweifel and Norris (1955) were the first to use this trinomial.

Habitat: Although occasionally encountered in the creosotebush flats of valley floors, *M. euryxanthus* in New Mexico is most commonly seen on rocky bajadas or in broad river valleys grown to mesquite. They are sometimes observed on the paved road in the rocky riparian area along Whitewater Creek near Glenwood.

Behavior: *Micruroides euryxanthus* is fossorial, generally found on the surface only after the summer rains have started. Although occasional specimens may be unearthed during construction, roadbuilding, or brush removal, it is largely nocturnal, and the majority of individuals are seen on the roadways at night. As spring approaches, *M. euryxanthus* may be seen abroad during daylight hours on overcast days. Seventeen records available from New Mexico indicate activity between April and September, although 82% of these records are July through September.

The observation that *M. euryxanthus* raises and moves the tail when agitated led Vitt and Hulse (1973) to suggest that the tail may serve as a head mimic and thus act as a decoy to potential predators. They found 14.3% of the 21 specimens examined had tail scars, suggesting that predation attempts directed at the head mimic had occurred. This species is known to exhibit "cloacal popping" which could add an auditory reinforcement to this tail display. Campbell and Lamar (1989) stated, "It may also bite, and fart loudly for its size, if agitated."

Reproduction: Generally two or three eggs are laid beneath stones or underground. Lowe et al. (1986) stated that egg laying is associated with summer rains in July–August, with hatching occurring in September. The newborn are 190–200 mm TL. Reproductive cycles, age and length at maturity, and courtship and mating behavior are unknown (Ernst, 1992a).

Food Habits: Adults seem to prefer *Leptotyphlops* spp. as prey (Vitt and Hulse, 1973), although they also feed on a variety of small, smooth-scaled, fossorial snakes including *Tantilla* spp., *Hypsiglena torquata*, and *Diadophis punctatus*. Several species of small lizards are also eaten. Captives are usually reluctant feeders but are known to eat ground skinks, *Scincella lateralis*, in addition to various small lizards and snakes. A captive ate pink mice on two occasions but disgorged them a few days later with little digestion evident.

Lowe et al. (1986) discussed the association of this species with *Leptotyphlops* spp. They noted that the ecological distribution of *Micruroides* is often correlated with that of blind snakes and suggested that the coral snake's small body size is selectively prey-directed towards blindsnakes.

Remarks: Some workers (McDowell, 1968; Smith et al., 1977; Savitsky, 1979) expressed doubts whether Elapidae is a natural group or merely a grade achieved independently by a number of higher snake lineages. McCarthy (1985) presented evidence to retain all proteroglyphous snakes, including *Micruroides*, in the family Elapidae.

Brown (1950) reported *M. euryxanthus* from El Paso County, Texas (MCZ 22645), and Wright and Wright (1957) provided a map depicting the range as widespread across the southwestern one-quarter of New Mexico, and into El Paso and Hudspeth counties of west Texas. Their map is inaccurate; only two specimens (AMNH 77456, UMMZ 7258) are known from east of the Continental Divide, both in New Mexico.

Although no human fatalities are attributed to the Arizona coral snake, it is a highly toxic species and should be regarded as dangerous. Russell (1967a) provided information on four cases treated by him during 1955–65.

Roze (1974) reviewed this species.

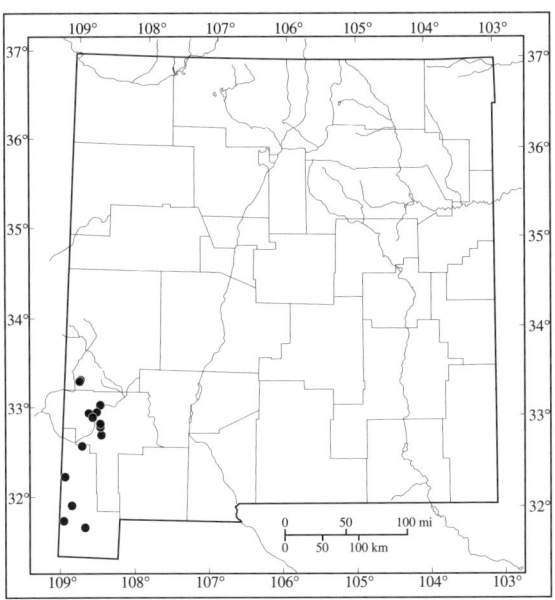

Distribution of Micruroides euryxanthus *in New Mexico.*

Vipers FAMILY VIPERIDAE

The family Viperidae is composed of 3 subfamilies, however only one, the Crotalinae, enters North America. The 140 or so members of this subfamily are commonly called "pit vipers" and are widely distributed through the Americas, parts of Asia, and barely into Europe. The common name refers to the deep pit on each side of the head which is lined with sensory organs and is used to detect heat of endothermic prey. These organs are extremely sensitive, capable of detecting temperature differences as little as 0.003°C (Shine, 1992). All pit vipers are venomous and venom delivery is through a highly evolved injection system, acting much like a hypodermic needle. In New Mexico, there are two genera and seven species of pit vipers; all are rattlesnakes in the genera *Crotalus* or *Sistrurus*, and all are characterized by the presence of a rattle at the tip of the tail. They are widespread in New Mexico and occur in a variety of habitats.

Rattlesnakes in New Mexico are increasingly threatened by unregulated commercial exploitation such as "rattlesnake roundups." These events support a commercial industry based on the value of rattlesnake skins for cowboy boots and other articles of fashion apparel. The potential impacts of this unregulated harvest have not been investigated, though most herpetologists feel that the practice is ecologically unsound and may lead to localized reductions of rattlesnake populations.

CROTALUS ATROX Baird and Girard, 1853 — *Western Diamondback Rattlesnake*

See Plate 117.

Type: Baird and Girard (1853a) described *Crotalus atrox* and gave the type locality as "Indianola" [Calhoun County, Texas]. The type specimen is USNM 7761. It was likely collected in 1851 by Mr. John H. Clark, an assistant to Lt. Col. J. D. Graham of the U.S. and Mexican Boundary Commission. See Remarks in the *Salvadora grahamiae* account for additional information. Paratypes are USNM 255 and USNM 7760 (Cochran, 1961).

Distribution: *Crotalus atrox* occurs from western Arkansas westward through eastern and south-central Oklahoma, Texas, central and southern New Mexico and Arizona, and possibly southern Nevada to southeastern California. It ranges southward in Mexico to northeastern Baja California and northern Sinaloa in the west, and northern Veracruz, Hidalgo, and Queŕetaro in the east (Ernst, 1992a). Disjunct populations are present in extreme southern Mexico (Klauber, 1956; Conant, pers. comm). In New Mexico, *C. atrox* occurs statewide except in the northern tier of counties and on the Mogollon Plateau. It ranges from 900 m to about 2350 m, and may reach 2500 m in Rio Arriba County.

Description: *Crotalus atrox* is the largest of all southwestern rattlesnakes. Individuals over 1.8 m TL are rare in New Mexico, although Texas boasts of specimens over 2.3 m TL. In general, adult females are slightly smaller than males; in a sample of 143 *C. atrox* from Arizona examined by Klauber (1972), males averaged 10.3% larger than females. Fitch and Pisani (1993) examined 1,011 specimens from the Oklahoma "Rattlesnake Roundups" and found males averaged about 10% longer than females. The tail is conspicuously black-and-white banded with bands of nearly equal width, thus the origin of the common name "coon-tail" rattler. Overall, the colors are muted with the background color variable, from light or dark gray or tan to reddish or pink. There are 24–45 brownish, light-bordered, diamond-shaped blotches on the back which tend to become obscure posteriorly. The postocular lateral stripe ends near or well in front of the corner of the mouth.

Males have 168–193 ventrals, 19–32 subcaudals, and 5–6 (3–8) black tail rings; females have 173–196 ventrals, 16–36 subcaudals, and 4 (2–6) black tail rings. Scale rows at midbody average 25 (23–29). There are 4–5 (3–8) small scales between the large supraoculars, 15–16 (12–18) upper labials, 16–17 (14–21) lower labials, 2–3 preoculars, and 3(2–6) postoculars. No prefrontals are present. The anal plate is undivided (Ernst, 1992a). Quinn (1979) examined 23 adult *C. atrox* from Oklahoma and found no overlap between sexes in the number of tail bands; 11 males had 5–6 tail bands, 12 females had 3–4 tail bands. However, Boyer (1958) examined 214 specimens from Oklahoma and found that males had 4–7 bands and females 3–6, thus the number of tail bands is not diagnostic in determining the sex of individual specimens.

Melanistic *Crotalus atrox* occur on the Pedro Armendariz lava field located in Socorro and Sierra counties in central New Mexico (Best and James, 1984).

Similar Species: In New Mexico, *C. atrox* can be easily distinguished from all but *C. scutulatus* by the presence of the black-and-white ringed "coon-tail." *Crotalus atrox* averages four or more small scales between the ocular scales while *C. scutulatus* rarely has more than two or three such scales. In addition, the white-and-black bands on the tail of *C. atrox* are generally of nearly equal width while the white bands on the tail of *C. scutulatus* are normally wider than the black bands, although this character varies among individuals. The scales that form the light-colored edges around the dorsal blotches of *C. atrox* are generally not uniformly light-colored, unlike those in *C. scutulatus*. This makes *C. scutulatus* appear more boldly patterned than *C. atrox*.

Systematics: *Crotalus atrox* is generally considered to be monotypic, consisting of one, wide-ranging, highly variable species. Lowe et al. (1986) however, considered *C. atrox* to be polytypic, assigning the name *C. a. atrox* to mainland individuals and *C. atrox tortugensis* to the population on Isla Tortuga in the Gulf of California.

Crotalus atrox was considered by Cope (1898 [1900]) to be a subspecies of *Crotalus adamanteus*, the eastern diamondback rattlesnake. He listed *C. adamanteus atrox* from Fort Wingate, N.M. The specimen, USNM 14819, was received by Dr. R. W. Shufeldt.

Klauber (1956, 1972) judged *Crotalus cinereus* to be the proper name for the western diamondback rattlesnake. However, Klauber (1952b) proposed that the International Commission on Zoological Nomenclature, in the interest of conserving widely used names, establish *atrox* as the accepted name for this snake and suppress *cinereus*. This proposal was favorably acted on by the Commission as Opinion 365, Nov. 16, 1955 (Klauber, 1972).

Habitat: *Crotalus atrox* inhabits a wide variety of habitats. It is primarily an inhabitant of dry or semi-arid lowland areas, usually found in brush-covered plains, dry washes, rocky outcrops, or desert foothills less than 1500 m elevation. Occasionally it reaches elevations over 1830 m, including specimens collected at 2135 m by C. M. Bogert near Otowi in Sandoval County. Klauber (1952a) reported specimens from 2135 m near Illescas, in western San Luis Potosí, Mexico.

Pough (1966) found *C. atrox* to be abundant in the mesquite-tarbush association in southeastern Arizona and southwestern New Mexico. [Pough's tarbush is *Acacia constricta* and not *Flourensia cernua* as used today.] Mendelson and Jennings (1992) found that *C. atrox* had increased in relative abundance in the same area, possibly with the succession of the semidesert grassland to a Chihuahuan desert scrub community.

Behavior: *Crotalus atrox* is quick to escape should it feel threatened, although it is equally quick to defend itself if escape is impossible, remaining in place and raising the head and anterior part of the body well above the ground. Envenomated bites are very serious and may be fatal even when treated.

Crotalus atrox frequently congregate at winter den sites on rocky hillsides. These dens are usually permanent structures and may include abandoned mines, small caves, or deep cracks in the rock that may also be occupied by other species (see Gregory, 1982; 1984 for a review of communal denning in snakes). In the Organ Mountains in Doña Ana County, Mark Doles and Doug Burkett (pers. comm.) have encountered *C. atrox*, *C. lepidus*, *C. molossus*, *Masticophis taeniatus*, and *Trimorphodon biscutatus* hibernating in the same den. Travis Perry (pers. comm.) found *C. atrox* and *C. molossus* occupying the same den in Hidalgo County, and Tom Moore (pers. comm.) found *Thamnophis elegans* in the same den in Otero County. We have never encountered *C. atrox* and *C. scutulatus* in the same den although they are syntopic in southwestern Hidalgo County.

In late April and May, adult *C. atrox* in southwestern Oklahoma studied by Landreth (1973) generally maintained a bimodal activity pattern with movements in the early morning and late afternoon, but not during the hot part of the day. During the hotter parts of the summer they tend to be most active during the cooler parts of the day. In the fall, snakes again assume the bimodal pattern of spring. The irregular, temperature-dependent winter activity is restricted to only the warmest part of the day when snakes may be seen basking at the mouth of the den site. Some of these snakes migrated in spring from dens to summer territories located 1–2 km away and returned to the same den in the fall. The distances moved each day by *C. atrox* in the summer were shorter than in the spring and usually consisted of trips from a hiding place to a feeding site near a small mammal run or burrow. Males tended to move farther than females, and individuals of both sexes moved most often during the spring. Landreth (1973) used radiotelemetry to demonstrate that the level of activity of *C. atrox* in Oklahoma varied seasonally. The movements of males studied averaged 102.4 m/day during the spring and 61.2 m/day during summer. Females averaged 62.4 m/day and 30.9 m/day. The mean home range size of eight *C. atrox* studied in the Sonoran Desert near Tucson, Arizona, with radiotelemetry by Beck (1995) was 5.4 hectares. Throughout the year, these snakes traveled a mean distance of 12.9 km, and traveled, on average, 50.8 m/day. The mean activity temperature was 29.5°C. *Crotalus atrox* showed seasonal changes in habitats and shelters used. Snakes used the rocky slopes during the winter, then moved onto creosotebush flats during the remainder of the year where they sought shelter under shrubs, in mammal burrows, and in woodrat mounds (Beck, 1991).

Reynolds (1982) found *C. atrox* to be active mainly from July to September, with most individuals observed during August. Price and LaPointe (1990) reported peak activity during July and August in Doña Ana County and attributed this to young-of-the-year. Specimen records from March through December are available from New Mexico. The majority were collected from April through September with peaks in May and August. Of 519 records, approximately 41% were collected during May and August.

Crotalus atrox will not hesitate to enter the water and has been observed crossing rivers and large reservoirs (our obs.).

Male *C. atrox* have often been observed in combat, although why these stylized performances take place is unclear. Lowe (1948) and Shaw (1948) first recognized this behavior as a territorial fight and not courtship as was once believed. Lowe and Norris (1950) suggested that the aggressive behavior on the part of snakes may be an expression of territorial defense, social or sexual domination, or sexual discrimination. Klauber (1972) believed the origin to be sexual impulse rather than territorial defense or social domination. Lowe et al. (1986) discussed the issues of limited resources and social dominance as stimuli that provoke this behavior.

Reproduction: Although mating in this viviparous species occasionally takes place at the den site, it most often occurs when the adults have left the dens and migrated to their summer ranges. Jacob et al. (1987) found males to be reproductively competent during their entire summer activity period, with testicular activity peaking in August. The young are 20.3–33 cm TL and are born during late July or August. In 33 broods from central and southern Texas, New Mexico, and Arizona, Fitch (1985) reported an average brood size of nine (2–21). Juvenal females from northwestern Texas examined by Tinkle (1962) matured in three years at approximately 900 mm SVL. The smallest sexually mature female was gravid at 800 mm SVL and the largest immature female was 936 mm SVL. Klauber (1937) reported one gravid *C. atrox* that was 742 mm SVL. The heaviest immature *C. atrox* examined by Tinkle (1962) weighed 450 gm, while the lightest mature snake weighed 320 gm. In general, adults weigh more than 500 gm, juveniles less than 400 gm. The largest and oldest snakes from Oklahoma that were measured by Fitch and Pisani (1993) were 10–15 years old.

Females may exhibit parental care of their young for a short time after birth. Price (1988) observed a gravid female that gave birth to at least eight young during mid August and then remained in the immediate vicinity with the young for at least six days. On several occasions, the young were observed in close contact with the female.

Food Habits: *Crotalus atrox* is a generalist in its food habits, taking a wide variety of vertebrate prey that is often ambushed as the snake lies in wait along well-used rodent trails (Thayer, 1988). It is a nocturnal hunter for most of the year, but forages more often during the day in spring and fall. Prey is detected by olfaction or infrared reception. In a study of the food habits of *C. atrox* in Texas, Beavers (1976) found that small mammals comprised the major diet by weight and frequency of occurrence. The mammals most commonly eaten included woodrats (*Neotoma*) and pocket mice (*Perognathus*), although harvest mice (*Reithrodontomys*), voles (*Microtus*), pygmy mice (*Baiomys*), kangaroo rats (*Dipodomys*), rabbits (*Sylvilagus*), pocket gophers (*Geomys*), and rock squirrels (*Spermophilus*) were also taken. Other vertebrates included various lizards and small passerine birds, including sparrows and towhees. In 80 *C. atrox* examined during the Oklahoma "Rattlesnake Roundups" that contained mammalian prey, Pisani and Stephenson (1991) reported the introduced Norway rat, *Rattus norvegicus*, as prey in 15% of the stomachs examined. Best and James (1984) reported a horned lark, a black-throated sparrow, feathers from a "sparrow-sized" bird, and body fragments of *Crotaphytus collaris* from four specimens they examined from Socorro County. A large number of incidental reports of the food habits of *C. atrox* are available, including those by Mitchell (1903), Ruthven (1907), Lewis (1950), Klauber (1972), and Lowe et al. (1986) who mention frogs and toads as prey items. See Ernst (1992a) for a review of the food habits of *C. atrox*. Beck (1995) reported that *C. atrox* can fulfill yearly maintenance energy requirements with a prey quantity equivalent to 93% of its body mass; a demand that can be met with 2 or 3 large meals.

Perez et al. (1978a, 1978b, 1979) reported the hispid cotton rat (*Sigmodon hispidus*), the gray woodrat (*Neotoma micropus*) and the opossum (*Didelphis virginiana*) to be resistant to the venom of *C. atrox*.

Remarks: A letter in the files of the New Mexico Department of Game and Fish describes *C. atrox* from San

Antonio Mountain. The locality given is in Rio Arriba County, west of US Hwy 285 and approximately 16 km south of the Colorado border. If the identification of this specimen is correct, the record represents a significant extension of the range in New Mexico as currently understood. Although *C. v. viridis* is known to range throughout this portion of northern New Mexico, the area should be investigated for the presence of *C. atrox*.

Logan et al. (1993), studying the ecology of mountain lions in the San Andres Mountains, reported a female mountain lion that was likely killed by a bite from a rattlesnake. The only rattlesnakes that occur in the area are *C. atrox*, *C. lepidus*, and *C. molossus*.

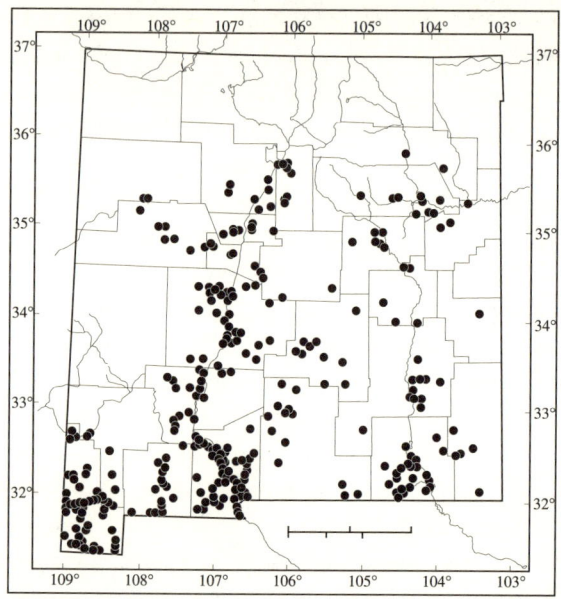

Distribution of Crotalus atrox *in New Mexico.*

CROTALUS LEPIDUS (Kennicott, 1861)   *Rock Rattlesnake*

See Plate 118A, 118B.

Type: Originally assigned the name *Caudisona lepida, Crotalus lepidus* was described by Kennicott (1861) from "Presidio del Norte and Eagle Pass, Texas." The type locality was restricted to Presidio (del Norte), Texas, by Smith and Taylor (1950b). This species was known at first only from two heads preserved in alcohol that were collected by the U.S. and Mexican Boundary Survey. These type specimens are apparently lost. The first complete individual (KU 2331) was collected at the head of Water Canyon in the Magdalena Mountains, about 20 miles west of Socorro, New Mexico, by Prof. F. H. Snow of the University of Kansas, in July 1881. This specimen was fully described by Cope (1883 [1884]).

Distribution: In the United States, *C. lepidus* occurs in isolated mountain ranges from southeastern Arizona and southern New Mexico to south-central Texas. The range in Mexico includes Coahuila and western Nuevo León, Chihuahua, and northeastern Sonora to southwestern Tamaulipas, western San Luis Potosí, Aguascalientes, northern Jalisco, Zacatecas, and southeastern Sinaloa. It probably also occurs in eastern Nayarit (Campbell and Lamar, 1989). In New Mexico, *C. lepidus* occurs in mountainous areas in Catron, Grant, Hidalgo, Luna, Socorro, Sierra, Eddy, Otero, and Doña Ana counties. It has been collected from about 1200 m in the Guadalupe Mountains (Eddy County) to the top of Animas Peak (Hidalgo County) at 2597 m. The specimen collected at Water Canyon in the Magdalena Mountains and described by Cope (1883 [1884]) represents the northernmost locality for *C. lepidus*. There are isolated records from the Guadalupe Mountains in Eddy and Otero counties. The broad desert flatlands that surround these areas serve as effective barriers to dispersal.

Description: *Crotalus lepidus* can be distinguished most readily by the conspicuous dark crossbars on the body. These crossbars may be evident throughout the length of the body or only posteriorly. The dark stripe from the eye to the corner of the mouth may or may not be evident. Of 209 specimens of *C. l. klauberi* measured by

Klauber (1952a), the longest was a 828 mm male from Chihuahua, Mexico. Tennant (1984) provided information on a 775 mm TL *C. l. lepidus* from Val Verde County, Texas.

Crotalus lepidus exhibits a striking amount of interpopulational and ontogenetic variation in color and pattern. The sexes may be strongly color dimorphic (Jacob and Altenbach, 1977), with males having a greenish ground color and the females tending to be grayish. Vincent (1982) described the color pattern variation found in *C. l. lepidus* from southwestern Texas. He examined the color pattern characteristics of 73 specimens and suggested differences in background color are the result of adaptation to dominant substrate colors.

There are 21–26 (usually 23) rows of keeled scales at midbody. Males have 147–172 ventrals and 20–33 subcaudals; females have 149–170 ventrals and 16–24 subcaudals. The male's tail is 7–10.1% of the total body length; the female's tail is 6–8.5%. There are 1–2 loreals, 3–4 preoculars, 3–5 postoculars, 12–15 upper labials, and 9–13 lower labials. The anal plate is undivided (Ernst, 1992a).

Similar Species: The distinctive background coloration and banded or mottled appearance of this small montane rattlesnake should serve to distinguish it from all other snakes in New Mexico. It can be distinguished from other sympatric rattlesnakes by the vertically divided upper preocular scale.

Systematics: Four subspecies of *C. lepidus* are currently recognized (Armstrong and Murphy, 1979). Two of these, *C. l. klauberi* and *C. l. lepidus*, occur in New Mexico. The nominotypical form is based on the description by Kennicott (1861). Gloyd (1936) described *C. l. klauberi* and gave the type locality as "Carr Canyon, Huachuca Mountains, Cochise County, Arizona." The type specimen is UMMZ 79895, an adult male collected 10 August 1930 by L. H. Clark. Gloyd (1936) differentiated *C. l. klauberi* from *C. l. lepidus* based on the absence of a postocular stripe, a dorsal pattern strongly contrasting with the ground color and equally distinct throughout the snake's length, less evident intermediate blotches, a lighter ventrum, and merged occipital spots, which are paired in *C. l. lepidus*. The specimen described by Cope (1883 [1884]) is now referred to as *C. lepidus klauberi* (Gloyd, 1940).

Habitat: *Crotalus lepidus* is usually found in steep, rugged, mountainous areas or desert canyons in the vicinity of rock outcroppings or talus slides. When found in pine-oak forests, it usually occurs in barren sunlit areas with numerous rocks and sparse vegetation. It is also occasionally found along rocky stream beds or on little-traveled roads which traverse these areas. Ernst (1992a) and Tennant (1984) suggested that *C. l. lepidus* may sometimes be found in low-lying desert habitat, where it occurs in barren rocky flats and mesquite grasslands. In such areas, it takes refuge under rocks or in animal burrows when disturbed.

Behavior: *Crotalus lepidus* is generally active during the morning hours and again just before sunset, although they may become nocturnal during the warmest months. Specimen records from New Mexico indicate that *C. lepidus* is active from March through October; of 69 records, 29% were collected during August. When disturbed, they are quick to retreat into the crevices between the rocks where they often betray their presence by continued rattling. Ritualized male combat was observed and described by Carpenter et al. (1976).

Reproduction: Copulation in captive specimens was observed on 26 June and 31 July with parturition occurring 363 and 312 days after copulation (Swinford, 1989). Williamson (1971) reported an adult female from Socorro County that had a brood of four young on 29 August. Broods of four and two, ranging in size from 193–240 mm SVL, born on 8 June and 24 June were reported by Swinford (1990). Armstrong and Murphy (1979) discussed three broods from southeastern Arizona born during mid- to late August with three, four, and five young produced. Mean lengths of four offspring from near Madera, Chihuahua, were TL 191 (181–196) mm and SVL 172 (164–177) mm. Lowe et al. (1986) reported a range of 2–8 young born to adults from Arizona, whereas Tennant (1984) gave a range of 2–5 offspring among several broods produced by captive adults. Klauber (1956) examined 14 litters with an average of 3.85 young.

An apparent natural hybrid between *C. willardi obscurus* and *C. lepidus klauberi* (NMSU 6610) from the

Peloncillo Mountains of southwest Hidalgo County was reported by Campbell et al. (1989).

Food Habits: Lizards, especially of the genus *Sceloporus*, and small rodents are the principal food items. Lowe et al. (1986) stated that small snakes and frogs may occasionally be taken. Swinford (1992) reported finding lizard scales (probably *Sceloporus undulatus* and *Urosaurus ornatus*) and insect parts in fecal samples examined from three wild-caught specimens. He suggested that the insect parts were ingested along with the lizards. Williamson (1971) described an instance of cannibalism in newly born captive specimens. Young of *C. lepidus klauberi* have bright yellow tail tips, which may serve as a lure to attract the lizards that are the primary diet of the juveniles (Kauffeld, 1943b).

Remarks: Russell (1983) reported on a *Crotalus lepidus klauberi* that lived at least 24.5 years in captivity. *Crotalus l. lepidus* is listed as endangered by the New Mexico Department of Game and Fish (NMGF, 1990).

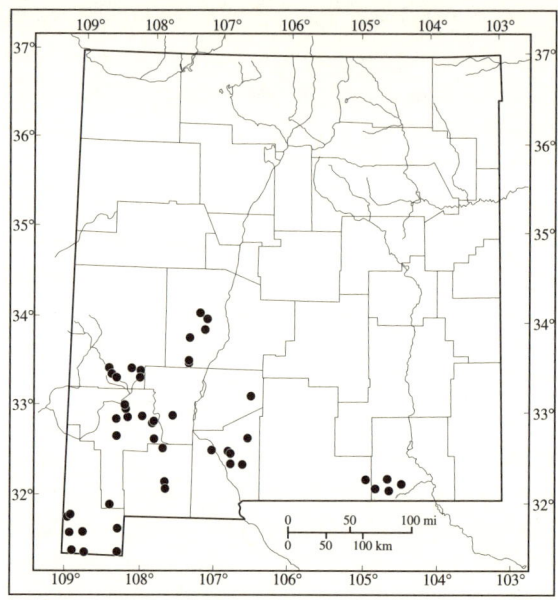

Distribution of Crotalus lepidus *in New Mexico.*

CROTALUS MOLOSSUS Baird & Girard, 1853   *Blacktail Rattlesnake*

See Plate 119.

Type: The type specimen, USNM 485, is a male collected by J. H. Clark from "Fort Webster, St. Rita del Cobre [Grant Co.], N[ew]. Mex[ico]." The date of collection is unknown (Baird and Girard, 1853a).

Distribution: *Crotalus molossus* is found from central and west-central Texas northwest through the southern two-thirds of New Mexico to northern and extreme western Arizona, and southward to the southern edge of the Mexican Plateau and Mesa del Sur, Oaxaca. It also occurs on the islands of Tiburón and San Esteban in the Gulf of California (Ernst, 1992a). In New Mexico, *C. molossus* generally avoids the eastern plains and grasslands. Specimens are commonly observed in the Gila Wilderness and range as far north as the Sandia Mountains near Placitas and along the Rio Puerco near Guadalupe in Sandoval County.

Description: Many consider this our most beautiful rattlesnake. The overall background color is variable, being gray, olive, or greenish-yellow with 20–41 brown or black dorsal rhombic blotches. The posterior dorsal blotches may form lateral bands that extend downward to the ventrals. Each scale is unicolored, with the pattern not cutting across individual scales. The tail is without banding and the black to dark-brown coloration of the tail does not extend forward onto the body. The top of the head and the muzzle are dark brown to black, and there is a distinct brown postocular stripe extending through the eye to the corner of the mouth. The young are similar to the adults in coloration but with dark crossbands often visible on the dark tail. Of 222 specimens from the United States measured by Klauber (1952a), the longest was a male 125.7 cm TL. Large adults may reach 130 cm TL (Tennant, 1984), although they average around 90 cm TL.

The keeled scales are usually in 27 (23–31) rows at midbody. Males have 166–199 ventrals, 22–30 subcaudals, and tails comprising 5.8–8.6% of the total body length. Females have 177–201 ventrals, 16–25 subcaudals, and tails that are 4.6–6.7% of the total length. There are 2–4 (1–9) loreals, 2–3 preoculars, 5 (3–7) postoculars, 17–18 (13–20) upper labials, and 17–18 (14–21) lower labials. The anal plate is single (Ernst, 1992a).

Similar Species: The solid-colored black or dark-brown tail and the light scales within the dark body blotches should distinguish this large rattlesnake from all others in New Mexico.

Systematics: Three subspecies of *C. molossus* are currently recognized (Price, 1980a); only the nominotypical form, *C. m. molossus* Baird and Girard, 1853, occurs in New Mexico.

Habitat: *Crotalus molossus* is an ecological generalist, occurring in a wide variety of habitats, including montane coniferous forests, talus slopes, rocky stream beds in riparian areas, lava flows on flat deserts, and occasionally arroyos in creosotebush desert flats. In a radiotelemetry study of *C. molossus* in southeastern Arizona, individual snakes were primarily rock dwellers, except during late summer and fall when they frequented arroyos and creosotebush flats (Beck, 1991). Armstrong and Murphy (1979) reported a specimen from the sand dunes just south of Ciudad Juárez, Chihuahua, Mexico. However, it is generally a semi-montane species, and is most commonly found in lower rocky areas of mountain foothills. In New Mexico, *C. molossus* ranges from approximately 1000–3150 m elevation.

Behavior: Although generally considered mild-mannered and quick to retreat, certain individuals will rattle vigorously and quickly assume a threatening pose when suddenly encountered in the field. Campbell and Lamar (1989) reported finding this species in trees several meters above the ground, possibly basking or foraging for nestling birds. In the Animas Mountains, we have observed *C. molossus* resting in the dense vegetation approximately one meter above ground.

Crotalus molossus has been collected in New Mexico during all months of the year except February and December. Of 178 New Mexico records, 45% of the snakes were found during July and August. Specimens collected during January were likely taken at the winter retreat where they may be observed basking on warm winter days. In the Sonoran Desert of southeastern Arizona, *C. molossus* (especially females) often forage and feed on diurnal small mammals throughout the winter months (Beck, 1991). The mean home range size of three *C. molossus* studied in the Sonoran Desert near Tucson, Arizona, with radiotelemetry by Beck (1995) was 3.5 hectares. Throughout the year these snakes traveled a mean distance of 15 km, and moved, on average, 42.9 m/day. The mean activity temperature was 29.5°C.

Reproduction: Bogert and Oliver (1945) described the hemipenis. Klauber (1972) stated that females mature at 703 mm. Copulation between an 888 mm SVL, 447 gm male and a 908 mm SVL, 462 gm female was observed on 26 July 1992 in Guadalupe Canyon in extreme southwestern Hidalgo County (B. R. Tomberlin, pers. comm.). Litter size is small, reported as 3–7 (Wright and Wright, 1957) or 3–16 (Lowe et al., 1986). Seventeen litters reported in the literature ranged from 3–16 young and averaged 6.7. Neonate *C. m. molossus* average 272 mm (229–315) TL, weigh 11–28 gm, and are born in July or August (Ernst, 1992a).

Food Habits: A wide variety of vertebrate prey items are taken. Reynolds and Scott (1982) examined 12 specimens from Chihuahua and found pocket mice (*Perognathus*), woodrats (*Neotoma*), and birds as prey items. Painter (1985) reported a green-tailed towhee from an individual from Grant County. Other mammals, including rock squirrels (*Spermophilus*), rabbits (*Sylvilagus*), chipmunks (*Tamias*), kangaroo rats (*Dipodomys*), gophers (*Geomys*), and deer mice (*Peromyscus*), are also eaten. Juveniles undoubtedly take a large number of lizards, primarily *Sceloporus*. Funk (1964) reported a 40 cm TL *Heloderma suspectum* eaten by an adult *C. molossus*. We expect that eating such prey is a very rare occurrence. Beck (1995) reported that *C. molossus* can fulfill yearly maintenance energy requirements with a prey quantity equivalent to 93% of their body mass; a demand that can be met with 2 or 3 large meals.

Remarks: Melanistic *C. molossus* have been reported from lava beds in southern and central New Mexico by Lewis (1949), Prieto and Jacobson (1968), and Best and James (1984). Price (1980a) reviewed this species.

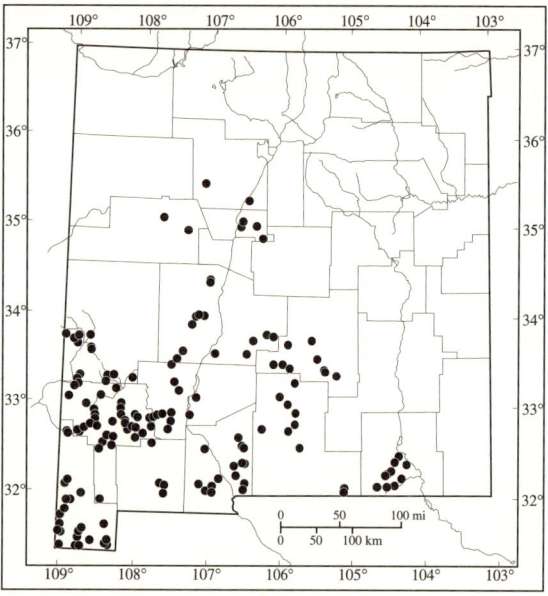

Distribution of Crotalus molossus in New Mexico.

CROTALUS SCUTULATUS (Kennicott, 1861) — *Mojave Rattlesnake*

See Plate 120.

Type: *Crotalus scutulatus* was originally described as *Caudisona scutulata* by Kennicott (1861). No type or type locality were designated, although the latter was restricted to "Wickenburg, Maricopa County, Arizona" by Smith and Taylor (1950b). Cope (1898 [1900]) referred to USNM 5021 as the type specimen of *C. scutulatus*, although USNM 5021 is actually a head of *C. viridis*. See Klauber (1972) for additional discussion.

Distribution: *Crotalus scutulatus* occurs from extreme southwestern Utah and southern Nevada into adjacent southern California, then southeast through Arizona to Trans-Pecos Texas and southward to the southern edge of the Mexican Plateau in Puebla and adjacent Veracruz (Ernst, 1992a). In New Mexico, *C. scutulatus* is most often encountered in the San Simon Valley of southwestern Hidalgo County. A single specimen has been taken in northern Hidalgo County near Virden and a small, little-known population occurs in the low desert regions in south-central Otero County. Elevation ranges from approximately 1200–1500 m. Gary Swinford (*in* Knight, 1985) reported a specimen (unverified by museum specimens) from near the Potrillo Mountains in southern Doña Ana County.

Description: *Crotalus scutulatus* is characterized by a black-and-white banded "coon-tail" with white bands that are usually wider than the black bands. There are usually two, rarely three, scales between the supraoculars. The ground color is variable, being light to dark gray, greenish, or brownish; some individuals may be relatively dark. There are 27–44 darker square blotches on the back that are bordered by light unicolored scales. The postocular lateral stripe passes above and beyond the corner of the mouth. Adults are usually 60–90 cm, although they may reach 140 cm TL.

There are usually 25 (21–29) keeled scale rows at midbody. Males have 165–190 ventrals, 21–29 subcaudals, and average 5 (3–8) dark tail rings; females have 167–192 ventrals, 15–25 subcaudals, and 3–4 (2–6) dark tail rings. Usually, one loreal, 2 preoculars, 2 postoculars, 14–15 (12–18) upper labials, and 15–16 (12–18) lower labials are present on each side of the head. The anal plate is entire (Ernst, 1992a).

Similar Species: *Crotalus scutulatus* is often confused with *C. atrox*, especially the distinctly marked young. These two species can be reliably separated only by counting the number of small scales between the supraocular scales; *C. scutulatus* usually has two, rarely three, while *C. atrox* averages more than four. In addition, the white bands on the tail of *C. scutulatus* are normally wider than the black bands, whereas the tail bands of *C. atrox* are normally of equal width, although this character varies widely among individuals. The postocular light stripe in *C. scutulatus* extends diagonally beyond the corner of the mouth; that of *C. atrox* ends well in front of the corner of the mouth. In New Mexico, *C. scutulatus* can be distinguished from all rattlesnakes, except *C. atrox*, by the presence of the black-and-white ringed "coon-tail."

Systematics: Although two subspecies are recognized (Gloyd, 1940), only the nominotypical form, *Crotalus s. scutulatus*, occurs in New Mexico. Gloyd (1940) was the first to use the trinomial. Knight (1985) suggested that the Continental Divide represents a barrier to gene flow, separating Chihuahuan and Sonoran populations to the east and west. Based on the considerable evolutionary divergence in the venoms of populations sampled by Knight (1985) he suggested the possibility of the existence of three subspecies of *C. scutulatus*.

Habitat: In New Mexico, *C. scutulatus* prefers lowland semi-arid grasslands or open areas of creosotebush and desert grasslands with dense growth of clump-forming grasses, such as *Sporobolus*. It is generally absent from rocky areas. It occurs sympatrically with *C. atrox* throughout much of its limited range in New Mexico. Pough (1966) felt as though this overlap was produced primarily by the penetration of *C. scutulatus* into less arid areas where *C. atrox* is more abundant.

Pough (1966) found *C. scutulatus* to be abundant in the *Ephedra* savanna (= Semidesert Grassland of Brown, 1982) in southeastern Arizona and southwestern New Mexico. However, Mendelson and Jennings (1992) found that *C. scutulatus* had decreased in relative abundance in the same area, possibly with the succession of the few remaining semidesert grasslands to a Chihuahuan desert scrub community.

Behavior: *Crotalus scutulatus* occupies only a small area of New Mexico and little is known of its habits. Most individuals encountered have been found on the roadways during warm summer nights. Of 73 specimens from New Mexico, nearly 60% were collected during August. In Chihuahua, Reynolds (1982) reported peak activity of *C. scutulatus* during July and August, with most snakes taken during August, approximately one month after the peak rainfall.

In an area of southwestern Hidalgo County where *C. atrox* and *C. scutulatus* are sympatric, we have monitored two hibernacula regularly for four years and have never observed *C. scutulatus* in the same hibernaculum with *C. atrox*. Individual *C. scutulatus*, however, are often observed basking at the entrance of abandoned rodent burrows or in soil crevices on warm fall days when numbers of *C. atrox* are gathered nearby at the hibernacula. Lowe et al. (1986) noted this behavior and stated that *C. scutulatus* usually does not aggregate in large numbers for the winter but generally spends the cold season singly or in pairs, often in woodrat burrows.

Individual temperament of *C. scutulatus* varies. Individuals encountered in the field are often excitable and will rattle and strike savagely; others only attempt to escape.

Reproduction: Very little information has been published on the reproductive biology of *C. scutulatus*. In northeastern Chihuahua, courtship was observed during August by Jacob et al. (1987), who found males to be reproductively active during their entire summer activity period, although testicular activity peaked during August. Van Devender and Lowe (1977) collected a gravid female from central Chihuahua that gave birth to eight young on 18 July. The offspring averaged 217.6 (204–221) mm SVL. Klauber (1972) recorded birth dates of *C. scutulatus* as 22 August and 18 and 20 September; 21 broods varied from 5–13 and averaged 8.6 young per brood.

Food Habits: *Crotalus scutulatus* feeds primarily on rodents and other small mammals. No studies of the food habits of *C. scutulatus* have been conducted in New Mexico, although in nearby Chihuahua, Reynolds and Scott (1982) found mammal remains in 91.7% of 48 specimens they examined. Kangaroo rats (*Dipodomys*) occurred in 39.6% of the stomachs containing food items, pocket mice (*Perognathus*) in 20.8%, white-footed mice (*Peromyscus*) in 16.5%, and ground squirrels (*Spermophilus*) in 10.4%. Other mammals included cottontail rabbits (*Sylvilagus*) and jackrabbits (*Lepus*). It is interesting they did not find pack rats (*Neotoma*) included in the diet. Klauber (1972) found mammalian remains in 21 specimens, lizard scales in two, and insect and centipede chitinous remains in a single individual. Reynolds (1978) noted insects and millipedes taken by young *C. scutulatus*. Tennant (1984) cited Earl Turner who observed a juvenal *C. scutulatus* feeding on *Coleonyx brevis* and who believed the juveniles regularly eat various lizards, including *Holbrookia*, *Sceloporus*, and *Uta*. He noted that these lizards are sometimes accepted by newly captured specimens in preference to laboratory mice. Dammann (1961) suggested that young *C. scutulatus* eat more lizards than do the young of *C. atrox*. Lowe et al. (1986) stated that *C. scutulatus* will take birds and an occasional frog or toad. Other food items reported in the diet of *C. scutulatus* include whiptail lizards, leaf-nosed snakes, spadefoot toads, and occasionally bird eggs (Ernst, 1992a and references therein).

As with most rattlesnakes, prey of *C. scutulatus* is selected on the basis of size. Those items that are either too large or too small are rejected, and potential prey that could harm the snake are not accepted (Reynolds and Scott, 1982). Reynolds (1978) discussed a captive adult *C. scutulatus* that retreated from a large packrat offered as food.

Remarks: The venom of *C. scutulatus* is very toxic, making it an extremely dangerous animal. Glenn and Straight (1978) and Glenn et al. (1983) identified two venom types, A and B. They discussed clinical differences and reported a significant geographical difference in the lethal toxicity. No significant differences in external morphology of the snakes producing these venom types could be established. The type A venom from California, Utah, southeastern Arizona, and southwestern New Mexico contains a major lethal toxin, Mojave toxin. Commercial antivenin seems to be less effective against venom A than venom B. Knight (1985) separated *C. s. scutulatus* into two distinct populations, Sonoran and Chihuahuan, based on variations found in venom proteins. He reported the Mojave toxin from Chihuahuan populations and, contrary to Glenn and Straight (1978) and Glenn et al. (1983), suggested that a bite victim may be expected to respond equally to a single antiserum, despite the geographic origin of the snake.

An intensive study of the natural history of *C. scutulatus* is needed as an expanding rural human population in southwestern New Mexico comes in contact with this species. Complete case histories involving human envenomation are rare, but the results of a bite by *C. scutulatus* are usually severe, with death occurring in a high percentage of untreated bites (Hardy, 1983, 1986; Russell, 1967b).

Jacob (1977) and Aird et al. (1989) reported hybridization between captive *C. scutulatus* and *C. atrox*. Glenn and Straight (1990) used venom characteristics as an indication of hybridization of *C. viridis* and *C. scutulatus* in southwestern New Mexico.

The presence of a population of *C. scutulatus* in the Potrillo Mountains (Knight, 1985) needs verification. Price (1982) reviewed this species.

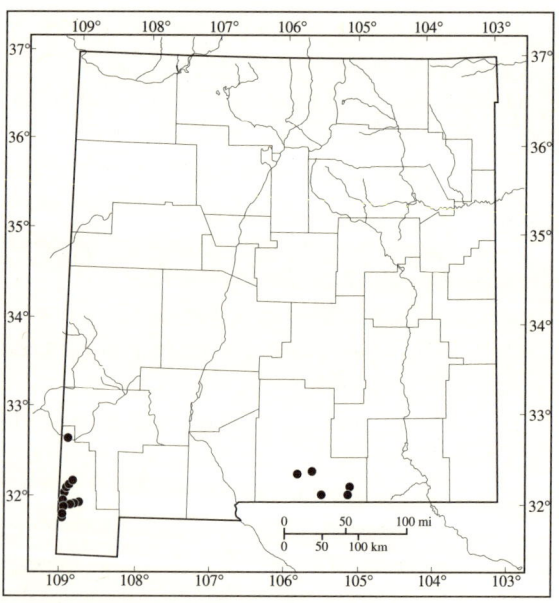

Distribution of Crotalus scutulatus *in New Mexico*.

CROTALUS VIRIDIS (Rafinesque, 1818)   *Western Rattlesnake*

See Plate 121.

Type: *Crotalus viridis* was originally described as *Crotalinus viridis* by Rafinesque (1818). No type specimen was designated. The type locality was stated as "the Upper Missouri [Valley]" although it was later restricted to "Gross, Boyd County, Nebraska" by Smith and Taylor (1950b).

Distribution: *Crotalus viridis* has the most extensive range of any North American rattlesnake. It ranges from south-central British Columbia, southeastern Alberta, and southwestern Saskatchewan southeastward through the United States to extreme western Iowa, Nebraska, and Kansas, and south to northern Baja California, northern Chihuahua, and northwestern Coahuila, Mexico. Elevations of occurrence range from near sea level to over 3300 m (Ernst, 1992a). In New Mexico, *C. viridis* is found statewide, where it occurs from 900 m to approximately 2600 m. One race is found in the higher elevations of the Gila Wilderness in Catron and Grant counties.

Description: This is the most widespread and variable rattlesnake in the United States and Canada. It is the only rattlesnake with more than two internasals in contact with the rostral. Coloration varies from jet black to light brown. Size of adults ranges from approximately 162.5 cm TL to barely 70 cm TL in the smallest subspecies. The species usually has 25–27 (21–29) scale rows at midbody, 158–196 ventrals, and 13–31 subcaudals. There are 1–3 loreals, 2 preoculars, 2–3 postoculars, 14–15 (10–19) upper labials, and 15–16 (11–20) lower labials. The anal plate is entire (Ernst, 1992a). The subspecies in New Mexico (see Systematics) are distinctive enough to warrant individual description.

Crotalus v. viridis may reach a total length of 152 cm. Ground color varies from light brown to a dusty yellowish tan. There are 35–55 dark brown dorsal blotches on the body (excluding the tail) that are narrowly bordered in white. They are mostly oval in shape at midbody but gradually elongate into narrow crossbands on the posterior third of the trunk. There are two rows of smaller blotches on the side, the uppermost larger and more diffuse, the lower ones darker with light-colored borders. Two oblique white lines are evident on the side of the head; the one behind the eye passes above the corner of the mouth. The belly is white and unspotted. *Crotalus viridis nuntius* is a small, dwarfed, and delicately brownish to pinkish replica of the prairie rattlesnake (Lowe et al., 1986). It rarely reaches 70 cm TL.

Maximum length of *Crotalus viridis cerberus* (Coues, 1875) is approximately 104 cm, although most adults encountered are less than 90 cm. This subspecies is characterized by a dark coloration varying from jet black without body blotches to a lighter blackish brown and black or dark grayish with body blotches. The blotches are large and usually poorly defined. In jet black individuals, there is often a series of yellow, cream, or gray light marks on the back corresponding to the interspace positions. The upper row of lateral blotches may be absent, or diffuse black or blackish brown. A dramatic color change occurs in dark individuals, which are jet black in daylight and brown to gray with dark blotches when foraging in darkness (Lowe et al., 1986).

Similar Species: This is the only rattlesnake in New Mexico with more than two internasals in contact with the rostral. It is most similar in appearance to *Sistrurus catenatus*, which possesses nine enlarged plates rather than numerous small scales on top of the head as in *C. viridis*. Both *C. atrox* and *C. scutulatus* have a sharply contrasting, black-and-white banded tail. Some individual *C. viridis* have this black-and-white banded tail although the bands tend to be somewhat obscure. *Crotalus molossus* has a solid colored black or dark-brown tail. *Heterodon nasicus* lacks a rattle and does not have elliptical pupils.

Systematics: Nine subspecies of *C. viridis* are currently recognized. Two, and possibly three, are found in New Mexico. In addition to the nominotypical subspecies, *C. v. cerberus* (Coues, 1875) also occurs in New Mexico. Coues gave the type locality as San Francisco Mountains, Coconino County, Arizona. The type specimens are represented by field numbers 509 (= ANSP 7085?) and 511 (= ANSP 7088). There were two additional syntypes, ANSP 7086–7087, whose present whereabouts are unknown (Klauber, 1972). *Crotalus v. nuntius* Klauber

(1935) was named from an adult male collected at Canyon Diablo, Coconino County, Arizona, by R. L. Borden on 9 August 1930. The holotype was LMK 3105, a specimen originally in the private collection of the late Laurence M. Klauber.

The zone of intergradation between *C. v. nuntius* and *C. v. viridis* lies somewhere in the Four Corners area of northwestern New Mexico. This zone cannot be exactly defined (Klauber, 1972), and until additional specimens are examined from this area, the inclusion of this small race in the herpetofauna of New Mexico is questionable. Ernst (1992a) described the range of this small (to 70 cm total length) subspecies as northeastern and north-central Arizona and adjacent extreme northwestern New Mexico.

Habitat: Although most commonly found on sandy or alkali soils of remnant grasslands, *Crotalus viridis* may also be encountered in open desert country, rocky hillsides, sparse, overgrazed shortgrass prairies, or into ponderosa pine forests to at least 2600 m.

Behavior: In New Mexico, although collection records are available for all months of the year, *C. v. viridis* is active mostly between March and November. Of 737 records examined from New Mexico, 40% were collected during July and August. Aldridge (1979a) suggested this extended activity period is due to flexible diel patterns, with *C. v. viridis* becoming diurnal in early spring and fall when air and substrate temperatures are below their normal activity range. During the summer, *C. viridis* is mostly nocturnal, seeking shelter in rodent burrows or rocky retreats during the hotter parts of the day. Individuals are often encountered on warm blacktop roads shortly after sunset.

During the winter months, *C. v. viridis* is known to hibernate in rodent burrows or communal rocky den sites that may be occupied by other species of snakes including *Pituophis melanoleucus*, *Masticophis flagellum*, *Coluber constrictor*, and *C. atrox*. Jacob and Painter (1980) monitored the body temperatures of hibernating *C. v. viridis* near Moriarty in Torrance County. They found active snakes in the den that often emerged to bask at the mouth of the hibernaculum as long as their body temperatures remained above 10°C. Within the 6-month period during which the snakes occupied the hibernaculum, their body temperatures ranged from 6–29°C, with a mean of 11.3°C. Increasing ground temperatures appeared to be responsible for emergence of these snakes in April.

David Duvall, his associates, and students have studied the ecology of *C. viridis* in Wyoming (see Duvall et al., 1990; Graves and Duvall, 1993 and references therein). They find that males and nonpregnant females make lengthy seasonal migrations. In the spring, individuals move away from the den site to foraging areas that contain large numbers of small rodents where the snakes become relatively sedentary. The pregnant females, however, move much shorter distances from the den site to perennially occupied localities, known as rookeries. At these sites they move little, exhibit reduced feeding, form aggregations with like females, and give birth in late summer. Although rookery use has been documented by many researchers working in diverse habitats, it is unknown if *C. viridis* in New Mexico uses such areas.

Reproduction: Throughout the wide range of *C. viridis*, litter size seems to be closely correlated with adult body size. The average number of young in 516 clutches listed in published accounts is 9.9 (Ernst, 1992a). In 17 broods of *C. v. viridis* from northeastern New Mexico, brood size ranged from 5–14 and averaged 9.5 (Fitch, 1985). Aldridge (1975) examined two specimens collected from Moriarty in early December and found sperm present in the seminiferous tubules. Aldridge (1979a) later examined 38 adult male *C. v. viridis* from central New Mexico and found that spermatogenesis peaked from mid- to late June and continued through August and September. Schuett et al. (1993) found that mating in *C. v. viridis* from Wyoming occurs from early through midsummer. Overwinter sperm storage thus becomes an obligatory component of the reproductive cycle in the Wyoming population. However, Holycross (*in* Schuett et al., 1993) documented mating in Nebraska sandhills populations soon after emergence from hibernacula in the early spring. He also suggested another mating period from mid- to late summer, based on observations of bisexual pairing. Not all female *C. v. viridis* reproduce annually; of 44 adult females from New Mexico examined by

Aldridge (1979b), only 73% showed signs of producing a litter during the year they were collected.

Food Habits: *Crotalus viridis* practices several feeding strategies, including active foraging around rodent burrows, sit-and-wait tactics along rodent trails, and carrion eating (Chiszar et al., 1990 and references therein). They depend primarily on rodents and other small mammals for their food supply, although their diet may include birds and their eggs, lizards, frogs, and insects. In Colorado, young of the year are known to eat earless lizards and fence lizards, whereas adults feed primarily on rodents (Klauber, 1972). Hamilton (1950) reported two *C. viridis* that had eaten *Phrynosoma douglassii* and Stabler (1948) found a spadefoot toad in the diet of another. Apparently, prey selection is size-limited and prey size increases as the snake grows. Ernst (1992a) provided an exhaustive list of prey items of *C. viridis*.

Remarks: Mello (1978) reported on the easternmost specimen of *C. v. cerberus* from "Grant Co: 37.0 km E and 12.5 km S Glenwood." This specimen was being maintained in a private collection and cannot be located for verification.

The longevity record for this species is a *C. v. viridis* kept 27 years, 9 months in captivity in Kansas. It was estimated to have been two years old when captured (Bailey et al., 1989).

Based on venom characteristics, Glenn and Straight (1990) reported hybridization between *C. v. viridis* and *C. s. scutulatus* in southwestern New Mexico.

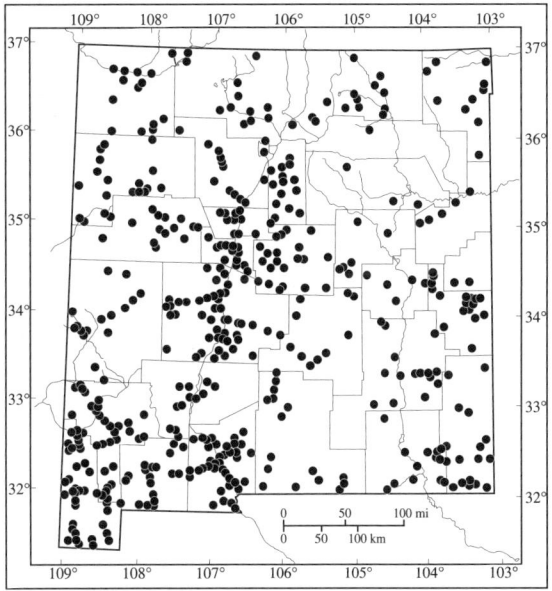

Distribution of Crotalus viridis *in New Mexico.*

CROTALUS WILLARDI Meek, 1905 — *Ridgenose Rattlesnake*

See Plate 122.

Type: The type specimen is FMNH 902. Meek (1905) presented the type locality for *Crotalus willardi* as "Tombstone, Arizona," however after correspondence with Frank C. Willard, the original collector, Swarth (1921) restricted the type locality to: "above Hamburg, middle branch of Ramsey Canyon, Huachuca Mountains (altitude about 7000 ft.), Cochise County, Arizona." Klauber (1949) accepted this revised type locality.

Distribution: *Crotalus willardi* ranges from southeastern Arizona and extreme southwestern New Mexico southward in the Sierra Madre Occidental to Zacatecas, Mexico. Elevational range is 1600–2750 m. In New Mexico, *C. willardi* is known only from southern Hidalgo County at approximately 1820–2590 m in the Animas Mountains and 1707–1890 m in the Peloncillo Mountains. Campbell et al. (1989) were the first to report *C. willardi* from the Peloncillo Mountains.

Description: *Crotalus willardi* is easily recognized by the unique ridge of upturned scales on the tip of the snout between the nostrils. The ground color is reddish brown, rust, yellowish brown, gray brown, or gray. There are 18–45 dorsal blotches that are deep chestnut or dark brown and are usually edged on the anterior and posterior margins with dark brown or black pigment. They are more or less subquadrate in shape and tend to merge with the ground color laterally. The pale interspaces between the middorsal blotches may form distinctive cream-colored, buff, or light brown crossbands. The top of the head is usually marked with a few dark brown or black spots. The belly becomes progressively darker posteriorly. The subcaudal scales are pinkish, reddish, or orangish with irregular dark markings (Campbell and Lamar, 1989).

The unique pattern of bright white facial stripes is characteristic of the species. There are two longitudinal stripes on the side of the face, one extending diagonally backward from the prenasal scale below the eye to the corner of the mouth, and a second extending backward along the upper and lower labials. A median white, vertical stripe extends downward from the rostral and mental scales (Ernst, 1992a).

The overall body coloration of specimens from New Mexico is normally various shades of gray, although brownish or reddish individuals occur. The facial pattern is usually indistinct. The posterior upper labials are pale with some dark peppering, and the postocular stripe is poorly defined. The rostral-medial stripe is absent.

Crotalus willardi has 23–31 rows of keeled scales at midbody. Males have 140–156 ventrals and 24–36 subcaudals; females have 146–160 ventrals and 21–32 subcaudals (Barker, 1992). Males have tails 9.1–11.5% of SVL; females only 7.9–9.8% of SVL. There are 1–3 loreals, 2–3 preoculars, 3–4 postoculars, and 13–14 (12–17) upper and lower labials (Ernst, 1992a).

Barker (1991) presented mensural characteristics for 39 *Crotalus willardi* from the Sierra San Luis in Chihuahua and Sonora, Mexico. In 24 males, mean TL was 436.4 mm (211–586); in 15 females mean TL was 408.4 (198–515). Select scale counts of 29 specimens from the Sierra San Luis, the Animas Mountains, and the Peloncillo Mountains examined by Barker (1992) are: dorsal scale rows at midbody 23–27 (25.2); rattle fringe scales 8–10 (8.4); ventrals (males) 147–151 (148.7), (females) 148–156 (152.4); subcaudals (males) 26–33 (29.7), females 21–30 (25.8). Andrew T. Holycross (pers. comm.) reported an adult male collected during June 1994 in the Animas Mountains that weighed 171 gm, and measured 558 SVL and 626 TL.

Young resemble the adults in most subspecies, although those of *C. willardi obscurus*, in contrast with their grayish parents, are dark brown, have yellow-orange pigment on their lips, and have mostly black tails. Young of *C. w. willardi* may have yellowish tails (Martin, 1975; Lowe et al., 1986), fading to an olive gray or brown soon after the second shed.

Similar Species: The pronounced canthal ridge and unique facial pattern will separate *C. willardi* from all other snakes in the United States. Although there are unverified reports of *C. atrox* and *C. viridis* from higher elevations of the Peloncillo Mountains, only *C. lepidus klauberi* and *C. molossus* are known to occur sympatrically with *C. willardi* in New Mexico and they both lack the canthal ridge.

Systematics: Five subspecies of *C. willardi* are currently recognized (Barker, 1992). Only *C. w. obscurus* occurs in New Mexico. The first *C. willardi* collected in New Mexico was taken from the Animas Mountains in Indian Creek Canyon near Animas Peak on 15 September 1957 by Robert A. Zeller (Bogert and Degenhardt, 1961). In the earlier literature, this form is referred to as *C. w. silus*, although Harris (1974) assigned the name *C. w. obscurus* to the population found in the Animas Mountains. A more formal description followed two years later when Harris and Simmons (1976) discussed the paleogeography and evolution of the entire *C. willardi* complex. However, the acceptance of *C. w. obscurus* as a valid subspecies remained in question until Barker (1992), studying morphology and biochemical characters of the entire complex, presented evidence that *C. w. obscurus* was indeed unique.

Habitat: *Crotalus willardi* is a montane generalist, occurring in New Mexico mostly above 1800 m elevation, although Barker (1991) suggested collectors have not investigated lower elevations where the species may also occur. Rado and Rowlands (1981) reported a specimen found at 1600 m in Arizona. Most specimens from New Mexico and northern Mexico have been collected in and along steep rocky canyons with intermittent streams or on exposed talus slopes. Of 13 collecting sites in the Sierra San Luis characterized by Barker (1991), SW-NW (225–315°) facing slopes seem to be favored; east facing slopes in the Animas Mountains are also often occupied (our obs.). Dominant vegetation in the narrow elevational band where *C. willardi obscurus* is known to exist includes various oaks, Apache and Chihuahua pine, alligator bark juniper, Arizona cypress, Arizona madrone, manzanita, and various grasses including *Sporobolus*, *Muhlenbergia*, and *Aristida* (Degenhardt, 1972; our obs.).

Behavior: Individuals are generally secretive and inconspicuous. They are more likely to rattle and attempt to escape than coil and strike. When encountered on talus, they will retreat deep into the talus, generally giving away their location by constant rattling. Individuals from the Sierra San Luis have been found hibernating 40–46 cm deep in talus and have been observed basking at air temperatures of 6–9°C (shade) and 26°C (sun).

Reproduction: Applegarth (1980) reported an average of 5.5 (2–9) young for 12 broods. Martin (1979) reported a brood of nine young, captive born to parents collected from the Animas Mountains, that averaged 205 (200–212) mm and weighed 7.35 (6.84–7.99) gm. An adult female, 49 cm TL, collected in the Peloncillo Mountains in April 1987, later had six young that averaged 21.5 (20–22) cm TL. Andy Holycross (pers. comm.) collected a gravid female from the Animas Mountains, Hidalgo County, that produced a brood of 8 young on 30 June–1 July. The best description of courtship and mating behavior in this species was reported by Tryon (1978) for a pair of captive *C. w. willardi* from Arizona.

An apparent natural hybrid between *C. willardi obscurus* and *C. lepidus klauberi* (NMSU 6610) was reported from the Peloncillo Mountains of southwestern New Mexico by Campbell et al. (1989).

Food Habits: Barker (1991) examined fecal samples collected from 12 wild caught *C. w. obscurus* from the Sierra San Luis of northern Sonora and Chihuahua, Mexico. Nine of these samples contained rodent hair, two contained lizard scales, and two contained body parts of the large centipede, *Scolopendra*. Applegarth (1980 and references therein) reported a wide variety of prey items taken by *C. willardi*, including various rodents, birds, lizards, snakes, and arthropods.

Remarks: The small New Mexico population of this unique montane rattlesnake could be negatively impacted by habitat destruction or by overzealous and irresponsible collectors. While investigating *C. willardi* in the Animas Mountains, Harris and Simmons (1975) reported encountering 15 collectors from six states during late August 1974, and they suggested that the species could not withstand such overexploitation. We observed an apparent decline in the population during the 1970s in the Animas Mountains. However, based on recent studies, it appears that the population is currently well protected and stable (A. T. Holycross, pers. comm; our unpbl. data). *Crotalus willardi obscurus* is considered a threatened species by the U.S. Fish and Wildlife Service (USDI, 1991) and an endangered species by the New Mexico Department of Game and Fish (NMGF, 1990).

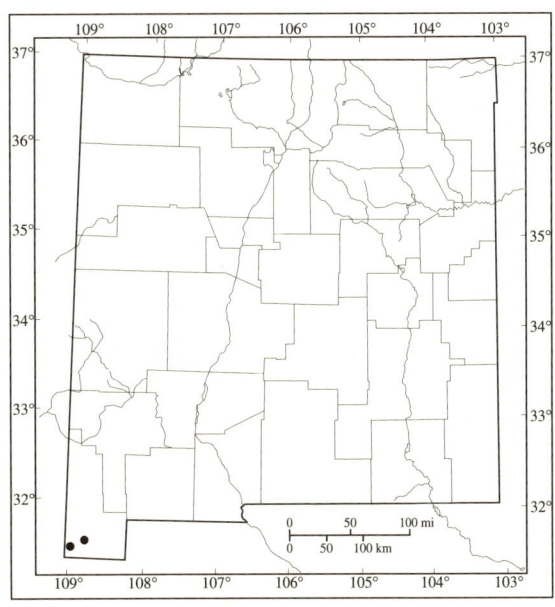

Distribution of Crotalus willardi *in New Mexico.*

SISTRURUS CATENATUS (Rafinesque, 1818)   *Massasauga*

See Plate 123.

Type: *Sistrurus catenatus* was first described as *Crotalinus catenatus* by Rafinesque (1818). He gave the type locality as "Prairies of the Upper Missouri." No holotype was designated.

Distribution: *Sistrurus catenatus* occurs largely as disjunct populations from west central New York and Pennsylvania to the Georgian Bay region of Ontario, then west across the central states to extreme southeastern Minnesota, southeastern Colorado, southeastern Arizona, and the Gulf Coast of Texas. Disjunct Mexican populations occur in the Cuatro Ciénegas Basin, Coahuila and near Aramberri, Nuevo León. Most of the range is below 1500 m elevation (Minton, 1983). In New Mexico, *S. catenatus* is found primarily in the central and eastern counties, most often in the shortgrass prairies in the central Rio Grande Valley and in the shinnery oak regions of Chaves and Lea counties. It occurs from approximately 925–2100 m elevation. Verified records are lacking from the northwestern counties and the Mogollon Rim in west-central New Mexico.

Description: *Sistrurus catenatus* is a small rattlesnake, with adults rarely over 55 cm TL in New Mexico. There are nine large scales on the top of the small head, and a broad dark lateral stripe on the face that is bordered on both sides by white. Dorsally there are 27–41 dark brown or blackish brown blotches that are generally wider than long. The venter is white or cream and is unmarked or with small dark flecks.

Sistrurus catenatus has 21–27 (usually 23) scale rows at midbody, 129–160 ventrals, and 19–34 subcaudals. Tail length is 10–12.5% of total length in males, but only 7.5–9.0% in females. There are two preoculars, 2–4 postoculars, 1–2 suboculars, 11–12 (9–14) upper labials, and 12–13 (10–16) lower labials. The anal plate is entire (Ernst, 1992a).

Similar Species: *Sistrurus catenatus* is often confused with the sympatric *Crotalus v. viridis*, although it is easily identified by the presence of nine large symmetrical scales on top of the head between the ocular scales. All other rattlesnakes in New Mexico lack these large scales and instead have numerous small scales on the top of the head. In addition, the presence of a small rattle will separate this species from similarly colored species, including *Heterodon nasicus* and *Hypsiglena torquata*.

Systematics: Three subspecies of *S. catenatus* are currently recognized, however only *S. c. edwardsii* occurs in New Mexico. This form was originally described by Baird and Girard (1853a) as *Crotalophorus edwardsii*. The type locality was listed as "Tamaulipas" (Mexico); the holotype, USNM 507, is presumed lost. USNM 506 from "Sonora, Mexico" collected by J. D. Graham and USNM 509 from "Tamaulipas, Mexico" collected by L. A. Edwards were designated as syntypes (Cochran, 1961). Minton (1983) illustrated a zone of intergradation between *S. c. edwardsii* and the western massasauga, *S. c. tergeminus* Say *in* James, 1823 in Union County, New Mexico, while Klauber (1972) illustrated a broader zone in the eastern plains of New Mexico. Until further study, we follow Conant and Collins (1991) and Ernst (1992a) and recognize only *S. c. edwardsii* as part of the herpetofauna of New Mexico. Individuals from the southeastern Colorado population are also thought to be intergrades between these subspecies (Maslin, 1965).

Habitat: In New Mexico, *Sistrurus* is an inhabitant of the desert grasslands or shortgrass prairies with sandy soil. They tend to avoid rocky habitat. They are common in the low growing, shrubby shinnery oak habitat in southeastern New Mexico where they are associated with pure stands of shinnery oak or a mixture of oak and various herbs and grasses.

Behavior: This small rattlesnake is often found tightly coiled in sandy soils as if resting or waiting for prey. Those encountered in this resting coil in the shinnery oak habitat of the Mescalero Sands in Chaves County are often reluctant to rattle or move until molested. Once disturbed they often quickly disappear into rodent burrows that they use for refugia. Such burrows may provide a humid microclimate that helps retard moisture loss from the snake's body. They have been observed resting in the extensive tunnels of kangaroo rat mounds. *Sistrurus catenatus* is primarily nocturnal, often found while crossing roadways at night where they sometimes stop, apparently utilizing heat retained in the pavement. It has been collected from March through November in New Mexico. Of 29 specimens with collecting dates available, 16 were collected during May and June.

Reproduction: Nothing is known of the reproductive biology of S. catenatus in New Mexico. Of 13 broods from Kansas, Oklahoma, and Texas reported by Fitch (1985), brood size averaged 5.9 (3–11). Klauber (1956) recorded four broods of *S. c. tergeminus* numbering three, four, nine, and nine. Greene and Oliver (1965) provided detailed information on brood size of six female *S. c. tergeminus* from Texas, including one female 694 mm TL with a litter of 11 young. In Arizona, Lowe et al. (1986) reported courtship and copulation in captive *S. c. edwardsii* from March into June. Known brood sizes are five and seven for two captive broods.

Food Habits: Greene and Oliver (1965) suggested that *S. c. tergeminus* is a rather opportunistic feeder on a wide variety of vertebrate prey. Stomach contents of 19 specimens examined from north-central Texas and south-central Kansas consisted entirely of vertebrate prey including frogs, lizards, snakes, shrews, and mice. They reported one specimen trying to devour a recently run-over hognose snake and another that ate a Texas horned lizard. A large centipede, *Scolopendra* sp., was discovered in an adult male (MSB 54904) found DOR near Albuquerque (Don Sias, pers. comm.). No detailed studies of the food habits of *S. c. edwardsii* are available from New Mexico.

Remarks: Chapman and Casto (1972) reported *S. c. edwardsii* from near Milnesand in Roosevelt County that was killed by a loggerhead shrike. Although not likely to be fatal, bites from this small rattlesnake may be extremely painful and may be accompanied by swelling, faintness, nausea, and a cold sweat. Andrew T. Holycross is currently investigating the status of an isolated population of *S. c. edwarsdii* in extreme southeastern Arizona, Cochise County. The apparent absence of this species in the grasslands of southwestern Hidalgo County is puzzling. Minton (1983) reviewed this species.

Distribution of Sistrurus catenatus in New Mexico.

AMPHIBIANS AND REPTILES OF QUESTIONABLE OCCURRENCE

The following species have been erroneously listed as part of the New Mexico herpetofauna, or have populations that occur within 40 km of the state line. Due to their close proximity to New Mexico's borders, some may eventually be found in this state.

AMPHIBIANS

Ambystoma rosaceum Taylor, 1941
Rosy Salamander

This common Mexican species (Taylor, 1941b) has been collected in Sonora, México, approximately 27 km south of the New Mexican border in the Sierra San Luis (Peter Warren, pers. comm.). *Ambystoma rosaceum* may be found breeding in stock ponds at higher elevations in southern Hidalgo County.

Pseudacris clarkii (Baird, 1854)
Spotted Chorus Frog

The inclusion of *Pseudacris clarkii* in the herpetofauna of New Mexico by Painter and Burkett (1991) is likely based on a misidentified specimen. A discriminant function analysis based on morphometric features of *P. clarkii* and *P. triseriata*, strongly indicated that the single specimen of *P. clarkii* from Colfax County (KU 78132) is actually referable to *P. triseriata* (Stuart, 1992a). The species has been reported to occur within 32.2 km of New Mexico in Bailey County, Texas (Tinkle and Knopf, 1964; Dixon, 1987). It may occur in New Mexico along the eastern boundary in Union, Colfax, Roosevelt, or Curry counties.

Spea intermontana (Cope, 1883 [1884])
Great Basin Spadefoot

The inclusion of *Spea intermontana* in the herpetofauna of New Mexico by Gehlbach (1965) is based on misidentified specimens, although it may be discovered in San Juan County in the extreme northwest corner of the state. Wiens and Titus (1991) suggest two species may actually be represented by the name *Spea intermontana*.

Rana tarahumarae Boulenger, 1917
Tarahumara Frog

The inclusion of *Rana tarahumarae* in New Mexico by Linsdale (1933), "one mile above the XSX Ranch, East Fork of the Gila River, Socorro County, New Mexico," Little and Keller (1937), "along Rio Grande near Mesilla Dam," and Wright and Wright (1949) is based on misidentified specimens of *R. catesbeiana* (Zweifel, 1968a). *Rana tarahumarae* is not considered a part of the herpetofauna of New Mexico.

REPTILES

Gopherus agassizii (Cooper, 1863)
Desert Tortoise

Hulse and Middendorf (1979) and Patterson (1982) reported *Gopherus agassizii* from the area around Portal and San Simon, Cochise County, Arizona. Hulse and Middendorf (1979) state, "It seems likely that the specimens seen by us [3 specimens] represent a large extension for *Gopherus agassizii* of approximately 108 km eastward, although we cannot discount the possibility of tourists in the area releasing the animals." David L. Hardy and Barney R. Tomberlin (pers. comm.) reported specimens that were collected in the foothills of the Chiricahua Mountains in Cochise County, Arizona and in the Peloncillo Mountains near Rodeo in Hidalgo County, New Mexico. On 4 September 1994, Dr. Ray Turner photographed an adult *G. agassizii* on the Gray Ranch in southern Hidalgo County as it crossed the road approximately 10 km east of Cloverdale. Cecil Schwalbe (pers. comm.) reported a population in Cochise County just north of the San Bernardino National Wildlife Refuge. Germano et al. (1994) do not include southeastern Arizona or southwestern New Mexico in their distribution map and discount all records from that area as being erroneous or resulting from introductions. Although numerous sightings exist from Hidalgo County and adjacent Cochise County, Arizona, the presence of a breeding population of *G. agassizii* in New Mexico is doubtful, yet uncertain. Specimens such as MSB 7619 from Santa Fe County are considered released or escaped captive animals. Fritts and Jennings (1994) discuss the distribution of *G. agassizii* in Mexico.

Malayemys subtrijuga (Schlegel and Müller, 1844)
Malayan Snail-eating Turtle

Price and Johnson (1978a) reported *Malayemys subtrijuga* from Elephant Butte, Sierra County. The species is native to Thailand and southern Vietnam south through the base of the Malay Peninsula to Sumatra and Java (King and Burke, 1986; Iverson, 1992) and is commonly seen in the pet trade. The specimen (NMSU 4241) reported by Price and Johnson (1978a) represented an escaped or released pet and the species is not considered part of the herpetofauna of New Mexico.

Terrapene carolina (Linnaeus, 1758)
Eastern Box Turtle

Although there are no literature reports of *T. carolina* from New Mexico, individuals are often found in or near metropolitan areas statewide. These occurrences are due to escaped or released backyard pets. *Terrapene carolina* is not considered part of the native herpetofauna of New Mexico.

Ophisaurus ventralis (Linnaeus, 1766)
Eastern Glass Lizard

Yarrow (1882) and Cope (1898 [1900]) recorded *Ophisaurus ventralis* from "between Arkansas River and Cimarron River, N.M." This record (USNM 5129) was included by Stejneger and Barbour (1917, 1923, 1933, 1939, 1943) in the first-fifth editions of the Checklist of North American Amphibians and Reptiles, although Schmidt (1953), removed it from the sixth edition. Van Denburgh (1924) suggested that the specimen was probably taken in Colorado. This is an eastern species and is not considered part of the herpetofauna of New Mexico or Colorado.

Sceloporus grammicus Wiegmann, 1828
Mesquite Lizard

Sceloporus grammicus has been collected in Sonora, México approximately 4.8–8 km south of the New Mexico border in the Sierra San Luis (UTACV-R 17365–68). Other specimens from Sonora (MSB 31246–53) have been brought to our attention by Hobart M. Smith (pers. comm.). *Sceloporus grammicus* may occur in the foothills of the northern Sierra San Luis or in the southern Animas Mountains of south-central Hidalgo County.

Sceloporus olivaceus Smith, 1934
Texas Spiny Lizard

Kennedy (1973), in his review of this species, mapped (with query) a specimen from the Capitan Mountains, Lincoln County. That specimen, USNM 44905, had earlier been referred to by Van Denburgh (1924) and Smith (1939). Degenhardt (*In* Kennedy, 1973) doubted the existence of the species in New Mexico. It is not considered part of the herpetofauna of New Mexico. USNM 44905 has been re-identified as *Sceloporus undulatus*.

Drymarchon corais melanurus Duméril, Bibron, and Duméril, 1854
Indigo Snake

Drymarchon corais melanurus was reported from New Mexico by Bailey (1928) who referred to it as the Mexican black snake. He included it in a brief listing of the nonpoisonous snakes from the Carlsbad Cave region without providing specimen records or the source for the record. We do not include this species in the herpetofauna of New Mexico.

Heterodon platirhinos Latreille, 1801
Eastern Hognose Snake

Heterodon platirhinos may occur in the eastern plains of New Mexico along the Canadian River drainage. Axtell (1983) reported a specimen (RWA 5931) that was "collected from a shady cottonwood grove on the sandy floodplain of Punta de Agua Creek in southwestern Hartley County, [Texas]." This locality is adjacent to New Mexico in the Texas Panhandle and is approximately 10–12 km E of the NM/TX state line.

Leptodeira septentrionalis (Kennicott *in* Baird, 1859)
Cat-eyed Snake

Noguchi (1909) erroneously included this species in the herpetofauna of New Mexico under the name *Sibon septentrionale* without providing any supporting data. This tropical species is not considered a part of the herpetofauna of New Mexico.

Opheodrys aestivus (Linnaeus, 1766)
Rough Green Snake

The rough green snake, under the name *Cyclophis aestivus*, has been recorded in New Mexico from the Cimarron River, Fort Bliss, and Old Fort Cobb (Cope, 1898 [1900]; Van Denburgh, 1924), although it is doubtful that these specimens are actually from New Mexico. Until evidence indicates otherwise, *Opheodrys aestivus* is not considered a part of the New Mexico herpetofauna.

Phyllorhynchus browni Stejneger, 1890
Pima Leafnose Snake

The inclusion of this species as a member of the herpetofauna of New Mexico by Wright and Wright (1952, 1957) is likely based on the misidentification of a specimen of *Gyalopion canum*. Medden (1927) reported a small snake collected in Hidalgo County at Granite Gap, a locality approximately 30.7 km north of Rodeo. The specimen was identified as *P. browni* based on the truncated snout and a figure in Cope (1898 [1900]). *Phyllorhynchus browni* is not considered to be a member of the herpetofauna of New Mexico.

Tantilla wilcoxi Stejneger, 1902 [1903]
Chihuahuan Black-headed Snake

Tantilla wilcoxi has been collected in Sonora, Mexico, approximately 4.8–8 km south of the New Mexico border in the Sierra San Luis. (UTACV-R 17805) It is expected to occur in the foothills of the northern Sierra San Luis or in the southern Animas or Peloncillo mountains of southern Hidalgo County.

Crotalus cerastes Hallowell, 1854
Sidewinder

Yarrow (1882) listed *Crotalus cerastes* from Fort Buchanan, New Mexico. This specimen, USNM 5022, was received from Dr. B. J. D. Irwin; no other information is provided. It is likely that it was taken in Arizona prior to the separation of New Mexico and Arizona with the division of the Territory of New Mexico. The sidewinder is not considered part of the herpetofauna of New Mexico although many individuals erroneously believe it occurs throughout the desert regions of southern New Mexico.

Crotalus pricei Van Denburgh, 1895
Twin-spotted Rattlesnake

Van Denburgh (1924) recorded this species as occurring at Steeple Rock, Grant County, New Mexico. Klauber (1934) reidentified the specimen, USNM 52273, as *Crotalus viridis* and stated that if *C. pricei* does occur in New Mexico it should be expected in the Animas or Peloncillo mountains. The population nearest to New Mexico occurs at about 2592 m elevation in the Chiricahua Mountains of Cochise County, Arizona. It is doubtful that this high-elevation species occurs here and it is not considered a part of the herpetofauna of New Mexico.

Crotalus tigris Kennicott *in* Baird, 1859
Tiger Rattlesnake

Cope (1898 [1900]) illustrated scale characteristics of USNM 471 from Sierra Verde, New Mexico. He stated that the locality is "situated on the boundary between Arizona and Sonora, nearly due S. of Baboquivari Peak, and about 50 miles northwest of Nogales." This locality lies well within the range of *C. tigris*. The specimen was taken in Arizona prior to the separation of New Mexico and Arizona with the division of the Territory of New Mexico. Currently the species is not considered part of the herpetofauna of New Mexico, although the locality of a specimen, MSB 56030, taken in Cochise County, Arizona, on 11 August 1993 by Chris Milensky is only 5.6 km west of the NM/AZ border (Painter and Milensky, 1993). During July 1994, Phil Rosen (pers. comm.) collected a specimen (UAZ 50299-PSV) 0.7 km W of the NM/AZ border in Guadalupe Canyon. It is reasonable to expect this species will be discovered in the rocky, desert scrub or canyon habitat in extreme southwestern Hidalgo County.

Wright and Wright (1952) included two species of snakes that are not considered by us to be part of the New Mexico herpetofauna: *Elaphe bairdi* and *Natrix* (= *Nerodia*) *rhombifera*. They provide no information on the source of these records.

LIST OF SCIENTIFIC AND COMMON PLANT NAMES USED

Family **Equisetaceae**
Equisetum spp.
Horsetail

Family **Ephedraceae**
Ephedra spp.
Ephedra, Joint fir, Mormon tea
Ephedra torreyana
Torrey joint fir
Ephedra trifurca
Longleaf ephedra

Family **Pinaceae**
Abies concolor
White fir
Abies lasiocarpa
Subalpine or Corkbark fir
Picea engelmannii
Engelmann spruce
Picea pungens
Blue spruce
Pinus edulis
Rocky Mountain piñon pine
Pinus engelmannii
Apache pine
Pinus flexilis
Limber pine
Pinus leiophylla
Chihuahua pine
Pinus ponderosa
Ponderosa pine
Pinus strobiformis
Southwestern white pine
Pseudotsuga menziesii
Douglas-fir

Family **Cupressaceae**
Cupressus arizonica
Arizona cypress

Juniperus spp.
Junipers, "cedars"
Juniperus deppeana
Alligator juniper
Juniperus erythrocarpa
Redberry juniper
Juniperus monosperma
One-seeded juniper
Juniperus osteosperma
Utah juniper
Juniperus scopulorum
Rocky Mountain juniper

Family **Gramineae**
Agropyron smithii
Western wheatgrass
Aristida spp.
Threeawn grass
Aristida adscensionis
Sixweeks threeawn
Aristida fendleriana
Fendler threeawn
Aristida longiseta
Red threeawn
Aristida ternipes
Spidergrass
Blepharoneuron tricholepis
Pine dropseed
Bothriochloa barbinodis
Cane bluestem
Bouteloua spp.
Grama grasses
Bouteloua barbata
Six-weeks grama
Bouteloua curtipendula
Sideoats grama

Bouteloua gracilis
Blue grama
Bouteloua hirsuta
Hairy grama
Bromus frondosus
Weeping brome
Bromus tectorum
Downy chess
Buchloe dactyloides
Buffalo grass
Cenchrus pauciflorus
Field sandbur
Danthonia spp.
Oatgrass
Deschampsia caespitosa
Tufted hairgrass
Distichlis stricta
Desert saltgrass
Erioneuron (= *Tridens*) *pulchellum*
Fluff grass
Festuca spp.
Fescue
Hilaria spp.
Galleta
Hilaria jamesii
Galleta
Hilaria mutica
Tobosa grass
Koeleria cristata
Junegrass
Muhlenbergia spp.
Muhly
Muhlenbergia torreyi
Ring muhly
Muhlenbergia virescens
Screwleaf muhly
Munroa squarrosa
False buffalograss

Oryzopsis hymenoides
Indian ricegrass
Panicum hallii
Hall's panic grass
Panicum obtusum
Vine mesquite
Poa spp.
Bluegrass
Scleropogon brevifolius
Burro grass
Setaria leucopila
Bristlegrass
Sporobolus spp.
Dropseed
Sporobolus airoides
Alkali sacaton
Sporobolus cryptandrus
Sand dropseed
Stipa spp.
Needlegrass

Family **Cyperaceae**
Carex spp.
Sedges
Cyperus spp.
Flatsedge
Eleocharis spp.
Spikerush
Scirpus spp.
Bulrush

Family **Juncaceae**
Juncus spp.
Rushes

Family **Liliaceae**
Dasylirion spp.
Sotol
Yucca baccata
Banana yucca
Yucca elata
Soaptree yucca

LIST OF PLANT NAMES

    Yucca glauca
    Small soapweed

Family   **Salicaceae**
    Populus spp.
    Cottonwood; Aspen
    Populus tremuloides
    Aspen

Family   **Juglandaceae**
    Juglans spp.
    Walnuts

Family   **Fagaceae**
    Quercus spp.
    Oaks
    Quercus arizonica
    Arizona white oak
    Quercus gambelii
    Gambel oak
    Quercus havardii
    Shinoak, Shinnery oak

Family   **Ulmaceae**
    Celtis pallida
    Desert hackberry
    Celtis reticulata
    Netleaf hackberry

Family   **Moraceae**
    Morus microphylla
    Texas mulberry

Family   **Chenopodiaceae**
    Atriplex spp.
    Saltbush
    Atriplex canescens
    Fourwing saltbush
    Artiplex confertifolia
    Shadscale
    Salsola kali
    Russian thistle
    Sarcobatus vermiculatus
    Greasewood

Family   **Platanaceae**
    Platanus wrightii
    Plane-tree, Sycamore

Family   **Rosaceae**
    Cercocarpus montanus
    Mountain mahogany
    Fallugia paradoxa
    Apache plume

Family   **Leguminosae**
    Acacia spp.
    Acacia
    Cercidium spp.
    Paloverde
    Krameria spp.
    Ratany, Krameria
    Mimosa spp.
    Mimosa, Catclaw
    Prosopis spp.
    Mesquite, Screwbean
    Prosopis glandulosa
    Mesquite
    Robinia neomexicana
    New Mexico locust
    Trifolium spp.
    Clover

Family   **Zygophyllaceae**
    Larrea tridentata
    Creosotebush

Family   **Euphorbiaceae**
    Jatropha dioica
    Rubber plant

Family   **Anacardiaceae**
    Rhus spp.
    Sumac
    Rhus microphylla
    Littleleaf sumac
    Rhus trilobata
    Skunkbush

Family   **Rhamnaceae**
    Condalia spp.
    Condalia

Family   **Tamaricaceae**
    Tamarix spp.
    Tamarisk; Saltcedar

Family   **Koeberliniaceae**
    Koeberlinia spinosa
    Allthorn

Family   **Cactaceae**
    Cereus giganteus
    Saguaro
    Opuntia spp.
    Prickly pear; cholla
    Opuntia leptocaulis
    Christmas cactus, tasajillo

Family   **Elaeagnaceae**
    Elaeagnus angustifolia
    Russian olive

Family   **Garryaceae**
    Garrya wrightii
    Wright silktassel

Family   **Ericaceae**
    Arbutus xalapensis
    Arizona madrone
    Arctostaphylos pungens
    Pointleaf manzanita

Family   **Fouquieriaceae**
    Fouquieria splendens
    Ocotillo, Devil's walkingstick

Family   **Labiatae**
    Monarda fistulosa
    Wild bergamot
    Poliomintha incana
    Hoary rosemary mint

Family   **Solanaceae**
    Lycium berlandieri
    Wolfberry

Family   **Bignoniaceae**
    Chilopsis linearis
    Desert willow

Family   **Compositae**
    Artemisia spp.
    Sagebrush
    Artemisia filifolia
    Sand sagebrush
    Artemisia tridentata
    Big sagebrush
    Ambrosia psilotachya
    Western ragweed
    Brickellia spp.
    Brickelbush
    Chrysothamnus nauseosus
    Rubber rabbitbrush
    Chrysothamnus pulchellus
    Southwest rabbitbrush
    Encelia farinosa
    Encelia
    Flourensia cernua
    Tarbush
    Franseria spp.
    Bursage
    Guiterrezia sarothrae
    Broom snakeweed
    Haplopappus laricifolius
    Turpentine bush
    Helianthus spp.
    Sunflowers

LIST OF MUSEUM SYMBOLIC CODES

AMNH—American Museum of Natural History
ANSP—Academy of Natural Sciences, Philadelphia
ASNHC—Angelo State University
ASU—Arizona State University
ASUC—Appalachian State University
ASUMZ—Arkansas State University Museum of Zoology
AUM—Auburn University Museum
BMNH—British Museum of Natural History
BS/FC—United States Fish & Wildlife Service
BWMC—Avila College
BYU—Brigham Young University
CA—Chicago Academy of Sciences
CACA—Carlsbad Caverns National Park
CAS—California Academy of Sciences
CM—Carnegie Museum
CU—Cornell University
DMNH—Dallas Museum of Natural History
EAL—Ernest A. Liner Private Collection
ENMU—Eastern New Mexico University
FMNH—Field Museum of Natural History
HSU—Humbolt State University
INHS—Illinois Natural History Survey
ISU—Indiana State University
JBI—John B. Iverson Private Collection
KU—University of Kansas Museum of Natural History
LACM—Los Angeles County Museum
LSUMZ—Louisiana State University Museum of Zoology
LSUS—Louisiana State University, Shreveport
LTU—Louisiana Tech University

MCZ—Museum of Comparative Zoology, Harvard University
MHP—Fort Hays State University
MNHN—Museum National d'Historie Naturelle, Paris
MPM—Milwaukee Public Museum
MSB—Museum of Southwestern Biology, University of New Mexico
MSUM—Michigan State University Museum
MSUMZ—Memphis State University Museum of Zoology
MVZ—Museum of Vertebrate Zoology, University California, Berkeley
MZUS—Universite de Strasbourg Musee de Zoologie
NCSM—North Carolina State Museum
NLU—Northeast Louisiana University
NMSU—New Mexico State University
NRM—Nuturhistoriska Riksmuseet
OMNH—University of Oklahoma
OS—Ottys Sanders Private Collection
OSUS—Oklahoma State University, Stillwater
OUVC—Ohio University Vertebrate Collection
PSM—University of Puget Sound Museum
RWA—Ralph W. Axtell Private Collection
SAM—Sherman A. Minton Private Collection
SDSNH—San Diego Natural History Museum
SIUC—Southern Illinois University, Carbondale
SJSU—San Jose State University
SMBU—Strecker Museum, Baylor University
SMF—Senckenberg Museum, Frankfurt
SREL—Savannah River Ecology Laboratory

SRSU—Sul Ross State University
TAIC—Texas A & I University
TCU—Texas Christian University
TCWC—Texas A & M University, Texas Cooperative Wildlife Coll.
TNHC—Texas Natural History Collection, Texas Memorial Museum
TTU—Texas Tech University Museum
TU—Tulane University
UADZ—University of Arkansas
UAZ—University of Arizona
UCC—University of Cincinnati
UCM—University of Colorado Museum of Natural History
UCS—University of Connecticut, Storrs
UCSB—University of California, Santa Barbara
UF-FSU—University of Florida, Florida State Museum
UIMNH—University of Illinois, Urbana-Champaign
UMMZ—University of Michigan Museum of Zoology
UMOC—University of Missouri Museum of Zoology
UN—University of Nebraska, Lincoln
UNCA—University of North Carolina, Asheville
USNM—National Museum of Natural History
UTA—University of Texas, Arlington
UTEP—University of Texas, El Paso
UWZM—University of Wisconsin, Madison
WNMU—Western New Mexico University
WTSU—West Texas A & M University
YPM—Peabody Museum, Yale University

GLOSSARY

A-1, A-2—abbreviations for first and second anterior labial tooth rows in tadpoles

A-2 gap—medial break in labial tooth row A-2

A-2 gap ratio—length of one section of row A-2/width of gap between sections regardless of which is larger; a number of 1 or larger indicates a gap that is equal to or shorter than the lateral sections

adolescent—not yet sexually mature

adpressed—in reference to the limbs, laid full length against the side of the body, with the hind legs pointing anteriorly and the forelegs pointing posteriorly

allopatric—occupying a geographic range separated from that of another population

alloploid—an organism produced from the combination of two or more sets of chromosomes from different species

alluvial—consisting of any material that has been carried or deposited by running water

amniotic sac—the membranous sac containing the developing embryo and its surrounding fluid

amplexus—the sexual embrace of frogs and toads; generally the male clasps the female from above with his forelimbs while fertilizing the eggs

anaerobic metabolism—the metabolic processes used by cells to produce energy without using oxygen

anal plate—the scale that covers the vent; usually referred to as single or divided

annelid—any of the members of the phylum Annelida, such as the earthworms and leeches

anterior—at or toward the head end of the body

anthropogenic—influenced or caused by the activities of humans

anus—the posterior end of the digestive tract through which waste is expelled

Arachnida—a class of Arthropoda allied to the insects characterized by eight legs, no wings or antennae, and two body regions and including spiders, ticks, mites, and scorpions

arboreal—pertaining to animals that live primarily in trees

arthropod—any member of the phylum Arthropoda, characterized by segmented bodies and jointed legs and including insects, arachnids, and crustaceans

atretic follicle—follicles that decline and disappear after only partly developing

auricular scales—scales bordering the ear opening anteriorly and projecting posteriorly over it

bajada—a gently sloping alluvial fan extending downward from the base of a mountain range into a basin

barbel—a short, fleshy appendage on the chin or throat of a turtle usually with a sensory function

basking—to rest in the direct rays of the sun, generally as thermoregulatory behavior or to help rid the body of parasites

beak—the horny mouthparts of a tadpole

bimodal activity period—a daily or seasonal time period in which there are two distinct peaks of activity divided by an interval in which activity is reduced or absent

biserial—occurring in two rows

boss—in spadefoots, a glandular or bony raised area on the midline of the head between the eyes

bridge—in turtles, the bony connection between the carapace and the plastron

canthus rostralis—the angle of the head from the tip of the snout to the anterior corner of the eye (in amphibians) or to the anterior corner of the eyebrow (in reptiles), which separates the dorsum of the head from the side of the snout

carapace—the top half of the shell of a turtle

carnivorous—a tendency to feed exclusively on animals and not plants

coelomic fat mass—a fat mass stored in the body cavity which is an important source of stored energy for reproductive growth

chinshields—in lizards, a paired series of enlarged scales on the chin diverging posteriorly from the mental or postmental scale; in snakes, one or two median pairs of elongate scales on the ventral surface of the head, posterior to the mental or to the first infralabial scales that may be in contact on the midventral line

chorus—an aggregation of singing male anurans

CL—carapace length; measured as a straight line from the anterior to the posterior edge of the carapace

cloaca—the chamber into which products from the digestive, reproductive, and urinary systems are passed, and which forms the common opening of these systems to the exterior of the body.

clone—genetically identical organisms descended from the same single parent by asexual processes

Coleoptera—an order of insects that includes the beetles and weevils; the largest order in the Animal Kingdom

commensalism—the relationship between two species in which one species benefits and the other is neither benefitted nor harmed

congener—a taxon that is a member of the same genus as another

coniferous—referring to the conifers, including pine, spruce, and fir

costal fold—the area between the costal grooves of salamanders

costal groove—vertical groove on the side of a salamander indicating position of a rib

cranial crests—the raised ridges on the heads of toads

crepuscular—active primarily at twilight and/or dawn

critical thermal maximum—the lowest temperature at which an organism, when heated, loses its ability to escape conditions that will result in death

critical thermal minimum—the highest temperature at which an organism, when cooled, loses its ability to escape conditions that will result in death

Crustacea—the class of arthropods consisting mainly of aquatic species such as crawfish and shrimp, also including terrestrial pillbugs

cycloid—an overlapping, rounded or oval-shaped scale with smooth edges

deciduous—falling off or shedding at a particular stage of growth

detritus—organic matter produced by the breakdown or decay of plants

dewlap—a vertical pendulous fold of skin on the throat of lizards

diploid—containing the normal complement of a single set of chromosomes from each parent in a body cell

disclimax—ecological succession that is maintained below climax for an extended time, usually because of human or animal disturbance or ecological effects

disjunct population—a population geographically separated from the main body of a species' range

diurnal—active primarily during daylight hours

dorsal—of or pertaining to the upper surface of the body

dorsum—the upper surface of the body

dorsolateral—pertaining to the area along the side of the back

ecotone—the transition zone between two distinct habitats

ectotherm—an animal whose body temperature is largely controlled by external environmental factors and that generates little or no internal body heat through its metabolism

edaphic—relating to the soil

egg tooth—deciduous tooth on the end of the premaxillary bone of squamates, or an analogous horny epidermal structure in other reptiles, which is used to break through the egg at hatching, and is lost shortly thereafter

electrophoretic analysis—a technique used to separate charged molecules by causing them to move in response to an electric field in proportion to their size; this procedure is often used in systematics to detect basic molecular differences between organisms

emarginate—refers to an actual indentation, not a fold, in the margin (lateral in the cases at hand) of the oral disc; disc may have to be opened in preserved specimens to see this emargination

endemic—restricted to a specific geographical region

endotherm—an organism able to regulate body temperature by means of internal, intrinsic sources of heat

ephemeral—existing for only a short time

epididymis—an elongate, coiled structure along the posterior surfaces of the testis that provides for the storage, transmission, and maturation of spermatozoa

erythrocyte—blood cells containing hemoglobin and specialized to carry oxygen

estivate—to spend a period of the warm season in a state of physiological inactivity

etymology—the origin and development of a word

eyes (dorsal)—eyes not included in a dorsal silhouette

eyes (lateral)—eyes included in a dorsal silhouette

fat body—in amphibians, a mass of fatty tissue attached to each genital gland; in reptiles, a distinct structure located in the urogenital region of the body cavity; an energy reserve mobilized for reproductive activities or maintenance during hibernation

femoral pores—a series of small openings in the scales (one per scale) on the underside of the thigh in some species of lizards

follicle—the structure that surrounds and nourishes each egg in the ovary of reptiles prior to ovulation

fossorial—living underground

friable—soils easily crumbled between the fingers

gene flow—movement of genes from one population of a species to another by interbreeding

gestation—the process of carrying and maintaining the young in the uterus from fertilization until birth

gonochoristic—from the Greek *gonos*, "procreation" and *chorizein*, "to separate"; the occurrence of two separate sexes in a species

gracile—gracefully slender

granular scales—tiny, rounded, smooth scales characteristic of certain lizards

gravid—pregnant or containing a mass of enlarged eggs almost ready to lay

gular fold—a well developed fold of skin across the posterior portion of the throat; seen in certain lizards and salamanders

haploid—the condition whereby certain cells of an organism are reduced to a single set of unpaired chromosomes

hemipenes—the paired, eversible, copulatory organ of male lizards and snakes

Hemiptera—the insect order containing the true bugs characterized by sucking mouthparts and including the stinkbugs and squash bugs

herbivorous—feeding primarily on plants

herpetology—the study of amphibians and reptiles

heterogeneous—dissimilar

hibernate—to spend the winter in a state of decreased physiological inactivity

hinge—a flexible joint between the different sections of the lower shell (plastron) in some turtles

holotype—the single specimen upon which the description of a new species or subspecies is based; the type-specimen

home range—the area in which the normal daily activities of an individual organism occurs

Homoptera—the insect order containing the sucking insects, characterized by membranous wings and including the cicadas, leafhoppers, aphids, and scale insects

hybrid—an individual resulting from a cross between two different species

Hymenoptera—the insect order containing the social insects, characterized by a clearly differentiated head, thorax, and abdomen and including the bees, wasps, and ants

infraspecific—the taxa below the rank of species

insectivore—feeding primarily on insects

intergradation—interbreeding of two closely related subspecies whose ranges overlap

interlabial—small scales that separate the chin shields from the infralabials in some lizards, notably whiptails

internasal—one or two pairs of scales on the median dorsal surface of the snout between the nasals

interoccipital spine—a small spine at the posterior medial margin of the headshield in horned lizards

intraspecific—referring to relationships between members of the same species

Isoptera—the insect order that contains the termites

jaw sheath—keratinized covering on jaws, usually having serrated edges for cutting material

juvenile (adj.—juvenal)—newly born or very young individual

keel—an elevated, longitudinal straight line, either sharp and well defined or broad and obtuse

labial teeth—keratinized teeth (not true teeth as in adult) arranged in transverse rows on the upper and lower labia of tadpoles

labile—liable to change; unstable

lamellae—in lizards, the transverse plates on the undersides of the digits

larvae—the immature form of an amphibian, usually with fins and gills

lateral—referring to the side

lateral fringe scales—the scales along the lateral margins of the body in horned lizards where the dorsal and ventral surfaces meet

lectotype—a specimen selected from the type series (syntypes) to become the holotype because the original author did not designate one; the remaining syntypes then become paralectotypes

lentic—refers to standing water

Lepidoptera—an insect order of scaly-winged insects characterized by a coiled sucking proboscis and including the butterflies, moths, and skippers

lineage—a sequence of ancestor-descendant populations with its own evolutionary tendencies and historical fate

lotic—refers to running water

lower labium—lower half of the oral disc in tadpoles

LTRF—(labial tooth row formula) the number of rows of teeth on the upper labium and lower labium written as a fraction, e.g. 2/3, 2/2, etc

malpais—the topography associated with an historic lava flow

marginal papillae—fleshy papillae on the margins of the oral disc

marginal scutes—in turtles, the scutes bordering the carapace

melanistic—black in color

melanophore—a cell containing black or dark pigment

mental scale—the azygous scale at the anterior edge of the lower jaw in lizards and snakes

meristic—pertaining to a change in the number of parts of an organism

mesic—a wet habitat, having only moderate moisture

mesoptychium—the middle and posterior throat area in lizards that is delimited by the pregular and gular folds

metamorphosis—a process whereby an amphibian undergoes a drastic change in body form and structure in transforming from the larval to the adult stage

metatarsal tubercle—a hardened, spadelike projection near the heel of a toad or spadefoot used in burrowing

midden—a nest of sticks and debris constructed by packrats of the genus *Neotoma*

mimic—to take on the appearance of another species in order to deceive a predator or prey organism

mitochondrial DNA (mtDNA)—the DNA of the mitochondria

monogamous—a condition when a male and female organism are paired one to one for at least one reproductive season

monotypic genus—a genus that contains only a single species

mucronate—terminated abruptly by a sharp terminal tip

mucus membrane—surfaces typically lining the oral or nasal cavities permanently kept moist by a layer of mucus

mutualism—a relationship between two species in which both benefit

neonate—a newborn

neotype—a specimen selected as type subsequent to the original description in cases where the primary types are known to be destroyed, lost, or unusable

nocturnal—being active primarily at night

nomenclature—a system of formally applying names to and classifying organisms

nominotypical—a taxon that contains the type specimen of, or bears the same name as, a subdivided higher taxon

nymph—the immature stage of various arthropods, usually similar to the adult state but may lack wings, some appendages, and mature genitalia

occipital spine—in horned lizards, the spines along the rear margin of the head shield on either side of the interoccipital spine

ocelli—eye-like markings or spots

olfaction—the sense of smell

omnivorous—a tendency to feed on both plants and animals

ontogenetic change—referring to changes in morphology, physiology, or other characteristics of an individual organism that occur with increasing growth and maturation

ontogeny—the developmental history of an individual organism

oogenesis—the growth and maturation of the ovum

oral disc—fleshy disc around the mouth with papillae in various patterns around the margin

oral flaps—fleshy hemispherical flaps that hang pendant over the mouth in microhylid tadpoles

oral papillae—in tadpoles, the small nipplelike projections that often form a fringe around the oral disc

Orthroptera—a insect order including the grasshoppers and crickets and characterized by chewing mouthparts and compound eyes

oviductal eggs—eggs that are positioned in the oviduct (the tube that leads from the ovary to the uterus)

oviparous—laying eggs that develop outside of the body

ovoviviparous—retaining eggs within the body almost until the time of hatching

oviposit—to lay eggs

ovulation—the discharge of an ovum from the ovary

panmixia—random mating between individuals within a population

paralectotype—a specimen or series of specimens in the syntypic series that are not the lectotype

parapatric—a population having a geographical range abutting that of another population

paratype—any specimen in the type series, exclusive of the holotype

PBT—see preferred body temperture

P-1, P-2, and P-3—abbreviations for first, second, and third posterior labial tooth rows in tadpoles

parotoid gland—in toads, a gland appearing as a swollen area on either side of the neck, above the level of the tympanum

parthenogenetic—from the Greek words which translate literally to "virgin birth"; refers to the species of all-female whiptail lizards that are able to reproduce without any involvement of males

phalanx—one of the finger or toe bones

photoperiod—the relative portion of light and darkness on a 24-hour basis

phylogenetic—the evolutionary relationships within and between groups of organisms

phylogeny—the evolutionary history of an individual taxon

physiography—the study of the natural features of the earth's surface, especially in its current aspects, including land formations, climate, and currents

pit vipers—any of the venomous snakes of the subfamily Crotalinae characterized by a heat-seeking organ between the eye and the nostril including the rattlesnakes, copperheads and cottonmouths

PL (plastron length)—the median length of the plastron along a straight line; occasionally this measurement is taken along the curved length of the plastron

plastron—the lower half of the shell of a turtle

polytypic—containing two or more taxa of potentially equal or subordinate rank

population—a group of organisms of the same species living in a specific geographic area

postantebrachial scales—in whiptail lizards, scales covering the posterior surface of the forearm

posterior—at or toward the rear or tail end of the body

postmental scale—the scale in certain lizards that lies in the midline directly behind the mental

postrictal scale—in horned lizards, the posteriormost scale in the postlabial series, situated near the anteroventral border of the ear opening

preanal scale—a scale situated in the pelvic region anterior to the anus

preferred body temperature—the temperature at which amphibians and reptiles will maintain themselves given the opportunity to select proper substrate, exposure to sunlight, and other thermal factors

pyramidal scales—conical scales projecting vertically from the dorsal body surface of horned lizards

recrudescence—the renewal of physiological activity in sex organs following a latent period

regenerated—the portion of new tissue or organ growth replacing that lost or damaged by injury

release call—a call, typically inaudible, given by a male anuran when clasped by another male during a breeding chorus

relict population—a group of organisms representing the surviving remnants of a population formerly more widespread in a given area

reticulated—having a network-like pattern

riparian—generally referring to the bank or floodplain of a river or stream

rostral—the scale forming the tip of the snout

saxicolous—living on or among rocks or boulders

scalation—referring to the size and arrangement of scales on reptiles

scute—a term usually applied to relatively large scales

seminiferous tubules—the convoluted tubules in the testis that are the sites of sperm production

sexual dimorphism—the condition of diagnostic morphological differences existing between sexes within a species

species—a group of organisms with a common ancestor that are on a unique phylogenetic trajectory

spermatogenesis—the process by which undifferentiated male germ cells develop into mature spermatozoa

spermatophore—a gelatinous mass, sometimes in the form of a stalklike structure consisting of a cone of jelly with a sperm cap on top, produced from the cloacal glands of male salamanders, that is used to hold a mass of spermatozoa

spermatozoa—mature sperm cells

spermiogenesis—the transformation of spermatids into spermatozoa

spiracle—an opening positioned at various places (left lateral side [sinistral], midventrally) where water exits after being passed over the gills and filtering system of tadpoles

spiracular tube—a tubelike projection from the side of a tadpole's head that contains the spiracle at its free end

Squamata—the order of reptiles containing the amphisbaenids, lizards, and snakes

subcaudals—the scales that cover the underside of the tail in snakes

submarginal papillae—fleshy papillae anywhere on the face of the oral disc except on the margins

subspecies—the taxonomic rank immediately below the species level; indicates a group of organisms geographically distinct from but able to interbreed with other populations of the same species where their ranges overlap

substrate—a term encompassing the biologically relevant characteristics of the ground upon which an organism lives

superciliary—any one of the series of scales lying along the outer edge of the supraoculars in lizards

superciliary spine—an elongate, pointed scale lying in the superciliary series of scales; in *Phrynosoma*, found at the posterior end of this series

supraocular scale—either of a pair of shields that lie above the eyes in snakes; the scales lying on the dorsum of the orbit in lizards

SVL—snout-vent length: in snakes and lizards, the distance between the tip of the snout to the posterior end of the scale covering the vent; in salamanders, the distance is measured to the posterior end of the elongated vent opening

sympatric—occupying the same geographic range, or having broadly overlapping ranges

syntopic—occupying the same specific habitat or microhabitat

syntype—any member of a type series for which no holotype was designated

tail autotomy—spontaneous or reflexive separation of the tail from the body of an organism

talus—accumulations of rock fragments below steep slopes or cliffs

taxon—any group of organisms that have been designated as belonging to a particular taxonomic rank

temporal spines—in horned lizards, the series of spines extending backwards and upwards from the posterior margin of the eye

terrestrial—living on the land

thermoregulate—to regulate the body's internal temperature by behavioral means

Thysanura—an order of wingless insects commonly called bristletails and silverfish

TL—total length; measured as a straight line from the tip of the snout to the tip of the tail

tooth row numbering—distal to proximal on upper labium; proximal to distal on lower labium; A-? —a given upper (anterior) row; P-? —a given lower (posterior) row

Tricoptera—an order of insects characterized by two pairs of hairy, scaled wings and long antennae and legs and including the caddisflies

triploid—an organism having three haploid chromosome sets in each body cell

tubercle—a small rounded protrusion

tympanum—the external ear drum in amphibians

type locality—the site at which the specimen that is used in the original description of a new species or subspecies was collected

type specimen—the specimen upon which the original name of a species or subspecies is based; cf. holotype, lectotype, neotype, etc.

uniserial—occurring in one row

upper labium—upper half of the oral disc

vascularized—richly supplied with blood vessels

vent—the external opening of the cloaca; the outlet for digestive, urinary, and reproductive tracts in amphibians and reptiles: in tadpoles, medial refers to the tube opening parallel with the midline; dextral refers to the tube opening to the right of the midline

ventral—referring to the underside

venter—the lower surface of the body

vermiculated—marked with wavy, irregular lines or streaks of color

versant—the slope of a mountain or mountain chain; hence the general slope, or declination, of a region

vestige—a degenerate or rudimentary organ, often more fully developed or functional in an earlier stage of development

vitellogenesis—the production and development of an egg yolk

xanthic—yellow or yellowish in color

xeric—dry

LITERATURE CITED

Abert, J. W. 1846. Report of the expedition led by Lieutenant Abert on the upper Arkansas and through the country of the Comanche Indians, in the fall of the year 1845. 29th Congress, 1st Sess., Sen. Exec. Doc. VIII (438):1–75.

———. 1848. Report of Lieut. J. W. Abert of his examination of New Mexico in the years 1846–1847, p. 417–546. *In* W. H. Emory, Notes of a military reconnaissance, from Fort Leavenworth, in Missouri, to San Diego, in California, including parts of the Arkansas, Del Norte, and Gila rivers. 30th Congress, 1st Sess., House of Rep. Exec. Doc. (41), Washington, D.C. 614 p. + 60 plates + 2 folded maps.

Adler, K. 1989. Herpetologists of the past, p. 5–141. *In* K. Adler (ed.), Contributions to the History of Herpetology. SSAR Contrib. Herpetol. (5). 202 p.

Adolph, S. C. 1990. Influence of behavioral thermoregulation on microhabitat use by two *Sceloporus* lizards. Ecology 71(1):315–327.

Agassiz, L. 1850. Lake Superior; its physical character, vegetation, and animals compared with those of other and similar regions. Gould, Kendall, and Lincoln, Boston. x + p. 9–428.

———. 1857. Contributions to the Natural History of the United States of America. Vol. 1. Little, Brown, and Co., Boston. 452 p.

Aird, S. D., L. J. Thirkhill, C. S. Seebart, and I. I. Kaiser. 1989. Venoms and morphology of western diamondback/mojave rattlesnake hybrids. J. Herpetol. 23(2):131–141.

Aldridge, R. D. 1975. Environmental control of spermatogenesis in the rattlesnake *Crotalus viridis*. Copeia 1975(3):493–496.

———. 1979a. Seasonal spermatogenesis in sympatric *Crotalus viridis* and *Arizona elegans* in New Mexico. J. Herpetol. 13(2):187–192.

———. 1979b. Female reproductive cycles of the snakes *Arizona elegans* and *Crotalus viridis*. Herpetologica 35(3):256–261.

Aleksiuk, M. 1976. Reptilian hibernation: evidence of adaptive strategies in *Thamnophis sirtalis parietalis*. Copeia 1976(1):170–178.

Alexander, C. E. and W. G. Whitford, 1968. Energy requirements of *Uta stansburiana*. Copeia 1968(4):678–683.

Altenbach, M. J. 1992. An annotated bibliography of the Jemez Mountains salamander *Plethodon neomexicanus*. Unpubl. Rep. New Mexico Dept. Game and Fish, Santa Fe. 24 p.

Altig, R. 1970. A key to the tadpoles of the continental United States and Canada. Herpetologica 26(2):180–207.

Anderson, C. L. 1992. Care and captive breeding of the desert glossy snake, *Arizona elegans eburnata*. New England Herpetol. Soc. Newsletter 14(1):1–3.

Anderson, J. D., D. D. Hassinger, and G. H. Dalrymple. 1971. Natural mortality of eggs and larvae of *Ambystoma t. tigrinum*. Ecology 52(6):1107–1112.

Anderson, R. A. and W. H. Karasov. 1981. Contrasts in energy intake and expenditure in sit-and-wait and widely foraging lizards. Oecologia 49(1):67–72.

Andersson, L. G. 1900. Catalogue of Linnaean type-specimens of Linnaeus's Reptilia in the Royal Museum in Stockholm. Bihang till Konglika. Vetenskaps-Akademiens. Handlingar, Stockholm. (4)26(1):1–29.

Andre, J. B. and J. A. MacMahon. 1980. Reproduction in three sympatric lizard species from west-central Utah. Great Basin Nat. 40(1):68–72.

Andrews, R. M. and B. R. Rose. 1994. Evolution of viviparity: constraints on egg retention. Physiol. Zool. 67(4):1006–1024.

Anton, T. G. 1994. Observation of predatory behavior in the regal ringneck snake (*Diadophis punctatus regalis*) under captive conditions. Bull. Chicago Herpetol. Soc. 29(5):95.

Applegarth, J. S. 1980. The ridge-nosed rattlesnake in New Mexico: A review of existing information and a search for suitable habitat on public lands. Unpubl. Rept. U.S. Bureau Land Management, Las Cruces. 140 p.

———. 1982. A survey of the softshell turtles (*Trionyx muticus* and *Trionyx spiniferus*) living in the Conchas Reservoir, San Miguel

County, New Mexico. Unpubl. Rept. U.S. Army Corps of Engineers, Albuquerque. 74 p.

Armstrong, B. L. and **J. B. Murphy.** 1979. The natural history of Mexican rattlesnakes. Univ. Kansas Mus. Nat. Hist. Spec. Publ. (5):1–88.

Arnberger, L. P. 1948. Gila monster swallows quail eggs whole. Herpetologica 4(6):209–210.

Asplund, K. K. 1964. Seasonal variation in the diet of *Urosaurus ornatus* in a riparian community. Herpetologica 20(2):91–94.

——— and **C. H. Lowe.** 1964. Reproductive cycles of the iguanid lizards *Urosaurus ornatus* and *Uta stansburiana* in southeastern Arizona. J. Morphol. 115(1):27–34.

Awbrey, F. T. 1972. "Mating call" of a *Bufo boreas* male. Copeia 1972(3):579–581.

Axtell, R. W. 1958. A monographic revision of the iguanid genus *Holbrookia*. Ph.D. Diss., Univ. Texas, Austin. viii + 222 p.

———. 1960. Orientation by *Holbrookia maculata* (Lacertilia, Iguanidae) to solar and reflected heat. Southwest. Nat. 5(1):47–48.

———. 1961a. *Eumeces epipleurotus* Cope, a revived name for the southwestern skink *Eumeces multivirgatus gaigei* Taylor. Texas J. Sci. 13(3):345–351.

———. 1961b. *Cnemidophorus inornatus*, the valid name for the little striped whiptail lizard, with the description of an annectant subspecies. Copeia 1961(2):148–158.

———. 1966. Geographic distribution of the unisexual whiptail *Cnemidophorus neomexicanus* (Sauria: Teiidae)—present and past. Herpetologica 22(4):241–253.

———. 1983. Range portrayal and reality: *Heterodon platyrhinos* distribution on the high plains of Texas and Oklahoma. J. Herpetol. 17(2):191–193.

———. 1986. *Coleonyx brevis*. *In* Interpretive Atlas of Texas Lizards (1):1–13. Privately printed. Southern Illinois University, Edwardsville.

———. 1987. *Sceloporus poinsettii*. *In* Interpretive Atlas of Texas Lizards (3):1–16. Privately printed. Southern Illinois University, Edwardsville.

———. 1988a. *Sceloporus graciosus*. *In* Interpretive Atlas of Texas Lizards (5):1–4. Privately printed. Southern Illinois University, Edwardsville.

———. 1988b. *Phrynosoma modestum*. *In* Interpretive Atlas of Texas Lizards (6):1–18. Privately printed. Southern Illinois University, Edwardsville.

———. 1994. *Cnemidophorus inornatus*. *In* Interpretive Atlas of Texas Lizards (14):1–17. Privately printed. Southern Illinois University, Edwardsville.

Bailey, B., **M. R. Terman**, and **R. Wall.** 1989. Noteworthy longevity in *Crotalus viridis viridis* (Rafinesque). Trans. Kansas Acad. Sci. 92:116–117.

Bailey, V. 1905. Biological survey of Texas. North Amer. Fauna (25):1–222.

———. 1913. Life zones and crop zones of New Mexico. North Amer. Fauna (35):1–100.

———. 1928. Animal Life of Carlsbad Caverns. Monogr. Amer. Soc. Mammal. (3):xiii + 195 p.

Baird, S. F. 1850. Revision of the North American tailed-Batrachia, with descriptions of new genera and species. J. Acad. Nat. Sci. Philadelphia 2nd Ser., Vol. 1, Pt. 4:281–294.

———. 1854. Descriptions of new genera and species of North American frogs. Proc. Acad. Nat. Sci. Philadelphia 7:59–62.

———. 1858 [1859]. Description of new genera and species of North American lizards in the Museum of the Smithsonian Institution. Proc. Acad. Nat. Sci. Philadelphia 10(16–19):253–256.

———. 1859. Reptiles of the Boundary, with notes by the naturalists of the Survey, p. 1–35. *In* William H. Emory, Report on the United States and Mexican Boundary Survey, made under the direction of the Secretary of the Interior. 34th Congress, 1st Sess., Sen. Exec. Doc. (108), Vol. II, Part II.

——— and **C. Girard.** 1851 [1852]. Reptiles, p. 336–353. *In* Howard Stansbury. Exploration and survey of the Valley of the Great Salt Lake of Utah, including a reconnaissance of a new route through the Rocky Mountains. U.S. 32nd Congress, Spec. Sess., Sen. Exec. Doc. 3. Lippincott, Grambo, and Co., Philadelphia. 487 p.

——— and ———. 1852a. Characteristics of some new reptiles in the Museum of the Smithsonian Institution. Parts I–III. Proc. Acad. Nat. Sci. Philadelphia 6:68–70, 125–129, 173.

——— and ———. 1852b. Description of new species of reptiles, collected by the U.S. Exploring Expedition under the command of Capt. Charles Wilkes, U.S.N. Part I. Proc. Acad. Nat. Sci. Philadelphia 6:174–177, 420–424.

——— and ———. 1853a. Catalogue of North American reptiles in the Museum of the Smithsonian Institution. Part 1. Serpentes. Smithsonian Institution, Washington, D.C. xvi + 172 p.

——— and ———. 1853b. Reptiles. *In* R. B. Marcy and G. B. McClellan (eds.), Exploration of the Red River of Louisiana in the year 1852. 32rd Congress, 2nd Sess., Sen. Exec. Doc. 8(54):217–244 + plates 1–11, Appendix F.

——— and ———. 1854. Reptiles. *In* R. B. Marcy and G. B. McClellan (eds.), Exploration of the Red River of Louisiana in the year 1852. 33rd Congress, 1st Sess., House of Rep. Exec. Doc.:188–215 + plates 1–11, Appendix F.

Baldridge, R. S. and **D. E. Wivagg.** 1992. Predation on imported fire ants by blind snakes. Texas J. Sci. 44(2):250–252.

Ballinger, R. E. 1973. Comparative demography of two viviparous iguanid lizards (*Sceloporus jarrovi* and *Sceloporus poinsetti*). Ecology 54(2):269–283.

———. 1974. Reproduction of the Texas horned lizard, *Phrynosoma cornutum*. Herpetologica 30(4):321–327.

———. 1976. Evolution of life history strategies: implications of recruitment in a lizard population following density manipulations. Southwest. Nat. 21(2):203–208.

———. 1977. Reproductive strategies: food availability as a source of proximal variation in a lizard. Ecology 58(3):628–635.

———. 1978. Reproduction, population structure, and effects of congeneric competition on the crevice spiny lizard, *Sceloporus poinsetti* (Iguanidae), in southwestern New Mexico. Southwest. Nat. 23(4):641–650.

———. 1979. Intraspecific variation in demography and life history of the lizard, *Sceloporus jarrovi*, along an altitudinal gradient in southeastern Arizona. Ecology 60(5):901–909.

———. 1980. Food limiting effects in populations of *Sceloporus jarrovi* (Iguanidae). Southwest. Nat. 25(4):554–557.

———. 1984. Survivorship of the lizard, *Urosaurus ornatus linearis*, in New Mexico. J. Herpetol. 18(4):480–481.

——— and **R. A. Ballinger.** 1979. Food resource utilization during periods of low and high food availability in *Sceloporus jarrovi* (Sauria: Iguanidae). Southwest. Nat. 24(2):347–363.

——— and **J. D. Congdon.** 1980. Food resource limitation of body growth rates in *Sceloporus scalaris* (Sauria: Iguanidae). Copeia 1980(4):921–923.

——— and ———. 1981. Population ecology and life history strategy of a montane lizard (*Sceloporus scalaris*) in southeastern Arizona. J. Nat. Hist. 15(2):213–222.

——— and **T. G. Hipp.** 1985. Reproduction in the collared lizard, *Crotaphytus collaris*, in west central Texas. Copeia 1985(4):976–980.

——— and **S. M. Jones.** 1985. Ecological disturbance in a sandhills prairie: impact and importance to the lizard community on Arapaho Prairie in western Nebraska. Prairie Nat. 17(2):91–100.

——— and **D. J. Ketels.** 1983. Male reproductive cycle of the lizard *Sceloporus virgatus*. J. Herpetol. 17(1):99–102.

——— and **C. O. McKinney.** 1967. Variation and polymorphism in the dorsal color pattern of *Uta stansburiana stejnegeri*. Amer. Midl. Nat. 77(2):476–483.

——— and ———. 1968. Occurrence of a patternless morph of *Cnemidophorus*. Herpetologica 24(3):264–265.

———, **M. E. Newlin,** and **S. J. Newlin.** 1977. Age-specific shift in the diet of the crevice spiny lizard, *Sceloporus poinsetti* in southwestern New Mexico. Amer. Midl. Nat. 97(2):482–484.

——— and **J. W. Nietfeldt.** 1989. Ontogenetic stages of reproductive maturity in the viviparous lizard, *Sceloporus jarrovi* (Iguanidae). J. Herpetol. 23(3):282–292.

———, ———, and **J. J. Krupa.** 1979. An experimental analysis of the role of the tail in attaining high running speed in *Cnemidophorus sexlineatus* (Reptilia: Squamata: Lacertilia). Herpetologica 35(2):114–116.

——— and **G. D. Schrank.** 1972. Reproductive potential of female whiptail lizards, *Cnemidophorus gularis gularis*. Herpetologica 28(3):217–222.

——— and **D. W. Tinkle.** 1972. Systematics and evolution of the genus *Uta* (Sauria: Iguanidae). Misc. Publ. Mus. Zool. Univ. Michigan (145):1–83.

———, **E. D. Tyler,** and **D. W. Tinkle.** 1972. Reproductive ecology of a west Texas population of the greater earless lizard, *Cophosaurus texanus*. Amer. Midl. Nat. 88(2):419–428.

Baltosser, W. H. and **T. L. Best.** 1990. Seasonal occurrence and habitat utilization by lizards in southwestern New Mexico. Southwest. Nat. 35(4):377–384.

Banicki, L. H. and **R. G. Webb.** 1982. Morphological variation of the Texas lyre snake (*Trimorphodon biscutatus vilkinsoni*) from the Franklin Mountains, West Texas. Southwest. Nat. 27(3):321–324.

Banta, B. H. 1960. Notes on the feeding of the western collared lizard, *Crotaphytus collaris baileyi* Stejneger. Wasmann J. Biol. 18(2):309–311.

———. 1971. The report of Captain Howard Stansbury and the original descriptions of some western North American amphibians and reptiles. Wasmann J. Biol. 29(2):169–184.

Barbault, R. 1977. Étude comparative des cycles journaliers d'activité des lézards *Cophosaurus texanus*, *Cnemidophorus scalaris*, *Cnemidophorus tigris* dans le désert de Mapimi (Mexique). Bull. Soc. Zool. France 102(2):159–168.

——— and **M. E. Maury.** 1981. Ecological organization of a Chihuahuan Desert lizard community. Oecologia 51(3):335–342.

———, **A. Ortega,** and **M. E. Maury.** 1985. Food partitioning and community organization in a mountain lizard guild of northern Mexico. Oecologia 65(4):550–554.

Barbour, T. 1921. A new lizard from Guaymas, Mexico. Proc. New England Zool. Club 7:79–80.

Barker, D. G. 1991. An investigation of the natural history of the New Mexico ridgenose rattlesnake, *Crotalus willardi obscurus*. Unpubl. Rept. New Mexico Dept. Game and Fish, Santa Fe. 100 p.

———. 1992. Variation, infraspecific relationships and biogeography of the ridgenose rattlesnake, *Crotalus willardi*, p. 89–105. *In* J. A. Campbell and E. D. Brodie, Jr. (eds.), Biology of the Pitvipers. Selva, Tyler, Texas. xi + 467 p.

Bauer, A. M. 1992. Lizards, p. 126–173. *In* H. G. Cogger and R. G. Zweifel (eds.), Reptiles and Amphibians. Smithmark Publ., Inc., New York. 240 p.

Baur, B. E. 1986. Longevity of horned lizards of the genus *Phrynosoma*. Bull. Maryland Herpetol. Soc. 22(3):149–151.

Baxter, G. T. and M. D. Stone. 1980. Amphibians and Reptiles of Wyoming. Bull. 16, Wyoming Game and Fish Dept., Cheyenne. 137 p.

Beavers, R. A. 1976. Food habits of the western diamondback rattlesnake, *Crotalus atrox*, in Texas. Southwest. Nat. 20(4):503–515.

Bechtel, H. B. and E. Bechtel. 1958. Reproduction in captive corn snakes, *Elaphe guttata guttata*. Copeia 1958(2):148–149.

Beck, D. D. 1990. Ecology and behavior of the Gila monster in southwestern Utah. J. Herpetol. 24(1):54–68.

———. 1991. Physiological and behavioral consequences of reptilian life in the slow lane: Ecology of beaded lizards and rattlesnakes. Ph.D. Diss., Univ. Arizona, Tucson. 181 p.

———. 1995. Ecology and energetics of three sympatric rattlesnake species in the Sonoran Desert. J. Herpetol. 29(2):211–223.

——— and C. H. Lowe, Jr. 1994. Resting metabolism of helodermatid lizards: allometric and ecological considerations. J. Comp. Physiol. B 164(2):124–129.

Behler, J. L. and F. W. King. 1979. The Audubon Society Field Guide to North American Reptiles and Amphibians. Alfred A. Knopf, New York. 743 p.

Belfit, S. C. and V. F. Belfit. 1985. Notes on the ecology of a population of *Eumeces obsoletus* (Scincidae) in New Mexico. Southwest. Nat. 30(4):612–614.

Bell, T. 1828 [1833]. Description of a new species of *Agama*, brought from the Columbia River by Mr. Douglass. Trans. Linn. Soc. London 16(1):105–107.

Beltz, E. and H. M. Smith. 1990. On the original description of *Masticophis flagellum piceus*. Bull. Chicago Herpetol. Soc. 25(5):87.

Bennion, R. S. and W. S. Parker. 1976. Field observations on courtship and aggressive behavior in desert striped whipsnakes, *Masticophis t. taeniatus*. Herpetologica 32(1):30–35.

Beringer, J. and T. R. Johnson. 1995. Natural history notes. *Rana catesbeiana* (Bullfrog). Diet. Herpetol. Rev. 26(2):98.

Berry, J. F. and C. M. Berry. 1984. A re-analysis of geographic variation and systematics in the yellow mud turtle, *Kinosternon flavescens* (Agassiz). Ann. Carnegie Mus. 53(7):185–206.

Best, T. L. and A. L. Gennaro. 1984. Feeding ecology of the lizard, *Uta stansburiana*, in southeastern New Mexico. J. Herpetol. 18(3):291–301.

——— and ———. 1985. Food habits of the western whiptail lizard (*Cnemidophorus tigris*) in southeastern New Mexico. Great Basin Nat. 45(3):527–534.

——— and H. C. James. 1984. Rattlesnakes (genus *Crotalus*) of the Pedro Armendariz lava field, New Mexico. Copeia 1984(1):213–215.

———, ———, and F. H. Best. 1983. Herpetofauna of the Pedro Armendariz lava field, New Mexico. Texas J. Sci. 35(3):245–255.

——— and G. S. Pfaffenberger. 1987. Age and sexual variation in the diet of collared Lizards (*Crotaphytus collaris*). Southwest. Nat. 32(4):415–426.

Beuchat, C. A. 1986. Reproductive influences on the thermoregulatory behavior of a live-bearing lizard. Copeia 1986(4):971–979.

———. 1989. Patterns and frequency of activity in a high altitude population of the iguanid lizard, *Sceloporus jarrovi*. J. Herpetol. 23(2):152–158.

Bezy, R. L., W. C. Sherbrooke, and C. H. Lowe. 1966. The rediscovery of *Eleutherodactylus latrans* in Arizona. Herpetologica 22(3):221–225.

Bickham, J. W. and J. A. MacMahon. 1972. Feeding habits of the western whiptail lizard, *Cnemidophorus tigris*. Southwest. Nat. 17(2):207–208.

Bishop, S. C. 1941. The Salamanders of New York. Bull. New York St. Mus. (324):1–365.

Bizer, J. R. 1978. Growth rates and size at metamorphosis of high elevation populations of *Ambystoma tigrinum*. Oecologia 34(2):175–184.

Black, J. H. and G. Sievert. 1989. A Field Guide to Amphibians of Oklahoma. Oklahoma Dept. Wildlife Cons. Oklahoma City. 80 p.

Blainville, H. M. D. de. 1835. Description de quelques espèces de reptiles de la Californie précédée de l'analyse d'un système général d'herpétologie et d'amphibiologie. Nouv. Ann. Mus. Nat. Hist. Nat. Paris 3(4):232–296.

Blair, A. P. 1947. Field observations on spadefoot toads. Copeia 1947(1):67.

———. 1955. Distribution, variation, and hybridization in a relict toad (*Bufo microscaphus*) in southwest Utah. Amer. Mus. Novitates (1722):1–38.

Blair, K. B., H. M. Smith, and D. Chiszar. 1994. Albinism and distributional records for *Lampropeltis triangulum* (Reptilia:Serpentes) in Panhandle Texas. Bull. Maryland Herpetol. Soc. 30(1):1–5.

Blair, W. F. 1955. Differentiation of mating calls in spadefoots, genus *Scaphiopus*. Texas J. Sci. 7(2):183–188.

———. 1956. Call differences as an isolation mechanism in southwestern toads (genus *Bufo*). Texas J. Sci. 8(1):87–106.

———. 1963. Evolutionary relationships of North American toads of the genus *Bufo*: a progress report. Evolution 17(1):1–16.

———. 1976. Some aspects of the biology of the ornate box turtle, *Terrapene ornata*. Southwest. Nat. 21(1):89–103.

——— (ed.). 1972. Evolution in the Genus *Bufo*. Univ. Texas Press, Austin. viii + 459 p.

LITERATURE CITED

———and A. P. Blair. 1941. Food habits of the collared lizard in northeastern Oklahoma. Amer. Midl. Nat. 26(1):230–232.

———and D. Pettus. 1954. The mating call and its significance in the Colorado River toad, (*Bufo alvarius* Girard). Texas J. Sci. 6(1):72–77.

Blanchard, F. N. 1921. A revision of the king snakes: genus *Lampropeltis*. Bull. U.S. Natl. Mus. (114):1–160.

———. 1924. A new snake of the genus *Arizona*. Occ. Pap. Mus. Zool. Univ. Michigan (150):1–5.

———. 1936 [1937]. Eggs and natural nests of the eastern ringneck snake, *Diadophis punctatus edwardsii*. Pap. Michigan Acad. Sci. Arts and Letters 22(1936):521–532.

———. 1942. The ringneck snakes, genus *Diadophis*. Bull. Chicago Acad. Sci. 7:1–144.

———and F. C. Blanchard. 1942. Mating of the garter snake *Thamnophis sirtalis sirtalis* (Linnaeus). Pap. Michigan Acad. Sci., Arts and Letters 27:334–336.

Blaney, R. M. 1973. *Lampropeltis*. Cat. Amer. Amphib. Rept.:150.1–150.2.

———. 1977. Systematics of the common kingsnake, *Lampropeltis getulus* (Linnaeus). Tulane Stud. Zool. Bot. 19(3–4):47–103.

———and P. J. Kimmich. 1973. Notes on the young of the Texas horned lizard, *Phrynosoma cornutum*. HISS News-J. 1(4):120.

Blaustein, A. R., P. D. Hoffman, D. G. Hokit, J. M. Kiesecker, S. C. Walls, and J. B. Hayes. 1994a. UV repair and resistance to solar UV-B in amphibian eggs: A link to population declines? Proc. Natl. Acad. Sci. USA. 91:1791–1795.

———, D. G. Hokit, R. K. O'Hara, and R. A. Holt. 1994b. Pathogenic fungus contributes to amphibian losses in the Pacific Northwest. Biol. Conserv. 67(3):251–254.

———and D. B. Wake. 1995. The puzzle of declining amphibian populations. Scientific American 272(4):52–57.

———, ———, and W. P. Sousa. 1994c. Amphibian declines: judging stability, persistence, and susceptibility of populations to local and global extinction. Conserv. Biol. 8(1):60–70.

Bloom, A. L. 1978. Geomorphology: a systematic analysis of Late Cenozoic landforms. Prentice-Hall, Inc., Englewood Cliffs, New Jersey. xvii + 510 p.

Bobyn, M. L. and R. J. Brooks. 1994. Interclutch and interpopulation variation in the effects of incubation conditions on sex, survival and growth of hatchling turtles (*Chelydra serpentina*). J. Zool., Lond. (1994) **233**, 233–257.

Bock, C. E., H. M. Smith, and J. H. Bock. 1990. The effect of livestock grazing upon abundance of the lizard, *Sceloporus scalaris*, in southeastern Arizona. J. Herpetol. 24(4):445–446.

Boucourt, M.-F. 1874. *In*: A. Duméril, M.-F. Bocourt, and F. Mocquard. Études sur les reptiles, p. 113–192. *In* Recherches zoologiques pour servir a l'histoire de la fauna de l'Amérique Centrale et du Mexique. Mission Scientifique au Mexique et dans l'Amérique Centrale, recherches Zool. part 3, sect. 1, livr. 3. Imprimerie au Nat., Paris.

———. 1879. *In*: A. Duméril, M.-F. Bocourt, and F. Mocquard. Études sur les reptiles, p. 361–440, *In* Recherches zoologiques pour servir a l'histoire de la fauna de l'Amérique Centrale et du Mexique. Mission Scientifique au Mexique et dans l'Amérique Centrale, recherches Zool. part 3, sect. 1. livr. 6. Imprimerie au Nat., Paris.

Boettger, O. 1883. Herpetologische Mittheilungen. I. Kurze Notizen uber Reptilien und Amphibien in der Heidelberger Universitats-Sammlung. Ber. Offenbach. Ver. Naturk. 22:147–152.

———. 1898. Katalog der Reptilien-Sammlung im Museum der Senckenbergischen Naturforschenden Gesellschaft in Frankfurt am Main. II. Teil (Schlangen). Senckenb. Naturforsch. Gesellsch. Frankfurt am Main. ix + 160 p.

Bogert, C. M. 1949. Thermoregulation and eccritic body temperatures in Mexican lizards of the genus *Sceloporus*. An. Inst. Biol. Univ. Mexico 20(1–2):415–426.

———. 1960. The influence of sound on the behavior of amphibians and reptiles, p. 37–320. *In* W. E. Lanyon and W. N. Tavolga (eds.), Animal Sounds and Communication. Publ. No. 7, A.I.B.S., Washington, D.C. xiii + 443 p.

———. 1962. Isolation mechanisms in toads of the *Bufo debilis* group in Arizona and western Mexico. Amer. Mus. Novitates (2100):1–37.

———and W. G. Degenhardt. 1961. An addition to the fauna of the United States, the Chihuahuan ridge-nosed rattlesnake in New Mexico. Amer. Mus. Novitates (2064):1–15.

———and M. R. del Campo. 1956. The Gila monster and its allies: The relationships, habits, and behavior of the lizards of the family Helodermatidae. Bull Amer. Mus. Nat. Hist. 109(1):1–238.

———and J. A. Oliver. 1945. A preliminary analysis of the herpetofauna of Sonora. Bull. Amer. Mus. Nat. Hist. 83(6):297–426.

Boulenger, G. A. 1882. Description of an apparently new species of lizard of the genus *Sceloporus*. Proc. Zool. Soc. London 1882:761–762.

———. 1885. Catalogue of the Lizards in the British Museum (Natural History). Second edition. Vol. 2. Trustees of the British Museum, London. xiii + 497 p. + pl. I–XXIV.

———. 1917. Description of new frogs of the genus *Rana*. Ann. Mag. Nat. Hist. 8(20):413–418.

Bowker, R. G., S. Damschroder, A. M. Sweet, and D. K. Anderson. 1986. Thermoregulatory behavior of the North American lizards *Cnemidophorus velox* and *Sceloporus undulatus*. Amphib.-Rept. 7:335–346.

Bowker, R. W. 1987. Life History Notes: *Elgaria kingi* (Arizona alligator lizard). Antipredator behavior. Herpetol. Rev. 18(4):73,75.

———. 1994. Natural History Notes: *Elgaria kingi* (Arizona alligator lizard). Size. Herpetol. Rev. 25(3):121.

Boyer, D. R. 1958. Sexual dimorphism in a population of western diamondback rattlesnakes. Herpetologica 13(3):213–221.

Boykin, K. and **N. Zucker.** 1993. Winter aggregation on a small rock cluster by the tree lizard *Urosaurus ornatus.* Southwest. Nat. 38(3):304–306.

Bragg, A. N. 1936. Notes on the breeding habits, eggs, and embryos of *Bufo cognatus* with a description of the tadpole. Copeia 1936(1):14–20.

———. 1937. A note on the metamorphosis of the tadpoles of *Bufo cognatus.* Copeia 1937(4):227–228.

———. 1941. Some observations on Amphibia at and near Las Vegas, New Mexico. Great Basin Nat. 2(3):109–117.

———. 1944. The spadefoot toads in Oklahoma with a summary of our knowledge of the group. Amer. Nat. 78(779):517–533.

———. 1945. The spadefoot toads in Oklahoma with a summary of our knowledge of the group. II. Amer. Nat. 79(780):52–72.

———. 1950a. Observations on the ecology and natural history of Anura. XVII. Adaptations and distribution in accordance with habits in Oklahoma, p. 59–100. *In* A. N. Bragg, A. O. Weese, H. A. Dundee, H. T. Fisher, A. Richards, and C. B. Clark (eds.), Researches on the Amphibia of Oklahoma. Univ. Oklahoma Press, Norman. 154 p.

———. 1950b. Size range in adults of the toad *Bufo cognatus.* Copeia 1950(2):153–154.

———. 1950c. Salientian breeding dates in Oklahoma, p. 35–38. *In* A. N. Bragg, A. O. Weese, H. A. Dundee, H. T. Fisher, A. Richards, and C. B. Clark (eds.), Researches on the Amphibia of Oklahoma. Univ. Oklahoma Press, Norman. 154 p.

———. 1955. The tadpole of *Bufo debilis debilis.* Herpetologica 11(3):211–212.

———. 1955–56. In quest of the spadefoots. New Mexico Quart. 25(4):345–358.

———. 1958 (1959). A melanistic tendency in the Great Plains toad, *Bufo cognatus.* Southwest. Nat. 3(1–4):229–230.

———. 1965. Gnomes of the Night. The Spadefoot Toads. Univ. Pennsylvania Press, Philadelphia. 127 p.

———and **H. A. Dundee.** 1949 [1950]. Reptiles collected in the vicinity of Las Vegas, New Mexico. Great Basin Nat. 9(3–4):55–57.

Brattstrom, B. H. 1953. An ecological restriction of the type locality of the western worm snake, *Leptotyphlops h. humilis.* Herpetologica 8(4):180–181.

———. 1965. Body temperatures of reptiles. Amer. Midland Nat. 73(2):376–422.

Breckenridge, W. J. 1970. Reptiles and Amphibians of Minnesota. 3rd ed. Univ. of Minnesota Press, Minneapolis. xiii + 202 p.

Brocchi, M. P. 1878 [1879]. Sur divers Batraciens anoures de l'Amérique Centrale. Bull Soc. Philo. Paris, ser. 7, 3(1):19–24.

Brodie, E. D., Jr. and **R. A. Altig.** 1967. Morphological variation in the Jemez Mountains salamander, *Plethodon neomexicanus.* Copeia 1967(3):670–672.

Brodie, E. D., III and **E. D. Brodie, Jr.** 1990. Tetrodotoxin resistance in garter snakes: An evolutionary response of predators to dangerous prey. Evolution 44(3):651–659.

Brophy, T. E. 1980. Food habits of sympatric larval *Ambystoma tigrinum* and *Notophthalmus viridescens.* J. Herpetol. 14(1):1–6.

Brown, A. E. 1901a. A new species of *Coluber* from western Texas. Proc. Acad. Nat. Sci. Philadelphia 53:292–295.

———. 1901b. A new species of *Ophibolus* from western Texas. Proc. Acad. Nat. Sci. Philadelphia 53:612–613.

Brown, B. C. 1950. An annotated check list of the Reptiles and Amphibians of Texas. Baylor Univ. Press, Waco. xii + 257 p.

Brown, D. E. 1982. Chihuahuan desertscrub, p. 169–179. *In* D. E. Brown (ed.), Biotic communities of the American Southwest–United States and Mexico. Desert Plants 4(1–4):1–342.

———and **N. B. Carmony.** 1991. Gila Monster. Facts and Folklore of America's Aztec Lizard. High-Lonesome Books, Silver City, New Mexico. ii + 127 p.

Brown, E. E. 1956. Nests and young of the six-lined racerunner *Cnemidophorus sexlineatus* Linnaeus. J. Elisha Mitchell Sci. Soc. 72(1):30–40.

Brown, H. A. 1976. The status of California and Arizona populations of the western spadefoot toads (genus *Scaphiopus*). Contrib. Sci. Nat. Hist. Mus. Los Angeles Co. (286):1–15.

Brown, L. E. 1974. Behavioral reactions of bullfrogs while attempting to eat toads. Southwest. Nat. 19(3):335–336.

———. 1992. *Rana blairi.* Cat. Amer. Amphib. Rept.:536.1–536.6.

———and **J. R. Pierce.** 1967. Male-male interactions and chorusing intensities of the Great Plains toad, *Bufo cognatus.* Copeia 1967(1):149–154.

Brown, T. L. and **R. V. Lucchino.** 1972. A record-sized specimen of the Texas horned lizard (*Phrynosoma cornutum*). Texas J. Sci. 24(3):353–354.

Brown, W. M. and **J. W. Wright.** 1979. Mitochondrial DNA analyses and the origin and relative age of parthenogenetic lizards (genus *Cnemidophorus*). Science 203:1247–1249.

Brown, W. S. and **W. S. Parker.** 1976. Movement ecology of *Coluber constrictor.* Copeia 1976(2):225–242.

———, ———, and **J. A. Elder.** 1974. Thermal and spatial relationships of two species of colubrid snakes during hibernation. Herpetologica 30(1):32–38.

Bugbee, R. E. 1942. Notes on animal occurrence and activity in the White Sands National Monument, New Mexico. Trans. Kansas Acad. Sci. 45(42):315–321.

Bull, J. J., R. C. Vogt, and C. J. McCoy. 1982. Sex determining temperatures in turtles: A geographic comparison. Evolution 36(2):326–332.

Bulova, S. J. 1994. Ecological correlates of population and individual variation in antipredator behavior of two species of desert lizards. Copeia 1994(4):980–992.

Bundy, R. E. 1951. New locality records of reptiles in New Mexico. Copeia 1951(4):314.

——— and J. Neess. 1958. Color variation in the round-tailed horned lizard, *Phrynosoma modestum*. Ecology 39(3):463–477.

Burger, W. L. 1950. New, revived, and reallocated names for North American whiptailed lizards, genus *Cnemidophorus*. Chicago Acad. Sci., Nat. Hist. Misc. (65):1–9.

Burke, R. L. 1994. Diurnal aggregation of banded geckos under field conditions. Southwest. Nat. 39(3):297–298.

Burkett, R. D. 1984. An ecological study of the cricket frog, *Acris crepitans*, p. 89–103. *In* R. A. Seigel, L. E. Hunt, J. L. Knight, L. Malaret, and N. L. Zuschlag (eds.), Vertebrate Ecology and Systematics—A Tribute to Henry S. Fitch. Univ. Kansas Mus. Nat. Hist. Spec. Publ. (10). 278 p.

Burkholder, G. L. and W. W. Tanner. 1974. Life history and ecology of the Great Basin sagebrush swift, *Sceloporus graciosus graciosus* Baird and Girard, 1852. Brigham Young Univ. Sci. Bull. Biol. Ser. 19(5):1–44.

Burns, T. A. 1970. Temperature of Yarrow's spiny lizard *Sceloporus jarrovi* at high altitudes. Herpetologica 26(1):9–16.

Bursey, C. R. and S. R. Goldberg. 1993. Diet of neonatal Yarrow's spiny lizard, *Sceloporus jarrovii* (Phrynosomatidae). Southwest. Nat. 38(4):381–383.

——— and ———. 1994. Growth of Yarrow's spiny lizard, *Sceloporus jarrovi* (Phrynosomatidae). Texas J. Sci. 46(1):13–20.

Burt, C. E. 1928. Insect food of Kansas lizards with notes on feeding habits. J. Kansas Entomol. Soc. 1(3):50–68.

———. 1931. A study of the teiid lizards of the genus *Cnemidophorus* with special reference to their phylogenetic relationships. Bull. U.S. Natl. Mus. (154):viii + 286 p.

———. 1935 [1936]. A key to the lizards of the United States and Canada. Trans. Kansas Acad. Sci. 38:255–305.

Bury, R. B. and J. A. Whelan. 1984. Ecology and management of the bullfrog. U.S. Dept. Interior, Fish Wildl. Serv., Resour. Publ. (155):1–23.

Cagle, F. R. 1950. Notes on *Holbrookia texana* in Texas. Copeia 1950(3):230.

Cagle, K. D., G. C. Packard, K. Miller, and M. J. Packard. 1993. Effects of microclimate in natural nests on development of embryonic painted turtles, (*Chrysemys picta*). Funct. Ecol. 7:653–660.

Cahn, A. R. 1937. The Turtles of Illinois. Illinois Biol. Monogr. 16(1–2). 218 p.

Caldwell, J. P. 1982. Disruptive selection: a tail color polymorphism in *Acris* tadpoles in response to differential predation. Can. J. Zool. 60(11):2818–2827.

Campbell, J. A. 1972. Reproduction in captive Trans-Pecos ratsnakes, *Elaphe subocularis*. Life history notes. *Elaphe subocularis* (Trans-Pecos Ratsnake). Reproduction. Herpetol. Rev. 4(4):129–130.

———, E. D. Brodie, Jr., D. G. Barker, and A. H. Price. 1989. An apparent natural hybrid rattlesnake and *Crotalus willardi* (Viperidae) from the Peloncillo Mountains of southwestern New Mexico. Herpetologica 45(3):344–349.

——— and W. W. Lamar. 1989. The Venomous Reptiles of Latin America. Cornell Univ. Press., Ithaca, New York. xii + 425 p.

Campbell, J. B. 1970a. New elevational records of the boreal toad (*Bufo boreas boreas*). Arctic and Alpine Res. 2(2):157–159.

———. 1970b. Hibernacula of a population of *Bufo boreas boreas* in the Colorado Front Range. Herpetologica 26(2):278–282.

———. 1970c. Food habits of the boreal toad, *Bufo boreas boreas*, in the Colorado front range. J. Herpetol. 4:83–85.

———. 1972. Reproduction and transformation of boreal toads in the Colorado Front Range. J. Colorado-Wyoming Acad. Sci. 7:114.

——— and W. G. Degenhardt. 1971. *Bufo boreas boreas* in New Mexico. Southwest. Nat. 16(1):219.

Campbell, H. 1953. Observations on snakes DOR in New Mexico. Herpetologica 9(3):157–160.

———. 1956. Snakes found dead on the roads of New Mexico. Copeia 1956(2):124–125.

——— and R. S. Simmons. 1961. Notes on the eggs and young of *Eumeces callicephalus* Bocourt. Herpetologica 17(3):212–213.

Carey, C. 1976. Thermal physiology and energetics of boreal toads, *Bufo boreas*. Ph.D. Diss., Univ. of Michigan, Ann Arbor. ix + 186 p.

———. 1978. Factors affecting body temperatures of toads. Oecologia 35(2):197–219.

———. 1987. Status of a breeding population of the western toad, *Bufo boreas boreas* at Lagunitas campground, New Mexico. Unpubl. Rept. New Mexico Dept. Game and Fish, Santa Fe. 23 p.

———. 1988. Physiological responses of Sacramento mountain salamanders to temperature and humidity. Unpubl. Rept. New Mexico Dept. Game and Fish, Santa Fe. 42 p.

———. 1993. Hypothesis concerning the causes of the disappearance of boreal toads from the mountains of Colorado. Conserv. Biol. 7(2):355–362.

Carpenter, C. C. 1960. Reproduction in Oklahoma *Sceloporus* and *Cnemidophorus*. Herpetologica 16(3):175–182.

———, J. C. Gillingham, and J. B. Murphy. 1976. The combat ritual of the rock rattlesnake (*Crotalus lepidus*). Copeia 1976(4):764–780.

——— and J. J. Krupa. 1989. Oklahoma herpetology: an annotated bibliography. Univ. Oklahoma Press, Norman. vii + 258 p.

Carr, A. F. 1952. Handbook of turtles: the turtles of the United States, Canada, and Baja California. Cornell Univ. Press, Ithaca, New York. xv + 542 p.

Censky, E. J. 1986. *Sceloporus graciosus*. Cat. Amer. Amphib. Rept.:386.1–386.4.

Chapel, W.L. 1939. Field notes on *Hyla wrightorum* Taylor. Copeia 1939(4):225–227.

Chapman, B. R. and S. D. Casto. 1972. Additional vertebrate prey of the loggerhead shrike. Wilson Bull. 84(4):496–497.

Chiasson, R. B. and C. H. Lowe. 1989. Ultrastructural scale patterns in *Nerodia* and *Thamnophis*. J. Herpetol. 23(2):109–118.

Chiszar, D., T. Melcer, R. Lee, C. W. Ratcliffe, and D. Duvall. 1990. Chemical cues used by prairie rattlesnakes (*Crotalus viridis*) to follow trails of rodent prey. J. Chem. Ecol. 16(1):79–86.

Chrapliwy, P. S., K. Williams, and H. M. Smith. 1961. Extensions of known range of certain amphibians and reptiles of Mexico. Herpetologica 12(2):121–124.

Christian, K. A. 1976. Ontogeny of the food niche of *Pseudacris triseriata*. M.S. thesis, Colorado St. Univ., Ft. Collins. viii + 70 p.

———, C. R. Tracy, and W. P. Porter. 1986. The effect of cold exposure during incubation of *Sceloporus undulatus* eggs. Copeia 1986(4):1012–1014.

Christiansen, J. L. 1969. Notes on hibernation of *Cnemidophorus neomexicanus* and *C. inornatus* (Sauria: Teiidae). J. Herpetol. 3(1–2):99–100.

———. 1971. Reproduction of *Cnemidophorus inornatus* and *Cnemidophorus neomexicanus* (Sauria, Teiidae) in northern New Mexico. Amer. Mus. Novitates (2442):1–48.

———, J. A. Cooper, J. W. Bickham, B. J. Galloway, and J. Springer. 1985. Aspects of the natural history of the yellow mud turtle, *Kinosternon flavescens* (Agassiz). Ann. Carnegie Mus. 53:185–206.

——— and W. G. Degenhardt. 1969. An unusual variant of the whiptail lizard, *Cnemidophorus gularis* (Sauria:Teidae), from New Mexico. Texas J. Sci. 21(1):95–97.

———, ———, and J. E. White. 1971. Habitat preferences of *Cnemidophorus inornatus* and *C. neomexicanus* with reference to conditions contributing to their hybridization. Copeia 1971(2):357–359.

——— and A. E. Dunham. 1972. Reproduction of the yellow mud turtle (*Kinosternon flavescens flavescens*) in New Mexico. Herpetologica 28(2):130–137.

——— and J. B. Iverson. 1993. *Kinosternon flavescens* (Agassiz, 1857). *In* P. C. H. Pritchard and A. G. J. Rhodin (eds.), The Conservation of Freshwater Turtles. IUCN/SSC Tortoise and Freshwater Turtle Specialist Group. Acct. No. 110, p. 1–4.

——— and E. O. Moll. 1973. Latitudinal reproductive variation within a single subspecies of painted turtle, *Chrysemys picta belli*. Herpetologica 29(2):152–163.

Chronic, H. 1987. Roadside geology of New Mexico. Mountain Press Publ. Co., Missoula, Montana. xiv + 257 p.

Clark, D. R. 1970. Loss of the left oviduct in the colubrid snake genus *Tantilla*. Herpetologica 26(1):130–133.

———. 1974. The western ribbon snake (*Thamnophis proximus*): ecology of a Texas population. Herpetologica 30(4):372–379.

——— and C. S. Lieb. 1973. Reproduction in the night snake (*Hypsiglena torquata*). Southwest. Nat. 18(2):248–252.

Clark, T. W., T. M. Campbell, III, D. G. Socha, and D. E. Casey. 1982. Prairie dog colony attributes and associated vertebrate species. Great Basin Nat. 42(4):572–582.

Clarke, R. F. 1965. An ethological study of the iguanid lizard genera *Callisaurus*, *Cophosaurus*, and *Holbrookia*. Emporia St. Res. Stud. 13(4):1–66.

Clarkson, R. W. and J. C. deVos, Jr. 1986. The bullfrog, *Rana catesbeiana* Shaw, in the lower Colorado River, Arizona-California. J. Herpetol. 20(1):42–49.

Cochran, D. M. 1961. Type specimens of reptiles and amphibians in the U.S. National Museum. Bull. U.S. Natl. Mus. (220):xv + 291 p.

Cockerell, T. D. A. 1896. Reptiles and batrachians of Mesilla Valley, New Mexico. Amer. Nat. 30:325–327.

Cohen, A. C. and J. L. Cohen. 1990. Ingestion of blister beetles by a Texas horned lizard. Southwest. Nat. 35(3):369.

Cole, C. J. 1962. Notes on the distribution and food habits of *Bufo alvarius* at the eastern edge of its range. Herpetologica 18(3):172–175.

———. 1963. Variation, distribution, and taxonomic status of the lizard, *Sceloporus undulatus virgatus* Smith. Copeia 1963(2):413–425.

———. 1968. *Sceloporus virgatus*. Cat. Amer. Amphib. Rept.:72.1–72.2.

———. 1983. Specific status of the North American fence lizards, *Sceloporus undulatus* and *Sceloporus occidentalis*, with comments on chromosome variation. Amer. Mus. Novitates (2768):1–13.

———, H. C. Dessauer, and G. F. Barrowclough. 1988. Hybrid origin of a unisexual species of whiptail lizard, *Cnemidophorus neomexicanus*, in western North America: new evidence and a review. Amer. Mus. Novitates (2905):1–38.

——— and L. M. Hardy. 1981. Systematics of North American colubrid snakes related to *Tantilla planiceps* (Blainville). Bull. Amer. Mus. Nat. Hist. 171(3):199–284.

——— and ———. 1983. *Tantilla hobartsmithi*. Cat Amer. Amphib. Rept.:318.1–318.2.

Collins, J. P. and J. E. Cheek. 1983. Effect of food and density on development of typical and cannibalistic salamander larvae in *Ambystoma tigrinum nebulosum*. Amer. Zool. 23(1):77–84.

——— and M. A. Lewis. 1979. Overwintering tadpoles and breeding season variation in the *Rana pipiens* complex in Arizona. Southwest. Nat. 24(2):371–373.

———, J. B. Mitton, and B. A. Pierce. 1980. *Ambystoma tigrinum*: a multispecies conglomerate? Copeia 1980(4):938–941.

Collins, J. T. 1982. Amphibians and reptiles in Kansas. 2nd ed. Univ. Kansas Mus. Nat. Hist. Publ. Educ. Ser. (8):xiii + 356 p.

———. 1989. New records of amphibians and reptiles in Kansas for 1989. Kansas Herpetol. Soc. Newsletter (78):16–21.

———. 1990. Standard common and current scientific names for North American amphibians and reptiles. Third edition. SSAR Herp. Circ. (19):v + 41 p.

———. 1991. Viewpoint: a new taxonomic arrangement for some North American amphibians and reptiles. Herpetol. Rev. 22(2):42–43.

Conant, R. 1949. Two new races of *Natrix erythrogaster*. Copeia 1949(1):1–15.

———. 1955a. Notes on *Natrix erythrogaster* from the eastern and western extremes of its range. Chicago Acad. Sci., Nat. Hist. Misc. (147):1–3.

———. 1955b. Notes on three Texas reptiles, including an addition to the fauna of the state. Amer. Mus. Novitates (1726):1–6.

———. 1963. Semiaquatic snakes of the genus *Thamnophis* from the isolated drainage system of the Río Nazas and the adjacent areas in Mexico. Copeia 1963(3):473–499.

———. 1965. Notes on reproduction in two natricine snakes from Mexico. Herpetologica 21(2):140–144.

———. 1968. Zoological exploration in Mexico—the route of Lieut. D. N. Couch in 1853. Amer. Mus. Novitates (2350):1–14.

———. 1969. A review of the water snakes of the genus *Natrix* in Mexico. Bull. Amer. Mus. Nat. Hist. 142(1):1–140.

———. 1975. A Field Guide to Reptiles and Amphibians of Eastern and Central North America. Second edition. Houghton Mifflin Co., Boston. xviii + 429 p. + 48 plates.

———. 1977(1978). Semiaquatic reptiles and amphibians of the Chihuahuan Desert and their relationships to drainage patterns of the region, p. 455–491. *In* R. H. Wauer and D. H. Riskind (eds.), Transactions of the Symposium on the Biological Resources of the Chihuahuan Desert Region, United States and Mexico. U.S. Dept. Interior, Natl. Park Serv., Trans. Proc. Ser. (3). xxii + 658 p.

——— and J. T. Collins. 1991. A Field Guide to Reptiles and Amphibians of Eastern and Central North America. Third edition. Houghton Mifflin Co., Boston. xx + 450 p. + 48 plates.

——— and C. J. Goin. 1948. A new subspecies of soft-shelled turtle from the central United States, with comments on the application of the name *Amyda*. Occ. Pap. Mus. Zool. Univ. Michigan (510):1–19.

Congdon, J. D., R. E. Ballinger, and K. A. Nagy. 1979. Energetics, temperature and water relations in winter aggregated *Sceloporus jarrovi* (Sauria: Iguanidae). Ecology 60(1):30–35.

———, G. L. Breitenbach, R. C. van Loben Sels, and D. W. Tinkle. 1987. Reproduction and nesting ecology of snapping turtles (*Chelydra serpentina*) in southeastern Michigan. Herpetologica 43(1):39–54.

——— and D. W. Tinkle. 1982. Energy expenditure in free-ranging sagebrush lizards (*Sceloporus graciosus*). Can. J. Zool. 60(6):1412–1416.

———, L. J. Vitt, and N. F. Hadley. 1978. Parental investment: comparative reproductive energetics in bisexual and unisexual lizards, genus *Cnemidophorus*. Amer. Nat. 112:509–521.

———, ———, and W. W. King. 1974. Geckos: adaptive significance and energetics of tail autotomy. Science 184:1379–1380.

Cook, F. R. 1964. Communal egg laying in the smooth green snake. Herpetologica 20(3):206.

Cooper, J. G. 1860. Report upon the reptiles collected on the survey, p. 292–306 + 11 Plates. *In* Chapter 4, Zoological Report. *In* Reports of explorations and surveys, to ascertain the most practicable and economical route for a railroad from the Mississippi River to the Pacific Ocean. Made under the direction of the Secretary of War, in 1853–5, according to acts of Congress of March 3, 1853, May 31, 1854, and August 5, 1854. Senate Exec. Doc. 36th Congress, 1st Sess. Vol. XII, Part III. 399 p.

———. 1863. Description of *Xerobates agassizii*. Proc. California Acad. Sci. 2:118–123.

Cooper, W. E., Jr., C. Caffrey, and L. J. Vitt. 1985a. Diel activity patterns in the banded gecko, *Coleonyx variegatus*. J. Herpetol. 19(2):308–311.

———, ———, and ———. 1985b. Aggregation in the banded gecko, *Coleonyx variegatus*. Herpetologica 41(3):342–350.

Cope, E. D. 1860. Catalog of the Colubridae in the museum of the Academy of Natural Sciences of Philadelphia, with notes and descriptions of new species. Part 2. Proc. Acad. Nat. Sci. Philadelphia 12(12–18):241–266.

———. 1861. Descriptions of reptiles from tropical America and Asia. Proc. Acad. Nat. Sci. Philadelphia 12(25–28):368–374.

——. 1863. On *Trachycephalus, Scaphiopus* and other American Batrachia. Proc. Acad. Nat. Sci. Philadelphia 15(2):43–54.

——. 1866a. Fourth contribution to the herpetology of tropical America. Proc. Acad. Nat. Sci. Philadelphia 18(2):123–132.

——. 1866b. On the structures and distribution of the genera of the arciferous Anura. J. Acad. Nat. Sci. Philadelphia, 2nd series, Vol. 6, Part 1:67–112.

——. 1866 [1867]. On the Reptilia and Batrachia of the Sonoran Province of the Nearctic region. Proc. Acad. Nat. Sci. Philadelphia 18(4):300–314.

——. 1869. On *Heloderma horridum*. Proc. Acad. Nat. Sci. Philadelphia 21(1):5–6.

——. 1875. Checklist of North American Batrachia and Reptilia; with a systematic list of higher groups, and an essay on geographical distribution. Based on the specimens contained in the U.S. National Museum. Bull. U.S. Natl. Mus. (1):1–104p.

——. 1878. A Texan cliff frog. Amer. Nat. 12:186.

——. 1880. On the zoological position of Texas. Bull. U.S. Natl. Mus. (17):1–51.

——. 1883a. A new snake from New Mexico. Amer. Nat. 17:1300–1301.

——. 1883b. Notes on the geographical distribution of Batrachia and Reptilia in western North America. Proc. Acad. Nat. Sci. Philadelphia 35:10–35.

——. 1885 [1886]. Thirteenth contribution to the herpetology of tropical America. Proc. Amer. Phil. Soc. 23(122):271–287.

——. 1892. A critical review of the characters and variations of the snakes of North America. Proc. U.S. Nat. Mus. 14:589–694.

——. 1896. The geographical distribution of Batrachia and Reptilia in North America. Amer. Nat. 30:886–902, 1003–1026.

——. 1898 [1900]. The crocodilians, lizards, and snakes of North America. Ann. Rept. U.S. Natl. Mus. 1898:153–1270.

Cordes, J. E., J. M. Walker, and **R. M. Abuhteba.** 1990. Genetic homogeneity in geographically remote populations of parthenogenetic *Cnemidophorus neomexicanus* (Sauria: Teiidae). Texas J. Sci. 42(3):303–305.

——, ——, **J. F. Scudday,** and **R. M. Abuhteba.** 1989. Distribution and habitat of the parthenogenetic whiptail lizard, *Cnemidophorus neomexicanus* (Sauria: Teiidae), in Texas. Texas J. Sci. 41(4):425–428.

Corn, P. S. 1993. Life History Notes: *Bufo boreas* (Boreal toad) Predation. Herpetol. Rev. 24(2):57.

——and **R. B. Bury.** 1986. Morphological variation and zoogeography of racers (*Coluber constrictor*) in the central Rocky Mountains. Herpetologica 42(2):258–264.

——and **J. C. Fogleman.** 1984. Extinction of montane populations of the northern leopard frog (*Rana pipiens*) in Colorado. J. Herpetol. 18(2):147–152.

——and **L. J. Livo.** 1989. Leopard frog and wood frog reproduction in Colorado and Wyoming. Northwest. Nat. 70(1):1–9.

——, **W. Stolzenburg,** and **R. B. Bury.** 1989. Acid precipitation studies in Colorado and Wyoming: interim report of surveys of montane amphibians and water chemistry. U.S. Fish Wildl. Serv. Biol. Rept. 80(40.26). 56 p.

Costanzo, J. P. 1989. A physiological basis for prolonged submergence in hibernating garter snakes *Thamnophis sirtalis*: evidence for an energy-sparing adaptation. Physiol. Zool. 62(2):580–592.

Cottam, W. P. 1937. Copulation in the western blue racer. Copeia 1937(4):229.

Coues, E. 1875. Synopsis of the reptiles and batrachians of Arizona, with critical and field notes, and an extensive synonymy, p. 585–633 + plates 16–25. *In* G. M. Wheeler, Report upon geographical and geological explorations and surveys west of the one hundredth meridian. Vol. V, Zoology. Engineer Dept., U.S. Army, Washington, D.C. 1021 p.

Cowles, R. B. 1946. Carrion eating by a snake. Herpetologica 3(4):121–122.

——and **C. M. Bogert.** 1935. Observations on the California lyre snake, *Trimorphodon vandenburghi* Klauber, with notes on the effectiveness of its venom. Copeia 1935(2):80–85.

——and ——. 1944. A preliminary study of the thermal requirements of desert reptiles. Bull. Amer. Mus. Nat. Hist. 83(5):261–296.

Cranston, T. 1989a. Natural history and captive husbandry of the western green rat snake. Vivarium 2(1):8–11, 23.

——. 1989b. Captive propagation and husbandry of the western green rat snake (*Senticolis triaspis intermedia*): the untold story, p. 81–85. *In* R. Gowen (ed.), Captive Propagation and Husbandry of Reptiles and Amphibians. Proceedings of the Fourth Northern California Herpetological Societies Conference on Captive Propagation and Husbandry of Reptiles and Amphibians.

Creel, G. C. 1963. Bat as a food item of *Rana pipiens*. Texas J. Sci. 15(1):104–106.

Creusere, F. M. and **W. G. Whitford.** 1976. Ecological relationships in a desert anuran community. Herpetologica 32(1):7–18.

——and ——. 1982. Temporal and spatial resource partitioning in a Chihuahuan Desert lizard community, p. 121–127. *In* N. J. Scott, Jr. (ed.), Herpetological communities. U.S. Dept. Interior, Fish Wildl. Serv., Wildl. Res. Rep. (13), Washington, D.C. iv + 239 p.

Crews, D. and **K. T. Fitzgerald.** 1980. "Sexual" behavior in parthenogenetic lizards (*Cnemidophorus*). Proc. Natl. Acad. Sci. USA 77(1):499–502.

——, **M. Grassman,** and **J. Lindzey.** 1986. Behavioral facilitation of reproduction in sexual and unisexual whiptail lizards. Proc. Natl. Acad. Sci. USA 83(24):9547–9550.

LITERATURE CITED

———and **M. C. Moore.** 1993. Psychobiology of reproduction of unisexual whiptail lizards, p. 257–282. *In* J. W. Wright and L. J. Vitt (eds.), Biology of Whiptail Lizards (genus *Cnemidophorus*). Oklahoma Mus. Nat. Hist., Norman. xiv + 417 p.

Criddle, S. 1937. Snakes from an anthill. Copeia 1937(2):142.

Cross, J. K. and **M. S. Rand.** 1979. Climbing activity in wild-ranging Gila monsters, *Heloderma suspectum*, (Helodermatidae). Southwest. Nat. 24(4):703–705.

Crowley, S. R. 1985. Thermal sensitivity of sprint-running in the lizard *Sceloporus undulatus*: support for a conservative view of thermal physiology. Oecologia 66(2):219–225.

———and **R. D. Pietruszka.** 1983. Aggressiveness and vocalization in the leopard lizard *(Gambelia wislizennii)*: the influence of temperature. Anim. Behav. 31(4):1055–1060.

Cuellar, O. 1966. Delayed fertilization in the lizard *Uta stansburiana*. Copeia 1966(3):549–552.

———. 1971. Reproduction and the mechanism of meiotic restitution in the parthenogenetic lizard *Cnemidophorus uniparens*. J. Morphol. 133(2):139–166.

———. 1976. Intraclonal histocompatibility in a parthenogenetic lizard: evidence of genetic homogeneity. Science 193:150–153.

———. 1977. Genetic homogeneity and speciation in the parthenogenetic lizards *Cnemidophorus velox* and *C. neomexicanus*: evidence from intraspecific histocompatibility. Evolution 31(1):24–31.

———. 1979. On the ecology of coexistence in parthenogenetic and bisexual lizards of the genus *Cnemidophorus*. Amer. Zool. 19(3):773–786.

———. 1981. Long-term analysis of reproductive periodicity in the lizard *Cnemidophorus uniparens*. Amer. Midl. Nat. 105(1):93–101.

———. 1993. Further observations on competition and natural history of coexisting parthenogenetic and bisexual whiptail lizards, p. 345–370. *In* J. W. Wright and L. J. Vitt (eds.), Biology of Whiptail Lizards (genus *Cnemidophorus*). Oklahoma Mus. Nat. Hist., Norman. xiv + 417 p.

———and **C. O. McKinney.** 1976. Natural hybridization between parthenogenetic and bisexual lizards: detection of uniparental source by skin grafting. J. Exp. Zool. 196(3):341–350.

———and **J. W. Wright.** 1992. Isogenicity in the unisexual lizard *Cnemidophorus velox*. C. R. Soc. Biogeogr. 68(4):157–160.

Cunningham, J. D. 1958. Alligator lizard notes. Herpetologica 14(3):180.

Curtis, L. 1950. Distribution of some Texas reptiles and amphibians. Field and Lab. 18(1):47.

Dall, W. H. 1915. Spencer Fullerton Baird, a biography. J. B. Lippincott Co., Philadelphia. xvi + 462 p.

Dalrymple, G. H. 1970. Caddis fly larvae feeding upon eggs of *Ambystoma t. tigrinum*. Herpetologica 26(1):128–129.

Dammann, A. E. 1961. Some factors affecting the distribution of sympatric species of rattlesnakes (genus *Crotalus*) in Arizona. Ph.D. Diss., Univ. Michigan, Ann Arbor. vi + 99 p.

Darevsky, I. S., L. A. Kupriyanova, and **T. Uzzell.** 1985. Parthenogenesis in reptiles, p. 411–526. *In* C. Gans and F. Billett (eds.), Biology of the Reptilia, Vol. 15, Development B. John Wiley and Sons, New York. x + 731 p.

Daudin, F. M. 1803. Histoire Naturelle, Générale et Particulière des Reptiles. Vol. 6. F. Dufart, Paris. 447 p + 10 plates.

David, P. 1994. Liste des reptiles actuels du monde. I. Chelonii. Dumerilia, 1:7–127.

Dawson, W. R. 1960. Physiological responses to temperature in the lizard *Eumeces obsoletus*. Physiol. Zool. 33(2):87–103.

———and **J. R. Templeton.** 1963. Physiological responses to temperature in the lizard *Crotaphytus collaris*. Physiol. Zool. 36(3):219–236.

De Carvalho, A. L. 1954. A preliminary synopsis of the genera of American microhylid frogs. Occ. Pap. Mus. Zool. Univ. Michigan (555):1–19.

Degenhardt, W. G. 1972. The ridge-nosed rattlesnake: an endangered species, p. 104–113. *In* Proceedings of the Symposium on Rare and Endangered Species of the Southwestern United States. New Mexico Dept. Game and Fish. Santa Fe, NM. vii + 167 p.

———. 1986. Geographic Distribution: *Gastrophryne olivacea* (Great Plains narrowmouth toad). Herpetol. Rev. 17(4):91.

———and **J. L. Christiansen.** 1974. Distribution and habitats of turtles in New Mexico. Southwest. Nat. 19(1):21–46.

———and **P. B. Degenhardt.** 1965. The host-parasite relationship between *Elaphe subocularis* (Reptilia: Colubridae) and *Aponomma elaphensis* (Acarina: Ixodidae). Southwest. Nat. 10(3):167–178.

———and **K. L. Jones.** 1972. A new sagebrush lizard, *Sceloporus graciosus*, from New Mexico and Texas. Herpetologica 28(3):212–217.

———and **G. E. Steele.** 1957. Additional specimens of *Trimorphodon vilkinsoni* from Texas. Copeia 1957(4):309–310.

Delson, J. and **W. G. Whitford.** 1973. Critical thermal maxima in several life history stages in desert and montane populations of *Ambystoma tigrinum*. Herpetologica 29(4):352–355.

DeMarco, V. G., R. W. Drenner, and **G. W. Ferguson.** 1985. Maximum prey size of an insectivorous lizard, *Sceloporus undulatus garmani*. Copeia 1985(4):1077–1080.

Densmore, L. D., III, J. W. Wright, and **W. M. Brown.** 1989a. Mitochondrial-DNA analyses and the origin and relative age of parthenogenetic lizards (genus *Cnemidophorus*). II. *C. neomexicanus* and the *C. tesselatus* complex. Evolution 43(5):943–957.

———, ———, and ———. 1989b. Mitochondrial-DNA analyses and the origin and relative age of parthenogenetic lizards (genus *Cnemidophorus*). IV. Nine *sexlineatus*-group unisexuals. Evolution 43(5):969–983.

Derickson, W. K. 1974. Lipid deposition and utilization in the sagebrush lizard, *Sceloporus graciosus*: its significance for reproduction and maintenance. Comp. Biochem. Physiol. 49A(2):267–273.

Deslippe, R. J. and **R. T. M'Closkey.** 1991. An experimental test of mate defense in an iguanid lizard (*Sceloporus graciosus*). Ecology 72(4):1218–1224.

———, ———, **S. P Dajczak,** and **C. P. Szpak.** 1990. Female tree lizards: oviposition and activity patterns during the breeding season. Copeia 1990(3):877–880.

Dessauer, H. C. and **C. J. Cole.** 1986. Clonal inheritance in parthenogenetic whiptail lizards: biochemical evidence. J. Heredity 77(1):8–12.

——— and ———. 1989. Diversity between and within nominal forms of unisexual teiid lizards, p. 49–71. *In* R. M. Dawley and J. P. Bogart (eds.), Evolution and Ecology of Unisexual Vertebrates. Bull. New York St. Mus. (466). iv + 302 p.

——— and ———. 1991. Genetics of whiptail lizards (Reptilia: Teiidae: *Cnemidophorus*) in a hybrid zone in southwestern New Mexico. Copeia 1991(3):622–637.

———, **W. Fox,** and **F. H. Pough.** 1962. Starch-gel electrophoresis of transferrins, esterases and other plasma proteins of hybrids between two subspecies of whiptail lizard (genus *Cnemidophorus*). Copeia 1962(4):767–774.

Devine, M. C. 1975. Copulatory plugs in snakes: Enforced chastity. Science 187:844–845.

Dial, B. E. 1978a. Aspects of the behavioral ecology of two Chihuahuan Desert geckos (Reptilia, Lacertilia, Gekkonidae). J. Herpetol. 12(2):209–216.

———. 1978b. The thermal ecology of two sympatric, nocturnal *Coleonyx* (Lacertilia: Gekkonidae). Herpetologica 34(2):194–201.

——— and **L. C. Fitzpatrick.** 1981. The energetic costs of tail autotomy to reproduction in the lizard *Coleonyx brevis* (Sauria: Gekkonidae). Oecologia 51(3):310–317.

——— and ———. 1982. Evaporative water loss in sympatric *Coleonyx* (Sauria: Gekkonidae). Comp. Biochem. Physiol. 71A(4):623–625.

———, **P. J. Weldon,** and **B. Curtis.** 1989. Chemosensory identification of snake predators (*Phyllorhynchus decurtatus*) by banded geckos (*Coleonyx variegatus*). J. Herpetol. 23(3):224–229.

Dick-Peddie, W. A. 1993. New Mexico Vegetation, Past, Present, and Future. Univ. New Mexico Press. Albuquerque. xxxii + 244 p. + folding map.

Dimmitt, M. A. and **R. Ruibal.** 1980a. Environmental correlates of emergence in spadefoot toads (*Scaphiopus*). J. Herpetol. 14(1):21–29.

——— and ———. 1980b. Exploitation of food resources by spadefoot toads (*Scaphiopus*). Copeia 1980(4):854–862.

Dixon, J. R. 1965. A taxonomic reevaluation of the night snake, *Hypsiglena ochrorhyncha*, and its relatives. Southwest. Nat. 10(2):125–131.

———. 1967. Aspects of the biology of the lizards of the White Sands, New Mexico. Los Angeles Co. Mus. Nat. Hist. Contrib. Sci. (129):1–22.

———. 1970a. *Coleonyx brevis*. Cat. Amer. Amphib. Rept.:88.1–88.2.

———. 1970b. *Coleonyx variegatus*. Cat. Amer. Amphib. Rept.:96.1–96.4.

———. 1987. Amphibians and Reptiles of Texas, With Keys, Taxonomic Synopses, Bibliography, and Distribution Maps. Texas A&M Univ. Press, College Station. xii + 434 p. + 13 plates.

——— and **R. H. Dean.** 1986. Status of the southern populations of the night snake (*Hypsiglena:* Colubridae) exclusive of California and Baja California. Southwest. Nat. 31(3):307–318.

——— and **R. R. Fleet.** 1976. *Arizona, A. elegans*. Cat. Amer. Amphib. Rept.:179.1–179.4.

——— and **P. A. Medica.** 1965. Noteworthy records of reptiles from New Mexico. Herpetologica 21(1):72–75.

——— and ———. 1966. Summer food of four species of lizards from the vicinity of White Sands, New Mexico. Los Angeles Co. Mus. Nat. Hist. Contrib. Sci. (121):1–6.

Dodson, S. I. and **V. E. Dodson.** 1971. The diet of *Ambystoma tigrinum* from western Colorado. Copeia 1971(4):614–624.

Donaldson, W., A. H. Price, and **J. Morse.** 1994. The current status and future prospects of the Texas horned lizard (*Phrynosoma cornutum*) in Texas. Texas J. Sci. 46(2):97–113.

Dott, R. H., Jr. and **R. L. Batten.** 1981. Evolution of the Earth. 3rd. ed. McGraw-Hill Book Co., Inc., New York. vii + 504 p.

Dorcas, M. E. and **K. D. Foltz.** 1991. Environmental effects on anuran advertisement calling. Amer. Zool. 31(5):111a.

Douglas, C. L. 1966. Amphibians and reptiles of Mesa Verde National Park, Colorado. Univ. Kansas Publ. Mus. Nat. Hist. 15(15):711–744.

Dowling, H. G. 1952. A taxonomic study of the ratsnakes, genus *Elaphe* Fitzinger. IV. A checklist of the American forms. Occ. Pap. Mus. Zool. Univ. Michigan (541):1–12.

———. 1960. A taxonomic study of the ratsnakes, genus *Elaphe* Fitzinger. VII. The *triaspis* section. Zoologica (New York) 45(2,6):53–80.

——— and **I. Fries.** 1987. A taxonomic study of the ratsnakes, VIII. A proposed new genus for *Elaphe triaspis* (Cope). Herpetologica 43(2):200–208.

——— and **R. M. Price.** 1988. A proposed new genus for *Elaphe subocularis* and *Elaphe rosaliae*. The Snake 20:52–63.

Drake, C. J. 1914. The food of *Rana pipiens* Schreber. Ohio Nat. 14:257–269.

Duellman, W. E. 1958. A preliminary analysis of the herpetofauna of Colima, Mexico. Occ. Pap. Mus. Zool. Univ. Michigan (589):1–22.

———. 1970. Hylid Frogs of Middle America. Monogr. Mus. Nat. Hist. Univ. Kansas (1). xi + 753 p. + 72 plates.

———. 1993. Amphibian species of the world: additions and corrections. Univ. Kansas Mus. Nat. Hist. Spec. Publ. (21):iii + 372 p.

——— and **L. Trueb.** 1986 [1985]. Biology of Amphibians. McGraw-Hill, Inc., New York. xix + 670 p.

——— and **R. G. Zweifel.** 1962. A synopsis of the lizards of the *sexlineatus* group (genus *Cnemidophorus*). Bull. Amer. Mus. Nat. Hist. 123(3):155–210.

Dugès, A. 1865. Du *Liophis janii*. Mem. Acad. Sci. Lett. Montpellier 6:32–33.

———. 1869. Catálogo de animales vertebrados observados en la República Mexicana. La Naturaleza 1:137–145.

Duméril, A. M. C., G. Bibron, and **A. H. A. Duméril.** 1854. Erpétologie générale ou histoire naturelle complète des reptiles. Librarie Encyclopedique de Roret, Vol. 7, pt. 2 Paris. xii + p. 781–1536.

Dundee, H. A. 1988. *Ambystoma tigrinum* locality records—be wary. Herpetol. Rev. 19(3):53.

———. 1995. Natural history notes. *Cnemidophorus gularis gularis* (Texas spotted whiptail). Maximum size. Herpetol. Rev. 26(2):100.

——— and **D. A. Rossman.** 1989. The Amphibians and Reptiles of Louisiana. Louisiana St. Univ. Press, Baton Rouge. xi + 300 p.

Dunham, A. E. 1981. Populations in a fluctuating environment: the comparative population ecology of the iguanid lizards *Sceloporus merriami* and *Urosaurus ornatus*. Misc. Publ. Mus. Zool. Univ. Michigan (158):1–62.

———. 1982. Demographic and life-history variation among populations of the iguanid lizard *Urosaurus ornatus*: implications for the study of life-history phenomena in lizards. Herpetologica 38(1):208–221.

Dunlap, K. D. and **J. W. Lang.** 1990. Offspring sex ratio varies with maternal size in the common garter snake, *Thamnophis sirtalis*. Copeia 1990(2):568–570.

Dunn, E. R. 1926. The Salamanders of the Family Plethodontidae. Smith College 50th Anniversary Publ., Northampton, Massachusetts. xiv + 441 p.

———. 1940. The races of *Ambystoma tigrinum*. Copeia 1940(3):154–162.

——— and **G. C. Wood.** 1939. Notes on eastern snakes of the genus *Coluber*. Notulae Naturae (Philadelphia) (5):1–4.

Dunson, W. A., R. L. Wyman, and **E. S. Corbett.** 1992. A symposium on amphibian declines and habitat acidification. J. Herpetol. 26(4):349–352.

Duvall, D., D. Chiszar, W. K. Hayes, J. K. Leonhardt, and **M. J. Goode.** 1990. Chemical and behavioral ecology of foraging in prairie rattlesnakes (*Crotalus viridis viridis*). J. Chem. Ecol. 16(1):87–101.

Dwyer, C. M. and **J. Hanken.** 1990. Limb skeletal variation in the Jemez Mountains salamander, *Plethodon neomexicanus*. Can. J. Zool. 68:1281–1287.

Easterla, D. A. 1975. Reproductive and ecological observations on *Tantilla rubra cucullata* from Big Bend National Park, Texas (Serpentes: Colubridae). Herpetologica 31(2):234–236.

Echternacht, A. C. 1967. Ecological relationships of two species of the lizard genus *Cnemidophorus* in the Santa Rita Mountains of Arizona. Amer. Midl. Nat. 78(2):448–459.

Edgren, R. A. 1952. A synopsis of the snakes of the genus *Heterodon*, with the diagnosis of a new race of *Heterodon nasicus* Baird and Girard. Nat. Hist. Misc. Chicago Acad. Sci. (112):1–4.

———. 1955. Possible thermo-regulatory burrowing in the lizard *Cnemidophorus sexlineatus*. Nat. Hist. Misc. Chicago Acad. Sci. (141):1–2.

Ellis, M. M. 1917. Amphibians and reptiles from the Pecos Valley. Copeia (43):39–40.

Ellis-Quinn, B. A. and **C. A. Simon.** 1989. Homing behavior of the lizard *Sceloporus jarrovi*. J. Herpetol. 23(2):146–152.

Emlen, S. T. 1968. Territoriality in the bullfrog, *Rana catesbeiana*. Copeia 1968(2):240–243.

Emory, W. H. 1848. Notes of a military reconnaissance, from Fort Leavenworth, in Missouri, to San Diego, in California, including parts of the Arkansas, Del Norte, and Gila rivers. 30th Congress, 1st Sess., House of Rep. Exec. Doc. (41):1–546.

———. 1857. Report of the United States and Mexican Boundary Survey, made under the direction of the Secretary of the Interior. Vol. I. 34th Congress, 1st Sess., House of Rep. Exec. Doc. (135). C. Wendell, printer, Washington, D.C. 174 p.

———. 1859. Report of the United States and Mexican Boundary Survey, made under the direction of the Secretary of the Interior. Vol. II. 34th Congress, 1st Sess., Senate Exec. Doc. (108), Washington, D.C.

Ernst, C. H. 1966. Overwintering of hatchling *Chelydra serpentina* in southeastern Pennsylvania. Philadelphia Herpetol. Soc. Bull. 14(1):8–9.

———. 1968. Evaporative water-loss relationships of turtles. J. Herpetol. 2(3–4):159–161.

———. 1971. *Chrysemys picta*. Cat. Amer. Amphib. Rept.: 106.1–106.4.

———. 1990a. Systematics, taxonomy, variation, and geographic distribution of the slider turtle, p. 57–67. *In* J. W. Gibbons (ed.), Life History and Ecology of the Slider Turtle. Smithsonian Inst. Press, Washington, D.C. xiv + 368 p.

———. 1990b. *Pseudemys gorzugi*. Cat. Amer. Amphib. Rept.:461.1–461.2.

———. 1992a. Venomous Reptiles of North America. Smithsonian Inst. Press, Washington, D.C. ix + 236 p.

———. 1992b. *Trachemys gaigeae*. Cat. Amer. Amphib. Rept.:538.1–538.4.

——— and **R. W. Barbour**. 1972. Turtles of the United States. Univ. Kentucky Press, Lexington. x + 347 p.

——— and ———. 1989. Turtles of the World. Smithsonian Inst. Press, Washington, D.C. xii + 313 p.

———, **J. W. Gibbons**, and **S. S. Novak**. 1988. *Chelydra*. Cat. Amer. Amphib. Rept.:419.1–419.4.

———, **J. E. Lovich**, and **R. W. Barbour**. 1994. Turtles of the United States and Canada. Smithsonian Inst. Press, Washington, D.C. xxxviii + 578 p.

——— and **J. F. McBreen**. 1991. *Terrapene*. Cat. Amer. Amphib. Rept.:511.1–511.6.

Estes, R., K. de Queiroz, and **J. Gauthier**. 1988. Phylogenetic relationships within Squamata, p. 119–281. *In* R. Estes and G. Pregill (eds.), Phylogenetic Relationships of Lizard Families: Essays Commemorating Charles L. Camp. Stanford Univ. Press, Stanford, California. xii + 631 p.

Ewert, M. A. and **J. M. Legler**. 1978. Hormonal induction of oviposition in turtles. Herpetologica 34(3):314–318.

——— and **C. E. Nelson**. 1991. Sex determination in turtles: Diverse patterns and some possible adaptive values. Copeia 1991(1):50–69.

Fellman, B. 1995. To eat or not to eat. National Wildlife 33(2):42–45.

Ferguson, D. E., J. P. McKowen, O. S. Bosarge, and **H. F. Landreth**. 1968. Sun compass orientation of bullfrogs. Copeia 1968(2):230–235.

Ferguson, G. W. 1970. Mating behavior of the side-blotched lizards of the genus *Uta* (Sauria: Iguanidae). Anim. Behav. 18(1):65–72.

——— and **T. Brockman**. 1980. Geographic differences of growth rate of *Sceloporus* lizards (Sauria: Iguanidae). Copeia 1980(2):259–264.

———, **K. L. Brown**, and **V. G. DeMarco**. 1982. Selective basis for the evolution of variable egg and hatchling size in some iguanid lizards. Herpetologica 38(1):178–188.

——— and **S. F. Fox**. 1984. Annual variation of survival advantage of large juvenile side-blotched lizards, *Uta stansburiana*: its causes and evolutionary significance. Evolution 38(2):342–349.

——— and **H. L. Snell**. 1986. Endogenous control of seasonal change of egg, hatchling, and clutch size of the lizard *Sceloporus undulatus garmani*. Herpetologica 42(2):185–191.

———, ———, and **A. J. Landwer**. 1990. Proximate control of variation of clutch, egg, and body size in a west-Texas population of *Uta stansburiana stejnegeri* (Sauria: Iguanidae). Herpetologica 46(2):227–238.

Ferguson, J. H. and **C. H. Lowe**. 1969. Evolutionary relationship in the *Bufo punctatus* group. Amer. Midl. Nat. 81(2):435–466.

Fernandez, P. J. and **J. P. Collins**. 1988. Effect of environment and ontogeny on color pattern variation in Arizona tiger salamanders (*Ambystoma tigrinum nebulosum* Hallowell). Copeia 1988(4):928–938.

Ferner, J. W. 1974. Home-range size and overlap in *Sceloporus undulatus erythrocheilus* (Reptilia: Iguanidae). Copeia 1974(2):332–337.

———. 1976. Notes on natural history and behavior of *Sceloporus undulatus erythrocheilus* in Colorado. Amer. Midl. Nat. 96(2):291–302.

Findley, J. S. 1959. A new station for the Sacramento mountain salamander in New Mexico. Southwest. Nat.4(3):155–156.

———. 1964. Verification of the occurrence of *Bufo microscaphus* Cope in New Mexico. Southwest. Nat. 9(2):107.

Fitch, H. S. 1955. Habits and adaptations of the Great Plains skink (*Eumeces obsoletus*). Ecol. Monogr. 25(1):59–83.

———. 1956a. An ecological study of the collared lizard (*Crotaphytus collaris*). Univ. Kansas Publ. Mus. Nat. Hist. 8(3):213–274.

———. 1956b. A field study of the Kansas ant-eating frog, *Gastrophryne olivacea*. Univ. Kansas Publ. Mus. Nat. Hist. 8(4):275–306.

———. 1958. Natural history of the six-lined racerunner (*Cnemidophorus sexlineatus*). Univ. Kansas Publ. Mus. Nat. Hist. 11(2):11–62.

———. 1960. Criteria for determining sex and breeding maturity in snakes. Herpetologica 16(1):49–51.

———. 1963. Natural history of the racer *Coluber constrictor*. Publ. Univ. Kansas Mus. Nat. Hist. 15(8):351–468.

———. 1965. An ecological study of the garter snake, *Thamnophis sirtalis*. Publ. Univ. Kansas Mus. Nat. Hist. 15(10):493–564.

———. 1970. Reproductive cycles of lizards and snakes. Univ. Kansas Mus. Nat. Hist. Misc. Publ. (52):1–247.

———. 1975. A demographic study of the ringneck snake (*Diadophis punctatus*) in Kansas. Univ. Kansas Mus. Nat. Hist. Misc. Publ. (62):1–53.

———. 1980. *Thamnophis sirtalis*. Cat. Amer. Amphib. Rept.:270.1–270.4.

LITERATURE CITED

———. 1983. *Thamnophis elegans.* Cat. Amer. Amphib. Rept.:320.1–320.4.

———. 1985. Variation in clutch and litter size in New World reptiles. Univ. Kansas Mus. Nat. Hist. Misc. Publ. (76):1–76.

———, **W. S. Brown,** and **W. S. Parker.** 1981. *Coluber mormon,* a species distinct from *C. constrictor.* Trans. Kansas Acad. Sci. 84(4):196–203.

———and **T. P. Maslin.** 1961. Occurrence of the garter snake, *Thamnophis sirtalis,* in the Great Plains and Rocky Mountains. Univ. Kansas Publ. Mus. Nat. Hist. 13(5):189–308.

———and **G. R. Pisani.** 1993. Life history traits of the western diamondback rattlesnake *(Crotalus atrox)* studied from roundup samples in Oklahoma. Occ. Pap. Mus. Nat. Hist. Univ. Kansas (156):1–24.

———and **W. W. Tanner.** 1951. Remarks concerning the systematics of the collared lizard, *(Crotaphytus collaris),* with a description of a new subspecies. Trans. Kansas Acad. Sci. 54(4):548–559.

Fitzgerald, L. A. 1986a. A preliminary status survey of *Thamnophis rufipunctatus* and *Thamnophis eques* in New Mexico. Unpubl. Rept., New Mexico Dept. Game and Fish, Santa Fe. 12 p.

———. 1986b. A comparison of the systematics and general biology of *Thamnophis rufipunctatus* and *Nerodia harteri.* Unpubl. Rept., U.S. Fish Wildl. Serv., Endangered Species Office, Albuquerque, New Mexico. 14 p.

———and **W. G. Degenhardt.** 1986. Geographic Distribution: *Bufo microscaphus* (Southwestern toad). Herpetol. Rev. 17(3):65.

Fleharty, E. D. 1967. Comparative ecology of *Thamnophis elegans, T. cyrtopsis,* and *T. rufipunctatus* in New Mexico. Southwest. Nat. 12(3):207–230.

Flores-Villela, O. 1993. Herpetofauna Mexicana. Spec. Publ. Carnegie Mus. Nat. Hist. (17):iv + 73 p.

Flury, A. 1950. A new kingsnake from Trans-Pecos Texas. Copeia 1950(3):215–217.

Force, E. R. 1930. The amphibians and reptiles of Tulsa County, Oklahoma and vicinity. Copeia 1930(2):25–39.

———. 1931. Habits and birth of young of the lined snake, *Tropidoclonion lineatum* (Hallowell). Copeia 1931(2):51–53.

———. 1936. Notes on the blind snake, *Leptotyphlops dulcis* (Baird and Girard) in northeastern Oklahoma. Proc. Oklahoma Acad. Sci. 16:24–26.

Ford, N. B. and **V. Cobb.** 1992. Timing of courtship in two colubrid snakes of the southern United States. Copeia 1992(2):573–577.

———and **J. P. Karges.** 1987. Reproduction in the checkered garter snake, *Thamnophis marcianus,* from southern Texas and northeastern Mexico: seasonality and evidence for multiple clutches. Southwest. Nat. 32(1):93–101.

———and **M. L. O'Bleness.** 1986. Species and sexual specificity of pheromone trails of the garter snake, *Thamnophis marcianus.* J. Herpetol. 20(2):259–262.

———and **R. A. Seigel.** 1989. Phenotypic plasticity in reproductive traits: evidence from a viviparous snake. Ecology 70(6):1768–1774.

Forester, D. C. 1973. Mating call as a reproductive isolating mechanism between *Scaphiopus bombifrons* and *S. hammondii.* Copeia 1973(1):60–67.

Forster, J. R. 1771. A catalogue of the animals of North America. *In* J. B. Bossu, Travels Through That Part of North America Formerly Called Louisiana. London Vol. 1. vii + 407 p.

Fouquette, M. J., Jr. 1954. Food competition among four sympatric species of garter snakes, genus *Thamnophis.* Texas J. Sci. 6(2):172–188.

———. 1968. Remarks on the type specimen of *Bufo alvarius* Girard. Great Basin Nat. 28(2):70–72.

———. 1970. *Bufo alvarius.* Cat. Amer. Amphib. Rept.:93.1–93.4.

Fowler, H. W. 1915 [1916]. Cold-blooded vertebrates from Florida, the West Indies, Costa Rica, and eastern Brazil. Proc. Acad. Nat. Sci. Philadelphia 67:244–269.

Fox, S. F. 1978. Natural selection on behavioral phenotypes of the lizard *Uta stansburiana.* Ecology 59(4):834–847.

———and **T. A. Baird.** 1992. The dear enemy phenomenon in the collared lizard, *Crotaphytus collaris,* with a cautionary note on experimental methodology. Anim. Behav. 44(4):780–782.

Fox, W. 1965. A comparison of the male urogenital systems of blind snakes, Leptotyphlopidae and Typhlopidae. Herpetologica 21(4):241–256.

———and **H. C. Dessauer.** 1962. The single right oviduct and other urogenital structures of female *Typhlops* and *Leptotyphlops.* Copeia 1962(3):590–597.

Frazer, N. B., J. L. Greene, and **J. W. Gibbons.** 1993. Temporal variation in growth rate and age at maturity of male painted turtles, *Chrysemys picta.* Amer. Midl. Nat. 130(2):314–324.

Freiburg, R. E. 1951. An ecological study of the narrow-mouthed toad *(Microhyla)* in northeastern Kansas. Trans. Kansas Acad. Sci. 54(3):374–386.

Fritts, T. H. and **R. D. Jennings.** 1994. Distribution, habitat use, and status of the desert tortoise in Mexico, p. 49–56. *In* R. B. Bury and D. J. Germano (eds.). Biology of North American tortoises. National Biological Survey, Fish and Wildlife Research (13):vi + 204 p.

Fritts, T. H., R. D. Jennings, and **N. J. Scott, Jr.** 1984. A review of the leopard frogs of New Mexico. Unpubl. Rept. New Mexico Dept. Game and Fish. Santa Fe. 157 p.

Frost, D. R. 1983a. Past occurrence of *Acris crepitans* (Hylidae) in Arizona. Southwest. Nat. 28(1):105.

———. 1983b. *Sonora semiannulata*. Cat. Amer. Amphib. Rept.:333.1–333.4.

———. (ed.). 1985. Amphibian Species of the World: A Taxonomic and Geographic Reference. Allen Press, Lawrence, Kansas. v + 732 p.

——— and **R. Etheridge.** 1989. A phylogenetic analysis and taxonomy of iguanian lizards (Reptilia: Squamata). Misc. Publ. Mus. Nat. Hist. Univ. Kansas (81):1–65.

——— and **D. M. Hillis.** 1990. Species in concept and practice: herpetological applications. Herpetologica 46(1):87–104.

———, **A. G. Kluge,** and ———. 1992. Species in contemporary herpetology: comments on phylogenetic inference and taxonomy. Herpetol. Rev. 23(2):46–54.

——— and **J. W. Wright.** 1988. The taxonomy of uniparental species, with special reference to parthenogenetic *Cnemidophorus* (Squamata: Teiidae). Syst. Zool. 37(2):200–209.

Frost, J. S. 1982. Functional genetic similarity between geographically separated populations of Mexican leopard frogs (*R. pipiens* complex). Syst. Zool. 31(1):57–67.

——— and **J. E. Platz.** 1983. Comparative assessment of modes of reproductive isolation among four species of leopard frogs (*Rana pipiens* complex). Evolution 37(1):66–78.

Funk, R. S. 1964. On the food of *Crotalus m. molossus*. Herpetologica 20(2):134.

Galbraith, D. A., C. A. Bishop, and **M. E. Obbard.** 1989. The influence of growth rate on age and body size at maturity in female snapping turtles (*Chelydra serpentina*). Copeia 1989(4):896–904.

Garman, S. 1883. North American reptilia. Part 1. Ophidia. Mem. Mus. Comp. Zool. 8(3):1–185.

———. 1887. Reptiles and batrachians from New Mexico and Texas. Bull. Essex Inst. 19:1–20.

Garrett, C. M. and **C. W. Painter.** 1992. Geographic Distribution: *Senticolis triaspis intermedia*. Herpetol. Rev. 23(4):124.

Garstka, W. R. 1982. Systematics of the *mexicana* species group of the colubrid genus *Lampropeltis*, with an hypothesis mimicry. Breviora (466):1–35.

Gates, G. O. 1956. A record length for the Arizona coral snake. Herpetologica 12(2):155.

———. 1957. A study of the herpetofauna in the vicinity of Wickenburg, Maricopa County, Arizona. Trans. Kansas Acad. Sci.60(4):403–418.

Gatten, R. E., Jr. 1985. Activity metabolism of lizards after thermal acclimation. J. Therm. Biol. 10(4):209–215.

Gebhardt, L. P., G. J. Stanton, and **S. De St. Jeor.** 1966. Transmission of WEE virus to snakes by infected *Culex tarsalis* mosquitoes. Proc. Soc. Exp. Biol. Med. 123(1):233–235.

Gehlbach, F. R. 1956. Annotated records of southwestern amphibians and reptiles. Trans. Kansas Acad. Sci. 59(3):364–372.

———. 1965. Herpetology of the Zuni Mountains region, northwestern New Mexico. Proc. U.S. Natl. Mus. 116(3505):243–332.

———. 1967a. *Ambystoma tigrinum*. Cat. Amer. Amphib. Rept.:52.1–52.4.

———. 1967b. *Lampropeltis mexicana*. Cat. Amer. Amphib. Rept.:55.1–55.2.

———. 1970. Death-feigning and erratic behavior in leptotyphlopid, colubrid, and elapid snakes. Herpetologica 26(1):24–34.

———. 1971. Lyre snakes of the *Trimorphodon biscutatus* complex: a taxonomic resume. Herpetologica 27(2):200–211.

———. 1974. Evolutionary relations of southwestern ringneck snakes (*Diadophis punctatus*). Herpetologica 30(2):140–148.

———. 1979. Biomes of the Guadalupe Escarpment: vegetation, lizards, and human impact, p. 427–439. *In* H. H. Genoways and R. J. Baker (eds.), Biological Investigations in the Guadalupe Mountains National Park, Texas. U.S. Dept. Interior, Natl. Park Serv., Proc. Trans. Ser. (4). xvii + 439 p.

——— and **J. K. Baker.** 1962. Kingsnakes allied with *Lampropeltis mexicana:* taxonomy and natural history. Copeia 1962(2):291–300.

——— and **C. J. McCoy, Jr.** 1965. Additional observations on variation and distribution of the gray-banded kingsnake, *Lampropeltis mexicana* (Garman). Herpetologica 21(1):35–38.

———, **J. F. Watkins,** and **J. C. Kroll.** 1971. Pheromone trail-following of typhlopid, leptotyphlopid, and colubrid snakes. Behavior 90(19):282–294.

Gennaro, A. L. 1972. Home range and movements of *Holbrookia maculata maculata* in eastern New Mexico. Herpetologica 28(2):165–168.

———. 1974. Growth, size, and age at sexual maturity of the lesser earless lizard, *Holbrookia maculata maculata* in eastern New Mexico. Herpetologica 30(1):85–90.

Germano, D. J., R. B. Bury, T. C. Esque, T. H. Fritts, and **P. A. Medica.** 1994. Range and habits of the desert tortoise, p. 73–84. *In* R. B. Bury and D. J. Germano (eds.). Biology of North American tortoises. National Biological Survey, Fish and Wildlife Research 13. vi + 204 p.

Gibbons, J. W. 1990. The slider turtle, p. 3–18. *In* J. W. Gibbons (ed.), Life History and Ecology of the Slider Turtle. Smithsonian Inst. Press, Washington, D.C. xiv + 368 p.

——— and **J. L. Greene.** 1979. X-ray photography: a technique to determine reproductive patterns in freshwater turtles. Herpetologica 35(1):86–89.

——— and ———. 1990. Reproduction in the slider and other species of turtles, p. 124–134. *In* J. W. Gibbons (ed.), Life History and Ecology of the Slider Turtle. Smithsonian Inst. Press. Washington, D.C. xiv + 368 p.

———, **S. S. Novak**, and **C. H. Ernst**. 1988. *Chelydra serpentina*. Cat. Amer. Amphib. Rept.:420.1–420.4.

——— and **R. D. Semlitsch**. 1987. Activity patterns, p. 396–421. *In* R. A. Seigel, J. T. Collins, and S. S. Novak (eds.), Snakes: Ecology and Evolutionary Biology. Macmillan Publ. Co., New York. xiv + 529 p.

Gillis, J. E. 1975. Characterization of a hybridizing complex of leopard frogs. Ph.D. Diss., Colorado St. Univ., Fort Collins. 136 p.

———. 1989. Selection for substrate reflectance-matching in two populations of red-chinned lizards (*Sceloporus undulatus erythrocheilus*) from Colorado. Amer. Midl. Nat. 121(1):197–200.

Gillis, R. and **R. E. Ballinger**. 1992. Reproductive ecology of red-chinned lizards (*Sceloporus undulatus erythrocheilus*) in southcentral Colorado: Comparisons with other populations of a wide-ranging species. Oecologia 89(2):236–243.

Girard, C. F. 1851. On a new American saurian reptile. Proc. Amer. Assoc. Adv. Sci. 4:200–202.

———. 1851 [1852]. A monographic essay on the genus *Phrynosoma*, p. 354–365. *In* H. Stansbury, Exploration and survey of the Valley of the Great Salt Lake of Utah, including a reconnaissance of a new route through the Rocky Mountains. U.S. 32nd Congress, Spec. Sess., Senate Exec. Doc. 3. Lippincott, Granbo, and Co., Philadelphia. 487 p.

———. 1854. A list of the North American bufonids, with diagnoses of new species. Proc. Acad. Nat. Sci Philadelphia 7:86–88.

———. 1858. Herpetology. *In* United States exploring expedition, during the years 1838, 1839, 1840, 1841, 1842, under the command of Charles Wilkes, U.S.N. J. B. Lippincott and Co., Philadelphia. Vol. 20, xvii + 496 p.

Glenn, J. L. and **R. C. Straight**. 1978. Mojave rattlesnake *Crotalus scutulatus scutulatus* venom: variation in toxicity with geographical origin. Toxicon 16(1):81–84.

——— and ———. 1990. Venom characteristics as an indicator of hybridization between *Crotalus viridis viridis* and *Crotalus scutulatus scutulatus* in New Mexico. Toxicon 28(7):857–862.

———, ———, **M. C. Wolfe**, and **D. L. Hardy**. 1983. Geographical variation in *Crotalus scutulatus scutulatus* (Mojave rattlesnake) venom properties. Toxicon 21(1):119–130.

Gloyd, H. K. 1936. The subspecies of *Crotalus lepidus*. Occ. Pap. Mus. Zool. Univ. Michigan (337):1–6.

———. 1940. The rattlesnakes, genera *Sistrurus* and *Crotalus*. A study in zoogeography and evolution. Spec. Publ. Chicago Acad. Sci. (4):1–270.

Goldberg, S. R. 1970. Seasonal ovarian histology of the ovoviviparous iguanid lizard *Sceloporus jarrovi* Cope. J. Morphol. 132(3):265–275.

———. 1971a. Reproductive cycle of the ovoviviparous iguanid lizard *Sceloporus jarrovi* Cope. Herpetologica 27(2):123–131.

———. 1971b. Reproduction in the short-horned lizard *Phrynosoma douglassi* in Arizona. Herpetologica 27(3):311–314.

———. 1975. Reproduction in the Arizona alligator lizard, *Gerrhonotus kingi*. Southwest. Nat. 20(3):412–413.

———. 1987. Reproductive cycle of the giant spotted whiptail, *Cnemidophorus burti stictogrammus*, in Arizona. Southwest. Nat. 32(4):510–511.

———. 1995. Reproduction in the western patchnose snake, *Salvadora hexalepis*, and the mountain patchnose snake, *Salvadora grahamiae* (Colubridae), from Arizona. Southwest. Nat. 40(1):119–120.

——— and **C. R. Bursey**. 1990. Winter feeding in the mountain spiny lizard, *Sceloporus jarrovi* (Iguanidae). J. Herpetol. 24(4):446–448.

——— and ———. 1992a. Helminths of the bunch grass lizard, *Sceloporus scalaris slevini* (Iguanidae). J. Helminthol. Soc. Washington 59(1):130–131.

——— and ———. 1992b. Gastrointestinal helminths of the southwestern earless lizard, *Cophosaurus texanus scitulus*, and the speckled earless lizard, *Holbrookia maculata approximans* (Phrynosomatidae). J. Helminthol. Soc. Washington 59(2):230–231.

——— and ———. 1992c. Prevalence of the nematode *Spauligodon giganticus* (Oxyurida: Pharyngodonidae) in neonatal Yarrow's spiny lizards, *Sceloporus jarrovii* (Sauria: Iguanidae). J. Parasitol. 78(3):539–541.

——— and ———. 1993. Duration of attachment of the chigger, *Eutrombicula lipovskyana* (Trombiculidae) in mite pockets of Yarrow's spiny lizard, *Sceloporus jarrovii* (Phrynosomatidae) from Arizona. J. Wildl. Dis. 29(1):142–144.

———, ———, and **N. Zucker**. 1993. Gastrointestinal helminths of the tree lizard, *Urosaurus ornatus* (Phrynosomatidae). J. Helminthol. Soc. Washington 60(1):118–121.

——— and **C. H. Lowe**. 1966. The reproductive cycle of the western whiptail lizard (*Cnemidophorus tigris*) in southern Arizona. J. Morphol. 118(4):543–548.

——— and **W. S. Parker**. 1975. Seasonal testicular histology of the colubrid snakes, *Masticophis taeniatus* and *Pituophis melanoleucus*. Herpetologica 31(3):317–322.

Good, D. A. and **J. W. Wright**. 1984. Allozymes and the hybrid origin of the parthenogenetic lizard *Cnemidophorus exsanguis*. Experientia 40(9):1012–1014.

Gorman, J. 1960. Treetoad studies, 1. *Hyla californiae*, new species. Herpetologica 16(3):214–222.

Gosner, K. L. 1960. A simplified table for staging anuran embryos and larvae with notes on identification. Herpetologica 16(3):183–190.

Graham, T. E. 1978. Preliminary notes on locomotor behavior in juvenile snapping turtles, *Chelydra serpentina*, under controlled conditions. Bull. Maryland Herpetol. Soc. 14(4):266–268.

———and **A. A. Graham.** 1991. Life History Notes: *Trionyx spiniferus spiniferus*. Burying behavior. Herpetol. Rev. 22(2):56–57.

Grant, K. P. and **L. E. Licht.** 1993. Acid tolerance of anuran embryos and larvae from central Ontario. J. Herpetol. 27(1):1–6.

Graves, B. M. and **D. Duvall.** 1990. Spring emergence patterns of wandering garter snakes and prairie rattlesnakes in Wyoming. J. Herpetol. 24(4):351–356.

———and———. 1993. Reproduction, rookery use, and thermoregulation in free-ranging, pregnant *Crotalus v. viridis*. J. Herpetol. 27(1):33–41.

Gray, J. E. 1831. Synopsis Reptilium or short descriptions of the species of reptiles. Part 1. Cataphracta, Tortoises, Crocodiles, and Enaliosaurians. Treuttel, Wurtz and Co.; G. B. Sowerby; W. Wood, London. viii + 85 p.

———. 1838. Catalogue of the slender-tongued saurians, with descriptions of many new genera and species. Part 2. Ann. Mag. Nat. Hist. (1)1(5):388–394.

———. 1845. Catalogue of the specimens of lizards in the collection of the British Museum. Taylor and Francis, London. xxviii + 289 p.

Green, J. 1825. Description of a new species of salamander. J. Acad. Nat. Sci. Philadelphia 5:116–118.

Greene, H. W. 1970. Beobachtungen zur fortpflanzung von *Sceloporus poinsettii* (Reptilia, Iguanidae). Salamandra 6(1/2):48–50.

———. 1984. Taxonomic status of the western racer, *Coluber constrictor mormon*. J. Herpetol. 18(2):210–211.

———and **G. V. Oliver, Jr.** 1965. Notes on the natural history of the western massasauga. Herpetologica 21(3):225–228.

Gregory, P. T. 1977. Life history parameters of the red-sided garter snake (*Thamnophis sirtalis parietalis*) in the interlake region of Manitoba. Natl. Mus. Canada Publ. Zool. 13:1–44.

———. 1982. Reptilian hibernation, p. 53–154. *In* C. Gans and F. H. Pough (eds.), Biology of the Reptilia, Vol. 13, Physiology D, Physiological Ecology. Academic Press, New York. xiii + 345 p.

———. 1984. Communal denning in snakes, p. 57–75. *In* R. A. Seigel, L. E. Hunt, J. L. Knight, L. Malaret, and N. L. Zuschlag (eds.), Vertebrate Ecology and Systematics—A Tribute to Henry S. Fitch. Special Publ No. 10. Mus. Nat. History, Univ. Kansas, Lawrence.

Grismer, L. L. 1988. Phylogeny, taxonomy, classification, and biogeography of eublepharid geckos, p. 369–469. *In* R. Estes and G. Pregill (eds.), Phylogenetic Relationships of the Lizard Families. Stanford Univ. Press, Stanford, California. xii + 631 p.

Grobman, A. B. 1941. A contribution to the knowledge of variation in *Opheodrys vernalis* (Harlan), with the description of a new subspecies. Univ. Michigan Mus. Zool. Misc. Publ. (50):1–38.

———. 1989. Clutch size and female length in *Opheodrys vernalis*. Herpetol. Rev. 20(4):84.

———. 1990. The effects of soil temperatures on emergence from hibernation of *Terrapene carolina* and *T. ornata*. Amer. Midl. Nat. 124(2):366–371.

———. 1992. Metamerism in the snake *Opheodrys vernalis*, with a description of a new subspecies. J. Herpetol. 26(2):176–186.

Günther. A. C. L. G. 1860. Description of *Leptodeira torquata*, a new snake from Central America. Ann. Mag. Nat. Hist. (3)5(27):169–171.

Guyer, C. and **A. D. Linder.** 1985. Thermal ecology and activity patterns of the short-horned lizard *(Phrynosoma douglassi)* and the sagebrush lizard *(Sceloporus graciosus)* in southeastern Idaho. Great Basin Nat. 45(4):607–614.

Hahn, D. E. 1968. A biogeographic analysis of the herpetofauna of the San Luis Valley, Colorado. M.S. Thesis, Louisiana St. Univ., Baton Rouge. vii + 103 p.

———. 1979a. Leptotyphlopidae, *Leptotyphlops*. Cat. Amer. Amphib. Rept.:230.1–230.4.

———. 1979b. *Leptotyphlops dulcis*. Cat. Amer. Amphib. Rept.:231.1–231.2.

———. 1979c. *Leptotyphlops humilis*. Cat. Amer. Amphib. Rept.:232.1–232.4.

Hahn, W. E. 1964. Seasonal changes in testicular and epididymal histology and spermatogenic rate in the lizard *Uta stansburiana stejnegeri*. J. Morphol. 115(3):447–460.

———and **D. W. Tinkle.** 1965. Fat body cycling and experimental evidence for its adaptive significance to ovarian follicle development in the lizard *Uta stansburiana*. J. Exp. Zool. 158(1):79–85.

Hakkila, M. 1994. An assessment of potential habitat and distribution of the gray-banded kingsnake (*Lampropeltis alterna*) in New Mexico. Unpubl. Rept. submitted to NM Dept. Game & Fish. Santa Fe, NM 12 pp + 3 maps.

Hall, R. J. 1971. Ecology of a population of the Great Plains Skink (*Eumeces obsoletus*). Univ. Kansas Sci. Bull. 49(7):357–388.

———. 1976. *Eumeces obsoletus*. Cat. Amer. Amphib. Rept.:186.1–186.2.

———and **H. S. Fitch.** 1971. Further observations on the demography of the Great Plains skink (*Eumeces obsoletus*). Trans. Kansas Acad. Sci. 74(1):93–98.

LITERATURE CITED

Hallowell, E. 1852. Descriptions of new species of reptiles inhabiting North America. Proc. Acad. Nat. Sci. Philadelphia 6:177–182.

———. 1852 [1853]. On a new genus and three new species of reptiles inhabiting North America. Proc. Acad. Nat. Sci. Philadelphia 6:206–209.

———. 1853. Reptiles, p. 106–147. *In* L. Sitgreaves. Report of an expedition down the Zuñi and Colorado Rivers in 1851. 32rd Congress, 2nd Sess., Sen. Exec. Doc. 59:1–198.

———. 1854. Descriptions of new reptiles from California. Proc. Acad. Nat. Sci. Philadelphia 7:91–97.

———. 1856 [1857?]. Notice of a collection of reptiles from Kansas and Nebraska, presented to the Academy of Natural Sciences by Dr. Hammond. Proc. Acad. Nat. Sci. Philadelphia 8:238–253.

———. 1857 [1858]. Description of several new North American reptiles. Proc. Acad. Nat. Sci. Philadelphia 9(14–16):215–216.

Halloy, M. and G. M. Burghardt. 1990. Ontogeny of fish capture and ingestion in four species of garter snakes (*Thamnophis*). Behavior 112(3–4):299–317.

Hamilton, W. J., Jr. 1947. Hibernation of the lined snake. Copeia 1947(1):209–210.

———. 1950. Food of the prairie rattlesnake. Herpetologica 6(2):34.

Hammerson, G. A. 1982. Amphibians and Reptiles in Colorado. Colorado Div. Wildl. Publ. DOW-M-I-27–82. vii + 131 p.

——— and H. M. Smith. 1991. The correct spelling of the name for the short-horned lizard of North America. Bull. Maryland Herpetol. Soc. 27(3):121–127.

———, L. Valentine, and L. J. Livo. 1991. Geographic Distribution: *Lampropeltis getula* (common kingsnake). Herpetol. Rev. 22(2):67.

Hanson, J. A. and J. L. Vial. 1956. Defensive behavior and effects of toxins in *Bufo alvarius*. Herpetologica 12(2):141–149.

Hardy, D. F. 1962. Ecology and behavior of the six-lined racerunner, *Cnemidophorus sexlineatus*. Univ. Kansas Sci. Bull. 43(1):1–73.

Hardy, D. L. 1983. Envenomation by the Mojave rattlesnake (*Crotalus scutulatus scutulatus*) in southern Arizona, U.S.A. Toxicon 21(1):111–118.

———. 1986. Fatal rattlesnake envenomation in Arizona: 1969–1984. Clinical Toxicol. 24(1):1–10.

Hardy, L. M. 1975. A systematic revision of the colubrid snake genus *Gyalopion*. J. Herpetol. 9(1):107–132.

———. 1976. *Gyalopion*, *G. canum*, *G. quadrangularis*. Cat. Amer. Amphib. Rept.:182.1–182.4.

——— and R. W. McDiarmid. 1969. The amphibians and reptiles of Sinaloa, Mexico. Univ. Kansas Publ. Mus. Nat. Hist. 18(3):39–252.

Harlan, R. 1825. Description of two new species of *Agama*. J. Acad. Nat. Sci. Philadelphia 4:296–304.

———. 1827. Genera of North American Reptilia, and a synopsis of the species. J. Acad. Nat. Sci. Philadelphia 6:7–37.

Harper, F. 1940. Some works of Bartram, Daudin, Latreille, and Sonnini, and their bearing upon North American herpetological nomenclature. Amer. Midl. Nat. 23(3):692–723.

———. 1947. A new cricket frog (*Acris*) from the middle western states. Proc. Biol. Soc. Washington 60:39–40.

———. 1955. The type locality of *Hyla triseriata* Wied. Proc. Biol. Soc. Washington 68:155–156.

Harris, A. H. 1959. Second record of the Arizona lyre snake in New Mexico. Southwest. Nat. 4(1):42–43.

——— (ed.). 1963. Ecological distribution of some vertebrates in the San Juan Basin, New Mexico. Mus. New Mexico Pap. Anthropol. (8):1–64.

Harris, D. M. 1985. Infralingual plicae: support for Boulenger's Teiidae (Sauria). Copeia 1985(3):560–565.

Harris, H. S., Jr. 1974. The New Mexican ridge-nosed rattlesnake. Natl. Parks Conserv. Mag. 48(3):22–24.

——— and R. S. Simmons. 1975. An endangered species, the New Mexican ridge-nosed rattlesnake. Bull. Maryland Herpetol. Soc. 11(1):1–7.

——— and ———. 1976. The paleogeography and evolution of *Crotalus willardi*, with a formal description of a new subspecies from New Mexico, United States. Bull. Maryland Herpetol. Soc. 12(1):1–22.

Harte, J. and E. Hoffman. 1989. Possible effects of acidic deposition on a Rocky Mountain population of the tiger salamander *Ambystoma tigrinum*. Conserv. Biol. 3(2):149–158.

Hartweg, N. 1939. A new American *Pseudemys*. Occ. Pap. Mus. Zool. Univ. Michigan (397):1–4.

———. 1940. Description of *Salvadora intermedia*, new species, with remarks on the *grahamiae* group. Copeia 1940(3):256–259.

———. 1944. Remarks on some Mexican snakes of the genus *Tantilla*. Occ. Pap. Mus. Zool. Univ. Michigan (486):1–9.

Hasson, O., R. Hibbard, and G. Ceballos. 1989. The pursuit deterrent function of tail-wagging in the zebra-tailed lizard (*Callisaurus draconoides*). Can. J. Zool. 67(5):1203–1209.

Hayes, M. P. and M. R. Jennings. 1986. Decline of ranid frog species in western North America: are bullfrogs (*Rana catesbeiana*) responsible? J. Herpetol. 20(4):490–509.

Heath, J. E. 1965. Temperature regulation and diurnal activity in horned lizards. Univ. California Publ. Zool. 64(3):97–136.

Heckrotte, C. 1962. The effect of the environmental factors in the locomotory activity of the plains garter snake (*Thamnophis radix radix*). Anim. Behav. 10(3–4):193–207.

Henderson, J. and **J. P. Harrington.** 1914. Ethnozoology of the Tewa Indians. Bull. Bureau Amer. Ethnology, Smithsonian Inst. (56):x + 76 p.

Hendricks, F. S. and **J. R. Dixon.** 1984. Population structure of *Cnemidophorus tigris* (Reptilia: Teiidae) east of the Continental Divide. Southwest. Nat. 29(1):137–140.

——— and ———. 1986. Systematics and biogeography of *Cnemidophorus marmoratus* (Sauria: Teiidae). Texas J. Sci. 38(4):327–402.

——— and ———. 1988. Regenerated tail frequencies in populations of *Cnemidophorus marmoratus* (Reptilia: Teiidae). Southwest. Nat. 33(1):121–124.

Hensley, M. M. 1949. Mammal diet of *Heloderma*. Herpetologica 5(6):152.

———. 1950. Results of a herpetological reconnaissance in extreme southwestern Arizona and adjacent Sonora, with a description of a new subspecies of the Sonoran whipsnake, *Masticophis bilineatus*. Trans. Kansas Acad. Sci. 53(2):270–288.

Herrick, C. L., J. Terry, and **H. N. Herrick, Jr.** 1899. Notes on a collection of lizards from New Mexico. Bull. Sci. Lab. Denison Univ. 11(6):117–148.

Hibbard, C. W. 1964. A brooding colony of the blind snake, *Leptotyphlops dulcis dissecta* Cope. Copeia 1964(1):222.

Hillis, D. M. 1988. Systematics of the *Rana pipiens* complex: puzzle and paradigm. Ann. Rev. Ecol. Syst. 19:39–63.

———, **J. S. Frost,** and **D. A. Wright.** 1983. Phylogeny and biogeography of the *Rana pipiens* complex: a biochemical evaluation. Syst. Zool. 32(2):132–143.

Hoff, C. L. and **F. R. Kay.** 1970. Herbivorous feeding in the lizard *Uta stansburiana steinegeri*. Southwest. Nat. 15(1):137–138.

Holland, R. L. 1965. A comparative study of morphology and plasma proteins of the blood in the lizard species *Cnemidophorus tesselatus* (Say) (Reptilia: Teiidae) from Colorado and New Mexico. M. S. Thesis, Univ. Colorado, Boulder. x + 116 p.

Holomuzki, J. R. 1986. Predator avoidance and diel patterns of microhabitat use by larval tiger salamanders. Ecology 67(3):737–748.

——— and **J. P. Collins.** 1987. Trophic dynamics of a top predator, *Ambystoma tigrinum nebulosum* (Caudata: Ambystomatidae) in a lentic community. Copeia 1987(4):949–957.

Holycross, A. T. and **C. Schwalbe.** 1995. Geographic Distribution. *Lampropeltis triangulum*. Herpetol. Rev. 26(1):46.

Hotton, N., III. 1955. A survey of adaptive relationships of dentition to diet in the North American Iguanidae. Amer. Midl. Nat. 53(1):88–114.

Hover, E. L. 1985. Differences in aggressive behavior between two throat color morphs in a lizard, *Urosaurus ornatus*. Copeia 1985 (4):933–940.

Howard, C. W. 1974. Comparative reproductive ecology of horned lizards (genus *Phrynosoma*) in southwestern United States and northern Mexico. J. Arizona Acad. Sci. 9(3):108–116.

Howland, J. M. 1992. Life history of *Cophosaurus texanus* (Sauria: Iguanidae): environmental correlates and interpopulational variation. Copeia 1992(1):82–93.

Huey, L. M. 1942. A vertebrate faunal survey of the Organ Pipe Cactus National Monument, Arizona. Trans. San Diego Soc. Nat. Hist. 9(32):353–376.

Hubbard, J. P., M. C. Conway, H. Campbell, G. Schmitt, and **M. D. Hatch.** 1978. Handbook of Species Endangered in New Mexico. New Mexico Dept. Game and Fish. Santa Fe. vi + 202 p.

Hughes, N. 1965. Comparison of frontoparietal bones of *Scaphiopus bombifrons* and *S. hammondi* as evidence of interspecific hybridization. Herpetologica 21(3):196–201.

Hulse, A. C. 1971. Fluorescence in *Leptotyphlops humilis* (Serpentes: Leptotyphlopidae). Southwest. Nat. 16(1):123–124.

———. 1973. Herpetofauna of the Fort Apache Indian Reservation, east central Arizona. J. Herpetol. 7(3):275–282.

———. 1974a. Food habits and feeding behavior in *Kinosternon sonoriense* (Chelonia: Kinosternidae). J. Herpetol. 8(3):195–199.

———. 1974b. An autecological study of *Kinosternon sonoriense* LeConte (Chelonia: Kinosternidae). Ph.D. Diss., Arizona St. Univ., Tempe. x + 105 p.

———. 1981. Ecology and reproduction of the parthenogenetic lizard *Cnemidophorus uniparens* (Teiidae). Ann. Carnegie Mus. 50(14):353–369.

——— and **G. A. Middendorf.** 1979. Notes on the occurrence of *Gopherus agassizi* (Testudinidae) in extreme eastern Arizona. Southwest. Nat. 24(3):545–546.

Humphrey, F. 1956. The gestation and incubation period of an Arizona kingsnake. Herpetologica 12(4):311.

Hunt, R. H. 1980. Toad sanctuary in a tarantula burrow. Natural History 89(3):48–53.

Hutchison, V. H., A. Vinegar, and **R. J. Kosh.** 1966. Critical thermal maxima in turtles. Herpetologica 22(1):32–41.

Inger, R. F. 1958. The vocal sac of the Colorado River toad (*Bufo alvarius* Girard). Texas J. Sci. 10(3):319–324.

Ingram, W., III and **W. W. Tanner.** 1971. A taxonomic study of *Crotaphytus collaris* between the Rio Grande and Colorado rivers. Brigham Young Univ. Sci. Bull. Biol. Ser. 13(2):1–29.

Irwin, L. N. 1965. Diel activity and social interaction of the lizard *Uta stansburiana stejnegeri*. Copeia 1965(1):99–101.

Iverson, J. B. 1976. *Kinosternon sonoriense*. Cat. Amer. Amphib. Rept.:176.1–176.2.

———. 1978. Distributional problems of the genus *Kinosternon* in the American southwest. Copeia 1978(3):476–479.

LITERATURE CITED

———. 1981. Biosystematics of the *Kinosternon hirtipes* species group (Testudines: Kinosternidae). Tulane Stud. Zool. Bot. 23(1):1–74.

———. 1990. Nesting and parental care in the mud turtle, *Kinosternon flavescens*. Can. J. Zool. 68(2):230–233.

———. 1991. Life history and demography of the yellow mud turtle, *Kinosternon flavescens*. Herpetologica 47(4):373–395.

———. 1992. A revised checklist with distribution maps of the turtles of the world. Privately printed, Richmond, Indiana. xiii + 363 p.

——— and **G. R. Smith**. 1993. Reproductive ecology of the painted turtle *(Chrysemys picta)* in the Nebraska Sandhills and across its range. Copeia 1993(1):1–21.

Jacob, J. S. 1977. An evaluation of the possibility of hybridization between the rattlesnakes *Crotalus atrox* and *C. scutulatus* in the southwestern United States. Southwest. Nat. 22(4):469–485.

——— and **J. S. Altenbach**. 1977. Sexual color dimorphism in *Crotalus lepidus klauberi* Gloyd (Reptilia, Serpentes, Viperidae). J. Herpetol. 11(1):81–84.

——— and **C. W. Painter**. 1980. Overwinter thermal ecology of *Crotalus viridis* in the north-central plains of New Mexico. Copeia 1980(4):799–805.

———, **S. R. Williams**, and **R. P. Reynolds**. 1987. Reproductive activity of male *Crotalus atrox* and *C. scutulatus* (Reptilia: Viperidae) in northeastern Chihuahua, Mexico. Southwest. Nat. 32(2):273–276.

Jacobson, E. R. and **W. G. Whitford**. 1971. Physiological responses to temperature in the patch-nosed snake, *Salvadora hexalepis*. Herpetologica 27(3):289–295.

James, E. 1823. An Account of an Expedition from Pittsburgh to the Rocky Mountains, performed in the Years 1819 and '20, by order of the Hon. J. C. Calhoun, Sec'y of War: under the command of Major Stephen H. Long. Two volumes. H. C. Carey, and I. Lea, Philadelphia. 503 + 442 p. + 98 p. atlas, 8 plates, 2 charts, and 2 folding maps.

Jameson, D. L. 1950. The development of *Eleutherodactylus latrans*. Copeia 1950(1):44–46.

———. 1954. Social patterns in the leptodactylid frogs *Syrrhophus* and *Eleutherodactylus*. Copeia 1954(1)36–38.

———. 1956. Duplicate feeding habits in snakes. Copeia 1956(1):54–55.

——— and **A. G. Flury**. 1949. The reptiles and amphibians of the Sierra Vieja range of southwestern Texas. Texas J. Sci. 1(2):54–79.

Jan, G. 1863. Eleno Sistematico Degli Ofidi Descritti e Diseganti per L'Iconografia Generale. Lombardi, Milan. viii + 143 p.

Jennings, M. R. and **M. P. Hayes**. 1994. Decline of native ranid frogs in the desert southwest, p. 183–211. *In* Brown, P. R. and J. W. Wright. eds. Herpetology of the North American Deserts. Proceedings of a Symposium. Southwest. Herpetol. Soc. Special Publ. No. 5. iv + 311 p.

Jennings, R. D. 1987. The status of *Rana berlandieri*, the Rio Grande leopard frog, and *Rana yavapaiensis*, the lowland leopard frog in New Mexico. Unpubl. Rept. New Mexico Dept. Game and Fish, Santa Fe. iv + 44 p.

———. 1988. Ecological studies of the Chiricahua leopard frog, *Rana chiricahuensis*, in New Mexico. Unpubl. Rept. New Mexico Dept. Game and Fish, Santa Fe. 29 p.

———. 1990. Activity and reproductive phenologies and their ecological correlates among populations of the Chiricahua leopard frog, *Rana chiricahuensis*. Unpubl. Rept. New Mexico Dept. Game and Fish, Santa Fe. 46 p.

——— and **N. J. Scott, Jr.** 1991. Global amphibian population declines: insights from leopard frogs in New Mexico. Unpubl. Rept. New Mexico Dept. Game and Fish, Santa Fe. 43 p.

——— and ———. 1993. Ecologically correlated morphological variation in tadpoles of the leopard frog, *Rana chiricahuensis*. J. Herpetol. 27(3):285–293.

Johnson, D. R. 1966. Diet and estimated energy assimilation of three Colorado lizards. Amer. Midl. Nat. 76(2):504–509.

Johnson, J. A. and **E. D. Brodie, Jr.** 1974. Defensive behavior of the western banded gecko, *Coleonyx variegatus*. Anim. Behav. 22(3):684–687.

Johnson, T. R. 1987. The Amphibians and Reptiles of Missouri. Missouri Dept. Conserv., Jefferson City. xi + 368 p.

Johnston, R. F. and **G. A. Schad**. 1959. Natural history of the salamander, *Aneides hardii*. Univ. Kansas Publ. Mus. Nat. Hist. 10(8):573–585.

Jones, J. P. 1926. The proper name for *Sceloporus consobrinus* Baird and Girard. Occ. Pap. Mus. Zool. Univ. Michigan (172):1–3.

Jones, K. B. 1981. Effects of grazing on lizard abundance and diversity in western Arizona. Southwest. Nat. 26(2):107–115.

———. 1983. Movement patterns and foraging ecology of Gila monsters (*Heloderma suspectum* Cope) in northwestern Arizona. Herpetologica 39(3):247–253.

———. 1990. Habitat use and predatory behavior of *Thamnophis cyrtopsis* (Serpentes: Colubridae) in a seasonally variable aquatic environment. Southwest. Nat. 35(2):115–122.

Jones, K. L. 1970. An ecological survey of the reptiles and amphibians of Chaco Canyon National Monument, San Juan County, New Mexico. M. S. Thesis, Univ. New Mexico, Albuquerque. 68 p.

———. 1978. Status of *Bufo boreas* in New Mexico project. Unpubl. Rept. New Mexico Dept. Game and Fish, Santa Fe. 9 p.

Jones, S. M. and R. E. Ballinger. 1987. Comparative life histories of *Holbrookia maculata* and *Sceloporus undulatus* in western Nebraska. Ecology 68(6):1828–1838.

——— and D. L. Droge. 1980. Home range size and spatial distributions of two sympatric lizard species (*Sceloporus undulatus, Holbrookia maculata*) in the sand hills of Nebraska. Herpetologica 36(2):127–132.

———, S. R. Waldschmidt, and M. A. Potvin. 1987. An experimental manipulation of food and water: growth and time-space utilization of hatchling lizards (*Sceloporus undulatus*). Oecologia 73(1):53–59.

Jones, T. R. and J. P. Collins. 1992. Analysis of a hybrid zone between subspecies of the tiger salamander (*Ambystoma tigrinum*) in central New Mexico, USA. J. Evol. Biol. 5(3):375–402.

Jordan, O. R. 1967. The occurrence of *Thamnophis sirtalis* and *T. radix* in the prairie-forest ecotone west of Itasca State Park, Minnesota. Herpetologica 23(4):303–308.

Jorgensen, C. D. and W. W. Tanner. 1963. The application of the density probability function to determine the home ranges of *Uta stansburiana stansburiana* and *Cnemidophorus tigris tigris*. Herpetologica 19(2):105–115.

Justus, J. T., M. Sandomir, T. Urquhart, and B. O. Ewan. 1977. Developmental rates of two species of toads from the desert southwest. Copeia 1977(3):592–594.

Kassing, E. F. 1961. A life history of the Great Plains ground snake, *Sonora episcopa episcopa* (Kennicott). Texas J. Sci. 13(2):185–203.

Kauffeld, C. F. 1943a. Field notes on some Arizona reptiles and amphibians. Amer. Midl. Nat. 29(2):342–359.

———. 1943b. Growth and feeding of new-born Price's and green rock rattlesnakes. Amer. Midl. Nat. 29(3):606–614.

———. 1948. Notes on a hook-nosed snake from Texas. Copeia 1948(4):301.

———. 1949. Arizona adventure. In Animaland. Staten Island Zool. Soc. 16(6):1–3.

———. 1954. Gila Monster. In Animaland. Staten Island Zool. Soc. 21(5):1 3.

———. 1969. Snakes: The Keeper and the Kept. Doubleday and Co., Inc., New York. xii + 248 p.

Kellogg, R. 1932. Mexican tailless amphibians in the United States National Museum. Bull. U.S. Natl. Mus. (160):iv + 224 p.

Kennedy, J. P. 1973. *Sceloporus olivaceus*. Cat. Amer. Amphib. Rept.:143.1–143.4.

Kennerly, C. B. R. 1856. Field notes and explanations, p. 5–17. *In* Report on the zoology of the expedition. *In* Reports of explorations and surveys, to ascertain the most practicable and economical route of a railroad from the Mississippi River to the Pacific Ocean. Vol. IV(VI, 1). 33rd Congress, 2nd Sess., House of Rep. Exec. Doc. (91), Washington, D.C.

Kennicott, R. 1859. Notes on *Coluber calligaster* of Say, and a description of new species of serpents in the collection of the North Western University of Evanston, Ill. Proc. Acad. Nat. Sci. Philadelphia 11(7–8):98–100.

———. 1860. Descriptions of new species of North American serpents in the Museum of the Smithsonian Institution, Washington. Proc. Acad. Nat. Sci. Philadelphia 12(22–24):328–338.

———. 1861. On three new forms of rattlesnakes. Proc. Acad. Nat. Sci. Philadelphia 13(11–17):206–208.

Kerfoot, W. C. 1968. Geographic variability of the lizard, *Sceloporus graciosus* Baird and Girard, in the eastern part of its range. Copeia 1968(1):139–152.

King, F. W. 1932. Herpetological records and notes from the vicinity of Tucson, Arizona, June and July, 1930. Copeia 1932(4):175–177.

——— and R. L. Burke (eds.). 1986. Crocodilian, Tuatara, and Turtle Species of the World: A Taxonomic and Geographic Reference. Assoc. Syst. Coll., Washington, D.C. xxii + 216 p.

King, O. M. 1960. Observations on Oklahoma toads. Southwest. Nat. 5(2):102–103.

Kingsbury, B. A. 1989. Factors influencing activity in *Coleonyx variegatus*. J. Herpetol. 23(4):399–404.

Klauber, L. M. 1931. Notes on the worm snakes of the southwest, with descriptions of two new subspecies. Trans. San Diego Soc. Nat. Hist. 6(28):333–352.

———. 1934. An addition to the fauna of New Mexico and a deletion. Copeia 1934(1):52.

———. 1935. A new subspecies of *Crotalus confluentus*, the prairie rattlesnake. Trans. San Diego Soc. Nat. Hist. 8(13):75–90.

———. 1937. A statistical study of the rattlesnakes. IV. The growth of the rattlesnake. Occ. Pap. San Diego Soc. Nat. Hist. (3):1–56.

———. 1939a. A new subspecies of the western worm snake. Trans. San Diego Soc. Nat. Hist. 9(14):67–68.

———. 1939b. Studies of reptile life in the arid southwest. Part I. Night collecting on the desert with ecological statistics. Bull. Zool. Soc. San Diego 14(1):6–64.

———. 1940a. The worm snakes of the genus *Leptotyphlops* in the United States and northern Mexico. Trans. San Diego Soc. Nat. Hist. 9(18):87–162.

———. 1940b. The lyre snakes (genus *Trimorphodon*) of the United States. Trans. San Diego Soc. Nat. Hist. 9(19):163–194.

———. 1941. The long-nosed snakes of the genus *Rhinocheilus*. Trans. San Diego Soc. Nat. Hist. 9(29):289–332.

LITERATURE CITED

———. 1945. The geckos of the genus *Coleonyx* with descriptions of new subspecies. Trans. San Diego Soc. Nat. Hist. 10(11):133–216.

———. 1946. The glossy snake, *Arizona*, with descriptions of new subspecies. Trans. San Diego Soc. Nat. Hist. 10(17):311–398.

———. 1947. Classification and ranges of the gopher snakes of the genus *Pituophis* in the western United States. Bull. Zool. Soc. San Diego 22:1–18.

———. 1948. Some misapplications of the Linnaean names applied to American snakes. Copeia 1948(1):1–14.

———. 1949. The subspecies of the ridge-nosed rattlesnake, *Crotalus willardi*. Trans. San Diego Soc. Nat. Hist. 11(8):121–140.

———. 1952a. Taxonomic studies of the rattlesnakes of mainland Mexico. Bull. Zool. Soc. San Diego (26):1–143.

———. 1952b. Proposed use of the plenary powers to preserve for the western diamond rattlesnake the trivial name "*atrox*" Baird and Girard, 1853 (as published in the combination "*Crotalus atrox*") by suppressing the trivial name "*cinereus*" Le Conte *in* Hallowell, 1852 (as published in combination "*Crotalus cinereus*") (Class Reptilia, Order Squamata). Bull. Zool. Nomen. 6:234–236.

———. 1956. Rattlesnakes: Their Habits, Life Histories, and Influence on Mankind. Two Vols. Univ. California Press, Berkeley. xxix + 708 p., 3 folding tables; xvii + p. 709–1476.

———. 1972. Rattlesnakes: Their Habits, Life Histories, and Influence on Mankind. 2nd ed. Two Vols. Univ. California Press, Berkeley. xxx + 704 p., 3 folding tables; xxviii + p. 741–1533.

Kluge, A. G. 1966. A new pelobatine frog from the Lower Miocene of South Dakota with a discussion of the evolution of the *Scaphiopus-Spea* complex. Contrib. Sci. Los Angeles Co. Mus. Nat. Hist. (133):1–26.

———. 1987. Cladistic relationships in the Gekkonoidea (Squamata, Sauria). Misc. Publ. Mus. Zool. Univ. Michigan (173):iv + 54 p.

Knight, R. A. 1985. Populational differences in venoms of *Crotalus scutulatus scutulatus*. M.S. Thesis, Sul Ross St. Univ., Alpine, Texas. vii + 38 p.

——— and D. Duerre. 1987. Notes on distribution, habitat, and sexual dimorphism of *Gerrhonotus kingii* (Lacertilia: Anguidae). Southwest. Nat. 32(2):283–285.

Knopf, G. N. 1966. Reproductive behavior and ecology of the unisexual lizard, *Cnemidophorus tesselatus* Say. Ph.D. Diss., Univ. Colorado, Boulder. 111 p.

Knowlton, G. F. 1938. Lizards in insect control. Ohio J. Sci. 38(5):235–238.

———. 1969 [1971]. Some insect food of Curlew Valley lizards. Proc. Utah Acad. Sci. Arts Lett. 41(2):160–161.

——— and W. L. Thomas. 1934. Insect food of Troutcreek lizards. Proc. Utah Acad. Sci. Arts Lett. 12:263–264.

Kofron, C. P. and A. A. Schreiber. 1987. Observations on aquatic turtles in a northeastern Missouri marsh. Southwest. Nat. 32(4):517–521.

Koster, W. J. 1940. The first record of the snake, *Hypsiglena*, from New Mexico. Herpetologica 2(2):30.

———. 1946a. Records of the snapping turtle from New Mexico. Copeia 1946(3):173.

———. 1946b. The robberfrog in New Mexico. Copeia 1946(3):173.

———. 1951. The distribution of the Gila monster in New Mexico. Herpetologica 7(3):97–101.

Krohmer, R. W. and R. D. Aldridge. 1985a. Male reproductive cycle of the lined snake (*Tropidoclonion lineatum*). Herpetologica 41(1):33–38.

——— and ———. 1985b. Female reproductive cycle of the lined snake (*Tropidoclonion lineatum*). Herpetologica 41(1):39–44.

Kroll, J. C. 1971. Combat behavior in male Great Plains ground snakes (*Sonora episcopa episcopa*). Texas J. Sci. 23(2):300.

Krupa, J. J. 1986a. Anuran breeding dates in central Oklahoma. Bull. Oklahoma Herpetol. Soc. 11(1–4):10–13.

———. 1986b. Multiple egg clutch production in the Great Plains toad. Prairie Nat. 18(3):151–152.

———. 1988. Fertilization efficiency of the Great Plains toad (*Bufo cognatus*). Copeia 1988(3):800–802.

———. 1989. Alternative mating tactics in the Great Plains toad. Anim. Behav. 37(6):1035–1043.

———. 1990a. Advertisement call variation in the Great Plains toad. Copeia 1990(3):884–886.

———. 1990b. *Bufo cognatus*. Cat. Amer. Amphib. Rept.:457.1–457.8.

———. 1991. Night chorus. Nebraskaland 69(3):8–15.

———. 1994. Breeding biology of the Great Plains toad in Oklahoma. J. Herpetol. 28(2):217–224.

———. 1995. Natural history notes. *Bufo woodhousii* (Woodhouse's toad). Fecundity. Herpetol. Rev. 26(3):142, 144.

Lacepède, De, B-G.-É. 1789. Historie naturelle des quadrupèdes ovipares et des serpens, Vol 2. Hôtel de thou. xx + 527 p.

Landreth, H. F. 1973. Orientation and behavior of the rattlesnake, *Crotalus atrox*. Copeia 1973(1):26–31.

Lang, J. W. 1969. Hibernation and movements of *Storeria occipitomaculata* in northern Minnesota. J. Herpetol. 3(3–4):196–197.

Lannoo, M. J. and **M. D. Bachmann.** 1984. Aspects of cannibalistic morphs in a population of *Ambystoma t. tigrinum* larvae. Amer. Midl. Nat. 112(1):103–110.

Lanza, B., S. Vanni, and **A. Nistri.** 1992. Salamanders and newts, p. 60–75. *In* H. G. Cogger and R. G. Zweifel (eds.), Reptiles and Amphibians. Smithmark Publ., Inc., New York. 240 p.

Lardner, P. J. 1969. Diurnal and seasonal locomotory activity in the Gila monster, *Heloderma suspectum* Cope. Ph.D. Diss., Univ. Arizona, Tucson. xiv + 99 p.

Larson, D. W. 1974. Maximum size for ambystomatid salamanders. Prairie Nat. 6(2):32.

Latreille, P. A. 1801. *In* C. S. Sonnini and P. A. Latreille. Histoire Naturelle des Reptiles, avec Figures Dessinées apres Nature. Vol. 2. Chez Deterville, Paris, 332 p. + 21 plates.

Lawson, P. A. and **D. M. Secoy.** 1991. The use of solar cues as migratory orientation guides by the plains garter snake, *Thamnophis radix.* Can. J. Zool. 69(10):2700–2702.

Lawson, R. and **H. C. Dessauer.** 1981. Electrophoretic evaluation of the colubrid genus *Elaphe* (Fitzinger). Isozyme Bull. 14:83.

LeConte, J. L. 1854. Descriptions of four new species of *Kinosternon.* Proc. Acad. Nat. Sci. Philadelphia 7:180–190.

Legler, J. M. 1955. Observations on the sexual behavior of captive turtles. Lloydia 18(2):95–99.

———. 1958 [1959]. The Texas slider (*Pseudemys floridana texana*) in New Mexico. Southwest. Nat. 3(1–4):230–231.

———. 1960a. Remarks on the natural history of the Big Bend slider, *Pseudemys scripta gaigae* Hartweg. Herpetologica 16(2):139–140.

———. 1960b. Natural history of the ornate box turtle, *Terrapene ornata ornata* Agassiz. Univ. Kansas Publ. Mus. Nat. Hist. 11(10):527–669.

———. 1990. The genus *Pseudemys* in Mesoamerica: taxonomy, distribution, and origins, p. 82–105. *In* J. W. Gibbons (ed.), Life History and Ecology of the Slider Turtle. Smithsonian Inst. Press, Washington, D.C. xiv + 368 p.

——— and **L. J. Sullivan.** 1979. The application of stomach-flushing to lizards and anurans. Herpetologica 35(2):107–110.

LeSueur, C. A. 1827. Note sur duex especes de tortues du genre *Trionyx* Gffr. St. H. Mem. Mus. Hist. Nat. Paris 15:257–268.

Leuck, B. E. 1982. Comparative burrow use and activity patterns of parthenogenetic and bisexual whiptail lizards (*Cnemidophorus*: Teiidae). Copeia 1982(2):416–424.

———. 1985. Comparative social behavior of bisexual and unisexual whiptail lizards (*Cnemidophorus*). J. Herpetol. 19(4):492–506.

———, **E. E. Leuck II,** and **R. T. B. Sherwood.** 1981. A new population of New Mexico whiptail lizards, *Cnemidophorus neomexicanus* (Teiidae). Southwest. Nat. 26(1):72–74.

Lewis, T. H. 1948. *Elaphe sclerotica* in New Mexico. Herpetologica 4(6):223.

———. 1949. Dark coloration in the reptiles of the Tularosa malpais, New Mexico. Copeia 1949(3):181–184.

———. 1950. The herpetofauna of the Tularosa Basin and Organ Mountains of New Mexico with notes on some ecological features of the Chihuahuan Desert. Herpetologica 6(1):1–10.

———. 1951. Dark coloration in the reptiles of the malpais of the Mexican border. Copeia 1951(4):311–312.

Lieb, C. S. 1985. Systematics and distribution of the skinks allied to *Eumeces tetragrammus* (Sauria: Scincidae). Contrib. Sci. Nat. Hist. Mus. Los Angeles Co. (357):1–19.

———. 1990. *Eumeces tetragrammus.* Cat. Amer. Amphib. Rept.:492.1–492.4.

Lindeman, P. V. 1991. Survivorship of overwintering hatchling painted turtles, *Chrysemys picta*, in northern Michigan. Canadian Field Nat. 105(2):263–266.

Linnaeus, C. 1758. Systema Naturae, Ed. 10. Laurenti Salvi, Stockholm. 823 p.

———. 1766. Systema Naturae, Ed. 12. Part 1. Laurenti Salvi, Stockholm. 532 p.

Linsdale, J. M. 1933. A specimen of *Rana tarahumarae* from New Mexico. Copeia 1933(4):222.

Little, E. L., Jr. 1940. Amphibians and reptiles of the Roosevelt Reservoir area, Arizona. Copeia 1940(4):260–265.

——— and **J. G. Keller.** 1937. Amphibians and reptiles of the Jornada Experimental Range, New Mexico. Copeia 1937(4):216–222.

Livezey, R. L. and **A. H. Wright.** 1947. A synoptic key to the salientian eggs of the United States. Amer. Midl. Nat. 37(1):179–222.

Livo, L. J. 1989. Life History Notes: *Eumeces obsoletus* (Great Plains skink). Escape behavior. Herpetol. Rev. 20(3):70–71.

Logan, K. A., L. A. Sweanor, J. F. Smith, J. L. Cashman, and **T. K. Ruth.** 1993. Ecology of mountain lions in a desert environment. 8th Ann. Unpubl. Rept. submitted by The Hornocker Wildlife Research Institute, Moscow, Idaho to New Mexico Dept. Game and Fish, Santa Fe. 49 p.

Long, D. R. 1985. Lipid utilization during reproduction in female *Kinosternon flavescens.* Herpetologica 41(1):58–65.

Lowe, C. H., Jr. 1948. Territorial behavior in snakes and the so-called courtship dance. Herpetologica 4(4):129–135.

———. 1950. The systematic status of the salamander *Plethodon hardii*, with a discussion of biogeographical problems in *Aneides.* Copeia 1950(2):92–99.

——. 1954. Normal field movements and growth rates of marked regal horned lizards (*Phrynosoma solare*). Ecology 35(3):420–421.

——. 1955a. The eastern limit of the Sonoran Desert in the United States with additions to the known herpetofauna of New Mexico. Ecology 36(2):343–345.

——. 1955b. A new species of whiptailed lizard (genus *Cnemidophorus*) from the Colorado Plateau of Arizona, New Mexico, Colorado, and Utah. Brevoria (47):1–7.

——. 1955c. The occurrence of the lizard *Cnemidophorus sexlineatus* in New Mexico. Copeia 1955(1):61–62.

——. 1955d. Generic status of the aquatic snake, *Thamnophis angustirostris*. Copeia 1955(4):307–309.

——. 1956. A new species and a new subspecies of whiptailed lizards (genus *Cnemidophorus*) of the inland southwest. Bull. Chicago Acad. Sci. 10(9):137–150.

——. 1964. The amphibians and reptiles of Arizona, p. 153–174. *In* C. H. Lowe (ed.), The Vertebrates of Arizona. Univ. Arizona Press, Tucson. 259 p.

——. 1966. The prairie lined racerunner. J. Arizona Acad. Sci. 4:44–45.

——, **C. J. Cole,** and **J. L. Patton.** 1967. Karyotype evolution and speciation in lizards (genus *Sceloporus*) during evolution of the North American Desert. Syst. Zool. 16(4):296–300.

——, **P. J. Lardner,** and **E. A. Halpern.** 1971. Supercooling in reptiles and other vertebrates. Comp. Biochem. Physiol. 39(1A):125–135.

—— and **K. S. Norris.** 1950. Aggressive behavior in male sidewinders, *Crotalus cerastes*, with a discussion of aggressive behavior and territoriality in snakes. Chicago Acad. Sci. Nat. Hist. Misc. (66):1–13.

—— and ——. 1956. A subspecies of the lizard *Sceloporus undulatus* from the White Sands of New Mexico. Herpetologica 12(2):125–127.

——, **C. R. Schwalbe,** and **T. B. Johnson.** 1986. The Venomous Reptiles of Arizona. Arizona Game and Fish Dept., Phoenix. ix + 115 p.

—— and **J. W. Wright.** 1964. Species of the *Cnemidophorus exsanguis* subgroup of whiptail lizards. J. Arizona Acad. Sci. 3:78–80.

—— and ——. 1966a. Evolution of parthenogenetic species of *Cnemidophorus* (whiptail lizards) in western North America. J. Ariz. Acad. Sci. 4(2):81–87.

—— and ——. 1966b. Chromosomes and karyotypes of cnemidophorine teiid lizards. Mamm. Chromosomes Newsl. (22):199–200.

—— and **R. G. Zweifel.** 1952. A new species of whiptailed lizard (genus *Cnemidophorus*) from New Mexico. Bull. Chicago Acad. Sci. 9(13):229–247.

Lynch, J. D. 1986. The definition of the middle American clade of *Eleutherodactylus* based on jaw musculature (Amphibia: Leptodactylidae). Herpetologica 42(2):248–258.

McAlister, W. H. 1961. The mechanics of sound production in North American *Bufo*. Copeia 1961(1):86–95.

——. 1962. Variation in *Rana pipiens* Schreber in Texas. Amer. Midl. Nat. 67(2):334–346.

McAllister, C. T. 1985. Food habits and feeding behavior of *Crotaphytus collaris collaris* (Iguanidae) from Arkansas and Missouri. Southwest. Nat. 30(4):597–600.

——. 1990a. Helminth parasites of unisexual and bisexual whiptail lizards (Teiidae) in North America. I. The Colorado checkered whiptail (*Cnemidophorus tesselatus*). J. Wildl. Dis. 26(1):139–142.

——. 1990b. Helminth parasites of unisexual and bisexual whiptail lizards (Teiidae) in North America. II. The New Mexico whiptail (*Cnemidophorus neomexicanus*). J. Wildl. Dis. 26(3):403–406.

——. 1990c. Helminth parasites of unisexual and bisexual whiptail lizards (Teiidae) in North America. IV. The Texas spotted whiptail (*Cnemidophorus gularis*). Texas J. Sci. 42(4):381–388.

——. 1992. Helminth parasites of unisexual and bisexual whiptail lizards (Teiidae) in North America. VIII. The Gila spotted whiptail (*Cnemidophorus flagellicaudus*), Sonoran spotted whiptail (*Cnemidophorus sonorae*), and Plateau striped whiptail (*Cnemidophorus velox*). Texas J. Sci. 44(2):233–239.

——, **J. E. Cordes,** and **J. M. Walker.** 1991. Helminth parasites of unisexual and bisexual whiptail lizards (Teiidae) in North America. VI. The gray-checkered whiptail (*Cnemidophorus dixoni*). Texas J. Sci. 43(3):309–314.

——, **S. J. Upton, C. M. Garrett, J. N. Stuart,** and **C. W. Painter.** 1993. Hemogregarines and *Sarcocystis* sp. (Apicomplexa) in a western green rat snake, *Senticolis triaspis intermedia* (Serpentes: Colubridae), from New Mexico. J. Helminthol. Soc. Washington 60(2):284–286.

MacBride, T. H. 1905. The Alamogordo Desert. Science (n.s.) 21:90–97.

McCarthy, C. J. 1985. Monophyly of elapid snakes (Serpentes: Elapidae). An assessment of the evidence. Zool. J. Linn. Soc. 83(1):79–83.

McClanahan, L., Jr. 1967. Adaptations of the spadefoot toad, *Scaphiopus couchi*, to desert environments. Comp. Biochem. Physiol. 20(1):73–99.

M'Closkey, R. T., K. A. Baia, and **R. W. Russell.** 1987. Defense of mates: a territory departure rule for male tree lizards following sex-ratio manipulation. Oecologia 73(1):28–31.

McCoy, C. J., Jr. 1964. The snake, *Tantilla yaquia*, in Arizona: an addition to the fauna of the United States. Copeia 1964(1):216–217.

———. 1967. Natural history notes on *Crotaphytus wislizeni* (Reptilia: Iguanidae) in Colorado. Amer. Midl. Nat. 77(1):138–146.

———. 1970. *Hemidactylus turcicus*. Cat. Amer. Amphib. Rept.:87.1–87.2.

———. 1975. Cave-associated snakes, *Elaphe guttata* in Oklahoma. Bull. Natl. Speleol. Soc. 37(2):41.

——— and F. R. Gehlbach. 1967. Cloacal hemorrhage and the defense display of the colubrid snake *Rhinocheilus lecontei*. Texas J. Sci. 19(4):349–352.

——— and G. A. Hoddenbach. 1966. Geographic variation in ovarian cycles and clutch size in *Cnemidophorus tigris* (Teiidae). Science 154:1671–1672.

McCranie, J. R. 1990. *Nerodia erythrogaster*. Cat. Amer. Amphib. Rept.:500.1–500.8.

——— and L. D. Wilson. 1987. The biogeography of the herpetofauna of the pine-oak woodlands of the Sierra Madre Occidental of Mexico. Milwaukee Pub. Mus. Contrib. Biol. Geol. (72):1–30.

McDiarmid, R. W. 1968. Variation, distribution and systematic status of the black-headed snake *Tantilla yaquia* Smith. Bull. So. California Acad. Sci. 67(3):159–177.

———. 1977. *Tantilla yaquia*. Cat. Amer. Amphib. Rept.:198.1–198.2.

McDowell, S. B. 1968. Affinities of the snakes usually called *Elaps lacteus* and *E. dorsalis*. Zool. J. Linn. Soc. London 47:561–578.

MacKay, W. P., S. J. Loring, T. M. Frost, and W. G. Whitford. 1990. Population dynamics of a playa community in the Chihuahuan Desert. Southwest. Nat. 35(4):393–402.

MacLean, W. P. 1974. Feeding and locomotor mechanisms of teiid lizards: functional morphology and evolution. Pap. Avul. Zool. Sao Paulo 27(15):179–213.

Mahmoud, I. Y. 1967. Courtship behavior and sexual maturity in four species of kinosternid turtles. Copeia 1967(2):314–325.

———. 1968. Feeding behavior in kinosternid turtles. Herpetologica 24(4):300–305.

———. 1969. Comparative ecology of the kinosternid turtles of Oklahoma. Southwest. Nat. 14(1):31–66.

Mahrt, J. L. 1989. Prevalence of malaria in populations of *Sceloporus jarrovi* (Reptilia: Iguanidae) in southeastern Arizona. Southwest. Nat. 34(3):436–438.

Maker, H. J., H. E. Dregne, V. G. Link, and J. U. Anderson. 1978. Soils of New Mexico. New Mexico St. Univ. Agri. Exp. Sta. Res. Rep. (285):1–132.

Maldonado-Koerdell, M. and I. L. Firschein. 1947. Notes on the ranges of some North American salamanders. Copeia 1947(2):140.

Malnate, E. V. 1971. A catalog of primary types in the herpetological collections of the Academy of Natural Sciences, Philadelphia (ANSP). Proc. Acad. Nat. Sci. Philadelphia 123(9):345–375.

Marchand, L. J. 1944. Notes on the courtship of Florida terrapin. Copeia 1944(3):191–192.

Martin, B. E. 1975. Notes on a brood of the Arizona ridge-nosed rattlesnake, *Crotalus willardi willardi*. Bull. Maryland Herpetol. Soc. 11(2):66–67.

———. 1979. A reproductive record for the New Mexican ridge-nosed rattlesnake (*Crotalus willardi obscurus*). Bull. Maryland Herpetol. Soc. 12(4):126–128.

Martins, E. P. 1991. Individual and sex differences in the use of the push-up display by the sagebrush lizard, *Sceloporus graciosus*. Anim. Behav. 41(3):403–416.

Maslin, T. P. 1956. *Sceloporus undulatus erythrocheilus* ssp. nov. (Reptilia, Iguanidae), from Colorado. Herpetologica 12(4):291–294.

———. 1962. All-female species of the lizard genus *Cnemidophorus*, Teiidae. Science 135:212–213.

———. 1965. The status of the rattlesnake *Sistrurus catenatus* (Crotalidae) in Colorado. Southwest. Nat. 10(1):31–34.

———, R. G. Beidleman, and C. H. Lowe, Jr. 1958. The status of the lizard *Cnemidophorus perplexus* Baird and Girard (Teiidae). Proc. U.S. Natl. Mus. 108(3406):331–345.

——— and D. M. Secoy. 1986. A checklist of the lizard genus *Cnemidophorus* (Teiidae). Contrib. Zool. Univ. Colorado Mus. (1):1–60.

Matthews, T. C. 1971. Genetic changes in a population of boreal chorus frogs (*Pseudacris triseriata*) polymorphic for color. Amer. Midl. Nat. 85(1):208–221.

Maury, M. E. 1995. Diet composition of the greater earless lizard (*Cophosaurus texanus*) in central Chihuahuan Desert. J. Herpetol. 29(2):266–272.

Mautz, W. J. 1982. Observations on an oviposition site of the side blotched lizard, *Uta stansburiana*. J. Herpetol. 16(3):331–332.

Mayhew, W. W. 1962. *Scaphiopus couchi* in California's Colorado Desert. Herpetologica 18(3):153–161.

Meade, G. P. 1932. Notes on the breeding habits of Say's king snake in captivity. Bull. Antivenin Inst. Amer. 5(3):70–71.

Mecham, J. S. 1956. The relationship between the ringneck snakes *Diadophis regalis* and *D. punctatus*. Copeia 1956(1):51–52.

———. 1957. The taxonomic status of some southwestern skinks of the *multivirgatus* group. Copeia 1957(2):111–123.

———. 1979. The biogeographical relationships of the amphibians and reptiles of the Guadalupe Mountains, p. 169–179. *In* H. H. Genoways and R. J. Baker (eds.), Biological

Investigations in the Guadalupe Mountains National Park, Texas. U.S. Dept. Interior, Natl. Park Serv., Proc. Trans. Ser. (4). xvii + 442 p.

———. 1980. *Eumeces multivirgatus*. Cat. Amer. Amphib. Rept.:241.1–241.2.

———, **M. J. Littlejohn, R. S. Oldham, L. E. Brown,** and **J. R. Brown.** 1973. A new species of leopard frog (*Rana pipiens* complex) from the plains of the central United States. Occ. Pap. Mus. Texas Tech. Univ. (18):1–11.

Medden, R. V. 1927. Notes on *Phyllorhynchus*. Copeia 1927(164):82–83.

Medica, P. A. 1965. Altitude record for *Gerrhonotus kingi*. Herpetologica 21(2):147–148.

———. 1967. Food habits, habitat preference, reproduction, and diurnal activity in four sympatric species of whiptail lizards (*Cnemidophorus*) in south central New Mexico. Bull. So. California Acad. Sci. 66(4):251–276.

———. 1975. *Rhinocheilus, Rhinocheilus lecontei*. Cat. Amer. Amphib. Rept.:175.1–175.4.

———, **F. B. Turner,** and **D. D. Smith.** 1973. Hormonal induction of color change in female leopard lizards, *Crotaphytus wislizenii*. Copeia 1973(4):658–661.

Meek, S. E. 1905. An annotated list of a collection of reptiles from southern California and northern Lower California. Field Mus. Zool. Ser. 7(1):1–19.

Mello, K. 1978. Geographic Distribution: *Crotalus viridis cerberus* (Arizona black rattlesnake). Herpetol. Rev. 9(1):22.

Mendelson, J. R., III and **W. B. Jennings.** 1992. Shifts in the relative abundance of snakes in a desert grassland. J. Herpetol. 26(1):38–45.

Merker, G. P. and **K. A. Nagy.** 1984. Energy utilization by free-ranging *Sceloporus virgatus* lizards. Ecology 65(2):575–581.

Mertens, R. 1967. Die herpetologische-sektion des natur-museums und forschungs-institutes Senckenberg in Frankfurt a. M. nebst einem verzeichnis ihrer typen. Senckenberg. Biol. 48(A):1–105.

——— and **L. Müller.** 1940. Die Amphibien und Reptilien Europas. Abhandl. Senckenberg. Naturf. Ges. (451):1–56.

Metcalf, A. L. and **E. L. Metcalf.** 1970. Observations on ornate box turtles (*Terrapene ornata ornata* Agassiz) Trans. Kansas Acad. Sci. 73(1):96–117.

——— and ———. 1985. Longevity in some ornate box turtles (*Terrapene ornata ornata*). J. Herpetol. 19(1):157–158.

Meyer, D. E. 1966. Drinking habits in the earless lizard, *Holbrookia maculata*, and in two species of horned lizards (*Phrynosoma*). Copeia 1966(1):126–128.

Meylan, P. A. 1987. The phylogenetic relationships of soft-shelled turtles (Family Trionychidae). Bull. Amer. Mus. Nat. Hist. 186(1):1–101.

Michel, L. 1976. Reproduction in a southwest New Mexican population of *Urosaurus ornatus*. Southwest. Nat. 21(3):281–299.

Middendorf, G. A., III and **W. C. Sherbrooke.** 1992. Canid elicitation of blood-squirting in a horned lizard (*Phrynosoma cornutum*). Copeia 1992(2):519–527.

——— and **C. A. Simon.** 1988. Thermoregulation in the iguanid lizard *Sceloporus jarrovi*: the influences of age, time, and light condition on body temperature and thermoregulatory behaviors. Southwest. Nat. 33(3):347–356.

Miller, D. J. 1979. A life history study of the gray-banded kingsnake, *Lampropeltis mexicana alterna* in Texas. Chihuahuan Desert Res. Inst. Cont. No. 87. Alpine, TX. 48 p.

Miller, K., G. F. Birchard, G. F. Packard, and **G. C. Packard.** 1989. Life History Notes: *Trionyx spiniferus*. Fecundity. Herpetol. Rev. 20(2):56.

Miller, R. R. 1946. The probable origin of the soft-shelled turtle in the Colorado River basin. Copeia 1946(1):46.

Milne, L. J. and **M. J. Milne.** 1950. Notes on the behavior of horned toads. Amer. Midl. Nat. 44(3):720–741.

Milstead, W. W. 1957. Some aspects of competition in natural populations of whiptail lizards (genus *Cnemidophorus*). Texas J. Sci. 9(4):410–447.

———. 1960. Relict species of the Chihuahuan Desert. Southwest. Nat. 5(2):75–88.

———. 1965. Changes in competing populations of whiptail lizards (*Cnemidophorus*) in southwestern Texas. Amer. Midl. Nat. 73(1):75–80.

——— and **D. W. Tinkle.** 1969. Interrelationships of feeding habits in a population of lizards in southwestern Texas. Amer. Midl. Nat. 81(2):491–499.

Minton, S. A. 1958 [1959]. Observations on amphibians and reptiles of the Big Bend region of Texas. Southwest. Nat. 3(1):28–54.

———. 1983. *Sistrurus catenatus*. Cat. Amer. Amphib. Rept.:332.1–332.2.

Mitchell, J. C. 1979. Ecology of southeastern Arizona whiptail lizards (*Cnemidophorus*: Teiidae): population densities, resource partitioning, and niche overlap. Can. J. Zool. 57(9):1487–1499.

———. 1984. Observations on the ecology and reproduction of the leopard lizard, *Gambelia wislizenii* (Iguanidae), in southeastern Arizona. Southwest. Nat. 29(4):509–511.

Mitchell, J. D. 1903. The poisonous snakes of Texas, with notes on their habits. Trans. Texas Acad. Sci. 5(1):21–48.

Mittleman, M. B. 1940. Two new lizards of the genus *Uta*. Herpetologica 2(2):33–38.

———. 1942. A summary of the iguanid genus *Urosaurus*. Bull. Mus. Comp. Zool. 91(2):103–181.

———. 1945. Type localities of two American turtles. Copeia 1945(3):171.

———. 1949. Geographic variation in Marcy's garter snake, *Thamnophis marcianus* (Baird and Girard). Bull. Chicago Acad. Sci. 8(10):235–249.

Mocquard, M. F. 1899. Reptiles et Batraciens recueillis au Mexique par M. Leon Diguet en 1896 et 1897. Bull. Soc. Philo. Paris, ser. 9, 1:154–169.

Moir, W. H. and **H. M. Smith.** 1970. Occurrence of an American salamander, *Aneides hardyi* (Taylor), in tundra habitat. Arc. Alp. Res. 2(2):155–156.

Moll, D. 1979. Subterranean feeding by the Illinois Mud Turtle, *Kinosternon flavescens spooneri*. J. Herpetol. 13(3):371–373.

Moll, E. O. 1973. Latitudinal and intersubspecific variation in reproduction of the painted turtle *Chrysemys picta*. Herpetologica 29(4):307–318.

———. 1979. Reproductive cycles and adaptations, p. 305–331. *In* M. Harless and H. Morelock (eds.), Turtles: Perspectives and Research. Wiley-Interscience, New York. xiv + 695 p.

Montanucci, R. R. 1967. Further studies on leopard lizards, *Crotaphytus wislizenii*. Herpetologica 23(2):119–126.

———. 1981. Habitat separation between *Phrynosoma douglassi* and *P. orbiculare* (Lacertilia: Iguanidae) in Mexico. Copeia 1981(1):147–153.

———. 1983. Breeding, captive care and longevity of the short-horned lizard *Phrynosoma douglassi*. Intl. Zoo Yrbk. 23:148–156.

———. 1987. A phylogenetic study of the horned lizards, genus *Phrynosoma*, based on skeletal and external morphology. Contrib. Sci. Nat. Hist. Mus. Los Angeles Co. (390):1–36.

——— and **B. E. Baur.** 1982. Mating and courtship-related behaviors of the short-horned lizard, *Phrynosoma douglassi*. Copeia 1982(4):971–974.

Moore, M. C. 1986. Elevated testosterone levels during nonbreeding-season territoriality in a fall-breeding lizard, *Sceloporus jarrovi*. J. Comp. Physiol. A 158:159–163.

———, **J. M. Whittier, A. J. Billy,** and **D. Crews.** 1985. Male-like behavior in an all-female lizard: relationship to ovarian cycle. Anim. Behav. 33(1):284–289.

Morafka, D. J. 1977. A Biogeographical Analysis of the Chihuahuan Desert Through its Herpetofauna. Biogeographica, vol. 9. Dr. W. Junk, Publ., The Hague. viii + 313 p.

Moritz, C. C., J. W. Wright, and **W. M. Brown.** 1989. Mitochondrial-DNA analyses and the origin and relative age of parthenogenetic lizards (genus *Cnemidophorus*). III. *C. velox* and *C. exsanguis*. Evolution 43(5):958–968.

Morreale, S. J. and **J. W. Gibbons.** 1986. Habitat suitability index models: slider turtle. U.S. Fish Wildl. Serv. Biol. Rept. 82(10.125):1–14.

Mosauer, W. 1932. The amphibians and reptiles of the Guadalupe Mountains of New Mexico and Texas. Occ. Pap. Mus. Zool. Univ. Michigan (246):1–18.

———. 1935. How fast can snakes travel? Copeia 1935(1):6–9.

Moyle, P. B. 1973. Effects of introduced bullfrogs, *Rana catesbeiana*, on the native frogs of the San Joaquin Valley, California. Copeia 1973(1):18–22.

Munger, J. C. 1984a. Home ranges of horned lizards (*Phrynosoma*): circumscribed and exclusive? Oecologia 62(3):351–360.

———. 1984b. Optimal foraging? Patch use by horned lizards (Iguanidae: *Phrynosoma*). Amer. Nat. 123(5):654–680.

———. 1984c. Long-term yield from harvester ant colonies: implications for horned lizard foraging strategy. Ecology 65(4):1077–1086.

———. 1986. Rate of death due to predation for two species of horned lizard, *Phrynosoma cornutum* and *P. modestum*. Copeia 1986(3):820–824.

Murphy, J. B., B. W. Tryon, and **B. J. Brecke.** 1978. An inventory of reproduction and social behavior in captive gray-banded kingsnakes, *Lampropeltis mexicana alterna* (Brown). Herpetologica 34(1):84–93.

Murphy, R. W. and **J. R. Ottley.** 1984. Distribution of amphibians and reptiles on islands in the Gulf of California. Ann. Carnegie Mus. 53(8):207–230.

Murray, K. F. 1957. Pleistocene climate and the fauna of Burnet Cave, New Mexico. Ecology 38(1):129–132.

Murray, L. T. 1939. Annotated list of amphibians and reptiles from the Chisos Mountains. Contrib. Baylor Univ. Mus. 24:4–16.

Musgrave, M. E. 1930. *Bufo alvarius*, a poisonous toad. Copeia 1930(4):96–98.

Mushinsky, H. R., J. J. Hebrard, and **D. S. Vodopich.** 1982. Ontogeny of water snake foraging ecology. Ecology 63(6):1624–1629.

——— and **K. H. Lotz.** 1980. Chemoreceptive responses of two sympatric water snakes to extracts of commonly ingested prey species: Ontogenetic and ecological considerations. J. Chem. Ecol. 6(3):523–535.

Neaves, W. B. 1969. Adenosine deaminase phenotypes among sexual and parthenogenetic lizards in the genus *Cnemidophorus* (Teiidae). J. Exp. Zool. 171(2):175–184.

———. 1971. Tetraploidy in a hybrid lizard of the genus *Cnemidophorus* (Teiidae). Breviora (381):1–24.

——— and **P. S. Gerald.** 1968. Lactate dehydrogenase isozymes in parthenogenetic teiid lizards (*Cnemidophorus*). Science 160:1004–1005.

——— and ———. 1969. Gene dosage at the lactate dehydrogenase locus in triploid and diploid teiid lizards. Science 164:557–558.

LITERATURE CITED

Nelson, C. E. 1972a. *Gastrophryne olivacea.* Cat. Amer. Amphib. Rept.:122.1–122.4.

———. 1972b. Systematic studies of the North American microhylid genus *Gastrophryne.* J. Herpetol. 6(2):111–137.

———. 1973. *Gastrophryne.* Cat. Amer. Amphib. Rept.:134.1–134.2.

——— and H. S. Cuellar. 1968. Anatomical comparison of tadpoles of the genera *Hypopachus* and *Gastrophryne* (Microhylidae). Copeia 1968(2):423–424.

Newlin, M. E. 1974. Reproduction, trophic ecology, and other aspects of natural history in the bunch grass lizard, *Sceloporus scalaris.* M.S. Thesis, Angelo State Univ., San Angelo, Texas. viii + 111 p.

———. 1976. Reproduction in the bunch grass lizard, *Sceloporus scalaris.* Herpetologica 32(2):171–184.

New Mexico Dept. Game and Fish. 1990. Amended listing of endangered wildlife in New Mexico. State Game Commission Reg. No. 682. 4 p.

Nickerson, M. A. and C. E. Mays. 1969 [1970]. A preliminary herpetofaunal analysis of the Graham (Pinaleno) Mountain region, Graham Co., Arizona with ecological comments. Trans. Kansas Acad. Sci. 72(4):492–505.

Nieuwolt, M. C. 1993. The ecology of movement and reproduction in the western box turtle in central New Mexico. Ph.D. Diss., Univ. New Mexico, Albuquerque. xiii + 84 p.

Noble, G. K. 1925. An outline of the relation of ontogeny to phylogeny within the Amphibia, I. Amer. Mus. Novitates (165):1–17.

Noguchi, H. 1909. Snake Venoms. An Investigation of Venomous Snakes With Special Reference to the Phenomena of Their Venoms. Carnegie Inst., Washington, D.C. 315 p.

Nolan, E. J. (ed.). 1913. An index to the scientific contents of the Journal and Proceedings of the Academy of Natural Sciences of Philadelphia (1812–1912). Acad. Nat. Sci. Philadelphia. xiv + 1419 p.

Norris, K. S. and R. G. Zweifel. 1950. Observations on the habits of the ornate box turtle, *Terrapene ornata* (Agassiz). Nat. Hist. Misc. Chicago Acad. Sci. (58):1–4.

Obbard, M. E. and R. J. Brooks. 1981. A radio-telemetry and mark-recapture study of activity in the common snapping turtle, (*Chelydra serpentina*). Copeia 1981(3):630–637.

Oldham, J. C. and H. M. Smith. 1991. The generic status of the smooth green snake, *Opheodrys vernalis.* Bull. Maryland Herpetol. Soc. 27(4):201–215.

Olson, D. H. 1989. Predation on breeding western toads (*Bufo boreas*). Copeia 1989(2):391–397.

Olson, R. E. 1959. Notes on some Texas herptiles. Herpetologica 15(1):48.

Ortega, A. and R. Barbault. 1986. Reproduction in the high elevation Mexican lizard *Sceloporus scalaris.* J. Herpetol. 20(1):111–114.

———, M. E. Maury, and R. Barbault. 1982. Spatial organization and habitat partitioning in a mountain lizard community of Mexico. Acta Oecologica/Oecol. Gener. 3(3):323–330.

Ortenburger, A. I. and R. D. Ortenburger. 1926. Field observations on some amphibians and reptiles of Pima County, Arizona. Proc. Oklahoma Acad. Sci. 6:101–121.

Osborn, H. F. 1931. Cope: Master Naturalist. Princeton Univ. Press. xvii + 740 p.

Pace, A. E. 1974. Systematic and biological studies of the leopard frogs (*Rana pipiens* complex) of the United States. Misc. Publ. Mus. Zool. Univ. Michigan (148):1–140.

Pack, H. J. 1922. Food habits of *Crotaphytus wislizenii* Baird and Girard. Proc. Biol. Soc. Washington 35:1–4.

———. 1923a. Food habits of *Crotaphytus collaris baileyi* (Stejneger). Proc. Biol. Soc. Washington 36:83–84.

———. 1923b. Food habits of *Cnemidophorus tessellatus tessellatus* (Say). Proc. Biol. Soc. Washington 36:85–90.

Pack, L. E., Jr. and W. W. Tanner. 1970. A taxonomic comparison of *Uta stansburiana* of the Great Basin and the Upper Colorado River Basin in Utah, with a description of a new subspecies. Great Basin Nat. 30(2):71–90.

Packard, G. C. and M. J. Packard. 1993a. Hatchling painted turtles (*Chrysemys picta*) survive exposure to subzero temperatures during hibernation by avoiding freezing. J. Comp. Physiol. B 163(2):147–152.

——— and ———. 1993b. Sources of variation in laboratory measurements of water relations of reptilian eggs and embryos. Physiol. Zool. 66(1):115–127.

———, ———, and G. F. Birchard. 1989. Sexual differentiation and hatching success by painted turtles incubating in different thermal and hydric environments. Herpetologica 45(4):385–392.

———, ———, and K. Miller. 1990. Life History Notes: *Chelydra serpentina* (Common snapping turtle). Fecundity. Herpetol. Rev. 21(4):92.

———, K. A. Ruble, and M. J. Packard. 1993. Hatchling snapping turtles overwintering in natural nests are inoculated by ice in frozen soil. J. Therm. Biol. 18(4):185–188.

Packard, M. J., G. C. Packard, and T. J. Boardman. 1980. Water balance of the eggs of a desert lizard (*Callisaurus draconoides*). Can. J. Zool. 58(11):2051–2058.

———, ———, and ———. 1982. Structure of eggshells and water relations of reptilian eggs. Herpetologica 38(1):136–155.

Painter, C. W. 1985. Herpetology of the Gila and San Francisco river drainages of southwestern New Mexico. Unpubl. Rept. New Mexico Dept. Game and Fish, Santa Fe. 333 p.

———. 1991a. Preliminary investigations of the distribution and natural history of the Rio Grande river cooter *(Pseudemys concinna gorzugi)* in New Mexico. Unpubl. Rept. New Mexico Dept. Game & Fish to Bur. Land Management, Carlsbad Res. Dist. 27 p.

———. 1991b. Interim Report: Results of the 1990 survey for the gray-checkered whiptail *(Cnemidophorus dixoni)* on Bureau of Land Management (BLM) lands in southwestern New Mexico. Unpubl. Rept. New Mexico Dept. Game & Fish. Santa Fe, NM. 20 p.

———. 1992. Interim Report: Results of the 1991 survey for the gray-checkered whiptail *(Cnemidophorus dixoni)* on Bureau of Land Management (BLM) lands in southwestern New Mexico. Unpubl. Rept. New Mexico Dept. Game & Fish. Santa Fe, NM. 32 p.

———. 1993a. Life History Notes: *Apalone spinifera emoryi* (Texas spiny softshell). Coloration. Herpetol. Rev. 24(4):148.

———. 1993b. Geographic Distribution: *Phrynosoma solare* (Regal horned lizard). Herpetol. Rev. 24(4):155.

———. 1995. Status of the bunch grass lizard, *Sceloporus scalaris slevini*, in the Animas Valley, Hidalgo County, New Mexico. Unpubl. Rept. New Mexico Dept. Game & Fish. Santa Fe, NM.

——— and **T. L. Brown**. 1991. Geographic Distribution: *Leptotyphlops dulcis dissectus*. Herpetol. Rev. 22(2):67.

——— and **R. D. Burkett**. 1991. Geographic Distribution: *Pseudacris clarkii* (Spotted chorus frog). Herpetol. Rev. 22(2):64.

———, **P. W. Hyder**, and **G. Swinford**. 1992. Three species new to the herpetofauna of New Mexico. Herpetol. Rev. 23(2):64.

——— and **C. M. Milensky**. 1993. Geographic Distribution: *Crotalus tigris* (Tiger rattlesnake). Herpetol. Rev. 24(4):155–156.

Parke, J. G. 1855. Report of explorations for that portion of a railroad route, near the thirty-second parallel of north latitude, lying between Dona Ana, on the Rio Grande, and Pima Villages, on the Gila, p. 1–28. *In* Reports of explorations and surveys, to ascertain the most practicable and economical route for a railroad from the Mississippi River to the Pacific Ocean. Vol. II(6). 33rd Congress, 2nd Sess., House of Rep. Exec. Doc. (91). Washington, D.C.

Parker, E. D., Jr. 1979a. Phenotypic consequences of parthenogenesis in *Cnemidophorus* lizards. II. Similarity of *C. tesselatus* to its sexual parental species. Evolution 33(4):1167–1179.

———. 1979b. Phenotypic consequences of parthenogenesis in *Cnemidophorus* lizards. I. Variability in parthenogenetic and sexual populations. Evolution 33(4):1150–1166.

——— and **R. K. Selander**. 1976. The organization of genetic diversity in the parthenogenetic lizard *Cnemidophorus tesselatus*. Genetics 84(4):791–805.

——— and ———. 1984. Low clonal diversity in the parthenogenetic lizard *Cnemidophorus neomexicanus* (Sauria: Teiidae). Herpetologica 40(3):245–252.

Parker, W. S. 1971. Ecological observations on the regal horned lizard *(Phrynosoma solare)* in Arizona. Herpetologica 27(3):333–338.

———. 1972a. Aspects of the ecology of a Sonoran Desert population of the western banded gecko, *Coleonyx variegatus* (Sauria, Eublepharinae). Amer. Midl. Nat. 88(1):209–224.

———. 1972b. Ecological study of the western whiptail lizard, *Cnemidophorus tigris gracilis*, in Arizona. Herpetologica 28(4):360–369.

———. 1973a. Natural history notes on the iguanid lizard *Urosaurus ornatus*. J. Herpetol. 7(1):21–26.

———. 1973b. Notes on reproduction of some lizards from Arizona, New Mexico, Texas, and Utah. Herpetologica 29(3):258–264.

———. 1974a. Home range, growth, and population density of *Uta stansburiana* in Arizona. J. Herpetol. 8(2):135–139.

———. 1974b. *Phrynosoma solare*. Cat. Amer. Amphib. Rept.:162.1–162.2.

———. 1976. Population estimates, age structure, and denning habits of whipsnakes, *Masticophis taeniatus*, in a northern Utah *Atriplex-Sarcobatus* community. Herpetologica 32(1):53–57.

———. 1982a. *Sceloporus magister*. Cat. Amer. Amphib. Rept.:290.1–290.4.

———. 1982b. *Masticophis taeniatus*. Cat. Amer. Amphib. Rept.:304.1–304.4.

——— and **W. S. Brown**. 1972. Telemetric study of movements and oviposition of two female *Masticophis t. taeniatus*. Copeia 1972(4):892–895.

——— and ———. 1973. Species composition and population changes in two complexes of snake hibernacula in northern Utah. Herpetologica 29(4):319–326.

——— and ———. 1980. Comparative ecology of two colubrid snakes, *Masticophis t. taeniatus* and *Pituophis melanoleucus deserticola*, in northern Utah. Milwaukee Pub. Mus. Publ. Biol. Geol. (7):1–104.

——— and **E. R. Pianka**. 1973. Notes on the ecology of the iguanid lizard *Sceloporus magister*. Herpetologica 29(2):143–152.

——— and ———. 1974. Further ecological observations on the western banded gecko, *Coleonyx variegatus*. Copeia 1974(2):528–531.

——— and ———. 1975. Comparative ecology of populations of the lizard *Uta stansburiana*. Copeia 1975(4):615–632.

——— and ———. 1976. Ecological observations on the leopard lizard *(Crotaphytus wislizeni)* in different parts of its range. Herpetologica 32(1):95–114.

Parmenter, R. R. 1980. Effects of food availability and water temperature on the feeding ecology of pond sliders *(Chrysemys s. scripta)*. Copeia 1980(3):503–514.

LITERATURE CITED

———. 1981. Digestive turnover rates in freshwater turtles: the influence of temperature and body size. Comp. Biochem. Physiol. 70A(2):235–238.

——— and **H. W. Avery**. 1990. The feeding ecology of the slider turtle, p. 257–266. *In* J. W. Gibbons (ed.), Life History and Ecology of the Slider Turtle. Smithsonian Inst. Press, Washington, D.C. xiv + 368 p.

Patterson, R. 1982. The distribution of the desert tortoise (*Gopherus agassizii*), p. 51–55. *In* R. B. Bury (ed.), North American tortoises: conservation and ecology. U.S. Fish Wildl. Serv., Wildl. Res. Rept. (12). Washington, D.C. vii + 126 p.

Paulissen, M. A. 1987a. Optimal foraging and intraspecific diet differences in the lizard *Cnemidophorus sexlineatus*. Oecologia 71(3):439–446.

———. 1987b. Diet of adult and juvenile six-lined racerunners, *Cnemidophorus sexlineatus* (Sauria: Teiidae). Southwest. Nat. 32(3):395–397.

———. 1988a. Ontogenetic comparison of body temperature selection and thermal tolerance of *Cnemidophorus sexlineatus*. J. Herpetol. 22(4):473–476.

———. 1988b. Ontogenetic and seasonal shifts in microhabitat use by the lizard *Cnemidophorus sexlineatus*. Copeia 1988(4):1021–1029.

———, **J. M. Walker, J. E. Cordes**, and **H. L. Taylor**. 1993. Diet of diploid and triploid populations of parthenogenetic whiptail lizards of the *Cnemidophorus tesselatus* complex (Teiidae) in southeastern Colorado. Southwest. Nat. 38(4):377–381.

Pechmann, J. H. K. and **H. M. Wilbur**. 1994. Putting declining amphibian populations in perspective: natural fluctuations and human impacts. Herpetologica 50(1):65–84.

Pedersen, S. C. 1993. Skull growth in cannibalistic tiger salamanders, *Ambystoma tigrinum*. Southwest. Nat. 38(4):316–324.

Perez, J. C., W. Haws, V. Garcia, and **B. Jennings**. 1978a. Resistance of warm-blooded animals to snake venoms. Toxicon 16(4):375–383.

———, ———, and **C. Hatch**. 1978b. Resistance of woodrats (*Neotoma micropus*) to *Crotalus atrox* venom. Toxicon 16(2):198–200.

———, **S. Pichyangkul**, and **V. E. Garcia**. 1979. The resistance of three species of warm-blooded animals to western diamondback rattlesnake (*Crotalus atrox*) venom. Toxicon 17(6):601–607.

Perkins, C. B. 1943. Notes on captive-bred snakes. Copeia 1943(2):108–112.

Perry-Richardson, J. J., C. W. Schofield, and **N. B. Ford**. 1990. Courtship of the garter snake, *Thamnophis marcianus*, with a description of a female behavior for coitus interruption. J. Herpetol. 24(1):76–78.

Peters, E. L. and **C. J. McCoy**. 1978. Geographic Distribution: *Bufo alvarius* (Colorado River toad). Herpetol. Rev. 9(3):107.

Peters, J. A. 1951. Studies on the lizard *Holbrookia texana* (Troschel) with descriptions of two new subspecies. Occ. Pap. Mus. Zool. Univ. Michigan (537):1–20.

Peterson, D. K. and **W. G. Whitford**. 1987. Foraging behavior of *Uta stansburiana* and *Cnemidophorus tigris* in two different habitats. Southwest. Nat. 32(4):427–433.

Pettus, D. and **G. M. Angleton**. 1967. Comparative reproductive biology of montane and piedmont chorus frogs. Evolution 21(3):500–507.

Pfennig, D. W. 1992. Polyphenism in spadefoot toad tadpoles as a locally adjusted evolutionarily stable strategy. Evolution 46(5):1408–1420.

———, **M. L. G. Loeb,** and **J. P. Collins**. 1991a. Pathogens as a factor limiting the spread of cannibalism in tiger salamanders. Oecologia 88(2):161–166.

———, **A. Mabry**, and **D. Orange**. 1991b. Environmental causes of correlations between age and size at metamorphosis in *Scaphiopus multiplicatus*. Ecology 72(6):2240–2248.

Phelan, J. P. and **K. G. Niessen**. 1989. Effect of satiation on activity patterns in *Sceloporus virgatus*. J. Herpetol. 23(4):424–426.

Phelan, R. L. and **B. H. Brattstrom**. 1955. Geographic variation in *Sceloporus magister*. Herpetologica 11(1):1–14.

Pianka, E. R. 1970. Comparative autecology of the lizard *Cnemidophorus tigris* in different parts of its geographic range. Ecology 51(4):703–720.

——— and **W. S. Parker**. 1972. Ecology of the iguanid lizard *Callisaurus draconoides*. Copeia 1972(3):493–508.

——— and ———. 1975. Ecology of horned lizards: a review with special reference to *Phrynosoma platyrhinos*. Copeia 1975(1):141–162.

Pierce, B. A. and **J. B. Mitton**. 1980. Patterns of allozyme variation in *Ambystoma tigrinum mavortium* and *A. t. nebulosum*. Copeia 1980(1):594–605.

Pierce, J. R. 1976. Distribution of two mating call types of the plains spadefoot, *Scaphiopus bombifrons*, in southwestern United States. Southwest. Nat. 20(4):578–582.

Pietruszka, R. D. 1986. Search tactics of desert lizards: how polarized are they? Anim. Behav. 34(6):1742–1758.

Pisani, G. R. and **B. R. Stephenson**. 1991. Food habits in Oklahoma *Crotalus atrox* in fall and early spring. Trans. Kansas Acad. Sci. 94(3–4):137–141.

Platt, D. R. 1969. Natural history of the hognose snakes *Heterodon platyrhinos* and *Heterodon nasicus*. Univ. Kansas Publ. Mus. Nat. Hist. 18(4):253–420.

———. 1983. *Heterodon*. Cat. Amer. Amphib. Rept.:315.1–315.2.

Platz, J. E. 1976. Biochemical and morphological variation of leopard frogs in Arizona. Copeia 1976(4):660–672.

———. 1988. *Rana yavapaiensis.* Cat. Amer. Amphib. Rept.:418.1–418.2.

———. 1989. Speciation within the chorus frog *Pseudacris triseriata*: morphometric and mating call analyses of the boreal and western subspecies. Copeia 1989(3):704–712.

———. 1991. *Rana berlandieri.* Cat. Amer. Amphib. Rept.:508.1–508.4.

———. 1993. *Rana subaquavocalis*, a remarkable new species of leopard frog (*Rana pipiens* complex) from southeastern Arizona that calls under water. J. Herpetol. 27(2):154–162.

——— and **D. C. Forester.** 1988. Geographic variation in mating call among the four subspecies of the chorus frog: *Pseudacris triseriata* (Wied). Copeia 1988(4):1062–1066.

——— and **J. S. Frost.** 1984. *Rana yavapaiensis*, a new species of leopard frog (*Rana pipiens* complex). Copeia 1984(4):940–948.

——— and **J. S. Mecham.** 1979. *Rana chiricahuensis*, a new species of leopard frog (*Rana pipiens* complex) from Arizona. Copeia 1979(3):383–390.

——— and ———. 1984. *Rana chiricahuensis.* Cat. Amer. Amphib. Rept.:347.1–347.2.

——— and **A. L. Platz.** 1973. *Rana pipiens* complex: hemoglobin phenotypes of sympatric and allopatric populations in Arizona. Science 179:1334–1336.

Plummer, M. V. 1976. Some aspects of nesting success in the turtle, *Trionyx muticus.* Herpetologica 32(4):353–359.

———. 1977a. Activity, habitat and population structure in the turtle, *Trionyx muticus.* Copeia 1977(3):431–440.

———. 1977b. Reproduction and growth in the turtle *Trionyx muticus.* Copeia 1977(3):440–447.

———. 1977c. Notes on the courtship and mating of the softshell turtle, *Trionyx muticus* (Reptilia, Testudines, Trionychidae). J. Herpetol. 11(1):90–92.

——— and **D. B. Farrar.** 1981. Sexual dietary differences in a population of *Trionyx muticus.* J. Herpetol. 15(2):175–179.

——— and **H. W. Shirer.** 1975. Movement patterns in a river population of the softshell turtle, *Trionyx muticus.* Occ. Pap. Mus. Nat. Hist. Univ. Kansas (43):1–26.

Pope, J. 1854 [1855]. Report of exploration of a route for the Pacific railroad, near the 32nd parallel of north latitude, from the Red River to the Rio Grande, p. 1–185. *In* reports of explorations and surveys, to ascertain the most practicable and economical route for a railroad from the Mississippi River to the Pacific Ocean. Vol. II(6). 33rd Congress, 2nd Sess., House of Rep. Exec. Doc. (91). Washington, D.C.

Pope, M. H. and **R. Highton.** 1980. Geographic genetic variation in the Sacramento mountain salamander, *Aneides hardii.* J. Herpetol. 14(4):343–346.

Porzer, L. M. 1981. Movement, behavior, and body temperature of the Gila monster (*Heloderma suspectum*) in Queen Creek, Pinal County, Arizona. M.S. Thesis, Arizona St. Univ., Tempe. x + 100 p.

Pough, F. H. 1961. Range extension of the New Mexican whiptail lizard *Cnemidophorus perplexus.* Herpetologica 17(4):270.

———. 1966. Ecological relationships of rattlesnakes in southeastern Arizona with notes on other species. Copeia 1966(4):676–683.

——— and **R. M. Andrews.** 1985. Use of anaerobic metabolism by free-ranging lizards. Physiol. Zool. 58(2):205–213.

Powell, G. L. and **A. P. Russell.** 1984. The diet of the eastern short-horned lizard (*Phrynosoma douglassi brevirostre*) in Alberta and its relationship to sexual size dimorphism. Can. J. Zool. 62(3):428–440.

Powers, J. H. 1903. The causes of acceleration and retardation in the metamorphosis of *Ambystoma tigrinum*: a preliminary report. Amer. Nat. 37:385–410.

———. 1907. Morphological variation and its causes in *Ambystoma tigrinum.* Univ. Nebraska Studies 7(3):197–273.

Price, A. H. 1980a. *Crotalus molossus.* Cat. Amer. Amphib. Rept.:242.1–242.2.

———. 1980b. Geographic Distribution: *Hemidactylus turcicus* (Mediterranean Gecko). Herpetol. Rev. 11(2):39.

———. 1980c. Geographic Distribution: *Thamnophis eques megalops* (Mexican garter snake). Herpetol. Rev. 11(2):39.

———. 1982. *Crotalus scutulatus.* Cat. Amer. Amphib. Rept.:291.1–291.2.

———. 1983. Annotated bibliography of the genus *Cnemidophorus* in New Mexico. Smithsonian Herp. Info. Serv. (58):1–92.

———. 1986a. *Cnemidophorus tesselatus.* Cat. Amer. Amphib. Rept.:398.1–398.2.

———. 1986b. Geographic Distribution: *Hylactophryne augusti latrans* (Eastern barking frog). Herpetol. Rev. 17(4):91.

———. 1987. Life History Notes: *Hypsiglena torquata jani* (Texas night snake). Behavior. Herpetol. Rev. 18(1):16.

———. 1988. Observations on maternal behavior and neonate aggregation in the western diamondback rattlesnake, *Crotalus atrox* (Crotalidae). Southwest. Nat. 33(3):370–373.

———. 1990. *Phrynosoma cornutum.* Cat. Amer. Amphib. Rept.:469.1–469.7.

———. 1992. Comparative behavior in lizards of the genus *Cnemidophorus* (Teiidae), with comments on the evolution of parthenogenesis in reptiles. Copeia 1992(2):323–331.

LITERATURE CITED

———and **D. G. Johnson**. 1978a. Geographic Distribution: *Malayemys subtrijuga* (Malayan snail-eating turtle). Herpetol. Rev. 9(3):107.

———and ———. 1978b. Geographic Distribution: *Hyla eximia* (Mountain treefrog). Herpetol. Rev. 10(1):142.

———and **J. L. LaPointe**. 1990. Activity patterns of a Chihuahuan Desert snake community. Ann. Carnegie Mus. 59(1):15–23.

———, ———, and **J. W. Atmar**. 1993. The ecology and evolutionary implications of competition and parthenogenesis in *Cnemidophorus*, p. 371–410. *In* J. W. Wright and L. J. Vitt (eds.), Biology of Whiptail Lizards (genus *Cnemidophorus*). Oklahoma Mus. Nat. Hist., Norman. xiv + 417 p.

———and **B. K. Sullivan**. 1988. *Bufo microscaphus*. Cat. Amer. Amphib. Rept.:415.1–415.3.

Price, R. M. 1990. *Bogertophis*. Cat. Amer. Amphib. Rept.:497.1–497.2.

———. 1991. *Senticolis*, *S. triaspis*. Cat. Amer. Amphib. Rept.:525.1–525.4.

Prieto, A. A. and **E. R. Jacobson**. 1968. A new locality for melanistic *Crotalus molossus molossus* in southern New Mexico. Herpetologica 24(4):339–340.

———and **W. G. Whitford**. 1971. Physiological responses to temperature in the horned lizards, *Phrynosoma cornutum* and *Phrynosoma douglassii*. Copeia 1971(3):498–504.

Punzo, F. 1974a. Comparative analysis of the feeding habits of two species of Arizona blind snakes, *Leptotyphlops h. humilis* and *Leptotyphlops d. dulcis*. J. Herpetol. 8(2):153–156.

———. 1974b. An analysis of the stomach contents of the gecko, *Coleonyx brevis*. Copeia 1974(3):779–780.

———. 1974c. A qualitative and quantitative study of the food items of the yellow mud turtle, *Kinosternon flavescens* (Agassiz). J. Herpetol. 8(3):269–271.

———. 1982. Tail autotomy and running speed in the lizards *Cophosaurus texanus* and *Una notata*. J. Herpetol. 16(3):329–331.

———. 1991. Feeding ecology of spadefooted toads (*Scaphiopus couchi* and *Spea multiplicata*) in western Texas. Herpetol. Rev. 22(3):79–80.

Pyburn, W. F. 1958. Size and movements of a local population of cricket frogs (*Acris crepitans*). Texas J. Sci. 10(3):325–342.

Quinn, H. R. 1979. Sexual dimorphism in tail pattern of Oklahoma snakes. Texas J. Sci. 31(2):157–160.

Rado, T. A. and **P. G. Rowlands**. 1981. A range extension and low elevation record for the Arizona ridgenose rattlesnake (*Crotalus w. willardi*). Herpetol. Rev. 12(1):15.

Rafinesque, C. S. 1818. Further account of discoveries in natural history in the western states. Amer. Month. Mag. Crit. Rev. 4:39–42.

———. 1832. Descriptions of two new genera of soft shell turtles of North America. Atlantic J. and Friend of Knowledge, Philadelphia 1(2):64–65.

Ramotnik, C. A. and **N. J. Scott, Jr.** 1988. Habitat requirements of New Mexico's endangered salamanders, p. 54–63. *In* R. C. Szaro, K. E. Severson, and D. R. Patton (eds.), Management of Amphibians, Reptiles, and Small Mammals in North America. USDA Forest Serv., Rocky Mountain Forest Range Exp. Sta., Gen. Tech. Rep. RM–166, Fort Collins, Colorado. 458 p.

Ramsey, L. W. 1948. Hibernation of *Holbrookia texana*. Herpetologica 4(6):223.

———. 1953. The lined snake, *Tropidoclonion lineatum* (Hallowell). Herpetologica 9(1):7–24.

———. 1956. Nesting of Texas horned lizards. Herpetologica 12(3):239–240.

Reagan, D. P. 1972. Ecology and distribution of the Jemez Mountains salamander, *Plethodon neomexicanus*. Copeia 1972(3):486–492.

Reese, R. W. 1969. The taxonomy and ecology of the tiger salamander (*Ambystoma tigrinum*) of Colorado. Ph.D. Diss., Univ. Colorado, Boulder. x + 154 p.

Reeve, W. L. 1952. Taxonomy and distribution of the horned lizards genus *Phrynosoma*. Univ. Kansas Sci. Bull. 34(Pt. 2, 14):817–960.

Regal, P. J. 1967. Voluntary hypothermia in reptiles. Science 155:1551–1553.

Reid, J. R. and **T. E. Lott**. 1963. Feeding of *Leptotyphlops dulcis dulcis* (Baird and Girard). Herpetologica 19(2):141–142.

Renaud, M. 1977. Polymorphic and polytypic variation in the Arizona treefrog (*Hyla wrightorum*). Ph.D. Diss., Arizona St. Univ., Tempe. 132 p.

Reuss, A. 1834. Zoologische Miscellen, Reptilien. Ophidier. Senckenberg Mus., Frankfurt am Main 1(2):127–162 + 3 plates.

Reynolds, R. P. 1978. Resource use, habitat selection, and seasonal activity of a Chihuahuan snake community. Ph.D. Diss., Univ. New Mexico, Albuquerque. xi + 85 p.

———. 1982. Seasonal incidence of snakes in northeastern Chihuahua, Mexico. Southwest. Nat. 27(2):161–166.

———and **N. J. Scott, Jr.** 1982. Use of a mammalian resource by a Chihuahuan snake community, p. 99–118. *In* N. J. Scott, Jr. (ed.), Herpetological Communities. U.S. Dept. Interior, Fish and Wildl. Serv., Wildl. Res. Rep. 13. iv + 239 p.

Reynolds, T. D. 1979. Response of reptile populations to different land management practices on the Idaho National Engineering Laboratory site. Great Basin Nat. 39(3):255–262.

Riemer, W. J. 1955. Comments on the distribution of certain Mexican toads. Herpetologica 11(1):17–23.

Roberts, L. A. 1968. Oxygen consumption in the lizard *Uta stansburiana*. Ecology 49(5):809–819.

Robinson, M. D. and **C. W. Romack.** 1973. The Mediterranean gecko (*Hemidactylus turcicus*), a species new to the herpetofauna of Arizona. J. Herpetol. 7(3):311–312.

Rogers, J. S. 1972. Discriminate function analysis of morphological relationships within the *Bufo cognatus* species group. Copeia 1972(2):381–383.

———. 1973. Protein polymorphism, genic heterozygosity and divergence in the toads *Bufo cognatus* and *B. speciosis*. Copeia 1973(2):322–330.

Rose, B. 1981. Factors affecting activity in *Sceloporus virgatus*. Ecology 62(3):706–716.

———. 1982. Lizard home ranges: methodology and functions. J. Herpetol. 16(3):253–269.

Rose, F. L. 1980. Turtles in arid and semi-arid regions. Bull. Ecol. Soc. Amer. 61(2):89.

——— and **D. Armentrout.** 1976. Adaptive strategies of *Ambystoma tigrinum* (Green) inhabiting the Llano Estacado of west Texas. J. Anim. Ecol. 45(3):713–729.

——— and **C. D. Barbour.** 1968. Ecology and reproductive cycles of the introduced gecko, *Hemidactylus turcicus*, in the southern United States. Amer. Midl. Nat. 79(1):159–168.

Rosen, P. C. 1987. Variation in female reproduction among populations of Sonoran mud turtles (*Kinosternon sonoriense*). M.S. Thesis, Arizona St. Univ., Tempe. viii + 80 p.

———. 1991. Comparative field study of thermal preferenda in garter snakes (*Thamnophis*). J. Herpetol. 25(3):301–312.

——— and **C. R. Schwalbe.** 1988. Status of the Mexican and narrow-headed garter snakes (*Thamnophis eques megalops* and *Thamnophis rufipunctatus*) in Arizona. Arizona Game and Fish Dept., Unpubl. Rep. submitted to U.S. Fish and Wildlife Service, Albuquerque, New Mexico. 69 p.

Rossman, D. A. 1963. The colubrid snake genus *Thamnophis*: a revision of the *sauritus* group. Bull. Florida St. Mus. 7(3):99–178.

———. 1970. *Thamnophis proximus*. Cat. Amer. Amphib. Rept.:98.1–98.3.

———. 1971. Systematics of the neotropical populations of *Thamnophis marcianus* (Serpentes:Colubridae). Occ. Pap. Mus. Zool. Louisiana State Univ. (41):1–13.

Roth, J. J., E. C. Roth, and **H. M. Smith.** 1991. The minimal transformation size in the salamander *Ambystoma tigrinum*. Bull. Chicago Herpetol. Soc. 26(12):269.

Routman, E. J. and **A. C. Hulse.** 1984. Ecology and reproduction of a parthenogenetic lizard, *Cnemidophorus sonorae*. J. Herpetol. 18(4):381–386.

Roze, J. A. 1974. *Micruroides, Micruroides euryxanthus*. Cat. Amer. Amphib. Rept.:98.1–98.3.

Ruby, D. E. 1977. Winter activity in Yarrow's spiny lizard, *Sceloporus jarrovi*. Herpetologica 33(3):322–333.

———. 1978. Seasonal changes in the territorial behavior of the iguanid lizard *Sceloporus jarrovi*. Copeia 1978(3):430–438.

———. 1986. Selection of home range site by females of the lizard, *Sceloporus jarrovi*. J. Herpetol. 20(3):466–469.

——— and **D. I. Baird.** 1994. Intraspecific variation in behavior: comparisons between populations at different altitudes of the lizard *Sceloporus jarrovi*. J. Herpetol. 28(1):70–78.

——— and **A. E. Dunham.** 1984. A population analysis of the ovoviviparous lizard *Sceloporus jarrovi* in the Pinaleno Mountains of southeastern Arizona. Herpetologica 40(4):425–436.

Ruibal, R. 1962. The adaptive value of bladder water in the toad, *Bufo cognatus*. Physiol. Zool. 35(3):218–223.

Rundquist, E. M. 1992. Gentle monsters: captive reproduction of Gila monsters. Captive Breeding 1(1):20–28.

Russell, F. E. 1967a. Bites by the Sonoran coral snake, *Micruroides euryxanthus*. Toxicon 5(1):39–42.

———. 1967b. Gel diffusion study on human sera following rattlesnake venom poisoning. Toxicon 5(2):147–148.

———. 1983. Snake Venom Poisoning. Scholium Intl., Great Neck, New York. xiv + 552 p., 2 plates.

——— and **C. M. Bogert.** 1980. Gila monster: its biology, venom and bite—a review. Toxicon 19(3):341–359.

Ruthven, A. G. 1907. A collection of reptiles and amphibians from southern New Mexico and Arizona. Bull. Amer. Mus. Nat. Hist. 23(23):483–603.

———. 1908. Variations and genetic relationships of the garter-snakes. Bull. U.S. Natl. Mus. (61):xii + 201 p.

Ryan, M. J. and **W. Wilczynski.** 1991. Evolution of intraspecific variation in the advertisement call of a cricket frog (*Acris crepitans*, Hylidae) Biol. J. Linnean Soc. 44(3):249–271.

Sabath, M. 1960. *Sceloporus g. graciosus* in southern New Mexico and Texas. Herpetologica 16(1):22.

——— and **R. Worthington.** 1959. Eggs and young of certain Texas reptiles. Herpetologica 15(1):31–32.

Sanders, O. 1987. Evolutionary hybridization and speciation in North American indigenous bufonids. Privately printed, Dallas, Texas. viii + 110 p.

——— and **H. M. Smith.** 1951. Geographic variation in toads of the *debilis* group of *Bufo*. Field and Lab. 19(4):141–160.

Sattler, P. W. 1980. Genetic relationships among selected species of North American *Scaphiopus*. Copeia 1980(4):605–610.

Savage, J. M. 1954. A revision of the toads of the *Bufo debilis* complex. Texas J. Sci. 6(1):83–112.

Savitzky, A. H. 1979. The origin of the New World proteroglyphous snakes and its bearing on the study of venom delivery systems in snakes. Ph.D. Diss., Univ. Kansas, Lawrence. 396 p.

Saxon, J. G. 1970. The biology of the lizard, *Cnemidophorus tesselatus*, and effects of pesticides upon the population in the Presidio Basin, Texas. Ph.D. Diss., Texas A&M Univ., College Station. 90 p.

Schad, G. A., R. H. Stewart, and **F. A. Harrington.** 1959. Geographical distribution and variation of the Sacramento Mountains salamander, *Aneides hardii*. Can. J. Zool. 37(3):299–303.

Schall, J. J. 1977. Thermal ecology of five sympatric species of *Cnemidophorus* (Sauria: Teiidae). Herpetologica 33(3):261–272.

———. 1978. Reproductive strategies in sympatric whiptail lizards (*Cnemidophorus*): two parthenogenetic and three bisexual species. Copeia 1978(1):108–116.

——— and **E. R. Pianka.** 1980. Evolution of escape behavior diversity. Amer. Nat. 115(4):551–566.

Schlegel, H. 1837. Essai sur la Physionomie des Serpens. J. Kips, J. Hz. and W. P. van Stockum, The Hague. Two parts, xxvii + 251 p. and xv + 606 p., Atlas, 21 plates, 3 maps, and 2 tables.

——— and **S. Müller.** 1844. Over de Schildpadden van den Indischen *In* C. J. Temminck (ed.), Verhandelingen over de Natuurlijke Geschiendenis der Nederlandsche Overzeesche Bezittingen door de Leden der Natuurkundige Commissie in Indië en Andere Schrijvers. 1839–44. Part 3. Zoologie, Schildpadden.

Schmidt, K. P. 1921. New species of North American lizards of the genera *Holbrookia* and *Uta*. Amer. Mus. Novitates (22):1–6.

———. 1922. A review of the North American genus of lizards *Holbrookia*. Bull. Amer. Mus. Nat. Hist. 46(12):709–725.

———. 1940. Notes on Texan snakes of the genus *Salvadora*. Field Mus. Nat. Hist. Zool. Ser. 24(12):143–150.

———. 1953. A Check List of North American Amphibians and Reptiles 6th ed. Amer. Soc. Ichthyol. Herpetol. by Univ. Chicago Press, Chicago. viii + 280 p.

Schmidt, P. J., W. C. Sherbrooke, and **J. O. Schmidt.** 1989. The detoxification of ant (*Pogonomyrmex*) venom by a blood factor in horned lizards (*Phrynosoma*). Copeia 1989(3):603–607.

Schneider, J. G. 1783. Allgemeine Naturgeschichte der Schildkröten, nebst einem systematischen Verseichnisse der einzelnen Arten und zwei Kupfern. Müller, Leipzig. xlviii + 364 p., 2 plates.

Schoepff, J. D. 1792–1801. Historia Testudinum iconobus Illustrata. [1792:1–32; 1793:33–88; 1794:89–136; 1801:137–160]. Palmii, Erlangen.

Schrank, G. D. and **R. E. Ballinger.** 1973. Male reproductive cycles in two species of lizards (*Cophosaurus texanus* and *Cnemidophorus gularis*). Herpetologica 29(3):289–293.

Schreber, H. 1782. Beitrag zur Naturgeschichte der Frosche. Der Naturforscher, Johann Jacob Gebaur, Halle 18(10):182–193.

Schubauer, J. P., J. W. Gibbons, and **J. R. Spotila.** 1990. Home range and movement patterns of slider turtles inhabiting Par Pond, p. 223–232. *In* J. W. Gibbons (ed.), Life History and Ecology of the Slider Turtle. Smithsonian Inst. Press, Washington, D.C. xiv + 368 p.

Schuett, G. W., P. A. Buttenhoff and **D. Duvall.** 1993. Corroborative evidence for the lack of spring-mating in certain populations of prairie rattlesnakes (*Crotalus viridis*). Herpetol. Nat. Hist. 1(1):101–106.

Schwalbe, C. R. and **P. C. Rosen.** 1988. Preliminary report on effect of bullfrogs on wetland herpetofaunas in southeastern Arizona, p. 166–173. *In* R. C. Szaro, K. E. Severson, and D. R. Patton (ed.), Management of amphibians, reptiles, and small mammals in North America. USDA Forest Serv., Rocky Mountain Forest Range Exp. Sta., Gen. Tech. Rept. RM–166, Fort Collins, Colorado. 458 p.

Schwartz, A. 1955. A clutch of eggs of *Aneides hardyi* (Taylor). Herpetologica 11(1):70.

Scott, N. J., Jr. 1992. Ranid frog survey of the Gray Ranch with recommendations for management of frog habitats. August 1990–September 1991. Unpubl. Rept. Gray Ranch, Animas, New Mexico. 12 p.

——— and **R. D. Jennings.** 1985. The tadpoles of five species of New Mexican leopard frogs. Occ. Pap. Mus. Southwest. Biol. (3):1–21.

——— and **R. W. McDiarmid.** 1984a. *Trimorphodon*. Cat Amer. Amphib. Rept.:352.1–351.2.

——— and ———. 1984b. *Trimorphodon biscutatus*. Cat. Amer. Amphib. Rept.:353.1–353.4.

——— and **C. A. Ramotnik.** 1992. Does the Sacramento mountain salamander require old-growth forests?, p. 170–178. *In* M. R. Kaufmann, W. H. Moir, and R. L Bassett (eds.), Old-growth forests in the Southwest and Rocky Mountain regions. Proceedings of a workshop. USDA Forest Serv., Rocky Mountain Forest Range Exp. Sta., Gen. Tech. Rept. RM–213, Fort Collins, Colorado. 201 p.

———, ———, **M. J. Altenbach,** and **B. E. Smith.** 1987. Distribution and ecological requirements of endemic salamanders in relation to forestry management. Summary of 1987 activities. Part 2: Santa Fe National Forest. Final Rept., USDA Forest Serv., Albuquerque, New Mexico. 33 p.

Scudday, J. F. 1971. The biogeography and some ecological aspects of the teiid lizards (*Cnemidophorus*) of Trans-Pecos Texas. Ph.D. Diss., Texas A&M Univ., College Station. 198 p.

———. 1973. A new species of lizard of the *Cnemidophorus tesselatus* group from Texas. J. Herpetol. 7(4):363–371.

———. 1977(1978). Some recent changes in the herpetofauna of the northern Chihuahuan Desert, p. 513–522. *In* R. H. Wauer and D. H. Riskind (eds.), Transactions of the Symposium on the Biological Resources of the Chihuahuan Desert Region, United

States and Mexico. U.S. Dept. Interior, Natl. Park Serv., Trans. Proc. Ser. (3). xxii + 658 p.

——— and **J. R. Dixon.** 1973. Diet and feeding behavior of teiid lizards from Trans-Pecos Texas. Southwest. Nat. 18(3):279–289.

Seidel, M. E. 1975a. Osmoregulation in the turtle *Trionyx spiniferus* from brackish and freshwater. Copeia 1975(1):124–128.

———. 1975b. Geographic Distribution: *Chrysemys scripta elegans* (Red-eared turtle). Herpetol. Rev. 6(4):116.

———. 1977. Respiratory metabolism of temperate and tropical American turtles (genus *Chrysemys*). Comp. Biochem. Physiol. 57a:297–298.

———. 1994. Morphometric analysis and taxonomy of cooters and red-bellied turtles in the North American genus *Pseudemys* (Emydidae). Chelonian Conservation and Biology 1(2):117–130.

——— and **S. L. Reynolds.** 1980. Aspects of evaporative water loss in the mud turtles *Kinosternon hirtipes* and *Kinosternon flavescens*. Comp. Biochem. Physiol. 67A(4):593–598.

Seigel, R. A. and **N. B. Ford.** 1987. Reproductive Ecology, p. 210–252. *In* R. A. Seigel, J. T. Collins, and S. S. Novak (eds.), Snakes: Ecology and Evolutionary Biology. Macmillan, New York. xiv + 529 p.

——— and ———. 1991. Phenotypic plasticity in the reproductive characteristics of an oviparous snake, *Elaphe guttata:* implications for life history studies. Herpetologica 47(3):301–307.

Selcer, K. W. 1986. Life history of a successful colonizer: the Mediterranean gecko, *Hemidactylus turcicus,* in southern Texas. Copeia 1986(4):956–962.

———. 1987. Seasonal variation in fatbody and liver mass of the introduced Mediterranean gecko, *Hemidactylus turcicus,* in Texas. J. Herpetol. 21(1):74–78.

———. 1990. Egg-size relationships in a lizard with fixed clutch size: variation in a population of the Mediterranean gecko. Herpetologica 46(1):15–21.

Semmler, R. C. 1979. Spatial and temporal activities of the yellow mud turtle, *Kinosternon flavescens,* in eastern New Mexico. M.S. Thesis, Univ. New Mexico, Albuquerque. 71 p.

Sena, A. P. 1978. Temperature relations and the critical thermal maximum of *Holbrookia maculata maculata* (Reptilia: Iguanidae). Southwest. Nat. 23(1):41–50.

———. 1985. The distribution and reproductive ecology of *Sceloporus graciosus arenicolous* in southeastern New Mexico. Final draft, Ph.D. Diss., Univ. New Mexico, Albuquerque. x + 81 p.

Sexton, O. J. 1979. Remarks on defensive behavior of hognose snakes, *Heterodon.* Herpetol. Rev. 10(3):86–87.

——— and **J. R. Bizer.** 1978. Life history patterns of *Ambystoma tigrinum* in montane Colorado. Amer. Midl. Nat. 99(1):101–118.

Shaffer, D. T., Jr. and **W. G. Whitford.** 1981. Behavioral responses of a predator, the round-tailed horned lizard, *Phrynosoma modestum* and its prey, honey pot ants, *Myrmecocystus* spp. Amer. Midl. Nat. 105(2):209–216.

Shannon, F. A. 1949. A western subspecies of *Bufo woodhousii* hitherto erroneously associated with *Bufo compactilis.* Bull. Chicago Acad. Sci. 8(15):301–312.

——— and **F. L. Humphrey.** 1963. Analysis of color pattern polymorphism in the snake *Rhinocheilus lecontei.* Herpetologica 19(3):153–160.

——— and **C. H. Lowe, Jr.** 1955. A new subspecies of *Bufo woodhousei* from the inland southwest. Herpetologica 11(3):185–190.

Shaw, C. E. 1948. The male combat "dance" of some crotalid snakes. Herpetologica 4(4):137–145.

———. 1950. The Gila monster in New Mexico. Herpetologica 6(2):37–39.

——— and **S. Campbell.** 1974. Snakes of the American West. A. E. Knopf, Inc., New York. xii + 332 p.

Shaw, G. 1802. General Zoology or Systematic Natural History. Vol III, Part I, Amphibia. Printed for the author by Thomas Davison, London.

Sheffield, S. R. and **N. Carter.** 1994. Natural history notes: *Phrynosoma cornutum* (Texas horned lizard). Arboreal behavior. Herpetol. Rev. 25(2):65,68.

Sherbrooke, W. C. 1981. Horned Lizards, Unique Reptiles of Western North America. Southwest Parks and Mon. Assoc., Globe, Arizona. 48 p.

———. 1987. Defensive head posture in horned lizards (*Phrynosoma:* Sauria: Iguanidae). Southwest. Nat. 32(4):512–515.

———. 1990. Rain-harvesting in the lizard, *Phrynosoma cornutum:* behavior and integumental morphology. J. Herpetol. 24(3):302–308.

———. 1991. Behavioral (predator-prey) interactions of captive grasshopper mice (*Onychomys torridus*) and horned lizards (*Phrynosoma cornutum* and *P. modestum*). Amer. Midl. Nat. 126(1):187–195.

——— and **S. K. Frost.** 1989. Integumental chromatophores of a color-change, thermoregulating lizard, *Phrynosoma modestum* (Iguanidae; Reptilia). Amer. Mus. Novitates (2943):1–14.

——— and **R. R. Montanucci.** 1988. Stone mimicry in the round-tailed horned lizard, *Phrynosoma modestum* (Sauria: Iguanidae). J. Arid Environ. 14(3):275–284.

Shields, L. M. and **R. G. Lindeborg.** 1956. Records of the spineless softshelled turtle and the snapping turtle from New Mexico. Copeia 1956(2):120–121.

Shine, R. 1992. Snakes, p. 174–211. *In* H. G. Cogger and R. G. Zweifel (eds.), Reptiles and Amphibians. Smithmark Publ., Inc., New York. 240 p.

Shockey, J. 1992. Ribbon snake envenomation. The Desert Monitor February–March 1992:15–16.

Simon, C. A. 1975. The influence of food abundance on territory size in the iguanid lizard *Sceloporus jarrovi*. Ecology 56(4):993–998.

———. 1976a. Lizard coexistence in four dimensions. Nat. Hist. 85(4):70–74.

———. 1976b. Size selection of prey by the lizard, *Sceloporus jarrovi*. Amer. Midl. Nat. 96(1):236–241.

——— and G. A. Middendorf. 1980. Spacing in juvenile lizards (*Sceloporus jarrovi*). Copeia 1980(1):141–146.

Simovich, M. A. 1994. The dynamics of a spadefoot toad (*Spea multiplicata* and *S. bombifrons*) hybridization system, p. 167–182. *In* Brown, P. R. and J. W. Wright (eds.) Herpetology of the North American Deserts. Proceedings of a Symposium. Southwest. Herpetol. Soc. Special Publ. (5) iv + 311 p.

Sitgreaves, L. 1853. Report of an expedition down the Zuñi and the Colorado Rivers, accompanied by maps, sketches, views, and illustrations. U.S. 32nd Congress, 2nd Sess., Senate Exec. Doc. (59):1–198.

Small, M. F., S. P. Tabor, and C. Fazzari. 1994. Life History Notes: *Masticophis flagellum* (western coachwhip). Foraging behavior. Herpetol. Rev. 25(1):28.

Smith, A. G. 1949. The subspecies of the plains garter snake, *Thamnophis radix*. Bull. Chicago Acad. Sci 8(14):285–300.

Smith, D.C. 1981. Competitive interactions of the striped plateau lizard (*Sceloporus virgatus*) and the tree lizard (*Urosaurus ornatus*). Ecology 62(3):679–687.

———. 1985. Home range and territory in the striped plateau lizard (*Sceloporus virgatus*). Anim. Behav. 33(2):417–427.

Smith, D. D. 1989. A comparison of food habits of sympatric *Cnemidophorus exsanguis* and *Cnemidophorus gularis* (Lacertilia, Teiidae). Southwest. Nat. 34(3):418–420.

———, P. A. Medica, and S. R. Sanborn. 1987. Ecological comparison of sympatric populations of sand lizards (*Cophosaurus texanus* and *Callisaurus draconoides*). Great Basin Nat. 47(2):175–185.

——— and W. W. Milstead. 1971. Stomach analyses of the crevice spiny lizard (*Sceloporus poinsetti*). Herpetologica 27(2):147–149.

Smith, G. R. and R. E. Ballinger. 1994a. Thermal ecology of *Sceloporus virgatus* from southeastern Arizona, with comparison to *Urosaurus ornatus*. J. Herpetol. 28(1):65–69.

——— and ———. 1994b. Temperature relationships in the high-altitude viviparous lizard *Sceloporus jarrovi*. Amer. Midl. Nat. 131(1):181–189.

——— and ———. 1994c. Variation in individual growth in the tree lizard, *Urosaurus ornatus*: effects of food and density. Acta Oecol. 15(3):317–324.

——— and ———. 1994d. Temporal and spatial variation in individual growth in the spiny lizard, *Sceloporus jarrovi*. Copeia 1994(4):1007–1013.

——— and ———. 1994e. Survivorship in a high-elevation population of *Sceloporus jarrovi* during a period of drought. Copeia 1994(4):1040–1042.

———, ———, and J. D. Congdon. 1993. Thermal ecology of the high-altitude bunch grass lizard, *Sceloporus scalaris*. Can. J. Zool. 71:2125–2155.

———, ———, and J. W. Nietfeldt. 1994. Elevational variation of growth rates in neonate *Sceloporus jarrovi*: an experimental evaluation. Funct. Ecol. 8:215–218.

Smith, H. M. 1934. The amphibians of Kansas. Amer. Midl. Nat. 15(4):377–528.

———. 1935. Notes on some Mexican lizards of the genus *Holbrookia*, with the description of a new species. Univ. Kansas Sci. Bull. 22(8):185–201.

———. 1937. A synopsis of the *scalaris* group of the lizard genus Sceloporus. Occ. Pap. Mus. Zool. Univ. Michigan, (36):1–8.

———. 1938a. Remarks on the status of the subspecies of *Sceloporus undulatus*, with descriptions of new species and subspecies of the *undulatus* group. Occ. Pap. Mus. Zool. Univ. Michigan (387):1–17.

———. 1938b. Notes on the snakes of the genus *Salvadora*. Univ. Kansas Sci. Bull. 25(12):229–237.

———. 1939. The Mexican and Central American lizards of the genus *Sceloporus*. Zool. Ser. Field Mus. Nat. Hist. (26):1–397.

———. 1941a. Notes on Mexican snakes of the genus *Elaphe*. Copeia 1941(3):132–136.

———. 1941b. Notes on the snake genus *Trimorphodon*. Proc. U.S. Natl. Mus. 91(3130):149–168.

———. 1942. A resume of Mexican snakes of the genus *Tantilla*. Zoologica 27(7):33–42.

———. 1943. The White Sands earless lizard. Zool. Ser. Field Mus. Nat. Hist. 24(30):339–344.

———. 1946. Handbook of Lizards: Lizards of the United States and of Canada. Comstock Publ. Co., Inc., Ithaca, New York. xxi + 557 p.

———. 1947. Subspecies of the Sonoran toad, (*Bufo compactilis* Wiegmann). Herpetologica 4(1):7–13.

———. 1949. Size maxima in terrestrial salamanders. Copeia 1949(1):71.

———. 1950. Handbook of Amphibians and Reptiles of Kansas. Misc. Publ. Univ. Kansas Mus. Nat. Hist. (2):1–336.

———. 1951. The identity of the ophidian name *Coluber eques* Reuss. Copeia 1951(2):138–140.

———. 1955. Effect of preservatives upon pattern in a Mexican garter snake. Herpetologica 11(3):165–168.

———. 1956. Handbook of Amphibians and Reptiles of Kansas. 2nd ed. Misc. Publ. Univ. Kansas Mus. Nat. Hist. (9):1–357.

———. 1957. Curious feeding habit of a blind snake, *Leptotyphlops*. Herpetologica 13(2):102.

———. 1965. Two new colubrid snakes from the United States and Mexico. J. Ohio Herpetol. Soc. 5(1):1–4.

———. 1978. A Guide to Field Identification. Amphibians of North America. Golden Press, New York. 160 p.

———, **E. L. Bell, J. S. Applegarth,** and **D. Chiszar.** 1992. Adaptive convergence in the lizard superspecies *Sceloporus undulatus*. Bull. Maryland Herpetol. Soc. 28(4):123–149.

———and **E. D. Brodie, Jr.** 1982. A Guide to Field Identification Reptiles of North America. Golden Press, New York. 240 p.

———and **W. L. Burger.** 1949. The identity of *Ameiva tesselata* Say. Bull. Chicago Acad. Sci. 8(13):277–284.

———and **D. Chiszar.** 1993. Apparent intergradation in Texas between the subspecies of the Texas blind snake (*Leptotyphlops dulcis*). Bull. Maryland Herpetol. Soc. 29(4):143–155.

———and ———. 1994. Variation in the lined snake (*Tropidoclonion lineatum*) in northern Texas. Bull. Maryland Herpetol. Soc. 30(1):6–14.

———, ———, and **J. A. Lemos-Espinal.** 1995a. A new subspecies of the polytypic lizard species *Sceloporus undulatus* (Sauria: Iguanidae) from northern Mexico. Texas J. Sci. 47(2):117–143.

———, ———, **J. R. Staley II,** and **K. Tepedelen.** 1994. Populational relationships in the corn snake *Elaphe guttata* (Reptilia: Serpentes). Texas J. Sci. 46(3):259–292.

———and **W. P. Hall.** 1974. Contributions to the concepts of reproductive cycles and the systematics of the *scalaris* group of the lizard genus *Sceloporus*. Great Basin Nat. 34(2):97–104.

———and **W. L. Necker.** 1943. Alfredo Dugés' types of Mexican reptiles and amphibians. An. Esc. Nac. Cien. Biol. 3(1,2):179–219.

———and **L. W. Ramsey.** 1952. A new turtle from Texas. Wasmann J. Biol. 10(1):45–54.

———and **R. W. Reese.** 1968. A record tiger salamander. Southwest. Nat. 13(3):370–372.

———and **R. B. Smith.** 1973. Synopsis of the Herpetofauna of Mexico. Vol. II. Analysis of the Literature Exclusive of the Mexican Axolotl. John Johnson, North Bennington, Vermont. xxxiii + 367 p.

———and ———. 1976. Synopsis of the Herpetofauna of Mexico. Vol. III. Source Analysis and Index for Mexican Reptiles. John Johnson, North Bennington, Vermont. Pages numbered by parts.

———and ———. 1979 [1980]. Synopsis of the Herpetofauna of Mexico. Volume VI. Guide to Mexican Turtles, Bibliographic Addendum III. John Johnson, North Bennington, Vermont. xvii + 1044 p.

———and ———. 1993. Synopsis of the Herpetofauna of Mexico. Volume VII. Bibliographic Addendum IV and Index, Bibliographic Addenda II–IV 1979–1991. Univ. Press of Colorado, Niwot. ix + 1082 p.

———, ———, and **H. L. Sawin.** 1977. A summary of snake classification. J. Herpetol. 11(2):115–121.

———and **E. D. Taylor.** 1945. An annotated checklist and key to the snakes of Mexico. Bull. U.S. Natl. Mus. (187):iv + 239 p.

———and ———. 1948. An annotated checklist and key to the Amphibia of Mexico. Bull. U.S. Natl. Mus. (194):iv + 118 p.

———and ———. 1950a. An annotated checklist and key to the reptiles of Mexico exclusive of the snakes. Bull. U.S. Natl. Mus. (199):v + 253 p.

———and ———. 1950b. Type localities of Mexican reptiles and amphibians. Univ. Kansas Sci. Bull. 33, Pt. 2(8):313–380.

———and **F. N. White.** 1955. Adrenal enlargement and its significance in the hognose snakes (*Heterodon*). Herpetologica 11(2):137–144.

———, **R. T. Zappalorti, A. R. Breisch,** and **D. L. McKinley.** 1995b. The type locality of the frog *Acris crepitans*. Herpetol. Rev. 26(1):14.

Smith, N. M. 1974. Observation of voice in the western collared lizard *Crotaphytus collaris bicinctores*. Great Basin Nat. 35(4):276.

Smith, P. W. 1956. The status, correct name, and geographic range of the boreal chorus frog. Proc. Biol. Soc. Washington 69:169–176.

———and **H. M. Smith.** 1962. The systematic and biogeographic status of two Illinois snakes. Occ. Pap. C. C. Adams Cent. Ecol. Stud. West. Michigan Univ. 5:1–10.

Snell, H. L., B. Gorum, and **A. Landwer.** 1993. Results of second years research on the effect of shinnery oak removal on the Dunes sagebrush lizard, *Sceloporus graciosus arenicolous*, in New Mexico. Unpubl. Rept., New Mexico Dept. Game and Fish, Santa Fe. 16 p.

Snider, A. T. and **J. K. Bowler.** 1992. Longevity of reptiles and amphibians in North American collections. Second edition. SSAR Herpetol. Circ. (21):iii + 40 p.

Snow, G. E. 1978. Largest reported tiger salamander. Bull. Maryland Herpetol. Soc. 14(2):89–90.

Snow, S. E. 1993. Establishing a breeding colony of Jo Daviess County, Illinois bullsnakes. Wisconsin Herpetol. Soc. March 1993:4–10.

Snyder, R. C. 1952. Quadrupedal and bipedal locomotion of lizards. Copeia 1952(2):64–70.

LITERATURE CITED

Sonnini, C. S. and **P. A. Latreille**. 1801. Histoire Naturelle des Reptiles, avec Figures dessinées apres Nature. Vol. 2. Chez Deterville, Paris, 332 p. + 21 plates.

Spencer, A. W. 1964. The relationship of dispersal and migration to gene flow in the boreal chorus frog. Ph.D. Diss., Colorado St. Univ., Fort Collins. vii + 116 p.

Spix, J. B. von. 1824. Animalia nova, sive species novae testudinum et ranarum, quos in itinere per Brasiliam, annis 1817–20 . . . collegit et descripsit. . . . Vol. 3. Munich. 37 p.

Spoecker, P. D. 1967a. Movements and seasonal activity cycles of the lizard *Uta stansburiana stejnegeri*. Amer. Midl. Nat. 77(2):484–494.

———. 1967b. Ectoparasites of a Mojave Desert population of the lizard *Uta stansburiana stejnegeri* Schmidt. Amer. Midl. Nat. 77(2):539–542.

Spotila, J. R., R. E. Foley, and E. A. Standora. 1990. Thermoregulation and climate space of the slider turtle, p. 288–298. *In* J. W. Gibbons (ed.), Life History and Ecology of the Slider Turtle. Smithsonian Inst. Press, Washington, D.C. xiv + 368 p.

Springer, S. 1928. An annotated list of the lizards of Lee's Ferry, Arizona. Copeia (169):100–104.

Sredl, M. J. and J. P. Collins. 1992. The interaction of predation, competition, and habitat complexity in structuring an amphibian community. Copeia 1992(3):607–614.

Stabler, R. M. 1948. Prairie rattlesnake eats a spadefoot toad. Herpetologica 4(5):168.

Stahnke, H. L. 1950. The food of the Gila monster. Herpetologica 6(1):103–106.

———. 1952. A note on the food of the Gila monster, *Heloderma suspectum* Cope. Herpetologica 8(2):64–65.

Staub, N. L. 1986. A status survey of the Sacramento mountain salamander, *Aneides hardii*, with an assessment of the impact of logging on salamander abundance. Unpubl. Rept. U.S. Fish Wildl. Serv., Albuquerque, New Mexico. 18 p.

———. 1993. Intraspecific agonistic behavior of the salamander *Aneides flavipunctatus* (Amphibia: Plethodontidae) with comparisons to other plethodontid species. Herpetologica 49(2):271–282.

Stebbins, R. C. 1948. Additional observations on home ranges and longevity in the lizard *Sceloporus graciosus*. Copeia 1948(1):20–22.

———. 1951. Amphibians of Western North America. Univ. California Press, Berkeley. ix + 539 p., 68 plates.

———. 1954. Amphibians and Reptiles of Western North America. McGraw-Hill Book Co., New York. xiv + 536 p.

———. 1958. A new alligator lizard from the Panamint Mountains, Inyo County, California. Amer. Mus. Novitates (1883):1–27.

———. 1962. Amphibians of Western North America. 2nd Printing. Univ. California Press, Berkeley. xiv + 536 p., 68 plates.

———. 1985. A Field Guide to Western Reptiles and Amphibians. Second edition. Houghton Mifflin Co., Boston. xvi + 336 p.

——— and W. J. Riemer. 1950. A new species of plethodontid salamander from the Jemez Mountains of New Mexico. Copeia 1950(2):73–80.

Stejneger, L. H. 1890a. Part V. Annotated list of reptiles and batrachians collected by Dr. C. Hart Merriam and Vernon Bailey on the San Francisco Mountain Plateau and Desert of the Little Colorado, Arizona, with descriptions of new species, p. 103–118. *In* Merriam, C. H. Results of a biological survey of the San Francisco Mountain region and Desert of the Little Colorado, Arizona. North Amer. Fauna (3). vii + 136 p., 13 pl., 5 maps.

———. 1890b. On a new genus and species of colubrine snakes from North America. Proc. U.S. Natl. Mus. 13(802):151–155.

———. 1893. Annotated list of the reptiles and batrachians collected by the Death Valley expedition in 1891, with descriptions of new species, p. 159–228. *In* The Death Valley expedition: a biological survey of parts of California, Nevada, Arizona, and Utah, Part II. North Amer. Fauna (7). 393 p., 14 pl., 5 maps.

———. 1902 [1903]. The reptiles of the Huachuca Mountains, Arizona. Proc. U.S. Natl. Mus. 25(1282):149–158.

———. 1922. Two geckos new to the fauna of the United States. Copeia (108):56.

——— and T. Barbour. 1917. A Check List of North American Amphibians and Reptiles. Harvard Univ. Press, Cambridge. iv + 5–125 p.

——— and ———. 1923. A Check List of North American Amphibians and Reptiles. 2nd ed. Harvard Univ. Press, Cambridge. x + 171 p.

——— and ———. 1933. A Check List of North American Amphibians and Reptiles. 3rd ed. Harvard Univ. Press, Cambridge. xiv + 185 p.

——— and ———. 1939. A Check List of North American Amphibians and Reptiles. 4th ed. Harvard Univ. Press, Cambridge. xvi + 207 p.

——— and ———. 1943. A Check List of North American Amphibians and Reptiles. 5th ed. Bull. Mus. Comp. Zool. 93(1):xix + 260 p.

Stevens, T. P. 1980. Notes on thermoregulation and reproduction in *Cnemidophorus flagellicaudus*. J. Herpetol. 14(4):418–420.

———. 1983. Reproduction in an upper elevation population of *Cnemidophorus inornatus* (Reptilia, Teiidae). Southwest. Nat. 28(1):9–20.

Stickel, W. H. 1943. The Mexican snakes of the genera *Sonora* and *Chionactis* with notes on the status of other colubrid genera. Proc. Biol. Soc. Washington 56(22):109–128.

Stone, W. 1911. On some collections of reptiles and batrachians from the western United States. Proc. Acad. Nat. Sci. Philadelphia 63:222–232.

——— and J. A. G. Rehn. 1903. On the terrestrial vertebrates of portions of southern New Mexico and western Texas. Proc. Acad. Nat. Sci. Philadelphia 55:16–34.

Storey, K. B., J. M. Storey, S. P. J. Brooks, T. A. Churchill, and R. J. Brooks. 1988. Hatchling turtles survive freezing during winter hibernation. Proc. Natl. Acad. Sci. USA 85(21):8350–8354.

Strecker, J. K., Jr. 1910. Notes on the robber frog (*Lithodytes latrans* Cope). Trans. St. Louis Acad. Sci. 19(5):73–82.

Stroud, C. P. 1949. A white spade-foot toad from the New Mexico white sands. Copeia 1949(3):232.

Stuart, J. N. 1988. Life History Notes: *Hypsiglena torquata jani* (Texas night snake). Behavior. Herpetol. Rev. 19(4):84–85.

———. 1991a. Partial albinism in a New Mexico population of *Bufo woodhousei*. Bull. Maryland Herpetol. Soc. 27(1):33.

———. 1991b. *Cnemidophorus exsanguis*. Cat. Amer. Amphib. Rept.:516.1–516.4.

———. 1992a. Status survey of the spotted chorus frog (*Pseudacris clarkii*) in New Mexico. Unpubl. Rept. New Mexico Dept. Game and Fish, Santa Fe. 13 p.

———. 1992b. Preliminary status survey of the Great Plains narrowmouth toad (*Gastrophryne olivacea*) in New Mexico. Unpubl. Rept. New Mexico Dept. Game and Fish, Santa Fe. 9 p.

———. 1995a. Natural History Notes: *Rana catesbeiana* (Bullfrog). Diet. Herpetol. Rev. 26(1):33.

———. 1995b. Notes on aquatic turtles of the Rio Grande drainage, New Mexico. Bull. Maryland Herpetol. Soc. 31(3):147–157.

——— and C. S. Clark. 1990. Life history notes. *Apalone spinifera emoryi* (Texas spiny softshell). Defensive behavior. Herpetol. Rev. 21(4):91–92.

——— and W. G. Degenhardt. 1986. Geographic Distribution: *Cnemidophorus uniparens* (desert grassland whiptail). Herpetol. Rev. 17(3):65.

——— and C. W. Painter. 1988. Geographic Distribution: *Chelydra serpentina serpentina*. Herpetol. Rev. 19(1):21.

——— and ———. 1993a. Life History Notes. *Rana catesbeiana* (Bullfrog). Cannibalism. Herpetol. Rev. 24(3):103.

——— and ———. 1993b. Notes on hibernation of the smooth green snake *Opheodrys vernalis*, in New Mexico. Bull. Maryland Herpetol. Soc. 29(3):140–142.

——— and ———. 1994. A review of the distribution and status of the boreal toad, *Bufo boreas boreas*, in New Mexico. Bull. Chicago Herpetol. Soc. 29(6):113–116.

———, ———, and B. C. Stearns. 1993. Life History Notes: *Trachemys gaigeae* (Big Bend slider). Maximum size. Herpetol. Rev. 24(1):32–33.

Sugg, D. W., L. A. Fitzgerald, and H. L. Snell. 1995. Growth rate, timing of reproduction, and size dimorphism in the southwestern earless lizard (*Cophosaurus texanus scitulus*). Southwest. Nat. 40(2):193–202.

Sullivan, B. K. 1984. Advertisement call variation and observations on breeding behavior of *Bufo debilis* and *B. punctatus*. J. Herpetol. 18(4):406–411.

———. 1986a. Hybridization between the toads *Bufo microscaphus* and *Bufo woodhousei* in Arizona: morphological variation. J. Herpetol. 20(1):11–21.

———. 1986b. Advertisement call variation in the Arizona tree frog, *Hyla wrightorum* Taylor, 1938. Great Basin Nat. 46(2):378–381.

———. 1989. Desert environments and the structure of anuran mating systems. J. Arid Environ. 17(2):175–183.

———. 1990. Natural hybridization between the Great Plains toad (*Bufo cognatus*) and the red-spotted toad (*Bufo punctatus*) from central Arizona. Great Basin Nat. 50(4):371–372.

———. 1992. Calling behavior of the southwestern toad (*Bufo microscaphus*). Herpetologica 48(4):383–389.

———. 1993. Distribution of the southwestern toad (*Bufo microscaphus*) in Arizona. Great Basin Nat. 53(4):402–406.

——— and T. Lamb. 1988. Hybridization between the toads *Bufo microscaphus* and *Bufo woodhousii* in Arizona: variation in release call and allozymes. Herpetologica 44(3):325–333.

——— and K. B. Malmos. 1994. Call variation in the Colorado River toad (*Bufo alvarius*): behavioral and phylogenetic implications. Herpetologica 50(2):146–156.

Swain, T. A. and H. M. Smith. 1978. Communal nesting in *Coluber constrictor* in Colorado (Reptilia: Serpentes). Herpetologica 32(2):175–177.

Swarth, H. S. 1921. The type locality of *Crotalus willardi* Meek. Copeia 1921(100):83.

Sweet, S. S. and W. S. Parker. 1990. *Pituophis melanoleucus*. Cat. Amer. Amphib. Rept.:474.1–474.8.

Swinford, G. W. 1989. Captive reproduction of the banded rock rattlesnake *Crotalus lepidus klauberi*, p. 99–110. *In* M. J. Uricheck (ed.), 13th International Herpetological Symposium on Captive Propagation and Husbandry. International Herpetological Symposium, Inc., Stanford, California. i + 214 p.

———. 1990. Collecting the banded rock rattlesnake (*Crotalus lepidus klauberi*) in the mountains of southern New Mexico. Purpose: captive breeding. The Vivarium 3(1):8–10.

———. 1992. Status of the mottled rock rattlesnake, *Crotalus lepidus lepidus* in southeastern New Mexico. Unpubl. Rept., New Mexico Dept. Game and Fish, Santa Fe. 17 p.

LITERATURE CITED

Tanner, D. L. 1975. Lizards of the New Mexican Llano Estacado and its adjacent river valleys. Stud. Nat. Sci. Eastern New Mexico Univ. 2(2):1–39.

Tanner, V. M. 1931. A synoptical study of Utah Amphibia. Proc. Utah Acad. Sci. Arts Lett. 8:159–198.

Tanner, W. W. 1944. A taxonomic study of the genus *Hypsiglena*. Great Basin Nat. 5(3–4):25–92.

———. 1949. Food of the wandering garter snake, *Thamnophis elegans vagrans* (Baird and Girard), in Utah. Herpetologica 5(4):85–86.

———. 1954. Herpetological notes concerning some reptiles of Utah and Arizona. Herpetologica 10(2):92–96.

———. 1955. A new *Sceloporus magister* from eastern Utah. Great Basin Nat. 15(1):32–34.

———. 1959. A new *Thamnophis* from western Chihuahua with notes on four other species. Herpetologica 15(4):165–172.

———. 1983. *Lampropeltis pyromelana*. Cat Amer. Amphib. Rept.:342.1–342.2.

———. 1985. Snakes of western Chihuahua. Great Basin Nat. 45(4):615–676.

———. 1987. Lizards and turtles of western Chihuahua. Great Basin Nat. 47(3):383–421.

———. 1989. Status of *Spea stagnalis* Cope (1875), *Spea intermontanus* Cope (1889), and a systematic review of *Spea hammondii* Baird (1839) (Amphibia: Anura). Great Basin Nat. 49(4):503–510.

———. 1990. *Thamnophis rufipunctatus*. Cat. Amer. Amphib. Rept.:505.1–505.2.

——— and B. H. Banta. 1963. The systematics of *Crotaphytus wislizeni*, the leopard lizards. Part I. A redescription of *Crotaphytus wislizeni wislizeni* Baird and Girard, and a description of a new subspecies from the upper Colorado River basin. Great Basin Nat. 23(3–4):129–148.

——— and D. Cox. 1981. Reproduction in the snake *Lampropeltis pyromelana*. Great Basin Nat. 41(3):314–316.

——— and J. E. Krogh. 1973. Ecology of *Sceloporus magister* at the Nevada Test Site, Nye County, Nevada. Great Basin Nat. 33(3):133–146.

——— and C. H. Lowe. 1989. Variations in *Thamnophis elegans* with descriptions of new subspecies. Great Basin Nat. 49(4):511–516.

Tanzer, E. C. 1970. Polymorphism in the *mexicana* complex of kingsnakes, with notes on their natural history. Herpetologica 26(4):419–428.

Taylor, E. H. 1931. Notes on two specimens of the rare snake *Ficimia cana* and the description of a new species of *Ficimia* from Texas. Copeia 1931(1):4–7.

———. 1933. Observations on the courtship of turtles. Univ. Kansas Sci. Bull. 21(6):269–271.

———. 1935. A taxonomic study of the cosmopolitan scincoid lizards of the genus *Eumeces* with an account of the distribution and relationships of its species. Univ. Kansas Sci. Bull. 23(1):1–643.

———. 1936 [1937]. Notes and comments on certain American and Mexican snakes of the genus *Tantilla*, with descriptions of new species. Trans. Kansas Acad. Sci. 39:335–348.

———. 1936 [1938]. Notes on the herpetological fauna of the Mexican state of Sonora. Univ. Kansas Sci. Bull. 24(19):475–503.

———. 1938a. On Mexican snakes of the genera *Trimorphodon* and *Hypsiglena*. Univ. Kansas Sci. Bull. 25(16):357–383.

———. 1938b. Frogs of the *Hyla eximia* group in Mexico, with descriptions of two new species. Univ. Kansas Sci. Bull. 25(19):421–445.

———. 1938 [1939]. New species of Mexican tailless Amphibia. Univ. Kansas Sci. Bull. 25(17):385–405.

———. 1941a. A new plethodontid salamander from New Mexico. Proc. Biol. Soc. Washington 54:77–79.

———. 1941b. Two new ambystomid salamanders from Chihuahua. Copeia 1941(3):143–146.

———. 1943. Herpetological novelties from Mexico. Univ. Kansas Sci. Bull. 29 (pt II, no. 8):343–360.

Taylor, H. L. 1965. Morphological variation in selected populations of the teiid lizards *Cnemidophorus velox* and *Cnemidophorus inornatus*. Univ. Colorado Stud. Ser. Biol. (21):1–27.

———, C. R. Cooley, R. A. Aguilar, and C. J. Obana. 1992. Factors affecting clutch size in the teiid lizards *Cnemidophorus tigris gracilis* and *C. t. septentrionalis*. J. Herpetol. 26(4):443–447.

———, C. Currie, and J. J. Baker. 1989. The mode of origin for males collected from natural clones of parthenogenetic lizards (*Cnemidophorus*): cytological evidence. J. Herpetol. 23(2):202–205.

———, L. A. Harris, G. L. Burkholder, and J. M. Walker. 1994. Relationship of clutch size to body size and elevation of habitat in the three subspecies of the teiid lizard, *Cnemidophorus tigris*. Copeia 1994(4):1047–1050.

——— and P. A. Medica. 1966. Natural hybridization of the bisexual teiid lizard *Cnemidophorus inornatus* and the unisexual *Cnemidophorus perplexus* in southern New Mexico. Univ. Colorado Stud. Ser. Biol. (22):1–9.

———, J. M. Walker, and P. A. Medica. 1967. Males of three normally parthenogenetic species of teiid lizards (genus *Cnemidophorus*). Copeia 1967(4):737–743.

Taylor, T. 1985. Life History Notes: *Eumeces callicephalus* (mountain skink). Reproduction. Herpetol. Rev. 16(1):27.

Tennant, A. 1984. The Snakes of Texas. Texas Monthly Press, Austin. 561 p.

Thayer, F. D., Jr. 1988. Life History Notes: *Crotalus atrox*: (Western diamondback rattlesnake). Hunting behavior. Herpetol. Rev. 19(2):35.

Thomas, R. A. and **J. R. Dixon.** 1976. A re-evaluation of the *Sceloporus scalaris* group (Sauria: Iguanidae). Southwest. Nat. 20(4):523–536.

Thompson, C. W. and **M. C. Moore.** 1991a. Throat colour reliably signals status in male tree lizards, *Urosaurus ornatus*. Anim. Behav. 42(5):745–753.

——— and ———. 1991b. Syntopic occurrence of multiple dewlap color morphs in male tree lizards, *Urosaurus ornatus*. Copeia 1991(2):493–503.

Thompson, D. B. and **T. R. Jones.** 1992. The occurrence of paedomorphic cave-dwelling tiger salamanders in central New Mexico, p. 3–6. *In* D. Belski (ed.), GYPKAP Report #2, 1988–1991. Southwest Region Natl. Speleol. Soc., Adobe Press, Albuquerque, New Mexico. 60 p.

Thompson, F. G. 1957. A new Mexican garter snake (genus *Thamnophis*) with notes on related forms. Occ. Pap. Mus. Zool. Univ. Michigan (584):1–10.

Tihen, J. A. 1937 [1938]. Additional distributional records of amphibians and reptiles in Kansas counties. Trans. Kansas Acad. Sci. 40:401–409.

Tilley, S. G. 1964. A quantitative study of the shrinkage in the digestive tract of the tiger salamander (*Ambystoma tigrinum*) during metamorphosis. J. Ohio Herpetol. Soc. 4:81–85.

Tinkham, E. R. 1971. The biology of the Gila monster, p. 387–413. *In* W. Bücherl and E. E. Buckley (eds.). Venomous animals and their venoms, vol. 2, Venomous Vertebrates. Academic Press, New York. 687 p.

Tinkle, D. W. 1957. Ecology, maturation and reproduction of *Thamnophis sauritus proximus*. Ecology 38(1):69–77.

———. 1959. Observations on the lizards *Cnemidophorus tigris*, *Cnemidophorus tessellatus* and *Crotaphytus wislizeni*. Southwest. Nat. 4(4):195–200.

———. 1961. Population structure and reproduction in the lizard *Uta stansburiana stejnegeri*. Amer. Midl. Nat. 66(1):206–234.

———. 1962. Reproductive potential and cycles in female *Crotalus atrox* from northwestern Texas. Copeia 1962(2):306–313.

———. 1965. Population structure and effective size of a lizard population. Evolution 19(4):569–573.

———. 1967. The life and demography of the side-blotched lizard, *Uta stansburiana*. Misc. Publ. Mus. Zool. Univ. Michigan (132):1–182.

———. 1969. Evolutionary implications of comparative population studies in the lizard *Uta stansburiana*, p. 133–154. *In* Systematic Biology, Natl. Acad. Sci. Publ. (1692), Washington, D.C. xiii + 632 p.

———. 1973. A population analysis of the sagebrush lizard, *Sceloporus graciosus* in southern Utah. Copeia 1973(2):284–296.

———. 1976. Comparative data on the population ecology of the desert spiny lizard, *Sceloporus magister*. Herpetologica 32(1):1–6.

——— and **R. E. Ballinger.** 1972. *Sceloporus undulatus*: a study of the intraspecific comparative demography of a lizard. Ecology 53(4):570–584.

——— and **A. E. Dunham.** 1983. Demography of the tree lizard, *Urosaurus ornatus* in central Arizona. Copeia 1983(3):585–598.

——— and ———. 1986. Comparative life histories of two syntopic sceloporine lizards. Copeia 1986(1):1–18.

———, ———, and **J. D. Congdon.** 1993. Life history and demographic variation in the lizard *Sceloporus graciosus*: a long-term study. Ecology 74(8):2413–2429.

——— and **N. F. Hadley.** 1975. Lizard reproductive effort: caloric estimates and comments on its evolution. Ecology 56(2):427–434.

——— and **G. N. Knopf.** 1964. Biologically significant distribution records for amphibians and reptiles in northwest Texas. Herpetologica 20(1):42–47.

———, **D. McGregor,** and **S. Dana.** 1962. Home range ecology of *Uta stansburiana stejnegeri*. Ecology 43(2):223–229.

———, **H. M. Wilbur,** and **S. G. Tilley.** 1970. Evolutionary strategies in lizard reproduction. Evolution 24(1):55–74.

——— and **D. W. Woodward.** 1967. Relative movements of lizards in natural populations as determined from recapture radii. Ecology 48(1):166–168.

Toliver, M. E. and **D. T. Jennings.** 1975. Food habits of *Sceloporus undulatus tristichus* Cope (Squamata: Iguanidae) in Arizona. Southwest. Nat. 20(1):1–11.

Townsend, C. H. T. 1893. On the life zones of the Organ Mountains and adjacent region in southern New Mexico, with notes on the fauna of the range. Science 22:313–315.

Trauth, S. E. 1985. Nest, eggs and hatchlings of the Mediterranean gecko, *Hemidactylus turcicus* (Sauria: Gekkonidae), from Texas. Southwest. Nat. 30(2):309–310.

———. 1987. Natural nests and egg clutches of the Texas spotted whiptail, *Cnemidophorus gularis gularis* (Sauria: Teiidae), from northcentral Texas. Southwest. Nat. 32(2):279–281.

———. 1992. A new subspecies of six-lined racerunner, *Cnemidophorus sexlineatus* (Sauria: Teiidae), from southern Texas. Texas J. Sci. 44(4):437–443.

Troschel, F. H. 1850 [1852]. *Cophosaurus texanus*, neue Eidechsen-gattung aus Texas. Arch. Naturges. (Berlin) 16(1):388–394, Pl. 6.

LITERATURE CITED

Tryon, B. W. 1976. Second generation reproduction and courtship behavior in the Trans-Pecos ratsnake, *Elaphe subocularis*. Herpetol. Rev. 7(4):156–157.

——. 1978. Reproduction in a pair of captive Arizona ridge-nosed rattlesnakes, *Crotalus willardi willardi* (Reptilia, Serpentes, Crotalidae). Bull. Maryland Herpetol. Soc. 14(2):83–88.

Tucker, J. K. and **M. E. Sullivan.** 1975. Unsuccessful attempts by bullfrogs to eat toads. Trans. Illinois St. Acad. Sci. 68:167.

Tuma, M. W. 1993. Life history notes. *Kinosternon flavescens* (yellow mud turtle). Multiple nesting. Herpetol. Rev. 24(1):31.

Ultsch, G. R. 1988. Blood gases, hematocrit, plasma ion concentrations, and acid-base status of musk turtles *(Sternotherus odoratus)* during simulated hibernation. Physiol. Zool. 61(1):78–94.

——, **C. V. Herbert,** and **D. C. Jackson.** 1984. The comparative physiology of diving in North American freshwater turtles, I: Submergence tolerance, gas exchange, and acid-base balance. Physiol. Zool 57(6):620–631.

United States Department of the Interior. 1991. Title 50-Wildlife and Fisheries. Part 17: Endangered and threatened wildlife and plants, Subpart B: Lists, endangered and threatened wildlife (CFR 17.11 and 17.12). U.S. Govt. Printing Office, Washington, D.C. 37 p.

——. 1994. Endangered and threatened wildlife and plants: animal candidate review for listing as endangered or threatened species. Federal Register 59(219):58982–59028.

United States Fish and Wildlife Service. 1986. Distribution and ecological requirements of endemic salamanders in relation to forestry management. Summary of 1986 activities. Part 2: Santa Fe National Forest. Unpubl. Rept. submitted to U.S. Forest Serv., Albuquerque, New Mexico. 40 p.

——. 1987. Distribution and ecological requirements of endemic salamanders in relation to forestry management. Summary of 1987 activities. Part 2: Santa Fe National Forest. Rept. submitted to U.S. Forest Service, Albuquerque, New Mexico. 40 p.

Vaeth, R. H. 1980. Observation of ophiophagy in the western hooknose snake, *Gyalopion canum*. Bull. Maryland Herpetol. Soc. 16(3):94–96.

Valett, B. B. and **D. L. Jameson.** 1961. The embryology of *Eleutherodactylus augusti latrans*. Copeia 1961(1):103–109.

Vance, T. 1978. A field key to the whiptail lizards (genus *Cnemidophorus*) Part I: The whiptails of the United States. Bull. Maryland Herpetol. Soc. 14(1):1–9.

——, **F. S. Hendricks, J. R. Dixon,** and **H. M. Smith.** 1991. The status of *Cnemidophorus tigris reticuloriens* Vance, 1978 (Reptilia: Lacertilia). Bull. Maryland Herpetol. Soc. 27(2):95–98.

Vance, V. J. 1973. Temperature preference and tolerance in the gecko, *Coleonyx variegatus*. Copeia 1973(3):615–617.

Van Denburgh, J. 1895. Description of a new rattlesnake (*Crotalus pricei*) from Arizona. Proc. California Acad. Sci. 2(5):856–857.

——. 1922a. The reptiles of western North America. Vol. I. Lizards. Occ. Pap. California Acad. Sci. (10):1–611.

——. 1922b. The reptiles of western North America. Vol. II. Snakes and turtles. Occ. Pap. California Acad. Sci. (10):617–1028.

——. 1924. Notes on the herpetology of New Mexico, with a list of species known from that state. Proc. California Acad. Sci. 13(12):189–230.

Van Devender, T. R. 1986. Climatic cadences and the composition of Chihuahuan Desert communities: the late Pleistocene packrat midden record, p. 285–299. *In* J. Diamond and T. J. Case (eds.), Community Ecology. Harper and Row, New York. xxii + 665 p.

——and **G. L. Bradley.** 1994. Late Quaternary amphibians and reptiles from Maravillas Canyon Cave, Texas, with discussion of the biogeography and evolution of the Chihuahuan Desert herpetofauna, p. 23–53. *In* P. R. Brown and J. W. Wright (eds.), Herpetology of the North American deserts: Proceedings of a symposium. Special Pub. 5, Southwest. Herpetol. Soc., Van Nuys, California. iv + 311 p.

——and **B. L. Everitt.** 1977. The latest Pleistocene and Recent vegetation of the Bishop's Cap, south-central New Mexico. Southwest. Nat. 22(3):337–352.

——and **C. W. Howard.** 1973. Notes on natural nests and hatching success in the regal horned lizard (*Phrynosoma solare*) in southern Arizona. Herpetologica 29(3):238–239.

——and **C. H. Lowe, Jr.** 1977. Amphibians and reptiles of Yepómera, Chihuahua, Mexico. J. Herpetol. 11(1):41–50.

Van Loben Sels, R. C. and **L. J. Vitt.** 1984. Desert lizard reproduction: seasonal and annual variation in *Urosaurus ornatus* (Iguanidae). Can. J. Zool. 62(9):1779–1787.

Vest, M. 1992. AMPHIBIAN HIGHLIGHTS Mountain treefrog, *Hyla eximia*. The Desert Monitor February/March:8–9.

Vial, J. L. 1957. A new size record for *Thamnophis marcianus* Baird and Girard. Copeia 1957(2):143.

Vincent, J. 1982. Color pattern variation in *Crotalus lepidus lepidus* (Viperidae) in southwest Texas. Southwest. Nat. 27(3):263–272.

Vinegar, M. B. 1972. The function of breeding coloration in the lizard, *Sceloporus virgatus*. Copeia 1972(4):660–664.

——. 1975a. Demography of the striped plateau lizard, *Sceloporus virgatus*. Ecology 56(1):172–182.

——. 1975b. Life history phenomena in two populations of the lizard *Sceloporus undulatus* in southwestern New Mexico. Amer. Midl. Nat. 93(2):388–402.

———. 1975c. Comparative aggression in *Sceloporus virgatus*, *S. undulatus consobrinus*, and *S. u. tristichus* (Sauria: Iguanidae). Anim. Behav. 23(2):279–286.

Vitt, L. J. 1975. Observations on reproduction in five species of Arizona snakes. Herpetologica 31(1):83–84.

———. 1977. Observations on clutch and egg size and evidence for multiple clutches in some lizards of southwestern United States. Herpetologica 33(3):333–338.

——— and **A. C. Hulse.** 1973. Observations on feeding habits and tail display of the Sonoran coral snake, *Micruroides euryxanthus*. Herpetologica 29(4):302–304.

——— and **R. D. Ohmart.** 1977a. Ecology and reproduction of lower Colorado River lizards: I. *Callisaurus draconoides* (Iguanidae). Herpetologica 33(2):214–222.

——— and ———. 1977b. Ecology and reproduction of Lower Colorado River lizards: II. *Cnemidophorus tigris* (Teiidae), with comparisons. Herpetologica 33(2):223–234.

———, **R. C. Van Loben Sels,** and **R. D. Ohmart.** 1981. Ecological relationships among arboreal desert lizards. Ecology 62(2):398–410.

Vogt, R. C. and **J. J. Bull.** 1982. Temperature controlled sex-determination in turtles: ecological and behavioral aspects. Herpetologica 38(1):156–164.

———, ———, **C. J. McCoy,** and **T. W. Houseal.** 1982. Incubation temperature influences sex determination in kinosternid turtles. Copeia 1982(2):480–482.

Vorhies, C. T. 1948. Food items of rattlesnakes. Copeia 1948(4):302–303.

Wagner, E. 1995. Ask the breeder: box turtles. Reptiles 2(4):10.

Wake, D. B. 1965. *Aneides hardii*. Cat. Amer. Amphib. Rept.:17.1–17.2.

———. 1991. Declining amphibian populations. Science 253:860.

Waldschmidt, S. R. 1979. The effect of statistically based models on home range size estimate in *Uta stansburiana*. Amer. Midl. Nat. 101(1):236–240.

———. 1983. The effect of supplemental feeding on home range size and activity patterns in the lizard *Uta stansburiana*. Oecologia 57(1–2):1–5.

——— and **C. R. Tracy.** 1983. Interactions between a lizard and its thermal environment: implications for sprint performance and space utilization in the lizard *Uta stansburiana*. Ecology 64(3):476–484.

Walker, J. M, J. E. Cordes, and **R. M. Abuhteba.** 1990. Hybridization between all-female *Cnemidophorus neomexicanus* and gonochoristic *C. sexlineatus* (Sauria: Teiidae). Amer. Midl. Nat. 123(2):404–408.

———, ———, **C. C. Cohn, H. L. Taylor, R. V. Kilambi,** and **R. L. Meyer.** 1994. Life history characteristics of three morphotypes in the parthenogenetic *Cnemidophorus dixoni* complex (Sauria: Teiidae) in Texas and New Mexico. Texas J. Sci. 46(1):27–33.

———, ———, **J. F. Scudday, R. V. Kilambi,** and **C. C. Cohn.** 1991. Activity, temperature, age, size, and reproduction in the parthenogenetic whiptail lizard *Cnemidophorus dixoni* in the Chinati Mountains in Trans-Pecos Texas. Amer. Midl. Nat. 126(2):256–268.

Ward, J. P. 1978. *Terrapene ornata*. Cat. Amer. Amphib. Rept.:217.1–217.4.

———. 1984. Relationships of chrysemyd turtles of North America (Testudines: Emydidae). Spec. Publ. Mus. Texas Tech. Univ. (21):1–50.

Wasserman, A. O. 1970a. *Scaphiopus couchii*. Cat. Amer. Amphib. Rept.:85.1–85.2.

———. 1970b. Chromosomal studies of the Pelobatidae (Salientia) and some instances of ploidy. Southwest. Nat. 15(2):239–248.

Watkins, J. F., F. R. Gehlbach, and **R. S. Baldridge.** 1967. Ability of the blind snake *Leptotyphlops dulcis* to follow pheromone trails of army ants, *Neivamyrmex nigrescens* and *N. opacithorax*. Southwest. Nat. 12(4):455–462.

Webb, R. G. 1960. Notes on some amphibians and reptiles from northern Mexico. Trans. Kansas Acad. Sci. 63(4):289–298.

———. 1962. North American recent soft-shelled turtles (Family Trionychidae). Univ. Kansas Publ. Mus. Nat. Hist. 13(10):429–611

———. 1966. Resurrected names for the Mexican populations of black-necked garter snakes, *Thamnophis cyrtopsis* (Kennicott). Tulane Stud. Zool. 13(2):55–70.

———. 1969. Survival adaptations of tiger salamanders (*Ambystoma tigrinum*) in the Chihuahuan desert, p. 143–147. *In* C. C. Hoff and M. L. Riedesel (eds.), Physiological Systems in Semiarid Environments. Univ. New Mexico Press, Albuquerque. ix + 239 p.

———. 1970a. *Gerrhonotus kingii*. Cat. Amer. Amphib. Rept.:97.1–97.4.

———. 1970b. Reptiles of Oklahoma. Univ. of Oklahoma Press, Norman. xi + 370 p.

———. 1973a. *Trionyx muticus*. Cat. Amer. Amphib. Rept.:139.1–139.2.

———. 1973b. *Trionyx spiniferus*. Cat. Amer. Amphib. Rept.:140.1–140.4.

———. 1976. A review of the garter snake *Thamnophis elegans* in Mexico. Contrib. Sci. Nat. Hist. Mus. Los Angeles Co. (284):1–13.

———. 1980. *Thamnophis cyrtopsis*. Cat. Amer. Amphib. Rept.:245.1–245.4.

LITERATURE CITED

———. 1988. Type and type locality of *Sceloporus poinsettii* Baird and Girard (Sauria: Iguanidae). Texas J. Sci. 40(4):407–415.

———. 1990a. *Trionyx*. Cat. Amer. Amphib. Rept.:487.1–487.7.

———. 1990b. Description of a new subspecies of *Bogertophis subocularis* (Brown) from northern Mexico (Serpentes: Colubridae). Texas J. Sci. 42(3):227–241.

——— and **R. W. Axtell.** 1986. Type and type-locality of *Sceloporus jarrovi* Cope, with travel-routes of Henry W. Henshaw in Arizona in 1873 and 1874. J. Herpetol. 20(1):32–41.

——— and **W. L. Roueche.** 1971. Life history aspects of the tiger salamander (*Ambystoma tigrinum mavortium*) in the Chihuahuan Desert. Great Basin Nat. 31(4):193–212.

Weese, A. O. 1917. An experimental study of the reactions of the horned lizard, *Phrynosoma modestum* Gir., a reptile of the semi-desert. Biol. Bull. 32:98–116.

———. 1919. Environmental reactions of *Phrynosoma*. Amer. Nat. 53:33–54.

Weigmann, D. L., M. Hakkila, K. Whitmore, and **R. A. Cole.** 1980. Survey of Sacramento mountain salamander habitat of the Cloudcroft and Mayhill districts of the Lincoln National Forest. Unpubl. Rept., Lincoln Natl. Forest, Alamogordo, New Mexico. 45 p.

Weinstein, S. A., C. F. DeWitt, and **L. A. Smith.** 1992. Variability of venom-neutralizing properties of serum from snakes of the colubrid genus *Lampropeltis*. J. Herpetol. 26(4):452–461.

Wells, K. D. 1977. The social behavior of anuran amphibians. Anim. Behav. 25(3):666–693.

Werler, J. E. 1951. Miscellaneous notes on the eggs and young of Texan and Mexican reptiles. Zoologica 36(1):37–48.

Whipple, A. W. 1854 [1856]. Itinerary, vii + 136 p. *In* reports of explorations and surveys, to ascertain the most practicable and economical route for a railroad from the Mississippi River to the Pacific Ocean. Vol. II(6). 33rd Congress, 2nd Sess., House of Rep. Exec. Doc. (91). Washington, D.C.

Whitaker, J. O., Jr., D. Rubin, and **J. R. Munsee.** 1977. Observations on food habits of four species of spadefoot toads, genus *Scaphiopus*. Herpetologica 33(4):468–475.

Whitford, W. G. 1968. Physiological responses to temperature and desiccation in the endemic New Mexico plethodontids, *Plethodon neomexicanus* and *Aneides hardii*. Copeia 1968(2):247–251.

——— and **M. Bryant.** 1979. Behavior of a predator and its prey: the horned lizard (*Phrynosoma cornutum*) and harvester ants (*Pogonomyrmex* spp.). Ecology 60(4):686–694.

——— and **F. M. Creusere.** 1977. Seasonal and yearly fluctuations in Chihuahuan Desert lizard communities. Herpetologica 33(1):54–65.

——— and **M. Massey.** 1970. Responses of a population of *Ambystoma tigrinum* to thermal and oxygen gradients. Herpetologica 26(3):372–376.

Whiting, M. J., J. R. Dixon, and **R. C. Murray.** 1993. Spatial distribution of a population of Texas horned lizards (*Phrynosoma cornutum*: Phrynosomatidae) relative to habitat and prey. Southwest. Nat. 38(2):150–154.

———, **J. C. Godwin,** and **M. K. Coldern.** 1991. Life history notes: *Cnemidophorus sexlineatus* (Six-lined racerunner) and *Cophosaurus texanus* (Texas earless lizard). Spider predation. Herpetol. Rev. 22(2):58.

———, **B. D. Greene, J. R. Dixon, A. L. Mercer,** and **C. C. Eckerman.** 1992. Observations on the foraging ecology of the western coachwhip snake, *Masticophis flagellum testaceus*. The Snake 24:157–160.

Wied-Neuwied, M. A. P. 1838. Reise in das innere Nörd-America in den Jahren 1832 bis 1834. Hoaelscher, Coblenz. (Smith and Smith [1973] state that "This work appears to have been issued in 20 parts, of which Hefte 1–6 [Vol. 1, p. 1–392] appeared in 1838, and the remainder of Vol. 1 in 1839; Hefte 15–17 [Vol. 2, p. 169–504] appeared in 1840, and the work was completed in 1841").

———. 1865. Verzeichniss der Reptilien, welche auf einer Reise im nordlichen America beobachtet wurden. Nova Acta Acad. Leopold. Carol. 32:1–146.

Wiegmann, A. F. A. 1828. Beiträge zur Amphibienkunde. Isis von Oken 21(3/4):364–383.

———. 1829. Ueber das Acaltetepon oder Temacuilcahuya des Hernandez, eine neue Gattung der Saurer, *Heloderma*, Isis von Oken 22:624–629.

Wiens, J. J. 1993. Phylogenetic systematics of the tree lizards (genus *Urosaurus*). Herpetologica 49(4):399–420.

——— and **T. A. Titus.** 1991. A phylogenetic analysis of *Spea* (Anura: Pelobatidae). Herpetologica 47(1):21–28.

Wiest, J. A., Jr. 1982. Anuran succession at temporary ponds in a post oak-savanna region of Texas, p. 39–47. *In* N. J. Scott, Jr. (ed.), Herpetological Communities. U.S. Dept. Interior, Fish Wildl. Serv., Wildl. Res. Rept. (13). iv + 239 p.

Williams J. L. (ed.). 1986. New Mexico in Maps. 2nd ed. Univ. New Mexico Press, Albuquerque. xvii + 409 p.

Williams, K. L. 1959. Nocturnal activity of some species of horned lizards, genus *Phrynosoma*. Herpetologica 15(1):43.

———. 1988. Systematics and Natural History of the American Milk Snake, *Lampropeltis triangulum*. 2nd revised ed. Milwaukee Public Mus., Milwaukee. x + 176 p.

———. 1994. *Lampropeltis triangulum*. Cat. Amer. Amphib. Rept.:594.1–594.10.

Williams, S. R. 1972. Reproduction and ecology of the Jemez Mountains salamander, *Plethodon neomexicanus*. M.S. Thesis, Univ. New Mexico, Albuquerque. 98 p.

―――. 1973. *Plethodon neomexicanus*. Cat. Amer. Amphib. Rept.:131.1–131.2.

―――. 1976. Comparative ecology and reproduction of the endemic New Mexico plethodontid salamanders, *Plethodon neomexicanus* and *Aneides hardii*. Ph.D. Diss., Univ. New Mexico, Albuquerque. 152 p.

―――. 1978. Comparative reproduction of the endemic New Mexico plethodontid salamanders, *Plethodon neomexicanus* and *Aneides hardii*. J. Herpetol. 12(4):471–476.

Williams, T. A. and **J. L. Christiansen.** 1981. The niches of two sympatric softshell turtles, *Trionyx muticus* and *Trionyx spiniferus*, in Iowa. J. Herpetol. 15(3):303–308.

Williamson, L. U., J. R. Spotila, and **E. A. Standora.** 1989. Growth, selected temperature, and CTM of young snapping turtles, *Chelydra serpentina*. J. Therm. Biol. 14:(1)33–39.

Williamson, M. A. 1971. An instance of cannibalism in *Crotalus lepidus* (Serpentes: Crotalidae). Herpetol. Rev. 3(1):18.

Wilson, B. S. 1991. Latitudinal variation in activity season mortality rates of the lizard *Uta stansburiana*. Ecol. Monogr. 61(4):393–414.

Wilson, L. D. 1970. The coachwhip snake, *Masticophis flagellum* (Shaw): Taxonomy and distribution. Tulane Stud. Zool. Bot. 16(2):31–99.

―――. 1973a. *Masticophis*. Cat. Amer. Amphib. Rept.:144.1–144.2.

―――. 1973b. *Masticophis flagellum*. Cat. Amer. Amphib. Rept.:145.1–145.4.

―――. 1978. *Coluber constrictor*. Cat. Amer. Amphib. Rept.:218.1–218.4.

―――. 1982. *Tantilla*. Cat. Amer. Amphib. Rept.:307.1–307.4.

Wislizenus, A., M. D. 1848. Memoir of a tour to northern Mexico, Connected with Col. Doniphan's expedition, in 1846 and 1847. 30th Congress, 1st Sess., Sen. Misc. Doc. (26). Washington, D.C. 141 p.

Woodbury, A. M. and **R. M. Hansen.** 1950. A snake den in Tintic Mountains, Utah. Herpetologica 6(3):66–70.

―――and **D. D. Parker.** 1956. A snake den in Cedar Mountains and notes on snakes and parasitic mites. Herpetologica 12(4):261–268.

Woodhouse, S. W. 1854. Report on natural history. *In* Report of an expedition down the Zuñi and Colorado Rivers in 1851 by L. Sitgreaves, 33rd Congress, 2nd Sess., Sen. Exec. Doc., p. 3–40.

Wagner, E. 1995. Ask the breeder: Box turtles. Reptiles 2(4):10.

Woodin, W. H. 1953. Notes on some reptiles from the Huachuca area of southeastern Arizona. Bull. Chicago Acad. Sci. 9(15):285–296.

Woodward, B. D. 1982a. Persistence and male mating success in *Bufo woodhousei*. Ecology 63(2):583–585.

―――. 1982b. Sexual selection and nonrandom mating patterns in desert anurans (*Bufo woodhousei, Scaphiopus couchi, S. multiplicatus* and *S. bombifrons*). Copeia 1982(2):351–355.

―――. 1982c. Tadpole competition in a desert anuran community. Oecologica 54(1):96–100.

―――. 1983. Predator-prey interactions and breeding-pond use by temporary-pond species in a desert anuran community. Ecology 64(6):1549–1555.

―――. 1984a. Arrival to and location of *Bufo woodhousei* in the breeding pond: effect on the operational sex ratio. Oecologia 62:240–244.

―――. 1984b. Operational sex ratios and sex biased mortality in *Scaphiopus* (Pelobatidae). Southwest. Nat. 29(2):232–233.

―――. 1987a. Clutch parameters and pond use in some Chihuahuan Desert anurans. Southwest. Nat. 32(1):13–19.

―――. 1987b. Intra- and interspecific variation in spadefoot toad (*Scaphiopus*) clutch parameters. Southwest. Nat. 32(1):127–131.

―――and **S. Mitchell.** 1985. The distribution of *Bufo boreas* in New Mexico. Unpubl. Rept., New Mexico Dept. Game and Fish, Santa Fe. 26 p.

―――and―――. 1990. Predation on frogs in breeding choruses. Southwest. Nat. 35(4):449–450.

Worthington, R. D. 1972. Density, growth rates and home range sizes of *Phrynosoma cornutum* in southern Doña Ana County, New Mexico. Herpetol. Rev. 4(4):128.

―――. 1980. *Elaphe subocularis*. Cat. Amer. Amphib. Rept.:268.1–268.2.

―――. 1982. Dry and wet year comparisons of clutch and adult body sizes of *Uta stansburiana stejnegeri*. J. Herpetol. 16(3):332–334.

―――and **E. R. Arvizo.** 1973. Density, growth, and home range of the lizard *Uta stansburiana stejnegeri* in southern Doña Ana County, New Mexico. Great Basin Nat. 33(2):124–128.

Wright, A. H and **A. A. Wright.** 1949. Handbook of Frogs and Toads of the United States and Canada. 3rd ed. Comstock Publ. Assoc., Ithaca, New York. xxi + 640 p.

―――and―――. 1952. List of the snakes of the United States and Canada by states and provinces. Amer. Midl. Nat. 48(3):574–603.

―――and―――. 1957. Handbook of Snakes of the United States and Canada, 2 vols. Comstock Publ. Assoc., Ithaca, New York. xviii + 564 p., ix + p. 565–1105.

LITERATURE CITED

Wright, J. W. 1963. *Cnemidophorus gularis* in New Mexico. Southwest. Nat. 8(1):56.

——. 1966. Variation in two sympatric whiptail lizards (*Cnemidophorus inornatus* and *C. velox*) in New Mexico. Southwest. Nat. 11(1):54–71.

——. 1968. Variation in three sympatric sibling species of whiptail lizards (genus *Cnemidophorus*). J. Herpetol. 1(1/4):1–20.

——. 1971. *Cnemidophorus neomexicanus*. Cat. Amer. Amphib. Rept.:109.1–109.3.

——. 1993. Evolution of the lizards in the genus *Cnemidophorus*, p. 27–81. *In* J. W. Wright and L. J. Vitt (eds.), Biology of Whiptail Lizards (genus *Cnemidophorus*). Oklahoma Mus. Nat. Hist. Norman. xiv + 417 p.

——. 1994. The North American deserts and species diversity in the lizards of the genus *Cnemidophorus*, p. 255–271. *In* P. R. Brown and J. W. Wright (eds.), Herpetology of the North American Deserts: Proceedings of a Symposium. Southwest. Herpetol. Soc. Spec. Publ. (5). iv + 311 p.

—— and **C. H. Lowe.** 1965. The rediscovery of *Cnemidophorus arizonae* Van Denburgh. J. Arizona Acad. Sci. 3(3):164–168.

—— and ——. 1967a. Evolution of the alloploid parthenospecies *Cnemidophorus tesselatus* (Say). Mamm. Chromosomes Newsl. 8(2):95–96.

—— and ——. 1967b. Hybridization in nature between parthenogenetic and bisexual species of whiptail lizards (genus *Cnemidophorus*). Amer. Mus. Novitates (2286):1–36.

—— and ——. 1968. Weeds, polyploids, parthenogenesis, and the geographical and ecological distribution of all-female species of *Cnemidophorus*. Copeia 1968(1):128–138.

—— and ——. 1993. Synopsis of the subspecies of the little striped whiptail lizard, *Cnemidophorus inornatus* Baird. J. Arizona-Nevada Acad. Sci. 27(1):129–157.

—— and **L. J. Vitt (eds.).** 1993. Biology of Whiptail Lizards (genus *Cnemidophorus*). Oklahoma Mus. Nat. Hist., Norman. xiv + 417 p.

Yarrow, H. C. 1875. Report upon the collections of batrachians and reptiles made in portions of Nevada, Utah, California, Colorado, New Mexico, and Arizona, during the years 1871, 1872, 1873, and 1874, p. 509–584 + plates 16–25. *In* Wheeler, Geo. M., Report upon geographical and geological explorations and surveys west of the one hundredth meridian. Vol. V, Zoology. Engineer Dept., U.S. Army, Washington, D.C.

——. 1882. Check list of North American Reptilia and Batrachia, with catalogue of specimens in U.S. National Museum. Bull. U.S. Natl. Mus. (24):v + 249 p.

Yntema, C. L. 1976. Effects of incubation temperature on sexual differentiation in the turtle, *Chelydra serpentina*. J. Morphol. 150(2):453–461.

Youngstrom, K. A. and **H. M. Smith.** 1936. Description of the larvae of *Pseudacris triseriata* and *Bufo woodhousii woodhousii* (Anura). Amer. Midl. Nat. 17(3):629–633.

Zucker, N. 1987. Behavior and movement patterns of the tree lizard *Urosaurus ornatus* (Sauria: Iguanidae) in semi-natural enclosures. Southwest. Nat. 32(3):321–333.

——. 1989. Dorsal darkening and territoriality in a wild population of the tree lizard, *Urosaurus ornatus*. J. Herpetol. 23(4):389–398.

—— and **W. Boecklen.** 1990. Variation in female throat coloration in the tree lizard (*Urosaurus ornatus*): relation to reproductive cycle and fecundity. Herpetologica 46(4):387–394.

Zug, G. R. 1993. Herpetology: an introductory biology of amphibians and reptiles. Academic Press, Inc., San Diego, California. xv + 527 p.

Zweifel, R. G. 1956. A survey of the frogs of the *augusti* group, genus *Eleutherodactylus*. Amer. Mus. Novitates (1813):1–35.

——. 1961. Larval development of the tree frogs *Hyla arenicolor* and *Hyla wrightorum*. Amer. Mus. Novitates (2056):1–19.

——. 1962a. Notes on the distribution and reproduction of the lizard *Eumeces callicephalus*. Herpetologica 18(1):63–65.

——. 1962b. Analysis of hybridization between two subspecies of the desert whiptail lizard, *Cnemidophorus tigris*. Copeia 1962(4):749–766.

——. 1965. Variation in and distribution of the unisexual lizard, *Cnemidophorus tesselatus*. Amer. Mus. Novitates (2235):1–49.

——. 1967. *Eleutherodactylus augusti*. Cat. Amer. Amphib. Rept.:41.1–41.4.

——. 1968a. *Rana tarahumarae*. Cat. Amer. Amphib. Rept.:66.1–66.2.

——. 1968b. Reproductive biology of anurans of the arid southwest, with emphasis on adaptation of embryos to temperature. Bull. Amer. Mus. Nat. Hist. 140(1):1–64.

——. 1970. Descriptive notes on larvae of toads of the *debilis* group, genus *Bufo*. Amer. Mus. Novitates (2407):1–13.

——. 1980. Aspects of the biology of a laboratory population of kingsnakes, p 141–152. *In* J. B Murphy and J. T. Collins (eds.), Reproductive biology and diseases of captive reptiles. SSAR Contrib. Herpetol. (1):x + 277 p.

——. 1992. Frogs and Toads, p. 76–105. *In* H. G. Cogger and R. G. Zweifel (eds.), Reptiles and Amphibians. Smithmark Publ., Inc., New York. 240 p.

—— and **K. S. Norris.** 1955. Contribution to the herpetology of Sonora, Mexico: descriptions of new subspecies of snakes (*Micruroides euryxanthus* and *Lampropeltis getulus*) and miscellaneous collecting notes. Amer. Midl. Nat. 54(1):230–249.

INDEX

New Mexico amphibians and reptiles are indexed under common and scientific names, others are indexed by common and/or scientific name. Page numbers in bold type indicate the location of the species account.

A

Acris 16, 65, 66
    Acris crepitans 11, 16, 32, **65–67**, map 67, 72, 294, 259
    Acris crepitans blanchardi 11, 65
    Acris gryllus blanchardi
      - see *Acris crepitans*
affinis - *Pituophis melanoleucus affinis*
Alligator lizard(s) 13, 250
alterna - *Lampropeltis alterna*
alvarius - *Bufo alvarius*
Ambystoma 19, 22, 326
    Ambystoma mavortia
      - see *Ambystoma tigrinum mavortium*
    Ambystoma nebulosum
      - see *Ambystoma tigrinum nebulosum*
    Ambystoma rosaceum 15, **359**
    Ambystoma tigrinum 15, 17, **20–23**, map 23, 70, 84
    Ambystoma tigrinum mavortium 11, 20, 22
    Ambystoma tigrinum nebulosum 11, 20, 22
AMBYSTOMATIDAE 11, 15, 19
Aneides 25
    Aneides hardii 4, 11, 15, 17, **25–27**, map 27, 28
ANGUIDAE 13, 125, 244
Anolis carolinensis 301

ANURA 11
Apalone - see *Trionyx*
approximans - *Holbrookia maculata approximans*
arenicolor - *Hyla arenicolor*
arenicolus - *Sceloporus arenicolus*
Arid land ribbon snake 14
Arizona
    Arizona elegans 254, **261–262**, amp 262, 265, 274, 276, 290, 295, 300, 304
    Arizona elegans elegans 13, 261
    Arizona elegans occidentalis 261
    Arizona elegans philipi 13, 261
Arizona alligator lizard 13
Arizona black rattlesnake 14
Arizona coral snake 14
Arizona garter snake 14
Arizona mountain kingsnake 13
Arizona tiger salamander 11
Arizona toad 11
Arizona zebratail lizard 12
arizonae - *Thamnophis elegans arizonae*
arizonae - *Cnemidophorus inornatus arizonae*
arnyi - *Diadophis punctatus arnyi*
Atomarchus multimaculatus
    - see *Thamnophis rufipunctatus*
atrox - *Crotalus atrox*
augusti - *Eleutherodactylus augusti*
auriceps - *Crotaphytus collaris auriceps*
australis - *Bufo woodhousii australis*

B

baileyi - *Crotaphytus collaris baileyi*
Banded rock rattlesnake 14

Barking frog **44–45**, map 45
Barred tiger salamander 11
bellii - *Chrysemys picta bellii*
berlandieri - *Rana berlandieri*
Big Bend patchnose snake 14, **299–301**, map 301
Big Bend slider 12, **107–109**, map 109
Big Bend tree lizard 12
bilineatus - *Masticophis bilineatus*
bimaculosus - *Sceloporus magister bimaculosus*
biscutatus - *Trimorphodon biscutatus*
Blackneck garter snake **312–314**, map 314
Blacktail rattlesnake 14, **346–348**, map 348
blairi - see *Lampropeltis alterna*
Blanchard's cricket frog 11
Bleached earless lizard 12
Blind snakes 13
Blotched water snake 14
bogerti - *Coleonyx variegatus bogerti*
Bogertophis 264
    Bogertophis subocularis 3, 253, **262–264**, map 264, 302, 333
    Bogertophis subocularis subocularis 14, 263
bombifrons - *Spea bombifrons*
Boreal chorus frog 11
Boreal toad 11
boreas - *Bufo boreas*
Box turtles 11, 99
brevis - *Coleonyx brevis*
Bufo 16, 46, 54, 56, 75, 321, 331
    Bufo alvarius 11, 16, 32, **47–48**, map 48

421

Bufo americanus 84
Bufo antecessor 61, 62
Bufo boreas 3, 5, 11, 16, 32, **49–51**, map 51, 56, 61
Bufo boreas boreas 51
Bufo cognatus 3, 11, 16, 32, 47, 48, **51–53**, map 53, 56, 59, 60, 72, 84
Bufo compactilis 52, 56, 60
Bufo compactilis speciosus - see *Bufo speciosus*
Bufo debilis 16, 31, 32, **53–55**, map 55, 58, 84
Bufo debilis insidior 11, 54
Bufo dorsalis 61
Bufo houstonensis 58
Bufo insidior - see *Bufo debilis*
Bufo kelloggi 54, 58
Bufo marinus 46
Bufo microscaphus 16, 32, **55–57**, map 57, 59, 60, 61
Bufo microscaphus microscaphus 4, 11, 56
Bufo punctatus 3, 4, 11, 16, 32, 52, 54, **56–59**, map 59
Bufo retiformis 54, 58
Bufo speciosus 3, 11, 16, 32, 52, 56, **59–60**, map 60, 61
Bufo valliceps 84
Bufo woodhousii 2, 16, 32, 47, 55, 56, 57, 59, 60, **61–63**, map 63, 84
Bufo woodhousii australis 11, 61
Bufo woodhousii woodhousii 11, 61
BUFONIDAE 11, 16, 31, 46, 84
Bullfrog 11, **83–85**, map 85
Bullsnake 14, **295–297**, map 297
Bunch grass lizard 12, **175–178**, map 178

C

callicephalus - *Eumeces tetragrammus callicephalus*
Callisaurus 143
    Callisaurus draconoides 127, 137, **139–141**, map 141, 142, 145, 158
    Callisaurus draconoides ventralis 12, 139, 141
canum - *Gyalopion canum*
Canyon spotted whiptail 205
Canyon tree lizard 12
Canyon treefrog 11, **67–69**, map 69
Cat-eyed snake 360

catenatus - *Sistrurus catenatus*
catesbeiana - *Rana catesbeiana*
CAUDATA 11
Caudisona - see *Crotalus*
celaenops - *Lampropeltis triangulum celaenops*
cephaloflavus - *Sceloporus magister cephaloflavus*
cereberus - *Crotalus viridis cereberus*
Checkered garter snake 14, **320–322**, map 322
Checkered whiptail 13, **212–215**, map 215
Chelydra 114, 115
    Chelydra serpentina 93, 95, **96–98**, map 98, 100, 103, 110, 114, 117, 120, 123
    Chelydra serpentina serpentina 11, 96
CHELYDRIDAE 11, 93, 95
Chihuahuan black-headed snake 360
Chihuahuan collared lizard 12
Chihuahuan spotted whiptail 13, 209
Chilopoma rufipunctatus - see *Thamnophis rufipunctatus*
Chiricahua leopard frog 11, **85–87**, map 87
chiricahuensis - *Rana chiricahuensis*
Chorus frog 71
Chrysemys 104, 109, 111, 114
    Chrysemys picta 94, **100–102**, map 102, 103, 108, 110, 113, 117, 123
    Chrysemys picta bellii 11, 100, 101
    Chrysemys picta dorsalis 100
Clark's spiny lizard **161–163**, map 163
clarkii - *Sceloporus clarkii*
Cnemidophorus 134, 137, 201, 202, key 203–204, 279,
    Cnemidophorus burti 4, 129, 159, **205–206**, map 206, 209, 212, 226
    Cnemidophorus burti stictogrammus 12, 205, 209, 210, 234
    Cnemidophorus costatus 209, 212, 226, 234
    Cnemidophorus dixoni 4, 13, 129, **207–208**, map 208, 213, 214, 226, 228
    Cnemidophorus exsanguis 4, 13, 130, 208, **209–211**, map 211, 212, 214, 216, 218, 221, 222, 226, 227
    Cnemidophorus flagellicaudus 4, 13, 130, 209, 210, **211–212**, map 212, 218, 226

Cnemidophorus grahamii 4, 13, 129, 207, 208, **212–215**, map 215, 222, 228
Cnemidophorus gularis 3, 129, 209, 214, **216–217**, map 217, 218, 224
Cnemidophorus gularis gularis 13, 216
Cnemidophorus inornatus 5, 130, 147, 149, 180, 182, 191, 209, 214, **218–221**, map 221, 219, 222, 224, 232, 234
Cnemidophorus inornatus arizonae 212, 226, 232
Cnemidophorus inornatus gypsi 13, 219
Cnemidophorus inornatus heptagrammus 13, 219
Cnemidophorus inornatus juniperus 13, 219, 221
Cnemidophorus inornatus llanuras 13, 219
Cnemidophorus marmoratus - see *Cnemidophorus tigris*
Cnemidophorus neomexicanus 13, 129, 191, 218, 221, **221–223**, map 223, 225
Cnemidophorus octolineatus - see *Cnemidophorus inornatus*
Cnemidophorus perplexus - see *Cnemidophorus neomexicanus*
Cnemidophorus sacki exsanguis 205
Cnemidophorus sacki stictogrammus 205
Cnemidophorus scalaris 209
Cnemidophorus septemvittatus 207, 209, 214, 279
Cnemidophorus sexlineatus 4, 130, 160, 209, 214, 218, 219, **223–225**, map 225, 232, 234
Cnemidophorus sexlineatus viridis 13, 216, 224
Cnemidophorus sonorae 4, 13, 130, 159, 205, 208, 209, 210, 211, 212, 218, **226–227**, map 227
Cnemidophorus tesselatus 222, - also see *Cnemidophorus grahamii*
Cnemidophorus tigris 2, 3, 5, 129, 136, 137, 158, 160, 164, 191, 193, 207, 208, 213, 214, 221, 222, 226, **227–231**, map 231, 264, 301
Cnemidophorus tigris gracilis 13, 229
Cnemidophorus tigris marmoratus 13, 207, 214, 222, 229
Cnemidophorus tigris reticuloriens 13, 228

INDEX

Cnemidophorus tigris septentrionalis 13, 228
Cnemidophorus uniparens 3, 5, 13, 130, 202, 208, 218, 219, 221, 222, 226, **231–233**, map 233, 234, 235
Cnemidophorus velox 13, 130, 164, 219, 222, 232, **233–235**, map 235
Coachwhip **289–291**, map 291
cognatus - Bufo cognatus
Coleonyx 258
    Coleonyx brevis 3, 12, 125, **195–196**, map 196, 197, 199, 264, 306, 350,
    Coleonyx variegatus, 125, 195, **197–199**, map 199
    Coleonyx variegatus bogerti 12, 197
    Coleonyx variegatus fasciatus 197
Collared lizards 12, 131, 132
collaris - Crotaphytus collaris
Colorado River toad 11, **47–48**, map 48
Colorado side-blotched lizard 12
Coluber 254
    Coluber constrictor 4, 134, **264–266**, map 266, 270, 276, 286, 289, 290, 295, 296, 315, 352
    Coluber constrictor flaviventris 13, 265, 266
    Coluber constrictor mormon 13, 265, 266
    Coluber proximus
      - see *Thamnophis proximus*
    Coluber triaspis see *Senticolis triaspis*
COLUBRIDAE 13, 252, 260
Colubrids 13, 260
Common garter snake **328–331**, map 331
Common kingsnake **280–282**, map 282
Common snapping turtle 11
consobrinus - Sceloporus undulatus consobrinus
constrictor - Coluber constrictor
Cophosaurus 140
    Cophosaurus texanus 3, 5, 127, 139, 140, **141–144**, map 144, 145, 214, 264, 279
    Cophosaurus texanus scitulus 12, 142
    Cophosaurus texanus texanus 142
Corn snake **269–271**, map 271
cornutum - Phrynosoma cornutum
Couch's spadefoot 11, **36–37**, map 37
couchii - Scaphiopus couchii
cowlesi - Sceloporus undulatus cowlesi

crepitans - Acris crepitans
Crevice spiny lizard 12, **173–179**, map 179
Crotalinus - see *Sistrurus* or *Crotalus*
Crotalus 276, 296, 340
    Crotalus atrox 3, 5, 14, 252, 261, 274, 290, 292, 296, 298, 300, 302, 333, **341–344**, map 344, 348, 349, 350, 351, 352, 354
    Crotalus atrox atrox 341
    Crotalus atrox tortugensis 341
    Crotalus adamanteus 341
    Crotalus cerastes 360
    Crotalus cinereus
      - see *Crotalus atrox*
    Crotalus lepidus 252, 283, 288, 302, 333, 342, **344–346**, map 346
    Crotalus lepidus klauberi 14, 344, 345, 346, 354, 355
    Crotalus lepidus lepidus 14, 345, 346
    Crotalus molossus 4, 252, 288, 292, 302, 333, 342, 344, **346–348**, map 348, 351
    Crotalus molossus molossus 14, 347, 354
    Crotalus pricei 361
    Crotalus scutulatus 3, 252, 341, 342, **348–350**, map 350, 351
    Crotalus scutulatus scutulatus 14, 349, 350, 353
    Crotalus tigris 361
    Crotalus viridis cerberus 14, 351, 353
    Crotalus viridis nuntius 14, 351, 352
    Crotalus viridis viridis 14, 344, 351, 352, 353, 356
    Crotalus viridis 2, 4, 5, 7, 252, 261, 265, 272, 274, 292, 296, 298, 315, 348, 350, **351–353**, map 353, 354, 361
    Crotalus willardi 4, 252, 283, 288, **353–355**, map 355
    Crotalus willardi obscurus 14, 245, 354, 355
    Crotalus willardi silus 354
    Crotalus willardi willardi 354, 355
CROTAPHYTIDAE 126, 131
Crotaphytus
    Crotaphytus collaris 2, 7, 126, **132–134**, map 134, 135, 144, 156, 166, 167, 182, 191, 264, 343
    Crotaphytus collaris auriceps 12, 133
    Crotaphytus collaris baileyi 12, 133

    Crotaphytus collaris collaris 12, 133
    Crotaphytus collaris fuscus 12, 133
cyrtopsis - Thamnophis cyrtopsis

D

debilis - Bufo debilis
Desert box turtle 11
Desert grassland whiptail 13, **231–233**, map 233
Desert kingsnake 13
Desert massasauga 14
Desert side-blotched lizard 12
Desert spiny lizard **170–172**, map 172
Desert striped whipsnake 14
Desert tortoise **359**
deserticola - Pituophis melanoleucus deserticola
diabolicus - Thamnophis proximus diabolicus
Diadophis 260, 267
    Diadophis punctatus 4, 254, 265, **267–269**, map 269, 272, 286, 315, 335, 339
    Diadophis punctatus arnyi 13, 267
    Diadophis punctatus regalis 13, 267, 268
Dicamptodon 19
dissectus - Leptotyphlops dulcis dissectus
dixoni - Cnemidophorus dixoni
dorsalis - Thamnophis sirtalis dorsalis
douglasii - Phrynosoma douglasii
draconoides - Callisaurus draconoides
Drymarchon corais melanurus 360
dulcis - Leptotyphlops dulcis
Dunes sagebrush lizard
    - see sand dune lizard

E

Earless lizards 12, 138
Eastern barking frog 11
Eastern box turtle **360**
Eastern collared lizard 12
Eastern glass lizard **360**
Eastern hognose snake **360**
Eastern marbled whiptail 13
Eastern yellowbelly racer 13
edwardsii - Sistrurus catenatus edwardsii
Elaphe 264, 265, 269, 295,
    Elaphe bairdi 361

Elaphe chlorosoma
- see *Senticolis triaspis*
Elaphe guttata 4, 253, **269–271**, map 271, 320
Elaphe guttata emoryi 13, 269, 270
Elaphe guttata guttata 270
Elaphe guttata meahllmorum 13, 269, 270
Elaphe laeta
- see *Elaphe guttata*
Elaphe obsoleta 290
Elaphe rosaliae 264
Elaphe subocularis
- see *Bogertophis subocularis*
Elaphe triaspis
- see *Senticollis triaspis*
ELAPIDAE 14, 252, 337
Elapids 14, 337
Elaps
- see *Micruroides*
elegans - *Holbrookia maculata elegans*
elegans - *Thamnophis elegans*
elegans - *Trachemys scripta elegans*
Eleutherodactylus 44
Eleutherodactylus augusti 3, 4, 15, 31, 43, **44–45**, map 45, 75, 270
Eleutherodactylus augusti cactorum 44
Eleutherodactylus augusti latrans 11, 44
Eleutherodactylus latrans
- see *Eleutherodactylus augusti*
Elgaria
Elgaria kingii 4, 125, **245–246**, map 246
Elgaria kingii nobilis 13, 245
elongatus - *Sceloporus undulatus elongatus*
emoryi - *Elaphe guttata emoryi*
emoryi - *Trionyx spiniferus emoryi*
EMYDIDAE 11, 99
Engystoma olivaceum
- see *Gastrophryne olivacea*
eques - *Thamnophis eques*
erythrocheilus - *Sceloporus undulatus erythrocheilus*
erythrogaster - *Natrix erythrogaster*
Eumeces 245, 279, 284
Eumeces callicephalus
- see *Eumeces tetragrammus*

Eumeces multivirgatus 126, **237–239**, map 239, 240, 242, 270, 286, 316
Eumeces gaigei
- see *Eumeces multivirgatus*
Eumeces multivirgatus epipleurotus 13, 238
Eumeces obsoletus 3, 13, 125, 160, 191, 238, **239–241**, map 241, 242
Eumeces tetragrammus 4, 126, 238, 240, **241–243**, map 243
Eumeces tetragrammus callicephalus 13, 242
euryxanthus - *Micruroides euryxanthus*
Eurycea multiplicata 27
Eutaenia
- see *Thamnophis*
Eutaenia megalops
- see *Thamnophis eques*
eximia - *Hyla eximia*
exsanguis - *Cnemidophorus exsanguis*

F

Fence lizard
- see Prairie lizard
Ficimia cana
- see *Gyalopion canum*
flagellicaudus - *Cnemidophorus flagellicaudus*
flagellum - *Masticophis flagellum*
flavescens - *Kinosternon flavescens*
flaviventris - *Coluber constrictor flaviventris*
Four-lined skink **241–243**, map 243
fuscus - *Crotaphytus collaris fuscus*

G

gaigeae - *Trachemys gaigeae*
Gambelia
Gambelia wislizenii 3, 126, 132, **135–137**, map 137, 147, 164, 172, 182, 191, 231
Gambelia wislizenii punctata 12, 136, 142
Gambelia wislizenii wislizenii 12, 136
garmani - *Sceloporus undulatus garmani*
Gastrophryne 75, 321, 331
Gastrophryne carolinensis 277
Gastrophryne olivacea 3, 11, 15, 31, **75–77**, map 77
Gastrophryne olivacea mazatlanensis 75, 76

Gastrophryne olivacea olivacea 75
Geckos 12, 194
GEKKONIDAE 12, 125, 194
Gerrhonotus
- see *Elgaria*
getula - *Lampropeltis getula*
Giant spotted whiptail 12
Gila monster **248–250**, map 250
Gila spotted whiptail 13, **211–212**, map 212
Glossy snake **261–262**, map 262
Gopher snake **295–297**, map 297
Gopherus
Gopherus agassizii 94, **359**
gorzugi - *Pseudemys gorzugi*
gracilis - *Cnemidophorus tigris gracilis*
graciosus - *Sceloporus graciosus*
grahamiae - *Salvadora grahamiae*
grahamii - *Cnemidophorus grahamii*
Gray-banded kingsnake 13, **278–279**, map 279
Gray checkered whiptail 13, **207–208**, map 208
Great Basin gopher snake 14
Great Basin spadefoot **359**
Great Plains narrowmouth toad 11, **75–77**, map 77
Great Plains rat snake 13
Great Plains skink 13, **239–241**, map 241
Great Plains toad 11, **51–53**, map 53
Greater earless lizard **141–144**, map 144
Green rat snake 14, **303–305**, map 305
Green toad **53–55**, map 55
Ground snake 14, **305–307**, map 307
gularis - *Cnemidophorus gularis*
guttata - *Elaphe guttata*
Gyalopion
Gyalopion canum 4, 13, 254, **271–273**, map 273, 274, 292, 360
gypsi - *Cnemidophorus inornatus gypsi*

H

hardii - *Aneides hardii*
hartwegi - *Trionyx spiniferus hartwegi*
haydeni - *Thamnophis radix haydeni*
Heloderma
Heloderma horridum 247, 248
Heloderma suspectum 126, **248–250**, map 250

INDEX

Heloderma suspectum suspectum 13, 248
HELODERMATIDAE 13, 126, 247
Hemidactylus
    Hemidactylus turcicus 12, 125, 195, 197, **199–200**, map 200
heptagrammus - Cnemidophorus inornatus heptagrammus
hernandesi - Phrynosoma douglasii hernandesi
Heterodon 260, 271, 272, 296
    Heterodon nasicus 4, 116, 252, **273–275**, map 275, 290, 296, 298, 300, 320, 351, 356
    Heterodon nasicus gloydi 273, 274
    Heterodon nasicus kennerlyi 13, 273
    Heterodon nasicus nasicus 13, 273, 275
    Heterodon platirhinos 252, 273, 275, **360**
hobartsmithi - Tantilla hobartsmithi
Holbrookia 350
    Holbrookia maculata 3, 5, 127, 137, 139, 142, **145–148**, map 148, 159, 160, 164, 182, 220
    Holbrookia maculata approximans 12, 146
    Holbrookia maculata elegans 12, 146
    Holbrookia maculata maculata 12, 146
    Holbrookia maculata ruthveni 12, 146
    Holbrookia texana
        - see *Cophosaurus texanus*
Hopi rattlesnake 14
Horned lizards 12, 138
humilis - Leptotyphlops humilis
Hyla 16, 64
    Hyla affinis 67
    Hyla arenicolor 11, 16, 33, **67–69**, map 69, 70, 72, 279
    Hyla eximia 4, 11, 16, 33, 68, **69–70**, map 71
    Hyla triseriata
        - see *Pseudacris triseriata*
    Hyla wrightorum 70
Hylactophryne augusti
    - see *Eleutherodactylus*
HYLIDAE 23, 31, 64
Hylodes augusti
    - see *Eleutherodactylus augusti*
Hylodes maculatus
    - see *Pseudacris triseriata*

Hypopachus 75
Hypsiglena 260, 261, 277
    Hypsiglena torquata 3, 144, 254, 265, **276–277**, map 277, 290, 298, 300, 306, 308, 332, 339, 356
    Hypsiglena torquata jani 13, 276, 277
    Hypsiglena torquata loreala 13, 276

I

IGUANIA 126, 131
Indigo snake **360**
inornatus - Cnemidophorus inornatus
intermedia - Senticolis triaspis intermedia

J

jani - Hypsiglena torquata jani
jarrovii - Sceloporus jarrovii
Jemez Mountains salamander 11, **27–29**, map 29
juniperus - Cnemidophorus inornatus juniperus

K

Kansas glossy snake 13
kennerlyi - Heterodon nasicus kennerlyi
kingii - Elgaria kingii
KINOSTERNIDAE 12, 94, 113
Kinosternon 100
    Kinosternon flavescens 4, 94, 100, **114–116**, map 116, 117, 120, 123, 275
    Kinosternon flavescens flavescens 12, 114
    Kinosternon hirtipes 115
    Kinosternon sonoriense 5, 85, 94, 114, **116–118**, map 118, 123
    Kinosternon sonoriense sonoriense 12, 117
klauberi - Crotalus lepidus klauberi

L

lambda - Trimorphodon biscutatus lambda
Lampropeltis 265, 269, 278, 298
    Lampropeltis alterna 13, 254, 263, **278–279**, map 279, 332
    Lampropeltis doliata
        - see *Lampropeltis triangulum*
    Lampropeltis getula 254, 265, **280–282**, map 282, 298, 320, 329
    Lampropeltis getula californiae 280, 281, 282
    Lampropeltis getula holbrookia 280

    Lampropeltis getula splendida 13, 280, 281
    Lampropeltis mexicana
        - see *Lampropeltis alterna*
    Lampropeltis pyromelana 4, **282–283**, map 283, 284, 288, 333, 338
    Lampropeltis pyromelana pyromelana 13, 254, 283
    Lampropeltis triangulum 254, 282, **284–285**, map 285, 296
    Lampropeltis triangulum celaenops 13, 284
    Lampropeltis triangulum gentilis 284
Lamprosaurus guttulatus
    - see *Eumeces multivirgatus*
lecontei - Rhinocheilus lecontei
Leopard frogs key 79
Leopard lizards 12, 131
lepidus - Crotalus lepidus
LEPTODACTYLIDAE 11, 15, 31, 43
Leptodeira septentrionalis **360**
LEPTOTYPHLOPIDAE 13, 252, 255
Leptotyphlops 256, 276, 305, 306, 308, 339
    Leptotyphlops dulcis 252, **255–257**, map 257, 258, 259, 277, 310, 335
    Leptotyphlops dulcis dissectus 13, 256, 257
    Leptotyphlops dulcis dulcis 256
    Leptotyphlops humilis 252, 256, 257, **258–259**, map 259
    Leptotyphlops humilis coahuilae 259
    Leptotyphlops humilis segregus 13, 258, 259
    Leptotyphlops myopicus
        - see *Leptotyphlops dulcis*
Lesser earless lizard **145–148**, map 148
levis - Urosaurus ornatus levis
linearis - Urosaurus ornatus linearis
lineatum - Tropidoclonion lineatum
Lined snake 14, **334–336**, map 336
Lined tree lizard 12
Liochlorophis xv, 286, 287
    Liochlorophis vernalis 4, 238, 254, 265, **286–287**, map 287, 289, 304, 315, 315, 325
    Liochlorophis vernalis blanchardi 13, 286
Little striped whiptail **218–221**, map 221
Little white whiptail 13

Lizards 12, key 125–130,
llanuras - Cnemidophorus inornatus llanuras
Longnose leopard lizard 12
Longnose snake **298–299**, map 299
loreala - Hypsiglena torquata loreala
Lowland leopard frog 11, **89–91**, map 91
Lungless salamanders 11, 24
Lyre snake **332–333**, map 333

M

Macroclemys temminckii 95
maculata - Holbrookia maculata
maculata - Pseudacris triseriata maculata
Madrean alligator lizard **245–246**, map 246
magister - Sceloporus magister
Malayan snail-eating turtle **359**
Malayemys subtrijuga **359**
Many-lined skink **237–239**, map 239
marcianus - Thamnophis marcianus
marmoratus - Cnemidophorus tigris marmoratus
Massasauga **356–357**, map 357
Masticophis 134, 225, 289, 312, 315, 320, 324, 329
    Masticophis bilineatus 254, 283, **288–289**, map 289, 291, 292, 302, 333
    Masticophis bilineatus bilineatus 13, 288
    Masticophis flagellum 144, 151, 172, 215, 231, 254, 274, **289–291**, map 291, 292, 295, 298, 300, 304, 352
    Masticophis flagellum piceus 13, 289, 291
    Masticophis flagellum testaceus 13, 289
    Masticophis taeniatus 3, 166, 182, 246, 254, 288, **291–293**, map 293, 296, 303, 318, 322, 342
    Masticophis taeniatus taeniatus 14, 292
mavortium - Ambystoma tigrinum mavortium
Mediterranean gecko 12, **199–200**, map 200
megalops - Thamnophis eques megalops
melanoleucus - Pituophis melanoleucus
Mesa Verde night snake 13
Mescalero sand dunes prairie lizard 12

Mesquite lizard **360**
Mexican garter snake 14, **317–319**, map 319
Mexican hognose snake 13
Microhyla
    - see *Gastrophryne*
MICROHYLIDAE 11, 15, 31, 74
microscaphus - Bufo microscaphus
Micruroides 339
    Micruroides euryxanthus 282, 284, 298, 337, **338–339**, map 339
    Micruroides euryxanthus euryxanthus 14, 338
Midland smooth softshell 12
Milk snake **284–285**, map 285
modestum - Phrynosoma modestum
Mojave rattlesnake 14, **384–350**, map 350
Mole salamanders 11
molossus - Crotalus molossus
mormon - Coluber constrictor mormon
Mottled rock rattlesnake 14
Mountain patchnose snake 14, **301–303**, map 303
Mountain short-horned lizard 12
Mountain skink 13
Mountain treefrog 11, **69–71**, map 71
Mud turtles 12, 113
multiplicata - Eurycea multiplicata
multivirgatus - Eumeces multivirgatus
muticus - Trionyx muticus
myopicus
    - see *Leptotyphlops dulcis dissectus*

N

Narrowhead garter snake 14, **326–328**, map 328
Narrowmouth toads 11, 74
nasicus - Heterodon nasicus
Natrix
    - see *Nerodia*
nebulosum - Ambystoma tigrinum nebulosum
neomexicanus - Plethodon neomexicanus
neomexicanus - Cnemidophorus neomexicanus
Nerodia 327
    Nerodia erythrogaster 252, **293–295**, map 295, 320
    Nerodia erythrogaster transversa 4, 14, 294

Nerodia rhombifera 361
New Mexico blind snake 13
New Mexico garter snake 14
New Mexico milk snake 13
New Mexico ridgenose rattlesnake 14
New Mexico spadefoot 11, **40–42**, map 42
New Mexico whiptail 13, **221–223**, map 223
Night snake **276–277**, map 277
nigriceps - Tantilla nigriceps
nobilis - Elgaria kingii nobilis
Northern cricket frog, **65–67**, map 67
Northern earless lizard 12
Northern leopard frog 11, **87–89**, map 89
Northern plateau lizard 12
Northern prairie lizard 12
Northern sagebrush lizard 12
Northern tree lizard 12
Northern whiptail 13
nuntius - Crotalus viridis nuntius

O

obscurus - Crotalus willardi obscurus
obsoletus - Eumeces obsoletus
olivacea - Gastrophryne olivacea
Ophibolus alternus
    - see *Lampropeltis alterna*
Opheodrys
    Opheodrys aestivus **360**
    Opheodrys vernalis
        - see *Liochlorophis vernalis*
Ophisaurus ventralis **360**
Orangeheaded spiny lizard 12
ornata - Terrapene ornata
Ornate box turtle 11, **104–107**, map 107
ornatus - Urosaurus ornatus

P

Painted turtle **100–102**, map 102
Painted Desert glossy snake 13
Pale leopard lizard 12
PELOBATIDAE 11, 16, 31, 35
philipi - Arizona elegans philipi
Phrynosoma 143, 138
    Phrynosoma cornutum 7, 12, 126, 136, **148–151**, map 151, 152, 155, 158, 160, 180, 191, 219
    Phrynosoma douglasii 4, 7, 127, 149, **151–154**, map 154, 155, 158, 353,

INDEX

Phrynosoma douglasii hernandesi 12, 152, 153
Phrynosoma douglasii ornatissimum 152, 153
Phrynosoma modestum 3, 12, 126, 134, 149, 152, **154–157**, map 157, 158, 160
Phrynosoma platirhinos 155
Phrynosoma solare 12, 127, 149, 152, 155, **157–159**, map 159
PHRYNOSOMATIDAE 12, 126, 138
Phyllorhynchus browni **360**
piceus - Masticophis flagellum piceus
picta - Chrysemys picta
Pima leafnose snake **360**
pipiens - Rana pipiens
Pituophis 261
　Pituophis catenifer
　　- see *Pituophis melanoleucus*
　Pituophis melanoleucus 2, 253, 265, 274, 276, 284, **295–297**, map 297, 300, 302, 352
　Pituophis melanoleucus affinis 14, 296
　Pituophis melanoleucus deserticola 14, 296
　Pituophis melanoleucus sayi 14, 296, 297
Plainbelly water snake **293–295**, map 295
Plains black-headed snake 14, **309–310**, map 310
Plains garter snake **324–326**, map 326
Plains hognose snake 13
Plains leopard frog 11, **81–83**, map 83
Plains spadefoot 11, **38–40**, map 40
Plains striped whiptail 13
Plateau striped whiptail 13
Plestiodon obsoletum
　- see *Eumeces obsoletus*
Plethodon hardii
　- see *Aneides hardii*
Plethodon neomexicanus 4, 11, 15, 17, 25, **27–29**, map 29
PLETHODONTIDAE 11, 15, 24
poinsettii - Sceloporus poinsetti
Prairie lined racerunner 13
Prairie rattlesnake 14
Prairie ringneck snake 13
proximus - Thamnophis proximus
Pseudacris 16, 277, 326
　Pseudacris clarkii 16, **359**

Pseudacris triseriata 3, 16, 33, 49, 65, 69, **71–73**, map 73, 359
Pseudacris triseriata maculata 11, 72
Pseudacris triseriata triseriata 11, 72
Pseudemys 109, 111, 114
Pseudemys concinna 103
Pseudemys concinna gorzugi 103
Pseudemys concinna texana 103
Pseudemys floridana 103
Pseudemys floridana texana 103, 104
Pseudemys gaigeae
　- see *Trachemys gaigeae*
Pseudemys gorzugi 11, 94, 100, **102–104**, map 104, 109, 114, 117, 123
Pseudemys texana 103
Pseudemys scripta
　- see *Trachemys scripta*
punctata - Gambelia wislizenii punctata
punctatus - Bufo punctatus
punctatus - Diadophis punctatus
pyromelana - Lampropeltis pyromelana

R

Racer **264–266**, map 266
radix - Thamnophis radix
Rana 16, 78, key 79
　Rana berlandieri 11, 16, 33, **79–81**, map 81, 82, 90, 116
　Rana blairi 3, 11, 16, 33, 79, 80, **81–83**, map 83, 88, 90
　Rana catesbeiana 11, 16, 22, 33, **83–85**, map 85, 319, 359
　Rana chiricahuensis 11, 16, 33, 79, **85–87**, map 87, 88, 90
　Rana montezumae 86
　Rana pipiens 11, 16, 33, 79, 82, 86, **87–89**, map 89, 314
　Rana subaquavocalis 86
　Rana tarahumarae 311, **359**
　Rana yavapaiensis 11, 16, 33, 79, 80, 82, 85, 86, **89–91**, map 91
RANIDAE 11, 16, 31, 78
Red coachwhip 13
Red-eared slider 12
Red-lipped prairie lizard 12
Red-spotted toad 11, **57–59**, map 59
Regal horned lizard 12
Regal ringneck snake 13
regalis - Diadophis punctatus regalis

Reticulate gila monster 13
reticuloriens - Cnemidophorus tigris reticuloriens
Rhinocheilus 261
　Rhinocheilus lecontei 3, 254, 274, 276, 282, 284, 290, **298–299**, map 299, 300, 306, 308
　Rhinocheilus lecontei clarus 298, 299
　Rhinocheilus lecontei lecontei 14, 298, 299
　Rhinocheilus lecontei tessellatus 14, 298
Rhyacotriton 19
Ridgenose rattlesnake **353–355**, map 355
Ringneck snake **267–269**, map 269
Rio Grande leopard frog 11, **79–81**, map 81
Rock rattlesnake **344–346**, map 346
Rosy salamander **359**
Rough green snake **360**
Roundtail horned lizard 12, **154–157**, map 157
rufipunctatus - Thamnophis rufipunctatus
ruthveni - Holbrookia maculata ruthveni

S

Sacramento mountain salamander 11, **25–27**, map 27
Salamander key 17
Salamanders 11
Sagebrush lizard **163–166**, map 166
Salvadora 225, 288, 291, 312, 315, 320, 324, 329
　Salvadora deserticola 14, 254, **299–301**, map 301, 302
　Salvadora grahamiae 4, 254, 257, 268, 272, 282, 292, 296, **301–303**, map 303, 306, 341
　Salvadora grahamiae grahamiae 14, 302
　Salvadora hexalepis 290, 300, 301
　Salvadora hexalepis deserticola
　　- see *Salvadora deserticola*
　Salvadora hexalepis hexalepis 300, 302
Sand dune lizard 12, **159–161**, map 161
SAURIA 12
sayi - Pituophis melanoleucus sayi
scalaris - Sceloporus scalaris
Scaphiopus 16, 35, 36, 38, 47, 49, 75, 277, 321, 331
　Scaphiopus couchii 3, 11, 16, 32, **36–39**, map 37, 38, 40, 41, 48, 277

Scaphiopus rectifrenis 36
Sceloporus 134, 135, 277, 346, 350
Sceloporus arenicolus 3, 12, 128, **159–161**, map 161, 164, 179
Sceloporus clarkii 4, 128, 159, **161–163**, map 163, 171
Sceloporus clarkii clarkii 12, 162
Sceloporus graciosus 3, 128, 137, 160, **163–166**, map 166, 176, 179, 180
Sceloporus graciosus graciosus 12, 164, 165
Sceloporus grammicus **360**
Sceloporus jarrovii 128, 162, **166–170**, map 170, 171, 173, 283
Sceloporus jarrovii jarrovii 12, 167
Sceloporus magister 3, 128, 158, 162, **170–172**, map 172, 191
Sceloporus magister bimaculosus, 12, 171
Sceloporus magister cephaloflavus 12, 171
Sceloporus olivaceus **360**
Sceloporus poinsettii 128, 162, 167, 171, **173–175**, map 175, 279
Sceloporus poinsettii poinsettii 12, 174
Sceloporus scalaris 3, 127, 160, **175–178**, map 178, 179
Sceloporus scalaris slevini 12, 164, 176
Sceloporus undulatus 2, 128, 137, 147, 149, 160, 164, 176, **178–182**, map 182, 191, 220, 279, 284, 301, 346, 360
Sceloporus undulatus consobrinus 12, 179, 180, 219
Sceloporus undulatus cowlesi 12, 179, 180
Sceloporus undulatus elongatus 12, 179
Sceloporus undulatus erythrocheilus 12, 179
Sceloporus undulatus garmani 12, 179
Sceloporus undulatus tedbrowni 12, 180
Sceloporus undulatus tristichus 12, 179
Sceloporus virgatus 4, 12, 128, 176, **183–185**, map 185
schmidti - Urosaurus ornatus schmidti
Scincella lateralis 339
SCINCIDAE 13, 125, 236
scitulus - Cophosaurus texanus scitulus
scripta - Trachemys scripta
scutulatus - Crotalus scutulatus
segregus - Leptotyphlops humilis segregus

semiannulata - Sonora semiannulata
Senticolis 264, 304
    Senticolis triaspis 4, 253, 286, 288, 289, **303–305**, map 305
    Senticolis triaspis intermedia 14, 304
SERPENTES 13
serpentina - Chelydra serpentina
septentrionalis - Cnemidophorus tigris septentrionalis
sexlineatus - Cnemidophorus sexlineatus
Short-horned lizard **151–154**, map 154
Sibon septentrionale
    - see *Leptodeira septentrionalis*
Side-blotched lizard(s) 12, 138, **189–193**, map 193
Sidewinder **360**
sirtalis - Thamnophis sirtalis
Sistrurus 340, 356
    Sistrurus catenatus 4, 252, 276, 351, **356–357**, map 357
    Sistrurus catenatus edwardsii 14, 356, 357
    Sistrurus catenatus tergeminus 356, 357
Six-lined racerunner **223–225**, map 225
Skinks 13
slevini - Sceloporus scalaris slevini
Slider **109–112**, map 112
Smooth green snake **286–287**, map 287
Smooth softshell turtle **120–122**, map 122
Snakes 12, 13, key 251– 254
Snapping turtle(s) 11, 95, **96–98**, map 98
Softshell turtles 12, 119
solare - Phrynosoma solare
Sonora 260, 277, 306
    Sonora episcopa
    - see *Sonora semiannulata*
    Sonora semiannulata 3, 14, 134, 196, 254, 256, 258, 272, 276, 298, **305–307**, map 307, 308, 309, 310
sonorae - Cnemidophorus sonorae
Sonoran gopher snake 14
Sonoran lyre snake 14
Sonoran mountain kingsnake **282–283**, map 283
Sonoran mud turtle 12, **116–118**, map 118
Sonoran spiny lizard 12
Sonoran spotted whiptail 13, **226–227**, map 227

Sonoran whipsnake 13, **288–289**, map 289
sonoriense - Kinosternon sonoriense
Southern plains rat snake 13
Southern plateau lizard 12
Southern prairie lizard 12
Southern whiptail 13
Southwestern black-headed snake 14, **307–309**, map 309
Southwestern earless lizard 12
Southwestern toad **55–57**, map 57
Southwestern Woodhouse's toad 11
Spadefoots 11, 35
Spea 16, 35, 38, 39, 47, 49, 75, 321, 331
    Spea bombifrons 3, 11, 16, 32, 36, **38–40**, map 40, 41, 72
    Spea hammondii
    - see *Spea multiplicata*
    Spea hammondii stagnalis
    - see *Spea multiplicata*
    Spea intermontana **359**
    Spea multiplicata 3, 11, 16, 32, 36, 38, 39, **40–42**, map 42
speciosus - Bufo speciosus
Speckled earless lizard 12
Spelerpes multiplicatus 27
spiniferus - Trionyx spiniferus
Spiny lizards 12, 138
Spiny softshell **122–124**, map 124
splendida - Lampropeltis getula splendida
Spotted chorus frog **359**
SQUAMATA 12
stansburiana - Uta stansburiana
stejnegeri - Uta stansburiana stejnegeri
Sternotherus odoratus 100
Striped plateau lizard 12, **183–185**, map 185
Striped whipsnake **291–293**, map 293
Storeria dekayi 329
Storeria occipitomaculata 287, 325
subocularis - Bogertophis subocularis
suspectum - Heloderma suspectum

T

taeniatus - Masticophis taeniatus
Tantilla 256, 258, 259, 260, 272, 276, 277, 302, 305, 306, 309, 310, 311, 320, 339
    Tantilla atriceps
    - see *Tantilla hobartsmithi*

INDEX

Tantilla bogerti
 - see *Tantilla yaquia*
Tantilla gracilis 310
Tantilla hobartsmithi 14, 254, **307–309**, map 309, 310, 311
Tantilla nigriceps 14, 254, 265, 307, 308, **309–310**, map 310
Tantilla nigriceps fumiceps 309
Tantilla nigriceps nigriceps 309
Tantilla planiceps
 - see *Tantilla hobartsmithi*
Tantilla utahensis 308
Tantilla wilcoxi **360**
Tantilla yaquia 14, 254, 307, 308, 309, **311–312**, map 312
Tarahumara frog **359**
Taricha granulosa 331
tedbrowni - *Sceloporus undulatus tedbrowni*
TEIIDAE 12, 126, 201
Terrapene 114
 Terrapene carolina 360
 Terrapene ornata 3, 4, 94, **104–107**, map 107, 114, 117
 Terrapene ornata luteola 11, 105, 106
 Terrapene ornata ornata 11, 105
tessellatus - *Rhinocheilus lecontei tessellatus*
testaceus - *Masticophis flagellum testaceus*
TESTUDINES 11
Testudinidae 94
tetragrammus - *Eumeces tetragrammus*
texanus - *Cophosaurus texanus*
Texas banded gecko 12, **195–196**, map 196
Texas blind snake **256–257**, map 257
Texas horned lizard 12, **148–151**, map 161
Texas longnose snake 14
Texas lyre snake 14
Texas night snake 13
Texas spiny lizard **360**
Texas spiny softshell 12
Texas spotted whiptail 12, **216–217**, map 217
Texas toad 11, **59–60**, map 60
Thamnophis 22, 288, 292, 312, 313, 317, 320, 321, 322, 327, 334

Thamnophis angustirostris
 - see *Thamnophis rufipunctatus*
Thamnophis cyrtopsis 4, 253, **312–314**, map 314, 315, 317, 320, 322, 324, 325, 329
Thamnophis cyrtopsis cyrtopsis 14, 312
Thamnophis dorsalis
 - see *Thamnophis cyrtopsis*
Thamnophis elegans 4, 29, 238, 253, 265, 286, 296, 312, 313, **314–317**, map 317, 322, 324, 325, 329, 331, 335, 342
Thamnophis elegans arizonae 14, 315, 316, 317
Thamnophis elegans errans 315
Thamnophis elegans vagrans 14, 315, 317
Thamnophis eques 4, 5, 85, 253, 313, 314, **317–319**, map 319, 320, 322, 324, 329
Thamnophis eques megalops 14, 318
Thamnophis marcianus 4, 62, 253, 312, 313, 315, 317, **320–322**, map 322, 324, 327, 329
Thamnophis marcianus marcianus 14, 320
Thamnophis macrostemma megalops
 - see *Thamnophis eques*
Thamnophis proximus 4, 253, 295, 318, **322–324**, map 324, 325
Thamnophis proximus diabolicus 14, 323
Thamnophis radix 4, 253, 265, 287, 313, 317, 320, 322, **324–326**, map 326, 329, 335
Thamnophis radix haydenii 14, 324
Thamnophis radix radix 325
Thamnophis rufipunctatus 4, 253, 313, 315, **326–328**, map 328
Thamnophis rufipunctatus rufipunctatus 14, 327
Thamnophis sirtalis 4, 253, 265, 287, 313, 315, 317, 320, 322, 324, 325, **328–331**, map 331
Thamnophis sirtalis dorsalis 14, 329, 330
Thamnophis sirtalis parietalis 330
Thamnophis subcarinatus megalops
 - see *Thamnophis eques*
Tiger rattlesnake **361**
Tiger salamander **20–23**, map 23
tigrinum - *Ambystoma tigrinum*

tigris - *Cnemidophorus tigris*
tigris - *Crotalus tigris*
Toads 11, 46
Toads and Frogs key 31–33
torquata - *Hypsiglena torquata*
Trachemys 100, 114, 115, 117
 Trachemys gaigeae 12, 94, 100, **107–109**, map 109, 110, 111, 114, 123
 Trachemys nebulosa 108, 110
 Trachemys scripta 94, 100, 103, 107, 108, **109–112**, map 112, 114, 120, 123
 Trachemys scripta elegans 12, 108, 110, 111
 Trachemys scripta scripta 111
 Trachemys scripta taylori 108
Trans-Pecos blind snake 13
Trans-Pecos rat snake 13, **262–264**, map 264
Trans-Pecos striped whiptail 13
Tree lizard(s) 12, 138, **185–189**, map 189
Treefrogs 11, 64
triangulum - *Lampropeltis triangulum*
triaspis - *Senticolis triaspis*
Trimorphodon 260
 Trimorphodon biscutatus 3, 254, 263, 276, 296, **332–333**, map 333, 342
 Trimorphodon biscutatus lambda 14, 332, 333
 Trimorphodon biscutatus vilkinsonii 14, 332, 333
TRIONYCHIDAE 12, 93, 119
Trionyx xv, 119, 120, 122, 123
 Trionyx muticus 93, 114, **120–122**, map 122, 123, 124
 Trionyx muticus muticus 12, 120
 Trionyx ocellatus 122
 Trionyx spiniferus 93, 100, 103, 108, 110, 114, 120, **122–124**, map 124
 Trionyx spiniferus emoryi 12, 123
 Trionyx spiniferus hartwegi 12, 123
triseriata - *Pseudacris triseriata*
tristicus - *Sceloporus undulatus tristicus*
Tropical frogs 11, 43
Tropidoclonion
 Tropidoclonion lineatum 14, 253, 265, 315, 320, **334–336**, map 336
 Tropidoclonion lineatum annectens 336
 Tropidoclonion lineatum lineatum 334, 336

Tropidoclonion lineatum mertensi 334, 336
Tropidoclonion lineatum texanum 336
True frogs 11, 78
Tucson banded gecko 12
turcicus - Hemidactylus turcicus
Turtles 11, key 93–94
Twin-spotted rattlesnake **361**
Twin-spotted spiny lizard 12
Typhlops 259

U

undulatus - Sceloporus undulatus
uniformis - Uta stansburiana uniformis
uniparens - Cnemidophorus uniparens
Urosaurus 138
 Urosaurus ornatus 4, 127, 158, 159, 180, **185–189**, map 189, 264, 268, 346
 Urosaurus ornatus levis 12, 186
 Urosaurus ornatus linearis 12, 186
 Urosaurus ornatus schmidti 12, 186
 Urosaurus ornatus wrighti 12, 186
Uta 277, 350
 Uta stansburiana 127, 136, 137, 139, 142, 145, 149, 160, 180, 186, **189–193**, map 193, 219, 264
 Uta stansburiana stansburiana 191
 Uta stansburiana stejnegeri 12, 190, 191
 Uta stansburiana uniformis 12, 190

V

vagrans - Thamnophis elegans vagrans
Variable skink 13
variegatus - Coleonyx variegatus
velox - Cnemidophorus velox
Venomous lizards 13, 253
ventralis - Callisaurus draconoides ventralis
vernalis - Liochlorophis vernalis
vilkinsonii - Trimorphodon biscutatus vilkinsonii
VIPERIDAE 14, 340
Vipers 14
virgatus - Sceloporus virgatus
 Virginia striatula 333
viridis - Cnemidophorus sexlineatus viridis
viridis - Crotalus viridis

W

Wandering garter snake 14
Water turtles 11
Western banded gecko **197–199**, map 199
Western blackneck garter snake 14
Western blind snake **258–259**, map 259
Western chorus frog 11
Western coachwhip 13
Western collared lizard 12
Western coral snake **338–339**, map 339
Western diamondback rattlesnake 14, **341–334**, map 344
Western earless lizard 12
Western green toad 11
Western hognose snake **273–275**, map 275
Western hooknose snake 13, **271–273**, map 273
Western longnose snake 14
Western marbled whiptail 13
Western painted turtle 11
Western plains garter snake 14
Western rattlesnake **351–353**, map 353
Western ribbon snake **322–324**, map 324
Western river cooter 11, **102–104**, map 104
Western smooth green snake 13
Western spiny softshell 12
Western terrestrial garter snake **314–317**, map 317
Western toad **49–51**, map 51
Western whiptail **227–231**, map 231
Western yellowbelly racer 13
Whiptails 12, 201, key 203– 204
White Sands prairie lizard 12
willardi - Crotalus willardi
wislizenii - Crotaphytus wislizenii
Woodhouse's toad 11, **61–63**, map 63
woodhousii - Bufo woodhousii
Woodland striped whiptail 13
wrighti - Urosaurus ornatus wrighti
wrightorum
 - see *Hyla eximia*

Y

Yaqui black-headed snake 14, **311–312**, map 312
yaquia - Tantilla yaquia
Yarrow's spiny lizard 12, **166–170**, map 170
yavapaiensis - Rana yavapaiensis
Yellow mud turtle 12, **114–116**, map 116
Yellowheaded collared lizard 12

Z

Zebratail lizard(s) 12, 138, **139–141**, map 141

CONVERSION TABLE

INTERNATIONAL SYSTEM OF UNITS (METRIC) TO ENGLISH UNITS OF MEASURE

cm x 0.394 = inches
m x 3.281 = feet
m x 1.094 = yards
km x 0.621 = miles
km2 x 0.386 = square miles
km2 x 247.097 = acres
m2 x 1.196 = square yards
ha x 2.471 = acres
l x 0.264 = gallons
kg x 2.205 = pounds
gm x 0.035 = ounces
degrees Celsius = 5/9 (degrees Fahrenheit - 32)

ENGLISH UNITS OF MEASURE TO INTERNATIONAL SYSTEM OF UNITS (METRIC)

inches x 25.4 = mm
inches x 2.54 = cm
feet x 0.305 = m
miles x 1.609 = km
acres x 0.405 = ha
gallons x 3.785 = l
ounces x 28.349 = gm
pounds x 0.454 = kg
degrees Fahrenheit = 9/5 (degrees Celsius) + 32